CONTINUITY EQUATION FOR ANY FLUID	$\rho_1 A_1 v_1 = \rho_2 A_2 v_2$	(6–4)
CONTINUITY EQUATION FOR LIQUIDS	$A_1 v_1 = A_2 v_2$	(6–5)
BERNOULLI'S EQUATION	$\dfrac{p_1}{\gamma} + z_1 + \dfrac{v_1^2}{2g} = \dfrac{p_2}{\gamma} + z_2 + \dfrac{v_2^2}{2g}$	(6–9)
TORRICELLI'S THEOREM	$v_2 = \sqrt{2gh}$	(6–16)
TIME REQUIRED TO DRAIN A TANK	$t_2 - t_1 = \dfrac{2(A_t/A_j)}{\sqrt{2g}}(h_1^{1/2} - h_2^{1/2})$	(6–26)
GENERAL ENERGY EQUATION	$\dfrac{p_1}{\gamma} + z_1 + \dfrac{v_1^2}{2g} + h_A - h_R - h_L = \dfrac{p_2}{\gamma} + z_2 + \dfrac{v_2^2}{2g}$	(7–3)
POWER ADDED TO A FLUID BY A PUMP	$P_A = h_A W = h_A \gamma Q$	(7–5)
PUMP EFFICIENCY	$e_M = \dfrac{\text{Power delivered to fluid}}{\text{Power put into pump}} = \dfrac{P_A}{P_I}$	(7–6)
POWER REMOVED FROM A FLUID BY A MOTOR	$P_R = h_R W = h_R \gamma Q$	(7–8)
MOTOR EFFICIENCY	$e_M = \dfrac{\text{Power output from motor}}{\text{Power delivered by fluid}} = \dfrac{P_O}{P_R}$	(7–9)
REYNOLDS NUMBER—CIRCULAR SECTIONS	$N_R = \dfrac{vD\rho}{\eta} = \dfrac{vD}{\nu}$	(8–1)
DARCY'S EQUATION FOR ENERGY LOSS	$h_L = f \times \dfrac{L}{D} \times \dfrac{v^2}{2g}$	(8–3)
HAGEN-POISEUILLE EQUATION	$h_L = \dfrac{32\eta L v}{\gamma D^2}$	(8–4)
FRICTION FACTOR FOR LAMINAR FLOW	$f = \dfrac{64}{N_R}$	(8–5)
FRICTION FACTOR FOR TURBULENT FLOW	$f = \dfrac{0.25}{\left[\log \left(\dfrac{1}{3.7(D/\epsilon)} + \dfrac{5.74}{N_R^{0.9}} \right) \right]^2}$	(8–7)
HAZEN-WILLIAMS FORMULA – U.S. CUSTOMARY UNITS	$v = 1.32\, C_h R^{0.63} s^{0.54}$	(8–8)

APPLIED FLUID MECHANICS

APPLIED FLUID MECHANICS

Seventh Edition

Robert L. Mott
University of Dayton

Joseph A. Untener
University of Dayton

PEARSON

Boston   Columbus   Indianapolis   New York   San Francisco   Upper Saddle River
Amsterdam   Cape Town   Dubai   London   Madrid   Milan   Munich   Paris   Montréal   Toronto
Delhi   Mexico City   São Paulo   Sydney   Hong Kong   Seoul   Singapore   Taipei   Tokyo

Editorial Director: Vernon R. Anthony
Acquisitions Editor: Lindsey Gill
Editorial Assistant: Nancy Kesterson
Director of Marketing: David Gesell
Senior Marketing Coordinator:
    Alicia Wozniak
Marketing Assistant: Les Roberts
Program Manager: Maren L. Beckman
Project Manager: Janet Portisch
Procurement Specialist:
    Deidra M. Skahill

Art Director: Jayne Conte
Cover Designer: Suzanne Duda
Cover Image: Fotolia
Media Project Manager: Leslie Brado
Full-Service Project Management:
    Mansi Negi/Aptara®, Inc.
Composition: Aptara®, Inc.
Printer/Binder: Courier Kendallville
Cover Printer: Lehigh/Phoenix
    Color Hagerstown
Text Font: 10/12 Minion

Credits and acknowledgments borrowed from other sources and reproduced, with permission, in this textbook appear on the appropriate page within text.

Microsoft® and Windows® are registered trademarks of the Microsoft Corporation in the U.S.A. and other countries. Screen shots and icons reprinted with permission from the Microsoft Corporation. This book is not sponsored or endorsed by or affiliated with the Microsoft Corporation.

Many of the designations by manufacturers and sellers to distinguish their products are claimed as trademarks. Where those designations appear in this book, and the publisher was aware of a trademark claim, the designations have been printed in initial caps or all caps.

Library of Congress Cataloging-in-Publication Data

Mott, Robert L.
    Applied fluid mechanics/Robert L. Mott, Joseph A. Untener. —
Seventh edition.
        pages cm
    Includes bibliographical references and index.
    ISBN-13: 978-0-13-255892-1
    ISBN-10: 0-13-255892-0
    1. Fluid mechanics.   I. Untener, Joseph A.   II. Title.
    TA357.M67 2015
    620.1'06—dc23

                        2013026227

10 9 8 7 6 5 4 3 2 1

ISBN 10:   0-13-255892-0
ISBN 13: 978-0-13-255892-1

BRIEF CONTENTS

1 The Nature of Fluids and the Study of Fluid Mechanics   1

2 Viscosity of Fluids   19

3 Pressure Measurement   38

4 Forces Due to Static Fluids   63

5 Buoyancy and Stability   93

6 Flow of Fluids and Bernoulli's Equation   117

7 General Energy Equation   154

8 Reynolds Number, Laminar Flow, Turbulent Flow, and Energy Losses Due to Friction   178

9 Velocity Profiles for Circular Sections and Flow in Noncircular Sections   205

10 Minor Losses   225

11 Series Pipeline Systems   264

12 Parallel and Branching Pipeline Systems   296

13 Pump Selection and Application   318

14 Open-Channel Flow   372

15 Flow Measurement   395

16 Forces Due to Fluids in Motion   418

17 Drag and Lift   432

18 Fans, Blowers, Compressors, and the Flow of Gases   450

19 Flow of Air in Ducts   470

Appendices   488

Answers to Selected Problems   516

Index   525

CONTENTS

Preface   xi

Acknowledgments   xv

1 The Nature of Fluids and the Study of Fluid Mechanics   1

The Big Picture   1

1.1   Objectives   3

1.2   Basic Introductory Concepts   3

1.3   The International System of Units (SI)   4

1.4   The U.S. Customary System   4

1.5   Weight and Mass   5

1.6   Temperature   6

1.7   Consistent Units in an Equation   6

1.8   The Definition of Pressure   8

1.9   Compressibility   10

1.10   Density, Specific Weight, and Specific Gravity   11

1.11   Surface Tension   14

References   15

Internet Resources   15

Practice Problems   15

Computer Aided Engineering Assignments   18

2 Viscosity of Fluids   19

The Big Picture   19

2.1   Objectives   20

2.2   Dynamic Viscosity   21

2.3   Kinematic Viscosity   22

2.4   Newtonian Fluids and Non-Newtonian Fluids   23

2.5   Variation of Viscosity with Temperature   25

2.6   Viscosity Measurement   27

2.7   SAE Viscosity Grades   32

2.8   ISO Viscosity Grades   33

2.9   Hydraulic Fluids for Fluid Power Systems   33

References   34

Internet Resources   35

Practice Problems   35

Computer Aided Engineering Assignments   37

3 Pressure Measurement   38

The Big Picture   38

3.1   Objectives   39

3.2   Absolute and Gage Pressure   39

3.3   Relationship between Pressure and Elevation   40

3.4   Development of the Pressure–Elevation Relation   43

3.5   Pascal's Paradox   45

3.6   Manometers   46

3.7   Barometers   51

3.8   Pressure Expressed as the Height of a Column of Liquid   52

3.9   Pressure Gages and Transducers   53

References   55

Internet Resources   55

Practice Problems   55

4 Forces Due to Static Fluids   63

The Big Picture   63

4.1   Objectives   65

4.2   Gases Under Pressure   65

4.3   Horizontal Flat Surfaces Under Liquids   66

4.4   Rectangular Walls   67

4.5   Submerged Plane Areas— General   69

4.6   Development of the General Procedure for Forces on Submerged Plane Areas   72

4.7   Piezometric Head   73

4.8   Distribution of Force on a Submerged Curved Surface   74

4.9   Effect of a Pressure above the Fluid Surface   78

4.10   Forces on a Curved Surface with Fluid Below It   78

4.11   Forces on Curved Surfaces with Fluid Above and Below   79

Practice Problems   80

Computer Aided Engineering Assignments   92

5 Buoyancy and Stability   93

The Big Picture   93

5.1   Objectives   94

5.2   Buoyancy   94

5.3   Buoyancy Materials   101

5.4   Stability of Completely Submerged Bodies   102

5.5   Stability of Floating Bodies   103

5.6   Degree of Stability   107

Reference   108

Internet Resources   108

Practice Problems   108

Stability Evaluation Projects   116

6 Flow of Fluids and Bernoulli's Equation   117

The Big Picture   117

6.1   Objectives   118

6.2   Fluid Flow Rate and the Continuity Equation   118

6.3   Commercially Available Pipe and Tubing   122

6.4   Recommended Velocity of Flow in Pipe and Tubing   124

6.5   Conservation of Energy—Bernoulli's Equation   127

6.6   Interpretation of Bernoulli's Equation   128

6.7   Restrictions on Bernoulli's Equation   129

6.8   Applications of Bernoulli's Equation   129

6.9   Torricelli's Theorem   137

6.10   Flow Due to a Falling Head   140

References   142

Internet Resources   142

Practice Problems   143

Analysis Projects Using Bernoulli's Equation and Torricelli's Theorem   153

7 General Energy Equation   154

The Big Picture   154

7.1   Objectives   155

7.2   Energy Losses and Additions   156

7.3   Nomenclature of Energy Losses and Additions   158

7.4   General Energy Equation   158

7.5   Power Required by Pumps   162

7.6   Power Delivered to Fluid Motors   165

Practice Problems   167

8 Reynolds Number, Laminar Flow, Turbulent Flow, and Energy Losses Due to Friction   178

The Big Picture   178

8.1   Objectives   181

8.2   Reynolds Number   181

8.3   Critical Reynolds Numbers   182

8.4   Darcy's Equation   183

8.5   Friction Loss in Laminar Flow   183

8.6   Friction Loss in Turbulent Flow   184

8.7   Use of Software for Pipe Flow Problems   190

8.8   Equations for the Friction Factor   194

8.9   Hazen–Williams Formula for Water Flow   195

8.10   Other Forms of the Hazen–Williams Formula   196

8.11   Nomograph for Solving the Hazen–Williams Formula   196

References   198

Internet Resources   198

Practice Problems   198

Computer Aided Engineering Assignments   204

9 Velocity Profiles for Circular Sections and Flow in Noncircular Sections   205

The Big Picture   205

9.1   Objectives   206

9.2   Velocity Profiles   207

9.3   Velocity Profile for Laminar Flow   207

9.4   Velocity Profile for Turbulent Flow   209

9.5   Flow in Noncircular Sections   212

9.6   Computational Fluid Dynamics   216

References   218

Internet Resources   218

Practice Problems   218

Computer Aided Engineering Assignments   224

10 Minor Losses   225

The Big Picture   225

10.1   Objectives   227

10.2   Resistance Coefficient   227

10.3   Sudden Enlargement   228

10.4   Exit Loss   231

10.5   Gradual Enlargement   231

10.6   Sudden Contraction   233

10.7   Gradual Contraction   236

10.8   Entrance Loss   237

10.9 Resistance Coefficients for Valves and Fittings 238

10.10 Application of Standard Valves 244

10.11 Pipe Bends 246

10.12 Pressure Drop in Fluid Power Valves 248

10.13 Flow Coefficients for Valves Using C_V 251

10.14 Plastic Valves 252

10.15 Using K-Factors in PIPE-FLO® Software 253

References 258

Internet Resources 258

Practice Problems 258

Computer Aided Analysis and Design Assignments 263

11 Series Pipeline Systems 264

The Big Picture 264

11.1 Objectives 265

11.2 Class I Systems 265

11.3 Spreadsheet Aid for Class I Problems 270

11.4 Class II Systems 272

11.5 Class III Systems 278

11.6 PIPE-FLO® Examples for Series Pipeline Systems 281

11.7 Pipeline Design for Structural Integrity 284

References 286

Internet Resources 286

Practice Problems 286

Computer Aided Analysis and Design Assignments 295

12 Parallel and Branching Pipeline Systems 296

The Big Picture 296

12.1 Objectives 298

12.2 Systems with Two Branches 298

12.3 Parallel Pipeline Systems and Pressure Boundaries in PIPE-FLO® 304

12.4 Systems with Three or More Branches—Networks 307

References 314

Internet Resources 314

Practice Problems 314

Computer Aided Engineering Assignments 317

13 Pump Selection and Application 318

The Big Picture 318

13.1 Objectives 319

13.2 Parameters Involved in Pump Selection 320

13.3 Types of Pumps 320

13.4 Positive-Displacement Pumps 320

13.5 Kinetic Pumps 326

13.6 Performance Data for Centrifugal Pumps 330

13.7 Affinity Laws for Centrifugal Pumps 332

13.8 Manufacturers' Data for Centrifugal Pumps 333

13.9 Net Positive Suction Head 341

13.10 Suction Line Details 346

13.11 Discharge Line Details 346

13.12 The System Resistance Curve 347

13.13 Pump Selection and the Operating Point for the System 350

13.14 Using PIPE-FLO® for Selection of Commercially Available Pumps 352

13.15 Alternate System Operating Modes 356

13.16 Pump Type Selection and Specific Speed 361

13.17 Life Cycle Costs for Pumped Fluid Systems 363

References 364

Internet Resources 365

Practice Problems 366

Supplemental Problem (PIPE-FLO® Only) 367

Design Problems 367

Design Problem Statements 368

Comprehensive Design Problem 370

14 Open-Channel Flow 372

The Big Picture 372

14.1 Objectives 373

14.2 Classification of Open-Channel Flow 374

14.3 Hydraulic Radius and Reynolds Number in Open-Channel Flow 375

14.4 Kinds of Open-Channel Flow 375

14.5 Uniform Steady Flow in Open Channels 376

14.6 The Geometry of Typical Open Channels 380

14.7 The Most Efficient Shapes for Open Channels 382

14.8 Critical Flow and Specific Energy 382

14.9 Hydraulic Jump 384

14.10 Open-Channel Flow Measurement 386

References 390

Digital Publications 390

Internet Resources 390

Practice Problems 391

Computer Aided Engineering Assignments 394

15 Flow Measurement 395

The Big Picture 395

15.1 Objectives 396

15.2 Flowmeter Selection Factors 396

15.3 Variable-Head Meters 397

15.4 Variable-Area Meters 404

15.5 Turbine Flowmeter 404

15.6 Vortex Flowmeter 404

15.7 Magnetic Flowmeter 406

15.8 Ultrasonic Flowmeters 408

15.9 Positive-Displacement Meters 408

15.10 Mass Flow Measurement 408

15.11 Velocity Probes 410

15.12 Level Measurement 414

15.13 Computer-Based Data Acquisition and Processing 414

References 415

Internet Resources 415

Review Questions 416

Practice Problems 416

Computer Aided Engineering Assignments 417

16 Forces Due to Fluids in Motion 418

The Big Picture 418

16.1 Objectives 419

16.2 Force Equation 419

16.3 Impulse–Momentum Equation 420

16.4 Problem-Solving Method Using the Force Equations 420

16.5 Forces on Stationary Objects 421

16.6 Forces on Bends in Pipelines 423

16.7 Forces on Moving Objects 426

Practice Problems 427

17 Drag and Lift 432

The Big Picture 432

17.1 Objectives 434

17.2 Drag Force Equation 434

17.3 Pressure Drag 435

17.4 Drag Coefficient 435

17.5 Friction Drag on Spheres in Laminar Flow 441

17.6 Vehicle Drag 441

17.7 Compressibility Effects and Cavitation 443

17.8 Lift and Drag on Airfoils 443

References 445

Internet Resources 446

Practice Problems 446

18 Fans, Blowers, Compressors, and the Flow of Gases 450

The Big Picture 450

18.1 Objectives 451

18.2 Gas Flow Rates and Pressures 451

18.3 Classification of Fans, Blowers, and Compressors 452

18.4 Flow of Compressed Air and Other Gases in Pipes 456

18.5 Flow of Air and Other Gases Through Nozzles 461

References 467

Internet Resources 467

Practice Problems 468

Computer Aided Engineering Assignments 469

19 Flow of Air in Ducts 470

The Big Picture 470

19.1 Objectives 472

19.2 Energy Losses in Ducts 472

19.3 Duct Design 477

19.4 Energy Efficiency and Practical Considerations in Duct Design 483

References 484

Internet Resources 484

Practice Problems 484

Appendices 488

Appendix A Properties of Water 488

Appendix B Properties of Common Liquids 490

Appendix C Typical Properties of Petroleum Lubricating Oils 492

Appendix D Variation of Viscosity with Temperature 493

Appendix E Properties of Air 496

Appendix F Dimensions of Steel Pipe 500

Appendix G Dimensions of Steel, Copper, and Plastic Tubing 502

Appendix H Dimensions of Type K Copper Tubing 505

Appendix I Dimensions of Ductile Iron Pipe 506

Appendix J Areas of Circles 507

Appendix K Conversion Factors 509

Appendix L Properties of Areas 511

Appendix M Properties of Solids 513

Appendix N Gas Constant, Adiabatic Exponent, and Critical Pressure Ratio for Selected Gases 515

Answers to Selected Problems 516

Index 525

PREFACE

INTRODUCTION

The objective of this book is to present the principles of fluid mechanics and the application of these principles to practical, applied problems. Primary emphasis is on fluid properties; the measurement of pressure, density, viscosity, and flow; fluid statics; flow of fluids in pipes and noncircular conduits; pump selection and application; open-channel flow; forces developed by fluids in motion; the design and analysis of heating, ventilation, and air conditioning (HVAC) ducts; and the flow of air and other gases.

Applications are shown in the mechanical field, including industrial fluid distribution, fluid power, and HVAC; in the chemical field, including flow in materials processing systems; and in the civil and environmental fields as applied to water and wastewater systems, fluid storage and distribution systems, and open-channel flow. This book is directed to anyone in an engineering field where the ability to apply the principles of fluid mechanics is the primary goal.

Those using this book are expected to have an understanding of algebra, trigonometry, and mechanics. After completing the book, the student should have the ability to design and analyze practical fluid flow systems and to continue learning in the field. Students could take other applied courses, such as those on fluid power, HVAC, and civil hydraulics, following this course. Alternatively, this book could be used to teach selected fluid mechanics topics within such courses.

APPROACH

The approach used in this book encourages the student to become intimately involved in learning the principles of fluid mechanics at seven levels:

1. Understanding concepts.
2. Recognizing how the principles of fluid mechanics apply to their own experience.
3. Recognizing and implementing logical approaches to problem solutions.
4. Performing the analyses and calculations required in the solutions.
5. Critiquing the design of a given system and recommending improvements.
6. Designing practical, efficient fluid systems.
7. Using computer-assisted approaches, both commercially available and self-developed, for design and analysis of fluid flow systems.

This multilevel approach has proven successful for several decades in building students' confidence in their ability to analyze and design fluid systems.

Concepts are presented in clear language and illustrated by reference to physical systems with which the reader should be familiar. An intuitive justification as well as a mathematical basis is given for each concept. The methods of solution to many types of complex problems are presented in step-by-step procedures. The importance of recognizing the relationships among what is known, what is to be found, and the choice of a solution procedure is emphasized.

Many practical problems in fluid mechanics require relatively long solution procedures. It has been the authors' experience that students often have difficulty in carrying out the details of the solution. For this reason, each example problem is worked in complete detail, including the manipulation of units in equations. In the more complex examples, a programmed instruction format is used in which the student is asked to perform a small segment of the solution before being shown the correct result. The programs are of the linear type in which one panel presents a concept and then either poses a question or asks that a certain operation be performed. The following panel gives the correct result and the details of how it was obtained. The program then continues.

The International System of Units (Système International d'Unités, or SI) and the U.S. Customary System of units are used approximately equally. The SI notation in this book follows the guidelines set forth by the National Institute of Standards and Technology (NIST), U.S. Department of Commerce, in its 2008 publication *The International System of Units (SI)* (NIST Special Publication 330), edited by Barry N. Taylor and Ambler Thompson.

COMPUTER-ASSISTED PROBLEM SOLVING AND DESIGN

Computer-assisted approaches to solving fluid flow problems are recommended only after the student has demonstrated competence in solving problems manually. They allow more comprehensive problems to be analyzed and give students tools for considering multiple design options while removing some of the burden of calculations. Also, many employers expect students to have not only the skill to use software, but the inclination to do so, and using software within the course effectively nurtures this skill. We recommend the following classroom learning policy.

> *Users of computer software must have solid understanding of the principles on which the software is based to ensure that analyses and design decisions are fundamentally sound. Software should be used only after mastering relevant analysis methods by careful study and using manual techniques.*

Computer-based assignments are included at the end of many chapters. These can be solved by a variety of techniques such as:

- The use of a spreadsheet such as Microsoft® Excel
- The use of technical computing software
- The use of commercially available software for fluid flow analysis

Chapter 11, Series Pipeline Systems, and Chapter 13, Pump Selection and Application, include example Excel spreadsheet aids for solving fairly complex system design and analysis problems.

New, powerful, commercially available software: A new feature of this 7th edition is the integration of the use of a major, internationally renowned software package for piping system analysis and design, called PIPE-FLO®, produced and marketed by Engineered Software, Inc. (often called ESI) in Lacey, Washington. As stated by ESI's CEO and president, along with several staff members, the methodology used in this textbook for analyzing pumped fluid flow systems is highly compatible with that used in their software. Students who learn well the principles and manual problem solving methods presented in this book will be well-prepared to apply them in industrial settings and they will also have learned the fundamentals of using PIPE-FLO® to perform the analyses of the kinds of fluid flow systems they will encounter in their careers. This skill should be an asset to students' career development.

Students using this book in classes will be informed about a unique link to the ESI website where a specially adapted version of the industry-scale software can be used. Virtually all of the piping analysis and design problems in this book can be set up and solved using this special version. The tools and techniques for building computer models of fluid flow systems are introduced carefully starting in Chapter 8 on energy losses due to friction in pipes and continuing through Chapter 13, covering minor losses, series pipeline systems, parallel and branching systems, and pump selection and application. As each new concept and problem-solving method is learned from this book, it is then applied to one or more example problems where students can develop their skills in creating and solving real problems. With each chapter, the kinds of systems that students will be able to complete expand in breadth and depth. New supplemental problems using PIPE-FLO® are in the book so students can extend and demonstrate their abilities in assignments, projects, or self-study. The integrated companion software,

PUMP-FLO®, provides access to catalog data for numerous types and sizes of pumps that students can use in assignments and to become more familiar with that method of specifying pumps in their future positions.

Students and instructors can access the special version of PIPE-FLO® at this site:

http://www.eng-software.com/appliedfluidmechanics

FEATURES NEW TO THE SEVENTH EDITION

The seventh edition continues the pattern of earlier editions in refining the presentation of several topics, enhancing the visual attractiveness and usability of the book, updating data and analysis techniques, and adding selected new material. The Big Picture begins each chapter as in the preceding two editions, but each has been radically improved with one or more attractive photographs or illustrations, a refined Exploration section that gets students personally involved with the concepts presented in the chapter, and brief Introductory Concepts that preview the chapter discussions. Feedback from instructors and students about this feature has been very positive. The extensive appendixes continue to be useful learning and problem-solving tools and several have been updated or expanded.

The following list highlights some of the changes in this edition:

- A large percentage of the illustrations have been upgraded in terms of realism, consistency, and graphic quality. Full color has been introduced enhancing the appearance and effectiveness of illustrations, graphs, and the general layout of the book.

- Many photographs of commercially available products have been updated and some new ones have been added.

- Most chapters include an extensive list of Internet resources that provide useful supplemental information such as commercially available products, additional data for problem solving and design, more in-depth coverage of certain topics, information about fluid mechanics software, and industry standards. The resources have been updated and many have been added to those in previous editions.

- The end-of-chapter references have been extensively revised, updated, and extended.

- Use of metric units has been expanded in several parts of the book. Two new Appendix tables have been added that feature purely metric sizes for steel, copper, and plastic tubing. Use of the metric DN-designations for standard Schedules 40 and 80 steel pipes have been more completely integrated into the discussions, example problems, and end-of chapter problems. Almost all metric-based problems use these new tables for pipe or tubing designations, dimensions, and flow areas. This should give students strong foundations on which to build a career in the global industrial scene in which they will pursue their careers.

- Many new, creative supplemental problems have been added to the end-of-chapter set of problems in several chapters to enhance student learning and to provide more variety for instructors in planning their courses.

- Graphical tools for selecting pipe sizes are refined in Chapter 6 and used in later chapters and design projects.

- The discussion of computational fluid mechanics included in Chapter 9 has been revised with attractive new graphics that are highly relevant to the study of pipe flow.

- The use of *K*-factors (resistance coefficients) based on the equivalent-length approach has been updated, expanded, and refined according to the latest version of the *Crane Technical Paper 410* (TP 410).

- Use of the flow coefficient C_V for evaluating the relationship between flow rate and pressure drop across valves has been expanded in Chapter 10 with new equations for use with metric units. It is also included in new parts of Chapter 13 that emphasize the use of valves as control elements.

- The section General Principles of Pipeline System Design has been refined in Chapter 11.

- Several sections in Chapter 13 on pump selection and application have been updated and revised to provide more depth, greater consistency with TP 410, a smoother development of relevant topics, and use of the PIPE-FLO® software.

INTRODUCING PROFESSOR JOSEPH A. UNTENER—NEW CO-AUTHOR OF THIS BOOK

We are pleased to announce that the seventh edition of *Applied Fluid Mechanics* has been co-authored by:

Robert L. Mott and Joseph A. Untener

Professor Untener has been an outstanding member of the faculty in the Department of Engineering Technology at the University of Dayton since 1987 when he was hired by Professor Mott. Joe's first course taught at UD was Fluid Mechanics, using the 2nd edition of this book, and he continues to include this course in his schedule. A gifted instructor, a strong leader, a valued colleague, and a wise counselor of students, Joe is a great choice for the task of preparing this book. He brings fresh ideas, a keen sense of style and methodology, and an eye for effective and attractive graphics. He initiated the major move toward integrating the PIPE-FLO® software into the book and managed the process of working with the leadership and staff of Engineered Software, Inc. His contributions should prove to be of great value to users of this book, both students and instructors.

DOWNLOAD INSTRUCTOR RESOURCES FROM THE INSTRUCTOR RESOURCE CENTER

This edition is accompanied by an Instructor's Solutions Manual and a complete Image Bank of all figures featured in the text. To access supplementary materials online, instructors need to request an instructor access code. Go to www.pearsonhighered.com/irc to register for an instructor access code. Within 48 hours of registering, you will receive a confirming email including an instructor access code. Once you have received your code, locate your text in the online catalog and click on the 'Instructor Resources' button on the left side of the catalog product page. Select a supplement, and a login page will appear. Once you have logged in, you can access the instructor material for all Pearson textbooks. If you have any difficulties accessing the site or downloading a supplement, please contact Customer Service at http://247pearsoned.custhelp.com/.

ACKNOWLEDGMENTS

We would like to thank all who helped and encouraged us in the writing of this book, including users of earlier editions and the several reviewers who provided detailed suggestions: William E. Cole, Northeastern University; Gary Crossman, Old Dominion University; Charles Drake, Ferris State University; Mark S. Frisina, Wentworth Institute of Technology; Dr. Roy A. Hartman, P. E., Texas A & M University; Dr. Greg E. Maksi, State Technical Institute at Memphis; Ali Ogut, Rochester Institute of Technology; Paul Ricketts, New Mexico State University; Mohammad E. Taslim, Northeastern University at Boston; Pao-lien Wang, University of North Carolina at Charlotte; and Steve Wells, Old Dominion University. Special thanks go to our colleagues of the University of Dayton, the late Jesse Wilder, David Myszka, Rebecca Blust, Michael Kozak, and James Penrod, who used earlier editions of this book in class and offered helpful suggestions. Robert Wolff, also of the University of Dayton, has provided much help in the use of the SI system of units, based on his long experience in metrication through the American Society for Engineering Education. Professor Wolff also consulted on fluid power applications. University of Dayton student, Tyler Runyan, provided significant input to this edition by providing student feedback on the text, rendering some illustrations, and generating solutions to problems using PIPE-FLO®. We thank all those from Engineered Software, Inc. (ESI), for their cooperation and assistance in incorporating the PIPE-FLO® software into this book. Particularly, we are grateful for the collaboration by Ray Hardee, Christy Bermensolo, and Buck Jones of ESI. We are grateful for the expert professional and personal service provided by the editorial and marketing staff of Pearson Education. Comments from students who used the book are also appreciated because the book was written for them.

Robert L. Mott and Joseph A. Untener

REVIEWERS

Eric Baldwin
Bluefield State College

Randy Bedington
Catawba Valley Community College

Chuck Drake
Ferris State

Ann Marie Hardin
Blue Mountain Community College

Francis Plunkett
Broome Community College

Mir Said Saidpour
Farmingdale State College-SUNY

Xiuling Wang
Calumet Purdue

THE NATURE OF FLUIDS AND THE STUDY OF FLUID MECHANICS

THE BIG PICTURE

As you begin the study of fluid mechanics, let's look at some fundamental concepts and look ahead to the major topics that you will study in this book. Try to identify where you have encountered either stationary or moving, pressurized fluids in your daily life. Consider the water system in your home, hotels, or commercial buildings. Think about how your car's fuel travels from the tank to the engine or how the cooling water flows through the engine and its cooling system. When enjoying time in an amusement park, consider how fluids are handled in water slides or boat rides. Look carefully at construction equipment to observe how pressurized fluids are used to actuate moving parts and to drive the machines. Visit manufacturing operations where automation equipment, material handling devices, and production machinery utilize pressurized fluids.

On a larger scale, consider the chemical processing plant shown in Fig. 1.1. Complex piping systems use pumps to transfer fluids from tanks and move them through various processing systems. The finished products may be stored in other tanks and then transferred to trucks or railroad cars to be delivered to customers.

Listed here are several of the major concepts you will study in this book:

- *Fluid mechanics* is the study of the behavior of fluids, either at rest (fluid statics) or in motion (fluid dynamics).

- Fluids can be either *liquids* or *gases* and they can be characterized by their physical properties such as density, specific weight, specific gravity, surface tension, and viscosity.

- Quantitatively analyzing fluid systems requires careful use of units for all terms. Both the SI metric system of units and the U.S. gravitational system are used in this book. Careful distinction between weight and mass is also essential.

- Fluid statics concepts that you will learn include the measurement of pressure, forces exerted on surfaces due to fluid pressure, buoyancy and stability of floating bodies.

- Learning how to analyze the behavior of fluids as they flow through circular pipes and tubes and through conduits with other shapes is important.

- We will consider the energy possessed by the fluid because of its velocity, elevation, and pressure.

- Accounting for energy losses, additions, or purposeful removals that occur as the fluid flows through the

FIGURE 1.1 Industrial and commercial fluid piping systems, like this one used in a chemical processing plant, involve complex arrangements requiring careful design and analysis. (*Source:* Nikolay Kazachok/Fotolia)

components of a fluid flow system enables you to analyze the performance of the system.

■ A flowing fluid loses energy due to friction as it moves along a conduit and as it encounters obstructions (like in a control valve) or changes its direction (like in a pipe elbow).

■ Energy can be added to a flowing fluid by pumps that create flow and increase the fluid's pressure.

■ Energy can be purposely removed by using it to drive a fluid motor, a turbine, or a hydraulic actuator.

■ Measurements of fluid pressure, temperature, and the fluid flow rate in a system are critical to understanding its performance.

Exploration

Now let's consider a variety of systems that use fluids and that illustrate some of the applications of concepts learned from this book. As you read this section, consider such factors as:

■ The basic function or purpose of the system

■ The kind of fluid or fluids that are in the system

■ The kinds of containers for the fluid or the conduits through which it flows

■ If the fluid flows, what causes the flow to occur? Describe the flow path.

■ What components of the system resist the flow of the fluid?

■ What characteristics of the fluid are important to the proper performance of the system?

1. In your home, you use water for many different purposes such as drinking, cooking, bathing, cleaning, and watering lawns and plants. Water also eliminates wastes from the home through sinks, drains, and toilets. Rain water, melting snow, and water in the ground must be managed to conduct it away from the home using gutters, downspouts, ditches, and sump pumps. Consider how the water is delivered to your home. What is the ultimate source of the water—a river, a reservoir, or natural groundwater? Is the water stored in tanks at some points in the process of getting it to your home? Notice that the water system needs to be at a fairly high pressure to be effective for its uses and to flow reliably through the system. How is that pressure created? Are there pumps in the system? Describe their function and how they operate. From where does each pump draw the water? To what places is the water delivered? What quantities of fluid are needed at the delivery points? What pressures are required? How is the flow of water controlled? What materials are used for the pipes, tubes, tanks, and other containers or conduits?

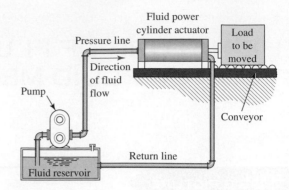

FIGURE 1.2 Typical piping system for fluid power.

As you study Chapters 6–13, you will learn how to analyze and design systems in which the water flows in a pipe or a tube. Chapter 14 discusses the cases of open-channel flow such as that in the gutters that catch the rain from the roof of your home.

2. In your car, describe the system that stores gasoline and then delivers it to the car's engine. How is the windshield washer fluid managed? Describe the cooling system and the nature of the coolant. Describe what happens when you apply the brakes, particularly as it relates to the hydraulic fluid in the braking system. The concepts in Chapters 6–13 will help you to describe and analyze these kinds of systems.

3. Consider the performance of an automated manufacturing system that is actuated by fluid power systems such as the one shown in Fig. 1.2. Describe the fluids, pumps, tubes, valves, and other components of the system. What is the function of the system? How does the fluid accomplish that function? How is energy introduced to the system and how is it dissipated away from the system?

4. Consider the kinds of objects that must float in fluids such as boats, jet skis, rafts, barges, and buoys. Why do they float? In what position or orientation do they float? Why do they maintain their orientation? The principles of buoyancy and stability are discussed in Chapter 5.

5. What examples can you think of where fluids at rest or in motion exert forces on an object? Any vessel containing a fluid under pressure should yield examples. Consider a swimming pool, a hydraulic cylinder, a dam or a retaining wall holding a fluid, a high-pressure washer system, a fire hose, wind during a tornado or a hurricane, and water flowing through a turbine to generate power. What other examples can you think of? Chapters 4, 16, and 17 discuss these cases.

6. Think of the many situations in which it is important to measure the flow rate of fluid in a system or the total quantity of fluid delivered. Consider measuring the gasoline that goes into your car so you can pay for just what you get. The water company wants to know how much water you use in a given month. Fluids often must be metered carefully into production processes in a factory. Liquid medicines and oxygen delivered to a patient in a hospital must be measured continuously for patient safety. Chapter 15 covers flow measurement.

There are many ways in which fluids affect your life. Completion of a fluid mechanics course using this book will help you understand how those fluids can be controlled. Studying this book will help you learn how to design and analyze fluid systems to determine the kind of components that should be used and their size.

1.1 OBJECTIVES

After completing this chapter, you should be able to:

1. Differentiate between a gas and a liquid.
2. Define *pressure*.
3. Identify the units for the basic quantities of time, length, force, mass, and temperature in the SI metric unit system and in the U.S. Customary unit system.
4. Properly set up equations to ensure consistency of units.
5. Define the relationship between force and mass.
6. Define *density*, *specific weight*, and *specific gravity* and the relationships among them.
7. Define *surface tension*.

1.2 BASIC INTRODUCTORY CONCEPTS

- **Pressure**   Pressure is defined as the amount of force exerted on a unit area of a substance or on a surface. This can be stated by the equation

$$p = \frac{F}{A} \qquad (1\text{--}1)$$

Fluids are subjected to large variations in pressure depending on the type of system in which they are used. Milk sitting in a glass is at the same pressure as the air above it. Water in the piping system in your home has a pressure somewhat greater than atmospheric pressure so that it will flow rapidly from a faucet. Oil in a fluid power system is typically maintained at high pressure to enable it to exert large forces to actuate construction equipment or automation devices in a factory. Gases such as oxygen, nitrogen, and helium are often stored in strong cylinders or spherical tanks under high pressure to permit rather large amounts to be held in a relatively small volume. Compressed air is often used in service stations and manufacturing facilities to operate tools or to inflate tires. More discussion about pressure is given in Chapter 3.

- **Liquids and Gases**   Fluids can be either liquids or gases.

When a liquid is held in a container, it tends to take the shape of the container, covering the bottom and the sides. The top surface, in contact with the atmosphere above it, maintains a uniform level. As the container is tipped, the liquid tends to pour out.

When a gas is held under pressure in a closed container, it tends to expand and completely fill the container. If the container is opened, the gas tends to expand more and escape from the container.

In addition to these familiar differences between gases and liquids, another difference is important in the study of fluid mechanics. Consider what happens to a liquid or a gas as the pressure on it is increased. If air (a gas) is trapped in a cylinder with a tight-fitting, movable piston inside it, you can compress the air fairly easily by pushing on the piston. Perhaps you have used a hand-operated pump to inflate a bicycle tire, a beach ball, an air mattress, or a basketball. As you move the piston, the volume of the gas is reduced appreciably as the pressure increases. But what would happen if the cylinder contained water rather than air? You could apply a large force, which would increase the pressure in the water, but the volume of the water would change very little. This observation leads to the following general descriptions of liquids and gases that we will use in this book:

1. Gases are readily compressible.
2. Liquids are only slightly compressible.

More discussion on compressibility is given later in this chapter. We will deal mostly with liquids in this book.

- **Weight and Mass**   An understanding of fluid properties requires a careful distinction between *mass* and *weight*. The following definitions apply:

Mass is the property of a body of fluid that is a measure of its inertia or resistance to a change in motion. It is also a measure of the quantity of fluid.

We use the symbol m for mass in this book.

Weight is the amount that a body of fluid weighs, that is, the force with which the fluid is attracted toward Earth by gravitation.

We use the symbol w for weight.

The relationship between weight and mass is discussed in Section 1.5 as we review the unit systems used in this book. You must be familiar with both the International System of Units, called *SI*, and the U.S. Customary System of units.

■ *Fluid Properties* The latter part of this chapter presents other fluid properties: *specific weight*, *density*, *specific gravity*, and *surface tension*. Chapter 2 presents an additional property, *viscosity*, which is a measure of the ease with which a fluid flows. It is also important in determining the character of the flow of fluids and the amount of energy that is lost from a fluid flowing in a system as discussed in Chapters 8–13.

1.3 THE INTERNATIONAL SYSTEM OF UNITS (SI)

In any technical work the units in which physical properties are measured must be stated. A system of units specifies the units of the basic quantities of length, time, force, and mass. The units of other terms are then derived from these.

The ultimate reference for the standard use of metric units throughout the world is the International System of Units (Système International d'Unités), abbreviated as SI. In the United States, the standard is given in the 2008 publication of the National Institute of Standards and Technology (NIST), U.S. Department of Commerce, *The International System of Units (SI)* (NIST Special Publication 330), edited by Barry N. Taylor and Ambler Thompson (see Reference 1). This is the standard used in this book.

The SI units for the basic quantities are

$$\text{length} = \text{meter (m)}$$
$$\text{time} = \text{second (s)}$$
$$\text{mass} = \text{kilogram (kg) or } \text{N·s}^2/\text{m}$$
$$\text{force} = \text{newton (N) or kg·m/s}^2$$

An equivalent unit for force is kg·m/s², as indicated above. This is derived from the relationship between force and mass,

$$F = ma$$

where a is the acceleration expressed in units of m/s². Therefore, the derived unit for force is

$$F = ma = \text{kg·m/s}^2 = \text{N}$$

Thus, a force of 1.0 N would give a mass of 1.0 kg an acceleration of 1.0 m/s². This means that either N or kg·m/s² can be used as the unit for force. In fact, some calculations in this book require that you be able to use both or to convert from one to the other.

Similarly, besides using the kg as the standard unit mass, we can use the equivalent unit N·s²/m. This can be derived again from $F = ma$:

$$m = \frac{F}{a} = \frac{\text{N}}{\text{m/s}^2} = \frac{\text{N·s}^2}{\text{m}}$$

Therefore, either kg or N·s²/m can be used for the unit of mass.

TABLE 1.1 SI unit prefixes

Prefix	SI symbol	Factor
terra	T	$10^{12} = 1\,000\,000\,000\,000$
giga	G	$10^9 = 1\,000\,000\,000$
mega	M	$10^6 = 1\,000\,000$
kilo	k	$10^3 = 1\,000$
milli	m	$10^{-3} = 0.001$
micro	μ	$10^{-6} = 0.000\,001$
nano	n	$10^{-9} = 0.000\,000\,001$
pico	p	$10^{-12} = 0.000\,000\,000\,001$

1.3.1 SI Unit Prefixes

Because the actual size of physical quantities in the study of fluid mechanics covers a wide range, prefixes are added to the basic quantities. Table 1.1 shows these prefixes. Standard usage in the SI system calls for only those prefixes varying in steps of 10^3 as shown. Results of calculations should normally be adjusted so that the number is between 0.1 and 10 000 times some multiple of 10^3.[*] Then the proper unit with a prefix can be specified. Note that some technical professionals and companies in Europe often use the prefix *centi*, as in centimeters, indicating a factor of 10^{-2}. Some examples follow showing how quantities are given in this book.

Computed Result	Reported Result
0.004 23 m	4.23×10^{-3} m, or 4.23 mm (millimeters)
15 700 kg	15.7×10^3 kg, or 15.7 Mg (megagrams)
86 330 N	86.33×10^3 N, or 86.33 kN (kilonewtons)

1.4 THE U.S. CUSTOMARY SYSTEM

Sometimes called the *English gravitational unit system* or the *pound-foot-second* system, the U.S. Customary System defines the basic quantities as follows:

$$\text{length} = \text{foot (ft)}$$
$$\text{time} = \text{second (s)}$$
$$\text{force} = \text{pound (lb)}$$
$$\text{mass} = \text{slug or lb-s}^2/\text{ft}$$

Probably the most difficult of these units to understand is the slug because we are more familiar with measuring in

---

[*] Because commas are used as decimal markers in many countries, we will not use commas to separate groups of digits. We will separate the digits into groups of three, counting both to the left and to the right from the decimal point, and use a space to separate the groups of three digits. We will not use a space if there are only four digits to the left or right of the decimal point unless required in tabular matter.

terms of pounds, seconds, and feet. It may help to note the relationship between force and mass,

$$F = ma$$

where a is acceleration expressed in units of ft/s^2. Therefore, the derived unit for mass is

$$m = \frac{F}{a} = \frac{\text{lb}}{\text{ft/s}^2} = \frac{\text{lb-s}^2}{\text{ft}} = \text{slug}$$

This means that you may use either slugs or lb-s^2/ft for the unit of mass. In fact, some calculations in this book require that you be able to use both or to convert from one to the other.

1.5 WEIGHT AND MASS

A rigid distinction is made between weight and mass in this book. Weight is a force and mass is the quantity of a substance. We relate these two terms by applying Newton's law of gravitation stated as *force equals mass times acceleration*, or

$$F = ma$$

When we speak of weight w, we imply that the acceleration is equal to g, the acceleration due to gravity. Then Newton's law becomes

⇨ **Weight–Mass Relationship**

$$w = mg \qquad\qquad (1–2)$$

In this book, we will use $g = 9.81$ m/s^2 in the SI system and $g = 32.2$ ft/s^2 in the U.S. Customary System. These are the standard values on Earth for g to three significant digits. To a greater degree of precision, we have the standard values $g = 9.806\ 65$ m/s^2 and $g = 32.1740$ ft/s^2. For high-precision work and at high elevations (such as aerospace operations) where the actual value of g is different from the standard, the local value should be used.

1.5.1 Weight and Mass in the SI Unit System

For example, consider a rock with a mass of 5.60 kg suspended by a wire. To determine what force is exerted on the wire, we use Newton's law of gravitation ($w = mg$):

$$w = mg = \text{mass} \times \text{acceleration due to gravity}$$

Under standard conditions, however, $g = 9.81$ m/s^2. Then, we have

$$w = 5.60 \text{ kg} \times 9.81 \text{ m/s}^2 = 54.9 \text{ kg·m/s}^2 = 54.9 \text{ N}$$

Thus, a 5.60 kg rock weighs 54.9 N.

We can also compute the mass of an object if we know its weight. For example, assume that we have measured the weight of a valve to be 8.25 N. What is its mass? We write

$$w = mg$$

$$m = \frac{w}{g} = \frac{8.25 \text{ N}}{9.81 \text{ m/s}^2} = \frac{0.841 \text{ N·s}^2}{\text{m}} = 0.841 \text{ kg}$$

1.5.2 Weight and Mass in the U.S. Customary Unit System

For an example of the weight–mass relationship in the U.S. Customary System, assume that we have measured the weight of a container of oil to be 84.6 lb. What is its mass? We write

$$w = mg$$

$$m = w/g = 84.6 \text{ lb}/32.2 \text{ ft/s}^2 = 2.63 \text{ lb-s}^2/\text{ft} = 2.63 \text{ slugs}$$

1.5.3 Mass Expressed as lbm (Pounds-Mass)

In the analysis of fluid systems, some professionals use the unit lbm (pounds-mass) for the unit of mass instead of the unit of slugs. In this system, an object or a quantity of fluid having a weight of 1.0 lb has a mass of 1.0 lbm. The pound-force is then sometimes designated lbf. It must be noted that the numerical equivalence of lbf and lbm applies *only* when the value of g is equal to the standard value.

This system is avoided in this book because it is not a coherent system. When one tries to relate force and mass units using Newton's law, one obtains

$$F = ma = \text{lbm}(\text{ft/s}^2) = \text{lbm-ft/s}^2$$

This is *not* the same as the lbf.

To overcome this difficulty, a conversion constant, commonly called g_c, is defined having both a numerical value and units. That is,

$$g_c = \frac{32.2 \text{ lbm}}{\text{lbf}/(\text{ft/s}^2)} = \frac{32.2 \text{ lbm-ft/s}^2}{\text{lbf}}$$

Then, to convert from lbm to lbf, we use a modified form of Newton's law:

$$F = m(a/g_c)$$

Letting the acceleration $a = g$, we find

$$F = m(g/g_c)$$

For example, to determine the weight of material in lbf that has a mass of 100 lbm, and assuming that the local value of g is equal to the standard value of 32.2 ft/s^2, we have

$$w = F = m\frac{g}{g_c} = 100 \text{ lbm} \frac{32.2 \text{ ft/s}^2}{\dfrac{32.2 \text{ lbm-ft/s}^2}{\text{lbf}}} = 100 \text{ lbf}$$

This shows that weight in lbf is numerically equal to mass in lbm *provided* $g = 32.2$ ft/s^2.

If the analysis were to be done for an object or fluid on the Moon, however, where g is approximately ⅙ of that on Earth, 5.4 ft/s^2, we would find

$$w = F = m\frac{g}{g_c} = 100 \text{ lbm} \frac{5.4 \text{ ft/s}^2}{\dfrac{32.2 \text{ lbm-ft/s}^2}{\text{lbf}}} = 16.8 \text{ lbf}$$

This is a dramatic difference.

In summary, because of the cumbersome nature of the relationship between lbm and lbf, we avoid the use of lbm in this book. Mass will be expressed in the unit of slugs when problems are in the U.S. Customary System of units.

1.6 TEMPERATURE

Temperature is most often indicated in °C (degrees Celsius) or °F (degrees Fahrenheit). You are probably familiar with the following values at sea level on Earth:

Water freezes at 0°C and boils at 100°C.

Water freezes at 32°F and boils at 212°F.

Thus, there are 100 Celsius degrees and 180 Fahrenheit degrees between the same two physical data points, and 1.0 Celsius degree equals 1.8 Fahrenheit degrees exactly. From these observations we can define the conversion procedures between these two systems as follows:

Given the temperature T_F in °F, the temperature T_C in °C is

$$T_C = (T_F - 32)/1.8$$

Given the temperature T_C in °C, the temperature T_F in °F is

$$T_F = 1.8T_C + 32$$

For example, given $T_F = 180°F$, we have

$$T_C = (T_F - 32)/1.8 = (180 - 32)/1.8 = 82.2°C$$

Given $T_C = 33°C$, we have

$$T_F = 1.8T_C + 32 = 1.8(33) + 32 = 91.4°F$$

In this book we will use the Celsius scale when problems are in SI units and the Fahrenheit scale when they are in U.S. Customary units.

1.6.1 Absolute Temperature

The Celsius and Fahrenheit temperature scales were defined according to arbitrary reference points, although the Celsius scale has convenient points of reference to the properties of water. The absolute temperature, on the other hand, is defined so the zero point corresponds to the condition where all molecular motion stops. This is called *absolute zero*.

In the SI unit system, the standard unit of temperature is the kelvin, for which the standard symbol is K and the reference (zero) point is absolute zero. Note that there is no degree symbol attached to the symbol K. The interval between points on the kelvin scale is the same as the interval used for the Celsius scale. Measurements have shown that the freezing point of water is 273.15 K above absolute zero. We can then make the conversion from the Celsius to the kelvin scale by using

$$T_K = T_C + 273.15$$

For example, given $T_C = 33°C$, we have

$$T_K = T_C + 273.15 = 33 + 273.15 = 306.15 \text{ K}$$

It has also been shown that absolute zero on the Fahrenheit scale is at −459.67°F. In some references you will find

another absolute temperature scale called the Rankine scale, where the interval is the same as for the Fahrenheit scale. Absolute zero is 0°R and any Fahrenheit measurement can be converted to °R by using

$$T_R = T_F + 459.67$$

Also, given the temperature in °F, we can compute the absolute temperature in K from

$$T_K = (T_F + 459.67)/1.8 = T_R/1.8$$

For example, given $T_F = 180°F$, the absolute temperature in K is

$$T_K = (T_F + 459.67)/1.8 = (180 + 459.67)/1.8$$
$$= (639.67°R)/1.8 = 355.37 \text{ K}$$

1.7 CONSISTENT UNITS IN AN EQUATION

The analyses required in fluid mechanics involve the algebraic manipulation of several terms. The equations are often complex, and it is extremely important that the results be dimensionally correct. That is, they must have their proper units. Indeed, answers will have the wrong numerical value if the units in the equation are not consistent. Table 1.2 summarizes standard and other common units for the quantities used in fluid mechanics.

A simple straightforward procedure called *unit cancellation* will ensure proper units in any kind of calculation, not only in fluid mechanics, but also in virtually all your technical work. The six steps of the procedure are listed below.

Unit-Cancellation Procedure

1. Solve the equation algebraically for the desired term.
2. Decide on the proper units for the result.
3. Substitute known values, including units.
4. Cancel units that appear in both the numerator and the denominator of any term.
5. Use conversion factors to eliminate unwanted units and obtain the proper units as decided in Step 2.
6. Perform the calculation.

This procedure, properly executed, will work for any equation. It is really very simple, but some practice may be required to use it. We are going to borrow some material from elementary physics, with which you should be familiar, to illustrate the method. However, the best way to learn how to do something is to do it. The following example problems are presented in a form called *programmed instruction*. You will be guided through the problems in a step-by-step fashion with your participation required at each step.

To proceed with the program you should cover all material under the heading Programmed Example Problem, using an opaque sheet of paper or a card. You should have a blank piece of paper handy on which to perform the requested operations. Then successively uncover one panel at a time down to the heavy line that runs across the page. The first panel presents a problem and asks you to perform

TABLE 1.2 Units for common quantities used in fluid mechanics in SI units and U.S. Customary units

Quantity	Basic Definition	Standard SI Units	Other Metric Units Often Used	Standard U.S. Units	Other U.S. Units Often Used
Length (L)	—	meter (m)	millimeter (mm); kilometer (km)	foot (ft)	inch (in); mile (mi)
Time	—	second (s)	hour (h); minute (min)	second (s)	hour (h); minute (min)
Mass (m)	Quantity of a substance	kilogram (kg)	$N \cdot s^2/m$	slug	$lb \cdot s^2/ft$
Force (F) or weight (w)	Push or pull on an object	newton (N)	$kg \cdot m/s^2$	pound (lb)	kip (1000 lb)
Pressure (p)	Force/area	N/m^2 or pascal (Pa)	kilopascals (kPa); bar	lb/ft^2 or psf	lb/in^2 or psi; kip/in^2 or ksi
Energy	Force times distance	$N \cdot m$ or Joule (J)	$kg \cdot m^2/s^2$	$lb \cdot ft$	$lb \cdot in$
Power (P)	Energy/time	watt (W) or $N \cdot m/s$ or J/s	kilowatt (kW)	lb·ft/s	horsepower (hp)
Volume (V)	L^3	m^3	liter (L)	ft^3	gallon (gal)
Area (A)	L^2	m^2	mm^2	ft^2	in^2
Volume flow rate (Q)	V/time	m^3/s	L/s; L/min; m^3/h	ft^3/s or cfs	gal/min (gpm); ft^3/min (cfm)
Weight flow rate (W)	w/time	N/s	kN/s; kN/min	lb/s	lb/min; lb/h
Mass flow rate (M)	M/time	kg/s	kg/hr	slugs/s	slugs/min; slugs/h
Specific weight (γ)	w/V	N/m^3 or $kg/m^2 \cdot s^2$		lb/ft^3	
Density (ρ)	M/V	kg/m^3 or $N \cdot s^2/m^4$		$slugs/ft^3$	

some operation or to answer a question. After doing what is asked, uncover the next panel, which will contain information that you can use to check your result. Then continue with the next panel, and so on through the program.

Remember, the purpose of this is to help you learn how to get correct answers using the unit-cancellation method. You may want to refer to the table of conversion factors in Appendix K.

PROGRAMMED EXAMPLE PROBLEM

Example Problem 1.1 Imagine you are traveling in a car at a constant speed of 80 kilometers per hour (km/h). How many seconds (s) would it take to travel 1.5 km?

For the solution, use the equation

$$s = vt$$

where s is the distance traveled, v is the speed, and t is the time. Using the unit-cancellation procedure outlined above, what is the first thing to do?

The first step is to solve for the desired term. Because you were asked to find time, you should have written

$$t = \frac{s}{v}$$

Now perform Step 2 of the procedure described above.

Step 2 is to decide on the proper units for the result, in this case time. From the problem statement the proper unit is seconds. If no specification had been given for units, you could choose any acceptable time unit such as hours.

Proceed to Step 3.

The result should look something like this:

$$t = \frac{s}{v} = \frac{1.5\,\text{km}}{80\,\text{km/h}}$$

For the purpose of cancellation it is not convenient to have the units in the form of a compound fraction as we have above. To clear this to a simple fraction, write it in the form

$$t = \frac{\dfrac{1.5\,\text{km}}{1}}{\dfrac{80\,\text{km}}{\text{h}}}$$

This can be reduced to

$$t = \frac{1.5\,\text{km·h}}{80\,\text{km}}$$

After some practice, equations may be written in this form directly. Now perform Step 4 of the procedure.

The result should now look like this:

$$t = \frac{1.5\,\cancel{\text{km}}\text{·h}}{80\,\cancel{\text{km}}}$$

This illustrates that units can be cancelled just as numbers can if they appear in both the numerator and the denominator of a term in an equation.

Now do Step 5.

The answer looks like this:

$$t = \frac{1.5\,\cancel{\text{km}}\text{·}\cancel{\text{h}}}{80\,\cancel{\text{km}}} \times \frac{3600\,\text{s}}{1\,\cancel{\text{h}}}$$

The equation in the preceding panel showed the result for time in hours after kilometer units were cancelled. Although hours is an acceptable time unit, our desired unit is seconds as determined in Step 2. Thus, the conversion factor 3600 s/1 h is required.

How did we know we have to multiply by 3600 instead of dividing?

The units determine this. Our objective in using the conversion factor was to eliminate the hour unit and obtain the second unit. Because the unwanted hour unit was in the numerator of the original equation, the hour unit in the conversion factor must be in the denominator in order to cancel.

Now that we have the time unit of seconds we can proceed with Step 6.

The correct answer is $t = 67.5\,\text{s}$.

1.8 THE DEFINITION OF PRESSURE

Pressure is defined as the amount of force exerted on a unit area of a substance. This can be stated by the equation

⇨ **Pressure**

$$p = \frac{F}{A} \qquad (1\text{–}3)$$

Two important principles about pressure were described by Blaise Pascal, a seventeenth-century scientist:

- Pressure acts uniformly in all directions on a small volume of a fluid.

- In a fluid confined by solid boundaries, pressure acts perpendicular to the boundary.

These principles, sometimes called *Pascal's laws*, are illustrated in Figs. 1.3 and 1.4.

Using Eq. (1–3) and the second of Pascal's laws, we can compute the magnitude of the pressure in a fluid if we know the amount of force exerted on a given area.

FIGURE 1.3 Pressure acting uniformly in all directions on a small volume of fluid.

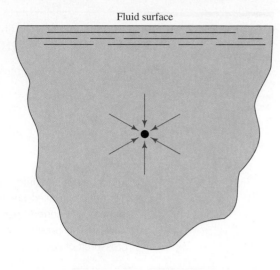

Fluid surface

FIGURE 1.4 Direction of fluid pressure on boundaries.

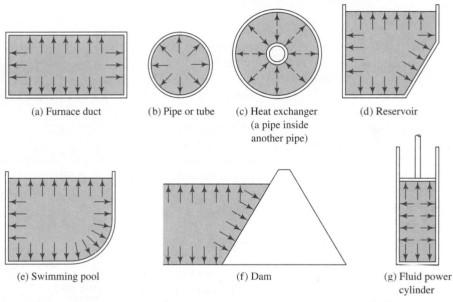

(a) Furnace duct

(b) Pipe or tube

(c) Heat exchanger (a pipe inside another pipe)

(d) Reservoir

(e) Swimming pool

(f) Dam

(g) Fluid power cylinder

Example Problem 1.2

Figure 1.5 shows a container of liquid with a movable piston supporting a load. Compute the magnitude of the pressure in the liquid under the piston if the total weight of the piston and the load is 500 N and the area of the piston is 2500 mm².

Solution

It is reasonable to assume that the entire surface of the fluid under the piston is sharing in the task of supporting the load. The second of Pascal's laws states that the fluid pressure acts perpendicular to the piston. Then, using Eq. (1–3), we have

$$p = \frac{F}{A} = \frac{500 \text{ N}}{2500 \text{ mm}^2} = 0.20 \text{ N/mm}^2$$

The standard unit of pressure in the SI system is the N/m², called the *pascal* (Pa) in honor of Blaise Pascal. The conversion can be made by using the factor 10^3 mm = 1 m. We have

$$p = \frac{0.20 \text{ N}}{\text{mm}^2} \times \frac{(10^3 \text{ mm})^2}{\text{m}^2} = 0.20 \times 10^6 \text{ N/m}^2 = 0.20 \text{ MPa}$$

Note that the pressure in N/mm² is numerically equal to pressure in MPa. It is not unusual to encounter pressure in the range of several megapascals (MPa) or several hundred kilopascals (kPa).

Pressure in the U.S. Customary System is illustrated in the following example problem.

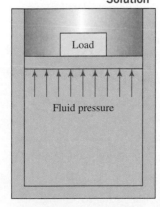

Load

Fluid pressure

FIGURE 1.5 Illustration of fluid pressure supporting a load.

Example Problem
1.3

A load of 200 pounds (lb) is exerted on a piston confining oil in a circular cylinder with an inside diameter of 2.50 inches (in). Compute the pressure in the oil at the piston. See Fig. 1.4.

Solution

To use Eq. (1–3), we must compute the area of the piston:

$$A = \pi D^2/4 = \pi(2.50 \text{ in})^2/4 = 4.91 \text{ in}^2$$

Then,

$$p = \frac{F}{A} = \frac{200 \text{ lb}}{4.91 \text{ in}^2} = 40.7 \text{ lb/in}^2$$

Although the standard unit for pressure in the U.S. Customary System is pounds per square foot (lb/ft²), it is not often used because it is inconvenient. Length measurements are more conveniently made in inches, and pounds per square inch (lb/in²), abbreviated psi, is used most often for pressure in this system. The pressure in the oil is 40.7 psi. This is a fairly low pressure; it is not unusual to encounter pressures of several hundred or several thousand psi.

The *bar* is another unit used by some people working in fluid mechanics and thermodynamics. The bar is defined as 10^5 Pa or 10^5 N/m². Another way of expressing the bar is $1 \text{ bar} = 100 \times 10^3$ N/m², which is equivalent to 100 kPa. Because atmospheric pressure near sea level is very nearly this value, the bar has a convenient point of physical reference. This, plus the fact that pressures expressed in bars yield smaller numbers, makes this unit attractive to some practitioners. You must realize, however, that the bar is not a part of the coherent SI system and that you must carefully convert it to N/m² (pascals) in problem solving.

1.9 COMPRESSIBILITY

Compressibility refers to the change in volume (V) of a substance that is subjected to a change in pressure on it. The usual quantity used to measure this phenomenon is the *bulk modulus of elasticity* or, simply, *bulk modulus, E*:

TABLE 1.3 Values for bulk modulus for selected liquids at atmospheric pressure and 68°F (20°C)

Liquid	Bulk Modulus	
	(psi)	(MPa)
Ethyl alcohol	130 000	896
Benzene	154 000	1 062
Machine oil	189 000	1 303
Water	316 000	2 179
Glycerin	654 000	4 509
Mercury	3 590 000	24 750

▷ **Bulk Modulus**

$$E = \frac{-\Delta p}{(\Delta V)/V} \qquad (1\text{–}4)$$

Because the quantities ΔV and V have the same units, the denominator of Eq. (1–4) is dimensionless. Therefore, the units for E are the same as those for the pressure.

As stated before, liquids are very slightly compressible, indicating that it would take a very large change in pressure to produce a small change in volume. Thus, the magnitudes of E for liquids, as shown in Table 1.3, are very high (see Reference 7). For this reason, *liquids will be considered incompressible in this book, unless stated otherwise.*

The term *bulk modulus* is not usually applied to gases, and the principles of thermodynamics must be applied to determine the change in volume of a gas with a change in pressure.

Example Problem
1.4

Compute the change in pressure that must be applied to water to change its volume by 1.0 percent.

Solution

The 1.0-percent volume change indicates that $\Delta V/V = -0.01$. Then, the required change in pressure is

$$\Delta p = -E[\Delta V/V] = [-316\,000 \text{ psi}][-0.01] = 3160 \text{ psi}$$

1.10 DENSITY, SPECIFIC WEIGHT, AND SPECIFIC GRAVITY

Because the study of fluid mechanics typically deals with a continuously flowing fluid or with a small amount of fluid at rest, it is most convenient to relate the mass and weight of the fluid to a given volume of the fluid. Thus, the properties of density and specific weight are defined as follows:

Density *is the amount of mass per unit volume of a substance.*

Therefore, using the Greek letter ρ (rho) for density, we write

▷ **Density**

$$\rho = m/V \qquad (1\text{–}5)$$

where V is the volume of the substance having a mass m. The units for density are kilograms per cubic meter (kg/m^3) in the SI system and slugs per cubic foot (slugs/ft^3) in the U.S. Customary System.

ASTM International, formerly the American Society for Testing and Materials, has published several standard test methods for measuring density that describe vessels having precisely known volumes called *pycnometers*. The proper filling, handling, temperature control, and reading of these devices are prescribed. Two types are the *Bingham pycnometer* and the *Lipkin bicapillary pycnometer*. The standards also call for the precise determination of the mass of the fluids in the pycnometers to the nearest 0.1 mg using an analytical balance. See References 3, 5, and 6.

Specific weight *is the amount of weight per unit volume of a substance.*

Using the Greek letter γ (gamma) for specific weight, we write

▷ **Specific Weight**

$$\gamma = w/V \qquad (1\text{–}6)$$

where V is the volume of a substance having the weight w. The units for specific weight are newtons per cubic meter (N/m^3) in the SI system and pounds per cubic foot (lb/ft^3) in the U.S. Customary System.

It is often convenient to indicate the specific weight or density of a fluid in terms of its relationship to the specific weight or density of a common fluid. When the term *specific gravity* is used in this book, the reference fluid is pure water at 4°C. At that temperature water has its greatest density. Then, specific gravity can be defined in either of two ways:

a. *Specific gravity* is the ratio of the density of a substance to the density of water at 4°C.

b. *Specific gravity* is the ratio of the specific weight of a substance to the specific weight of water at 4°C.

These definitions for specific gravity (sg) can be shown mathematically as

▷ **Specific Gravity**

$$sg = \frac{\gamma_s}{\gamma_w @ 4°C} = \frac{\rho_s}{\rho_w @ 4°C} \qquad (1\text{–}7)$$

where the subscript s refers to the substance whose specific gravity is being determined and the subscript w refers to water. The properties of water at 4°C are constant, having the following values:

$$\gamma_w @ 4°C = 9.81 \text{ kN/m}^3 \qquad \gamma_w @ 4°C = 62.4 \text{ lb/ft}^3$$
$$\text{or}$$
$$\rho_w @ 4°C = 1000 \text{ kg/m}^3 \qquad \rho_w @ 4°C = 1.94 \text{ slugs/ft}^3$$

Therefore, the mathematical definition of specific gravity can be written as

$$sg = \frac{\gamma_s}{9.81 \text{ kN/m}^3} = \frac{\rho_s}{1000 \text{ kg/m}^3} \text{ or}$$
$$sg = \frac{\gamma_s}{62.4 \text{ lb/ft}^3} = \frac{\rho_s}{1.94 \text{ slugs/ft}^3} \qquad (1\text{–}8)$$

This definition holds regardless of the temperature at which the specific gravity is being determined.

The properties of fluids do, however, vary with temperature. In general, the density (and therefore the specific weight and the specific gravity) decreases with increasing temperature. The properties of water at various temperatures are listed in Appendix A. The properties of other liquids at a few selected temperatures are listed in Appendices B and C. See Reference 9 for more such data.

You should seek other references, such as References 8 and 10, for data on specific gravity at specified temperatures if they are not reported in the appendix and if high precision is desired. One estimate that gives reasonable accuracy for petroleum oils, as presented more fully in References 8 and 9, is that the specific gravity of oils decreases approximately 0.036 for a 100°F (37.8°C) rise in temperature. This applies for nominal values of specific gravity from 0.80 to 1.00 and for temperatures in the range from approximately 32°F to 400°F (0°C to 204°C).

Some industry sectors prefer modified definitions for specific gravity. Instead of using the properties of water at 4°C (39.2°F) as the basis, the petroleum industry and others use water at 60°F (15.6°C). This makes little difference for typical design and analysis. Although the density of water at 4°C is 1000.00 kg/m^3, at 60°F it is 999.04 kg/m^3. The difference is less than 0.1 percent. References 3, 4, 6–8, and 10 contain more extensive tables of the properties of water at temperatures from 0°C to 100°C (32°F to 212°F).

Specific gravity in the Baumé and API scales is discussed in Section 1.10.2. We will use water at 4°C as the basis for specific gravity in this book.

The ASTM also refers to the property of specific gravity as *relative density*. See References 3–6.

1.10.1 Relation Between Density and Specific Weight

Quite often the specific weight of a substance must be found when its density is known and vice versa. The conversion from one to the other can be made using the following equation:

▷ **γ-ρ Relation**

$$\gamma = \rho g \qquad (1\text{–}9)$$

where g is the acceleration due to gravity. This equation can be justified by referring to the definitions of density and specific gravity and by using the equation relating mass to weight, $w = mg$.

The definition of specific weight is

$$\gamma = \frac{w}{V}$$

By multiplying both the numerator and the denominator of this equation by g we obtain

$$\gamma = \frac{wg}{Vg}$$

But $m = w/g$. Therefore, we have

$$\gamma = \frac{mg}{V}$$

Because $\rho = m/V$, we get

$$\gamma = \rho g$$

The following problems illustrate the definitions of the basic fluid properties presented above and the relationships among the various properties.

Example Problem 1.5

Calculate the weight of a reservoir of oil if it has a mass of 825 kg.

Solution

Because $w = mg$, and using $g = 9.81 \text{ m/s}^2$, we have

$$w = 825 \text{ kg} \times 9.81 \text{ m/s}^2 = 8093 \text{ kg·m/s}^2$$

Substituting the newton for the unit kg·m/s^2, we have

$$w = 8093 \text{ N} = 8.093 \times 10^3 \text{ N} = 8.093 \text{ kN}$$

Example Problem 1.6

If the reservoir from Example Problem 1.5 has a volume of 0.917 m^3, compute the density, the specific weight, and the specific gravity of the oil.

Solution

Density:

$$\rho_o = \frac{m}{V} = \frac{825 \text{ kg}}{0.917 \text{ m}^3} = 900 \text{ kg/m}^3$$

Specific weight:

$$\gamma_o = \frac{w}{V} = \frac{8.093 \text{ kN}}{0.917 \text{ m}^3} = 8.83 \text{ kN/m}^3$$

Specific gravity:

$$sg_o = \frac{\rho_o}{\rho_w @ 4°C} = \frac{900 \text{ kg/m}^3}{1000 \text{ kg/m}^3} = 0.90$$

Example Problem 1.7

Glycerin at 20°C has a specific gravity of 1.263. Compute its density and specific weight.

Solution

Density:

$$\rho_g = (sg)_g(1000 \text{ kg/m}^3) = (1.263)(1000 \text{ kg/m}^3) = 1263 \text{ kg/m}^3$$

Specific weight:

$$\gamma_g = (sg)_g(9.81 \text{ kN/m}^3) = (1.263)(9.81 \text{ kN/m}^3) = 12.39 \text{ kN/m}^3$$

Example Problem	A pint of water weighs 1.041 lb. Find its mass.
1.8	

Solution Because $w = mg$, the mass is

$$m = \frac{w}{g} = \frac{1.041 \text{ lb}}{32.2 \text{ ft/s}^2} = \frac{1.041 \text{ lb-s}^2}{32.2 \text{ ft}}$$

$$= 0.0323 \text{ lb-s}^2/\text{ft} = 0.0323 \text{ slugs}$$

Remember that the units of slugs and lb-s^2/ft are the same.

Example Problem	One gallon of mercury has a mass of 3.51 slugs. Find its weight.
1.9	

Solution Using $g = 32.2$ ft/s^2 in Equation 1–2,

$$w = mg = 3.51 \text{ slugs} \times 32.2 \text{ ft/s}^2 = 113 \text{ slug-ft/s}^2$$

This is correct, but the units may seem confusing because weight is normally expressed in pounds. The units of mass may be rewritten as lb-s^2/ft, and we have

$$w = mg = 3.51 \frac{\text{lb-s}^2}{\text{ft}} \times \frac{32.2 \text{ ft}}{\text{s}^2} = 113 \text{ lb}$$

1.10.2 Specific Gravity in Degrees Baumé or Degrees API

The reference temperature for specific gravity measurements on the Baumé or American Petroleum Institute (API) scale is 60°F rather than 4°C as defined before. To emphasize this difference, the API or Baumé specific gravity is often reported as

$$\text{Specific gravity } \frac{60°}{60°} \text{F}$$

This notation indicates that both the reference fluid (water) and the oil are at 60°F.

Specific gravities of crude oils vary widely depending on where they are found. Those from the western U.S. range from approximately 0.87 to 0.92. Eastern U.S. oil fields produce oil of about 0.82 specific gravity. Mexican crude oil is among the highest at 0.97. A few heavy asphaltic oils have sg > 1.0. (See Reference 7.)

Most oils are distilled before they are used to enhance their quality of burning. The resulting gasolines, kerosenes, and fuel oils have specific gravities ranging from about 0.67 to 0.98.

The equation used to compute specific gravity when the degrees Baumé are known is different for fluids lighter than water and fluids heavier than water. For liquids heavier than water,

$$sg = \frac{145}{145 - \text{deg Baumé}} \tag{1–10}$$

Thus, to compute the degrees Baumé for a given specific gravity, use

$$\text{deg Baumé} = 145 - \frac{145}{sg} \tag{1–11}$$

For liquids lighter than water,

$$sg = \frac{140}{130 + \text{deg Baumé}} \tag{1–12}$$

$$\text{deg Baumé} = \frac{140}{sg} - 130 \tag{1–13}$$

The API has developed a scale that is slightly different from the Baumé scale for liquids lighter than water. The formulas are

$$sg = \frac{141.5}{131.5 + \text{deg API}} \tag{1–14}$$

$$\text{deg API} = \frac{141.5}{sg} - 131.5 \tag{1–15}$$

Degrees API for oils may range from 10 to 80. Most fuel grades will fall in the range of API 20 to 70, corresponding to specific gravities from 0.93 to 0.70. Note that the heavier oils have the lower values of degrees API. Reference 9 contains useful tables listing specific gravity as a function of degrees API.

ASTM Standards D 287 and D 6822 (References 2 and 4, respectively) describe standard test methods for determining API gravity using a *hydrometer*. Figure 1.6 is a sketch of a typical hydrometer incorporating a weighted glass bulb with a smaller-diameter stem at the top that is designed to float upright in the test liquid. Based on the principles of buoyancy (see Chapter 5), the hydrometer rests at a position that is dependent on the density of the liquid. The stem is marked with a calibrated scale from which the direct reading of density, specific gravity, or API gravity can be made. Because of the importance of temperature to an accurate measurement of density, some hydrometers, called *thermohydrometers*, have a built-in precision thermometer.

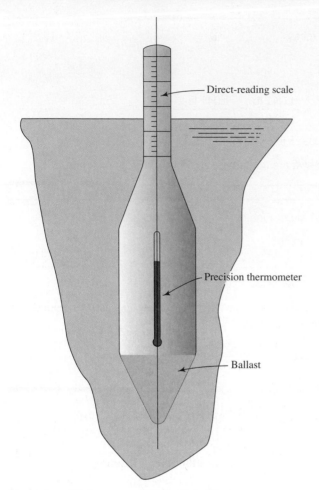

Direct-reading scale

Precision thermometer

Ballast

FIGURE 1.6   Hydrometer with built-in thermometer (thermohydrometer).

$$\text{Surface tension} = \frac{\text{work}}{\text{area}} = \frac{\text{N·m}}{\text{m}^2} = \text{N/m}$$

$$\text{Or: Surface tension} = \frac{\text{work}}{\text{area}} = \frac{\text{ft·lb}}{\text{ft}^2} = \text{lb/ft}$$

Surface tension is also the reason that water droplets assume a nearly spherical shape. In addition, the phenomenon of capillarity depends on the surface tension. The surface of a liquid in a small-diameter tube will assume a curved shape that depends on the surface tension of the liquid. Mercury will form virtually an extended bulbous shape. The surface of water, however, will settle into a depressed cavity with the liquid seeming to climb the walls of the tube by a small amount. Adhesion of the liquid to the walls of the tube contributes to this behavior.

The movement of liquids within small spaces depends on this capillary action. *Wicking* is the term often used to describe the rise of a fluid from a liquid surface into a woven material. The movement of liquids within soils is also affected by surface tension and the corresponding capillary action.

Table 1.4 gives the surface tension of water at atmospheric pressure at various temperatures. The SI units used here are mN/m, where 1000 mN = 1.0 N. Similarly, U.S. Customary units are mlb/ft, where 1000 mlb = 1.0 lb force. Table 1.5 gives values for a variety of common liquids also at atmospheric pressure at selected temperatures.

1.11 SURFACE TENSION

You can experiment with the surface tension of water by trying to cause an object to be supported on the surface when you would otherwise predict it would sink. For example, it is fairly easy to place a small needle on a still water surface so that it is supported by the surface tension of the water. Note that it is not significantly supported by buoyancy. If the needle is submerged, it will readily sink to the bottom.

Then, if you place a very small amount of dishwashing detergent in the water when the needle is supported, it will almost immediately sink. The detergent lowers the surface tension dramatically.

Surface tension acts somewhat like a film at the interface between the liquid water surface and the air above it. The water molecules beneath the surface are attracted to each other and to those at the surface. Quantitatively, surface tension is measured as the work per unit area required to move lower molecules to the surface of the liquid. The resulting units are force per unit length, such as N/m or lb/ft. These units can be found as follows:

TABLE 1.4 Surface tension of water			
Temperature (°F)	Surface Tension (mlb/ft)	Temperature (°C)	Surface Tension (mN/m)
32	5.18	0	75.6
40	5.13	5	74.9
50	5.09	10	74.2
60	5.03	20	72.8
70	4.97	30	71.2
80	4.91	40	69.6
90	4.86	50	67.9
100	4.79	60	66.2
120	4.67	70	64.5
140	4.53	80	62.7
160	4.40	90	60.8
180	4.26	100	58.9
200	4.12		
212	4.04		

Source: Adapted with permission from data from *CRC Handbook of Chemistry and Physics*, CRC Press LLC, Boca Raton, FL. (Reference 10)

Notes: Values taken at atmospheric pressure
1.0 lb = 1000 mlb; 1.0 N = 1000 mN

TABLE 1.5 Surface tension of some common liquids

Surface Tension at Stated Temperature

Liquid	10°C (mN/m)	50°F (mlb/ft)	25°C (mN/m)	77°F (mlb/ft)	50°C (mN/m)	122°F (mlb/ft)	75°C (mN/m)	167°F (mlb/ft)	100°C (mN/m)	212°F lb/ft)
Water	74.2	5.08	72.0	4.93	67.9	4.65	63.6	4.36	58.9	4.04
Methanol	23.2	1.59	22.1	1.51	20.1	1.38				
Ethanol	23.2	1.59	22.0	1.51	19.9	1.36				
Ethylene glycol			48.0	3.29	45.8	3.14	43.5	2.98	41.3	2.83
Acetone	24.57	1.68	22.72	1.56	19.65	1.35				
Benzene			28.2	1.93	25.0	1.71	21.8	1.49		
Mercury	488	33.4	485	33.2	480	32.9	475	32.5	470	32.2

Source: Adapted with permission from data from *CRC Handbook of Chemistry and Physics*, CRC Press LLC, Boca Raton, FL. (Reference 10)
Notes: Values taken at atmospheric pressure 1.0 lb = 1000 mlb; 1.0 N = 1000 mN

REFERENCES

1. Taylor, Barry N., and Ambler Thompson, eds. 2008. *The International System of Units (SI)* (NIST Special Publication 330). Washington, DC: National Institute of Standards and Technology, U.S. Department of Commerce.

2. ASTM International. 2006. *Standard D 287-92(2006): Standard Test Method for API Gravity of Crude Petroleum and Petroleum Products (Hydrometer Method)*. West Conshohocken, PA: Author.

3. _____. 2007. *Standard D 1217-93(2007): Standard Test Method for Density and Relative Density (Specific Gravity) of Liquids by Bingham Pycnometer*. West Conshohocken, PA: Author.

4. _____. 2008. *Standard D 6822-02(2008): Standard Test Method for Density, Relative Density (Specific Gravity), or API Gravity of Crude Petroleum and Liquid Petroleum Products by Thermohydrometer Method*. West Conshohocken, PA: Author.

5. _____. 2007. *Standard D 1480-07: Standard Test Method for Density and Relative Density (Specific Gravity) of Viscous Materials by Bingham Pycnometer*. West Conshohocken, PA: Author.

6. _____. 2007. *Standard D 1481-02(2007): Standard Test Method for Density and Relative Density (Specific Gravity) of Viscous Materials by Lipkin Bicapillary Pycnometer*. West Conshohocken, PA: Author.

7. Avallone, Eugene A., Theodore Baumeister, and Ali Sadegh, eds. 2007. *Marks' Standard Handbook for Mechanical Engineers*, 11th ed. New York: McGraw-Hill.

8. Bolz, Ray E., and George L. Tuve, eds. 1973. *CRC Handbook of Tables for Applied Engineering Science*, 2nd ed. Boca Raton, FL: CRC Press.

9. Heald, C. C., ed. 2002. *Cameron Hydraulic Data*, 19th ed. Irving, TX: Flowserve. [Previous editions were published by Ingersoll-Dresser Pump Co., Liberty Corner, NJ.]

10. Haynes, William H., ed. 2011. *CRC Handbook of Chemistry and Physics*, 92nd ed. Boca Raton, FL: CRC Press.

INTERNET RESOURCES

1. **Hydraulic Institute (HI):** HI is a nonprofit association serving the pump industry. It provides product standards in North America and worldwide.

2. **ASTM International:** ASTM establishes standards in a variety of fields, including fluid mechanics. Many ASTM standards are cited in this book for testing methods and fluid properties.

3. **Flow Control Network:** The website for *Flow Control Magazine* is a source of information on fluid flow technology, applications of fluid mechanics, and products that measure, control, and contain liquids, gases, and powders. It also includes links to standards organizations important to the fluids industry.

4. **GlobalSpec:** A searchable database of suppliers of a wide variety of technical products, including pumps, flow control, and flow measurement.

PRACTICE PROBLEMS

Conversion Factors

1.1 Convert 1250 millimeters to meters.
1.2 Convert 1600 square millimeters to square meters.
1.3 Convert 3.65×10^3 cubic millimeters to cubic meters.
1.4 Convert 2.05 square meters to square millimeters.
1.5 Convert 0.391 cubic meters to cubic millimeters.
1.6 Convert 55.0 gallons to cubic meters.
1.7 An automobile is moving at 80 kilometers per hour. Calculate its speed in meters per second.
1.8 Convert a length of 25.3 feet to meters.
1.9 Convert a distance of 1.86 miles to meters.
1.10 Convert a length of 8.65 inches to millimeters.
1.11 Convert a distance of 2580 feet to meters.
1.12 Convert a volume of 480 cubic feet to cubic meters.
1.13 Convert a volume of 7390 cubic centimeters to cubic meters.
1.14 Convert a volume of 6.35 liters to cubic meters.

1.15 Convert 6.0 feet per second to meters per second.

1.16 Convert 2500 cubic feet per minute to cubic meters per second.

Note: In all Practice Problems sections in this book, the problems will use both SI and U.S. Customary System units. In most problems, units are consistent and in the same system. It is expected that solutions are completed in the given unit system.

Consistent Units in an Equation

A body moving with constant velocity obeys the relationship $s = vt$, where s = distance, v = velocity, and t = time.

1.17 A car travels 0.50 km in 10.6 s. Calculate its average speed in m/s.

1.18 In an attempt at a land speed record, a car travels 1.50 km in 5.2 s. Calculate its average speed in km/h.

1.19 A car travels 1000 ft in 14 s. Calculate its average speed in mi/h.

1.20 In an attempt at a land speed record, a car travels 1 mi in 5.7 s. Calculate its average speed in mi/h.

A body starting from rest with constant acceleration moves according to the relationship $s = \frac{1}{2}at^2$, where s = distance, a = acceleration, and t = time.

1.21 If a body moves 3.2 km in 4.7 min with constant acceleration, calculate the acceleration in m/s^2.

1.22 An object is dropped from a height of 13 m. Neglecting air resistance, how long would it take for the body to strike the ground? Use $a = g = 9.81$ m/s^2.

1.23 If a body moves 3.2 km in 4.7 min with constant acceleration, calculate the acceleration in ft/s^2.

1.24 An object is dropped from a height of 53 in. Neglecting air resistance, how long would it take for the body to strike the ground? Use $a = g = 32.2$ ft/s^2.

The formula for kinetic energy is $KE = \frac{1}{2}mv^2$, where m = mass and v = velocity.

1.25 Calculate the kinetic energy in N·m of a 15-kg mass if it has a velocity of 1.20 m/s.

1.26 Calculate the kinetic energy in N·m of a 3600-kg truck moving at 16 km/h.

1.27 Calculate the kinetic energy in N·m of a 75-kg box moving on a conveyor at 6.85 m/s.

1.28 Calculate the mass of a body in kg if it has a kinetic energy of 38.6 N·m when moving at 31.5 km/h.

1.29 Calculate the mass of a body in g if it has a kinetic energy of 94.6 mN·m when moving at 2.25 m/s.

1.30 Calculate the velocity in m/s of a 12-kg object if it has a kinetic energy of 15 N·m.

1.31 Calculate the velocity in m/s of a 175-g body if it has a kinetic energy of 212 mN·m.

1.32 Calculate the kinetic energy in ft-lb of a 1-slug mass if it has a velocity of 4 ft/s.

1.33 Calculate the kinetic energy in ft-lb of an 8000-lb truck moving at 10 mi/h.

1.34 Calculate the kinetic energy in ft-lb of a 150-lb box moving on a conveyor at 20 ft/s.

1.35 Calculate the mass of a body in slugs if it has a kinetic energy of 15 ft-lb when moving at 2.2 ft/s.

1.36 Calculate the weight of a body in lb if it has a kinetic energy of 38.6 ft-lb when moving at 19.5 mi/h.

1.37 Calculate the velocity in ft/s of a 30-lb object if it has a kinetic energy of 10 ft-lb.

1.38 Calculate the velocity in ft/s of a 6-oz body if it has a kinetic energy of 30 in-oz.

One measure of a baseball pitcher's performance is earned run average or ERA. It is the average number of earned runs allowed if all the innings pitched were converted to equivalent nine-inning games. Therefore, the units for ERA are runs per game.

1.39 If a pitcher has allowed 39 runs during 141 innings, calculate the ERA.

1.40 A pitcher has an ERA of 3.12 runs/game and has pitched 150 innings. How many earned runs has the pitcher allowed?

1.41 A pitcher has an ERA of 2.79 runs/game and has allowed 40 earned runs. How many innings have been pitched?

1.42 A pitcher has allowed 49 earned runs during 123 innings. Calculate the ERA.

The Definition of Pressure

1.43 Compute the pressure produced in the oil in a closed cylinder by a piston exerting a force of 2500 lb on the enclosed oil. The piston has a diameter of 3.00 in.

1.44 A hydraulic cylinder must be able to exert a force of 8700 lb. The piston diameter is 1.50 in. Compute the required pressure in the oil.

1.45 Compute the pressure produced in the oil in a closed cylinder by a piston exerting a force of 12.0 kN on the enclosed oil. The piston has a diameter of 75 mm.

1.46 A hydraulic cylinder must be able to exert a force of 38.8 kN. The piston diameter is 40 mm. Compute the required pressure in the oil.

1.47 The hydraulic lift for an automobile service garage has a cylinder with a diameter of 8.0 in. What pressure must the oil have to be able to lift 6000 lb?

1.48 A coining press is used to produce commemorative coins with the likenesses of all the U.S. presidents. The coining process requires a force of 18 000 lb. The hydraulic cylinder has a diameter of 2.50 in. Compute the required oil pressure.

1.49 The maximum pressure that can be developed for a certain fluid power cylinder is 20.5 MPa. Compute the force it can exert if its piston diameter is 50 mm.

1.50 The maximum pressure that can be developed for a certain fluid power cylinder is 6000 psi. Compute the force it can exert if its piston diameter is 2.00 in.

1.51 The maximum pressure that can be developed for a certain fluid power cylinder is 5000 psi. Compute the required diameter for the piston if the cylinder must exert a force of 20 000 lb.

1.52 The maximum pressure that can be developed for a certain fluid power cylinder is 15.0 MPa. Compute the required diameter for the piston if the cylinder must exert a force of 30 kN.

1.53 A line of fluid power cylinders has a range of diameters in 1.00-in increments from 1.00 to 8.00 in. Compute the force that could be exerted by each cylinder with a fluid pressure of 500 psi. Draw a graph of the force versus cylinder diameter.

1.54 A line of fluid power cylinders has a range of diameters in 1.00-in increments from 1.00 to 8.00 in. Compute the pressure required by each cylinder if it must exert a force of 5000 lb. Draw a graph of the pressure versus cylinder diameter.

1.55 Determine your weight in newtons. Then, compute the pressure in pascals that would be created on the oil in a 20-mm-diameter cylinder if you stood on a piston in the cylinder. Convert the resulting pressure to psi.

1.56 For the pressure you computed in Problem 1.55, compute the force in newtons that could be exerted on a piston with 250-mm diameter. Then, convert the resulting force to pounds.

Bulk Modulus

1.57 Compute the pressure change required to cause a decrease in the volume of ethyl alcohol by 1.00 percent. Express the result in both psi and MPa.

1.58 Compute the pressure change required to cause a decrease in the volume of mercury by 1.00 percent. Express the result in both psi and MPa.

1.59 Compute the pressure change required to cause a decrease in the volume of machine oil by 1.00 percent. Express the result in both psi and MPa.

1.60 For the conditions described in Problem 1.59, assume that the 1.00-percent volume change occurred in a cylinder with an inside diameter of 1.00 in and a length of 12.00 in. Compute the axial distance the piston would travel as the volume change occurs.

1.61 A certain hydraulic system operates at 3000 psi. Compute the percentage change in the volume of the oil in the system as the pressure is increased from zero to 3000 psi if the oil is similar to the machine oil listed in Table 1.4.

1.62 A certain hydraulic system operates at 20.0 MPa. Compute the percentage change in the volume of the oil in the system if the oil is similar to the machine oil listed in Table 1.4.

1.63 A measure of the stiffness of a linear actuator system is the amount of force required to cause a certain linear deflection. For an actuator that has an inside diameter of 0.50 in and a length of 42.0 in and that is filled with machine oil, compute the stiffness in lb/in.

1.64 Repeat Problem 1.63 but change the length of the cylinder to 10.0 in. Compare the results.

1.65 Repeat Problem 1.63 but change the cylinder diameter to 2.00 in. Compare the results.

1.66 Using the results of Problems 1.63–1.65, generate a statement about the general design approach to achieving a very stiff system.

Force and Mass

1.67 Calculate the mass of a can of oil if it weighs 610 N.

1.68 Calculate the mass of a tank of gasoline if it weighs 1.35 kN.

1.69 Calculate the weight of 1 m^3 of kerosene if it has a mass of 825 kg.

1.70 Calculate the weight of a jar of castor oil if it has a mass of 450 g.

1.71 Calculate the mass of 1 gal of oil if it weighs 7.8 lb.

1.72 Calculate the mass of 1 ft^3 of gasoline if it weighs 42.0 lb.

1.73 Calculate the weight of 1 ft^3 of kerosene if it has a mass of 1.58 slugs.

1.74 Calculate the weight of 1 gal of water if it has a mass of 0.258 slug.

1.75 Assume that a man weighs 160 lb (force).
  a. Compute his mass in slugs.
  b. Compute his weight in N.
  c. Compute his mass in kg.

1.76 In the United States, hamburger and other meats are sold by the pound. Assuming that this is 1.00-lb force, compute the mass in slugs, the mass in kg, and the weight in N.

1.77 The metric ton is 1000 kg (mass). Compute the force in newtons required to lift it.

1.78 Convert the force found in Problem 1.77 to lb.

1.79 Determine your weight in lb and N and your mass in slugs and kg.

Density, Specific Weight, and Specific Gravity

1.80 The specific gravity of benzene is 0.876. Calculate its specific weight and its density in SI units.

1.81 Air at 16°C and standard atmospheric pressure has a specific weight of 12.02 N/m^3. Calculate its density.

1.82 Carbon dioxide has a density of 1.964 kg/m^3 at 0°C. Calculate its specific weight.

1.83 A certain medium lubricating oil has a specific weight of 8.860 kN/m^3 at 5°C and 8.483 kN/m^3 at 50°C. Calculate its specific gravity at each temperature.

1.84 At 100°C mercury has a specific weight of 130.4 kN/m^3. What volume of the mercury would weigh 2.25 kN?

1.85 A cylindrical can 150 mm in diameter is filled to a depth of 100 mm with a fuel oil. The oil has a mass of 1.56 kg. Calculate its density, specific weight, and specific gravity.

1.86 Glycerin has a specific gravity of 1.258. How much would 0.50 m^3 of glycerin weigh? What would be its mass?

1.87 The fuel tank of an automobile holds 0.095 m^3. If it is full of gasoline having a specific gravity of 0.68, calculate the weight of the gasoline.

1.88 The density of muriatic acid is 1200 kg/m^3. Calculate its specific weight and its specific gravity.

1.89 Liquid ammonia has a specific gravity of 0.826. Calculate the volume of ammonia that would weigh 22.0 N.

1.90 Vinegar has a density of 1080 kg/m^3. Calculate its specific weight and its specific gravity.

1.91 Methyl alcohol has a specific gravity of 0.789. Calculate its density and its specific weight.

1.92 A cylindrical container is 150 mm in diameter and weighs 2.25 N when empty. When filled to a depth of 200 mm with a certain oil, it weighs 35.4 N. Calculate the specific gravity of the oil.

1.93 A storage vessel for gasoline (sg = 0.68) is a vertical cylinder 10 m in diameter. If it is filled to a depth of 6.75 m, calculate the weight and mass of the gasoline.

1.94 What volume of mercury (sg = 13.54) would weigh the same as 0.020 m^3 of castor oil, which has a specific weight of 9.42 kN/m^3?

1.95 A rock has a specific gravity of 2.32 and a volume of 1.42×10^{-4} m^3. How much does it weigh?

1.96 The specific gravity of benzene is 0.876. Calculate its specific weight and its density in U.S. Customary System units.

1.97 Air at 59°F and standard atmospheric pressure has a specific weight of 0.0765 lb/ft^3. Calculate its density.

1.98 Carbon dioxide has a density of 0.003 81 slug/ft^3 at 32°F. Calculate its specific weight.

1.99 A certain medium lubricating oil has a specific weight of 56.4 lb/ft^3 at 40°F and 54.0 lb/ft^3 at 120°F. Calculate its specific gravity at each temperature.

1.100 At 212°F mercury has a specific weight of 834 lb/ft^3. What volume of the mercury would weigh 500 lb?

1.101 One gallon of a certain fuel oil weighs 7.50 lb. Calculate its specific weight, its density, and its specific gravity.

1.102 Glycerin has a specific gravity of 1.258. How much would 50 gal of glycerin weigh?

1.103 The fuel tank of an automobile holds 25.0 gal. If it is full of gasoline having a density of 1.32 slugs/ft^3, calculate the weight of the gasoline.

1.104 The density of muriatic acid is 1.20 g/cm^3. Calculate its density in slugs/ft^3, its specific weight in lb/ft^3, and its specific gravity. (Note that specific gravity and density in g/cm^3 are numerically equal.)

1.105 Liquid ammonia has a specific gravity of 0.826. Calculate the volume in cm^3 that would weigh 5.0 lb.

1.106 Vinegar has a density of 1.08 g/cm^3. Calculate its specific weight in lb/ft^3.

1.107 Alcohol has a specific gravity of 0.79. Calculate its density both in $slugs/ft^3$ and g/cm^3.

1.108 A cylindrical container has a 6.0-in diameter and weighs 0.50 lb when empty. When filled to a depth of 8.0 in with a certain oil, it weighs 7.95 lb. Calculate the specific gravity of the oil.

1.109 A storage vessel for gasoline (sg = 0.68) is a vertical cylinder 30 ft in diameter. If it is filled to a depth of 22 ft, calculate the number of gallons in the tank and the weight of the gasoline.

1.110 How many gallons of mercury (sg = 13.54) would weigh the same as 5 gal of castor oil, which has a specific weight of 59.69 lb/ft^3?

1.111 A rock has a specific gravity of 2.32 and a volume of 8.64 in^3. How much does it weigh?

Supplemental Problems

Use explicit and careful unit analysis to set up and solve the fluid problems below:

1.112 A village of 75 people desires a tank to store a 3-day supply of water. If the average daily usage per person is 1.7 gal, determine the required size of the tank in cubic feet.

1.113 A cylindrical tank has a diameter of 38 in with its axis vertical. Determine the depth of fluid in the tank when it is holding 85 gal of fluid.

1.114 What is the required rate, in N/min, to empty a tank containing 80 N of fluid in 5 s?

1.115 An empty tank measuring 1.5 m by 2.5 m on the bottom is filled at a rate of 60 L/min. Determine the time required for the fluid to reach a depth of 25 cm.

1.116 A tank that is 2 ft in diameter and 18 in tall is to be filled with a fluid in 90 s. Determine the required fill rate in gal/min.

1.117 A standard pump design can be upgraded to higher efficiency for an additional capital investment of $17,000. What is the period for payback if the upgrade saves 7500 $/year?

1.118 What is the annual cost to run a 2 HP system if it is to run continuously and the cost for energy is 0.10 $/kW-hr?

For Problems 1.119 to 1.121: A piston/cylinder arrangement like the one shown in Fig. 1.7 is used to pump liquid. It moves a volume of liquid equal to its displacement, which is the area of the piston face times the length of the stroke, for each revolution of the crank. Perform the following calculations.

1.119 Determine the displacement, in liters, for one revolution of a pump with a 75-mm diameter piston and 100 mm-stroke.

1.120 Determine the flow rate, in m^3/hr, for another pump that has a displacement of 2.2 L/revolution is run at 80 revolutions/min (rpm).

1.121 At what speed, in rpm, does a single cylinder pump with a 1.0-in diameter piston and a 2.5-in stroke need to be run to provide 20 gal/min of flow?

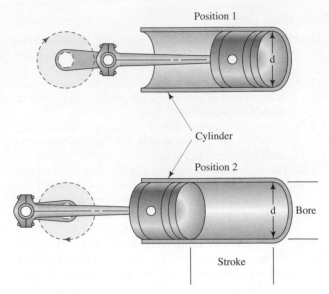

FIGURE 1.7

COMPUTER AIDED ENGINEERING ASSIGNMENTS

1. Write a program that computes the specific weight of water for a given temperature using the data from Appendix A. Such a program could be part of a more comprehensive program to be written later. The following options could be used:

   a. Enter the table data for specific weight as a function of temperature into an array. Then, for a specified temperature, search the array for the corresponding specific weight. Interpolate temperatures between values given in the table.

   b. Include data in both SI units and U.S. Customary System units.

   c. Include density.

   d. Include checks in the program to ensure that the specified temperature is within the range given in the tables (i.e., above the freezing point and below the boiling point).

   e. Instead of using the table look-up approach, use a curve-fit technique to obtain equations of the properties of water versus temperature. Then compute the desired property value for any specified temperature.

2. Use a spreadsheet to display the values of specific weight and density of water from Appendix A. Then create curve-fit equations for specific weight versus temperature and density versus temperature using the *Trendlines* feature of the spreadsheet chart. Add this equation to the spreadsheet to produce computed values of specific weight and density for any given temperature. Compute the percent difference between the table values and the computed values. Display graphs for specific weight versus temperature and density versus temperature on the spreadsheet showing the equations used.

VISCOSITY OF FLUIDS

The ease with which a fluid flows through pipes or pours from a container is an indication of its viscosity. Fluids that flow and pour easily have relatively low viscosity while those that pour or flow more slowly have high viscosity. Think about some of the fluids you encounter frequently and recall how easily or slowly they pour:

■ **Liquids:** Water, milk, juices, sodas, vinegar, syrup for waffles and pancakes, cooking oil, chocolate syrup for ice cream sundaes, mouthwash, shampoo, hair conditioner, liquid soap or detergent, jams and jellies, paint, varnish, sunscreen, insect repellants, gasoline, kerosene, motor oil, household and shop lubricants, cleaning fluids in spray bottles or those that are poured out, windshield washer fluids, weed control liquids, refrigerants in the liquid state, liquid chemicals and mixtures in a factory, oil-hydraulic fluids used in fluid power systems for automated machinery, and many others.

■ **Gases:** The air you breathe; air flow through a forced air heating, ventilating and air conditioning (HVAC) system in your home, office, or school; air flow drawn over a car's radiator to keep the coolant at an effective temperature, refrigerants in a gaseous state; natural gas used for home heating, cooking, or hot water heating; compressed air used in pneumatic actuation and control systems in a factory; steam, chemical vapors, atomized spray cleaners, and non-stick spray cooking oils.

■ **High viscosity fluids and semi-solids:** Catsup, mustard, salad dressing, peanut butter, apple butter, mayonnaise, face cream, ointments, tooth paste, artist paint, adhesives, sealants, grease, tar, wax, and liquid polymers.

Considering the *liquids*, you likely noticed that water pours easily and rapidly from a faucet, garden hose, or a bucket. However, oils, syrups, and shampoos pour much more slowly as illustrated in Fig. 2.1, which shows oil being poured from a cup. Water has a relatively low viscosity while oil has a relatively high viscosity. We also say that *oil is more viscous than water*. A good exercise is to add any other liquid you can think of to the list above and then arrange the complete list in the approximate order of how easily they flow, that is, put them in the order from the less viscous to the more viscous.

Gases are also fluids, although they behave much differently from liquids as explained in general in Chapter 1. We don't often think of *pouring* a gas because it moves freely unless confined into a container. However, there are many situations in which gases are *flowing* through pipes, tubing, ductwork, or conduits of other shapes. Consider the high pressure air you put in the tires of your car, bicycle, or motorcycle; the heated or cooled air delivered by an HVAC system; the compressed air delivered throughout a factory to drive automation devices; the movement

FIGURE 2.1 Lubricating oil with a relatively high viscosity pouring from a cup. (*Source:* runique/Fotolia)

of refrigerants in their gaseous state through the tubing in a refrigeration or air conditioning system; or the flow of chemical vapors in a distillation process of a petroleum refining plant. You must consider the viscosity of these gases when designing the flow systems.

High viscosity fluids and semi-solids are those which do not readily pour at all. Think about trying to pour catsup and mustard onto your sandwich. You typically have to shake the bottle, beat on the bottom, or squeeze it to get it on the bread. Other such fluids listed above behave similarly although, given enough time, all of them conform to their containers. These fluids behave very differently as compared with the more normal liquids described earlier and their behavior is described later in this chapter.

You have likely noticed that cold viscous fluids pour more slowly than when they are warm. Examples are motor oil, lubricating oil, and syrups. This phenomenon is due to the fact that a liquid's viscosity typically increases as the temperature drops. You will see data to support this observation in this chapter.

Internet resource 9 states the definition of viscosity as:

Viscosity is the internal friction of a fluid, caused by molecular attraction, which makes it resist a tendency to flow.

The internal friction, in turn, causes energy losses to occur as the fluid flows through pipes or other conduits. You will use the property of viscosity in Chapters 8 and 9 when we predict the energy lost from a fluid as it flows in a pipe, a tube, or a conduit of some other shape. Then in Chapters 10–13, it continues to be an important factor in designing and analyzing fluid flow systems. Also, in Chapter 13 on Pump Selection and Application, we show that the performance of a pump is affected by the fluid's viscosity. On a more general basis, viscosity measurement is often used as a measure of product quality and consistency. It can be sensed by the customer when the viscosity of a food product such as syrup is either too high (thick) or too low (thin). In materials processing, viscosity can often affect the mixing of constituents or chemical reactions.

It is important for you to learn how to define fluid viscosity, the units used for it, what industry standards apply to viscosity measurement of fluids such as engine oils and lubricants, and to become familiar with some of the commercially available instruments used to measure it.

Exploration

Now perform some experiments to demonstrate the wide range of viscosities for different kinds of fluids at different temperatures.

- Obtain samples of three different fluids with noticeably different viscosities. Examples are water, oil (cooking or lubricating), liquid detergent or other kinds of cleaning fluid, and foods that are fluids (e.g., tomato juice or catsup).
- Put some of each kind of fluid in the refrigerator and keep some at room temperature.
- Obtain a small, disposable container to use for a test cup and make a small hole in its bottom.
- For each fluid at both room and refrigerated temperature, pour the same amount into the test cup while holding your finger over the hole to keep the fluid in.
- Uncover the hole and allow the fluid to drain out while measuring the time to empty the cup.
- Compare the times for the different fluids at each temperature and the amount of change in time between the two temperatures.

Discuss your results with your fellow students and your instructor.

This chapter describes the physical nature of viscosity, defines both dynamic viscosity and kinematic viscosity, discusses the units for viscosity, and describes several methods for measuring the viscosity of fluids. Standards for testing and classifying viscosities for lubricants, developed by SAE International, ASTM International, International Standards Organization (ISO), and the Coordinating European Council (CEC) are also discussed.

2.1 OBJECTIVES

After completing this chapter, you should be able to:

1. Define *dynamic viscosity*.

2. Define *kinematic viscosity*.

3. Identify the units of viscosity in both the SI system and the U.S. Customary System.

4. Describe the difference between a *Newtonian fluid* and a *non-Newtonian fluid*.

5. Describe the methods of viscosity measurement using the *rotating-drum viscometer*, the *capillary-tube viscometer*, the *falling-ball viscometer*, and the *Saybolt Universal viscometer*.

6. Describe the variation of viscosity with temperature for both liquids and gases and define *viscosity index*.

7. Identify several types of commercially available viscometers.

8. Describe the viscosity of lubricants using the SAE viscosity grades and the ISO viscosity grades.

2.2 DYNAMIC VISCOSITY

As a fluid moves, a shear stress develops in it, the magnitude of which depends on the viscosity of the fluid. *Shear stress*, denoted by the Greek letter τ (tau), can be defined as the force required to slide one unit area layer of a substance over another. Thus, τ is a force divided by an area and can be measured in the units of N/m² (Pa) or lb/ft². In fluids such as water, oil, alcohol, or other common liquids the magnitude of the shearing stress is directly proportional to the change of velocity between different positions in the fluid.

Figure 2.2 illustrates the concept of velocity change in a fluid by showing a thin layer of fluid between two surfaces, one of which is stationary while the other is moving. A fundamental condition that exists when a real fluid is in contact with a boundary surface is that the fluid has the same velocity as the boundary. In Fig. 2.2, then, the fluid in contact with the lower surface has a zero velocity and that in contact with the upper surface has the velocity v. If the distance between the two surfaces is small, then the rate of change of velocity with position y is linear. That is, it varies in a straight-line manner. The *velocity gradient* is a measure of the velocity change and is defined as $\Delta v / \Delta y$. This is also called the *shear rate*.

The fact that the shear stress in the fluid is directly proportional to the velocity gradient can be stated mathematically as

$$\tau = \eta(\Delta v / \Delta y) \tag{2-1}$$

where the constant of proportionality η (the Greek letter eta) is called the *dynamic viscosity* of the fluid. The term *absolute viscosity* is sometimes used.

You can gain a physical feel for the relationship expressed in Eq. (2–1) by stirring a fluid with a rod. The action of stirring causes a velocity gradient to be created in the fluid. A greater force is required to stir cold oil having a high viscosity (a high value of η) than is required to stir water, which has a low viscosity. This is an indication of the higher shear stress in the cold oil. The direct application of Eq. (2–1) is used in some types of viscosity measuring devices as will be explained later.

2.2.1 Units for Dynamic Viscosity

Many different unit systems are used to express viscosity. The systems used most frequently are described here for dynamic viscosity and in the next section for kinematic viscosity.

The definition of dynamic viscosity can be derived from Eq. (2–1) by solving for η:

⮥ **Dynamic Viscosity**

$$\eta = \frac{\tau}{\Delta v / \Delta y} = \tau\left(\frac{\Delta y}{\Delta v}\right) \tag{2-2}$$

The units for η can be derived by substituting the SI units into Eq. (2–2) as follows:

$$\eta = \frac{N}{m^2} \times \frac{m}{m/s} = \frac{N \cdot s}{m^2}$$

Because Pa is another name for N/m², we can also express η as

$$\eta = Pa \cdot s$$

This is the standard unit for dynamic viscosity as stated in official documents of the National Institute for Standards and Technology (NIST), ASTM International, SAE International, ISO, and the Coordinating European Council (CEC). See Internet resources 1–4 of this chapter and Reference 1 from Chapter 1.

Sometimes, when units for η are being combined with other terms—especially density—it is convenient to express η in terms of kg rather than N. Because 1 N = 1 kg·m/s², η can be expressed as

$$\eta = N \times \frac{s}{m^2} = \frac{kg \cdot m}{s^2} \times \frac{s}{m^2} = \frac{kg}{m \cdot s}$$

Thus, N·s/m², Pa·s, or kg/m·s may all be used for η in the SI system.

Table 2.1 lists the dynamic viscosity units in the three most widely used systems. The dimensions of force multiplied by time divided by length squared are evident in each system. The units of poise and centipoise are listed here because much published data are given in these units. They are part of the obsolete metric system called cgs, derived from its base units of centimeter, dyne, gram, and second. Summary tables listing many conversion factors are included in Appendix K. Also, Internet resource 14 contains online conversion calculators for units of both dynamic and kinematic viscosity along with large lists of viscosity conversion factors.

Dynamic viscosities of common industrial liquids, such as those listed in Appendices A–D and in Section 2.7, range from approximately 1.0×10^{-4} Pa·s to 60.0 Pa·s. Because

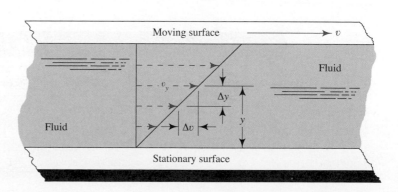

FIGURE 2.2 Velocity gradient in a moving fluid.

TABLE 2.1 Units for dynamic viscosity, η (Greek letter eta)	
Unit System	**Dynamic Viscosity (η) Units**
International System (SI)	$N{\cdot}s/m^2$, Pa·s, or kg/(m·s)
U.S. Customary System	lb·s/ft^2 or slug/(ft·s)
cgs system (obsolete)	poise = dyne·s/cm^2 = g/(cm·s) = 0.1 Pa·s centipoise = poise/100 = 0.001 Pa·s = 1.0 mPa·s

of this common range, many sources of fluid property data and the scales of viscosity-measurement instruments are listed in a more convenient unit of mPa·s, where

$$1.0 \text{ mPa·s} = 1.0 \times 10^{-3} \text{ Pa·s}$$

Note that the older unit of centipoise is numerically equivalent to mPa·s. Then the range given above expressed in mPa·s is

$$1.0 \times 10^{-4} \text{ Pa·s} = 0.10 \times 10^{-3} \text{ Pa·s} = 0.10 \text{ mPa·s}$$

to

$$60.0 \text{ Pa·s} = 60\,000 \times 10^{-3} \text{ Pa·s} = 60\,000 \text{ mPa·s}$$

Note that the value of 60 000 mPa·s is for engine-lubricating oil at extremely low temperatures as indicated in Section 2.7 in the discussion of SAE viscosity ratings for engine oils. This is the maximum dynamic viscosity accepted under cold starting conditions to ensure that the oil is able to flow into the engine's oil pump.

2.3 KINEMATIC VISCOSITY

Many calculations in fluid mechanics involve the ratio of the dynamic viscosity to the density of the fluid. As a matter of convenience, the kinematic viscosity ν (the Greek letter nu) is defined as

⇨ **Kinematic Viscosity**

$$\nu = \eta/\rho \qquad (2\text{–}3)$$

Because η and ρ are both properties of the fluid, ν is also a property. It is an unfortunate inconvenience that the Greek letter ν and the lower case v ('vee') in this text look very similar. Use care with these terms.

2.3.1 Units for Kinematic Viscosity

We can derive the SI units for kinematic viscosity by substituting the previously developed units for η and ρ:

$$\nu = \frac{\eta}{\rho} = \eta\left(\frac{1}{\rho}\right)$$

$$\nu = \frac{\text{kg}}{\text{m·s}} \times \frac{\text{m}^3}{\text{kg}}$$

$$\nu = \text{m}^2/\text{s}$$

Table 2.2 lists the kinematic viscosity units in the three most widely used systems. The basic dimensions of length squared divided by time are evident in each system. The obsolete units of stoke and centistoke are listed because published data often employ these units. Appendix K lists conversion factors.

Kinematic viscosities of common industrial liquids, such as those listed in Appendices A–D and in Section 2.7, range from approximately 1.0×10^{-7} m^2/s to 7.0×10^{-2} m^2/s. More convenient values are often reported in mm^2/s, where

$$1.0 \times 10^6 \text{ mm}^2/\text{s} = 1.0 \text{ m}^2/\text{s}$$

Note that the older unit of centistoke is numerically equivalent to mm^2/s. Then the range listed above expressed in mm^2/s is

$$1.0 \times 10^{-7} \text{ m}^2/\text{s} = (0.10 \times 10^{-6} \text{ m}^2/\text{s})(10^6 \text{ mm}^2/1.0 \text{ m}^2)$$
$$= 0.10 \text{ mm}^2/\text{s}$$

to

$$7.0 \times 10^{-2} \text{ m}^2/\text{s} = (70\,000 \times 10^{-6} \text{ m}^2/\text{s})(\text{mm}^2/1.0 \text{ m}^2)$$
$$= 70\,000 \text{ mm}^2/\text{s}$$

Again the very large value is for extremely cold engine oil.

TABLE 2.2 Units for kinematic viscosity, ν (Greek letter nu)	
Unit System	**Kinematic Viscosity (ν) Units**
International System (SI)	m^2/s
U.S. Customary System	ft^2/s
cgs system (obsolete)	stoke = cm^2/s = 1×10^{-4} m^2/s centistoke = stoke/100 = 1×10^{-6} m^2/s = 1 mm^2/s

2.4 NEWTONIAN FLUIDS AND NON-NEWTONIAN FLUIDS

The study of the deformation and flow characteristics of substances is called *rheology*, which is the field from which we learn about the viscosity of fluids. One important distinction is between a *Newtonian fluid* and a *non-Newtonian fluid*. Any fluid that behaves in accordance with Fig. 2.2 and Eq. (2–1) is called a *Newtonian fluid*. The viscosity η is a function only of the condition of the fluid, particularly its temperature. The magnitude of the velocity gradient $\Delta v / \Delta y$ has no effect on the magnitude of η. Most common fluids such as water, oil, gasoline, alcohol, kerosene, benzene, and glycerin are classified as Newtonian fluids. See Appendices A–E for viscosity data for water, several other Newtonian fluids, air, and other gases. See also Reference 12 that contains numerous tables and charts of viscosity data for petroleum oil and other common fluids. Internet resource 19 also lists many useful values for viscosities for oils. Most fluids considered in later chapters of this book are Newtonian.

In contrast to the behavior of Newtonian fluids, a fluid that does not behave in accordance with Eq. (2–1) is called a *non-Newtonian fluid*. The difference between the two is shown in Fig. 2.3. The viscosity of the non-Newtonian fluid is dependent on the velocity gradient in addition to the condition of the fluid.

Note that in Fig. 2.3(a), the slope of the curve for shear stress versus the velocity gradient is a measure of the *apparent viscosity* of the fluid. The steeper the slope, the higher is the apparent viscosity. Because Newtonian fluids have a linear relationship between shear stress and velocity gradient, the slope is constant and, therefore, the viscosity is constant. The slopes of the curves for non-Newtonian fluids vary and Fig. 2.3(b) shows how viscosity changes with velocity gradient.

Two major classifications of non-Newtonian fluids are *time-independent* and *time-dependent* fluids. As their name implies, time-independent fluids have a viscosity at any given shear stress that does not vary with time. The viscosity of time-dependent fluids, however, changes with time.

Three types of time-independent fluids can be defined as:

- *Pseudoplastic* The plot of shear stress versus velocity gradient lies above the straight, constant sloped line for Newtonian fluids, as shown in Fig. 2.3. The curve begins steeply, indicating a high apparent viscosity. Then the slope decreases with increasing velocity gradient. Examples of such fluids are blood plasma, molten polyethylene, latexes, syrups, adhesives, molasses, and inks.

- *Dilatant Fluids* Again referring to Fig. 2.3, the plot of shear stress versus velocity gradient or dilatant fluids lies below the straight line for Newtonian fluids. The curve begins with a low slope, indicating a low apparent viscosity. Then, the slope increases with increasing velocity gradient. Examples of dilatant fluids are slurries with high concentrations of solids such as corn starch in ethylene glycol, starch in water, and titanium dioxide, an ingredient in paint.

- *Bingham Fluids* Sometimes called *plug-flow fluids*, Bingham fluids require the development of a significant level of shear stress before flow will begin, as illustrated in Fig. 2.3. Once flow starts, there is an essentially linear slope to the curve indicating a constant apparent viscosity. Examples of Bingham fluids are chocolate, catsup, mustard, mayonnaise, toothpaste, paint, asphalt, some greases, and water suspensions of fly ash or sewage sludge.

2.4.1 Time-Dependent Fluids

Time-dependent fluids are very difficult to analyze because apparent viscosity varies with time as well as with velocity gradient and temperature. Examples of time-dependent fluids are some crude oils at low temperatures, printer's ink, liquid nylon and other polymer solutions, some jellies, flour dough, some kinds of greases, and paints. Figure 2.4 shows two types of time-dependent fluids where in each case the temperature is held constant. The vertical axis is the apparent dynamic viscosity, η, and the horizontal axis is time. The left part of the curves show stable viscosity when the shear rate is not changing. Then, when the shear rate changes, the

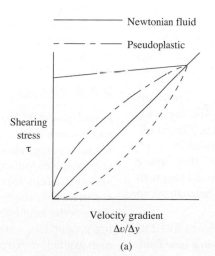
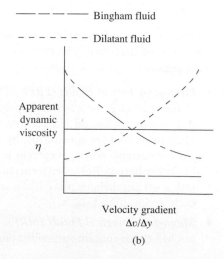

FIGURE 2.3 Newtonian and non-Newtonian fluids.

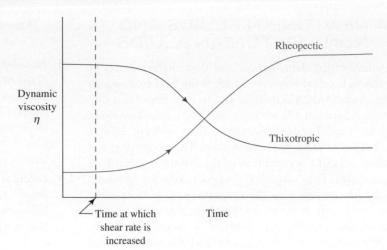

FIGURE 2.4 Behavior of time-dependent fluids.

apparent viscosity changes, either increasing or decreasing depending on the type of fluid described here.

- **Thixotropic Fluids.** A fluid that exhibits *thixotropy* whereby the apparent viscosity decreases with time as shear rate remains constant. This is the most common type of time-dependent fluid.

- **Rheopectic Fluids.** A fluid that exhibits *rheopexy* whereby the viscosity increases with time. *Rheopectic fluids* are quite rare.

2.4.2 Actively Adjustable Fluids

Other types of fluids of more recent development are those for which the rheological properties, particularly the viscosity and stiffness, can be changed actively by varying an electric current or by changing the magnetic field around the material. Adjustments can be made manually or by computer control rapidly and they are reversible. Applications include shock absorbers for vehicles where harder or softer rides can be selected, to adjust for varying loads on the vehicle, or when increased damping is used to reduce bounce and jounce when operating on rough, off-road locations; adjusting the motion of truck drivers' seats; active control of clutches; tuning of engine mounts to minimize vibration; providing adjustable damping in buildings and bridges to resist earthquakes; in various prosthetic devices for handicapped people; and computer controlled Braille displays. Two types are described here. Refer to Internet resource 5 for more details.

- **Electrorheological Fluids (ERF)** These are suspensions of fine particles, such as starch, polymers, and ceramics, in a nonconducting oil, such as mineral oil or silicone oil. Fluid properties are controllable by the application of an electric current. When no current is applied, they behave like other liquids. But when a current is applied they turn into a gel and behave more like a solid. The change can occur in less than $1/1000$ s.

- **Magnetorheological Fluids (MRF)** Similar to ERF fluids, MR fluids contain suspended particles in a base fluid.

However, in this case, the particles are fine iron powders. The base fluid can be a petroleum oil, silicone oil, or water. When there is no magnetic field present, the MRF behaves much like other fluids, with a viscosity that ranges from 0.2 Pa·s to 0.3 Pa·s at 25°C. The presence of a magnetic field can cause the MRF to become virtually solid such that it can withstand a shear stress of up to 100 kPa. The change can be controlled electronically quite rapidly.

2.4.3 Nanofluids

Nanofluids are those that contain extremely small, nano-scale particles (less than 100 nm in diameter) in base fluids such as water, ethylene glycol coolants, oil and synthetic lubricants, biological fluids, and polymer solutions. The nanoparticle materials can be metals such as aluminum and copper, silicon carbide, aluminum dioxide, copper oxide, graphite, carbon nanotubes and several others. The nanoparticles have far higher surface to volume ratios than conventional fluids, mixtures, or suspensions, leading to enhanced thermal conductivity and other physical properties. One major use of nanofluids is to enhance the overall performance of fluids used to cool electronic devices. Used in lubrication applications, improved flow characteristics can be obtained while maintaining lubricity and carrying heat away from critical surfaces. Bio-medical, drug delivery, and environmental control applications are also being researched and developed. See Reference 16.

2.4.4 Viscosity of Liquid Polymers

Liquid polymers are the subject of much industrial study and research because of their importance in product design, manufacturing, lubrication, and health care. They are decidedly non-Newtonian, and we need a variety of additional viscosity terminology to describe their behavior. See Internet resources 6, 7, and 9–12 for commercial equipment used to characterize liquid polymers, either in the laboratory or during production; some are designed to sample the polymer melt just before extrusion or injection into a die.

Five additional viscosity factors are typically measured or computed for polymers:

1. *Relative viscosity*
2. *Inherent viscosity*
3. *Reduced viscosity*
4. *Specific viscosity*
5. *Intrinsic viscosity* (also called *limiting viscosity number*)

A solvent is added to the liquid polymer prior to performing some of these tests and making the final calculations. Examples of polymer/solvent combinations are as follows:

1. Nylon in formic acid
2. Nylon in sulfuric acid
3. Epoxy resins in methanol
4. Cellulose acetate in acetone and methylene chloride
5. Polycarbonate in methylene chloride

The concentration (*C*) of polymer, measured in grams per 100 mL, must be known. The following calculations are then completed:

Relative Viscosity, η_{rel}. The ratio of the viscosities of the polymer solution and of the pure solvent at the same temperature

Inherent Viscosity, η_{inh}. The ratio of the natural logarithm of the relative viscosity and the concentration *C*

Specific Viscosity, η_{spec}. The relative viscosity of the polymer solution minus 1

Reduced Viscosity, η_{red}. The specific viscosity divided by the concentration

Intrinsic Viscosity, η_{intr}. The ratio of the specific viscosity to the concentration, extrapolated to zero concentration. The relative viscosity is measured at several concentrations and the resulting trend line of specific viscosities is extrapolated to zero concentration. Intrinsic viscosity is a measure of the molecular weight of the polymer or the degree of polymerization.

Testing procedures for liquid polymers must be carefully chosen because of their non-Newtonian nature. Figure 2.3(a) shows that the apparent viscosity changes as the velocity gradient changes, and the rate of shearing within the fluid also changes as the velocity gradient changes. Therefore, it is important to control the shear rate, also called the strain rate, in the fluid during testing. Reference 13 includes an extensive discussion of the importance of controlling the shear rate and the types of rheometers that are recommended for different types of fluids.

Many liquid polymers and other non-Newtonian fluids exhibit time-dependent viscoelastic characteristics in addition to basic viscosity. Examples are extruded plastics, adhesives, paints, coatings, and emulsions. For these materials, it is helpful to measure their elongational behavior to control manufacturing processes or application procedures. This

type of testing is called *extensional rheometry*. See Internet resource 11.

2.5 VARIATION OF VISCOSITY WITH TEMPERATURE

You are probably familiar with some examples of the variation of fluid viscosity with temperature. Engine oil is generally quite difficult to pour when it is cold, indicating that it has a high viscosity. As the temperature of the oil is increased, its viscosity decreases noticeably.

All fluids exhibit this behavior to some extent. Appendix D gives two graphs of dynamic viscosity versus temperature for many common liquids. Notice that viscosity is plotted on a logarithmic scale because of the large range of numerical values. To check your ability to interpret these graphs, a few examples are listed in Table 2.3.

Gases behave differently from liquids in that the viscosity increases as the temperature increases. Also, the general magnitude of the viscosities and the amount of change is generally smaller than that for liquids.

2.5.1 Viscosity Index

A measure of how greatly the viscosity of a fluid changes with temperature is given by its viscosity index, sometimes referred to as *VI*. This is especially important for lubricating oils and hydraulic fluids used in equipment that must operate at wide extremes of temperature.

A fluid with a high viscosity index exhibits a small change in viscosity with temperature. A fluid with a low viscosity index exhibits a large change in viscosity with temperature.

Typical curves for oils with *VI* values of 50, 100, 150, 200, 250, and 300 are shown in Fig. 2.5 on chart paper created especially for viscosity index that results in the curves being straight lines. Viscosity index is determined by measuring the kinematic viscosity of the sample fluid at 40°C and 100°C (104°F and 212°F) and comparing these values with those of certain reference fluids that were assigned *VI* values of 0 and 100. Standard ASTM D 2270 gives the complete method. See Reference 3.

TABLE 2.3	Selected values of viscosity read from Appendix D	
Fluid	**Temperature (°C)**	**Dynamic Viscosity (N·s/m² or Pa·s)**
Water	20	1.0×10^{-3}
Water	70	4.0×10^{-4}
Gasoline	20	3.1×10^{-4}
Gasoline	62	2.0×10^{-4}
SAE 30 oil	20	3.5×10^{-1}
SAE 30 oil	80	1.9×10^{-2}

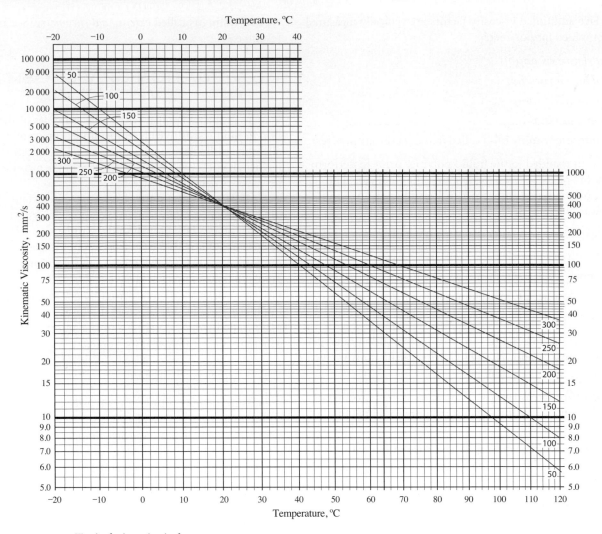

FIGURE 2.5 Typical viscosity index curves.

The general form of the equation for calculating the viscosity index for a type of oil that has a *VI* value up to and including 100 is given by the following formula. All kinematic viscosity values are in the unit of mm²/s:

$$VI = \frac{L - U}{L - H} \times 100 \qquad (2\text{–}4)$$

where

U = Kinematic viscosity at 40°C of the test oil

L = Kinematic viscosity at 40°C of a standard oil of 0 *VI* having the same viscosity at 100°C as the test oil

H = Kinematic viscosity at 40°C of a standard oil of 100 *VI* having the same viscosity at 100°C as the test oil

The values of L and H can be found from a table in ASTM Standard D 2270 for oils with kinematic viscosities between 2.0 mm²/s and 70.0 mm²/s at 100°C. This range encompasses most of the practical oils used as fuels or for lubrication.

For oils with $VI > 100$, ASTM Standard D 2270 gives an alternate method of computing *VI* that also depends on obtaining values from the table in the standard.

Look closer at the *VI* curves in Fig. 2.5. They are plotted for the special case where each type of oil has the same value of kinematic viscosity of 400 mm²/s at 20°C (68°F), approximately at room temperature. Table 2.4 gives the kinematic viscosity for six types of oil having different values of viscosity index (VI) at −20°C (−4°F), 20°C (68°F), and 100°C (212°F).

Notice the huge range of the values. The *VI* 50 oil has a very high viscosity at the cold temperature, and it may be difficult to make it flow to critical surfaces for lubrication. Conversely, at the hot temperature, the viscosity has decreased to such a low value that it may not have adequate lubricating ability. The amount of variation is much less for the types of oil with high viscosity indexes.

Lubricants and hydraulic fluids with a high *VI* should be used in engines, machinery, and construction equipment used outdoors where temperatures vary over wide ranges. In a given day the oil could experience the −20°C to +100°C range illustrated.

The higher values of *VI* are obtained by blending selected oils with high paraffin content or by adding special polymers that increase *VI* while maintaining good lubricating properties, and good performance in engines, pumps, valves, and actuators.

TABLE 2.4	Viscosity readings of types of oil with a variety of viscosity index (*VI*) values at three different temperatures		
Viscosity Index *VI*	Kinematic Viscosity ν (mm²/s)		
	At –20°C	At 20°C	At 100°C
50	47 900	400	9.11
100	21 572	400	12.6
150	9985	400	18.5
200	5514	400	26.4
250	3378	400	37.1
300	2256	400	51.3

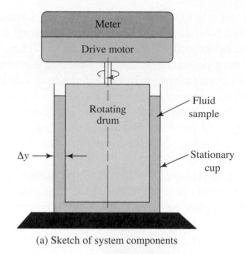

(a) Sketch of system components

2.6 VISCOSITY MEASUREMENT

Procedures and equipment for measuring viscosity are numerous. Some employ fundamental principles of fluid mechanics to indicate viscosity in its basic units. Others indicate only relative values for viscosity, which can be used to compare different fluids. In this section we will describe several common methods used for viscosity measurement. Devices for characterizing the flow behavior of liquids are called *viscometers* or *rheometers*. You should become familiar with the numerous suppliers of viscosity measurement instruments and systems. Some are designed for laboratory use while others are designed to be integral with production processes to maintain quality control and to record data for historical documentation of product characteristics. Inter net resources 6–14 are examples of such suppliers.

ASTM International, ISO, and CEC generate standards for viscosity measurement and reporting. See Internet resources 1, 3, and 4 along with References 1–11 for ASTM standards pertinent to the discussion in this section. Specific standards are cited in the sections that follow. Another important standards-setting organization is SAE International that defines and publishes many standards for fuels and lubricants. See Internet resource 2 and References 14 and 15. More discussion of SAE standards is included in Section 2.7. The German standards organization, DIN, also develops and publishes standards that are cited by some manufacturers of viscometers. (See www.din.de.)

2.6.1 Rotating-Drum Viscometer

The apparatus shown in Fig. 2.6(a) measures dynamic viscosity, η, by its definition given in Eq. (2–2), which we can write in the form

$$\eta = \tau / (\Delta v / \Delta y)$$

The outer cup is held stationary while the motor in the meter drives the rotating drum. The space Δy between the rotating drum and the cup is small. The part of the fluid in contact with the outer cup is stationary, whereas the fluid in contact with the surface of the inner drum is moving

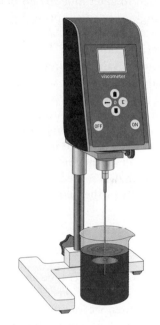

FIGURE 2.6 Rotating-drum viscometer.

with a velocity equal to the surface speed of the drum. Therefore, a known velocity gradient $\Delta v / \Delta y$, is set up in the fluid. The fluid viscosity causes a shearing stress τ in the fluid that exerts a drag torque on the rotating drum. The meter senses the drag torque and indicates viscosity directly on the display. Special consideration is given to the fluid in contact with the bottom of the drum because its velocity varies from zero at the center to the higher value at the outer diameter. Different models for the style of tester shown in Fig. 2.6(b), and different rotors for each tester allow measurement of a wide range of viscosity levels. This kind of tester can be used for a variety of fluids such as paint, ink, food, petroleum products, cosmetics, and adhesives. The tester is battery operated and can be either mounted on a stand as shown or hand held for in-plant operation. See Internet resources 5–14.

A variant of the rotating-drum viscometer, called a *cold-cranking simulator,* is described in Reference 5 and is often used in testing engine oils for their ability to start in cold temperatures. In this apparatus a universal motor drives a rotor, which is closely fitted inside a stator. The rotor speed is related to the viscosity of the test oil that fills the space between the stator and the rotor because of the viscous drag produced by the oil. Speed measurement is correlated to viscosity in mPa·s by reference to a calibration chart prepared by running a set of at least five standard calibration oils of known viscosity on the particular apparatus being used. The resulting data are used by engine designers and users to ensure the proper operation of the engine at cold temperatures.

SAE International specifies that the pumpability viscosity requirements for engine oils be determined using the methods described in Reference 9. A small rotary viscometer is used, and the oil is cooled to very low temperatures as described later in Section 2.7. It is also recommended that Reference 7 be used to determine the borderline pumping temperature of engine oils when specifying new oil formulations.

A novel design called the Stabinger viscometer employs a variation on the rotating-drum principle. The apparatus includes a small tube with a light cylindrical rotor suspended inside. Magnetic forces are used to maintain the rotor in position. The outer tube is rotated at a constant, specified speed, and viscous drag causes the internal rotor to rotate at a speed that is dependent on the fluid viscosity. A small magnet on the rotor creates a rotating magnetic field that is sensed outside the outer tube. The dynamic viscosity of the fluid can be computed from the simple equation

$$\eta = \frac{K}{(n_2/n_1 - 1)}$$

where n_2 is the speed of the outer tube and n_1 is the speed of the internal rotor. K is a calibration constant provided by the instrument manufacturer. See Internet resource 13.

Other designs for rotary viscometers employ a paddle-type rotor mounted to a small-diameter shaft that is submerged in the test fluid. As with other rotary styles of viscometers, the measurement is based on the torque required to drive the paddle at a fixed speed while submerged in the test fluid. See Internet resources 6 and 9.

2.6.2 Capillary Tube Viscometer

Figure 2.7 shows two reservoirs connected by a long, small-diameter tube called a *capillary tube*. As the fluid flows through the tube with a constant velocity, some energy is lost from the system, causing a pressure drop that can be measured by using manometers. The magnitude of the pressure drop is related to the fluid viscosity by the following equation, which is developed in Chapter 8:

$$\eta = \frac{(p_1 - p_2)D^2}{32vL} \qquad (2\text{--}5)$$

In Eq. (2–5), D is the inside diameter of the tube, v is the fluid velocity, and L is the length of the tube between points 1 and 2 where the pressure difference is measured.

2.6.3 Standard Calibrated Glass Capillary Viscometers

References 1 and 2 describe the use of standard glass capillary viscometers to measure the kinematic viscosity of transparent and opaque liquids. Figures 2.8 and 2.9 show 2 of the 17 types of viscometers discussed in the standards. Other capillary viscometers are available that are integrated units having temperature control and automatic sequencing of small samples of fluid through the device. See Fig. 2.10 and Internet resource 12.

In preparation for the viscosity test, the viscometer tube is charged with a specified quantity of test fluid. After stabilizing the test temperature, suction is used to draw fluid through the bulb and slightly above the upper timing mark. The suction is removed and the fluid is allowed to flow by gravity. The working section of the tube is the capillary below the lower timing mark indicated in the figures. The time required for the leading edge of the meniscus to pass from the upper timing mark to the lower timing mark is recorded. The kinematic viscosity is computed by multiplying the flow time by the calibration constant of the viscometer supplied by the vendor. The viscosity unit used in these tests is the centistoke (cSt), which is equivalent to mm²/s. This value must be multiplied by 10^{-6} to obtain the

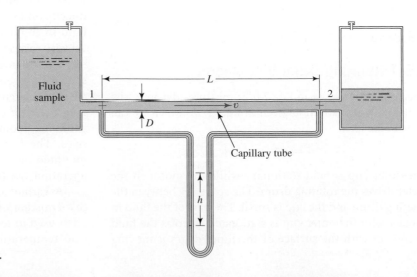

FIGURE 2.7 Capillary-tube viscometer.

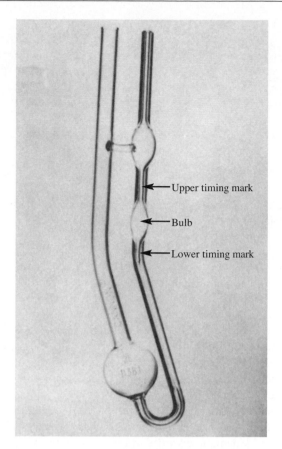

FIGURE 2.8 Cannon–Fenske routine viscometer.
(*Source:* Fisher Scientific)

FIGURE 2.10 Automated multi-range capillary viscometer.
(*Source:* Precision Scientific Petroleum Instruments Company)

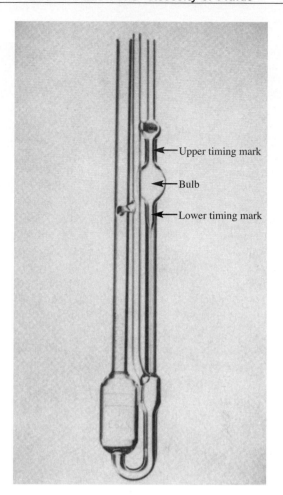

FIGURE 2.9 Ubbelohde viscometer. (*Source:* Fisher Scientific,
Pittsburgh, PA)

SI standard unit of m^2/s, which is used for calculations in
this book.

2.6.4 Falling-Ball Viscometer

As a body falls in a fluid under the influence of gravity only,
it will accelerate until the downward force (its weight) is just
balanced by the buoyant force and the viscous drag force acting upward. Its velocity at that time is called the *terminal
velocity, v*. The falling-ball viscometer sketched in Fig. 2.11
uses this principle by causing a spherical ball to fall freely
through the fluid and measuring the time required for the
ball to drop a known distance. Thus, the velocity can be calculated. Figure 2.12 shows a free-body diagram of the ball,
where w is the weight of the ball, F_b is the buoyant force, and
F_d is the viscous drag force on the ball, discussed more fully
in Chapter 17. When the ball has reached its terminal velocity, it is in equilibrium. Therefore, we have

$$w - F_b - F_d = 0. \qquad (2\text{–}6)$$

If γ_s is the specific weight of the sphere, γ_f is the specific
weight of the fluid, V is the volume of the sphere, and D is
the diameter of the sphere, we have

$$w = \gamma_s V = \gamma_s \pi D^3/6 \qquad (2\text{–}7)$$
$$F_b = \gamma_f V = \gamma_f \pi D^3/6 \qquad (2\text{–}8)$$

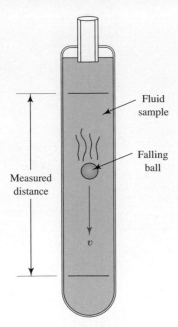

FIGURE 2.11 Falling-ball viscometer.

For very viscous fluids and a small velocity, the drag force on the sphere is

$$F_d = 3\pi\eta vD \qquad (2\text{-}9)$$

Equation (2–6) then becomes

$$\eta = \frac{(\gamma_s - \gamma_f)D^2}{18v} \qquad (2\text{-}10)$$

For visual timing of the descent of the ball, the fluid must be transparent so we can observe the falling ball and time its travel. However, some commercially available falling-ball viscometers have automatic sensing of the position of the ball, so that opaque fluids can be used. Some falling-ball viscometers employ a tube that is slightly inclined to the vertical so that the motion is a combination of rolling and sliding. Calibration between time of travel and viscosity is provided by the manufacturer. Several types and sizes of balls are available to enable the viscometer to be used for fluids with a wide range of viscosities, typically 0.5 mPa·s to 10^5 mPa·s. Balls are made from stainless steel, nickel–iron alloy, and glass. See Internet resources 9 and 12.

2.6.5 Saybolt Universal Viscometer

The ease with which a fluid flows through a small-diameter orifice is an indication of its viscosity. This is the principle on which the Saybolt viscometer is based. The fluid sample is placed in an apparatus similar to that sketched in Fig. 2.13, in which the external vessel maintains a constant temperature of the test fluid. After steady flow from the orifice is established, the time required to collect 60 mL of the fluid is measured. The resulting time is reported as the viscosity of the fluid in Saybolt Universal seconds (SUS). Because the measurement is not based on the basic definition of viscosity, the results are only relative. However, they do serve to compare the viscosities of different fluids. The advantage of this

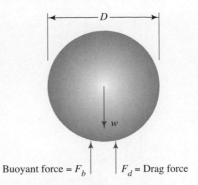

FIGURE 2.12 Free-body diagram of a ball in a falling-ball viscometer.

procedure is that it is simple and requires relatively unsophisticated equipment. See Internet resources 8, 10, and 11.

The use of the Saybolt viscometer is covered by ASTM Standard D 88 (Reference 10). However, this standard recommends that other methods be used for viscosity measurement, such as those listed in References 1 and 2 describing the use of glass capillary viscometers. Furthermore, it is recommended that kinematic viscosity data be reported in the proper SI unit, mm²/s.

ASTM Standard 2161 (Reference 11) describes the preferred conversion methods between viscosity measured in SUS and kinematic viscosity in mm²/s. However, the introduction to the standard states that the use of the Saybolt viscometer is now obsolete in the petroleum industry. Other industries may continue to use it because of historical data and because it is an easy method to use. Figure 2.14 shows a graph of SUS versus kinematic viscosity ν in mm²/s for a fluid temperature of 100°F. The curve is straight above $\nu = 75$ mm²/s, following the equation

$$\text{SUS} = 4.632\nu \qquad (2\text{-}11)$$

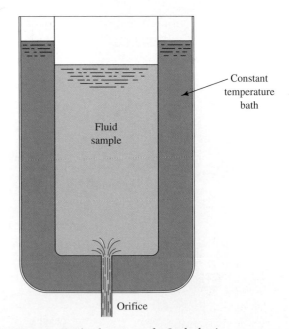

FIGURE 2.13 Basic elements of a Saybolt viscometer.

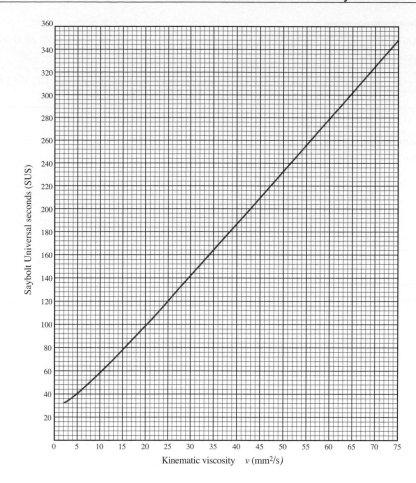

FIGURE 2.14 Kinematic viscosity ν in SUS versus ν in mm²/s at 100°F.

Kinematic viscosity ν (mm²/s)

For a fluid temperature of 210°F, the equation for the straight-line portion is

$$SUS = 4.664\nu \qquad (2\text{–}12)$$

These equations can be used down to approximately $\nu = 50$ mm²/s with an error of less than 0.5 percent and down to approximately $\nu = 38$ mm²/s with an error of less than 1.0 percent (<1.0 SUS).

The SUS value for any other temperature t in degrees Fahrenheit can be found by multiplying the SUS value for 100°F by the factor A shown in Fig. 2.15. The factor A can be computed from

$$A = 6.061 \times 10^{-5}t + 0.994$$
$$\text{(rounded to three decimal places)} \qquad (2\text{–}13)$$

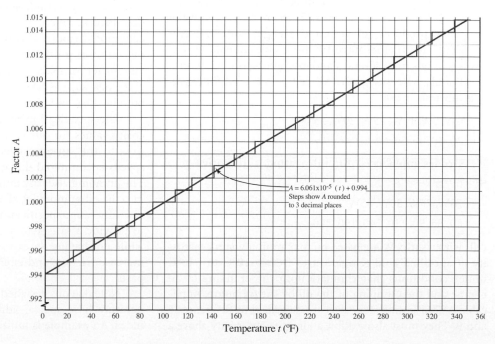

FIGURE 2.15 Factor A versus temperature t in degrees Fahrenheit used to determine the kinematic viscosity in SUS for any temperature.

$A = 6.061 \times 10^{-5}\ (t) + 0.994$
Steps show A rounded to 3 decimal places

Temperature t (°F)

**Example Problem
2.1**

Given that a fluid at 100°F has a kinematic viscosity of 30.0 mm²/s, determine the equivalent SUS value at 100°F.

Solution

Because $\nu < 75 \text{ mm}^2/\text{s}$, use Fig. 2.14 to find $\nu = 141.5$ SUS.

**Example Problem
2.2**

Given that a fluid at 100°F has a kinematic viscosity of 220 mm²/s, determine the equivalent SUS value at 100°F.

Solution

Because $\nu > 75 \text{ mm}^2/\text{s}$, use Eq. (2–11):

$$SUS = 4.632\nu = 4.632(220) = 1019 \text{ SUS}$$

**Example Problem
2.3**

Given that a fluid at 260°F has a kinematic viscosity of 145 mm²/s, determine its kinematic viscosity in SUS at 260°F.

Solution

Use Eq. (2–13) to compute the factor A:

$$A = 6.061 \times 10^{-5}t + 0.994 = 6.061 \times 10^{-5}(260) + 0.994 = 1.010$$

Now find the kinematic viscosity at 100°F using Eq. (2–11):

$$SUS = 4.632\nu = 4.632(145) = 671.6 \text{ SUS}$$

Finally, multiply this value by A to get the SUS value at 260°F:

$$SUS = A(671.6) = 1.010(671.6) = 678 \text{ SUS}$$

2.7 SAE VISCOSITY GRADES

SAE International has developed rating systems for engine oils (Reference 14) and automotive gear lubricants (Reference 15) which indicate the viscosity of the oils at specified temperatures. Note the ASTM testing standards listed as References 1–11. Internet resources 15–18 are representative of the many producers of automotive engine oils and gear lubricants. Internet resource 19 offers tables of data for viscosities of oils from several standards-setting organizations.

Popular viscosity grades for engine oils used for crankcase lubrication are:

0W, 5W, 10W, 15W, 20W, 25W

20, 30, 40, 50, 60

Grades often used for lubricating automotive gear transmissions are:

70W, 75W, 80W, 85W

80, 85, 90, 110, 140, 190, 250

Oils with a suffix W are based on maximum dynamic viscosity at specified cold temperatures from −10°C to −40°C under conditions that simulate both the cranking of an engine and the pumping of the oil by the oil pump. Applicable ASTM testing standards are described in References 5 and 9. They must also exhibit a kinematic viscosity above a specified minimum at 100°C using a glass capillary viscometer as described in Reference 1. Those without the suffix W are rated for viscosity at high temperatures by two different methods described in Reference 14, the kinematic viscosity under low-shear-rate conditions at 100°C and the dynamic viscosity under high-shear-rate conditions at 150°F as described in References 1 and 8. The ratings simulate the conditions in journal bearings and for sliding surfaces. Internet resource 6 offers the Ravenfield Tapered Plug HTHS viscometer for making such measurements. Multiviscosity-grade oils, such as SAE 10W-40, must meet the standards at both the low- and high-temperature conditions.

The specification of maximum low-temperature viscosity values for oils is related to the ability of the oil to flow to the surfaces needing lubrication at the engine speeds encountered during starting at cold temperatures. The pumping viscosity indicates the ability of the oil to flow into the oil pump inlet of an engine. The high-temperature viscosity range specifications relate to the ability of the oil to provide a satisfactory oil film to carry expected loads while not having an excessively high viscosity that would increase friction and energy losses generated by moving parts.

Note that oils designed to operate at wide ranges of temperature have special additives to increase the viscosity index. An example is multiviscosity engine oil (e.g. 5W-40) that must meet stringent low-temperature viscosity limits while maintaining a sufficiently high viscosity at higher engine operating temperatures for effective lubrication. In addition, automotive hydraulic system oils that must operate with similar performance in cold and warm climates and machine-tool hydraulic system oils that must operate outdoors as well as indoors must have high viscosity indexes. Achieving a high viscosity index in oil often calls for the blending of polymeric materials with the petroleum. The resulting blend may exhibit non-Newtonian characteristics, particularly at the lower temperatures.

See also Appendix C for typical properties of petroleum lubricating oils used in engines, gear drives, hydraulic systems, and machine tool applications.

2.8 ISO VISCOSITY GRADES

Lubricants used in industrial applications must be available in a wide range of viscosities to meet the needs of production machinery, bearings, gear drives, electrical machines, fans and blowers, fluid power systems, mobile equipment, and many other devices. The designers of such systems must ensure that the lubricant can withstand the temperatures to

be experienced while providing sufficient load-carrying ability. The result is a need for a wide range of viscosities.

To meet these requirements and still have an economical and manageable number of options, ASTM Standard D 2422 (Reference 4) defines a set of 20 ISO viscosity grades. The standard designation includes the prefix ISO VG followed by a number representing the nominal kinematic viscosity in mm²/s (cSt) for a temperature of 40°C. Table 2.5 gives the data. The maximum and minimum values are ±10 percent from the nominal. Although the standard is voluntary, the intent is to encourage producers and users of lubricants to agree on the specification of viscosities from the list. This system is gaining favor throughout world markets. The CEC develops lubricant performance standards for many European countries and that have been adopted by others throughout the world. See Internet resource 3. Internet resources 15–18 include examples of the many companies that provide oils and lubricants for the automotive and industrial markets. Internet resource 19 provides comparisons between ISO grades and some others.

2.9 HYDRAULIC FLUIDS FOR FLUID POWER SYSTEMS

Fluid power systems use fluids under pressure to actuate linear or rotary devices used in construction equipment, industrial automation systems, agricultural equipment, aircraft hydraulic systems, automotive braking systems, and many others. Fluid power includes both air-type systems, commonly called *pneumatics*, and liquid-type systems, usually referred to as hydraulic systems. This section will deal with liquid-type systems.

There are several types of hydraulic fluids in common use, including

- Petroleum oils
- Water–glycol fluids
- High water-based fluids (HWBF)
- Silicone fluids
- Synthetic oils

The primary characteristics of such fluids for operation in fluid power systems are

- Adequate viscosity for the purpose
- High lubricating capability, sometimes called lubricity
- Cleanliness
- Chemical stability at operating temperatures
- Noncorrosiveness with the materials used in fluid power systems
- Inability to support bacteria growth
- Ecologically acceptable
- High bulk modulus (low compressibility)

You should examine carefully the environment in which the fluid power system is to be used and select a fluid that is optimal for the application. Trade-offs will typically be

TABLE 2.5 ISO viscosity grades

Grade ISO VG	Kinematic Viscosity at 40°C (cSt) or (mm²/s)		
	Nominal	Minimum	Maximum
2	2.2	1.98	2.40
3	3.2	2.88	3.52
5	4.6	4.14	5.06
7	6.8	6.12	7.48
10	10	9.00	11.0
15	15	13.5	16.5
22	22	19.8	24.2
32	32	28.8	35.2
46	46	41.4	50.6
68	68	61.2	74.8
100	100	90.0	110
150	150	135	165
220	220	198	242
320	320	288	352
460	460	414	506
680	680	612	748
1000	1000	900	1100
1500	1500	1350	1650
2200	2200	1980	2420
3200	3200	2880	3520

Source: Reprinted with permission from ASTM Standard 2422. Copyright ASTM. (See Reference 4.)

required so that the combination of properties is acceptable. Suppliers of components, particularly pumps and valves, should be consulted for appropriate fluids to use with their products. Internet resources 15–18 provide information and data from representative suppliers of hydraulic fluids in automotive, construction, and general industrial machinery applications.

Viscosity is one of the most important properties because it relates to lubricity and the ability of the fluid to be pumped and to flow through the tubing, piping, actuators, valves, and other control devices found in fluid power systems.

Common industrial fluid power systems require fluids with viscosities in the range of ISO grades 32, 46, or 68. See Table 2.5 for the kinematic viscosity ranges for such fluids. In general, the ISO grade number is the nominal kinematic viscosity in the unit of mm^2/s.

Special care is needed when extreme temperatures are encountered. Consider the case of the fluid power system on a piece of construction equipment that is kept outdoors throughout the year. In winter, the temperature may range to $-20°F$ ($-29°C$). When starting the system at that temperature you must consider the ability of the fluid to flow into the intake ports of the pumps, through the piping systems, and through the control valves. The fluid viscosity may be greater than $800\ mm^2/s$. Then, when the system has warmed to approximately $150°F$ ($66°C$), the fluid viscosity may be as low as $15\ mm^2/s$. The performance of the pumps and valves is likely to be remarkably different under this range of conditions. Also, as you will learn in Chapter 8, the very nature of the flow may change as the viscosities change. At the cold temperatures the fluid flow will likely be laminar, whereas at the higher temperatures with the decreased viscosities the flow may be turbulent. Hydraulic fluids for operation at these ranges of temperatures should have a high viscosity index, as described earlier in this chapter.

Petroleum oils may be very similar to the automotive engine oils discussed earlier in this chapter. SAE 10W and SAE 20W-20 are appropriate. However, several additives are required to inhibit the growth of bacteria, to ensure compatibility with seals and other parts of fluid power components, to improve its antiwear performance in pumps, and to improve the viscosity index. Suppliers of hydraulic fluids should be consulted for recommendations of specific formulations. Some of the additives used to improve viscosity are polymeric materials, and they may change the flow characteristics dramatically under certain high-pressure conditions that may occur within valves and pumps. The oils may behave as non-Newtonian fluids.

Silicone fluids are desirable when high temperatures are to be encountered, as in work near furnaces, hot processes, and some vehicle braking systems. These fluids exhibit very high thermal stability. Compatibility with the pumps and valves of the system must be checked.

High water-based fluids (HWBF) are desirable where fire resistance is needed. Water-in-oil emulsions contain approximately 40% oil blended in water with a significant variety and quantity of additives to tailor the fluid properties to the application. A different class of fluids, called oil-in-water emulsions, contains 90–95 percent water with the balance being oil and additives. Such emulsions typically appear to be milky white because the oil is dispersed in the form of very small droplets.

Water–glycol fluids are also fire resistant, containing approximately 35–50 percent water, with the balance being any of several glycols along with additives suitable for the environment in which the system is to be operated.

REFERENCES

1. ASTM International. 2011. D445-11A: *Standard Test Method for Kinematic Viscosity of Transparent and Opaque Liquids*, West Conshohocken, PA: Author. DOI: 10.1520/D445-11A, www.astm.org.

2. _____ . 2007. D446-07: *Standard Specifications for Glass Capillary Kinematic Viscometers*. West Conshohocken, PA: Author. DOI: 10.1520/D446-07, www.astm.org.

3. _____. 2010. D2270-10: *Standard Practice for Calculating Viscosity Index from Kinematic Viscosity at 40 and 100°C*. West Conshohocken, PA: Author. DOI: 10.1520/D2270-10, www.astm.org.

4. _____2007. D2422-07: *Standard Classification of Industrial Lubricants by Viscosity System*. West Conshohocken, PA: Author. DOI: 10.1520/D2422-07, www.astm.org.

5. _____ .2010. D5293-10e1: *Standard Test Method for Apparent Viscosity of Engine Oils and Base Stocks Between -5 and $-35°C$ Using the Cold-Cranking Simulator*. West Conshohocken, PA: Author. DOI: 10.1520/D5293-10, www.astm.org.

6. _____. 2009. D2983-09: *Standard Test Method for Low-Temperature Viscosity of Automotive Fluid Lubricants Measured by Brookfield Viscometer*. West Conshohocken, PA: Author. DOI: 10.1520/D2983-09, www.astm.org.

7. _____ . 2007. D3829-07: *Standard Test Method for Predicting the Borderline Pumping Temperature of Engine Oil*. West Conshohocken, PA: Author. DOI: 10.1520/D3829-07, www.astm.org.

8. _____ . 2010. D4683-10: *Standard Test Method for Measuring Viscosity of New and Used Engine Oils at High-Shear Rate and High Temperature by Tapered Bearing Simulator Viscometer at 150°C*. West Conshohocken, PA: Author. DOI: 10.1520/D4683-10, www.astm.org.

9. _____ . 2008. D4684-08: *Standard Test Method for Determination of Yield Stress and Apparent Viscosity of Engine Oils at Low Temperature*. West Conshohocken, PA: Author. DOI: 10.1520/D4684-08, www.astm.org.

10. _____ . 2007. D88-07: *Standard Test Method for Saybolt Viscosity*. West Conshohocken, PA: Author. DOI: 10.1520/D88-07, www.astm.org.

11. _____ . 2010. D2161-10: *Standard Practice for Conversion of Kinematic Viscosity to Saybolt Universal Viscosity or to Saybolt Furol Viscosity*. West Conshohocken, PA: Author. DOI: 10.1520/D2161-10, www.astm.org.

12. Heald, C. C., ed. 2002. *Cameron Hydraulic Data*, 19th ed. Irving, TX: Flowserve. (Earlier editions were published by Ingersoll-Dresser Pump Co., Liberty Corner, NJ.)

13. Schramm, Gebhard. 2002. *A Practical Approach to Rheology and Rheometry.* Karlsruhe, Germany: Thermo Haake.

14. SAE International (SAE). 2009. *SAE Standard J300: Engine Oil Viscosity Classification.* Warrendale, PA: Author.

15. _____ . 2005. *SAE Standard J306: Automotive Gear Lubricant Viscosity Classification.* Warrendale, PA: Author.

16. Votz, Sebastian. 2010. *Thermal Nanosystems and Nanomaterials.* New York: Springer Publishing.

INTERNET RESOURCES

1. **ASTM International:** Develops and publishes standards for testing procedures and properties of numerous kinds of materials, including fluids.

2. **SAE International:** The engineering society for advancing mobility—land, sea, air, and space. Publisher of numerous industry standards including viscosity of lubricants and fuels.

3. **ISO (International Organization for Standardization):** ISO is the world's largest developer of voluntary International Standards.

4. **The Coordinating European Council (CEC):** Developer of fluid performance test methods used extensively in Europe and widely throughout the world. Represents the motor, oil, petroleum additive, and allied industries in performance evaluation of transportation fuels, lubricants, and other fluids.

5. **Lord Corporation:** Producer of a wide variety of vibration mounts and damping devices, including magneto-rheological fluids and their applications. From the home page, select *Products & Solutions* and then *Magneto-Rheological (MR)*.

6. **Cannon Instrument Company:** Producer of many types of viscometers and other instruments for measuring fluid properties.

7. **Fisher Scientific:** Supplier of numerous instruments and materials for laboratory and scientific uses, including viscometers under the Fisher brand and many others.

8. **Kohler Instrument Company:** A leading producer and supplier of petroleum and petrochemical instrumentation worldwide, including manual and automated petroleum ASTM testing equipment for viscosity, density, and tribology friction and wear properties.

9. **Brookfield Engineering Laboratories:** The world's leading manufacturer of viscosity-measuring equipment for laboratory and process control applications.

10. **Malvern Instruments Ltd.:** The company designs, manufactures, and sells materials-characterization instruments, including rheometers, viscometers, and particle analysis devices.

11. **Thermo Scientific Corporation:** Producer of many types of measurement equipment for industry and scientific laboratories and production operations. The Haake Division produces several types of viscometers and rheometers including the falling ball and rotary types. Part of ThremoFisher Scientific Inc.

12. **PAC L.P.:** PAC is a leading global provider of advanced analytical instruments for laboratories and online process applications in industries such as refinery, petrochemical, biofuels, environmental, food & beverage, and pharmaceutical. Search on *paclp*. PAC consists of several product lines featuring viscosity measurement and testing of other fluid properties, such as Cambridge Viscosity, ISL, PetroSpec, Walter Herzog, and PSPI - Precision Scientific.

13. **Anton Paar:** Manufacturer of instruments for measuring viscosity, density, concentration, and other properties of fluids.

14. **Cole-Parmer Company:** Cole-Parmer is a leading global source of laboratory and industrial fluid handling products, instrumentation, equipment, and supplies, including viscometers, pumps, flowmeters, and other fluid mechanics related products. The site includes viscosity conversion calculators for both dynamic and kinematic viscosity along with lists of viscosity conversion factors.

15. **Wynn's USA:** From the home page, select *C.A.M.P.*, then *Products & Equipment.* Wynn's is a producer and distributor of automotive lubricant products including engine oil, transmission fluid, brake fluid, and general purpose lubricants. A division of Illinois Tool Works, Inc.

16. **Mobil Industrial Lubricants:** Producer of a wide range of industrial hydraulic oils and other industrial lubricants. The site includes a product search feature related to specific applications.

17. **Castrol Limited:** Producer of industrial and automotive oils and lubricants for construction, machinery, and general industrial hydraulic systems.

18. **CITGO Petroleum Corporation:** Producer of a full range of engine oils, hydraulic fluids, lubricants, and greases for the automotive, construction, and general industrial markets.

19. **Tribology-ABC:** Part of *Engineering-ABC*, a website with a huge set of data helpful in many kinds of engineering calculations. From the home page, select the letter *V*, then select *Viscosity* to connect to the page listing basic definitions of viscosity terms, ISO viscosity grades, AGMA viscosity classifications for gear oils, SAE viscosity grades for engine and automotive gear oils and a comparison of all of these classifications.

PRACTICE PROBLEMS

2.1 Define *shear stress* as it applies to a moving fluid.

2.2 Define *velocity gradient*.

2.3 State the mathematical definition for *dynamic viscosity*.

2.4 Which would have the greater dynamic viscosity, a cold lubricating oil or fresh water? Why?

2.5 State the standard units for dynamic viscosity in the SI system.

2.6 State the standard units for dynamic viscosity in the U.S. Customary System.

2.7 State the equivalent units for *poise* in terms of the basic quantities in the cgs system.

2.8 Why are the units of poise and centipoise considered obsolete?

2.9 State the mathematical definition for *kinematic viscosity*.

2.10 State the standard units for kinematic viscosity in the SI system.

2.11 State the standard units for kinematic viscosity in the U.S. Customary System.

2.12 State the equivalent units for *stoke* in terms of the basic quantities in the cgs system.

2.13 Why are the units of stoke and centistoke considered obsolete?

2.14 Define a *Newtonian fluid*.

2.15 Define a *non-Newtonian fluid*.

2.16 Give five examples of Newtonian fluids.

2.17 Give four examples of the types of fluids that are non-Newtonian.

Appendix D gives dynamic viscosity for a variety of fluids as a function of temperature. Using this appendix, give the value of the viscosity for the following fluids:

2.18 Water at 40°C.

2.19 Water at 5°C.

2.20 Air at 40°C.

2.21 Hydrogen at 40°C.

2.22 Glycerin at 40°C.

2.23 Glycerin at 20°C.

2.24 Water at 40°F.

2.25 Water at 150°F.

2.26 Air at 40°F.

2.27 Hydrogen at 40°F.

2.28 Glycerin at 60°F.

2.29 Glycerin at 110°F.

2.30 Mercury at 60°F.

2.31 Mercury at 210°F.

2.32 SAE 10 oil at 60°F.

2.33 SAE 10 oil at 210°F.

2.34 SAE 30 oil at 60°F.

2.35 SAE 30 oil at 210°F.

2.36 Define *viscosity index* (*VI*).

2.37 If you want to choose a fluid that exhibits a small change in viscosity as the temperature changes, would you choose one with a high *VI* or a low *VI*?

2.38 Which type of viscosity measurement method uses the basic definition of dynamic viscosity for direct computation?

2.39 In the *rotating-drum viscometer,* describe how the velocity gradient is created in the fluid to be measured.

2.40 In the *rotating-drum viscometer,* describe how the magnitude of the shear stress is measured.

2.41 What measurements must be taken to determine dynamic viscosity when using a *capillary tube viscometer*?

2.42 Define the term *terminal velocity* as it applies to *a falling-ball viscometer.*

2.43 What measurements must be taken to determine dynamic viscosity when using the *falling-ball viscometer*?

2.44 Describe the basic features of the *Saybolt Universal viscometer.*

2.45 Are the results of the Saybolt viscometer tests considered to be direct measurements of viscosity?

2.46 Does the Saybolt viscometer produce data related to a fluid's dynamic viscosity or kinematic viscosity?

2.47 Which type of viscometer is prescribed by SAE for measurements of viscosity of oils at 100°C?

2.48 Describe the difference between an SAE 20 oil and an SAE 20W oil.

2.49 What grades of SAE oil are suitable for lubricating the crankcases of engines?

2.50 What grades of SAE oil are suitable for lubricating gear-type transmissions?

2.51 If you were asked to check the viscosity of an oil that is described as SAE 40, at what temperatures would you make the measurements?

2.52 If you were asked to check the viscosity of an oil that is described as SAE 10W, at what temperatures would you make the measurements?

2.53 How would you determine the viscosity of an oil labeled SAE 5W-40 for comparison with SAE standards?

2.54 The viscosity of a lubricating oil is given as 500 SUS at 100°F. Calculate the viscosity in m^2/s and ft^2/s.

2.55 Using the data from Table 2.5, report the minimum, nominal, and maximum values for viscosity for ISO grades VG 10, VG 65, VG 220, and VG 1000.

2.56 Convert a dynamic viscosity measurement of 4500 cP into Pa·s and $lb·s/ft^2$.

2.57 Convert a kinematic viscosity measurement of 5.6 cSt into m^2/s and ft^2/s.

2.58 The viscosity of an oil is given as 80 SUS at 100°F. Determine the viscosity in m^2/s.

2.59 Convert a viscosity measurement of 6.5×10^{-3} Pa·s into the units of $lb·s/ft^2$.

2.60 An oil container indicates that it has a viscosity of 0.12 poise at 60°C. Which oil in Appendix D has a similar viscosity?

2.61 In a falling-ball viscometer, a steel ball 1.6 mm in diameter is allowed to fall freely in a heavy fuel oil having a specific gravity of 0.94. Steel weighs 77 kN/m^3. If the ball is observed to fall 250 mm in 10.4 s, calculate the viscosity of the oil.

2.62 A capillary tube viscometer similar to that shown in Fig. 2.7 is being used to measure the viscosity of an oil having a specific gravity of 0.90. The following data apply:
Tube inside diameter = 2.5 mm = D
Length between manometer taps = 300 mm = L
Manometer fluid = mercury
Manometer deflection = 177 mm = h
Velocity of flow = 1.58 m/s = v
Determine the viscosity of the oil.

2.63 In a falling-ball viscometer, a steel ball with a diameter of 0.063 in is allowed to fall freely in a heavy fuel oil having a specific gravity of 0.94. Steel weighs 0.283 lb/in^3. If the ball is observed to fall 10.0 in in 10.4 s, calculate the dynamic viscosity of the oil in $lb·s^2/ft$.

2.64 A capillary type viscometer similar to that shown in Fig. 2.7 is being used to measure the viscosity of an oil having a specific gravity of 0.90. The following data apply:
Tube inside diameter = 0.100 in = D
Length between manometer taps = 12.0 in = L
Manometer fluid = mercury
Manometer deflection = 7.00 in = h
Velocity of flow = 4.82 ft/s = v
Determine the dynamic viscosity of the oil in $lb·s^2/ft$.

2.65 A fluid has a kinematic viscosity of 15.0 mm^2/s at 100°F. Determine its equivalent viscosity in SUS at that temperature.

2.66 A fluid has a kinematic viscosity of 55.3 mm^2/s at 100°F. Determine its equivalent viscosity in SUS at that temperature.

2.67 A fluid has a kinematic viscosity of 188 mm^2/s at 100°F. Determine its equivalent viscosity in SUS at that temperature.

2.68 A fluid has a kinematic viscosity of 244 mm^2/s at 100°F. Determine its equivalent viscosity in SUS at that temperature.

2.69 A fluid has a kinematic viscosity of 153 mm^2/s at 40°F. Determine its equivalent viscosity in SUS at that temperature.

2.70 A fluid has a kinematic viscosity of 205 mm^2/s at 190°F. Determine its equivalent viscosity in SUS at that temperature.

2.71 An oil is tested using a Saybolt viscometer and its viscosity is 6250 SUS at 100°F. Determine the kinematic viscosity of the oil in mm²/s at that temperature.

2.72 An oil is tested using a Saybolt viscometer and its viscosity is 438 SUS at 100°F. Determine the kinematic viscosity of the oil in mm²/s at that temperature.

2.73 An oil is tested using a Saybolt viscometer and its viscosity is 68 SUS at 100°F. Determine the kinematic viscosity of the oil in mm²/s at that temperature.

2.74 An oil is tested using a Saybolt viscometer and its viscosity is 176 SUS at 100°F. Determine the kinematic viscosity of the oil in mm²/s at that temperature.

2.75 An oil is tested using a Saybolt viscometer and its viscosity is 4690 SUS at 80°C. Determine the kinematic viscosity of the oil in mm²/s at that temperature.

2.76 An oil is tested using a Saybolt viscometer and its viscosity is 526 SUS at 40°C. Determine the kinematic viscosity of the oil in mm²/s at that temperature.

2.77 Convert all of the kinematic viscosity data in Table 2.5 for ISO viscosity grades from mm²/s (cSt) to SUS.

COMPUTER AIDED ENGINEERING ASSIGNMENTS

1. Write a program to convert viscosity units from any given system to another system using the conversion factors and techniques from Appendix K.

2. Write a program to determine the viscosity of water at a given temperature using data from Appendix A. This program could be joined with the one you wrote in Chapter 1, which used other properties of water. Use the same options described in Chapter 1.

3. Use a spreadsheet to display the values of kinematic viscosity and dynamic viscosity of water from Appendix A. Then create curve-fit equations for both types of viscosity versus temperature using the *Trendlines* feature of the spreadsheet chart. Display graphs for both viscosities versus temperature on the spreadsheet showing the equations used.

PRESSURE MEASUREMENT

Review the definition of pressure from Chapter 1:

$$p = F/A \qquad (3\text{–}1)$$

Pressure equals force divided by area.

> *The standard unit for pressure in the SI system is N/m^2, called the pascal (Pa).*
> *Other convenient SI units for fluid mechanics are the kPa, MPa, and bar.*
> *The standard unit for pressure in the U.S. Customary System is lb/ft^2.*
> *A convenient U.S. unit for fluid mechanics is lb/in^2, often called psi.*

In this chapter you will learn about commonly used methods of measuring and reporting values for pressure in a fluid and pressure difference between two points in a fluid system. Important concepts and terms include *absolute pressure, gage pressure, the relationship between pressure and changes in elevation within the fluid, the standard atmosphere, and Pascal's paradox.* You will also learn about several types of pressure measurement devices and equipment such as *manometers, barometers, pressure gages, and pressure transducers.*

Exploration

Think about situations where you observed pressure being measured or reported and try to recall the magnitude of the pressure, how it was measured, the units in which the pressure was reported, and the type of equipment that generated the pressure. Perhaps you have been in a scene like that shown in Fig. 3.1!

What examples of pressure measurement can you recall? Here are a few to get you started.

- Have you measured the pressure in tires for automobiles or bicycles?
- Have you observed the pressure reading on a steam or hot water boiler?
- Have you measured the pressure in a water supply system or observed places where the pressure was particularly low or high?
- Have you seen pressure gages mounted on pumps or at key components of hydraulic or pneumatic fluid power systems?
- Have you heard weather reports giving the pressure of the atmosphere, sometimes called the barometric pressure?
- Have you experienced increased pressure on your body as you dive deeper into water?
- Have you gone scuba diving?
- Have you seen movies in which submarines or undersea research vehicles are used?
- Have you visited places (like Denver, Colorado or gone mountain climbing) or flown at high altitudes where the air pressure is significantly lower than when you were on the ground and nearer to sea level?

Discuss these situations and others you can recall among your fellow students and with the course instructor.

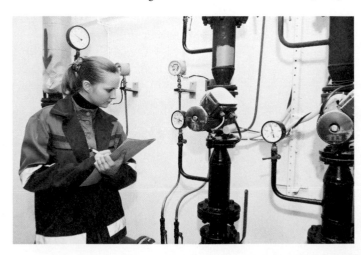

FIGURE 3.1 Knowing how to read and interpret pressure measurements in a laboratory, commercial building systems, and industrial processes is an important skill. (*Source:* Kadmy/Fotolia)

3.1 OBJECTIVES

After completing this chapter, you should be able to:

1. Define the relationship between *absolute pressure, gage pressure*, and *atmospheric pressure.*

2. Describe the degree of variation of atmospheric pressure near Earth's surface.

3. Describe the properties of air at standard atmospheric pressure.

4. Describe the properties of the atmosphere at elevations from sea level to 30 000 m (about 100 000 ft).

5. Define the relationship between a change in elevation and the change in pressure in a fluid.

6. Describe how a *manometer* works and how it is used to measure pressure.

7. Describe a U-tube manometer, a differential manometer, a well-type manometer, and an inclined well-type manometer.

8. Describe a *barometer* and how it indicates the value of the local atmospheric pressure.

9. Describe various types of pressure gages and pressure transducers.

3.2 ABSOLUTE AND GAGE PRESSURE

When making calculations involving pressure in a fluid, you must make the measurements relative to some reference pressure.

Normally the reference pressure is that of the atmosphere, and the resulting measured pressure is called gage pressure.

Pressure measured relative to a perfect vacuum is called absolute pressure.

It is extremely important for you to know the difference between these two ways of measuring pressure and to be able to convert from one to the other. A simple equation relates the two pressure-measuring systems:

▷ **Absolute and Gage Pressure**

$$p_{abs} = p_{gage} + p_{atm} \qquad (3\text{–}2)$$

where

$$p_{abs} = \text{Absolute pressure}$$
$$p_{gage} = \text{Gage pressure}$$
$$p_{atm} = \text{Atmospheric pressure}$$

Figure 3.2 shows a graphical interpretation of this equation. Some basic concepts may help you understand the equation and the graphic display in the figure:

1. A perfect vacuum is the lowest possible pressure. Therefore, an absolute pressure will always be positive.

2. A gage pressure above atmospheric pressure is positive.

3. A gage pressure below atmospheric pressure is negative, sometimes called *vacuum.*

4. Gage pressure will be indicated in the units of Pa(gage) or psig.

5. Absolute pressure will be indicated in the units of Pa(abs) or psia.

6. The magnitude of the atmospheric pressure varies with location and with climatic conditions. The barometric pressure as broadcast in weather reports is an indication of the continually varying atmospheric pressure.

7. The range of normal variation of atmospheric pressure near Earth's surface is approximately from 95 kPa(abs) to 105 kPa(abs), or from 13.8 psia to 15.3 psia.

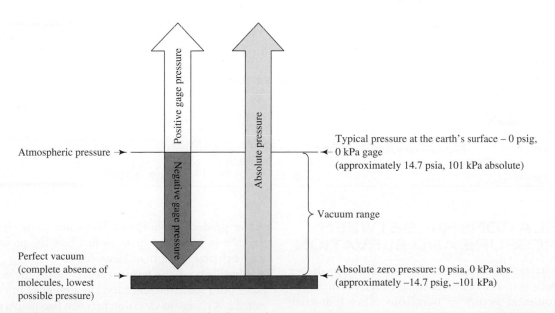

FIGURE 3.2 Comparison between absolute and gage pressures.

8. At sea level, the standard atmospheric pressure is 101.3 kPa(abs), or 14.69 psia.

9. In this book we will assume the atmospheric pressure to be 101 kPa(abs), or 14.7 psia, unless the prevailing atmospheric pressure is given.

Example Problem 3.1

Express a pressure of 155 kPa(gage) as an absolute pressure. The local atmospheric pressure is 98 kPa(abs).

Solution

$$p_{abs} = p_{gage} + p_{atm}$$
$$p_{abs} = 155 \text{ kPa(gage)} + 98 \text{ kPa(abs)} = 253 \text{ kPa(abs)}$$

Notice that the units in this calculation are kilopascals (kPa) for each term and are consistent. The indication of gage or absolute is for convenience and clarity.

Example Problem 3.2

Express a pressure of 225 kPa(abs) as a gage pressure. The local atmospheric pressure is 101 kPa(abs).

Solution

$$p_{abs} = p_{gage} + p_{atm}$$

Solving algebraically for p_{gage} gives

$$p_{gage} = p_{abs} - p_{atm}$$
$$p_{gage} = 225 \text{ kPa(abs)} - 101 \text{ kPa(abs)} = 124 \text{ kPa(gage)}$$

Example Problem 3.3

Express a pressure of 10.9 psia as a gage pressure. The local atmospheric pressure is 15.0 psia.

Solution

$$p_{abs} = p_{gage} + p_{atm}$$
$$p_{gage} = p_{abs} - p_{atm}$$
$$p_{gage} = 10.9 \text{ psia} - 15.0 \text{ psia} = -4.1 \text{ psig}$$

Notice that the result is negative. This can also be read "4.1 psi below atmospheric pressure" or "4.1 psi vacuum."

Example Problem 3.4

Express a pressure of −6.2 psig as an absolute pressure.

Solution

$$p_{abs} = p_{gage} + p_{atm}$$

Because no value was given for the atmospheric pressure, we will use $p_{atm} = 14.7$ psia:

$$p_{abs} = -6.2 \text{ psig} + 14.7 \text{ psia} = 8.5 \text{ psia}$$

3.3 RELATIONSHIP BETWEEN PRESSURE AND ELEVATION

You are probably familiar with the fact that as you go deeper in a fluid, such as in a swimming pool, the pressure increases. This phenomenon occurs in numerous other industrial, transportation, aerospace, appliance, commercial, and con- sumer product applications. There are many situations in which it is important to know just how the pressure varies with a change in depth or elevation.

In this book the term *elevation* means the vertical dis- tance from some reference level to a point of interest and is called *z*. A *change* in elevation between two points is called *h*.

FIGURE 3.3 Illustration of reference level for elevation.

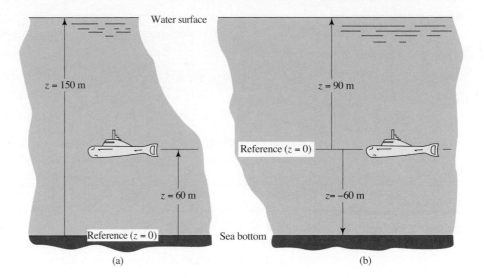

(a)   (b)

Elevation will always be measured positively in the upward direction. In other words, a higher point has a larger elevation than a lower point.

The reference level can be taken at any level, as illustrated in Fig. 3.3, which shows a submarine under water. In part (a) of the figure the sea bottom is taken as reference, whereas in part (b) the position of the submarine is the reference level. Because fluid mechanics calculations usually consider differences in elevation, it is advisable to choose the lowest point of interest in a problem as the reference level to eliminate the use of negative values for z. This will be especially important in later work in Chapters 6–13.

The change in pressure in a homogeneous liquid at rest due to a change in elevation can be calculated from

▷ **Pressure–Elevation Relationship**

$$\Delta p = \gamma h \qquad (3\text{–}3)$$

where

Δp = Change in pressure
γ = Specific weight of liquid
h = Change in elevation

Some general conclusions from Eq. (3–3) will help you to apply it properly:

1. The equation is valid only for a homogeneous liquid at rest.
2. Points on the same horizontal level have the same pressure.
3. The change in pressure is directly proportional to the specific weight of the liquid.
4. Pressure varies linearly with the change in elevation or depth.
5. A decrease in elevation causes an increase in pressure. (This is what happens when you go deeper in a swimming pool.)
6. An increase in elevation causes a decrease in pressure.

Equation (3–3) does not apply to gases because the specific weight of a gas changes with a change in pressure. However, it requires a large change in elevation to produce a significant change in pressure in a gas. For example, an increase in elevation of 300 m (about 1000 ft) in the atmosphere causes a decrease in pressure of only 3.4 kPa (about 0.5 psi). *In this book we assume that the pressure in a gas is uniform unless otherwise specified.*

**Example Problem
3.5**

Calculate the change in water pressure from the surface to a depth of 5 m.

Solution

Use Eq. (3–3), $\Delta p = \gamma h$, and let $\gamma = 9.81$ kN/m^3 for water and $h = 5$ m. Then we have

$$\Delta p = (9.81 \text{ kN/m}^3)(5.0 \text{ m}) = 49.05 \text{ kN/m}^2 = 49.05 \text{ kPa}$$

If the surface of the water is exposed to the atmosphere, the pressure there is 0 Pa(gage). Descending in the water (decreasing elevation) produces an increase in pressure. Therefore, at 5 m the pressure is 49.05 kPa(gage).

Example Problem 3.6

Calculate the change in water pressure from the surface to a depth of 15 ft.

Solution

Use Eq. (3–3), $\Delta p = \gamma h$, and let $\gamma = 62.4$ lb/ft^3 for water and $h = 15$ ft. Then we have

$$\Delta p = \frac{62.4 \text{ lb}}{\text{ft}^3} \times 15 \text{ ft} \times \frac{1 \text{ ft}^2}{144 \text{ in}^2} = 6.5\frac{\text{lb}}{\text{in}^2}$$

If the surface of the water is exposed to the atmosphere, the pressure there is 0 psig. Descending in the water (decreasing elevation) produces an increase in pressure. Therefore, at 15 ft the pressure is 6.5 psig.

Example Problem 3.7

Figure 3.4 shows a tank of oil with one side open to the atmosphere and the other side sealed with air above the oil. The oil has a specific gravity of 0.90. Calculate the gage pressure at points A, B, C, D, E, and F and the air pressure in the right side of the tank.

Solution

Point A At this point, the oil is exposed to the atmosphere, and therefore

$$p_A = 0 \text{ Pa(gage)}$$

Point B The change in elevation between point A and point B is 3.0 m, with B lower than A. To use Eq. (3–3) we need the specific weight of the oil:

$$\gamma_{\text{oil}} = (sg)_{\text{oil}}(9.81 \text{ kN/m}^3) = (0.90)(9.81 \text{ kN/m}^3) = 8.83 \text{ kN/m}^3$$

Then, we have

$$\Delta p_{A-B} = \gamma h = (8.83 \text{ kN/m}^3)(3.0 \text{ m}) = 26.5 \text{ kN/m}^2 = 26.5 \text{ kPa}$$

Now, the pressure at B is

$$p_B = p_A + \Delta p_{A-B} = 0 \text{ Pa(gage)} + 26.5 \text{ kPa} = 26.5 \text{ kPa(gage)}$$

Point C The change in elevation from point A to point C is 6.0 m, with C lower than A. Then, the pressure at point C is

$$\Delta p_{A-C} = \gamma h = (8.83 \text{ kN/m}^3)(6.0 \text{ m}) = 53.0 \text{ kN/m}^2 = 53.0 \text{ kPa}$$
$$p_C = p_A + \Delta p_{A-C} = 0 \text{ Pa(gage)} + 53.0 \text{ kPa} = 53.0 \text{ kPa(gage)}$$

Point D Because point D is at the same level as point B, the pressure is the same. That is, we have

$$p_D = p_B = 26.5 \text{ kPa(gage)}$$

Point E Because point E is at the same level as point A, the pressure is the same. That is, we have

$$p_E = p_A = 0 \text{ Pa(gage)}$$

Point F The change in elevation between point A and point F is 1.5 m, with F higher than A. Then, the pressure at F is

$$\Delta p_{A-F} = -\gamma h = (-8.83 \text{ kN/m}^3)(1.5 \text{ m}) = -13.2 \text{ kN/m}^2 = -13.2 \text{ kPa}$$
$$p_F = p_A + \Delta p_{A-F} = 0 \text{ Pa(gage)} + (-13.2 \text{ kPa}) = -13.2 \text{ kPa(gage)}$$

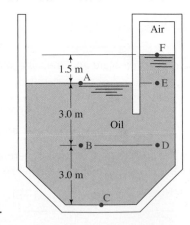

FIGURE 3.4 Tank for Example Problem 3.7.

Air Pressure   Because the air in the right side of the tank is exposed to the surface of the oil, where $p_F = -13.2$ kPa, the air pressure is also -13.2 kPa, (gage), or 13.2 kPa below atmospheric pressure.

3.3.1 Summary of Observations from Example Problem 3.7

The results from Problem 3.7 are summarized below and they illustrate general conclusions that can be applied when using Eq. (3–3):

a. The pressure increases as the depth in the fluid increases. This result can be seen from $p_C > p_B > p_A$.

b. Pressure varies linearly with a change in elevation; that is, p_C is two times greater than p_B, and C is at twice the depth of B.

c. Pressure on the same horizontal level is the same. Note that $p_E = p_A$ and $p_D = p_B$.

d. The decrease in pressure from E to F occurs because point F is at a higher elevation than point E. Note that p_F is negative; that is, it is below the atmospheric pressure that exists at A and E.

3.4 DEVELOPMENT OF THE PRESSURE–ELEVATION RELATION

In Section 3.3 we introduced Eq. (3–3) as the relationship between a change in elevation in a liquid, h, and a change in pressure, Δp, stated as,

$$\Delta p = \gamma h \qquad (3–3)$$

where γ is the specific weight of the liquid. This section presents the basis for this equation.

Figure 3.5 shows a body of static fluid with a specific weight, γ, and a small cylindrical volume of the fluid some where below the surface. The actual shape of the volume is arbitrary.

Because the entire body of fluid is stationary and in equilibrium, the small cylinder of the fluid is also in equilibrium. From physics, we know that for a body in static equilibrium, the sum of forces acting on it in all directions must be zero.

First consider the forces acting in the horizontal direction. A thin ring around the cylinder is shown at some arbitrary elevation in Fig. 3.6. The vectors acting on the ring represent the horizontal forces exerted on it by the fluid pressure. Recall from previous work that the pressure at any horizontal level in a static fluid is the same. Also recall that the pressure at a boundary, and therefore the force due to the pressure, acts perpendicular to the boundary. We can then conclude that the horizontal forces are completely balanced around the sides of the cylinder.

Now consider Fig. 3.7. The forces acting on the cylinder in the vertical direction are shown. The following concepts are illustrated in the figure:

1. The fluid pressure at the level of the bottom of the cylinder is called p_1.

2. The fluid pressure at the level of the top of the cylinder is called p_2.

3. The elevation difference between the top and the bottom of the cylinder is called dz, where $dz = z_2 - z_1$.

4. The pressure change that occurs in the fluid between the level of the bottom and the top of the cylinder is called dp. Therefore, $p_2 = p_1 + dp$.

5. The area of the top and bottom of the cylinder is called A.

6. The volume of the cylinder is the product of the area A and the height of the cylinder dz. That is, $V = A(dz)$.

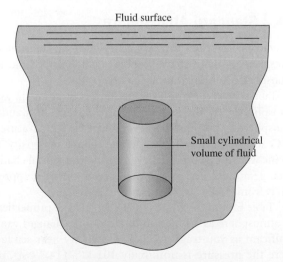

FIGURE 3.5 Small volume of fluid within a body of static fluid.

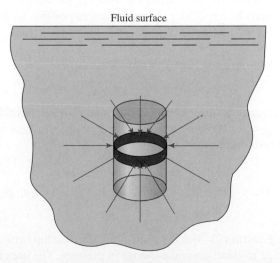

FIGURE 3.6 Pressure forces acting in a horizontal plane on a thin ring.

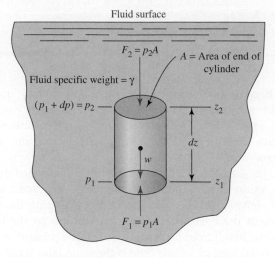

FIGURE 3.7 Forces acting in the vertical direction.

7. The weight of the fluid within the cylinder is the product of the specific weight of the fluid γ and the volume of the cylinder. That is, $w = \gamma V = \gamma A(dz)$. The weight is a force acting on the cylinder in the downward direction through the centroid of the cylindrical volume.

8. The force acting on the bottom of the cylinder due to the fluid pressure p_1 is the product of the pressure and the area A. That is, $F_1 = p_1 A$. This force acts vertically upward, perpendicular to the bottom of the cylinder.

9. The force acting on the top of the cylinder due to the fluid pressure p_2 is the product of the pressure and the area A. That is, $F_2 = p_2 A$. This force acts vertically downward, perpendicular to the top of the cylinder. Because $p_2 = p_1 + dp$, another expression for the force F_2 is

$$F_2 = (p_1 + dp)A \qquad (3\text{–}4)$$

Now we can apply the principle of static equilibrium, which states that the sum of the forces in the vertical direction must be zero. Using upward forces as positive, we get

$$\sum F_v = 0 = F_1 - F_2 - w \qquad (3\text{–}5)$$

Substituting from Steps 7–9 gives

$$p_1 A - (p_1 + dp)A - \gamma(dz)A = 0 \qquad (3\text{–}6)$$

Notice that the area A appears in all terms on the left side of Eq. (3–6). It can be eliminated by dividing all terms by A. The resulting equation is

$$p_1 - p_1 - dp - \gamma(dz) = 0 \qquad (3\text{–}7)$$

Now the p_1 term can be cancelled out. Solving for dp gives

$$dp = -\gamma(dz) \qquad (3\text{–}8)$$

Equation (3–8) is the controlling relationship between a change in elevation and a change in pressure. The use of Eq. (3–8), however, depends on the kind of fluid. Remember that the equation was developed for a very small element of the fluid. The process of integration extends Eq. (3–8) to large changes in elevation, as follows:

$$\int_{p_1}^{p_2} dp = \int_{z_1}^{z_2} -\gamma(dz) \qquad (3\text{–}9)$$

Equation (3–9) develops differently for liquids and for gases because the specific weight is constant for liquids and it varies with changes in pressure for gases.

3.4.1 Liquids

A liquid is considered to be incompressible. Thus, its specific weight γ is a constant. This allows γ to be taken outside the integral sign in Eq. (3–9). Then,

$$\int_{p_1}^{p_2} dp = -\gamma \int_{z_1}^{z_2} (dz) \qquad (3\text{–}10)$$

Completing the integration process and applying the limits gives

$$p_2 - p_1 = -\gamma(z_2 - z_1) \qquad (3\text{–}11)$$

For convenience, we define $\Delta p = p_2 - p_1$ and $h = z_1 - z_2$. Equation (3–11) becomes

$$\Delta p = \gamma h$$

which is identical to Eq. (3–3). The signs for Δp and h can be assigned at the time of use of the formula by recalling that pressure increases as depth in the fluid increases and vice versa.

3.4.2 Gases

Because a gas is compressible, its specific weight changes as pressure changes. To complete the integration process called for in Eq. (3–9), you must know the relationship between the change in pressure and the change in specific weight. The relationship is different for different gases, but a complete discussion of those relationships is beyond the scope of this text and requires the study of thermodynamics.

3.4.3 Standard Atmosphere

Appendix E describes the properties of air in the *standard atmosphere* as defined by the U.S. National Oceanic and Atmospheric Administration (NOAA).

Tables E1 and E2 give the properties of air at standard atmospheric pressure as temperature varies. The standard atmosphere is taken at sea level and at a temperature of $15°\,C$ as listed below Table E1. The change in density and specific weight is substantial even within the typical changes in temperature experienced in temperate climates, approximately from $-30°\,C$ $(-22°\,F)$ to $40°\,C$ $(104°\,F)$.

Table E3 and the graphs in Fig. E1 give the properties of the atmosphere as a function of elevation. Changes can be significant as you travel from a coastal city near sea level, where the pressure is nominally 101 kPa (14.7 psi), to a mountain town that may be at an altitude of 3000 m (9850 ft) or more, where the pressure is only about 70 kPa (10 psi), a

FIGURE 3.8 Illustration of Pascal's paradox.

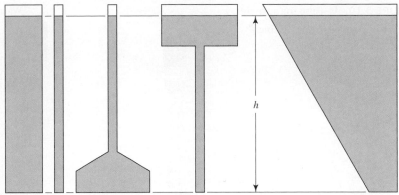

Pressure is the same at the bottom of all containers if the same fluid is in all containers.

reduction of about 31 percent. The density of air decreases by approximately 26 percent. Commercial aircraft often fly at 10 000 m (32 800 ft) or higher, where the pressure is approximately 27 kPa (4.0 psi), requiring pressurization of the fuselage. Here the air density is only about 0.4 kg/m³, compared with 1.23 kg/m³ at sea level, dramatically affecting the lift forces on the aircraft's wings.

3.5 PASCAL'S PARADOX

In the development of the relationship $\Delta p = \gamma h$, the shape and size of the small volume of fluid does not affect the result. The change in pressure depends only on the change in elevation and the type of fluid, not on the shape or size of the fluid container. Therefore, all the containers shown in Fig. 3.8 would have the same pressure at the bottom, even though they contain vastly different amounts of fluid. This observation is called *Pascal's paradox*, in honor of Blaise

Pascal, the seventeenth-century scientist who contributed much to the world's knowledge of the behavior of fluids.

This phenomenon is useful when a consistently high pressure must be produced on a system of interconnected pipes and tanks. Water systems for cities often include water towers located on high hills, as shown in Fig. 3.9. Besides providing a reserve supply of water, the primary purpose of such tanks is to maintain a sufficiently high pressure in the water system for satisfactory delivery of the water to residential, commercial, and industrial users.

In industrial or laboratory applications, a standpipe containing a static liquid can be used to create a stable pressure on a particular process or system. It is positioned at a high elevation relative to the system and is connected to the system by pipes. Raising or lowering the level of the fluid in the standpipe changes the pressure in the system. Standpipes are sometimes placed on the roofs of buildings to maintain water pressure in local fire-fighting systems.

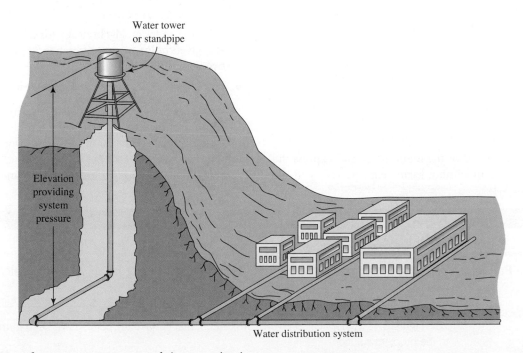

FIGURE 3.9 Use of a water tower or a standpipe to maintain water system pressure.

FIGURE 3.10 U-tube manometer.
(*Source:* Dwyer Instruments, Inc.)

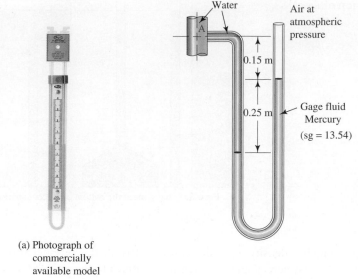

(a) Photograph of
commercially
available model

3.6 MANOMETERS

This and following sections describe several types of pressure-measurement devices. The first is the *manometer*, which uses the relationship between a change in pressure and a change in elevation in a static fluid, $\Delta p = \gamma h$ (see Sections 3.3 and 3.4). Photographs of commercially available manometers are shown in Figs. 3.10, 3.13, and 3.14 (see Internet resource 1).

The simplest kind of manometer is the U-tube (Fig. 3.10). One end of the U-tube is connected to the pressure that is to be measured and the other end is left open to the atmosphere. The tube contains a liquid called the *gage fluid*, which does not mix with the fluid whose pressure is to be measured. Typical gage fluids are water, mercury, and colored light oils.

Under the action of the pressure to be measured, the gage fluid is displaced from its normal position. Because the fluids in the manometer are at rest, the equation $\Delta p = \gamma h$ can be used to write expressions for the changes in pressure that occur throughout the manometer. These expressions can then be combined and solved algebraically for the desired pressure. Because manometers are used in many real situations such as those included in Chapters 6–13, you should learn the following step-by-step procedure.

Procedure for Writing the Equation for A Manometer

1. Start from one end of the manometer and express the pressure there in symbol form (e.g., p_A refers to the pressure at point A). If one end is open as shown in Fig. 3.10, the pressure is atmospheric pressure, taken to be zero gage pressure.

2. Add terms representing changes in pressure using $\Delta p = \gamma h$, proceeding from the starting point and including each column of each fluid separately.

3. When the movement from one point to another is downward, the pressure increases and the value of Δp is added. Conversely, when the movement from one point to the next is upward, the pressure decreases and Δp is subtracted.

4. Continue this process until the other end point is reached. The result is an expression for the pressure at that end point. Equate this expression to the symbol for the pressure at the final point, giving a complete equation for the manometer.

5. Solve the equation algebraically for the desired pressure at a given point or the difference in pressure between two points of interest.

6. Enter known data and solve for the desired pressure.

Working several practice problems will help you to apply this procedure correctly. The following problems are written in the programmed instruction format. To work through the program, cover the material below the heading "Programmed Example Problems" and then uncover one panel at a time.

PROGRAMMED EXAMPLE PROBLEMS

Example Problem 3.8 Using Fig. 3.10, calculate the pressure at point A. Perform Step 1 of the procedure before going to the next panel.

Figure 3.11 is identical to Fig. 3.10(b) except that certain key points have been numbered for use in the problem solution.

FIGURE 3.11 U-tube manometer.

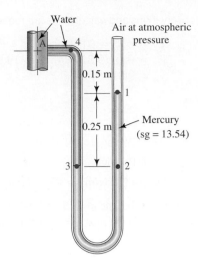

The only point for which the pressure is known is the surface of the mercury in the right leg of the manometer, point 1, and we can call that pressure p_1. Now, how can an expression be written for the pressure that exists within the mercury, 0.25 m below this surface at point 2?

The expression is

$$p_1 + \gamma_m(0.25\,\text{m})$$

The term $\gamma_m(0.25\,\text{m})$ is the change in pressure between points 1 and 2 due to a change in elevation, where γ_m is the specific weight of mercury, the gage fluid. This pressure change is added to p_1 because there is an increase in pressure as we descend in a fluid.

So far we have an expression for the pressure at point 2 in the right leg of the manometer. Now write the expression for the pressure at point 3 in the left leg.

This is the expression:

$$p_1 + \gamma_m(0.25\,\text{m})$$

Because points 2 and 3 are on the same level in the same fluid at rest, their pressures are equal.

Continue and write the expression for the pressure at point 4.

$$p_1 + \gamma_m(0.25\,\text{m}) - \gamma_w(0.40\,\text{m})$$

where γ_w is the specific weight of water. Remember, there is a decrease in pressure between points 3 and 4, so this last term must be subtracted from our previous expression.

What must you do to get an expression for the pressure at point A?

Nothing. Because points A and 4 are on the same level, their pressures are equal.

Now perform Step 4 of the procedure.

By setting the previous relationship to the pressure at point A, you should now have

$$p_1 + \gamma_m(0.25\,\text{m}) - \gamma_w(0.40\,\text{m}) = p_A$$

or, expressed as the equation for the pressure at point A,

$$p_A = p_1 + \gamma_m(0.25\,\text{m}) - \gamma_w(0.40\,\text{m})$$

Be sure to write the complete equation for the pressure at point A. Now do Steps 5 and 6.

Several observations and calculations are required here:

$$p_1 = p_{atm} = 0\,Pa(gage)$$
$$\gamma_m = (sg)_m(9.81\,kN/m^3) = (13.54)(9.81\,kN/m^3) = 132.8\,kN/m^3$$
$$\gamma_w = 9.81\,kN/m^3$$

Then, we have

$$p_A = p_1 + \gamma_m(0.25\,m) - \gamma_w(0.40\,m)$$
$$= 0\,Pa(gage) + (132.8\,kN/m^3)(0.25\,m) - (9.81\,kN/m^3)(0.40\,m)$$
$$= 0\,Pa(gage) + 33.20\,kN/m^2 - 3.92\,kN/m^2$$
$$p_A = 29.28\,kN/m^2 = 29.28\,kPa(gage)$$

Remember to include the units in your calculations. Review this problem to be sure you understand every step before going to the next panel for another problem.

Example Problem 3.9

Calculate the difference in pressure between points A and B in Fig. 3.12 and express it as $p_B - p_A$.
     This type of manometer is called a *differential manometer* because it indicates the difference between the pressure at two points but not the actual value of either one. Do Step 1 of the procedure to write the equation for the manometer.

We could start either at point A or point B. Let's start at A and call the pressure there p_A. Now write the expression for the pressure at point 1 in the left leg of the manometer.

You should have

$$p_A + \gamma_o(33.75\,in)$$

where γ_o is the specific weight of the oil. Note the use of the complete change in elevation from point A to point 1.
     What is the pressure at point 2?

It is the same as the pressure at point 1 because the two points are on the same level. Go on to point 3 in the manometer.

FIGURE 3.12 Differential manometer.

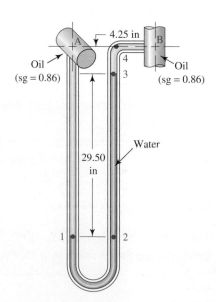

The expression should now look like this:

$$p_A + \gamma_o(33.75 \text{ in}) - \gamma_w(29.5 \text{ in})$$

Now write the expression for the pressure at point 4.

This is the desired expression:

$$p_A + \gamma_o(33.75 \text{ in}) - \gamma_w(29.5 \text{ in}) - \gamma_o(4.25 \text{ in})$$

This is also the expression for the pressure at B because points 4 and B are on the same level. Now do Steps 4–6 of the procedure.

Our final expression should be the complete manometer equation

$$p_A + \gamma_o(33.75 \text{ in}) - \gamma_w(29.5 \text{ in}) - \gamma_o(4.25 \text{ in}) = p_B$$

or, solving for the requested form of the differential pressure $p_B - p_A$,

$$p_B - p_A = \gamma_o(33.75 \text{ in}) - \gamma_w(29.5 \text{ in}) - y_o(4.25 \text{ in})$$

The known values are

$$\gamma_o = (sg)_o(62.4 \text{ lb/ft}^3) = (0.86)(62.4 \text{ lb/ft}^3) = 53.7 \text{ lb/ft}^3$$
$$\gamma_w = 62.4 \text{ lb/ft}^3$$

In this case it may help to simplify the expression before substituting known values. Because two terms are multiplied by γ_o they can be combined as follows:

$$p_B - p_A = \gamma_o(29.5 \text{ in}) - \gamma_w(29.5 \text{ in})$$

Factoring out the common term gives

$$p_B - p_A = (29.5 \text{ in})(\gamma_o - \gamma_w)$$

This looks simpler than the original equation. The difference between p_B and p_A is a function of the *difference* between the specific weights of the two fluids.

The pressure at B, then, is

$$p_B - p_A = (29.5 \text{ in})(53.7 - 62.4)\frac{\text{lb}}{\text{ft}^3} \times \frac{1 \text{ ft}^3}{1728 \text{ in}^3}$$
$$= \frac{(29.5)(-8.7)\text{lb/in}^2}{1728}$$
$$p_B - p_A = -0.149 \text{ lb/in}^2$$

The negative sign indicates that the magnitude of p_A is greater than that of p_B. Notice that using a gage fluid with a specific weight very close to that of the fluid in the system makes the manometer very sensitive. A large displacement of the column of gage fluid is caused by a small differential pressure and this allows a very accurate reading.

Figure 3.13 shows another type of manometer, the *well-type manometer*. When a pressure is applied to a well-type manometer, the fluid level in the well drops a small amount while the level in the right leg rises a larger amount in proportion to the ratio of the areas of the well and the tube. A scale is placed alongside the tube so that the deflection can be read directly. The scale is calibrated to account for the small drop in the well level.

The *inclined well-type manometer*, shown in Fig. 3.14, has the same features as the well-type manometer but offers a greater sensitivity by placing the scale along the inclined tube. The scale length is increased as a function of the angle of inclination of the tube, θ. For example, if the angle θ in Fig. 3.14(b) is 15°, the ratio of scale length L to manometer deflection h is

$$\frac{h}{L} = \sin \theta$$

or

$$\frac{L}{h} = \frac{1}{\sin \theta} = \frac{1}{\sin 15°} = \frac{1}{0.259} = 3.86$$

The scale would be calibrated so that the deflection could be read directly.

FIGURE 3.13   Well-type manometer.
(*Source:* Dwyer Instruments, Inc.)

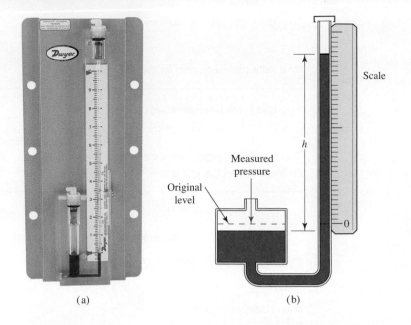

(a)   (b)

FIGURE 3.14   Inclined well-type
manometer. (*Source:* Dwyer Instruments, Inc.)

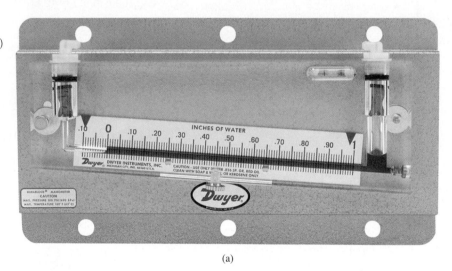

(a)

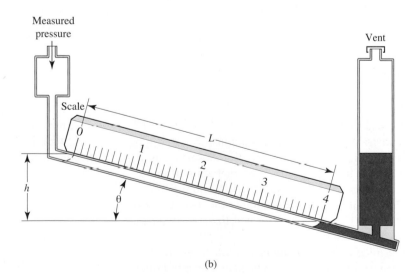

(b)

3.7 BAROMETERS

The device for measuring the atmospheric pressure is called a *barometer.* A simple type is shown in Fig. 3.15. It consists of a long tube closed at one end that is initially filled completely with mercury. The open end is then submerged under the surface of a container of mercury and allowed to come to equilibrium, as shown in Fig. 3.15. A void is produced at the top of the tube that is very nearly a perfect vacuum, containing mercury vapor at a pressure of only 0.17 Pa at 20° C. By starting at this point and writing an equation similar to those for manometers, we get

$$0 + \gamma_m h = p_{atm}$$

or

$$p_{atm} = \gamma_m h \qquad (3\text{--}12)$$

Because the specific weight of mercury is approximately constant, a change in atmospheric pressure will cause a change in the height of the mercury column. This height is often reported as the barometric pressure. To obtain true atmospheric pressure it is necessary to multiply h by γ_m.

Precision measurement of the atmospheric pressure with a mercury manometer requires that the specific weight of the mercury be adjusted for changes in temperature. In this book, we will use the values given in Appendix K.

In SI units

$$\gamma = 133.3 \text{ kN/m}^3$$

In U.S. Customary System units

$$\gamma = 848.7 \text{ lb/ft}^3$$

The atmospheric pressure varies from time to time, as reported on weather reports. The atmospheric pressure also varies with altitude. A decrease of approximately 1.0 in of mercury occurs per 1000 ft of increase in altitude. In SI units, the decrease is approximately 85 mm of mercury per 1000 m. See also Appendix E for variations in atmospheric pressure with altitude.

The development of the barometer dates to the early seventeenth century, with the Italian scientist Evangelista Torricelli publishing his work in 1643. Figure 3.15(b) shows a style of scientific barometer in which the atmospheric pressure acts directly on the surface of the mercury in the container at the bottom, called a cistern. The overall length of the barometer is 900 mm (36 in) and the mercury tube has an inside diameter of 7.7 mm (0.31 in). The readings are taken at the top of the mercury column, as shown in Fig. 3.15(c), using a vernier scale that allows reading to 0.1 millibar (mb), where 1.0 bar equals 100 kPa, approximately the normal atmospheric pressure. Thus, normal atmospheric pressure is approximately 1000 mb. The unit of mb is sometimes reported as hPa (hectopascal), which is equal to 100 Pa. Scales are also available in

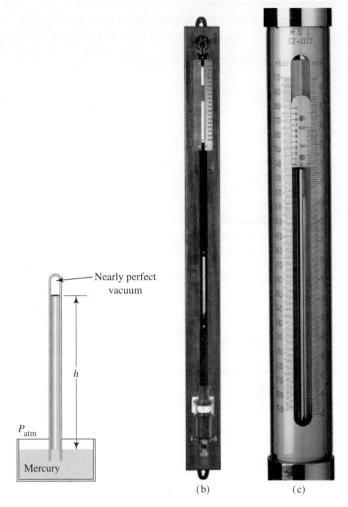

(b)   (c)

FIGURE 3.15   Barometers. (*Source:* Russell Scientific Instruments, Ltd.)

mmHg and inHg. See Internet resource 6 for several other styles of mercury barometers used in laboratories and meteorological offices. Care must be exercised in their use because of the environmental hazard posed by the mercury. Scale ranges on commercial barometers are approximately as follows:

870–1100 mb

650–825 mmHg

25.5–32.5 inHg

A more popular form of barometer is called the *aneroid barometer,* introduced around the year 1840 by Lucien Vidie in France. This mechanical instrument gives the barometric pressure reading using a pointer on a circular dial, as seen on barometers available for home use. The mechanism incorporates a flexible sealed vacuum chamber that changes height as the local atmospheric pressure on the outside changes. The movement acts through a linkage to drive the pointer. See Internet resource 6.

Example Problem   A news broadcaster reports that the barometric pressure is 772 mm of mercury. Calculate the atmo-
3.10   spheric pressure in kPa(abs).

Solution   In Eq. (3–12),

$$p_{atm} = \gamma_m h$$
$$\gamma_m = 133.3 \text{ kN/m}^3$$
$$h = 0.772 \text{ m}$$

Then we have

$$p_{atm} = (133.3 \text{ kN/m}^3)(0.772 \text{ m}) = 102.9 \text{ kN/m}^2 = 102.9 \text{ kPa(abs)}$$

Example Problem   The standard atmospheric pressure is 101.325 kPa. Calculate the height of a mercury column equiva-
3.11   lent to this pressure.

Solution   We begin with Eq. (3–12), $p_{atm} = \gamma_m h$, and write

$$h = \frac{p_{atm}}{\gamma_m} = \frac{101.325 \times 10^3 \text{ N}}{\text{m}^2} \times \frac{\text{m}^3}{133.3 \times 10^3 \text{ N}} = 0.7600 \text{ m} = 760.0 \text{ mm}$$

Example Problem   A news broadcaster reports that the barometric pressure is 30.40 in of mercury. Calculate the pressure
3.12   in psia.

Solution   In Eq. (3–12), set

$$\gamma_m = 848.7 \text{ lb/ft}^3$$
$$h = 30.40 \text{ in}$$

Then we have

$$p_{atm} = \frac{848.7 \text{ lb}}{\text{ft}^3} \times 30.40 \text{ in} \times \frac{1 \text{ ft}^3}{1728 \text{ in}^3} = 14.93 \text{ lb/in}^2$$
$$p_{atm} = 14.93 \text{ psia}$$

Example Problem   The standard atmospheric pressure is 14.696 psia. Calculate the height of a mercury column equiva-
3.13   lent to this pressure.

Solution   Write Eq. (3–12) as

$$h = \frac{p_{atm}}{\gamma_m} = \frac{14.696 \text{ lb}}{\text{in}^2} \times \frac{\text{ft}^3}{848.7 \text{ lb}} \times \frac{1728 \text{ in}^3}{\text{ft}^3} = 29.92 \text{ in}$$

3.8 PRESSURE EXPRESSED AS THE HEIGHT OF A COLUMN OF LIQUID

When measuring pressures in some fluid flow systems, such as air flow in heating ducts, the magnitude of the pressure reading is often small. Manometers are sometimes used to measure these pressures, and their readings are given in units such as *inches of water* (inH$_2$O or inWC for *inches of water column*) rather than the conventional units of psi or Pa.

To convert from such units to those needed in calculations, the pressure–elevation relationship must be used. For example, a pressure of 1.0 inH$_2$O expressed in psi units is given from $p = \gamma h$ as

$$p = \frac{62.4 \text{ lb}}{\text{ft}^3} (1.0 \text{ inH}_2\text{O}) \frac{1 \text{ ft}^3}{1728 \text{ in}^3} = 0.0361 \text{ lb/in}^2$$

$$= 0.0361 \text{ psi}$$

Then we can use this as a conversion factor,

$$1.0 \text{ inH}_2\text{O} = 0.0361 \text{ psi}$$

FIGURE 3.16 Bourdon tube pressure gage.

(a) Front view

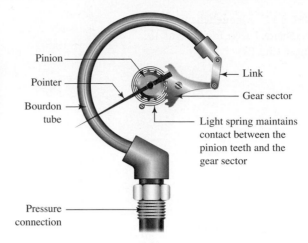

(b) Internal parts showing the Bourdon tube and the indicator mechanism

Converting this to Pa, using 6895 Pa = 1.00 psi from Appendix K, gives

$$1.0 \, \text{inH}_2\text{O} = 0.0361 \, \text{psi} \times 6895 \, \text{Pa}/1.00 \, \text{psi} = 249.0 \, \text{Pa}$$

Similarly, somewhat higher pressures are measured with a mercury manometer. Using $\gamma = 133.3 \, \text{kN/m}^3$ or $\gamma = 848.7 \, \text{lb/ft}^3$, we can develop the conversion factors,

$$1.0 \, \text{inHg} = 1.0 \, \text{in of mercury} = 0.491 \, \text{psi}$$
$$1.0 \, \text{mmHg} = 1.0 \, \text{mm of mercury} = 0.01934 \, \text{psi}$$
$$1.0 \, \text{mmHg} = 1.0 \, \text{mm of mercury} = 133.3 \, \text{Pa}$$

Remember that the temperature of the gage fluid can affect its specific weight and, therefore, the accuracy of these factors. See Appendix K for other conversion factors for pressure.

3.9 PRESSURE GAGES AND TRANSDUCERS

There are many needs for measuring pressure as outlined in the Big Picture section of this chapter. For those situations where only a visual indication is needed at the site where the pressure is being measured, a *pressure gage* is most often used. In other cases there is a need to measure pressure at one point and display the value at another. The general term for such a measurement device is *pressure transducer*, meaning that the sensed pressure causes an electrical signal to be generated that can be transmitted to a remote location such as a central control station where it is displayed digitally. Alternatively, the signal can be part of an automatic control system. Some manufacturers of transducers that are also configured to transmit the signal to remote sites simply call such devices *transmitters*. Some pressure gages and transducers employ integral switches that can emit audible signals and/or actuate process operations at set pressure values. This section describes some of the many types of pressure gages, transducers, and transmitters.

3.9.1 Pressure Gages

A widely used pressure-measuring device is the *Bourdon tube pressure gage** (Figure 3.16). The pressure to be measured is applied to the inside of a hollow tube with a flattened oval cross section, which is normally formed into a segment of a circle or a spiral as shown in part (b) of the figure. The increased pressure inside the tube causes the spiral to be opened somewhat. The movement of the end of the tube is transmitted through a linkage that causes a pointer to rotate.

The scale of the gage normally reads zero when the gage is open to atmospheric pressure and is calibrated in pascals (Pa) or other units of pressure above zero. Therefore, this type of gage reads *gage pressure* directly. Some gages are capable of reading pressures below atmospheric. Internet resource 2 shows a variety of gage styles.

Figure 3.17 shows a pressure gage using an actuation means called *Magnehelic*®[†]. The pointer is attached to a helix, made from a material having a high magnetic permeability that is supported in sapphire bearings. A leaf spring is driven up and down by the motion of a flexible diaphragm, not shown in the figure. At the end of the spring, the C-shaped element contains a strong magnet placed in close proximity to the outer surface of the helix. As the leaf spring moves up and down, the helix rotates to follow the magnet, moving the pointer. Note that there is no physical contact between the magnet and the helix. Calibration of the gage is accomplished by adjusting the length of the spring at its clamped end. See Internet resource 1 for supplier information.

3.9.2 Pressure Transducers and Transmitters

Figure 3.18 shows an example of a *pressure transducer*. The pressure to be measured is introduced through the pressure

---

*Note that two spellings, *gage* and *gauge*, are often used interchangeably.

[†]Magnehelic is a registered trade name of Dwyer Instruments, Inc., Michigan City, IN.

FIGURE 3.17 Magnehelic pressure gage. (*Source:* Dwyer Instruments, Inc., Michigan City, IN)

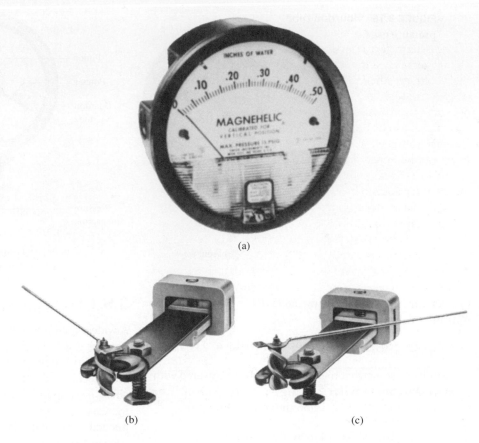

(a)

(b)   (c)

FIGURE 3.18 Strain gage pressure transducer.

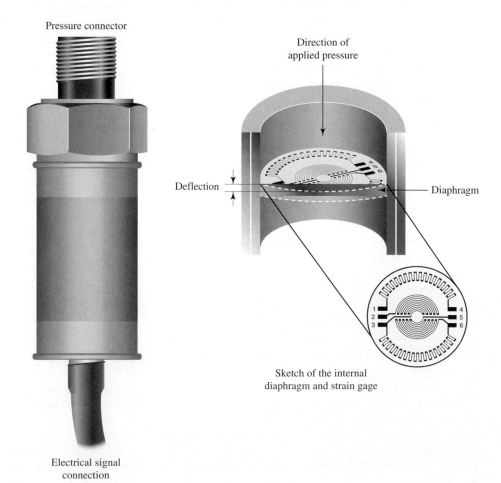

Pressure connector

Direction of applied pressure

Deflection

Diaphragm

Electrical signal connection

Sketch of the internal diaphragm and strain gage

port and acts on a sensing element that generates a signal proportional to the applied pressure. The sensing element can be a strain gage bonded to a diaphragm that is deformed by the pressure. As the strain gages sense the deformation of the diaphragm, their resistance changes. Passing an electrical current through the gages and connecting them into a network, called a *Wheatstone bridge*, causes a change in electrical voltage to be produced. The readout device is typically a digital voltmeter, calibrated in pressure units. The strain gages can be either thin metal foil or silicon. See Internet resources 1–5 and 7 for some commercially available transducers and transmitters.

Other transducers employ crystals, such as quartz and barium titanate that exhibit a *piezoelectric effect* in which the electrical charge across the crystal varies with stress in the crystal. Causing a pressure to exert a force, either directly or indirectly, on the crystal leads to a voltage change related to the pressure change. See References 2, 8, and 9 for additional detail about these sensing devices.

REFERENCES

1. Avallone, Eugene A., Theodore Baumeister, and Ali Sadegh, eds. 2007. *Marks' Standard Handbook for Mechanical Engineers*, 11th ed. New York: McGraw-Hill.

2. Busse, Donald W. 1987 (March). Quartz Transducers for Precision under Pressure. *Mechanical Engineering Magazine* 109(5):52–56.

3. CAPT (Center for the Advancement of Process Technology). 2010. *Instrumentation*, Upper Saddle River, NJ: Pearson/Prentice Hall.

4. Gillum, Donald R. 2009. *Industrial Pressure, Level, and Density Measurement*, 2nd ed. Research Triangle Park, NC: ISA—The International Society of Automation.

5. Holman, Jack P. 2012. *Experimental Methods for Engineers*, 8th ed. New York: McGraw-Hill.

6. Kutz, Myer. 2013. *Handbook of Measurement in Science and Engineering*, New York: John Wiley & Sons.

7. Walters, Sam. 1987 (March). Inside Pressure Measurement. *Mechanical Engineering Magazine* 109(5):41–47.

8. Worden, Roy D. 1987 (March). Designing a Fused-Quartz Pressure Transducer. *Mechanical Engineering Magazine* 109(5):48–51.

9. Vives, Antonio Arnau. 2010. *Piezoelectric Transducers and Applications*, New York: Springer Publishing.

INTERNET RESOURCES

1. **Dwyer Instruments, Inc.:** Manufacturer of instruments for measuring pressure, flow, air velocity, level, temperature, and humidity. Also provides valves, data acquisition systems, and combustions testing.

2. **Ametek U.S. Gauge, Inc.:** Manufacturer of a wide variety of pressure gages and transducers using solid-state and strain gage technology. Also provides level sensors, pressure transmitters, and pneumatic controllers.

3. **Ametek Power Instruments:** Manufacturer of sensors, instruments, and monitoring systems for the power generation, transmission, distribution, nuclear, oil, and petrochemical markets, including pressure transducers, temperature sensors, and transmitters.

4. **Honeywell Sensing & Control:** From the home page, select *Products & Information*, then *Sensors*, for information about several lines of strain gage-type pressure transducers, digital pressure gages, and digital pressure indicators, along with a variety of sensors for mechanical loads, vibration, motion, and temperature. Part of the Honeywell Sensing and Control unit of Honeywell International, Inc.

5. **Cooper Controls—Polaron Components Limited:** From the home page, select *Products*, then *Pressure Transducers* to learn more about Polaron pressure sensors, vibration monitors, motors, motion sensors, switches and other devices.

6. **Russell Scientific Instruments:** Manufacturer of precision barometers, thermometers, and other scientific instruments for industrial, meteorological, household, and other uses.

7. **Rosemount, Inc.:** From the home page, select *Product Quick Links* to learn more about industrial pressure transducers and transmitters. Rosemount also produces sensors for temperature, flow, and level. Part of Emerson Process Management.

PRACTICE PROBLEMS

Absolute and Gage Pressure

3.1 Write the expression for computing the pressure in a fluid.
3.2 Define *absolute pressure*.
3.3 Define *gage pressure*.
3.4 Define *atmospheric pressure*.
3.5 Write the expression relating gage pressure, absolute pressure, and atmospheric pressure.

State whether statements 3.6–3.10 are (or can be) true or false. For those that are false, explain why.

3.6 The value for the absolute pressure will always be greater than that for the gage pressure.
3.7 As long as you stay on the surface of Earth, the atmospheric pressure will be 14.7 psia.
3.8 The pressure in a certain tank is −55.8 Pa(abs).
3.9 The pressure in a certain tank is −4.65 psig.
3.10 The pressure in a certain tank is −150 kPa(gage).
3.11 If you were to ride in an open-cockpit airplane to an elevation of 4000 ft above sea level, what would the atmospheric pressure be if it conforms to the standard atmosphere?
3.12 The peak of a certain mountain is 13 500 ft above sea level. What is the approximate atmospheric pressure?
3.13 Expressed as a gage pressure, what is the pressure at the surface of a glass of milk?

Problems 3.14–3.33 require that you convert the given pressure from gage to absolute pressure or from absolute to gage pressure as indicated. The value of the atmospheric pressure is given.

Problem	Given Pressure	Patm	Express Result As:
3.14	583 kPa(abs)	103 kPa(abs)	Gage pressure
3.15	157 kPa(abs)	101 kPa(abs)	Gage pressure
3.16	30 kPa(abs)	100 kPa(abs)	Gage pressure
3.17	74 kPa(abs)	97 kPa(abs)	Gage pressure
3.18	101 kPa(abs)	104 kPa(abs)	Gage pressure
3.19	284 kPa(gage)	100 kPa(abs)	Absolute pressure
3.20	128 kPa(gage)	98.0 kPa(abs)	Absolute pressure
3.21	4.1 kPa(gage)	101.3 kPa(abs)	Absolute pressure
3.22	−29.6 kPa (gage)	101.3 kPa(abs)	Absolute pressure
3.23	−86.0 kPa (gage)	99.0 kPa(abs)	Absolute pressure
3.24	84.5 psia	14.9 psia	Gage pressure
3.25	22.8 psia	14.7 psia	Gage pressure
3.26	4.3 psia	14.6 psia	Gage pressure
3.27	10.8 psia	14.0 psia	Gage pressure
3.28	14.7 psia	15.1 psia	Gage pressure
3.29	41.2 psig	14.5 psia	Absolute pressure
3.30	18.5 psig	14.2 psia	Absolute pressure
3.31	0.6 psig	14.7 psia	Absolute pressure
3.32	−4.3 psig	14.7 psia	Absolute pressure
3.33	−12.5 psig	14.4 psia	Absolute pressure

Pressure–Elevation Relationship

3.34 If milk has a specific gravity of 1.08, what is the pressure at the bottom of a milk can 550 mm deep?

3.35 The pressure in an unknown fluid at a depth of 4.0 ft is measured to be 1.820 psig. Compute the specific gravity of the fluid.

3.36 The pressure at the bottom of a tank of propyl alcohol at 25° C must be maintained at 52.75 kPa(gage). What depth of alcohol should be maintained?

3.37 When you dive to a depth of 12.50 ft in seawater, what is the pressure?

3.38 A water storage tank is on the roof of a factory building and the surface of the water is 50.0 ft above the floor of the factory. If a pipe connects the storage tank to the floor level and the pipe is full of static water, what is the pressure in the pipe at floor level?

3.39 An open tank contains ethylene glycol at 25°C. Compute the pressure at a depth of 3.0 m.

3.40 For the tank of ethylene glycol described in Problem 3.39, compute the pressure at a depth of 12.0 m.

3.41 Figure 3.19 shows a diagram of the hydraulic system for a vehicle lift. An air compressor maintains pressure above the oil in the reservoir. What must the air pressure be if the pressure at point A must be at least 180 psig?

3.42 Figure 3.20 shows a clothes washing machine. The pump draws fluid from the tub and delivers it to the disposal sink. Compute the pressure at the inlet to the pump when the water is static (no flow). The soapy water solution has a specific gravity of 1.15.

3.43 An airplane is flying at 10.6 km altitude. In its nonpressurized cargo bay is a container of mercury 325 mm deep.

FIGURE 3.19 Vehicle lift for Problem 3.41.

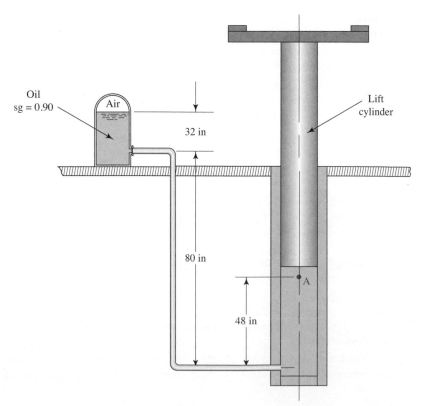

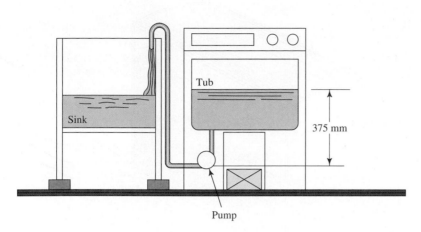

FIGURE 3.20 Washing machine for Problem 3.42.

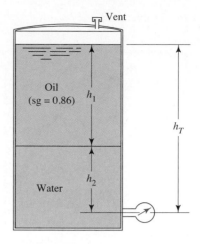

FIGURE 3.22 Problems 3.48–3.50.

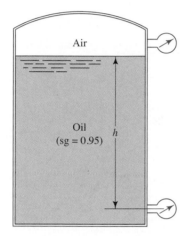

FIGURE 3.21 Problems 3.44–3.47.

The container is vented to the local atmosphere. What is the absolute pressure at the surface of the mercury and at the bottom of the container? Assume the conditions of the standard atmosphere prevail for pressure. Use sg = 13.54 for the mercury.

3.44 For the tank shown in Fig. 3.21, determine the reading of the bottom pressure gage in psig if the top of the tank is vented to the atmosphere and the depth of the oil h is 28.50 ft.

3.45 For the tank shown in Fig. 3.21, determine the reading of the bottom pressure gage in psig if the top of the tank is sealed, the top gage reads 50.0 psig, and the depth of the oil h is 28.50 ft.

3.46 For the tank shown in Fig. 3.21, determine the reading of the bottom pressure gage in psig if the top of the tank is sealed, the top gage reads −10.8 psig, and the depth of the oil h is 6.25 ft.

3.47 For the tank shown in Fig. 3.21, determine the depth of the oil h if the reading of the bottom pressure gage is 35.5 psig, the top of the tank is sealed, and the top gage reads 30.0 psig.

3.48 For the tank in Fig. 3.22, compute the depth of the oil if the depth of the water is 2.80 m and the gage at the bottom of the tank reads 52.3 kPa(gage).

3.49 For the tank in Fig. 3.22, compute the depth of the water if the depth of the oil is 6.90 m and the gage at the bottom of the tank reads 125.3 kPa(gage).

3.50 Figure 3.22 represents an oil storage drum that is open to the atmosphere at the top. Some water was accidentally pumped into the tank and settled to the bottom as shown in the figure. Calculate the depth of the water h_2 if the pressure gage at the bottom reads 158 kPa(gage). The total depth $h_T = 18.0$ m.

3.51 A storage tank for sulfuric acid is 1.5 m in diameter and 4.0 m high. If the acid has a specific gravity of 1.80, calculate the pressure at the bottom of the tank. The tank is open to the atmosphere at the top.

3.52 A storage drum for crude oil (sg = 0.89) is 32 ft deep and open at the top. Calculate the pressure at the bottom.

3.53 The greatest known depth in the ocean is approximately 11.0 km. Assuming that the specific weight of the water is constant at 10.0 kN/m^3, calculate the pressure at this depth.

3.54 Figure 3.23 shows a closed tank that contains gasoline floating on water. Calculate the air pressure above the gasoline.

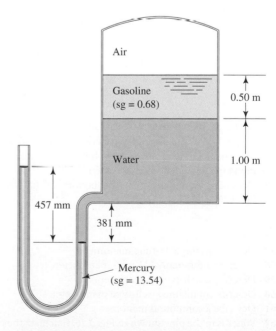

FIGURE 3.23 Problem 3.54.

FIGURE 3.24   Problem 3.55.

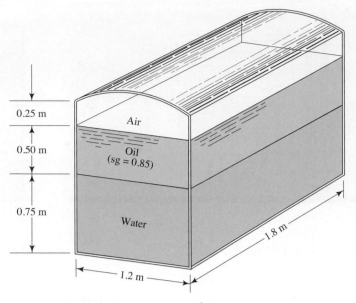

FIGURE 3.25   Problem 3.56.

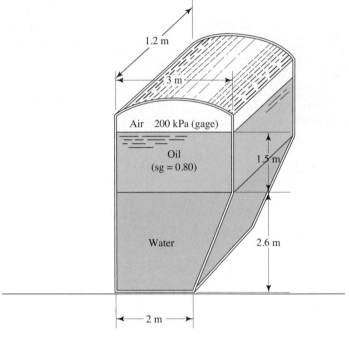

3.55 Figure 3.24 shows a closed container holding water and oil. Air at 34 kPa below atmospheric pressure is above the oil. Calculate the pressure at the bottom of the container in kPa(gage).

3.56 Determine the pressure at the bottom of the tank in Fig. 3.25.

Manometers

3.57 Describe a simple U-tube manometer.

3.58 Describe a differential U-tube manometer.

3.59 Describe a well-type manometer.

3.60 Describe an inclined well-type manometer.

3.61 Describe a compound manometer.

3.62 Water is in the pipe shown in Fig. 3.26. Calculate the pressure at point A in kPa(gage).

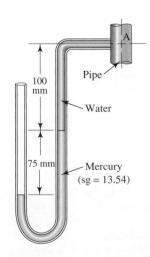

FIGURE 3.26   Problem 3.62.

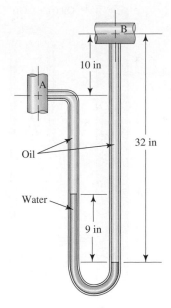

FIGURE 3.27 Problem 3.63.

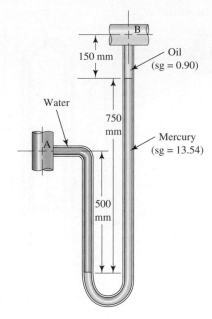

FIGURE 3.29 Problem 3.65.

3.63 For the differential manometer shown in Fig. 3.27, calculate the pressure difference between points A and B. The specific gravity of the oil is 0.85.

3.64 For the manometer shown in Fig. 3.28, calculate $(p_A - p_B)$.

3.65 For the manometer shown in Fig. 3.29, calculate $(p_A - p_B)$.

3.66 For the manometer shown in Fig. 3.30, calculate $(p_A - p_B)$.

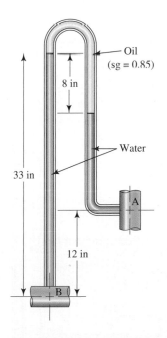

FIGURE 3.28 Problem 3.64.

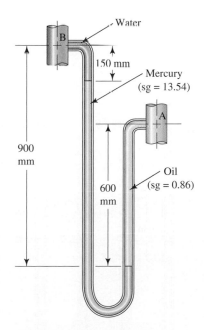

FIGURE 3.30 Problem 3.66.

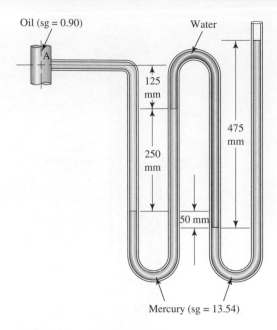

Oil (sg = 0.90)

Water

125 mm

475 mm

250 mm

50 mm

Mercury (sg = 13.54)

FIGURE 3.31 Problem 3.67.

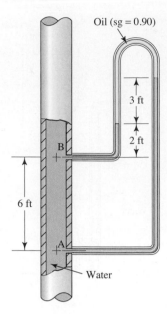

Oil (sg = 0.90)

3 ft

2 ft

B

6 ft

A

Water

FIGURE 3.33 Problem 3.69.

3.67 For the compound manometer shown in Fig. 3.31, calculate the pressure at point A.

3.68 For the compound differential manometer in Fig. 3.32, calculate $(p_A - p_B)$.

3.69 Figure 3.33 shows a manometer being used to indicate the difference in pressure between two points in a pipe. Calculate $(p_A - p_B)$.

3.70 For the well-type manometer in Fig. 3.34, calculate p_A.

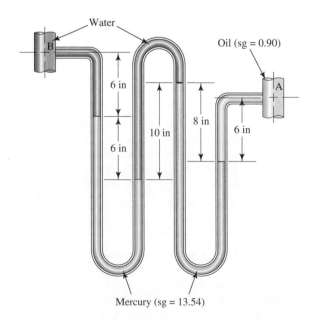

Water

B

Oil (sg = 0.90)

A

6 in

8 in

6 in

10 in

6 in

Mercury (sg = 13.54)

FIGURE 3.32 Problem 3.68.

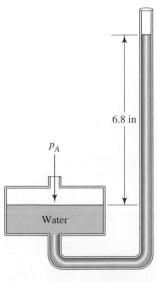

6.8 in

p_A

Water

FIGURE 3.34 Problem 3.70.

FIGURE 3.35 Problem 3.71.

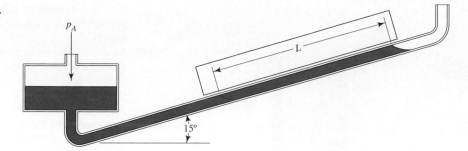

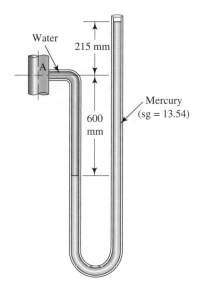

3.71 Figure 3.35 shows an inclined well-type manometer in which the distance L indicates the movement of the gage fluid level as the pressure p_A is applied above the well. The gage fluid has a specific gravity of 0.87 and L = 115 mm. Neglecting the drop in fluid level in the well, calculate p_A.

3.72 **a.** Determine the gage pressure at point A in Fig. 3.36.
  b. If the barometric pressure is 737 mm of mercury, express the pressure at point A in kPa(abs).

FIGURE 3.36 Problem 3.72.

Barometers

3.73 What is the function of a barometer?

3.74 Describe the construction of a barometer.

3.75 Why is mercury a convenient fluid to use in a barometer?

3.76 If water were to be used instead of mercury in a barometer, how high would the water column be?

3.77 What is the barometric pressure reading in inches of mercury corresponding to 14.696 psia?

3.78 What is the barometric pressure reading in millimeters of mercury corresponding to 101.325 kPa(abs)?

3.79 Why must a barometric pressure reading be corrected for temperature?

3.80 By how much would the barometric pressure reading decrease from its sea-level value at an elevation of 1250 ft?

3.81 Denver, Colorado, is called the "Mile-High City" because it is located at an elevation of approximately 5200 ft. Assuming that the sea-level pressure is 101.3 kPa(abs),

what would be the approximate atmospheric pressure in Denver?

3.82 The barometric pressure is reported to be 28.6 in of mercury. Calculate the atmospheric pressure in psia.

3.83 A barometer indicates the atmospheric pressure to be 30.65 in of mercury. Calculate the atmospheric pressure in psia.

3.84 What would be the reading of a barometer in inches of mercury corresponding to an atmospheric pressure of 14.2 psia?

3.85 A barometer reads 745 mm of mercury. Calculate the barometric pressure reading in kPa(abs).

Pressure Expressed as the Height of a Column of Liquid

3.86 The pressure in a heating duct is measured to be 5.37 in H_2O. Express this pressure in psi and Pa.

3.87 The pressure in a ventilation duct at the inlet to a fan is measured to be −3.68 inH_2O. Express this pressure in psi and Pa.

3.88 The pressure in an air conditioning duct is measured to be 3.24 mmHg. Express this pressure in Pa and psi.

3.89 The pressure in a compressed natural gas line is measured to be 21.6 mmHg. Express this pressure in Pa and psi.

3.90 The pressure in a vacuum chamber is −68.2 kPa. Express this pressure in mmHg.

3.91 The pressure in a vacuum chamber is −12.6 psig. Express this pressure in inHg.

3.92 The performance of a fan is rated at a pressure differential of 12.4 inWC. Express this pressure in psi and Pa.

3.93 The pressure of a pressure blower is rated at a pressure differential of 115 inWC. Express this pressure in psi and Pa.

Supplemental Problems

3.94 A passive solar water heater is to be installed on the roof of a multi-story building. The heater tank is open to atmospheric pressure and is mounted 16 m above ground level. In the static (non-flowing) state, what gage pressure, in kPa, must the plumbing line be designed to withstand if it is connected all the way down to ground level?

3.95 The elevated tank similar to the one shown in Fig. 3.37 is part of a water delivery system to be built for a small village. Find the required elevation of the tank if a minimum gage pressure of 160 kPa is required at the outlet when the water is static (no flow). Note that the level calculated will establish the height for the bottom of the tank when it is nearly empty. When the level of water is higher, the outlet pressure will increase.

FIGURE 3.37   Problem 3.95.

3.96 In the "eye" of a hurricane, pressure can sometimes drop from normal atmospheric pressure all the way down to 11 psia. What would be the height reading, in inches, of a mercury barometer there?

3.97 A concrete form used to pour a basement wall is to hold wet concrete mix (sg = 2.6) during construction. The wall is to be 3 m high, 10 m long, and 150 mm thick. What pressure does the wet concrete exert at the bottom of the form?

3.98 An environmental instrumentation package is to be designed to be lowered into the Mariana Trench to a depth of 11 km into the Pacific Ocean. If the case is to be watertight at that depth in sea water, what pressure must it be designed to withstand?

3.99 A scuba diver will descend "one and a half atmospheres" into a fresh water lake. Calculate the depth of the dive. Note that "an atmosphere" is a measure sometimes used by divers to indicate a depth in water that results in a pressure increase equivalent to one standard atmospheric pressure.

3.100 An inclined manometer similar to the one shown in Figure 3.14 is used for sensitive pressure measurement. It is inclined at an angle of 25 degrees above the horizontal and uses red gage fluid with a specific gravity of 0.826. How far apart should the marks along the inclined tube be to indicate a pressure of "one inch of water"?

3.101 A meteorologist reports a "high pressure system" with barometric pressure of 790 mm of mercury and then later in the year a "low pressure system" with a pressure of 738 mm of mercury. What is the total difference in atmospheric pressure, in kPa?

3.102 What is the pressure, in psig, at the bottom of a swimming pool that is 10 ft deep?

3.103 If air has a constant specific weight of 0.075 lb/ft³, what pressure difference would result when driving from the base to the top of Pike's Peak, if the climb for the trip is 8400 ft?

FORCES DUE TO STATIC FLUIDS

Recall that pressure is force divided by the area on which it acts, $p = F/A$. We are now concerned with the force produced by the pressure in a fluid that acts on the walls of containers. When the pressure is uniform over all the area of interest, the force is simply $F = pA$. When the pressure varies over the surface of interest, other methods must be used to account for this variation before we can compute the magnitude of the resultant force on the surface. The location of the resultant force, called the *center of pressure*, must also be located so that an analysis of the effects of that force can be done.

Look at the photo in Fig. 4.1 showing some children admiring exotic fish in an aquarium. It is essential that the design and fabrication of the aquarium ensures that the glass will not break and that the children are safe. Here the water pressure increases linearly with the depth of the fluid as discussed in Chapter 3.

Exploration

Identify several examples where the force exerted by a fluid on the surfaces that contain it may be of importance. Discuss your examples among your fellow students and with the course instructor, addressing these questions:

- How does the force act on its container? Does the pressure vary at different points in the fluid? If so, how does it vary?
- How is the design of the container affected by the force created by the fluid pressure?

- What would be the consequence if the forces exceeded the ability of the container to withstand them? How would the container fail?

This chapter will help you discover the principles governing the generation of forces due to fluid acting on plane (flat) or curved surfaces. Some of the solution procedures will be for special cases such as flat horizontal surfaces, surfaces containing gases, or rectangular walls exposed to the free surface of the fluid. Other cases cover more general situations where pressure variations must be considered and where both the magnitude and the location of the resultant force must be computed.

Introductory Concepts

Here we consider the effects of fluid pressure acting on plane (flat) and curved surfaces in applications such as those shown in Fig. 4.2. In each case, the fluid exerts a force on the surface of interest that acts perpendicular to the surface, considering the basic definition of pressure, $p = F/A$, and the corresponding form, $F = pA$.

We apply these equations directly only when the pressure is uniform over the entire area of interest. An example is when the fluid is a gas for which we consider the pressure to be equal throughout the gas because of its low specific weight. In addition, if the change in depth is small, the variation is often neglected. For example, the pressure acting on the piston in the fluid power actuator in Fig. 4.2(a) can be considered approximately constant if

FIGURE 4.1 An aquarium scene where forces due to fluid pressure must be considered.
(*Source:* Iuliia Sokolovska/Fotolia)

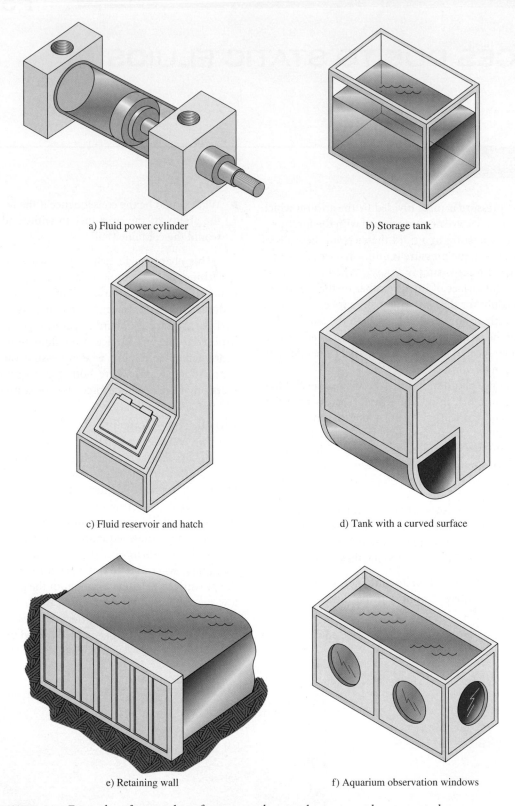

a) Fluid power cylinder

b) Storage tank

c) Fluid reservoir and hatch

d) Tank with a curved surface

e) Retaining wall

f) Aquarium observation windows

FIGURE 4.2 Examples of cases where forces on submerged areas must be computed.

the fluid is either air, as in a pneumatic fluid power system, or for oil in a hydraulic system. Another example of the use of $F = pA$ is the exertion of liquid pressure on a flat, horizontal surface as on the bottom of the tanks in Fig. 4.2(b), (c), and (f).

In other cases where the surface of interest is vertical, inclined, or curved, we must take into account the variation of pressure with depth; special analysis approaches are developed in this chapter. You should review Chapter 3 on the subjects of absolute and gage pressure, the variation of pressure with elevation, and piezometric head. We will show methods of computing the *resultant force* on the surface and the location of the *center of pressure* where the resultant force can be assumed to act when computing the effect of the distributed force.

Consider the side walls of the tanks, the hatch in the inclined wall of the fluid reservoir, the retaining wall, and the aquarium windows. The retaining wall is an example of a special case that we call *rectangular walls*, for which the pressure varies linearly from zero (gage) at the top surface of the fluid to some larger pressure at the bottom of the wall. The fluid reservoir hatch and the aquarium windows require a more general approach because no part of the area of interest involves the zero pressure.

4.1 OBJECTIVES

After completing this chapter, you should be able to:

1. Compute the force exerted on a plane area by a pressurized gas.

2. Compute the force exerted by any static fluid acting on a horizontal plane area.

3. Compute the resultant force exerted on a rectangular wall by a static liquid.

4. Define the term *center of pressure*.

5. Compute the resultant force exerted on any submerged plane area by a static liquid.

6. Show the vector representing the resultant force on any submerged plane area in its proper location and direction.

7. Visualize the distribution of force on a submerged curved surface.

8. Compute the total resultant force on the curved surface.

9. Compute the direction in which the resultant force acts and show its line of action on a sketch of the surface.

10. Include the effect of a pressure head over the liquid on the force on a plane or curved surface.

4.2 GASES UNDER PRESSURE

Figure 4.3 shows a pneumatic cylinder of the type used in automated machinery. The air pressure acts on the piston face, producing a force that causes the linear movement of the rod. The pressure also acts on the end of the cylinder, tending to pull it apart. This is the reason for the four tie rods between the end caps of the cylinder. The distribution of pressure within a gas is very nearly uniform. Therefore, we can calculate the force on the piston and the cylinder ends directly from $F = pA$.

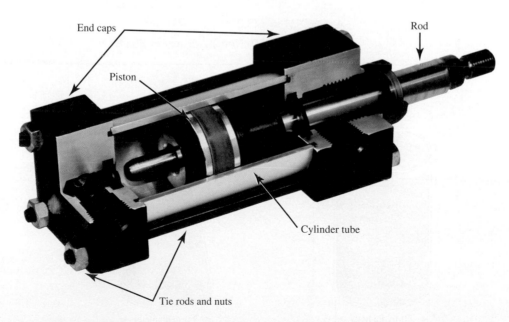

FIGURE 4.3   Fluid power cylinder. (*Source:* Norgren, Inc.)

Example Problem   If the cylinder in Fig. 4.3 has an internal diameter of 2 in and operates at a pressure of 300 psig, calcu-
4.1   late the force on the ends of the cylinder.

Solution

$$F = pA$$

$$A = \frac{\pi D^2}{4} = \frac{\pi (2 \text{ in})^2}{4} = 3.14 \text{ in}^2$$

$$F = \frac{300 \text{ lb}}{\text{in}^2} \times 3.14 \text{ in}^2 = 942 \text{ lb}$$

Notice that gage pressure was used in the calculation of force instead of absolute pressure. The additional force due to atmospheric pressure acts on both sides of the area and is thus balanced. If the
pressure on the outside surface is not atmospheric, then all external forces must be considered to
determine a net force on the area.

4.3 HORIZONTAL FLAT SURFACES UNDER LIQUIDS

Figure 4.4(a) shows a cylindrical drum containing oil and
water. The pressure in the water at the bottom of the drum is
uniform across the entire area because it is a horizontal plane
in a fluid at rest. Again, we can simply use $F = pA$ to calculate the force on the bottom.

Example Problem   If the drum in Figure 4.4(a) is open to the atmosphere at the top, calculate the force on the bottom.
4.2

Solution   To use $F = pA$ we must first calculate the pressure at the bottom of the drum p_B and the area of
the bottom:

$$p_B = p_{atm} + \gamma_o(2.4 \text{ m}) + \gamma_w(1.5 \text{ m})$$
$$\gamma_o = (sg)_o(9.81 \text{ kN/m}^3) = (0.90)(9.81 \text{ kN/m}^3) = 8.83 \text{ kN/m}^3$$
$$p_B = 0 \text{ Pa(gage)} + (8.83 \text{ kN/m}^3)(2.4 \text{ m}) + (9.81 \text{ kN/m}^3)(1.5 \text{ m})$$
$$= (0 + 21.2 + 14.7) \text{ kPa} = 35.9 \text{ kPa(gage)}$$
$$A = \pi D^2/4 = \pi(3.0 \text{ m})^2/4 = 7.07 \text{ m}^2$$
$$F = p_B A = (35.9 \text{ kN/m}^2)(7.07 \text{ m}^2) = 253.8 \text{ kN}$$

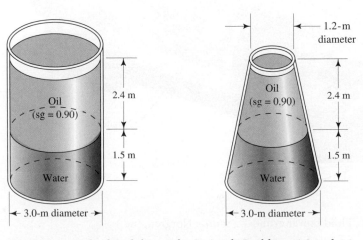

FIGURE 4.4 Cylindrical drums for Example Problems 4.2 and 4.3.

Example Problem **4.3**	Would there be any difference between the force on the bottom of the drum in Fig. 4.4(a) and that on the bottom of the cone-shaped container in Fig. 4.4(b)?
Solution	The force would be the same because the pressure at the bottom is dependent only on the depth and specific weight of the fluid in the container. The total weight of fluid is not the controlling factor. Recall Pascal's paradox in Section 3.4.
	Comment: The force computed in these two example problems is the force exerted by the fluid on the inside bottom of the container. Of course, when designing the support structure for the container, the total weight of the container and the fluids must be considered. For the structural design, the cone-shaped container will be lighter than the cylindrical drum.

4.4 RECTANGULAR WALLS

The retaining walls shown in Figs. 4.2(e) and 4.5 are typical examples of rectangular walls exposed to a pressure varying from zero on the surface of the fluid to a maximum at the bottom of the wall. The force due to the fluid pressure tends to overturn the wall or break it at the place where it is fixed to the bottom.

The actual force is distributed over the entire wall, but for the purpose of analysis it is desirable to determine the resultant force and the place where it acts, called the *center of pressure*. That is, if the entire force were concentrated at a single point, where would that point be and what would the magnitude of the force be?

Figure 4.6 shows the pressure distribution on the vertical retaining wall. As indicated by the equation $\Delta p = \gamma h$,

the pressure varies linearly (in a straight-line manner) with depth in the fluid. The lengths of the dashed arrows represent the magnitude of the fluid pressure at various points on the wall. Because of this linear variation in pressure, the total resultant force can be calculated from the equation

$$F_R = p_{avg} \times A \qquad (4\text{–}1)$$

where p_{avg} is the average pressure and A is the total area of the wall. But the average pressure is at the middle of the wall and can be calculated from the equation

$$p_{avg} = \gamma(h/2) \qquad (4\text{–}2)$$

where h is the total depth of the fluid.

Therefore, we have

FIGURE 4.5 Rectangular walls.

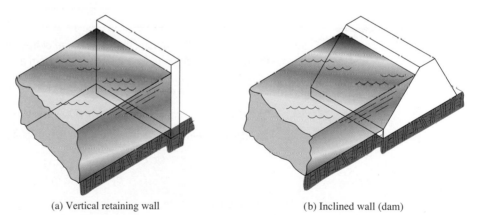

(a) Vertical retaining wall   (b) Inclined wall (dam)

FIGURE 4.6 Vertical rectangular wall.

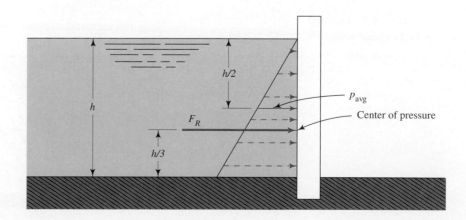

▷ **Resultant Force on a Rectangular Wall**

$$F_R = \gamma(h/2)A \qquad (4\text{–}3)$$

The pressure distribution shown in Fig. 4.6 indicates that a greater portion of the force acts on the lower part of the wall than on the upper part. The center of pressure is at the centroid of the pressure distribution triangle, one third of the distance from the bottom of the wall. The resultant force F_R acts perpendicular to the wall at this point.

The procedure for calculating the magnitude of the resultant force due to fluid pressure and the location of the center of pressure on a rectangular wall such as those shown in Fig. 4.5 is listed below. The procedure applies whether the wall is vertical or inclined.

Procedure for Computing the Force on a Rectangular Wall

1. Calculate the magnitude of the resultant force F_R from

$$F_R = \gamma(h/2)A$$

where

γ = Specific weight of the fluid

h = Total depth of the fluid

A = Total area of the wall

2. Locate the center of pressure at a vertical distance of $h/3$ from the bottom of the wall.

3. Show the resultant force acting at the center of pressure perpendicular to the wall.

Example Problem 4.4

In Fig. 4.6, the fluid is gasoline (sg = 0.68) and the total depth is 12 ft. The wall is 40 ft long. Calculate the magnitude of the resultant force on the wall and the location of the center of pressure.

Solution   **Step 1.**

$$F_R = \gamma(h/2)A$$
$$\gamma = (0.68)(62.4\ \text{lb/ft}^3) = 42.4\ \text{lb/ft}^3$$
$$A = (12\ \text{ft})(40\ \text{ft}) = 480\ \text{ft}^2$$
$$F_R = \frac{42.4\ \text{lb}}{\text{ft}^3} \times \frac{12\ \text{ft}}{2} \times 480\ \text{ft}^2 = 122{,}000\ \text{lb}$$

Step 2. The center of pressure is at a distance of

$$h/3 = 12\ \text{ft}/3 = 4\ \text{ft}$$

from the bottom of the wall.

Step 3. The force F_R acts perpendicular to the wall at the center of pressure as shown in Fig. 4.6.

Example Problem 4.5

Figure 4.7 shows a dam 30.5 m long that retains 8 m of fresh water and is inclined at an angle θ of 60°. Calculate the magnitude of the resultant force on the dam and the location of the center of pressure.

Solution   **Step 1.**

$$F_R = \gamma(h/2)A$$

To calculate the area of the dam we need the length of its face, called L in Fig. 4.7:

$$\sin\theta = h/L$$
$$L = h/\sin\theta = 8\ \text{m}/\sin 60° = 9.24\ \text{m}$$

Then, the area of the dam is

$$A = (9.24\ \text{m})(30.5\ \text{m}) = 281.8\ \text{m}^2$$

FIGURE 4.7 Inclined rectangular wall.

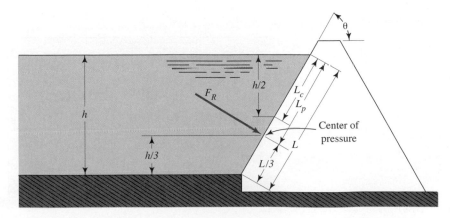

Now we can calculate the resultant force:

$$F_R = \gamma(h/2)A = \frac{9.81 \text{ kN}}{\text{m}^3} \times \frac{8 \text{ m}}{2} \times 281.8 \text{ m}^2$$

$$= 11\,060 \text{ kN} = 11.06 \text{ MN}$$

Step 2. The center of pressure is at a vertical distance of

$$h/3 = 8 \text{ m}/3 = 2.67 \text{ m}$$

from the bottom of the dam, or measured from the bottom of the dam along the face of the dam, the center of pressure is at

$$L/3 = 9.24 \text{ m}/3 = 3.08 \text{ m}$$

Measured along the face of the dam we define

L_p = Distance from the free surface of the fluid to the center of pressure
$L_p = L - L/3$
$L_p = 9.24 \text{ m} - 3.08 \text{ m} = 6.16 \text{ m}$

We show F_R acting at the center of pressure perpendicular to the wall.

4.5 SUBMERGED PLANE AREAS—GENERAL

The procedure we will discuss in this section applies to problems dealing with plane areas, either vertical or inclined, that are completely submerged in the fluid. As in previous problems, the procedure will enable us to calculate the magnitude of the resultant force on an area and the location of the center of pressure where we can assume the resultant force to act.

Figure 4.8 shows a tank that has a rectangular window in an inclined wall. The standard dimensions and symbols used

in the procedure described later are shown in the figure and defined as follows:

F_R — Resultant force on the area due to the fluid pressure

— The *center of pressure* of the area is the point at which the resultant force can be considered to act

— The *centroid* of the area is the point at which the area would be balanced if suspended from that point; it is equivalent to the center of gravity of a solid body

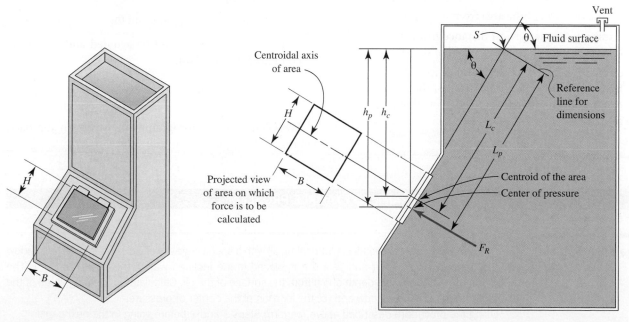

FIGURE 4.8 Force on a submerged plane area.

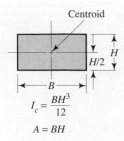

$$I_c = \frac{BH^3}{12}$$

$$A = BH$$

FIGURE 4.9 Properties of a rectangle.

θ Angle of inclination of the area

h_c Depth of fluid from the free surface to the centroid of the area

L_c Distance from the level of the free surface of the fluid to the centroid of the area, measured along the angle of inclination of the area

L_p Distance from the level of the free surface of the fluid to the center of pressure of the area, measured along the angle of inclination of the area

h_p Vertical distance from the free surface to the center of pressure of the area

B, H Dimensions of the area

Figure 4.9 shows the location of the centroid of a rectangle. Other shapes are described in Appendix L.

The following procedure will help you calculate the magnitude of the resultant force on a submerged plane area due to fluid pressure and the location of the center of pressure.

Procedure for Computing the Force on a Submerged Plane Area

1. Identify the point where the angle of inclination of the area of interest intersects the level of the free surface of the fluid. This may require the extension of the angled surface or the fluid surface line. Call this point S.

2. Locate the centroid of the area from its geometry.

3. Determine h_c as the *vertical* distance from the level of the free surface down to the centroid of the area.

4. Determine L_c as the *inclined* distance from the level of the free surface down to the centroid of the area. This is the distance from S to the centroid. Note that h_c and L_c are related by

$$h_c = L_c \sin \theta$$

5. Calculate the total area A on which the force is to be determined.

6. Calculate the resultant force from

⮞ **Resultant Force on a Submerged Plane Area**

$$F_R = \gamma h_c A \qquad (4\text{--}4)$$

where γ is the specific weight of the fluid. This equation states that the resultant force is the product of the pressure at the centroid of the area and the total area.

7. Calculate I_c, the moment of inertia of the area about its centroidal axis.

8. Calculate the location of the center of pressure from

⮞ **Location of Center of Pressure**

$$L_p = L_c + \frac{I_c}{L_c A} \qquad (4\text{--}5)$$

Notice that the center of pressure is always below the centroid of an area. In some cases it may be of interest to calculate only the difference between L_p and L_c from

$$L_p - L_c = \frac{I_c}{L_c A} \qquad (4\text{--}6)$$

9. Sketch the resultant force F_R acting at the center of pressure, perpendicular to the area.

10. Show the dimension L_p on the sketch in a manner similar to that used in Fig. 4.8.

11. Draw the dimension lines for L_c and L_p from a reference line drawn through point S and perpendicular to the angle of inclination of the area.

12. If it is desired to compute the vertical depth to the center of pressure, h_p, either of two methods can be used. If the distance L_p has already been computed, use

$$h_p = L_p \sin \theta$$

Alternatively, Step 8 could be avoided and h_p can be computed directly from

$$h_p = h_c + \frac{I_c \sin^2 \theta}{h_c A}$$

We will now use the programmed instruction approach to illustrate the application of this procedure.

PROGRAMMED EXAMPLE PROBLEM

Example Problem 4.6   The tank shown in Fig. 4.8 contains a lubricating oil with a specific gravity of 0.91. A rectangular window with the dimensions $B = 4$ ft and $H = 2$ ft is placed in the inclined wall of the tank ($\theta = 60°$). The centroid of the window is at a depth of 5 ft from the surface of the oil. Calculate (a) the magnitude of the resultant force F_R on the window and (b) the location of the center of pressure.

Using the procedure described above, perform Steps 1 and 2 before going to the next panel.

FIGURE 4.10 Rectangular window for Example Problem 4.6.

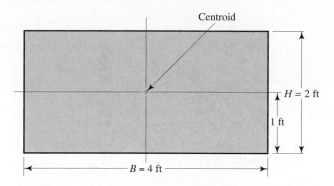

Point S is shown in Fig. 4.8.

The area of interest is the rectangular window sketched in Fig. 4.10. The centroid is at the intersection of the axes of symmetry of the rectangle.

Now, for Step 3, what is the distance h_c?

From the problem statement we know that $h_c = 5$ ft, the vertical depth from the free surface of the oil to the centroid of the window.

Now calculate L_c. See Step 4.

The terms L_c and h_c are related in this case by

$$h_c = L_c \sin \theta$$

Therefore, we have

$$L_c = h_c/\sin \theta = 5 \text{ ft}/\sin 60° = 5.77$$

Both h_c and L_c will be needed for later calculations.

Go on to Step 5.

Because the area of the rectangle is BH,

$$A = BH = (4 \text{ ft})(2 \text{ ft}) - 8 \text{ ft}^2$$

Now do Step 6.

In the equation $F_R = \gamma h_c A$ we need the specific weight of the oil:

$$\gamma_o = (sg)_o(62.4 \text{ lb/ft}^3) = (0.91)(62.4 \text{ lb/ft}^3)$$
$$= 56.8 \text{ lb/ft}^3$$

Then we have

$$F_R = \gamma_o h_c A = \frac{56.8 \text{ lb}}{\text{ft}^3} \times 5 \text{ ft} \times 8 \text{ ft}^2 = 2270 \text{ lb}$$

The next steps concern the location of the center of pressure. Go on to Step 7.

Fig. 4.9 shows the equation for I_c for a rectangle. Using $B = 4$ ft and $H = 2$ ft, we find,

$$I_c = BH^3/12 = (4 \text{ ft})(2 \text{ ft})^3/12 = 2.67 \text{ ft}^4$$

Now we have all the data necessary to do Step 8.

Because $I_c = 2.67 \text{ ft}^4$, $L_c = 5.77$ ft, and $A = 8 \text{ ft}^2$,

$$L_p = L_c + \frac{I_c}{L_c A} = 5.77 \text{ ft} + \frac{2.67 \text{ ft}^4}{(5.77 \text{ ft})(8 \text{ ft}^2)}$$
$$L_p = 5.77 \text{ ft} + 0.058 \text{ ft} = 5.828 \text{ ft}$$

The result is $L_p = 5.828$ ft.

This means that the center of pressure is 0.058 ft (or 0.70 in) below the centroid of the window. Steps 9–11 are already completed in Fig. 4.8. Be sure you understand how the dimension L_p is drawn from the reference line.

4.6 DEVELOPMENT OF THE GENERAL PROCEDURE FOR FORCES ON SUBMERGED PLANE AREAS

Section 4.5 showed the use of the principles for computing the resultant force on a submerged plane area and for finding the location of the center of pressure. Equation (4–4) gives the resultant force, and Eq. (4–6) gives the distance between the centroid of the area of interest and the center of pressure. Figure 4.8 illustrates the various terms. This section shows the development of those relationships.

4.6.1 Resultant Force

The *resultant force* is defined as the summation of the forces on small elements of interest. Figure 4.11 illustrates the concept using the same rectangular window used in Fig. 4.8. Actually, the shape of the area is arbitrary. On any small area dA, there exists a force dF acting perpendicular to the area owing to the fluid pressure p. But the magnitude of the pressure at any depth h in a static liquid of specific weight γ is $p = \gamma h$. Then, the force is

$$dF = p(dA) = \gamma h(dA) \qquad (4\text{–}7)$$

Because the area is inclined at an angle θ, it is convenient to work in the plane of the area, using y to denote the position on the area at any depth h. Note that

$$h = y \sin \theta \qquad (4\text{–}8)$$

where y is measured from the level of the free surface of the fluid along the angle of inclination of the area. Then,

$$dF = \gamma(y \sin \theta)(dA) \qquad (4\text{–}9)$$

The summation of forces over the entire area is accomplished by the mathematical process of integration,

$$F_R = \int_A dF = \int_A \gamma(y \sin \theta)(dA) = \gamma \sin \theta \int_A y(dA)$$

From mechanics we learn that $\int y(dA)$ is equal to the product of the total area times the distance to the centroid of the area from the reference axis. That is,

$$\int_A y(dA) = L_c A$$

Then, the resultant force F_R is

$$F_R = \gamma \sin \theta (L_c A) \qquad (4\text{–}10)$$

Now we can substitute $h_c = L_c \sin \theta$, finding

$$F_R = \gamma h_c A \qquad (4\text{–}11)$$

This is the same form as Eq. (4–4). Because each of the small forces dF acted perpendicular to the area, the resultant force also acts perpendicular to the area.

4.6.2 Center of Pressure

The *center of pressure* is that point on an area where the resultant force can be assumed to act so as to have the same effect as the distributed force over the entire area due to fluid pressure. We can express this effect in terms of the moment of a force with respect to an axis through S perpendicular to the page.

See Fig. 4.11. The moment of each small force dF with respect to this axis is

$$dM = dF \cdot y$$

FIGURE 4.11 Development of the general procedure for forces on submerged plane areas.

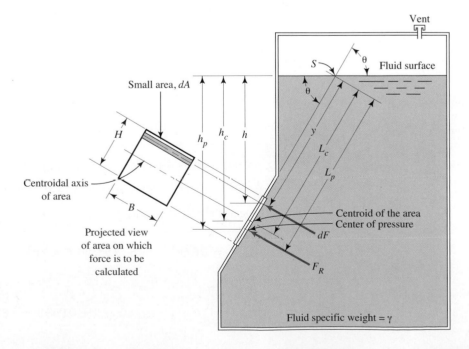

But $dF = \gamma(y \sin \theta)(dA)$. Then,

$$dM = y[\gamma(y \sin \theta)(dA)] = \gamma \sin \theta(y^2 dA)$$

The moment of all the forces on the entire area is found by integrating over the area. Now, if we assume that the resultant force F_R acts at the center of pressure, its moment with respect to the axis through S is $F_R L_p$. Then,

$$F_R L_p = \int \gamma \sin \theta(y^2 dA) = \gamma \sin \theta \int (y^2 dA)$$

Again from mechanics, we learn that $\int (y^2 dA)$ is defined as the moment of inertia I of the entire area with respect to the axis from which y is measured. Then,

$$F_R L_p = \gamma \sin \theta(I)$$

Solving for L_p gives

$$L_p = \frac{\gamma \sin \theta(I)}{F_R}$$

Substituting for F_R from Eq. (4–10) gives

$$L_p = \frac{\gamma \sin \theta(I)}{\gamma \sin \theta(L_c A)} = \frac{I}{L_c A} \qquad (4\text{–}12)$$

A more convenient expression can be developed by using the transfer theorem for moment of inertia from mechanics. That is,

$$I = I_c + AL_c^2$$

where I_c is the moment of inertia of the area of interest with respect to its own centroidal axis and L_c is the distance from the reference axis to the centroid. Equation (4–12) then becomes

$$L_p = \frac{I}{L_c A} = \frac{I_c + AL_c^2}{L_c A} = \frac{I_c}{L_c A} + L_c \qquad (4\text{–}13)$$

Rearranging gives the same form as Eq. (4–6):

$$L_p - L_c = \frac{I_c}{L_c A}$$

We now continue the development by creating an expression for the vertical depth to the center of pressure h_p. Starting from Eq. (4–13), note the following relationships:

$$h_p = L_p \sin \theta$$
$$L_c = h_c / \sin \theta$$

Then,

$$h_p = L_p \sin \theta = \sin \theta \left[\frac{h_c}{\sin \theta} + \frac{I_c}{(h_c / \sin \theta)A} \right]$$

$$h_p = h_c + \frac{I_c \sin^2 \theta}{h_c A}$$

4.7 PIEZOMETRIC HEAD

In all the problems demonstrated so far, the free surface of the fluid was exposed to the ambient pressure, where $p = 0$ (gage). Therefore, our calculations for pressure within the fluid were also gage pressures. It was appropriate to use gage pressures for computing the magnitude of the net force on the areas of interest because the ambient pressure also acts outside the area.

A change is required in our procedure if the pressure above the free surface of the fluid is different from the ambient pressure outside the area. A convenient method would be to use the concept of *piezometric head*, in which the actual pressure above the fluid, p_a, is converted into an equivalent depth of the fluid, h_a, that would create the same pressure (Fig. 4.12):

⤷ **Piezometric Head**

$$h_a = p_a / \gamma \qquad (4\text{–}14)$$

FIGURE 4.12 Illustration of piezometric head for Example Problem 4.7.

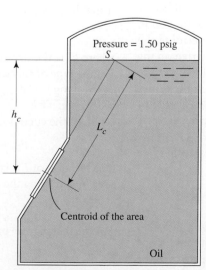

(a) Tank from Fig. 4.8 with pressure above the oil

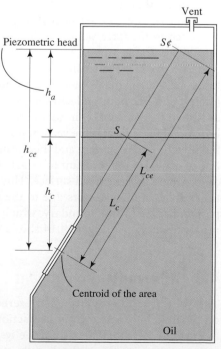

(b) Tank showing piezometric head equivalent to pressure above the oil

This depth is added to any depth h below the free surface to obtain an *equivalent depth*, h_e. That is,

$$h_e = h + h_a \qquad (4\text{--}15)$$

Then, h_e can be used in any calculation requiring a depth to compute pressure. For example, in Fig. 4.12, the equivalent depth to the centroid is

$$h_{ce} = h_c + h_a$$

Example Problem 4.7 Repeat Example Problem 4.6, except consider that the tank shown in Fig. 4.8 is sealed at its top and that there is a pressure of 1.50 psig above the oil.

Solution Several calculations in the solution to Example Problem 4.6 used the depth to the centroid, h_c, given to be 5.0 ft below the surface of the oil. With the pressure above the oil, we must add the piezometric head h_a from Eq. (4–14). Using $\gamma = 56.8 \, \text{lb/ft}^3$, we get

$$h_a = \frac{p_a}{\gamma} = \frac{1.5 \, \text{lb}}{\text{in}^2} \frac{144 \, \text{in}^2}{\text{ft}^2} \frac{\text{ft}^3}{56.8 \, \text{lb}} = 3.80 \, \text{ft}$$

Then, the equivalent depth to the centroid is

$$h_{ce} = h_c + h_a = 5.00 \, \text{ft} + 3.80 \, \text{ft} = 8.80 \, \text{ft}$$

The resultant force is then

$$F_R = \gamma h_{ce} A = (56.8 \, \text{lb/ft}^3)(8.80 \, \text{ft})(8.0 \, \text{ft}^2) = 4000 \, \text{lb}$$

Compare this with the value of 2270 lb found before for the open tank.

The center of pressure also changes because the distance L_c changes to L_{ce} as follows:

$$L_{ce} = h_{ce}/\sin \theta = 8.80 \, \text{ft}/\sin 60° = 10.16$$

$$L_{pe} - L_{ce} = \frac{I_c}{L_{ce}A} = \frac{2.67 \, \text{ft}^4}{(10.16 \, \text{ft})(8 \, \text{ft}^2)} = 0.033 \, \text{ft}$$

The corresponding distance from Example Problem 4.6 was 0.058 ft.

4.8 DISTRIBUTION OF FORCE ON A SUBMERGED CURVED SURFACE

Figure 4.13 shows a tank holding a liquid with its top surface open to the atmosphere. Part of the left wall is vertical, and the lower portion is a segment of a cylinder. Here we are interested in the force acting on the curved surface due to the fluid pressure.

One way to visualize the total force system involved is to isolate the volume of fluid directly above the surface of interest as a free body and show all the forces acting on it, as shown in Fig. 4.14. Our goal here is to determine the horizontal force F_H and the vertical force F_V exerted on the fluid by the curved surface and their resultant force F_R. The line of action of the resultant force acts through the center of curvature of the curved surface. This is because each of the individual force vectors due to the fluid pressure acts perpendicular to the boundary, which is then along the radius of curvature. Figure 4.14 shows the resulting force vectors.

4.8.1 Horizontal Component

The vertical solid wall at the left exerts horizontal forces on the fluid in contact with it in reaction to the forces due to the fluid pressure. This part of the system behaves in the same manner as the vertical walls studied earlier. The

resultant force F_1 acts at a distance $h/3$ from the bottom of the wall.

The force F_{2a} on the right side of the upper part to a depth of h is equal to F_1 in magnitude and acts in the opposite direction. Thus, they have no effect on the curved surface.

By summing forces in the horizontal direction, you can see that F_H must be equal to F_{2b} acting on the lower part of the right side. The area on which F_{2b} acts is the *projection* of the curved surface onto a vertical plane.

The magnitude of F_{2b} and its location can be found using the procedures developed for plane surfaces. That is,

$$F_{2b} = \gamma h_c A \qquad (4\text{--}16)$$

where h_c is the depth to the centroid of the projected area. For the type of surface shown in Fig. 4.14, the projected area is a rectangle. Calling the height of the rectangle s, you can see that $h_c = h + s/2$. Also, the area is sw, where w is the width of the curved surface. Then,

$$F_{2b} = F_H = \gamma sw(h + s/2) \qquad (4\text{--}17)$$

The location of F_{2b} is the center of pressure of the projected area. Again using the principles developed earlier, we get

$$h_p - h_c = I_c/(h_c A)$$

For the rectangular projected area, however,

$$I_c = ws^3/12$$

$$A = sw$$

FIGURE 4.13 Tank with a curved surface containing a static fluid.

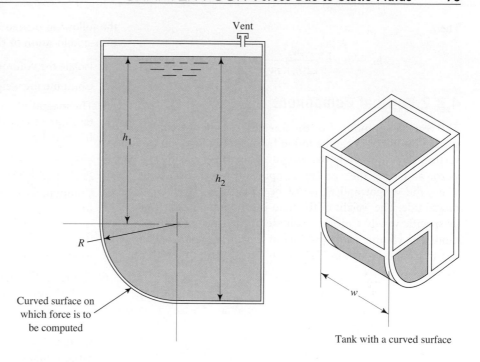

Curved surface on which force is to be computed

Tank with a curved surface

FIGURE 4.14 Free-body diagram of a volume of fluid above the curved surface.

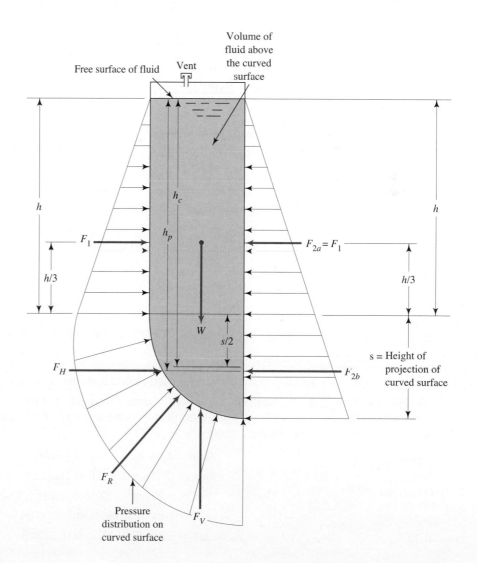

Free surface of fluid

Vent

Volume of fluid above the curved surface

Pressure distribution on curved surface

s = Height of projection of curved surface

Then,

$$h_p - h_c = \frac{ws^3}{12(h_c)(sw)} = \frac{s^2}{12h_c} \quad (4\text{–}18)$$

4.8.2 Vertical Component

The vertical component of the force exerted by the curved surface on the fluid can be found by summing forces in the vertical direction. Only the weight of the fluid acts downward, and only the vertical component F_V acts upward. Then, the weight and F_V must be equal to each other in magnitude. The weight of the fluid is simply the product of its specific weight times the volume of the isolated body of fluid. The volume is the product of the cross-sectional area of the volume shown in Fig. 4.14 and the length of interest, w. That is,

$$F_V = \gamma(\text{volume}) = \gamma Aw \quad (4\text{–}19)$$

4.8.3 Resultant Force

The total resultant force F_R is

$$F_R = \sqrt{F_H^2 + F_V^2} \quad (4\text{–}20)$$

The resultant force acts at an angle ϕ relative to the horizontal found from

$$\phi = \tan^{-1}(F_V/F_H) \quad (4\text{–}21)$$

4.8.4 Summary of the Procedure for Computing the Force on a Submerged Curved Surface

Given a curved surface submerged beneath a static liquid similar to the configuration shown in Fig. 4.13, we can use the following procedure to compute the magnitude, direction, and location of the resultant force on the surface.

1. Isolate the volume of fluid above the surface.

2. Compute the weight of the isolated volume.

3. The magnitude of the vertical component of the resultant force is equal to the weight of the isolated volume. It acts in line with the centroid of the isolated volume.

4. Draw a projection of the curved surface onto a vertical plane and determine its height, called s.

5. Compute the depth to the centroid of the projected area from

$$h_c = h + s/2$$

where h is the depth to the top of the projected area.

6. Compute the magnitude of the horizontal component of the resultant force from

$$F_H = \gamma sw(h + s/2) = \gamma swh_c$$

7. Compute the depth to the line of action of the horizontal component from

$$h_p = h_c + s^2/(12h_c)$$

8. Compute the resultant force from

$$F_R = \sqrt{F_V^2 + F_H^2}$$

9. Compute the angle of inclination of the resultant force relative to the horizontal component from

$$\phi = \tan^{-1}(F_V/F_H)$$

10. Show the resultant force acting on the curved surface in such a direction that its line of action passes through the center of curvature of the surface.

Example Problem 4.8

For the tank shown in Fig. 4.13, the following dimensions apply:

$$h_1 = 3.00 \text{ m}$$
$$h_2 = 4.50 \text{ m}$$
$$w = 2.50 \text{ m}$$
$$\gamma = 9.81 \text{kN/m}^3 \text{ (water)}$$

Compute the horizontal and vertical components of the resultant force on the curved surface and the resultant force itself. Show these force vectors on a sketch.

Solution

Using the steps outlined above:

1. The volume above the curved surface is shown in Fig. 4.15.

2. The weight of the isolated volume is the product of the specific weight of the water times the volume. The volume is the product of the area times the length w. The total area is the sum of a rectangle and a quarter circle.

$$\text{Area} = A_1 + A_2 = h_1 \cdot R + \tfrac{1}{4}(\pi R^2)$$
$$\text{Area} = (3.00 \text{ m})(1.50 \text{ m}) + \tfrac{1}{4}[\pi(1.50 \text{ m})^2] = 4.50 \text{ m}^2 + 1.767 \text{ m}^2$$
$$\text{Area} = 6.267 \text{ m}^2$$
$$\text{Volume} = \text{area} \cdot w = (6.267 \text{ m}^2)(2.50 \text{ m}) = 15.67 \text{ m}^3$$
$$\text{Weight} = \gamma V = (9.81 \text{ kN/m}^3)(15.67 \text{ m}^3) = 153.7 \text{ kN}$$

FIGURE 4.15 Isolated volume above the curved surface for Example Problem 4.8.

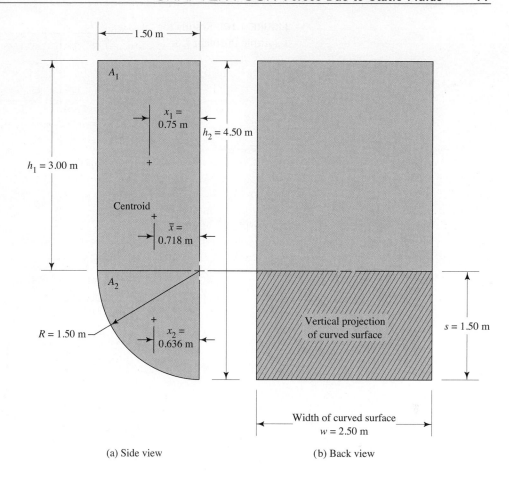

(a) Side view   (b) Back view

3. Then, $F_V = 153.7$ kN, acting upward through the centroid of the volume. The location of the centroid is found using the composite-area technique. Refer to Fig. 4.15 for the data. Each value should be obvious except x_2, the location of the centroid of the quadrant. From Appendix L,

$$x_2 = 0.424R = 0.424(1.50 \text{ m}) = 0.636 \text{ m}$$

Then, the location of the centroid for the composite area is

$$\bar{x} = \frac{A_1x_1 + A_2x_2}{A_1 + A_2} = \frac{(4.50)(0.75) + (1.767)(0.636)}{4.50 + 1.767} = 0.718 \text{ m}$$

4. The vertical projection of the curved surface is shown in Fig. 4.15. The height s equals 1.50 m.
5. The depth to the centroid of the projected area is

$$h_c = h_1 + s/2 = 3.00 \text{ m} + (1.50 \text{ m})/2 = 3.75 \text{ m}$$

6. The magnitude of the horizontal force is

$$F_H = \gamma s w(h_1 + s/2) = \gamma s w h_c$$
$$F_H = (9.81 \text{ kN/m}^3)(1.50 \text{ m})(2.50 \text{ m})(3.75 \text{ m}) = 138.0 \text{ kN}$$

7. The depth to the line of action of the horizontal component is found from

$$h_p = h_c + s^2/(12h_c)$$
$$h_p = 3.75 \text{ m} + (1.50)^2/[(12)(3.75)] = 3.80 \text{ m}$$

8. The resultant force is computed from

$$F_R = \sqrt{F_V^2 + F_H^2}$$
$$F_R = \sqrt{(153.7 \text{ kN})^2 + (138.0 \text{ kN})^2} = 206.5 \text{ kN}$$

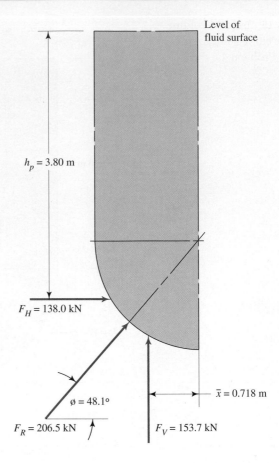

FIGURE 4.16 Results for
Example Problem 4.8.

9. The angle of inclination of the resultant force relative to the horizontal is computed from

$$\phi = \tan^{-1}(F_V/F_H)$$

$$\phi = \tan^{-1}(153.7/138.0) = 48.1°$$

10. The horizontal component, the vertical component, and the resultant force are shown in Fig. 4.16. Note that the line of action of F_R is through the center of curvature of the surface. Also note that the vertical component is acting through the centroid of the volume of liquid above the surface. The horizontal component is acting through the center of pressure of the projected area at a depth h_p from the level of the free surface of the fluid.

4.9 EFFECT OF A PRESSURE ABOVE THE FLUID SURFACE

In the preceding discussion of force on a submerged curved surface, the magnitude of the force was directly dependent on the depth of the static fluid above the surface of interest. If an additional pressure exists above the fluid or if the fluid itself is pressurized, the effect is to add to the actual depth a depth of fluid h_a equivalent to p/γ. This is the same procedure, called *piezometric head*, used in Section 4.7. The new equivalent depth is used to compute both the vertical and horizontal forces.

4.10 FORCES ON A CURVED SURFACE WITH FLUID BELOW IT

To this point, problems have considered curved surfaces supporting a fluid above. An important concept presented for such problems was that the vertical force on the curved surface was equal to the weight of the fluid above the surface.

Now, consider the type of curved surface shown in Fig. 4.17, in which the fluid is restrained below the surface. Fluid pressure on such a surface causes forces that tend to push it upward and to the right. The surface and its connections

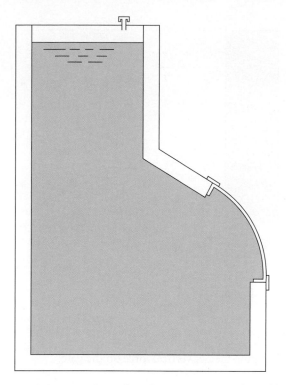

FIGURE 4.17 Curved surface restraining a liquid below it.

then would have to exert reaction forces downward and to the left on the contained fluid.

The pressure in the fluid at any point is dependent on the depth of fluid to that point from the level of the free surface. This situation is equivalent to having the curved surface supporting a volume of liquid *above* it, except for the direction of the force vectors. Figure 4.18 shows that we can visualize an imaginary volume of fluid extending from the surface of interest to the level of the free surface or to the piezometric line if the fluid is under an additional pressure. Then, as before, the horizontal component of the force exerted by the curved surface on the fluid is the force on the projection of the curved surface on a vertical plane. The vertical component is equal to the weight of the imaginary volume of fluid above the surface.

4.11 FORCES ON CURVED SURFACES WITH FLUID ABOVE AND BELOW

Figure 4.19 shows a semicylindrical gate projecting into a tank containing an oil. The force due to fluid pressure would have a horizontal component acting to the right on the gate. This force acts on the projection of the surface on a vertical plane and is computed in the same manner as used in Section 4.7.

In the vertical direction, the force on the top of the gate would act downward and would equal the weight of the oil above the gate. However, there is also a force acting upward on the bottom surface of the gate equal to the total weight of the fluid, both real and imaginary, above that surface. The net vertical force is the difference between the two forces, equal to the weight of the semicylindrical volume of fluid displaced by the gate itself (Fig. 4.20).

FIGURE 4.18 Forces exerted by a curved surface on the fluid.

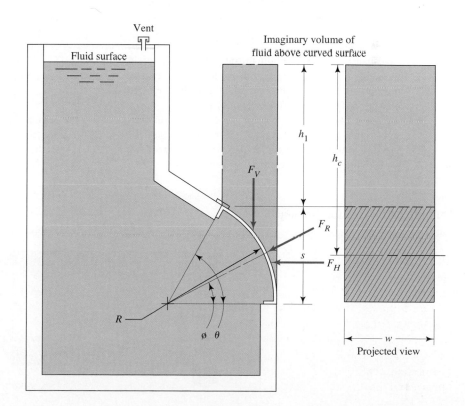

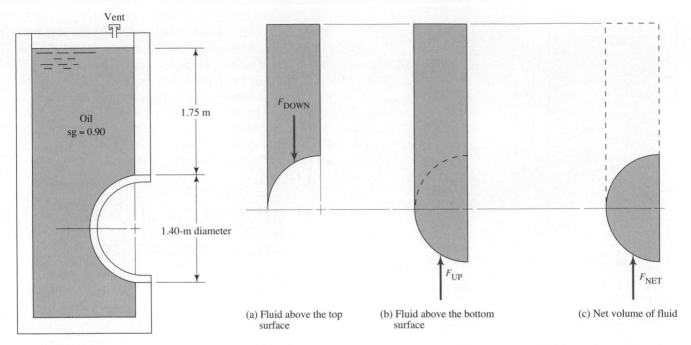

FIGURE 4.19 Semicylindrical gate.

(a) Fluid above the top surface

(b) Fluid above the bottom surface

(c) Net volume of fluid

FIGURE 4.20 Volumes used to compute the net vertical force on the gate.

PRACTICE PROBLEMS

Forces Due to Gas Pressure

4.1 Figure 4.21 shows a vacuum tank with a flat circular observation window in one end. If the pressure in the tank is 0.12 psia when the barometer reads 30.5 in of mercury, calculate the total force on the window.

4.2 The flat left end of the tank shown in Fig. 4.21 is secured with a bolted flange. If the inside diameter of the tank is 30 in and the internal pressure is raised to +14.4 psig, calculate the total force that must be resisted by the bolts in the flange.

4.3 An exhaust system for a room creates a partial vacuum in the room of 1.20 in of water relative to the atmospheric pressure outside the room. Compute the net force exerted on a 36- by 80-in door to this room.

4.4 A piece of 14-in Schedule 40 pipe is used as a pressure vessel by capping its ends. Compute the force on the caps if the pressure in the pipe is raised to 325 psig. See Appendix F for the dimensions of the pipe.

4.5 A pressure relief valve is designed so that the gas pressure in the tank acts on a piston with a diameter of 30 mm. How much spring force must be applied to the outside of the piston to hold the valve closed under a pressure of 3.50 MPa?

4.6 A gas-powered cannon shoots projectiles by introducing nitrogen gas at 20.5 MPa into a cylinder having an inside diameter of 50 mm. Compute the force exerted on the projectile.

4.7 The egress hatch of a manned spacecraft is designed so that the internal pressure in the cabin applies a force to help maintain the seal. If the internal pressure is 34.4 kPa(abs) and the external pressure is a perfect vacuum, calculate the force on a square hatch 800 mm on a side.

Forces on Horizontal Flat Surfaces under Liquids

4.8 A tank containing liquid ammonia at 77°F has a flat horizontal bottom. A rectangular door, 24 in by 18 in, is installed in the bottom to provide access for cleaning. Compute the force on the door if the depth of ammonia is 12.3 ft.

FIGURE 4.21 Tank for Problems 4.1 and 4.2.

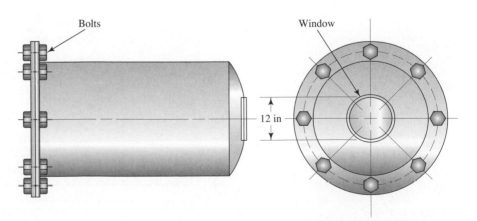

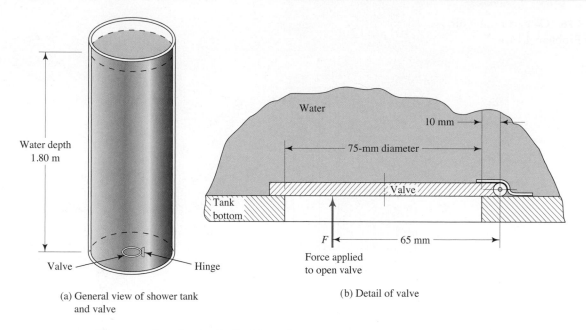

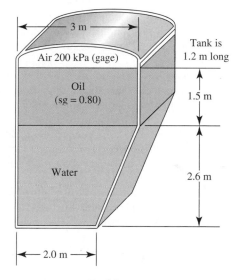

(a) General view of shower tank and valve

(b) Detail of valve

FIGURE 4.22 Shower tank and valve for Problem 4.10.

4.9 The bottom of a laboratory vat has a hole in it to allow the liquid mercury to pour out. The hole is sealed by a rubber stopper pushed in the hole and held by friction. What force tends to push the 0.75-in-diameter stopper out of the hole if the depth of the mercury is 28.0 in?

4.10 A simple shower for remote locations is designed with a cylindrical tank 500 mm in diameter and 1.800 m high as shown in Fig. 4.22. The water flows through a flapper valve in the bottom through a 75-mm-diameter opening. The flapper must be pushed upward to open the valve. How much force is required to open the valve?

4.11 Calculate the total force on the bottom of the closed tank shown in Fig. 4.23 if the air pressure is 52 kPa(gage).

4.12 If the length of the tank in Fig. 4.24 is 1.2 m, calculate the total force on the bottom of the tank.

4.13 An observation port in a small submarine is located in a horizontal surface of the sub. The shape of the port is shown in Fig. 4.25. Compute the total force acting on the port when the pressure inside the sub is 100 kPa(abs) and the sub is operating at a depth of 175 m in seawater.

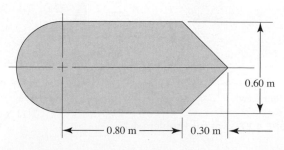

FIGURE 4.24 Problem 4.12.

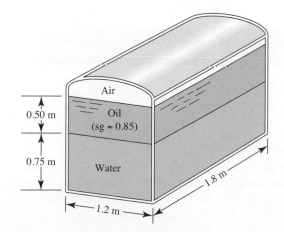

FIGURE 4.23 Problem 4.11.

FIGURE 4.25 Port for Problem 4.13.

FIGURE 4.26 Gate in a reservoir wall for Problem 4.14.

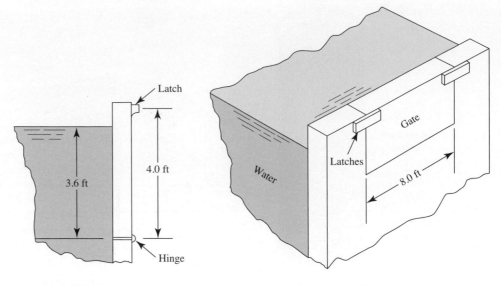

Forces on Rectangular Walls

4.14 A rectangular gate is installed in a vertical wall of a reservoir, as shown in Fig. 4.26. Compute the magnitude of the resultant force on the gate and the location of the center of pressure. Also compute the force on each of the two latches shown.

4.15 A vat has a sloped side, as shown in Fig. 4.27. Compute the resultant force on this side if the vat contains 15.5 ft of glycerin. Also compute the location of the center of pressure and show it on a sketch with the resultant force.

4.16 The wall shown in Fig. 4.28 is 20 ft long. (a) Calculate the total force on the wall due to water pressure and locate the center of pressure; (b) calculate the moment due to this force at the base of the wall.

4.17 If the wall in Fig. 4.29 is 4 m long, calculate the total force on the wall due to the oil pressure. Also determine the location of the center of pressure and show the resultant force on the wall.

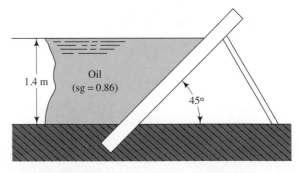

FIGURE 4.28 Problem 4.16.

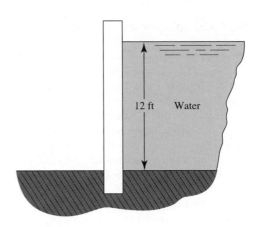

FIGURE 4.29 Problem 4.17.

Forces on Submerged Plane Areas

For each of the cases shown in Figs. 4.30–4.41, compute the magnitude of the resultant force on the indicated area and the location of the center of pressure. Show the resultant force on the area and clearly dimension its location.

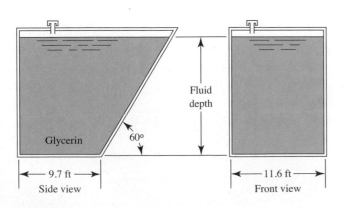

FIGURE 4.27 Vat for Problem 4.15.

4.18 Refer to Fig. 4.30.

4.20 Refer to Fig. 4.32.

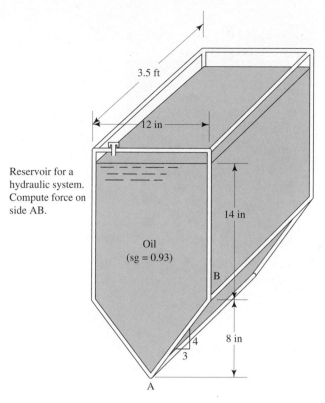

Reservoir for a hydraulic system. Compute force on side AB.

FIGURE 4.30 Problem 4.18.

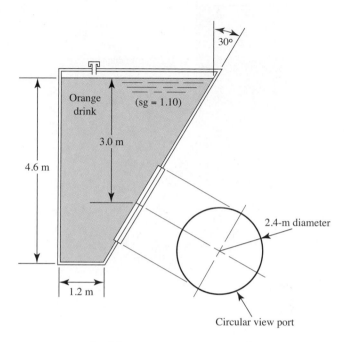

FIGURE 4.32 Problems 4.20, 4.36, 4.37, and 4.44.

4.19 Refer to Fig. 4.31.

4.21 Refer to Fig. 4.33.

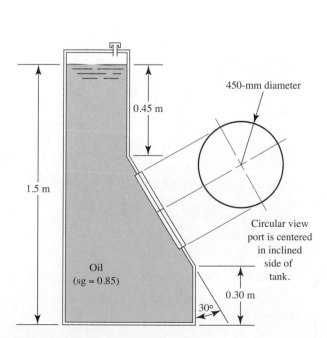

FIGURE 4.31 Problems 4.19 and 4.43.

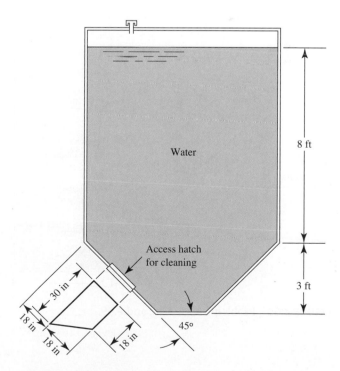

FIGURE 4.33 Problem 4.21.

4.22 Refer to Fig. 4.34.

4.24 Refer to Fig. 4.36.

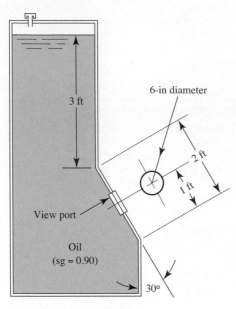

FIGURE 4.34 Problem 4.22.

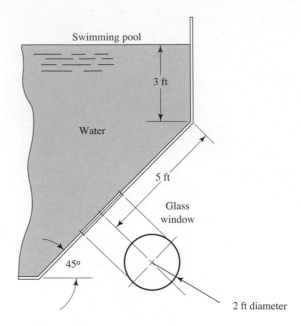

FIGURE 4.36 Problem 4.24.

4.23 Refer to Fig. 4.35.

4.25 Refer to Fig. 4.37.

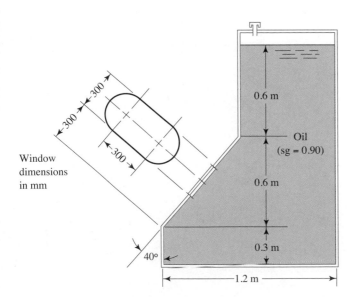

FIGURE 4.35 Problems 4.23, 4.38, and 4.39.

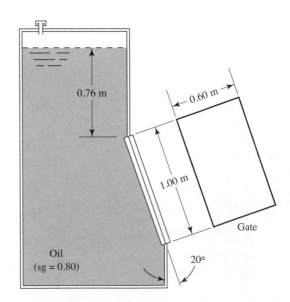

FIGURE 4.37 Problem 4.25.

4.26 Refer to Fig. 4.38.

FIGURE 4.38 Problems 4.26 and 4.45.

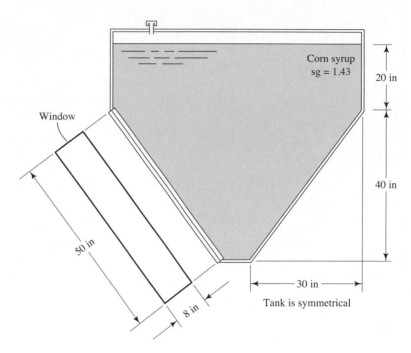

4.27 Refer to Fig. 4.39.

4.28 Refer to Fig. 4.40.

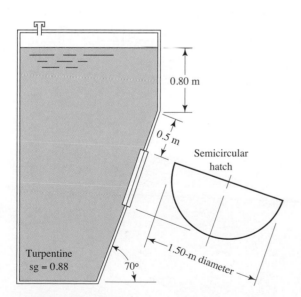

FIGURE 4.39 Problem 4.27.

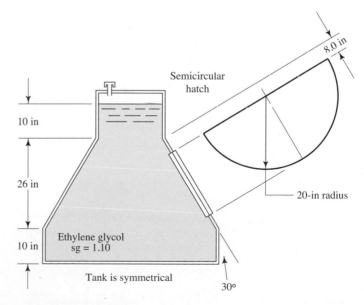

FIGURE 4.40 Problems 4.28 and 4.46.

4.29 Refer to Fig. 4.41.

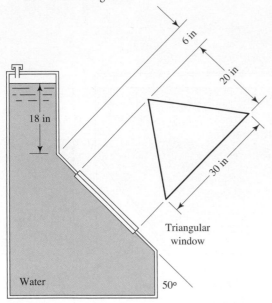

4.30 Figure 4.42 shows a gasoline tank filled into the filler pipe. The gasoline has a specific gravity of 0.67. Calculate the total force on each flat end of the tank and determine the location of the center of pressure.

4.31 If the tank in Fig. 4.42 is filled just to the bottom of the filler pipe with gasoline (sg = 0.67), calculate the magnitude and location of the resultant force on the flat end.

4.32 If the tank in Fig. 4.42 is only half full of gasoline (sg = 0.67), calculate the magnitude and location of the resultant force on the flat end.

4.33 For the water tank shown in Fig. 4.43, compute the magnitude and location of the total force on the vertical back wall.

4.34 For the water tank shown in Fig. 4.43, compute the magnitude and location of the total force on each vertical end wall.

4.35 For the water tank shown in Fig. 4.43, compute the magnitude and location of the total force on the inclined wall.

4.36 For the orange-drink tank shown in Fig. 4.32, compute the magnitude and location of the total force on each vertical end wall. The tank is 3.0 m long.

FIGURE 4.41 Problem 4.29.

FIGURE 4.42 Problems 4.30–4.32.

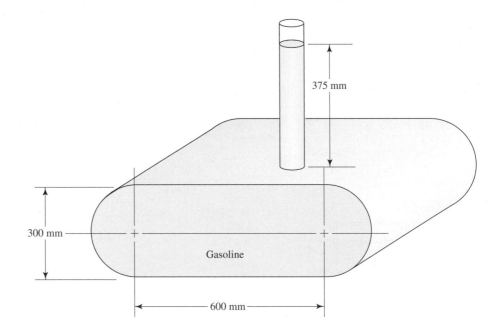

FIGURE 4.43 Problems 4.33–4.35.

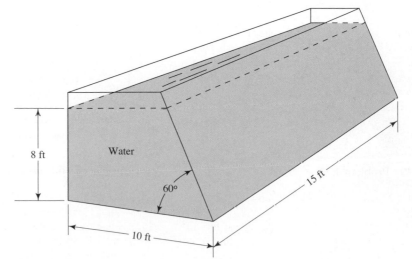

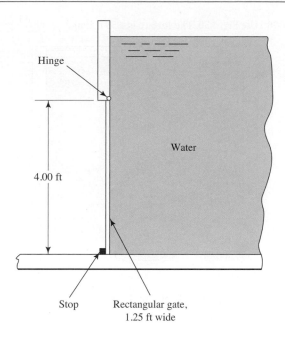

FIGURE 4.44 Problem 4.40.

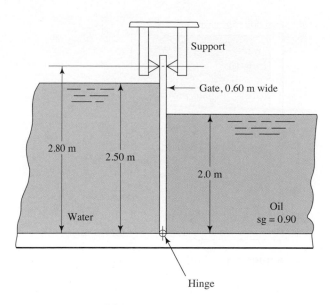

FIGURE 4.45 Problem 4.41.

4.37 For the orange-drink tank shown in Fig. 4.32, compute the magnitude and location of the total force on the vertical back wall. The tank is 3.0 m long.

4.38 For the oil tank shown in Fig. 4.35, compute the magnitude and location of the total force on each vertical end wall. The tank is 1.2 m long.

4.39 For the oil tank shown in Fig. 4.35, compute the magnitude and location of the total force on the vertical back wall. The tank is 1.2 m long.

4.40 Figure 4.44 shows a rectangular gate holding water behind it. If the water is 6.00 ft deep, compute the magnitude and location of the resultant force on the gate. Then compute the forces on the hinge at the top and on the stop at the bottom.

4.41 Figure 4.45 shows a gate hinged at its bottom and held by a simple support at its top. The gate separates two fluids. Compute the net force on the gate due to the fluid on each side. Then compute the force on the hinge and on the support.

4.42 Figure 4.46 shows a tank of water with a circular pipe connected to its bottom. A circular gate seals the pipe opening to prohibit flow. To drain the tank, a winch is used to pull the gate open. Compute the amount of force that the winch cable must exert to open the gate.

Piezometric Head

4.43 Repeat Problem 4.19 (Fig. 4.31), except that the tank is now sealed at the top with a pressure of 13.8 kPa above the oil.

4.44 Repeat Problem 4.20 (Fig. 4.32), except that the tank is now sealed at the top with a pressure of 25.0 kPa above the fluid.

4.45 Repeat Problem 4.26 (Fig. 4.38), except that the tank is now sealed at the top with a pressure of 2.50 psig above the fluid.

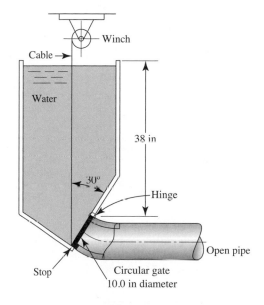

FIGURE 4.46 Problem 4.42.

4.46 Repeat Problem 4.28 (Fig. 4.40), except that the tank is now sealed at the top with a pressure of 4.0 psig above the fluid.

Forces on Curved Surfaces

General Note for Problems 4.47–4.54. For each problem, one curved surface is shown restraining a body of static fluid. Compute the magnitude of the horizontal component of the force and compute the vertical component of the force exerted by the fluid on that surface. Then compute the magnitude of the resultant force and its direction. Show the resultant force acting on the curved surface. In each case the surface of interest is a portion of a cylinder with the length of the surface given in the problem statement.

4.47 Use Fig. 4.47. The surface is 2.00 m long.

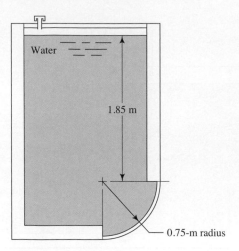

Water

1.85 m

0.75-m radius

FIGURE 4.47 Problems 4.47 and 4.55.

4.48 Use Fig. 4.48. The surface is 2.50 m long.

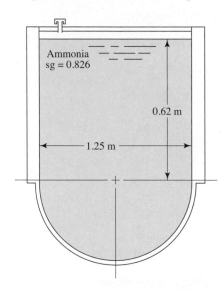

Ammonia
sg = 0.826

0.62 m

1.25 m

FIGURE 4.48 Problems 4.48 and 4.56.

4.49 Use Fig. 4.49. The surface is 5.00 ft long.

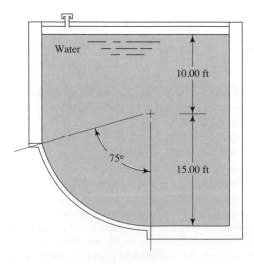

Water

10.00 ft

75°

15.00 ft

FIGURE 4.49 Problem 4.49.

4.50 Use Fig. 4.50. The surface is 4.50 ft long.

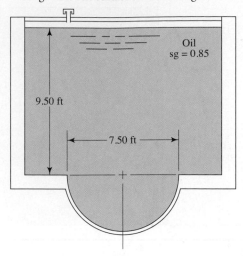

Oil
sg = 0.85

9.50 ft

7.50 ft

FIGURE 4.50 Problem 4.50.

4.51 Use Fig. 4.51. The surface is 4.00 m long.

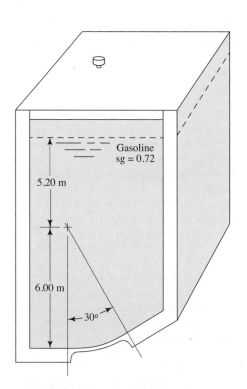

Gasoline
sg = 0.72

5.20 m

6.00 m

30°

FIGURE 4.51 Problem 4.51.

4.52 Use Fig. 4.52. The surface is 1.50 m long.

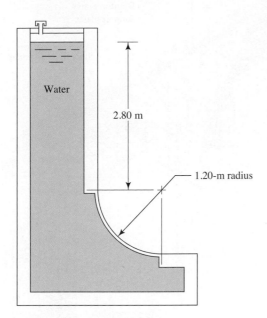

FIGURE 4.52 Problem 4.52.

4.53 Use Fig. 4.53. The surface is 1.50 m long.

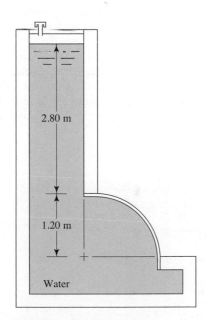

FIGURE 4.53 Problem 4.53.

4.54 Use Fig. 4.54. The surface is 60 in long.

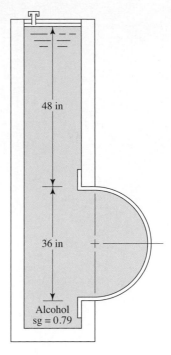

FIGURE 4.54 Problem 4.54.

4.55 Repeat Problem 4.47 using Fig. 4.47, except that there is now 7.50 kPa air pressure above the fluid.

4.56 Repeat Problem 4.48 using Fig. 4.48, except that there is now 4.65 kPa air pressure above the fluid.

Supplemental Problems

4.57 The tank in Fig. 4.55 has a view port in the inclined side. Compute the magnitude of the resultant force on the panel. Show the resultant force on the door clearly and dimension its location.

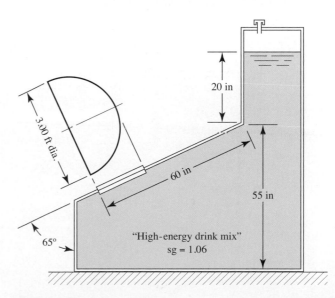

FIGURE 4.55 Problem 4.57.

FIGURE 4.56 Problem 4.58.

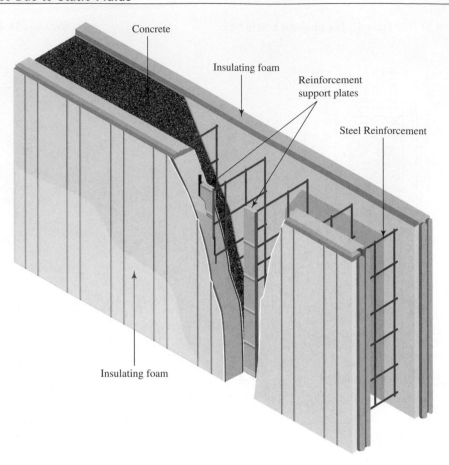

4.58 Insulated concrete forms (ICFs) are becoming more and more common for a variety of reasons including the desire to build more energy efficient "green" structures. Instead of using temporary forms like lumber to hold poured concrete in place until it has cured, an ICF is essentially a rigid lightweight foam container for a poured wall that is left in place permanently, providing an added layer of insulation. See the basic layout of the system in Fig. 4.56. The form, of course, needs to provide adequate strength to contain the wet concrete

(sg = 2.4) until it cures. A leak or blowout at the bottom of the form would have a catastrophic effect on construction. What is the maximum pressure that the form needs withstand if the wall is to be 4.5 in thick, 35 ft wide, and 14 ft tall?

4.59 Locks are installed in rivers to allow boats to pass safely around a dam and through the associated change in water level. The doors of the lock must hold back the water as the boat passes from one level to the next. In Fig. 4.57, the rightmost pair of doors is shown separating

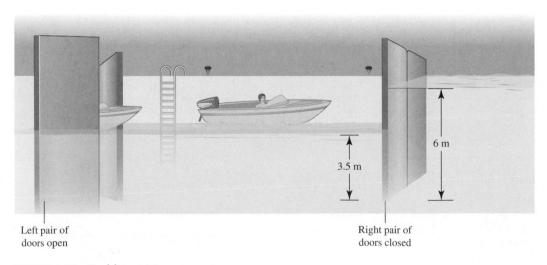

Left pair of
doors open

Right pair of
doors closed

FIGURE 4.57 Problem 4.59.

FIGURE 4.58   Problem 4.60.

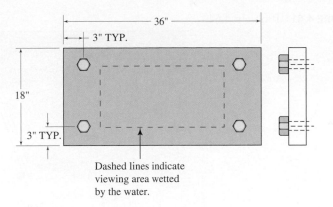

Dashed lines indicate
viewing area wetted
by the water.

FIGURE 4.59   Problem 4.62.

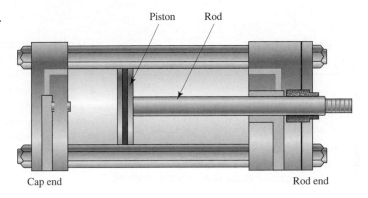

the high side, at a depth of 6 m, from the low side at a
depth of 3.5 m. The lock is 8 m wide. Calculate the
resulting forces on this pair of doors, and draw a free-body
diagram of this pair of doors showing the opposing
pressure distributions.

4.60 When a dam is installed in a river that has salmon, an
alternative path of travel for spawning is critical to the
survival of the salmon. It is desired to install a series of
observation windows so that the public can watch the
salmon "run" through this bypass during the season. The
design for observation windows is shown in Fig. 4.58 and
the wetted portion of the window will be a rectangle of 10
in by 28 in at the center of the window. Four bolts in the
holes shown in the Figure hold the window in place. If the
windows are to be installed with their centers at a depth of
3 ft below the surface of the water, determine the force on
each of the four bolts.

4.61 A wealthy eccentric is interested in having an entire interior wall of his home converted to a seawater aquarium.
Determine the total force exerted on the wall and location
of its resultant if the wall is 3 m high and 7 m long.

4.62 A pneumatic cylinder like the one shown in Fig. 4.59 is
used to push a box on an automated packaging machine.
It will be plumbed into the existing compressed air supply
in the plant that provides 60 psig. To extend the cylinder,
air is supplied to the "cap end", pushing on the full face
of the piston. To retract, air is supplied to the "rod end"
pushing on the piston, except for where the rod attaches.
Calculate the force available to extend and to retract the
cylinder if the piston bore has a diameter of 3 in and the
diameter of the rod is 7/8 in.

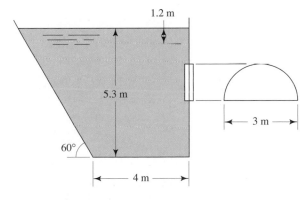

FIGURE 4.60   Problem 4.63.

4.63 Determine the magnitude and the location of the force of
water pushing on the semi-circular window shown in Fig.
4.60. Show the pressure distribution and the resultant force.

4.64 For the hinged gate shown in Fig. 4.61, determine the
magnitude, direction, and location of the force of the
fluid acting on it. Complete a free-body diagram and determine the force that the gate exerts on the stop.

4.65 A large holding tank has an 8-in-diameter drain hole on
the bottom horizontal surface of the tank. A stopper is
lightly placed to cover the hole and a cable is attached to
that stopper. When the tank needs to be drained, a person
on a walkway over the tank is to pull on that cable. If the
tank is filled to a depth of 30 ft with sea water, and the
walkway is 8 ft above the fluid surface, how much force is
required to remove the stopper?

FIGURE 4.61 Problem 4.64.

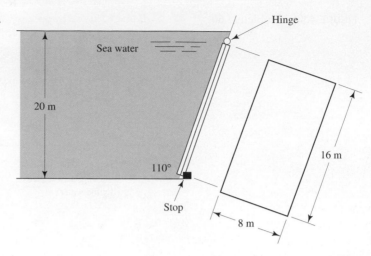

COMPUTER AIDED ENGINEERING ASSIGNMENTS

1. Write a program to solve Problem 4.41 with any combination of data for the variables in Fig. 4.45, including the depth on either side of the gate and the specific gravity of the fluids.

2. Write a program to solve Problem 4.42 (Fig. 4.46) with any combination of data, including the size of the gate, the depth of the fluid, the specific gravity of the fluid, and the angle of inclination of the gate.

3. Write a program to solve curved surface problems of the type shown in Figs. 4.47–4.51 for any combination of variables, including the depth of the fluid, the angular size of the curved segment, the specific gravity of the fluid, and the radius of the surface.

4. For Program 1, cause the depth h to vary over some specified range, giving the output for each value.

BUOYANCY AND STABILITY

Whenever an object is floating in a fluid or when it is completely submerged in the fluid, it is subjected to a *buoyant force* that tends to lift it upward, helping to support it. *Buoyancy* is the tendency of a fluid to exert a supporting force on a body placed in the fluid. You need to understand the concept of buoyancy and make calculations to determine the net forces exerted on objects immersed in fluids or the position of an object when it is floating. You also need to learn about the *stability* of floating or submerged bodies to ensure that they will stay in the preferred orientation even when subjected to external forces that tend to tip them over.

Stability refers to the ability of a body in a fluid to return to its original position after being tilted about a horizontal axis.

Consider the two photographs shown in Fig. 5.1(a) and (b). The boats must float safely under any expected condition of loading while also remaining stable against the forces of the wind acting on the sails or wave action against the hull. The scuba diver will typically tend to float but carefully measured weights are added to allow him or her to swim at whatever depth is desired. Discarding the weights is one way to rise to the surface.

(a)

(b)

FIGURE 5.1 Have you experienced the fun activities shown in these two scenes? Both depend on the principles of buoyancy and stability—the focus of this chapter. (a) Racing sail boats. (*Source:* synto/Fotolia) (b) A scuba diver exploring sea life. (*Source:* Richard Carey/Fotolia LLC)

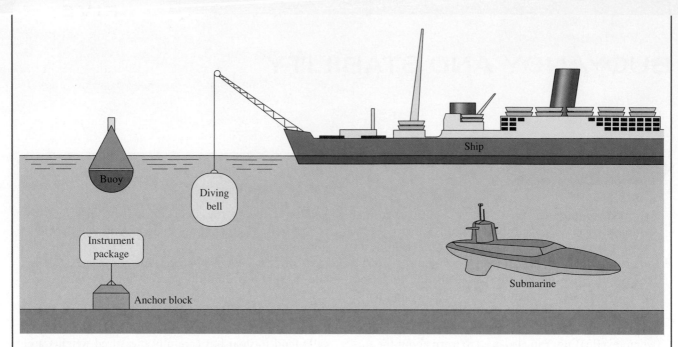

FIGURE 5.2 Examples of types of buoyancy problems.

Exploration

When you lie still in a swimming pool, you will float even though you are almost completely submerged. Wearing a life-vest or holding a buoyant cushion helps. Where else have you observed objects floating in water or other fluids? Examples are any kind of boat, a jet ski, a buoy, a hollow plastic ball, an air mattress, a water toy, or a wooden stick. Where have you seen objects completely submerged in a fluid? Examples are a submarine, dishes in a sink, and a scuba diver. Write down at least five other situations where you observed or felt the tendency of a fluid to support something. Describe whether the object tends to float or sink and its desired orientation. Discuss your observations with your fellow students and with the course instructor.

Introductory Concepts

The objects shown in Fig. 5.2 show different floating tendencies. The buoy and the ship are obviously designed to float and to maintain a specified orientation. The diving bell would tend to sink unless supported by the cable

from the crane on the ship. The instrument package tends to float and must be restrained by the cable attached to a heavy anchor block on the sea bottom. However, the submarine is designed to be able to adjust its ballast to hover at any depth (a condition called *neutral buoyancy*), dive deeper, or rise to the surface and float. Consider any kind of boat, raft, or other floating objects that are expected to maintain a particular orientation when placed in a fluid. How can it be designed to ensure that it will float at a desired level and be stable when given some angular displacement? Why is a canoe more likely to tip over than a large boat with a broad beam when you stand up or move around in it?

This chapter provides the fundamental principles of both buoyancy and stability to help you develop the ability to analyze and design devices that will operate effectively while floating or submerged in a fluid. You will learn how to calculate the magnitude of the buoyant force, to determine the position of a floating body in a fluid, and to calculate the degree of stability of a submerged or floating object.

5.1 OBJECTIVES

After completing this chapter, you should be able to:

1. Write the equation for the buoyant force.
2. Analyze the case of bodies floating on a fluid.
3. Use the principle of static equilibrium to solve for the forces involved in buoyancy problems.
4. Define the conditions that must be met for a body to be stable when completely submerged in a fluid.
5. Define the conditions that must be met for a body to be stable when floating on a fluid.
6. Define the term *metacenter* and compute its location.

5.2 BUOYANCY

A body in a fluid, whether floating or submerged, is buoyed up by a force equal to the weight of the fluid displaced.

The buoyant force acts vertically upward through the centroid of the displaced volume.

These principles were discovered by the Greek scholar Archimedes, and the buoyant force can be defined mathematically as follows:

▷ **Buoyant Force**

$$F_b = \gamma_f V_d \qquad (5\text{--}1)$$

where

F_b = Buoyant force
γ_f = Specific weight of the fluid
V_d = Displaced volume of the fluid

When a body is floating freely, it displaces a sufficient volume of fluid to just balance its own weight.

The analysis of problems dealing with buoyancy requires the application of the equation of static equilibrium in the vertical direction, $\Sigma F_v = 0$, assuming the object is at rest in the fluid. The following procedure is recommended for all problems, whether they involve floating or submerged bodies.

Procedure for Solving Buoyancy Problems

1. Determine the objective of the problem solution. Do you want to determine a force, a weight, a volume, or a specific weight?

2. Draw a free-body diagram of the object in the fluid. Show all forces that act on the free body in the vertical direction, including the weight of the body, the buoyant force, and all external forces. If the direction of some force is not known, assume the most probable direction and show it on the free body.

3. Write the equation of static equilibrium in the vertical direction, $\Sigma F_v = 0$, assuming the positive direction to be upward.

4. Solve for the desired force, weight, volume, or specific weight, remembering the following concepts:
   a. The buoyant force is calculated from $F_b = \gamma_f V_d$.
   b. The weight of a solid object is the product of its total volume and its specific weight; that is, $w = \gamma V$.
   c. An object with an average specific weight less than that of the fluid will tend to float because $w < F_b$ with the object submerged.
   d. An object with an average specific weight greater than that of the fluid will tend to sink because $w > F_b$ with the object submerged.
   e. *Neutral buoyancy* occurs when a body stays in a given position wherever it is submerged in a fluid. An object whose average specific weight is equal to that of the fluid is neutrally buoyant.

PROGRAMMED EXAMPLE PROBLEMS

Example Problem 5.1

A cube 0.50 m on a side is made of bronze having a specific weight of 86.9 kN/m³. Determine the magnitude and direction of the force required to hold the cube in equilibrium when completely submerged (a) in water and (b) in mercury. The specific gravity of mercury is 13.54.

Solution

Consider part (a) first. Imagine the cube of bronze submerged in water. Now do Step 1 of the procedure.

On the assumption that the bronze cube will not stay in equilibrium by itself, some external force is required. The objective is to find the magnitude of this force and the direction in which it would act—that is, up or down.

Now do Step 2 of the procedure before looking at the next panel.

The free body is simply the cube itself. There are three forces acting on the cube in the vertical direction, as shown in Fig. 5.3:

- The weight of the cube *w*, acting downward through its center of gravity
- The buoyant force F_b, acting upward through the centroid of the displaced volume
- The externally applied supporting force F_e

Part (a) of Fig. 5.3 shows the cube as a three-dimensional object with the three forces acting along a vertical line through the centroid of the volume. This is the preferred visualization of the free-body diagram. However, for most problems it is suitable to use a simplified two-dimensional sketch as shown in part (b).

How do we know to draw the force F_e in the upward direction?

FIGURE 5.3   Free-body diagram of a cube.

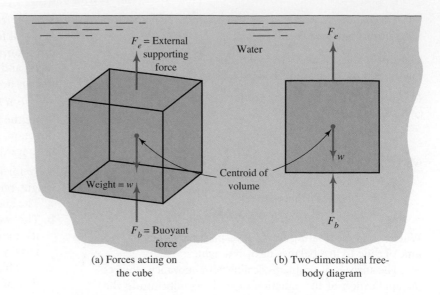

(a) Forces acting on
the cube

(b) Two-dimensional free-
body diagram

We really do not know for certain. However, experience should indicate that without an external force the solid bronze cube would tend to sink in water. Therefore, an upward force seems to be required to hold the cube in equilibrium. If our choice is wrong, the final result will indicate that to us.

Now, assuming that the forces are as shown in Fig. 5.3, go on to Step 3.

The equation should look as follows (assume that positive forces act upward):

$$\Sigma F_v = 0$$

$$F_b + F_e - w = 0 \tag{5–2}$$

As a part of Step 4, solve this equation algebraically for the desired term.

You should now have

$$F_e = w - F_b \tag{5–2}$$

because the objective is to find the external force.

How do we calculate the weight of the cube w?

Item b under Step 4 of the procedure indicates that $w = \gamma_B V$, where γ_B is the specific weight of the bronze cube and V is its total volume. For the cube, because each side is 0.50 m, we have

$$V = (0.50 \text{ m})^3 = 0.125 \text{ m}^3$$

and

$$w = \gamma_B V = (86.9 \text{ kN/m}^3)(0.125 \text{ m}^3) = 10.86 \text{ kN}$$

There is another unknown on the right side of Eq. (5–3). How do we calculate F_b?

Check Step 4a of the procedure if you have forgotten. Write

$$F_b = \gamma_f V_d$$

In this case γ_f is the specific weight of the water (9.81 kN/ m³), and the displaced volume V_d is equal to the total volume of the cube, which we already know to be 0.125 m³. Then, we have

$$F_b = \gamma_f V_d = (9.81 \text{ kN/m}^3)(0.125 \text{ m}^3) = 1.23 \text{ kN}$$

Now we can complete our solution for F_e.

The solution is

$$F_e = w - F_b = 10.86 \text{ kN} - 1.23 \text{ kN} = 9.63 \text{ kN}$$

FIGURE 5.4 Two possible free-body diagrams.

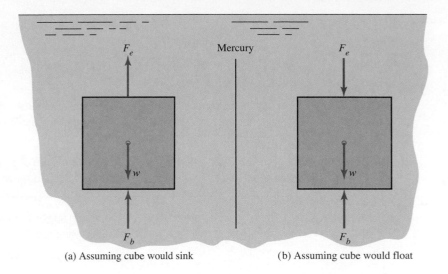

(a) Assuming cube would sink   (b) Assuming cube would float

Result Part (a)   Notice that the result is positive. This means that our assumed direction for F_e was correct. Then the solution to the problem is that an upward force of 9.63 kN is required to hold the block of bronze in equilibrium under water.

What about part (b) of the problem, where the cube is submerged in mercury? Our objective is the same as before—to determine the magnitude and direction of the force required to hold the cube in equilibrium.

Now do Step 2 of the procedure.

Either of the two free-body diagrams is correct as shown in Fig. 5.4, depending on the assumed direction for the external force F_e. The solution for the two diagrams will be carried out simultaneously so you can check your work regardless of which diagram looks like yours, and to demonstrate that either approach will yield the correct answer.

Now do Step 3 of the procedure.

The following are the correct equations of equilibrium. Notice the differences and relate them to the figures:

$$F_b + F_e - w = 0 \quad | \quad F_b - F_e - w = 0$$

Now, solve for F_e.

You should now have

$$F_e = w - F_b \quad | \quad F_e = F_b - w$$

Because the magnitudes of w and F_b are the same for each equation, they can now be calculated.

As in part (a) of the problem, the weight of the cube is

$$w = \gamma_B V = (86.9 \text{ kN/m}^3)(0.125 \text{ m}^3) = 10.86 \text{ kN}$$

For the buoyant force F_b, you should have

$$F_b = \gamma_m V = (sg)_m (9.81 \text{ kN/m}^3)(V)$$

where the subscript m refers to mercury. We then have

$$F_b = (13.54)(9.81 \text{ kN/m}^3)(0.125 \text{ m}^3) = 16.60 \text{ kN}$$

Now go on with the solution for F_e.

The correct answers are

$$F_e = w - F_b$$
$$= 10.86\,\text{kN} - 16.60\,\text{kN}$$
$$= -5.74\,\text{kN}$$

$$F_e = F_b - w$$
$$= 16.60\,\text{kN} - 10.86\,\text{kN}$$
$$= +5.74\,\text{kN}$$

Notice that both solutions yield the same numerical value, but they have opposite signs. The negative sign for the solution on the left means that the assumed direction for F_e in Fig. 5.4(a) was wrong. Therefore, both approaches give the same result.

Result Part (b)   The required external force is a downward force of 5.74 kN.

How could you have reasoned from the start that a downward force would be required?

Items c and d of Step 4 of the procedure suggest that the specific weight of the cube and the fluid be compared. In this case we have the following results:

$$\text{For the bronze cube } \gamma_B = 86.9\,\text{kN/m}^3$$
$$\text{For the fluid (mercury) } \gamma_m = (13.54)(9.81\,\text{kN/m}^3)$$
$$= 132.8\,\text{kN/m}^3$$

Comment   Because the specific weight of the cube is less than that of the mercury, it would tend to float without an external force. Therefore, a downward force, as pictured in Fig. 5.4(b), would be required to hold it in equilibrium under the surface of the mercury.

This example problem is concluded.

Example Problem 5.2   A certain solid metal object weighs 60 lb when measured in the normal manner in air, but it has such an irregular shape that it is difficult to calculate its volume by geometry. Use the principle of buoyancy to calculate its volume and specific weight.

Solution   It is given that the weight of the object is 60 lb. Now, using a setup similar to that in Fig. 5.5, we find its apparent weight while submerged in water to be 46.5 lb. Using these data and the procedure for analyzing buoyancy problems, we can find the volume of the object.

Now apply Step 2 of the procedure and draw the free-body diagram of the object while it is suspended in the water.

The free-body diagram of the object while it is suspended in the water should look like Fig. 5.6. In this figure what are the two forces F_e and w?

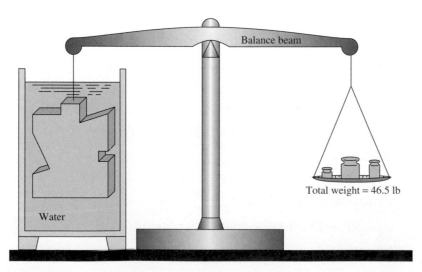

FIGURE 5.5 Metal object suspended in a fluid.

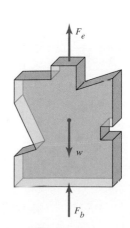

FIGURE 5.6 Free-body diagram.

We know that $w = 60$ lb, the weight of the object in air, and $F_e = 46.5$ lb, the supporting force exerted by the balance shown in Fig. 5.5.

Now do Step 3 of the procedure.

Using $\Sigma F_v = 0$, we get

$$F_b + F_e - w = 0$$

Our objective is to find the total volume V of the object. How can we get V into this equation?

We use this equation from Step 4a,

$$F_b = \gamma_f V$$

where γ_f is the specific weight of the water, 62.4 lb/ft^3.

Substitute this into the preceding equation and solve for V.

You should now have

$$F_b + F_e - w = 0$$
$$\gamma_f V + F_e - w = 0$$
$$\gamma_f V = w - F_e$$
$$V = \frac{w - F_e}{\gamma_f}$$

Now we can put in the known values and calculate V.

Result   The result is $V = 0.216$ ft^3. This is how it is done:

$$V = \frac{w - F_e}{\gamma_f} = (60 - 46.5)\text{lb}\left(\frac{\text{ft}^3}{62.4\ \text{lb}}\right) = \frac{13.5\ \text{ft}^3}{62.4} = 0.216\ \text{ft}^3$$

Comment   Now that the volume of the object is known, the specific weight of the material can be found.

$$\gamma = \frac{w}{V} = \frac{60\ \text{lb}}{0.216\ \text{ft}^3} = 278\ \text{lb/ft}^3$$

This is approximately the specific weight of a titanium alloy.

The next two problems are worked out in detail and should serve to check your ability to solve buoyancy problems. After reading the problem statement, you should complete the solution yourself before reading the panel on which a correct solution is given. Be sure to read the problem carefully and use the proper units in your calculations. Although there is more than one way to solve some problems, it is possible to get the correct answer by the wrong method. If your method is different from that given, be sure yours is based on sound principles before assuming it is correct.

Example Problem 5.3   A cube 80 mm on a side is made of a rigid foam material and floats in water with 60 mm of the cube below the surface. Calculate the magnitude and direction of the force required to hold it completely submerged in glycerin, which has a specific gravity of 1.26.

Complete the solution before looking at the next panel.

Solution   First calculate the weight of the cube, then the force required to hold the cube submerged in glycerin. Use the free-body diagrams in Fig. 5.7: (a) a cube floating on water and (b) a cube submerged in glycerin.

FIGURE 5.7 Free-body diagrams.

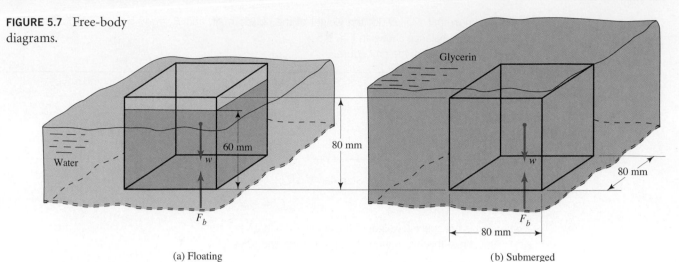

(a) Floating

(b) Submerged

From Fig. 5.7(a), we have

$$\Sigma F_v = 0$$

$$F_b - w = 0$$

$$w = F_b = \gamma_f V_d$$

$$V_d = (80\text{ mm})(80\text{ mm})(60\text{ mm}) = 384 \times 10^3 \text{ mm}^3$$

(submerged volume of cube)

$$w = \left(\frac{9.81 \times 10^3\text{ N}}{\text{m}^3}\right)(384 \times 10^3 \text{ mm}^3)\left(\frac{1\text{ m}^3}{(10^3\text{ mm})^3}\right)$$

$$= 3.77\text{ N}$$

From Fig. 5.7(b), we have

$$\Sigma F_v = 0$$

$$F_b - F_e - w = 0$$

$$F_e = F_b - w = \gamma_f V_d - 3.77\text{ N}$$

$$V_d = (80\text{ mm})^3 = 512 \times 10^3 \text{ mm}^3$$

(total volume of cube)

$$\gamma_f = (1.26)(9.81\text{ kN/m}^3) = 12.36\text{ kN/m}^3$$

$$F_e = \gamma_f V_d - 3.77\text{ N}$$

$$= \left(\frac{12.36 \times 10^3\text{ N}}{\text{m}^3}\right)(512 \times 10^3 \text{ mm}^3)\left(\frac{1\text{ m}^3}{(10^3\text{mm})^3}\right) - 3.77\text{ N}$$

$$F_e = 6.33\text{ N} - 3.77\text{ N} = 2.56\text{ N}$$

Result A downward force of 2.56 N is required to hold the cube submerged in glycerin.

Example Problem 5.4 A brass cube 6 in on a side weighs 67 lb. We want to hold this cube in equilibrium under water by attaching a light foam buoy to it. If the foam weighs 4.5 lb/ft³, what is the minimum required volume of the buoy? Complete the solution before looking at the next panel.

Solution Calculate the minimum volume of foam to hold the brass cube in equilibrium.

Notice that the foam and brass in Fig. 5.8 are considered as parts of a single system and that there is a buoyant force on each. The subscript F refers to the foam and the subscript B refers to the brass. No external force is required.

The equilibrium equation is

$$\Sigma F_v = 0$$

$$0 = F_{b_B} + F_{b_F} - w_B - w_F \qquad (5\text{--}4)$$

FIGURE 5.8 Free-body diagram for brass and foam together.

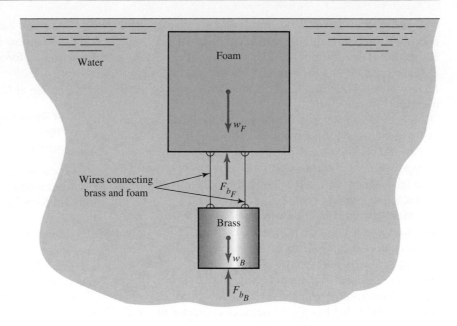

$$w_B = 67 \text{ lb} \quad \text{(given)}$$

$$F_{b_B} = \gamma_f V_{d_B} = \left(\frac{62.4 \text{ lb}}{\text{ft}^3}\right)(6 \text{ in})^3\left(\frac{\text{ft}^3}{1728 \text{ in}^3}\right) = 7.8 \text{ lb}$$

$$w_F = \gamma_F V_F$$

$$F_{b_F} = \gamma_f V_F$$

Substitute these quantities into Eq. (5–4):

$$F_{b_B} + F_{b_F} - w_B - w_F = 0$$

$$7.8 \text{ lb} + \gamma_f V_F - 67 \text{ lb} - \gamma_F V_F = 0$$

Solve for V_F, using $\gamma_f = 62.4 \text{ lb/ft}^3$ and $\gamma_F = 4.5 \text{ lb/ft}^3$:

$$\gamma_f V_F - \gamma_F V_F = 67 \text{ lb} - 7.8 \text{ lb} = 59.2 \text{ lb}$$

$$V_F(\gamma_f - \gamma_F) = 59.2 \text{ lb}$$

$$V_F = \frac{59.2 \text{ lb}}{\gamma_f - \gamma_F} = \frac{59.2 \text{ lb ft}^3}{(62.4 - 4.5) \text{ lb}}$$

$$V_F = 1.02 \text{ ft}^3$$

Result   This means that if 1.02 ft³ of foam were attached to the brass cube, the combination would be in equilibrium in water without any external force. It would be neutrally buoyant.
This completes the programmed example problems.

5.3 BUOYANCY MATERIALS

The design of floating bodies often requires the use of lightweight materials that offer a high degree of buoyancy. In addition, when a relatively heavy object must be moved while submerged in a fluid, it is often desirable to add buoyancy to facilitate mobility. The buoyancy material should typically have the following properties:

- Low specific weight and density
- Little or no tendency to absorb the fluid
- Compatibility with the fluid in which it will operate

- Ability to be formed to appropriate shapes
- Ability to withstand fluid pressures to which it will be subjected
- Abrasion resistance and damage tolerance
- Attractive appearance

Foam materials are popular for buoyancy applications. They are made up of a continuous network of closed, hollow cells that contain air or other light gases to yield the low specific weight. The closed cells also ensure that the fluid is not absorbed. The following tests are performed to evaluate the performance of foams: density, tensile strength, tensile elongation, tear strength, compression set, compressive deflection, thermal stability, thermal conductivity, and water absorption. The details of the tests are prescribed in ASTM D 3575, *Standard Test Methods for Flexible Cellular Materials Made from Olefin Polymers*. Other standards apply to other materials.

The specific weights of buoyancy foams range from approximately 2.0 lb/ft^3 to 40 lb/ft^3. This is often reported as density, taking the unit lb to be pound-mass. Compressive strengths generally increase with density. Applications in a deep sea environment call for the denser, stiffer, and heavier foams.

Materials used include urethane, polyethylene, olefin polymers, vinyl chloride polymers, extruded polystyrene, and sponge or expanded rubber. Undersea applications often employ syntactic foam materials made up of tiny hollow spheres embedded in a surrounding plastic such as fiberglass, polyester, epoxy, or vinyl ester resins to produce a composite material that has good buoyancy characteristics with abrasion resistance and low fluid absorption. See Internet resources 1–5.

The forms in which buoyancy materials are commercially available include planks (approximately 50 mm × 500 mm × 2750 mm or 2 in × 20 in × 110 in), billets (175 mm × 500 mm × 1200 mm or 7 in × 20 in × 48 in), cylinders, and hollow cylinders. Specially fabricated products can be made in almost limitless forms using molds or foaming in place. Two-part pourable urethane is available wherein two liquids, a polyether polyol and a polyfunctional isocyanite, are mixed at the point of use. The mixture expands rapidly producing the familiar, closed-cell foam structure. See Internet resources 3 and 5.

5.4 STABILITY OF COMPLETELY SUBMERGED BODIES

A body in a fluid is considered stable if it will return to its original position after being rotated a small amount about a horizontal axis. Two familiar examples of bodies completely submerged in a fluid are submarines and weather balloons.

It is important for these kinds of objects to remain in a specific orientation despite the action of currents, winds, or maneuvering forces.

▷ **Condition of Stability for Submerged Bodies**

The condition for stability of bodies completely submerged in a fluid is that the center of gravity of the body must be below the center of buoyancy.

The center of buoyancy of a body is at the centroid of the displaced volume of fluid, and it is through this point that the buoyant force acts in a vertical direction. The weight of the body acts vertically downward through the center of gravity.

The sketch of an undersea research vehicle shown in Fig. 5.9 has a stable configuration due to its shape and the location of equipment within the structure. An example is the *Alvin* deep submergence vehicle, owned by the U.S. Navy and operated by the Woods Hole Oceanographic Institution. See Internet resources 6 and 7. It can operate at depths down to 4.50 km (14 700 ft), where the pressure is 45.5 MPa (6600 psi). The overall length is 7.1 m (23.3 ft), the beam (width) is 2.6 m (8.5 ft), and the height is 3.7 m (12.0 ft). Its three-person crew pilots the vehicle and performs scientific observations from inside a spherical titanium pressure hull having a diameter of 2.08 m (82 in). When loaded, its weight is approximately 165 kN (37 000 lb), depending on the weight of the crew and experimental equipment. The design places heavier equipment such as batteries, descent weights, pressure vessels, variable ballast spheres, and motor controls in the lower part of the structure. Much of the upper structure is filled with light syntactic foam to provide buoyancy. This causes the center of gravity (cg) to be lower than the center of buoyancy (cb), achieving stability. In one configuration, the center of gravity is located 1.34 m (4.40 ft) above the bottom and the center of buoyancy is at 1.51 m (4.94 ft).

Figure 5.9(a) shows the approximate cross-sectional shape of the vehicle with the cg and the cb shown in their respective positions along the vertical centerline of the hull. Figure 5.9(b) shows the hull with some angular displacement with the total weight *w* acting vertically downward through

FIGURE 5.9 Stability of a submerged submarine.

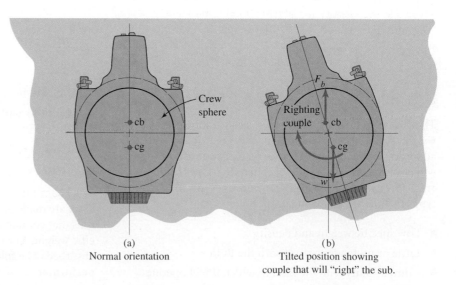

(a)
Normal orientation

(b)
Tilted position showing
couple that will "right" the sub.

FIGURE 5.10 Method of finding the metacenter.

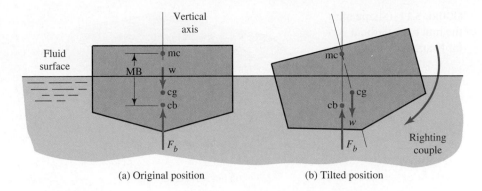

(a) Original position   (b) Tilted position

the cg and the buoyant force F_b acting vertically upward through the cb. Because their lines of action are now offset, these forces create a *righting couple* that brings the vehicle back to its original orientation, demonstrating stability.

If the cg is above the cb, the couple created when the body is tilted would produce an *overturning couple* that would cause it to capsize. Solid, homogeneous objects have the cg and cb coincident and they exhibit *neutral stability* when completely submerged, meaning that they tend to stay in whatever position they are placed.

5.5 STABILITY OF FLOATING BODIES

The condition for the stability of floating bodies is different from that for completely submerged bodies; the reason is illustrated in Fig. 5.10, which shows the approximate cross section of a ship's hull. In part (a) of the figure, the floating body is at its equilibrium orientation and the center of gravity (cg) is above the center of buoyancy (cb). A vertical line through these points will be called the *vertical axis* of the body. Figure 5.10(b) shows that if the body is rotated slightly, the center of buoyancy shifts to a new position because the geometry of the displaced volume has changed. The buoyant force and the weight now produce a righting couple that tends to return the body to its original orientation. Thus, the body is stable.

To state the condition for stability of a floating body, we must define a new term, *metacenter*. The metacenter (mc) is defined as the intersection of the vertical axis of a body when in its equilibrium position and a vertical line through the new position of the center of buoyancy when the body is rotated slightly. This is illustrated in Fig. 5.10(b).

▷ **Condition of Stability for Floating Bodies**

A floating body is stable if its center of gravity is below the metacenter.

It is possible to determine analytically if a floating body is stable by calculating the location of its metacenter. The distance to the *meta*center from the center of *buoy*ancy is called *MB* and is calculated from

$$MB = I/V_d \qquad (5\text{--}5)$$

In this equation, V_d is the displaced volume of fluid and I is the *least* moment of inertia of a horizontal section of the body taken at the surface of the fluid. *If the distance MB places the metacenter above the center of gravity, the body is stable.*

Procedure for Evaluating the Stability of Floating Bodies

1. Determine the position of the floating body, using the principles of buoyancy.

2. Locate the center of buoyancy, cb; compute the distance from some reference axis to cb, called y_{cb}. Usually, the bottom of the object is taken as the reference axis.

3. Locate the center of gravity, cg; compute y_{cg} measured from the same reference axis.

4. Determine the shape of the area at the fluid surface and compute the *smallest* moment of inertia I for that shape.

5. Compute the displaced volume V_d.

6. Compute $MB = I/V_d$.

7. Compute $y_{mc} = y_{cb} + MB$.

8. If $y_{mc} > y_{cg}$, the body is stable.

9. If $y_{mc} < y_{cg}$, the body is unstable.

PROGRAMMED EXAMPLE PROBLEMS

Example Problem 5.5   Figure 5.11(a) shows a flatboat hull that, when fully loaded, weighs 150 kN. Parts (b)–(d) show the top, front, and side views of the boat, respectively. Note the location of the center of gravity, cg. Determine whether the boat is stable in fresh water.

Solution   First, find out whether the boat will float.

This is done by finding how far the boat will sink into the water, using the principles of buoyancy stated in Section 5.1. Complete that calculation before going to the next panel.

FIGURE 5.11 Shape of the hull for a flatboat for Example Problem 5.5.

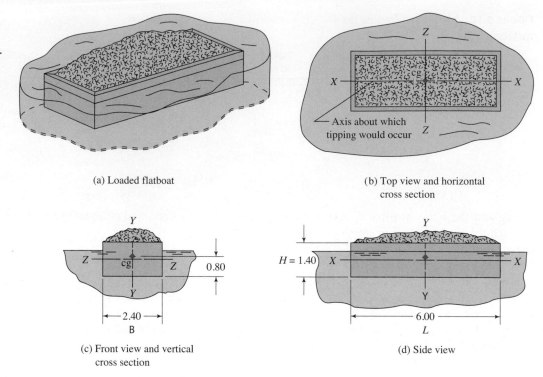

(a) Loaded flatboat

(b) Top view and horizontal cross section

(c) Front view and vertical cross section

(d) Side view

The depth of submergence or draft of the boat is 1.06 m, as shown in Fig. 5.12, found by the following method:

Equation of equilibrium:

$$\Sigma F_v = 0 = F_b - w$$
$$w = F_b$$

Submerged volume:

$$V_d = B \times L \times X$$

Buoyant force:

$$F_b = \gamma_f V_d = \gamma_f \times B \times L \times X$$

Then, we have

$$w = F_b = \gamma_f \times B \times L \times X$$

$$X = \frac{w}{B \times L \times \gamma_f} = \frac{150 \text{ kN}}{(2.4 \text{ m})(6.0 \text{ m})} \times \frac{\text{m}^3}{(9.81 \text{ kN})} = 1.06 \text{ m}$$

It floats with 1.06 m submerged. Where is the center of buoyancy?

FIGURE 5.12 Free-body diagram of boat hull.

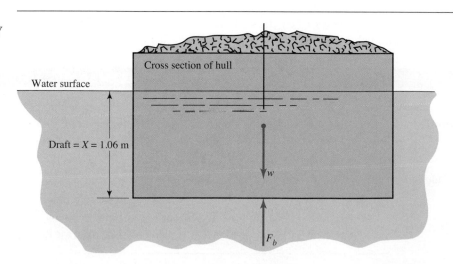

FIGURE 5.13 Location of center of buoyancy and center of gravity.

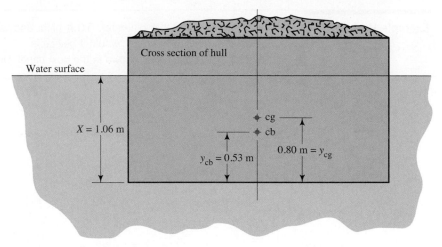

It is at the center of the displaced volume of water. In this case, as shown in Fig. 5.13, it is on the vertical axis of the boat at a distance of 0.53 m from the bottom. That is half of the draft, X. Then $y_{cb} = 0.53$ m.

Because the center of gravity is above the center of buoyancy, we must locate the metacenter to determine whether the boat is stable. Using Eq. (5–5), calculate the distance MB and show it on the sketch.

The result is MB = 0.45 m, as shown in Fig. 5.14. Here is how it is done:

$$MB = I/V_d$$
$$V_d = L \times B \times X = (6.0\,\text{m})(2.4\,\text{m})(1.06\,\text{m}) = 15.26\,\text{m}^3$$

The moment of inertia I is determined about the axis X–X in Fig. 5.11(b) because this would yield the smallest value for I. See parts (c) and (d) of Fig. 5.11 for the dimensions L and B.

$$I = \frac{LB^3}{12} = \frac{(6.0\,\text{m})(2.4\,\text{m})^3}{12} = 6.91\,\text{m}^4$$

Then, the distance from the center of buoyancy to the metacenter is

$$MB - I/V_d = 6.91\,\text{m}^4/15.26\,\text{m}^3 = 0.45\,\text{m}$$

The position of the metacenter is found from

$$y_{mc} = y_{cb} + MB = 0.53\,\text{m} + 0.45\,\text{m} = 0.98\,\text{m}$$

Is the boat stable?

Result   Yes, it is. Because the metacenter is above the center of gravity, as shown in Fig. 5.14, the boat is stable. That is, for $y_{cg} = 0.80$ m and $y_{mc} = 0.98$ m, $y_{mc} > y_{cg}$. Now, read the next panel for another problem.

FIGURE 5.14 Location of the metacenter.

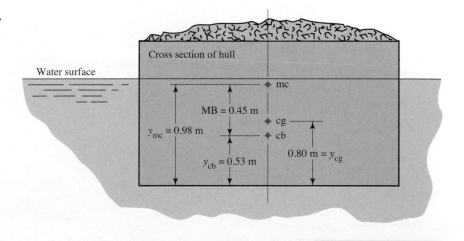

Example Problem 5.6   A solid cylinder is 3.0 ft in diameter, 6.0 ft high, and weighs 1550 lb. If the cylinder is placed in oil (sg = 0.90) with its axis vertical, would it be stable?

The complete solution is shown in the next panel. Do this problem and then look at the solution.

Solution   Position of cylinder in oil (Fig. 5.15):

$$V_d = \text{submerged volume} = AX = \frac{\pi D^2}{4}(X)$$

Equilibrium equation:

$$\Sigma F_v = 0$$

$$w = F_b = \gamma_o V_d = \gamma_o \frac{\pi D^2}{4}(X)$$

$$X = \frac{4w}{\pi D^2 \gamma_o} = \frac{(4)(1550 \text{ lb}) \text{ ft}^3}{(\pi)(3.0 \text{ ft})^2(0.90)(62.4 \text{ lb})} = 3.90 \text{ ft}$$

The center of buoyancy cb is at a distance $X/2$ from the bottom of the cylinder.

$$y_{cb} = X/2 = 3.90 \text{ ft}/2 = 1.95 \text{ ft}$$

The center of gravity cg is at $H/2 = 3.0$ ft from the bottom of the cylinder, assuming the material of the cylinder is of uniform specific weight. The position of the metacenter mc, using Eq. (5–5), is

$$MB = I/V_d$$

$$I = \frac{\pi D^4}{64} = \frac{\pi(3.0 \text{ ft})^4}{64} = 3.98 \text{ ft}^4$$

$$V_d = AX = \frac{\pi D^2}{4}(X) = \frac{\pi(3.0 \text{ ft})^2}{4}(3.90 \text{ ft}) = 27.6 \text{ ft}^3$$

$$MB = I/V_d = 3.98 \text{ ft}^4/27.6 \text{ ft}^3 = 0.144 \text{ ft}$$

$$y_{mc} = y_{cb} + MB = 1.95 \text{ ft} + 0.14 \text{ ft} = 2.09 \text{ ft}$$

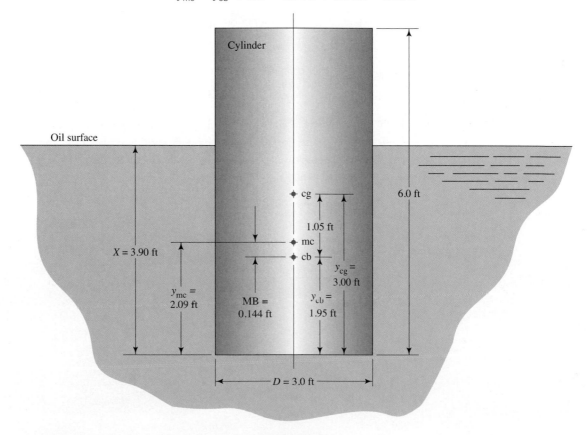

FIGURE 5.15   Complete solution for Example Problem 5.6.

Result   In summary, y_{cg} = 3.00 ft and y_{mc} = 2.09 ft. Because the metacenter is below the center of gravity ($y_{mc} < y_{cg}$), the cylinder is not stable in the position shown. It would tend to fall to one side until it reached a stable orientation, probably with the axis horizontal or nearly so.
This completes the programmed instruction.

The conditions for stability of bodies in a fluid can be summarized as follows.

■ *Completely submerged bodies are stable if the center of gravity is below the center of buoyancy.*

■ *Floating bodies are stable if the center of gravity is below the metacenter.*

5.6 DEGREE OF STABILITY

Although the limiting case of stability has been stated as any design for which the metacenter is above the center of gravity, some objects can be more stable than others. One measure of relative stability is called the *metacentric height*,

defined as the distance to the metacenter above the center of gravity and called *MG*. An object with a larger metacentric height is more stable than one with a smaller value.

Refer to Fig. 5.16. The metacentric height is labeled MG. Using the procedures discussed in this chapter, we can compute MG from

$$MG = y_{mc} - y_{cg} \qquad (5\text{–}6)$$

Reference 1 states that small seagoing vessels should have a minimum value of MG of 1.5 ft (0.46 m). Large ships should have MG > 3.5 ft (1.07 m). The metacentric height should not be too large, however, because the ship may then exhibit the uncomfortable rocking motions that cause seasickness.

Example Problem 5.7   Compute the metacentric height for the flatboat hull described in Example Problem 5.5.

Solution   From the results of Example Problem 5.5,

$$y_{mc} = 0.98 \text{ m from the bottom of the hull}$$
$$y_{cg} = 0.80 \text{ m}$$

Then, the metacentric height is

$$MG = y_{mc} - y_{cg} = 0.98 \text{ m} - 0.80 \text{ m} = 0.18 \text{ m}$$

5.6.1 Static Stability Curve

Another measure of the stability of a floating object is the amount of offset between the line of action of the weight of the object acting through the center of gravity and that of

the buoyant force acting through the center of buoyancy. Earlier, in Fig. 5.10, it was shown that the product of one of these forces and the amount of the offset produces the righting couple that causes the object to return to its original position and thus to be stable.

FIGURE 5.16 Degree of stability as indicated by the metacentric height and the righting arm.

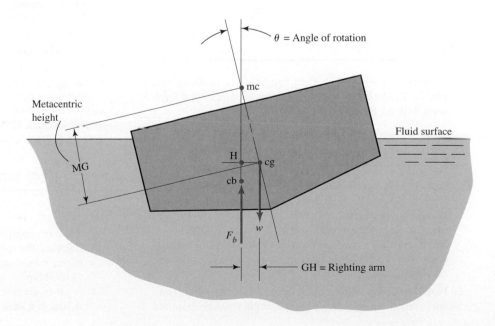

FIGURE 5.17 Static stability curve for a floating body.

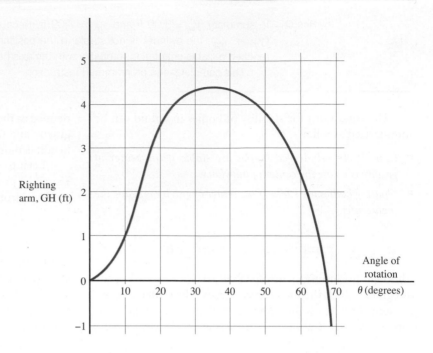

Figure 5.16 shows a sketch of a boat hull in a rotated position with the weight and the buoyant force shown. A horizontal line drawn through the center of gravity intersects the line of action of the buoyant force at point *H*. The horizontal distance, GH, is called the *righting arm* and is a measure of the magnitude of the righting couple. The distance GH varies as the angle of rotation varies, and Fig. 5.17 shows a characteristic plot of the righting arm versus the angle of rotation for a ship. Such a plot is called a *static stability curve*. As long as the value of GH remains positive, the ship is stable. Conversely, when GH becomes negative, the ship is unstable and it will overturn. Note that the object for which the example data in Fig. 5.17 apply, the object would become unstable at an angle of rotation of about 68 degrees. Also, because of the steep slope of the curve after about 50 degrees, that represents a reasonable recommended limit for rotation.

REFERENCE

1. Avallone, Eugene A., Theodore Baumeister, and Ali Sadegh, eds. 2007. *Marks' Standard Handbook for Mechanical Engineers,* 11th ed. New York: McGraw-Hill.

INTERNET RESOURCES

1. **Sealed Air Protective Packaging:** From the home page select *Protective Packaging*, then *Browse by Product Type*, then *Foam Packaging* for product information for many types of foam materials for industrial, packaging, and marine applications. Several formulations of polyethylene foams are used as buoyancy components. Brand names include Cellu-Cushion®, Ethafoam®, and Stratocell®.

2. **Flotation Technologies:** Manufacturer of deep-water buoyancy systems, specializing in high-strength syntactic foam and polyurethane elastomer products used for buoys, floats, instrument collars, and other forms applied to surface flotation or subsurface buoyancy to 6000 m (20 000 ft) depth.

3. **Marine Foam.com:** From the home page select *Floatation Foams*. They are a provider of marine and buoyancy products under the Marine Foam and Buoyancy Foam names, along with pourable urethane foam.

4. **Cuming Corporation:** A provider of syntactic foams and insulation equipment for the offshore oil and gas industries, including buoys and floats. A sister company to Flotation Technologies.

5. **U.S. Composites, Inc.:** A distributor of composite materials for the marine, automotive, aerospace, and art communities, including urethane foam, fiberglass, epoxy, carbon fiber composites, Kevlar, and others.

6. **National Undersea Research Program (NURP):** The federal government agency that sponsors undersea research. Part of the National Oceanographic and Atmospheric Administration (NOAA). See also Internet resource 7.

7. **Woods Hole Oceanographic Institute:** A research organization that performs both undersea and surface-based projects, including the operation of several U.S. Navy-owned deep submergence vehicles, such as the *Alvin* human-occupied submersible vehicle (HOV), the *Jason* remotely-operated undersea vehicle (ROV), and the *Sentry* autonomous undersea vehicle (AUV). See also Internet resource 6.

PRACTICE PROBLEMS

Buoyancy

5.1 The instrument package shown in Fig. 5.18 weighs 258 N. Calculate the tension in the cable if the package is completely submerged in seawater having a specific weight of 10.05 kN/m³.

5.2 A 1.0-m-diameter hollow sphere weighing 200 N is attached to a solid concrete block weighing 4.1 kN. If the

FIGURE 5.18   Problem 5.1.

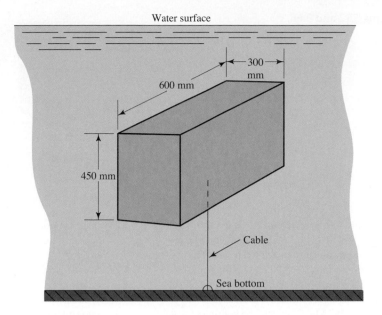

concrete has a specific weight of 23.6 kN/m³, will the two objects together float or sink in water?

5.3 A certain standard steel pipe has an outside diameter of 168 mm, and a 1 m length of the pipe weighs 277 N. Would the pipe float or sink in glycerin (sg = 1.26) if its ends are closed?

5.4 A cylindrical float has a 10-in diameter and is 12 in long. What should be the specific weight of the float material if it is to have 9/10 of its volume below the surface of a fluid with a specific gravity of 1.10?

5.5 A buoy is a solid cylinder 0.3 m in diameter and 1.2 m long. It is made of a material with a specific weight of 7.9 kN/m³. If it floats upright, how much of its length is above the water?

5.6 A float to be used as a level indicator is being designed to float in oil, which has a specific gravity of 0.90. It is to be a cube 100 mm on a side, and is to have 75 mm submerged in the oil. Calculate the required specific weight of the float material.

5.7 A concrete block with a specific weight of 23.6 kN/m³ is suspended by a rope in a solution with a specific gravity of 1.15. What is the volume of the concrete block if the tension in the rope is 2.67 kN?

5.8 Figure 5.19 shows a pump partially submerged in oil (sg = 0.90) and supported by springs. If the total weight of the pump is 14.6 lb and the submerged volume is 40 in³, calculate the supporting force exerted by the springs.

5.9 A steel cube 100 mm on a side weighs 80 N. We want to hold the cube in equilibrium under water by attaching a light foam buoy to it. If the foam weighs 470 N/m³, what is the minimum required volume of the buoy?

5.10 A cylindrical drum is 2 ft in diameter, 3 ft long, and weighs 30 lb when empty. Aluminum weights are to be placed inside the drum to make it neutrally buoyant in fresh water. What volume of aluminum will be required if it weighs 0.100 lb/in³?

5.11 If the aluminum weights described in Problem 5.10 are placed outside the drum, what volume will be required?

5.12 Figure 5.20 shows a cube floating in a fluid. Derive an expression relating the submerged depth X, the specific weight of the cube, and the specific weight of the fluid.

5.13 A hydrometer is a device for indicating the specific gravity of liquids. Figure 5.21 shows the design for a hydrometer in which the bottom part is a hollow cylinder with a 1.00-in diameter, and the top is a tube with a 0.25-in diameter. The empty hydrometer weighs 0.020 lb. What weight of steel balls should be added to make the hydrometer float in the position shown in fresh water? (Note that this is for a specific gravity of 1.00.)

5.14 For the hydrometer designed in Problem 5.13, what will be the specific gravity of the fluid in which the hydrometer would float at the top mark?

5.15 For the hydrometer designed in Problem 5.13, what will be the specific gravity of the fluid in which the hydrometer would float at the bottom mark?

5.16 A buoy is to support a cone-shaped instrument package, as shown in Fig. 5.22. The buoy is made from a uniform material having a specific weight of 8.00 lb/ft³. At least 1.50 ft of the buoy must be above the surface of the seawater for safety and visibility. Calculate the maximum allowable weight of the instrument package.

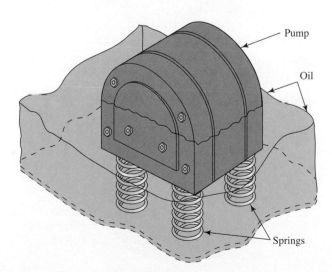

FIGURE 5.19   Problem 5.8.

FIGURE 5.20 Problems 5.12 and 5.60.

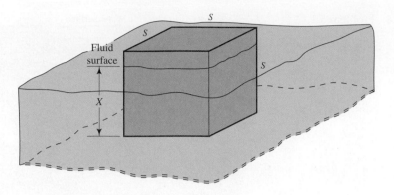

5.17 A cube has side dimensions of 18.00 in. It is made of steel having a specific weight of 491 lb/ft³. What force is required to hold it in equilibrium under fresh water?

5.18 A cube has side dimensions of 18.00 in. It is made of steel having a specific weight of 491 lb/ft³. What force is required to hold it in equilibrium under mercury?

5.19 A ship has a mass of 292 Mg. Compute the volume of seawater it will displace when floating.

5.20 An iceberg has a specific weight of 8.72 kN/m³. What portion of its volume is above the surface when in seawater?

5.21 A cylindrical log has a diameter of 450 mm and a length of 6.75 m. When the log is floating in fresh water with its long axis horizontal, 110 mm of its diameter is above the surface. What is the specific weight of the wood in the log?

5.22 The cylinder shown in Fig. 5.23 is made from a uniform material. What is its specific weight?

5.23 If the cylinder from Problem 5.22 is placed in fresh water at 95°C, how much of its height would be above the surface?

5.24 A brass weight is to be attached to the bottom of the cylinder described in Problems 5.22 and 5.23, so that the cylinder will be completely submerged and neutrally buoyant in water at 95°C. The brass is to be a cylinder with the same diameter as the original cylinder shown in Fig. 5.24. What is the required thickness of the brass?

5.25 For the cylinder with the added brass (described in Problem 5.24), what will happen if the water were cooled to 15°C?

5.26 For the composite cylinder shown in Fig. 5.25, what thickness of brass is required to cause the cylinder to float in the position shown in carbon tetrachloride at 25°C?

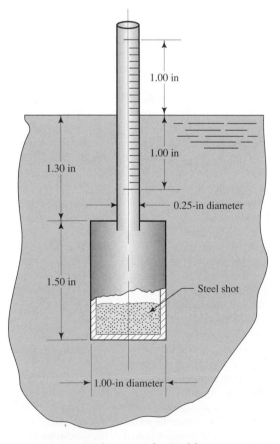

FIGURE 5.21 Hydrometer for Problems 5.13–5.15.

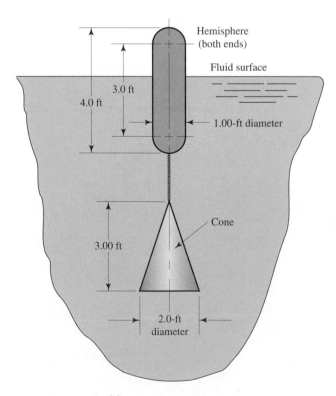

FIGURE 5.22 Problem 5.16.

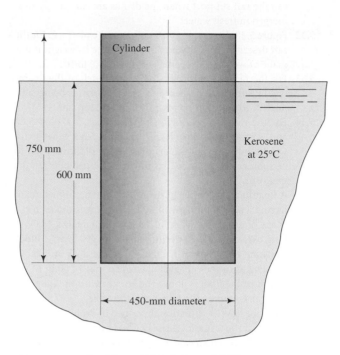

FIGURE 5.23 Problems 5.22–5.25 and 5.52.

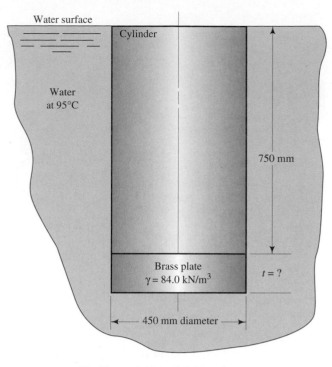

FIGURE 5.24 Problems 5.24 and 5.25.

5.27 A vessel for a special experiment has a hollow cylinder for its upper part and a solid hemisphere for its lower part, as shown in Fig. 5.26. What must be the total weight of the vessel if it is to sit upright, submerged to a depth of 0.75 m, in a fluid having a specific gravity of 1.16?

5.28 A light foam cup similar to a disposable coffee cup has a weight of 0.05 N, a uniform diameter of 82.0 mm, and a length of 150 mm. How much of its height would be submerged if it were placed in water?

5.29 A light foam cup similar to a disposable coffee cup has a weight of 0.05 N. A steel bar is placed inside the cup. The bar has a specific weight of 76.8 kN/m³, a diameter of 38.0 mm, and a length of 80.0 mm. How much of the height of the cup will be submerged if it is placed in water? The cup has a uniform diameter of 82.0 mm and a length of 150 mm.

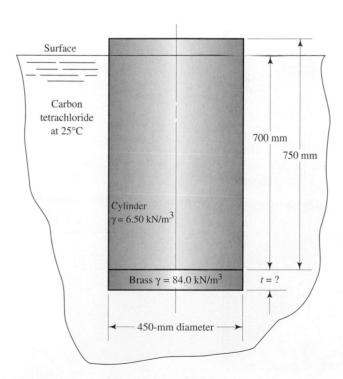

FIGURE 5.25 Problems 5.26 and 5.53.

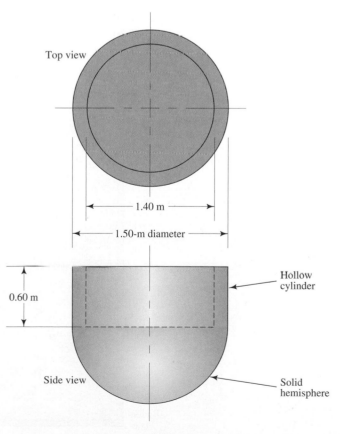

FIGURE 5.26 Problems 5.27 and 5.48.

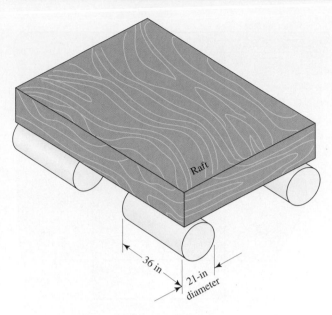

FIGURE 5.27   Problems 5.31, 5.33, and 5.34.

5.30 Repeat Problem 5.29, but consider that the steel bar is fastened outside the bottom of the cup instead of being placed inside.

5.31 Figure 5.27 shows a raft made of four hollow drums supporting a platform. Each drum weighs 30 lb. How much total weight of the platform and anything placed on it

can the raft support when the drums are completely submerged in fresh water?

5.32 Figure 5.28 shows the construction of the platform for the raft described in Problem 5.31. Compute its weight if it is made of wood of a specific weight of 40.0 lb/ft³.

5.33 For the raft shown in Fig. 5.27, how much of the drums will be submerged when only the platform is being supported? Refer to Problems 5.31 and 5.32 for data.

5.34 For the raft and platform shown in Figs. 5.27 and 5.28 and described in Problems 5.31 and 5.32, what extra weight will cause all the drums and the platform itself to be submerged? Assume that no air is trapped beneath the platform.

5.35 A float in an ocean harbor is made from a uniform foam having a specific weight of 12.00 lb/ft³. It is made in the shape of a rectangular solid 18.00 in square and 48.00 in long. A concrete (specific weight = 150 lb/ft³) block weighing 600 lb in air is attached to the float by a cable. The length of the cable is adjusted so that 14.00 in of the height of the float is above the surface with the long axis vertical. Compute the tension in the cable.

5.36 Describe how the situation described in Problem 5.35 will change if the water level rises by 18 in during high tide.

5.37 A cube 6.00 in on a side is made from aluminum having a specific weight of 0.100 lb/in³. If the cube is suspended on a wire with half its volume in water and the other half in oil (sg = 0.85), what is the tension in the wire?

5.38 A solid cylinder with its axis horizontal sits completely submerged in a fluid on the bottom of a tank. Compute

FIGURE 5.28   Raft construction for Problems 5.32 and 5.34.

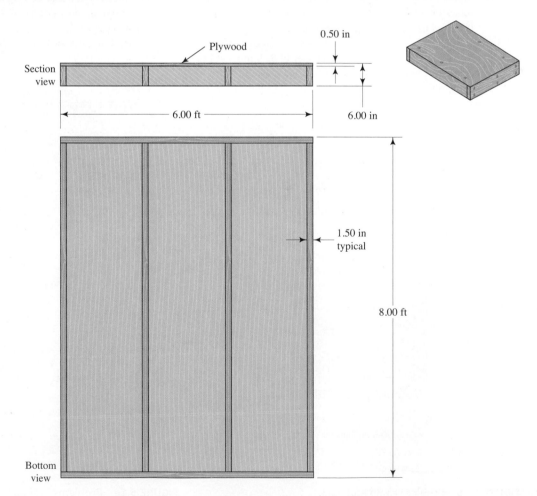

FIGURE 5.29 Problem 5.41.

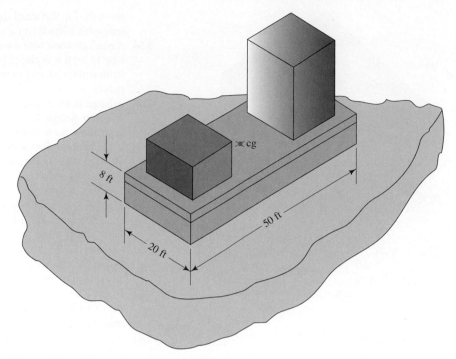

the force exerted by the cylinder on the bottom of the tank for the following data: $D = 6.00$ in, $L = 10.00$ in, $\gamma_c = 0.284$ lb/in^3 (steel), $\gamma_f = 62.4$ lb/ft^3.

Stability

5.39 A cylindrical block of wood is 1.00 m in diameter and 1.00 m long and has a specific weight of 8.00 kN/m^3. Will it float in a stable manner in water with its axis vertical?

5.40 A container for an emergency beacon is a rectangular shape 30.0 in wide, 40.0 in long, and 22.0 in high. Its center of gravity is 10.50 in above its base. The container weighs 250 lb. Will the box be stable with the 30.0-in by 40.0-in side parallel to the surface in plain water?

5.41 The large platform shown in Fig. 5.29 carries equipment and supplies to offshore installations. The total weight of the system is 450 000 lb, and its center of gravity is even with the top of the platform, 8 ft from the bottom. Will the platform be stable in seawater in the position shown?

5.42 Will the cylindrical float described in Problem 5.4 be stable if placed in the fluid with its axis vertical?

5.43 Will the buoy described in Problem 5.5 be stable if placed in the water with its axis vertical?

5.44 Will the float described in Problem 5.6 be stable if placed in the oil with its top surface horizontal?

5.45 A closed, hollow, empty drum has a diameter of 24.0 in, a length of 48.0 in, and a weight of 70.0 lb. Will it float stably if placed upright in water?

5.46 Figure 5.30 shows a river scow used to carry bulk materials. Assume that the scow's center of gravity is at its centroid and that it floats with 8.00 ft submerged. Determine the minimum width that will ensure stability in seawater.

5.47 Repeat Problem 5.46, except assume that crushed coal is added to the scow so that the scow is submerged to a depth of 16.0 ft and its center of gravity is raised to 13.50 ft from the bottom. Determine the minimum width for stability.

5.48 For the vessel shown in Fig. 5.26 and described in Problem 5.27, assume that it floats with just the entire hemisphere submerged and that its center of gravity is 0.65 m from the top. Is it stable in the position shown?

5.49 For the foam cup described in Problem 5.28, will it float stably if placed in the water with its axis vertical?

5.50 Referring to Problem 5.29, assume that the steel bar is placed inside the cup with its long axis vertical. Will the cup float stably?

5.51 Referring to Problem 5.30, assume that the steel bar is fastened to the bottom of the cup with the long axis of the bar horizontal. Will the cup float stably?

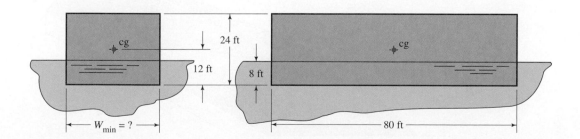

FIGURE 5.30 Problems 5.46 and 5.47.

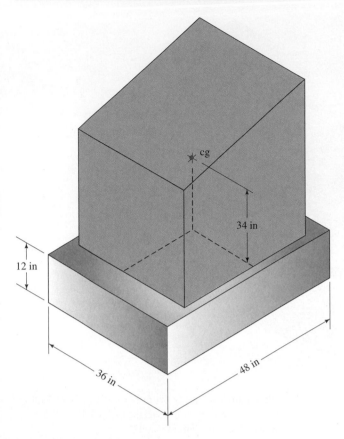

FIGURE 5.31 Problem 5.55.

5.52 Will the cylinder shown in Fig. 5.23 and described in Problem 5.22 be stable in the position shown?

5.53 Will the cylinder together with the brass plate shown in Fig. 5.25 and described in Problem 5.26 be stable in the position shown?

5.54 A proposed design for a part of a seawall consists of a rectangular solid weighing 3840 lb with dimensions of 8.00 ft by 4.00 ft by 2.00 ft. The 8.00-ft side is to be vertical. Will this object float stably in seawater?

5.55 A platform is being designed to support some water pollution testing equipment. As shown in Fig. 5.31, its base is 36.00 in wide, 48.00 in long, and 12.00 in high. The entire system weighs 130 lb, and its center of gravity is 34.0 in above the top surface of the platform. Is the proposed system stable when floating in seawater?

5.56 A block of wood with a specific weight of 32 lb/ft^3 is 6 by 6 by 12 in. If it is placed in oil (sg = 0.90) with the 6 by 12-in surface parallel to the surface of the oil, would it be stable?

5.57 A barge is 60 ft long, 20 ft wide, and 8 ft deep. When empty, it weighs 210 000 lb, and its center of gravity is 1.5 ft above the bottom. Is it stable when floating in water?

5.58 If the barge in Problem 5.57 is loaded with 240 000 lb of loose coal having an average density of 45 lb/ft^3, how much of the barge would be below the water? Is it stable?

5.59 A piece of cork having a specific weight of 2.36 kN/m^3 is shaped as shown in Fig. 5.32. (a) To what depth will it sink in turpentine (sg = 0.87) if placed in the orientation shown? (b) Is it stable in this position?

5.60 Figure 5.20 shows a cube floating in a fluid. (a) Derive an expression for the depth of submergence X that would ensure that the cube is stable in the position shown. (b) Using the expression derived in (a), determine the required distance X for a cube 75 mm on a side.

5.61 A boat is shown in Fig. 5.33. Its geometry at the water line is the same as the top surface. The hull is solid. Is the boat stable?

5.62 (a) If the cone shown in Fig. 5.34 is made of pine wood with a specific weight of 30 lb/ft^3, will it be stable in the position shown floating in water? (b) Will it be stable if it is made of teak wood with a specific weight of 55 lb/ft^3?

5.63 Refer to Fig. 5.35. The vessel shown is to be used for a special experiment in which it will float in a fluid having a specific gravity of 1.16. It is required that the top surface of the vessel is 0.25 m above the fluid surface.

   a. What should be the total weight of the vessel and its contents?

   b. If the contents of the vessel have a weight of 5.0 kN, determine the required specific weight of the material from which the vessel is made.

   c. The center of gravity for the vessel and its contents is 0.40 m down from the rim of the open top of the cylinder. Is the vessel stable?

5.64 A golf club head is made from aluminum having a specific weight of 0.100 lb/in^3. In air it weighs 0.500 lb. What would be its apparent weight when suspended in cool water?

FIGURE 5.32 Problem 5.59.

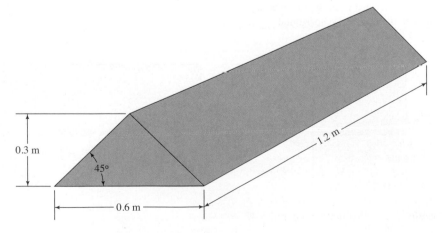

FIGURE 5.33 Problem 5.61.

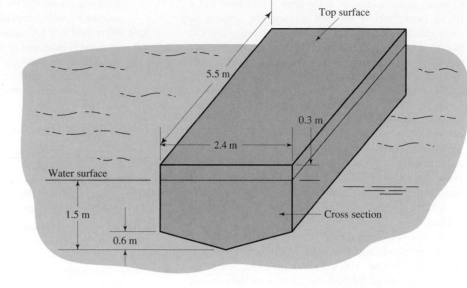

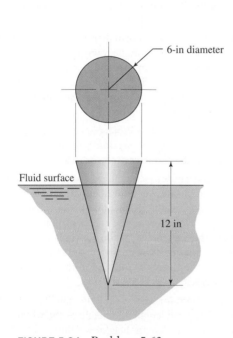

FIGURE 5.34 Problem 5.62.

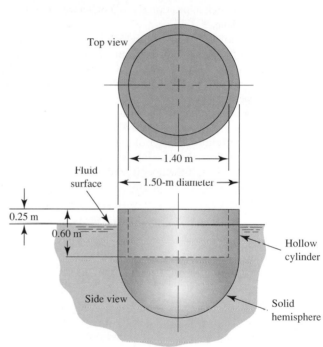

FIGURE 5.35 Problem 5.63.

Supplemental Problems

5.65 Wetsuits are prohibited in some triathlons due to the added buoyancy they provide the swimmer, essentially holding a greater portion of the body above the water and decreasing the power required to swim. If a given suit is made of 1.8 square yards of the material, and it is 0.25 in thick and has a specific weight of 38 lb/ft³, what net buoyant effect would aid the swimmer in seawater?

5.66 A cylinder that is 500 mm in diameter and 2.0 m long has a specific weight of 535 N/m³. It is held down into position with a cable attached to the sea floor. At this location, the sea is 500 m deep and the cylinder is to be held in a fully submerged position just 3 m above the sea floor. Find the resulting tension in the cable.

5.67 The diving bell shown in Fig. 5.2 weighs 72 kN and has a volume of 6.5 m³. Find the tension in the cable when the

sub is (a) hanging above the water, and (b) once the sub is lowered into the sea water. When released from the cable, will the bell tend to sink or float?

5.68 A hot air balloon is needed to lift a load with a mass of 125 kg from the earth's surface. If the ambient air is 20°C and the air in the balloon can be heated to 110°C, determine the required diameter of the balloon if it is approximated to be a sphere. Also explain why the load is to be carried **below** the balloon. Note the specific weight of air at various temperatures is available in the Appendix E.

5.69 A scuba diver with wet suit, tank, and gear has a mass of 78 kg. The diver and gear displace a total volume of 82.5 L of sea water. The diver would like to add enough lead weights to become neutrally buoyant for a dive. How much lead (sg = 11.35) should be added to the weight belt to achieve neutral buoyancy?

5.70 Concrete is to be poured for a large foundation, but a round passageway is required through the concrete to carry utilities through it. A lightweight plastic tube will be placed horizontally in the empty form to keep this duct way open and then wet concrete (sg = 2.6) will be poured into the form and around the tube. Since this tube of negligible weight will be buoyed up while the concrete mix is wet, it must be tethered down to keep it from floating up though the wet concrete mix. If this tube is 120 cm in diameter, how much force, per meter of length, must be provided to tether it down and hold it in place until the concrete sets?

5.71 Does steel float? It has a specific gravity of 7.85, so certainly not in water, but look through the Appendix B and see if there is a fluid in which a solid steel cube will float. Explain.

5.72 A toy castle is to be molded from polyethylene to be used as a decoration in aquariums. Polyethylene has a specific weight of 9125 N/m^3. Will it sink naturally or need to be weighted to stay at the bottom? Determine the required force to keep it in position if it is molded with 125 cm^3 of polyethylene.

5.73 An undersea camera (Figure 5.36) is to hang from a float in the ocean, allowing it to take constant video of undersea life. The camera is relatively heavy; it weighs 40 pounds and has a volume of only 0.3 ft^3. It is to hang 2 ft below the float on a pair of wires, allowing it take video at this constant depth. The float will be cylindrical, have a specific gravity, sg = 0.15, have a thickness of 6 in. What is the **minimum** diameter of the float? Note that for the minimum float, the waterline will be coincident with the top of the float. If the float is made of a larger diameter,

what will happen? If the float is made of a smaller diameter, what would happen?

5.74 Work problem 5.73 again, but this time the camera is to sit atop the float, out of the water. What minimum diameter float is required for this arrangement? If the float is made of a larger diameter, what will happen? If the float is made of a smaller diameter, what would happen?

STABILITY EVALUATION PROJECTS

Note: The following projects can be done using spreadsheets or calculation software. Plotting key results, such as for metacentric height, may be added as a feature to any project.

1. Write a program for evaluating the stability of a circular cylinder placed in a fluid with its axis vertical. Call for input data for diameter, length, and weight (or specific weight) of the cylinder; location of the center of gravity; and specific weight of the fluid. Solve the position of the cylinder when it is floating, the location of the center of buoyancy, and the metacenter. Compare the location of the metacenter with the center of gravity to evaluate stability.

2. For any cylinder of a uniform density floating in any fluid and containing a specified volume, vary the diameter from a small value to larger values in selected increments. Then compute the required height of the cylinder to obtain the specified volume. Finally, evaluate the stability of the cylinder if it is placed in the fluid with its axis vertical.

3. For the results found in Project 2, compute the metacentric height (as described in Section 5.6). Plot the metacentric height versus the diameter of the cylinder.

4. Write a program for evaluating the stability of a rectangular block placed in a fluid in a specified orientation. Call for input data for length, width, height, and weight (or specific weight) of the block; location of the center of gravity; and specific weight of the fluid. Solve for the position of the block when it is floating, the location of the center of buoyancy, and the metacenter. Compare the location of the metacenter with the center of gravity to evaluate stability.

5. Write a program for determining the stability of a rectangular block with a given length and height as the width varies. Call for input data for length, height, weight (or specific weight), and fluid specific weight. Vary the width in selected increments from small values to larger values, and compute the range of widths for which the metacentric height is positive, that is, for which the design would be stable. Plot a graph of metacentric height versus width.

6 in

2 ft

FIGURE 5.36 Problem 5.73.

FLOW OF FLUIDS AND BERNOULLI'S EQUATION

This chapter begins the study of *fluid dynamics*. While Chapters 1–5 dealt mostly with fluids at rest, this chapter deals with moving fluids, and primary emphasis is placed on the flow of fluids through pipes or tubes. Here you will learn several fundamental principles and following chapters 7–13 continue to build on those foundations. Your ultimate goal is to build your knowledge and skills that are needed to design and analyze the performance of pumped piping systems as they are applied in industrial applications and certain products.

You will learn how to analyze the effects of pressure, flow rate, and elevation of the fluid on the behavior of a fluid flow system. A fundamental concept used to analyze and design fluid flow systems is *Bernoulli's principle*, providing a way to account for three important types of energy possessed by fluids. Applications of the principle range from explaining how a chimney works to how an aircraft can fly and how fluids flow through pipes and tubes. Bernoulli's principle is used widely, including the design of an aesthetically pleasing fountain like the one shown in Fig. 6.1.

Introductory Concepts

Three measures of fluid flow rate are commonly used in fluid flow analysis:

- *Volume flow rate Q* is the volume of fluid flowing past a given section per unit time.
- *Weight flow rate W* is the weight of fluid flowing past a given section per unit time.

- *Mass flow rate M* is the mass of fluid flowing past a given section per unit time.

You will learn how to relate these terms to each other at various points in a system using the *principle of continuity*.

You must also learn to account for three kinds of energy possessed by the fluid at any point of interest within a fluid flow system:

- *kinetic energy* due to the motion of the fluid
- *potential energy* due to the elevation of the fluid
- *flow energy*, energy content based on the pressure in the fluid and its specific weight

Bernoulli's equation, based on the principle of conservation of energy, is the fundamental tool developed in this chapter for tracking changes in these three kinds of energy in a system. Later chapters will add additional terms to permit the analysis of many more kinds of energy losses from and additions to the fluid.

Exploration

Where have you observed fluids being transported through pipes and tubes? Try to identify five different systems and describe each, giving:

- The kind of fluid flowing
- The purpose of the system
- The kind of pipe or tube used and the material from which it was made

FIGURE 6.1 We admire attractive fountains with many jets of water reaching high into the air. How do they do that? This chapter will help you understand how. (*Source:* Vitas/Fotolia)

- The size of the pipe or tube and whether there were any changes in the size
- Any changes in the elevation of the fluid
- Information about the pressure in the fluid at any point

As an example, consider the cooling system for an automotive engine.

- The fluid, called a *coolant*, is a blend of water, an antifreeze component such as ethylene glycol, and other additives to inhibit corrosion and to ensure long life of the fluid and the system components.
- The purpose of the system is to have the coolant pick up heat from the engine block and deliver it to the car's radiator, where it is taken away by the flow of air through the finned coils. The coolant temperature may reach 125°C (257°F) as it leaves the engine. The main functional elements of the system are the water pump, the radiator typically mounted in front of the engine, and the cooling passages within the engine.
- Many kinds of conduits are used to carry the fluid, including:
  - Rigid hollow circular steel or copper tubes connect the radiator to the water pump and to the engine block. The tubes are typically small, with an inside diameter of approximately 10 mm (0.40 in).

- The fluid flows from the water pump and then through passages within the engine that are quite complex in shape. They are cast into the engine block to place the coolant around the cylinders where the heat of combustion moves through the metal cylinder wall into the fluid circulating through the engine.
- The fluid travels from the engine block to the radiator through a larger rubber hose, having an inside diameter about 40 mm (1.6 in).
- The fluid typically enters the top of the radiator where a manifold distributes it to a series of narrow rectangular channels within the radiator.
- At the bottom of the radiator, the fluid collects and is drawn out by the suction side of the pump.
- The elevation difference from the bottom of the radiator to the top of the engine is about 500 mm (20 in). The fluid is pressurized to about 100 kPa (15 psi) throughout the system to raise its boiling point to allow it to carry away much heat while remaining liquid.
- The pump causes the flow and raises the pressure of the fluid from its inlet to the outlet, so it can overcome flow resistances throughout the system.

Now describe the systems you discovered and discuss them with your fellow students and with the course instructor. Keep a record of the systems used here because you will be asked to reconsider them in Chapters 7–13.

Looking Ahead

In this chapter you will begin to learn how to analyze the behavior and performance of fluid flow systems. You will lay the foundation for learning many other aspects of fluid flow that you will study in the following chapters where you will be analyzing and designing systems for moving a desired amount of fluid from a source point to a desired destination, including the specification of pipes, valves, fittings, and a suitable pump.

6.1 OBJECTIVES

After completing this chapter, you should be able to:

1. Define *volume flow rate*, *weight flow rate*, and *mass flow rate* and their units.
2. Define *steady flow* and the *principle of continuity*.
3. Write the continuity equation, and use it to relate the volume flow rate, area, and velocity of flow between two points in a fluid flow system.
4. Describe five types of commercially available pipe and tubing: steel pipe, ductile iron pipe, steel tubing, copper tubing, and plastic pipe and tubing.
5. Specify the desired size of pipe or tubing for carrying a given flow rate of fluid at a specified velocity.
6. State recommended velocities of flow and typical volume flow rates for various types of systems.
7. Define *potential energy*, *kinetic energy*, and *flow energy* as they relate to fluid flow systems.

8. Apply the principle of conservation of energy to develop *Bernoulli's equation*, and state the restrictions on its use.
9. Define the terms *pressure head*, *elevation head*, *velocity head*, and *total head*.
10. Apply Bernoulli's equation to fluid flow systems.
11. Define *Torricelli's theorem*, and apply it to compute the flow rate of fluid from a tank and the time required to empty a tank.

6.2 FLUID FLOW RATE AND THE CONTINUITY EQUATION

6.2.1 Fluid Flow Rate

The quantity of fluid flowing in a system per unit time can be expressed by the following three different terms:

Q The *volume flow rate* is the volume of fluid flowing past a section per unit time.

TABLE 6.1 Flow rates—Definitions and units

Symbol	Name	Definition	SI Units	U.S. Customary System Units
Q	Volume flow rate	$Q = Av$	m^3/s	ft^3/s
W	Weight flow rate	$W = \gamma Q$	N/s	lb/s
		$W = \gamma Av$		
M	Mass flow rate	$M = \rho Q$	kg/s	slugs/s
		$M = \rho Av$		

W The *weight flow rate* is the weight of fluid flowing past a section per unit time.

M The *mass flow rate* is the mass of fluid flowing past a section per unit time.

The most fundamental of these three terms is the volume flow rate Q, which is calculated from

⇨ **Volume Flow Rate**

$$Q = Av \qquad (6\text{--}1)$$

where A is the area of the section and v is the average velocity of flow. The units of Q can be derived as follows, using SI units for illustration:

$$Q = Av = m^2 \times m/s = m^3/s$$

The weight flow rate W is related to Q by

⇨ **Weight Flow Rate**

$$W = \gamma Q \qquad (6\text{--}2)$$

where γ is the specific weight of the fluid. The SI units of W are then

$$W = \gamma Q = N/m^3 \times m^3/s = N/s$$

The mass flow rate M is related to Q by

⇨ **Mass Flow Rate**

$$M = \rho Q \qquad (6\text{--}3)$$

where ρ is the density of the fluid. The SI units of M are then

$$M = \rho Q = kg/m^3 \times m^3/s = kg/s$$

Table 6.1 summarizes these three types of fluid flow rates and gives the standard units in both the SI system and the U.S. Customary System. Because both the cubic meter per second and cubic foot per second are very large flow rates, other units are frequently used, such as liters per minute (L/min), cubic meters per hour (m³/h), and gallons per minute (gal/min or gpm; this text will use gal/min). Useful conversions are

⇨ **Conversion Factors for Volume Flow Rates**

$$1.0 \text{ L/min} = 0.060 \text{ m}^3/\text{h}$$
$$1.0 \text{ m}^3/\text{s} = 60\ 000 \text{ L/min}$$
$$1.0 \text{ gal/min} = 3.785 \text{ L/min}$$
$$1.0 \text{ gal/min} = 0.2271 \text{ m}^3/\text{h}$$
$$1.0 \text{ ft}^3/\text{s} = 449 \text{ gal/min}$$

Table 6.2 lists typical volume flow rates for different kinds of systems. Following are example problems illustrating the conversion of units from one system to another that are often required in problem solving to ensure consistent units in equations.

TABLE 6.2 Typical volume flow rates for a variety of types of systems

Type of system	Flow rate (m³/h)	(L/min)	(gal/min)
Reciprocating pumps—heavy fluids and slurries	0.90–7.5	15–125	4–33
Industrial oil hydraulic systems	0.60–6.0	10–100	3–30
Hydraulic systems for mobile equipment	6.0–36	100–600	30–150
Centrifugal pumps in chemical processes	2.4–270	40–4500	10–1200
Flood control and drainage pumps	12–240	200–4000	50–1000
Centrifugal pumps handling mine wastes	2.4–900	40–15 000	10–4000
Centrifugal fire-fighting pumps	108–570	1800–9500	500–2500

Example Problem   Convert a flow rate of 30 gal/min to ft³/s.
6.1

Solution   The flow rate is

$$Q = 30 \text{ gal/min}\left(\frac{1.0 \text{ ft}^3/\text{s}}{449 \text{ gal/min}}\right) = 6.68 \times 10^{-2} \text{ft}^3/\text{s} = 0.0668 \text{ ft}^3/\text{s}$$

Example Problem   Convert a flow rate of 600 L/min to m³/s.
6.2

Solution

$$Q = 600 \text{ L/min}\left(\frac{1.0 \text{ m}^3/\text{s}}{60\ 000 \text{ L/min}}\right) = 0.010 \text{ m}^3/\text{s}$$

Example Problem   Convert a flow rate of 30 gal/min to L/min.
6.3

Solution

$$Q = 30 \text{ gal/min}\left(\frac{3.785 \text{ L/min}}{1.0 \text{ gal/min}}\right) = 113.6 \text{ L/min}$$

6.2.2 The Continuity Equation

The method of calculating the velocity of flow of a fluid in a closed pipe system depends on the *principle of continuity*. Consider the pipe in Fig. 6.2. A fluid is flowing from section 1 to section 2 at a constant rate. That is, the quantity of fluid flowing past any section in a given amount of time is constant. This is referred to as *steady flow*. Now if there is no fluid added, stored, or removed between section 1 and section 2, then the mass of fluid flowing past section 2 in a given amount of time must be the same as that flowing

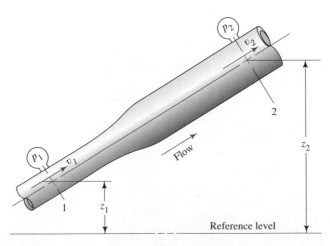

FIGURE 6.2   Portion of a fluid distribution system showing variations in velocity, pressure, and elevation.

past section 1. This can be expressed in terms of the mass flow rate as

$$M_1 = M_2$$

or, because $M = \rho A v$, we have

⮑ **Continuity Equation for Any Fluid**

$$\rho_1 A_1 v_1 = \rho_2 A_2 v_2 \qquad (6\text{–}4)$$

Equation (6–4) is a mathematical statement of the principle of continuity and is called the *continuity equation*. It is used to relate the fluid density, flow area, and velocity of flow at two sections of the system in which there is steady flow. It is valid for all fluids, whether gas or liquid.

If the fluid in the pipe in Fig. 6.2 is a liquid that can be considered incompressible, then the terms ρ_1 and ρ_2 in Eq. (6–4) are equal and they can be cancelled from Eq. (6–4). The equation then becomes

⮑ **Continuity Equation for Liquids**

$$A_1 v_1 = A_2 v_2 \qquad (6\text{–}5)$$

or, because $Q = Av$, we have

$$Q_1 = Q_2$$

Equation (6–5) is the continuity equation as applied to liquids; it states that for steady flow the volume flow rate is the same at any section. It can also be used for gases at low velocity, that is, less than 100 m/s, with little error.

Example Problem 6.4

In Fig. 6.2 the inside diameters of the pipe at sections 1 and 2 are 50 mm and 100 mm, respectively. Water at 70°C is flowing with an average velocity of 8.0 m/s at section 1. Calculate the following:

a. Velocity at section 2
b. Volume flow rate
c. Weight flow rate
d. Mass flow rate

Solution

a. Velocity at section 2.
From Eq. (6–5) we have

$$A_1 v_1 = A_2 v_2$$

$$v_2 = v_1 \left(\frac{A_1}{A_2} \right)$$

$$A_1 = \frac{\pi D_1^2}{4} = \frac{\pi (50 \text{ mm})^2}{4} = 1963 \text{ mm}^2$$

$$A_2 = \frac{\pi D_2^2}{4} = \frac{\pi (100 \text{ mm})^2}{4} = 7854 \text{ mm}^2$$

Then the velocity at section 2 is

$$v_2 = v_1 \left(\frac{A_1}{A_2} \right) = \frac{8.0 \text{ m}}{\text{s}} \times \frac{1963 \text{ mm}^2}{7854 \text{ mm}^2} = 2.0 \text{ m/s}$$

Notice that for steady flow of a liquid, as the flow area increases, the velocity decreases. This is independent of pressure and elevation.

b. Volume flow rate Q.
From Table 6.1, $Q = Av$. Because of the principle of continuity we could use the conditions either at section 1 or at section 2 to calculate Q. At section 1 we have

$$Q = A_1 v_1 = 1963 \text{ mm}^2 \times \frac{8.0 \text{ m}}{\text{s}} \times \frac{1 \text{ m}^2}{(10^3 \text{ mm})^2} = 0.0157 \text{ m}^3/\text{s}$$

c. Weight flow rate W.
From Table 6.1, $W = \gamma Q$. At 70°C the specific weight of water is 9.59 kN/m^3. Then the weight flow rate is

$$W = \gamma Q = \frac{9.59 \text{ kN}}{\text{m}^3} \times \frac{0.0157 \text{ m}^3}{\text{s}} = 0.151 \text{ kN/s}$$

d. Mass flow rate M.
From Table 6.1, $M = \rho Q$. At 70°C the density of water is 978 kg/m^3. Then the mass flow rate is

$$M = \rho Q = \frac{978 \text{ kg}}{\text{m}^3} \times \frac{0.0157 \text{ m}^3}{\text{s}} = 15.36 \text{ kg/s}$$

Example Problem 6.5

At one section in an air distribution system, air at 14.7 psia and 100°F has an average velocity of 1200 ft/min and the duct is 12 in square. At another section, the duct is round with a diameter of 18 in, and the velocity is measured to be 900 ft/min. Calculate (a) the density of the air in the round section and (b) the weight flow rate of air in pounds per hour. At 14.7 psia and 100°F, the density of air is 2.20×10^{-3} slugs/ft^3 and the specific weight is 7.09×10^{-2} lb/ft^3.

Solution

According to the continuity equation for gases, Eq. (6–4), we have

$$\rho_1 A_1 v_1 = \rho_2 A_2 v_2$$

Then, we can calculate the area of the two sections and solve for ρ_2:

$$\rho_2 = \rho_1\left(\frac{A_1}{A_2}\right)\left(\frac{v_1}{v_2}\right)$$

$$A_1 = (12 \text{ in})(12 \text{ in}) = 144 \text{ in}^2$$

$$A_2 = \frac{\pi D_2^2}{4} = \frac{\pi(18 \text{ in})^2}{4} = 254 \text{ in}^2$$

a. Then, the density of the air in the round section is

$$\rho_2 = (2.20 \times 10^{-3} \text{ slugs/ft}^3)\left(\frac{144 \text{ in}^2}{254 \text{ in}^2}\right)\left(\frac{1200 \text{ ft/min}}{900 \text{ ft/min}}\right)$$

$$\rho_2 = 1.66 \times 10^{-3} \text{ slugs/ft}^3$$

b. The weight flow rate can be found at section 1 from $W = \gamma_1 A_1 v_1$. Then, the weight flow rate is

$$W = \gamma_1 A_1 v_1$$

$$W = (7.09 \times 10^{-2} \text{ lb/ft}^3)(144 \text{ in}^2)\left(\frac{1200 \text{ ft}}{\text{min}}\right)\left(\frac{1 \text{ ft}^2}{144 \text{ in}^2}\right)\left(\frac{60 \text{ min}}{\text{h}}\right)$$

$$W = 5100 \text{ lb/h}$$

6.3 COMMERCIALLY AVAILABLE PIPE AND TUBING

We will describe several widely used types of standard pipe and tubing in this section. Data are given in Appendices F–I in either U.S. units or metric units for actual outside diameter, inside diameter, wall thickness, and flow area for selected sizes and types. Many more are commercially available. Refer to References 2–5 and Internet resources 2–15. You can see that the dimensions are listed in inches (in) and millimeters (mm) for outside diameter, inside diameter, and wall thickness. The flow areas are listed in square feet (ft^2) and square meters (m^2) to help you maintain consistent units in calculations. Data for inside diameters are also given in ft for the U.S. Customary System for unit consistency.

Specifying piping and tubing for a particular application is the responsibility of the designer and it has a significant impact on the cost, life, safety, and performance of the system. For many applications, codes and standards must be followed as established by U.S. governmental agencies or organizations such as the following:

American Water Works Association (AWWA)

American Fire Sprinkler Association (AFSA)

National Fire Protection Association (NFPA)

ASTM International (ASTM) [Formerly American Society for Testing and Materials]

NSF International (NSF) [Formerly National Sanitation Foundation]

International Association of Plumbing and Mechanical Officials (IAPMO)

Standards for various international organizations should also be consulted, such as:

International Organization for Standardization (ISO)

British Standards (BS)   *European Standards (EN)*

German Standards (DIN)   *Japanese Standards (JIS)*

6.3.1 Steel Pipe

General-purpose pipe lines are often constructed of steel pipe. Standard sizes are designated by the Nominal Pipe Size (NPS) and schedule number. The nominal size is merely the standard designation and it is not used for calculations. Schedule numbers are related to the permissible operating pressure of the pipe and to the allowable stress of the steel in the pipe. The range of schedule numbers is from 10 to 160, with the higher numbers indicating a heavier wall thickness. Because all schedules of pipe of a given nominal size have the same outside diameter, the higher schedules have a smaller inside diameter. The most complete series of steel pipe available are Schedules 40 and 80. Data for these two schedules are given in SI units and in U.S. Customary System units in Appendix F. Refer to *ANSI/ASME Standard B31.1: Power Piping* for a method of computing the minimum acceptable wall thickness for pipes. See Reference 1.

Nominal Pipe Sizes in Metric Units   Because of the long experience with manufacturing standard pipe according to the standard NPS sizes and schedule numbers, they continue to be used often even when the piping system is specified in metric units. In such cases, the DN set of equivalents has been established by the International Standards

Organization (ISO). The symbol DN is used to designate the nominal diameter (*diametre nominel*) in mm. Appendix F shows the DN designation alongside the NPS designation. For example, a DN 50-mm Schedule 40 steel pipe has the same dimensions as a 2-in Schedule 40 steel pipe.

6.3.2 Steel Tubing

Standard steel tubing is used in fluid power systems, condensers, heat exchangers, engine fuel systems, and industrial fluid processing systems. Standard inch sizes are designated by outside diameter and wall thickness in inches. Standard sizes from $\frac{1}{8}$ in to 2 in for several wall thickness gauges are tabulated in Appendix G.1. Other diameters and wall thicknesses are available. Data from Appendix G.1 can be used for metric problems by selecting the equivalent metric converted data listed in the table.

Designers working on all-metric systems should specify tubing made to convenient metric dimensions. Appendix Table G.2 shows data for a sample set of outside diameters and wall thicknesses. Many more choices are available. See Internet resource 13.

6.3.3 Copper Tubing

The Copper Development Association (CDA) develops standards for copper tubing made to U.S. Customary unit sizes. There are six types of CDA copper tubing offered, and the choice of which one to use depends on the application, considering the environment, fluid pressure, and fluid properties. See Internet resource 3 for details on all types and sizes available. Tube dimensions are given in the section called *Properties*. The following are brief descriptions of typical uses:

1. Type K: Used for water service, fuel oil, natural gas, and compressed air.

2. Type L: Similar to Type K, but with a smaller wall thickness.

3. Type M: Similar to Types K and L, but with smaller wall thicknesses; preferred for most water services and heating applications at moderate pressures.

4. Type DWV: Drain, waste, and vent uses in plumbing systems.

5. Type ACR: Air conditioning, refrigeration, natural gas, liquefied petroleum (LP) gas, and compressed air.

6. Type OXY/MED: Used for oxygen or medical gas distribution, compressed medical air, and vacuum applications. Available in sizes similar to Types K and L, but with special processing for increased cleanliness.

Copper tubing is available in either a soft, annealed condition or hard drawn. Drawn tubing is stiffer and stronger, maintains a straight form, and can carry higher pressures. Annealed tubing is easier to form into coils and other special shapes. Nominal or standard sizes for Types K, L, M, and DWV are all $\frac{1}{8}$ in less than the actual outside diameter. The wall thicknesses are different for each type so that the inside diameter and flow areas vary. This system of dimensions is

sometimes referred to as *Copper Tube Sizes* (*CTS*). The nominal size for Type ACR tubing is equal to the outside diameter. Appendix H gives data for dimensions of Type K tubing, including outside diameter, inside diameter, wall thickness, and flow area in both U.S. and SI units.

Copper tubing is also available made to convenient SI metric dimensions, and sample data are included in Appendix G.2. See Internet resource 13 for data for a more complete set of available sizes.

6.3.4 Ductile Iron Pipe

Water, gas, and sewage lines are often made of ductile iron pipe because of its strength, ductility, and relative ease of handling. It has replaced cast iron in many applications. Standard fittings are supplied with the pipe for convenient installation above or below ground. Several classes of ductile iron pipe are available for use in systems with a range of pressures. Appendix I lists the dimensions of cement lined, Class 150 pipe for 150 psi (1.03 MPa) service in nominal sizes from 4 to 48 in. Actual inside and outside diameters are larger than nominal sizes. Other internal linings and coatings are available. Internet resource 4 gives data for all sizes, linings, coatings, and classes. Appendix I gives data for a sample of commercially available ductile iron pipe. In a manner similar to steel pipe, the designation for ductile iron pipe is a nominal inch-size that is only approximately equal to the inside diameter. Actual data from the tables must be used in problem solving. For convenience, the inch-based data are converted to equivalent metric data in the appendix table.

6.3.5 Plastic Pipe and Tubing

Plastic pipe and tubing are being used in a wide variety of applications where their light weight, ease of installation, corrosion and chemical resistance, and very good flow characteristics present advantages. Examples are water and gas distribution, sewer and wastewater, oil and gas production, irrigation, mining, and many industrial applications. Numerous types of plastic are used such as polyethylene (PE), cross-linked polyethylene (PEX), polyamide (PA), polypropylene (PP), polyvinyl chloride (PVC), chlorinated polyvinyl chloride (CPVC), polyvinylidene fluoride (PVDF), food-grade vinyl, and nylon. See Internet resources 6–9 and 14.

Because some plastic pipe and tubing serve the same markets as metals for which special size standards have been common for decades, many plastic products conform to existing standards of NPS, Ductile Iron Pipe Sizes (DIPS), or CTS. Specific manufacturers' data for outside diameter (OD), inside diameter (ID), wall thickness, and flow area should be confirmed.

Plastic pipe is also made to convenient metric sizes. Appendix G.3 lists examples of commercially available sizes of PVC plastic pipes. Many more sizes can be found at Internet resource 14. In addition to dimensions and flow area, Appendix G.3 lists the pressure ratings for the given sizes.

Commonly used pressure ratings include 6 bar (87 psi), 10 bar (145 psi), and 16 bar (232 psi). Note the relationship between diameter, wall thickness, and pressure ratings in the table.

Other plastic piping systems use the Standard Inside Dimension Ratio (SIDR) or Standard Dimension Ratio (SDR). The SIDR system is based on the ratio of the average specified inside diameter to the minimum specified wall thickness (ID/t). It is used where the ID is critical to the application. The ID remains constant and the OD changes with wall thickness to accommodate different pressures and structural and handling considerations. The SDR is based on the ratio of the average specified outside diameter to the minimum specified wall thickness (OD/t). The OD remains constant and the ID and wall thickness change. The SDR system is useful because the pressure rating of the pipe is directly related to this ratio. For example, for plastic pipe with a hydrostatic design stress rating of 1250 psi (11 MPa), the pressure ratings for different SDR ratings are as follows:

SDR	Pressure Rating
26	50 psi (345 kPa)
21	62 psi (427 kPa)
17	80 psi (552 kPa)
13.5	100 psi (690 kPa)

These pressure ratings are for water at 73°F (23°C). In general, plastic pipe and tubing can be found rated up to 250 psi (1380 kPa). See Internet resource 6.

6.3.6 Hydraulic Hose

Reinforced flexible hose is used extensively in hydraulic fluid power systems and other industrial applications where flow lines must flex in service. Hose materials include butyl rubber, synthetic rubber, silicone rubber, thermoplastic elastomers, and nylon. Braided reinforcement may be made from steel wire, Kevlar, polyester, and fabric. Industrial applications include steam, compressed air, chemical transfer, coolants, heaters, fuel transfer, lubricants, refrigerants, paper stock, power steering fluids, propane, water, foods, and beverages. SAE International Standard J517, *Hydraulic Hose,* defines many standard types and sizes according to their pressure rating and flow capacity. Sizes include inside diameters of 3/16, 1/4, 5/16, 3/8, 1/2, 5/8, 3/4, 1, 1¼, 1½, 2, 2½, 3, 3½, and 4 in. Pressure ratings range from 35 psig to over 10 000 psig (240 kPA to 69 MPa) to cover high-pressure fluid power and hydraulic jacking applications to low-pressure suction and return lines and low-pressure fluid transfer applications. See Internet resources 11 and 12.

6.4 RECOMMENDED VELOCITY OF FLOW IN PIPE AND TUBING

Many factors affect the selection of a satisfactory velocity of flow in fluid systems. Some of the important ones are the type of fluid, the length of the flow system, the type of pipe or tube, the pressure drop that can be tolerated, the devices (e.g., pumps, valves, etc.) that may be connected to the pipe or tube, the temperature, the pressure, and the noise.

When we discussed the continuity equation in Section 6.2, we learned that the velocity of flow increases as the area of the flow path decreases. Therefore, smaller tubes will cause higher velocities and larger tubes will provide lower velocities. Later we will explain that energy losses and the corresponding pressure drop increase dramatically as the flow velocity increases. For this reason it is desirable to keep the velocities low. Because larger pipes and tubes are more costly, however, some limits are necessary.

Figure 6.3 provides very rough guidance for specifying pipe sizes as a function of volume flow rate for typical pumped fluid distribution systems. The data were abstracted from an analysis of the rated volume flow rate for many commercially available centrifugal pumps operating near their best efficiency point and observing the size of the suction and discharge connections. In general, the flow velocity is kept lower in suction lines providing flow into a pump to ensure proper filling of the suction inlet passages. The lower velocity also helps to limit energy losses in the suction line, keeping the pressure at the pump inlet relatively high to ensure that pure liquid enters the pump. Lower pressures may cause a damaging condition called *cavitation* to occur, resulting in excessive noise, significantly degraded performance, and rapid erosion of pump and impeller surfaces. Cavitation is discussed more fully in Chapter 13.

Note that specifying one size larger or one size smaller than indicated by the lines in Fig. 6.3 will not affect the performance of the system very much. In general, you should favor the larger pipe size to achieve a lower velocity unless there are difficulties with space, cost, or compatibility with a given pump connection.

The resulting flow velocities from the recommended pipe sizes in Fig. 6.3 are generally lower for the smaller pipes and higher for the larger pipes, as shown for the following data.

Volume Flow Rate		Suction Line			Discharge Line		
		Pipe	Velocity		Pipe	Velocity	
gal/min	m³/h	Size (in)	ft/s	m/s	Size (in)	ft/s	m/s
10	2.3	1	3.7	1.1	¾	6.0	1.8
100	22.7	2½	6.7	2.0	2	9.6	2.9
500	114	5	8.0	2.4	3½	16.2	4.9
2000	454	8	12.8	3.9	6	22.2	6.8

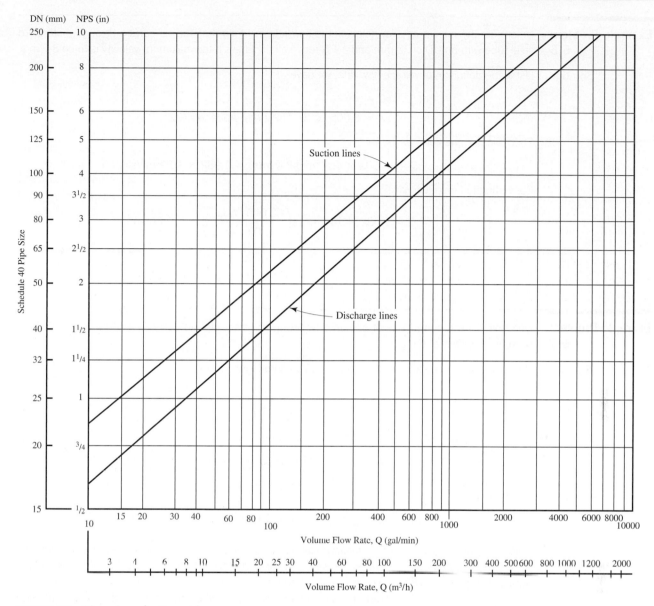

FIGURE 6.3 Pipe-size selection aid.

It is the responsibility of the system designer to specify the final pipe sizes that will yield reasonably optimum performance considering energy losses, pressures at critical points in the system, pump power required, and life cycle cost.

Data for volume flow rate in Fig. 6.3 are given in gal/min for the U.S. Customary System and in m^3/h for the SI system because most manufacturers rate their pumps in such units. Conversions to the standard units of ft^3/s and m^3/s must be done before using the flow rates in calculations in this book.

Recommended Flow Velocities for Specialized Systems
The data in Fig. 6.3 apply to general fluid distribution systems. You are advised to seek other information about industry practice in specific fields for which you are designing fluid flow systems.

For example, recommended flow velocities for fluid power systems are as follows (see Internet resources 11 and 15):

| Type of Service | Recommended Range of Velocity | |
	ft/s	m/s
Suction lines	2–4	0.6–1.2
Return lines	4–13	1.5–4.0
Discharge lines	7–25	2.1–7.6

The suction line delivers the hydraulic fluid from the reservoir to the intake port of the pump. A discharge line carries the high-pressure fluid from the pump outlet to working components such as actuators or fluid motors. A return line carries fluid from actuators, pressure relief valves, or fluid motors back to the reservoir.

The U.S. Army Corps of Engineers manual *Liquid Process Piping* recommends that for normal liquid service applications, the flow velocity should be in the range of 1.2 m/s to 3.0 m/s (3.9 ft/s to 9.8 ft/s). Specific applications may allow greater velocities. See Reference 6.

Example Problem 6.6

Determine the maximum allowable volume flow rate in L/min that can be carried through a standard steel tube with an OD of 32 mm and a 1.5-mm-wall thickness if the maximum velocity is to be 3.0 m/s.

Solution

Using the definition of volume flow rate, we have

$$Q = Av$$
$$A = 6.605 \times 10^{-4} \, m^2 \text{ (from Appendix G.2)}$$

Then, we find the flow rate

$$Q = (6.605 \times 10^{-4} \, m^2)(3.0 \text{ m/s}) = 1.982 \times 10^{-3} \, m^3/s$$

Converting to L/min, we have

$$Q = 1.982 \times 10^{-3} \, m^3/s \left(\frac{60\,000 \text{ L/min}}{1.0 \text{ m}^3/s} \right) = 119 \text{ L/min}$$

Example Problem 6.7

Determine the required size of standard Schedule 40 steel pipe to carry 192 m³/h of water with a maximum velocity of 6.0 m/s.

Solution

Because Q and v are known, the required area can be found from

$$Q = Av$$
$$A = Q/v$$

First, we must convert the volume flow rate to the units of m³/s:

$$Q = 192 \text{ m}^3/h(1 \text{ h}/3600 \text{ s}) = 0.0533 \text{ m}^3/s$$

Then, we have

$$A = \frac{Q}{v} = \frac{0.0533 \text{ m}^3/s}{6.0 \text{ m/s}} = 0.008\,88 \text{ m}^2 = 8.88 \times 10^{-3} m^2$$

This must be interpreted as the *minimum* allowable area because any smaller area would produce a velocity higher than 6.0 m/s. Therefore, we must look in Appendix F for a standard DN pipe with a flow area just larger than $8.88 \times 10^{-3} m^2$. A standard DN 125-mm Schedule 40 steel pipe, with a flow area of $1.291 \times 10^{-2} m^2$ is required. The actual velocity of flow when this pipe carries 0.0533 m³/s of water is

$$v = \frac{Q}{A} = \frac{0.0533 \text{ m}^3/s}{1.291 \times 10^{-2} m^2} = 4.13 \text{ m/s}$$

If the next-smaller pipe (a DN 100-mm Schedule 40 pipe) is used, the velocity is

$$v = \frac{Q}{A} = \frac{0.0533 \text{ m}^3/s}{8.213 \times 10^{-3} m^2} = 6.49 \text{ m/s (too high)}$$

Example Problem 6.8

A pumped fluid distribution system is being designed to deliver 400 gal/min of water to a cooling system in a power generation plant. Use Fig. 6.3 to make an initial selection of Schedule 40 pipe sizes for the suction and discharge lines for the system. Then compute the actual average velocity of flow for each pipe.

Solution

Entering Fig. 6.3 at $Q = 400$ gal/min, we select the following:

Suction pipe, 4-in Schedule 40: $A_s = 0.08840 \text{ ft}^2$ (from Appendix F)

Discharge pipe, 3-in Schedule 40: $A_d = 0.05132 \text{ ft}^2$ (from Appendix F)

The actual average velocity of flow in each pipe is

$$v_s = \frac{Q}{A_s} = \frac{400 \text{ gal/min}}{0.08840 \text{ ft}^2} \frac{1 \text{ ft}^3/s}{449 \text{ gal/min}} = 10.08 \text{ ft/s}$$

$$v_d = \frac{Q}{A_d} = \frac{400 \text{ gal/min}}{0.05132 \text{ ft}^2} \frac{1 \text{ ft}^3/s}{449 \text{ gal/min}} = 17.36 \text{ ft/s}$$

Comment   Although these pipe sizes and velocities should be acceptable in normal service, there are situations where lower velocities are desirable to limit energy losses in the system. Compute the velocities resulting from selecting the next-larger standard Schedule 40 pipe size for both the suction and discharge lines:

Suction pipe, 5-in Schedule 40: $A_s = 0.1390 \text{ ft}^2$ (from Appendix F)

Discharge pipe, 3 1/2-in Schedule 40: $A_d = 0.06868 \text{ ft}^2$ (from Appendix F)

The actual average velocity of flow in each pipe is

$$v_s = \frac{Q}{A_s} = \frac{400 \text{ gal/min}}{0.1390 \text{ ft}^2} \frac{1 \text{ ft}^3/s}{449 \text{ gal/min}} = 6.41 \text{ ft/s}$$

$$v_d = \frac{Q}{A_d} = \frac{400 \text{ gal/min}}{0.06868 \text{ ft}^2} \frac{1 \text{ ft}^3/s}{449 \text{ gal/min}} = 12.97 \text{ ft/s}$$

If the pump connections were the 4-in and 3-in sizes from the initial selection, a gradual reducer and gradual enlargement could be designed to connect these pipes to the pump.

6.5 CONSERVATION OF ENERGY—BERNOULLI'S EQUATION

The analysis of a pipeline problem such as that illustrated in Fig. 6.2 accounts for all the energy within the system. In physics you learned that energy can be neither created nor destroyed, but it can be transformed from one form into another. This is a statement of the law of *conservation of energy*.

There are three forms of energy that are always considered when analyzing a pipe flow problem. Consider an element of fluid, as shown in Fig. 6.1, inside a pipe in a flow system. It is located at a certain elevation z, has a velocity v, and pressure p. The element of fluid possesses the following forms of energy:

1. *Potential Energy.* Due to its elevation, the potential energy of the element relative to some reference level is

$$\text{PE} = wz \qquad (6\text{–}6)$$

where w is the weight of the element.

2. *Kinetic Energy.* Due to its velocity, the kinetic energy of the element is

$$\text{KE} = wv^2/2g \qquad (6\text{–}7)$$

3. *Flow Energy.* Sometimes called *pressure energy* or *flow work*, this represents the amount of work necessary to move the element of fluid across a certain section against the pressure p. Flow energy is abbreviated FE and is calculated from

$$\text{FE} = wp/\gamma \qquad (6\text{–}8)$$

Equation (6–8) can be derived as follows. Figure 6.5 shows the element of fluid in the pipe being moved across a section. The force on the element is pA, where p is the pressure at the section and A is the area of the section. In moving the element across the section, the force moves a distance L equal to the length of the element. Therefore, the work done is

$$\text{Work} = pAL = pV$$

where V is the volume of the element. The weight of the element w is

$$w = \gamma V$$

where γ is the specific weight of the fluid. Then, the volume of the element is

$$V = w/\gamma$$

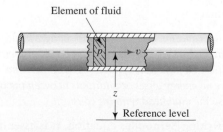

Element of fluid

z

Reference level

FIGURE 6.4   Element of a fluid in a pipe.

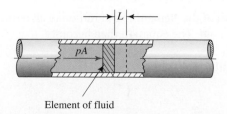

L

pA

Element of fluid

FIGURE 6.5   Flow energy.

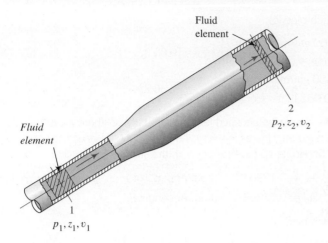

FIGURE 6.6 Fluid elements used in Bernoulli's equation.

and we have

$$\text{Work} = pV = pw/\gamma$$

which is called flow energy in Eq. (6–8).

The total amount of energy of these three forms possessed by the element of fluid is the sum E,

$$E = \text{FE} + \text{PE} + \text{KE}$$
$$E = wp/\gamma + wz + wv^2/2g$$

Each of these terms is expressed in units of energy, which are Newton-meters (N $\cdot$ m) in the SI unit system and foot-pounds (ft-lb) in the U.S. Customary System.

Now consider the element of fluid in Fig. 6.6, which moves from section 1 to section 2. The values for p, z, and v are different at the two sections. At section 1, the total energy is

$$E_1 = \frac{wp_1}{\gamma} + wz_1 + \frac{wv_1^2}{2g}$$

At section 2, the total energy is

$$E_2 = \frac{wp_2}{\gamma} + wz_2 + \frac{wv_2^2}{2g}$$

If no energy is added to the fluid or lost between sections 1 and 2, then the principle of conservation of energy requires that

$$E_1 = E_2$$
$$\frac{wp_1}{\gamma} + wz_1 + \frac{wv_1^2}{2g} = \frac{wp_2}{\gamma} + wz_2 + \frac{wv_2^2}{2g}$$

The weight of the element w is common to all terms and can be divided out. The equation then becomes

⇨ **Bernoulli's Equation**

$$\frac{p_1}{\gamma} + z_1 + \frac{v_1^2}{2g} = \frac{p_2}{\gamma} + z_2 + \frac{v_2^2}{2g} \qquad (6\text{–}9)$$

This is referred to as *Bernoulli's equation.*

6.6 INTERPRETATION OF BERNOULLI'S EQUATION

Each term in Bernoulli's equation, Eq. (6–9), resulted from dividing an expression for energy by the weight of an element of the fluid. Therefore,

Each term in Bernoulli's equation is one form of the energy possessed by the fluid per unit weight of fluid flowing in the system.

The units for each term are "energy per unit weight." In the SI system the units are N·m/N and in the U.S. Customary System the units are lb·ft/lb.

Notice, however, that the force (or weight) unit appears in both the numerator and the denominator and it can be cancelled. The resulting unit is simply the meter (m) or foot (ft) and it can be interpreted to be a height. In fluid flow analysis, the terms are typically expressed as "head," referring to a height above a reference level. Specifically,

p/γ is called the pressure head.

z is called the elevation head.

$v^2/2g$ is called the velocity head.

The sum of these three terms is called the total head.

Because each term in Bernoulli's equation represents a height, a diagram similar to that shown in Fig. 6.7 helps visualize the relationship among the three types of energy. As the fluid moves from point 1 to point 2, the magnitude of each term may change in value. If no energy is lost from or added to the fluid, however, the total head remains at a constant level. Bernoulli's equation is used to determine how the values of pressure head, elevation head, and velocity head change as the fluid moves through the system.

In Fig. 6.7 you can see that the velocity head at section 2 will be less than that at section 1. This can be shown by the continuity equation,

$$A_1 v_1 = A_2 v_2$$
$$v_2 = v_1(A_1/A_2)$$

Because $A_1 < A_2$, v_2 must be less than v_1. And because the velocity is squared in the velocity head term, $v_2^2/2g$ is much less than $v_1^2/2g$.

Typically, when the section size expands, as it does in Fig. 6.7, the pressure head increases because the velocity head decreases. That is how Fig. 6.7 is constructed. However, the actual change is also affected by the change in the elevation head; in this case the elevation head increased between points 1 and 2.

In summary,

Bernoulli's equation accounts for the changes in elevation head, pressure head, and velocity head between two points in a fluid flow system. It is assumed that there are no energy losses or additions between the two points, so the total head remains constant.

When writing Bernoulli's equation, it is essential that the pressures at the two reference points be expressed both as

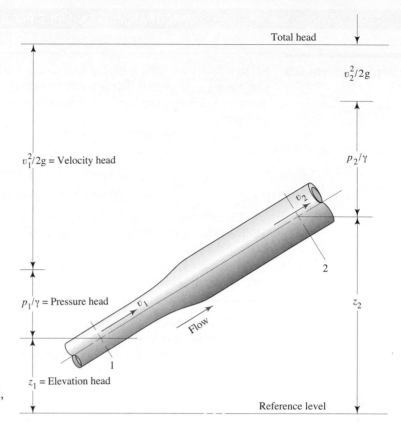

FIGURE 6.7 Pressure head, elevation head, velocity head, and total head.

absolute pressures or both as gage pressures. That is, they must both have the same reference pressure. In most problems it will be convenient to use gage pressure because parts of the fluid system exposed to the atmosphere will then have zero pressure. Also, most pressures are measured by a gage relative to the local atmospheric pressure.

6.7 RESTRICTIONS ON BERNOULLI'S EQUATION

Although Bernoulli's equation is applicable to a large number of practical problems, there are several limitations that must be understood to apply it properly.

1. It is valid only for incompressible fluids because the specific weight of the fluid is assumed to be the same at the two sections of interest.

2. There can be no mechanical devices between the two sections of interest that would add energy to or remove energy from the system, because the equation states that the total energy in the fluid is constant.

3. There can be no heat transferred into or out of the fluid.

4. There can be no energy lost due to friction.

In reality no system satisfies all these restrictions. However, there are many systems for which only a negligible error will result when Bernoulli's equation is used. Also, the use of this equation allows a rapid calculation if only a rough estimate is required. In Chapter 7, limitations 2 and 4 will be eliminated by expanding Bernoulli's equation into the *general energy equation*.

6.8 APPLICATIONS OF BERNOULLI'S EQUATION

We will present several programmed example problems to illustrate the use of Bernoulli's equation. Although it is not possible to cover all types of problems with a certain solution method, the general approach to problems of fluid flow is described here.

Procedure for Applying Bernoulli's Equation

1. Decide which items are known and what is to be found.

2. Decide which two sections in the system will be used when writing Bernoulli's equation. One section is chosen for which many data values are known. The second is usually the section at which something is to be calculated.

3. Write Bernoulli's equation for the two selected sections in the system. It is important that the equation is written *in the direction of flow*. That is, the flow must proceed *from* the section on the left side of the equation *to* that on the right side.

4. Be explicit when labeling the subscripts for the pressure head, elevation head, and velocity head terms in Bernoulli's equation. You should note where the reference points are on a sketch of the system.

5. Simplify the equation, if possible, by cancelling terms that are zero or those that are equal on both sides of the equation.

6. Solve the equation algebraically for the desired term.

7. Substitute known quantities and calculate the result, being careful to use consistent units throughout the calculation.

PROGRAMMED EXAMPLE PROBLEM

Example Problem 6.9   In Fig. 6.7, water at 10°C is flowing from section 1 to section 2. At section 1, which is 25 mm in diameter, the gage pressure is 345 kPa and the velocity of flow is 3.0 m/s. Section 2, which is 50 mm in diameter, is 2.0 m above section 1. Assuming there are no energy losses in the system, calculate the pressure p_2.

List the items that are known from the problem statement before looking at the next panel.

$$D_1 = 25 \text{ mm} \qquad v_1 = 3.0 \text{ m/s} \qquad z_2 - z_1 = 2.0 \text{ m}$$
$$D_2 = 50 \text{ mm} \qquad p_1 = 345 \text{ kPa(gage)}$$

The pressure p_2 is to be found. In other words, we are asked to calculate the pressure at section 2, which is different from the pressure at section 1 because there is a change in elevation and flow area between the two sections.

We are going to use Bernoulli's equation to solve the problem. Which two sections should be used when writing the equation?

In this case, sections 1 and 2 are the obvious choices. At section 1, we know p_1, v_1, and z_1. The unknown pressure p_2 is at section 2.

Now write Bernoulli's equation. [See Eq. (6–9).]

It should look like this:

$$\frac{p_1}{\gamma} + z_1 + \frac{v_1^2}{2g} = \frac{p_2}{\gamma} + z_2 + \frac{v_2^2}{2g}$$

The three terms on the left refer to section 1 and the three on the right refer to section 2.

Solve for p_2 in terms of the other variables.

The algebraic solution for p_2 could look like this:

$$\frac{p_1}{\gamma} + z_1 + \frac{v_2^2}{2g} = \frac{p_2}{\gamma} + z_2 + \frac{v_2^2}{2g}$$

$$\frac{p_2}{\gamma} = \frac{p_1}{\gamma} + z_1 + \frac{v_1^2}{2g} - z_2 - \frac{v_2^2}{2g}$$

$$p_2 = \gamma \left(\frac{p_1}{\gamma} + z_1 + \frac{v_1^2}{2g} - z_2 - \frac{v_2^2}{2g} \right)$$

This is correct. However, it is convenient to group the elevation heads and velocity heads together so you are considering differences in the values of like quantities. Also, because $\gamma(p_1/\gamma) = p_1$, the final solution for p_2 should be

$$p_2 = p_1 + \gamma \left(z_1 - z_2 + \frac{v_1^2 - v_2^2}{2g} \right) \qquad \textbf{(6–10)}$$

Are the values of all the terms on the right side of this equation known?

Everything was given except γ, v_2, and g. Of course, $g = 9.81 \text{ m/s}^2$. Because water at 10°C is flowing in the system, $\gamma = 9.81 \text{ kN/m}^3$. How can v_2 be determined?

The continuity equation is used:

$$A_1 v_1 = A_2 v_2$$
$$v_2 = v_1 (A_1/A_2)$$

Calculate v_2 now.

You should have $v_2 = 0.75$ m/s, found from

$$A_1 = \pi D_1^2/4 = \pi(25 \text{ mm})^2/4 = 491 \text{ mm}^2$$
$$A_2 = \pi D_2^2/4 = \pi(50 \text{ mm})^2/4 = 1963 \text{ mm}^2$$
$$v_2 = v_1(A_1/A_2) = 3.0 \text{ m/s}(491 \text{ mm}^2/1963 \text{ mm}^2) = 0.75 \text{ m/s}$$

Now substitute the known values into Eq. (6–10).

$$p_2 = 345 \text{ kPa} + \frac{9.81 \text{ kN}}{\text{m}^3}\left(-2.0 \text{ m} + \frac{(3.0 \text{ m/s})^2 - (0.75 \text{ m/s})^2}{2(9.81 \text{ m/s}^2)}\right)$$

Notice that $z_1 - z_2 = -2.0$ m. Neither z_1 nor z_2 is known, but it is known that z_2 is 2.0 m greater than z_1. Therefore, the difference $z_1 - z_2$ must be negative.

Now complete the calculation for p_2.

The final answer is $p_2 = 329.6$ kPa. This is 15.4 kPa less than p_1. The details of the solution are

$$p_2 = 345 \text{ kPa} + \frac{9.81 \text{ kN}}{\text{m}^3}\left(-2.0 \text{ m} + \frac{(9.0 - 0.563)\text{m}^2/\text{s}^2}{2(9.81)\text{m/s}^2}\right)$$

$$p_2 = 345 \text{ kPa} + \frac{9.81 \text{ kN}}{\text{m}^3}(-2.0 \text{ m} + 0.43 \text{ m})$$

$$p_2 = 345 \text{ kPa} - 15.4 \text{ kN/m}^2 = 345 \text{ kPa} - 15.4 \text{ kPa}$$

$$p_2 = 329.6 \text{ kPa}$$

The pressure p_2 is a gage pressure because it was computed relative to p_1, which was also a gage pressure. In later problem solutions, we will assume the pressures to be gage unless otherwise stated.

6.8.1 Tanks, Reservoirs, and Nozzles Exposed to the Atmosphere

Figure 6.8 shows a fluid flow system in which a siphon draws fluid from a tank or reservoir and delivers it through a nozzle at the end of the pipe. Note that the surface of the tank (point A) and the free stream of fluid exiting the nozzle (section F) are not confined by solid boundaries and are exposed to the prevailing atmosphere. Therefore, the pressure at those sections is zero gage pressure. We then use the following rule:

When the fluid at a reference point is exposed to the atmosphere, the pressure is zero and the pressure head term can be cancelled from Bernoulli's equation.

The tank from which the fluid is being drawn can be assumed to be quite large compared to the size of the flow area inside the pipe. Now, because $v = Q/A$, the velocity at the surface of such a tank will be very small. Furthermore, when we use the velocity to compute the velocity head, $v^2/2g$, we *square* the velocity. The process of squaring a small number much less than 1.0 produces an even smaller number. For these reasons, we adopt the following rule:

The velocity head at the surface of a tank or reservoir is considered to be zero and it can be cancelled from Bernoulli's equation.

FIGURE 6.8 Siphon for Example Problem 6.10.

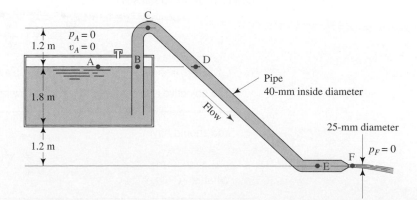

6.8.2 When Both Reference Points Are in the Same Pipe

Also notice in Fig. 6.8 that several points of interest (points B–E) are inside the pipe, which has a uniform flow area. Under the conditions of steady flow assumed for these problems, the velocity will be the same throughout the pipe. Then the following rule applies when steady flow occurs:

When the two points of reference for Bernoulli's equation are both inside a pipe of the same size, the velocity head terms on both sides of the equation are equal and can be cancelled.

6.8.3 When Elevations Are Equal at Both Reference Points

Similarly, the following rule applies when the reference points are on the same level:

When the two points of reference for Bernoulli's equation are both at the same elevation, the elevation head terms z_1 and z_2 are equal and can be cancelled.

The four observations made in Sections 6.8.1–6.8.3 enable the simplification of Bernoulli's equation and make the algebraic manipulations easier. Example Problem 6.10 uses these observations.

PROGRAMMED EXAMPLE PROBLEM

Example Problem 6.10

Figure 6.8 shows a siphon that is used to draw water from a swimming pool. The tube that makes up the siphon has an ID of 40 mm and terminates with a 25-mm diameter nozzle. Assuming that there are no energy losses in the system, calculate the volume flow rate through the siphon and the pressure at points B–E.

The first step in this problem solution is to calculate the volume flow rate Q, using Bernoulli's equation. The two most convenient points to use for this calculation are A and F. What is known about point A?

Point A is the free surface of the water in the pool. Therefore, $p_A = 0$ Pa. Also, because the surface area of the pool is very large, the velocity of the water at the surface is very nearly zero. Therefore, we will assume $v_A = 0$.

What do we know about point F?

Point F is in the free stream of water outside the nozzle. Because the stream is exposed to atmospheric pressure, the pressure $p_F = 0$ Pa. We also know that point F is 3.0 m below point A.

Now write Bernoulli's equation for points A and F.

You should have

$$\frac{p_A}{\gamma} + z_A + \frac{v_A^2}{2g} = \frac{p_F}{\gamma} + z_F + \frac{v_F^2}{2g}$$

Taking into account the information in the previous two panels, how can we simplify this equation?

Because $p_A = 0$ Pa, $p_F = 0$ Pa, and v_A is approximately zero, we can cancel them from the equation. What remains is

$$\cancel{\frac{p_A}{\gamma}}^{0} + z_A + \cancel{\frac{v_A^2}{2g}}^{0} = \cancel{\frac{p_F}{\gamma}}^{0} + z_F + \frac{v_F^2}{2g}$$

$$z_A = z_F + \frac{v_F^2}{2g}$$

The objective is to calculate the volume flow rate, which depends on the velocity. Solve for v_F now.

You should have

$$v_F = \sqrt{(z_A - z_F)2g}$$

What is $z_A - z_F$?

From Fig. 6.8 we see that $z_A - z_F = 3.0$ m. Notice that the difference is positive because z_A is greater than z_F. We can now solve for the value of v_F.

The result is

$$v_F = \sqrt{(3.0\text{ m})(2)(9.81\text{ m/s}^2)} = \sqrt{58.9}\text{ m/s} = 7.67\text{ m/s}$$

Now, how can Q be calculated?

Using the definition of volume flow rate, $Q = Av$, compute the volume flow rate.

The result is

$$Q = A_F v_F$$
$$v_F = 7.67\text{ m/s}$$
$$A_F = \pi(25\text{ mm})^2/4 = 491\text{ mm}^2$$
$$Q = 491\text{ mm}^2 \left(\frac{7.67\text{ m}}{\text{s}}\right)\left(\frac{1\text{ m}^2}{10^6\text{ mm}^2}\right) = 3.77 \times 10^{-3}\text{ m}^3/\text{s}$$

The first part of the problem is now complete. Now use Bernoulli's equation to determine p_B. What two points should be used?

Points A and B are the best. As shown in the previous panels, using point A allows the equation to be simplified greatly, and because we are looking for p_B, we must choose point B.

Write Bernoulli's equation for points A and B, simplify it as before, and solve for p_B.

Here is one possible solution procedure:

$$\cancel{\frac{p_A}{\gamma}}^{0} + z_A + \cancel{\frac{v_A^2}{2g}}^{0} = \frac{p_B}{\gamma} + z_B + \frac{v_B^2}{2g}$$

Because $p_A = 0$ Pa and $v_A = 0$, we have

$$z_A = \frac{p_B}{\gamma} + z_B + \frac{v_B^2}{2g}$$
$$p_B = \gamma[(z_A - z_B) - v_B^2/2g] \qquad\qquad \textbf{(6–11)}$$

What is $z_A - z_B$?

It is zero. Because the two points are on the same level, their elevations are the same. Can you find v_B?

We can calculate v_B by using the continuity equation:

$$Q = A_B v_B$$
$$v_B = Q/A_B$$

The area of a 40-mm diameter tube can be found in Appendix J. Complete the calculation for v_B.

The result is

$$v_B = Q/A_B$$
$$Q = 3.77 \times 10^{-3}\text{ m}^3/\text{s}$$
$$A_B = 1.257 \times 10^{-3}\text{ m}^2 \quad \text{(from Appendix J)}$$
$$v_B = \frac{3.77 \times 10^{-3}\text{ m}^3}{\text{s}} \times \frac{1}{1.257 \times 10^{-3}\text{ m}^2} = 3.00\text{ m/s}$$

We now have all the data we need to calculate p_B from Eq. (6–11).

The pressure at point B is

$$p_B = \gamma[(z_A - z_B) - v_B^2/2g]$$

$$\frac{v_B^2}{2g} = \frac{(3.00)^2\, m^2}{s^2} \times \frac{s^2}{(2)(9.81)\, m} = 0.459\ m$$

$$p_B = (9.81\ kN/m^3)(0 - 0.459\ m)$$

$$p_B = -4.50\ kN/m^2$$

$$p_B = -4.50\ kPa$$

The negative sign indicates that p_B is 4.50 kPa below atmospheric pressure. Notice that when we deal with fluids in motion, the concept that points on the same level have the same pressure does *not* apply as it does with fluids at rest.

The next three panels present the solutions for the pressures p_C, p_D, and p_E, which can be found in a manner very similar to that used for p_B. Complete the solution for p_C before looking at the next panel.

The answer is $p_C = -16.27$ kPa. We use Bernoulli's equation.

$$\cancel{\frac{p_A}{\gamma}}^{0} + z_A + \cancel{\frac{v_A^2}{2g}}^{0} = \frac{p_C}{\gamma} + z_C + \frac{v_C^2}{2g}$$

Because $p_A = 0$ and $v_A = 0$, the pressure at point C is found by the following process.

$$z_A = \frac{p_C}{\gamma} + z_C + \frac{v_C^2}{2g}$$

$$p_C = \gamma[(z_A - z_C) - v_C^2/2g]$$

$$z_A - z_C = -1.2\ m \quad (\text{negative, because } z_C \text{ is greater than } z_A)$$

$$v_C = v_B = 3.00\ m/s \quad (\text{because } A_C = A_B)$$

$$\frac{v_C^2}{2g} = \frac{v_B^2}{2g} = 0.459\ m$$

$$p_C = (9.81\ kN/m^3)(-1.2\ m - 0.459\ m)$$

$$p_C = -16.27\ kN/m^2$$

$$p_C = -16.27\ kPa$$

Complete the calculation for p_D before looking at the next panel.

The answer is $p_D = -4.50$ kPa. This is the same as p_B because the elevation and the velocity at points B and D are equal. Solution by Bernoulli's equation would prove this.
Now find p_E.

The pressure at point E is 24.93 kPa. We use Bernoulli's equation:

$$\cancel{\frac{p_A}{\gamma}}^{0} + z_A + \cancel{\frac{v_A^2}{2g}}^{0} = \frac{p_E}{\gamma} + z_E + \frac{v_E^2}{2g}$$

Because $p_A = 0$ and $v_A = 0$, we have

$$z_A = \frac{p_E}{\gamma} + z_E + \frac{v_E^2}{2g}$$

$$p_E = \gamma[(z_A - z_E) - v_E^2/2g]$$

$$z_A - z_E = +3.0\ m$$

$$v_E = v_B = 3.00\ m/s$$

$$\frac{v_E^2}{2g} = \frac{v_B^2}{2g} = 0.459\ m$$

$$p_E = (9.81\ kN/m^3)(3.0\ m - 0.459\ m)$$

$$p_E = 24.93\ kN/m^2$$

$$p_E = 24.93\ kPa$$

Summary of the Results of Example Problem 6.10

1. The velocity of flow from the nozzle, and therefore the volume flow rate delivered by the siphon, depends on the elevation difference between the free surface of the fluid and the outlet of the nozzle.

2. The pressure at point B is below atmospheric pressure even though it is on the same level as point A, which is exposed to the atmosphere. In Eq. (6–11), Bernoulli's equation shows that the pressure head at B is decreased by the amount of the velocity head. That is, some of the energy is converted to kinetic energy, resulting in a lower pressure at B.

3. When steady flow exists, the velocity of flow is the same at all points where the tube size is the same.

4. The pressure at point C is the lowest in the system because point C is at the highest elevation.

5. The pressure at point D is the same as that at point B because both are on the same elevation and the velocity head at both points is the same.

6. The pressure at point E is the highest in the system because point E is at the lowest elevation.

6.8.4 Venturi Meters and Other Closed Systems with Unknown Velocities

Figure 6.9 shows a device called a *venturi meter* that can be used to measure the velocity of flow in a fluid flow system. A more complete description of the venturi meter is given in Chapter 15. However, the analysis of such a device is based on the application of Bernoulli's equation. The reduced-diameter section at B causes the velocity of flow to increase there with a corresponding decrease in the pressure. It will be shown that the velocity of flow is dependent on the *difference* in pressure between points A and B. Therefore, a differential manometer as shown is convenient to use.

We will also show in the solution to the following problem that we must combine the continuity equation with Bernoulli's equation to solve for the desired velocity of flow.

Example Problem 6.11 The venturi meter shown in Fig. 6.9 carries water at 60°C. The inside dimensions are machined to the sizes shown in the figure. The specific gravity of the gage fluid in the manometer is 1.25. Calculate the velocity of flow at section A and the volume flow rate of water.

Solution The problem solution will be shown in the steps outlined at the beginning of this section but the programmed technique will not be used.

FIGURE 6.9 Venturi meter system for Example Problem 6.11.

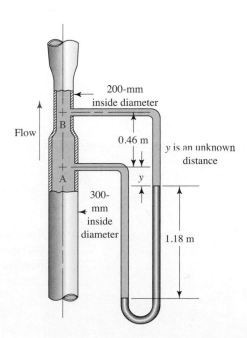

1. *Decide what items are known and what is to be found.* The elevation difference between points A and B is known. The manometer allows the determination of the difference in pressure between points A and B. The sizes of the sections at A and B are known.

    The velocity is not known at any point in the system and the velocity at point A was specifically requested.

2. *Decide on sections of interest.* Points A and B are the obvious choices.

3. *Write Bernoulli's equation* between points A and B:

$$\frac{p_A}{\gamma} + z_A + \frac{v_A^2}{2g} = \frac{p_B}{\gamma} + z_B + \frac{v_B^2}{2g}$$

The specific weight γ is for water at 60°C, which is 9.65 kN/m^3 (Appendix A).

4. *Simplify the equation*, if possible, by eliminating terms that are zero or terms that are equal on both sides of the equation. No simplification can be done here.

5. *Solve the equation algebraically for the desired term.* This step will require significant effort. First, note that *both* of the velocities are unknown. However, we can find the difference in pressures between A and B and the elevation difference is known. Therefore, it is convenient to bring both pressure terms and both elevation terms onto the left side of the equation in the form of differences. Then the two velocity terms can be moved to the right side. The result is

$$\frac{p_A - p_B}{\gamma} + (z_A - z_B) = \frac{v_B^2 - v_A^2}{2g} \tag{6-12}$$

6. *Calculate the result.* Several steps are required. The elevation difference is

$$z_A - z_B = -0.46 \text{ m} \tag{6-13}$$

The value is negative because B is higher than A. This value will be used in Eq. (6–12) later.

    The pressure-head difference term can be evaluated by writing the equation for the manometer. We will use γ_g for the specific weight of the gage fluid, where

$$\gamma_g = 1.25(\gamma_w \text{ at } 4°C) = 1.25(9.81 \text{ kN/m}^3) = 12.26 \text{ kN/m}^3$$

A new problem occurs here because the data in Fig. 6.9 do not include the vertical distance from point A to the level of the gage fluid in the right leg of the manometer. We will show that this problem will be eliminated by simply calling this unknown distance y or any other variable name.

    Now we can write the manometer equation starting at A:

$$p_A + \gamma(y) + \gamma(1.18 \text{ m}) - \gamma_g(1.18 \text{ m}) - \gamma(y) - \gamma(0.46 \text{ m}) = p_B$$

Note that the two terms containing the unknown y variable can be cancelled out.

    Solving for the pressure difference $p_A - p_B$, we find

$$p_A - p_B = \gamma(0.46 \text{ m} - 1.18 \text{ m}) + \gamma_g(1.18 \text{ m})$$
$$p_A - p_B = \gamma(-0.72 \text{ m}) + \gamma_g(1.18 \text{ m})$$

Notice in Eq. (6–12), however, that we really need $(p_A - p_B)/\gamma$. If we divide both sides of the above equation by γ, we get the desired term:

$$\frac{p_A - p_B}{\gamma} = -0.72 \text{ m} + \frac{\gamma_g(1.18 \text{ m})}{\gamma}$$

$$= -0.72 \text{ m} + \frac{12.26 \text{ kN/m}^3(1.18 \text{ m})}{9.65 \text{ kN/m}^3}$$

$$(p_A - p_B)/\gamma = -0.72 \text{ m} + 1.50 \text{ m} = 0.78 \text{ m} \tag{6-14}$$

The entire left side of Eq. (6–12) has now been evaluated. Note, however, that there are still *two* unknowns on the right side, v_A and v_B. We can eliminate one unknown by finding another independent equation that relates these two variables. A convenient equation is the *continuity equation*,

$$A_A v_A = A_B v_B$$

Solving for v_B in terms of v_A, we obtain

$$v_B = v_A(A_A/A_B)$$

The areas for the 200-mm and 300-mm diameter sections can be found in Appendix J. Then,

$$v_B = v_A(7.069 \times 10^{-2}/3.142 \times 10^{-2}) = 2.25v_A$$

But we need v_B^2:

$$v_B^2 = 5.06\, v_A^2$$

Then,

$$v_B^2 - v_A^2 = 5.06\, v_A^2 - v_A^2 = 4.06\, v_A^2 \qquad \textbf{(6–15)}$$

We can now take these results, the elevation head difference [Eq. (6–13)] and the pressure head difference [Eq. (6–14)], back into Eq. (6–12) and complete the solution. Equation (6–12) becomes

$$0.78\ \text{m} - 0.46\ \text{m} = 4.06\, v_A^2/2g$$

Solving for v_A gives

$$v_A = \sqrt{\frac{2g(0.32\ \text{m})}{4.06}} = \sqrt{\frac{2(9.81\ \text{m/s}^2)(0.32\ \text{m})}{4.06}}$$

$$v_A = 1.24\ \text{m/s}$$

The problem statement also asked for the volume flow rate, which can be computed from

$$Q = A_A v_A = (7.069 \times 10^{-2}\,\text{m}^2)(1.24\ \text{m/s}) = 8.77 \times 10^{-2}\ \text{m}^3/\text{s}$$

This example problem is completed.

6.9 TORRICELLI'S THEOREM

In the siphon analyzed in Example Problem 6.10, it was observed that the velocity of flow from the siphon depends on the elevation difference between the free surface of the fluid and the outlet of the siphon. A classic application of this observation is shown in Fig. 6.10. Fluid

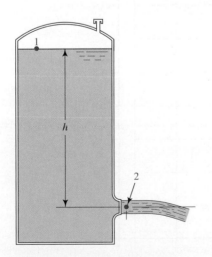

FIGURE 6.10 Flow from a tank.

is flowing from the side of a tank through a smooth, rounded nozzle. To determine the velocity of flow from the nozzle, write Bernoulli's equation between a reference point on the fluid surface and a point in the jet issuing from the nozzle:

$$\frac{p_1}{\gamma} + z_1 + \frac{v_1^2}{2g} = \frac{p_2}{\gamma} + z_2 + \frac{v_2^2}{2g}$$

However, $p_1 = p_2 = 0$, and v_1 is approximately zero. Thus,

$$\cancel{\frac{p_1}{\gamma}}^{0} + z_1 + \cancel{\frac{v_1^2}{2g}}^{0} = \cancel{\frac{p_2}{\gamma}}^{0} + z_2 + \frac{v_2^2}{2g}$$

Then, solving for v_2 gives

$$v_2 = \sqrt{2g(z_1 - z_2)}$$

Letting $h = (z_1 - z_2)$, we have

▷ **Torricelli's Theorem**

$$v_2 = \sqrt{2gh} \qquad \textbf{(6–16)}$$

Equation (6–16) is called *Torricelli's theorem* in honor of Evangelista Torricelli, who discovered it approximately in the year 1645. See Internet resource 1.

Example Problem
6.12

For the tank shown in Fig. 6.10, compute the velocity of flow from the nozzle for a fluid depth h of 3.00 m.

Solution

This is a direct application of Torricelli's theorem:

$$v_2 = \sqrt{2gh} = \sqrt{(2)(9.81 \text{ m/s}^2)(3.0 \text{ m})} = 7.67 \text{ m/s}$$

Example Problem
6.13

For the tank shown in Fig. 6.10, compute the velocity of flow from the nozzle and the volume flow rate for a range of depth from 3.0 m to 0.50 m in steps of 0.50 m. The diameter of the jet at the nozzle is 50 mm.

Solution

The same procedure used in Example Problem 6.12 can be used to determine the velocity at any depth. So, at $h = 3.0$ m, $v_2 = 7.67$ m/s. The volume flow rate is computed by multiplying this velocity by the area of the jet:

$$A_j = 1.963 \times 10^{-3} \text{m}^2 \text{ (from Appendix J)}$$

Then,

$$Q = A_j v_2 = (1.963 \times 10^{-3} \text{m}^2)(7.67 \text{ m/s}) = 1.51 \times 10^{-2} \text{m}^3/2$$

Using the same procedure, we compute the following data:

Depth h (m)	v_2(m/s)	Q (m³/s)
3.0	7.67	1.51×10^{-2}
2.5	7.00	1.38×10^{-2}
2.0	6.26	1.23×10^{-2}
1.5	5.42	1.07×10^{-2}
1.0	4.43	0.87×10^{-2}
0.5	3.13	0.61×10^{-2}

Figure 6.11 is a plot of velocity and volume flow rate versus depth.

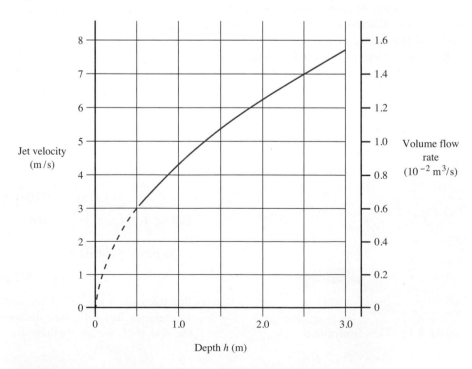

FIGURE 6.11 Jet velocity and volume flow rate versus fluid depth.

FIGURE 6.12 Vertical jet.

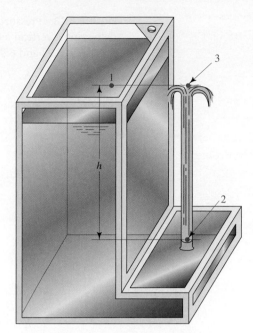

Another interesting application of Torricelli's theorem is shown in Fig. 6.12, in which a jet of fluid is shooting upward. If no energy losses occur, the jet will reach a height equal to the elevation of the free surface of the fluid in the tank. Of course, at this height the velocity in the stream is zero. This can be demonstrated using Bernoulli's equation. First obtain an expression for the velocity of the jet at point 2:

$$\frac{\cancel{p_1}^{0}}{\gamma} + z_1 + \frac{\cancel{v_1^2}^{0}}{2g} = \frac{\cancel{p_2}^{0}}{\gamma} + z_2 + \frac{v_2^2}{2g}$$

This is an identical situation to that encountered in the initial development of Torricelli's theorem. Then, as in Eq. (6–16),

$$v_2 = \sqrt{2gh}$$

Now, write Bernoulli's equation between point 2 and point 3 at the level of the free surface of the fluid, but in the fluid stream:

$$\frac{\cancel{p_2}^{0}}{\gamma} + z_2 + \frac{v_2^2}{2g} = \frac{\cancel{p_3}^{0}}{\gamma} + z_3 + \frac{v_3^2}{2g}$$

However, $p_2 = p_3 = 0$. Then, solving for v_3, we have

$$v_3 = \sqrt{v_2^2 + 2g(z_2 - z_3)}$$

From Eq. (6–16), $v_2^2 = 2gh$. Also, $(z_2 - z_3) = -h$. Then,

$$v_3 = \sqrt{2gh + 2g(-h)} = 0$$

This result verifies that the stream just reaches the height of the free surface of the fluid in the tank.

To make a jet go higher (as with some decorative fountains, for example), a greater pressure can be developed above the fluid in the reservoir, or a pump can be used to develop a higher pressure.

**Example Problem
6.14**

Using a system similar to that shown in Fig. 6.13, compute the required air pressure above the water to cause the jet to rise 40.0 ft from the nozzle. The depth $h = 6.0$ ft.

Solution

First, use Bernoulli's equation to obtain an expression for the velocity of flow from the nozzle as a function of the air pressure.

$$\frac{p_1}{\gamma} + z_1 + \frac{\cancel{v_1^2}^{0}}{2g} = \frac{\cancel{p_2}^{0}}{\gamma} + z_2 + \frac{v_2^2}{2g}$$

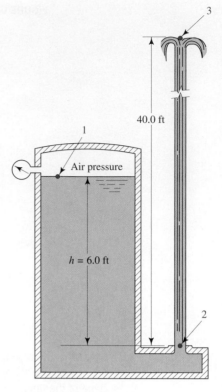

FIGURE 6.13 Pressurized tank delivering a vertical jet. Also used for Problems 6.93 and 6.94.

Here, we can see that $v_1 = 0$ and $p_2 = 0$. Solving for v_2 gives

$$v_2 = \sqrt{2g[(p_1/\gamma) + (z_1 - z_2)]}$$

As before, letting $h = (z_1 - z_2)$, we have

$$v_2 = \sqrt{2g[(p_1/\gamma) + h]} \qquad \qquad \textbf{(6–17)}$$

This is similar to Torricelli's theorem. It was shown above that for $v = \sqrt{2gh}$, the jet rises to a height h. By analogy, the pressurized system would cause the jet to rise to a height of $[(p_1/\gamma) + h]$. Then, in this problem, if we want a height of 40.0 ft and $h = 6.0$ ft,

$$p_1/\gamma + h = 40.0 \text{ ft}$$

$$p_1/\gamma = 40.0 \text{ ft} - h = 40.0 \text{ ft} - 6.0 \text{ ft} = 34.0 \text{ ft}$$

and

$$p_1 = \gamma(34.0 \text{ ft})$$

$$p_1 = (62.4 \text{ lb/ft}^3)(34.0 \text{ ft})(1 \text{ ft}^2)/(144 \text{ in}^2)$$

$$p_1 = 14.73 \text{ psig}$$

In Chapter 4, we defined the pressure head p/γ in such applications as the *piezometric head*. Then the *total head* above the nozzle is $p_1/\gamma + h$.

6.10 FLOW DUE TO A FALLING HEAD

As stated earlier, most problems considered in this book are for situations in which the flow rate is constant. However, in Section 6.90, it was shown that the flow rate depends on the pressure head available to cause the flow. The results of Example Problem 6.13, plotted in Fig. 6.11, show that the velocity and volume flow rate issuing from an orifice in a tank decrease in a nonlinear manner as the fluid flows from the tank and the depth of the fluid decreases.

In this section, we will develop a method for computing the time required to empty a tank, considering the variation of velocity as the depth decreases. Figure 6.14 shows a tank with a smooth, well-rounded nozzle in the bottom through which fluid is discharging. For a given depth of fluid h, Torricelli's theorem tells us that the velocity of flow in the jet is

$$v_j = \sqrt{2gh}$$

The volume flow rate through the nozzle is $Q = A_j v_j$ in such units as cubic meters per second (m³/s) or cubic feet per

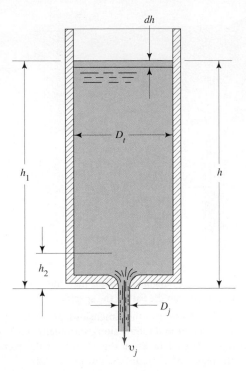

FIGURE 6.14 Flow from a tank with falling head. Also for Problems 6.95–6.106.

second (ft^3/s). In a small amount of time dt, the volume of fluid flowing through the nozzle is

$$\text{Volume flowing} = Q(dt) = A_j v_j (dt) \qquad (6-18)$$

Meanwhile, because fluid is leaving the tank, the fluid level is decreasing. During the small time increment dt, the fluid level drops a small distance dh. Then, the volume of fluid removed from the tank is

$$\text{Volume removed} = -A_t dh \qquad (6-19)$$

These two volumes must be equal. Then,

$$A_j v_j (dt) = -A_t dh \qquad (6-20)$$

Solving for the time dt, we have

$$dt = \frac{-(A_t/A_j)}{v_j} dh \qquad (6-21)$$

From Torricelli's theorem, we can substitute $v_j = \sqrt{2gh}$. Then,

$$dt = \frac{-(A_t/A_j)}{\sqrt{2gh}} dh \qquad (6-22)$$

Rewriting to separate the terms involving h gives

$$dt = \frac{-(A_t/A_j)}{\sqrt{2g}} h^{-1/2} dh \qquad (6-23)$$

The time required for the fluid level to fall from a depth h_1 to a depth h_2 can be found by integrating Eq. (6–23):

$$\int_{t_1}^{t_2} dt = \frac{-(A_t/A_j)}{\sqrt{2g}} \int_{h_1}^{h_2} h^{-1/2} dh \qquad (6-24)$$

$$t_2 - t_1 = \frac{-(A_t/A_j)}{\sqrt{2g}} \frac{\left(h_2^{1/2} - h_1^{1/2}\right)}{\frac{1}{2}} \qquad (6-25)$$

We can reverse the two terms involving h and remove the minus sign. At the same time clearing the $\frac{1}{2}$ from the denominator, we get

▷ **Time Required to Drain a Tank**

$$t_2 - t_1 = \frac{2(A_t/A_j)}{\sqrt{2g}} \left(h_1^{1/2} - h_2^{1/2}\right) \qquad (6-26)$$

Equation (6–26) can be used to compute the time required to drain a tank from h_1 to h_2.

Example Problem 6.15

For the tank shown in Fig. 6.14, find the time required to drain the tank from a level of 3.0 m to 0.50 m. The tank has a diameter of 1.50 m and the nozzle has a diameter of 50 mm.

Solution

To use Eq. (6–26), the required areas are

$$A_t = \pi (1.50 \text{ m})^2 / 4 = 1.767 \text{ m}^2$$
$$A_j = \pi (0.05 \text{ m})^2 / 4 = 0.001963 \text{ m}^2$$

The ratio of these two areas is required:

$$\frac{A_t}{A_j} = \frac{1.767 \text{ m}^2}{0.001963 \text{ m}^2} = 900$$

Now, in Eq. (6–26),

$$t_2 - t_1 = \frac{2(A_t/A_j)}{\sqrt{2g}} (h_1^{1/2} - h_2^{1/2})$$

$$t_2 - t_1 = \frac{2(900)}{\sqrt{2(9.81 \text{ m/s}^2)}} [(3.0 \text{ m})^{1/2} - (0.5 \text{ m})^{1/2}]$$

$$t_2 - t_1 = 417 \text{ s}$$

This is equivalent to 6 min and 57 s.

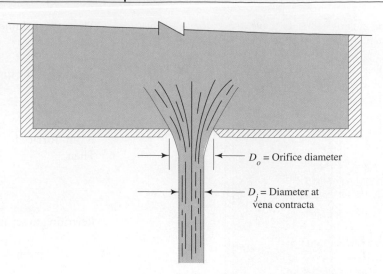

FIGURE 6.15 Flow through a sharp-edged orifice.

D_o = Orifice diameter

D_j = Diameter at vena contracta

6.10.1 Draining a Pressurized Tank

If the tank in Fig. 6.14 is sealed with a pressure above the fluid, the piezometric head p/γ should be added to the actual liquid depth before completing the calculations called for in Eq. (6–25).

6.10.2 Effect of the Type of Nozzle

The development of Eq. (6–26) assumes that the diameter of the jet of fluid flowing from the nozzle is the same as the diameter of the nozzle itself. This is very nearly true for the well-rounded nozzles depicted in Figs. 6.10, 6.12, and 6.14. However, if the nozzle is sharper, the minimum diameter of the jet is significantly smaller than the diameter of the opening. For example, Fig. 6.15 shows the flow from a tank through a sharp-edged orifice. The proper area to use for A_j in Eq. (6–26) is that at the smallest diameter. This point, called the *vena contracta*, occurs slightly outside the orifice. For this sharp-edged orifice, $A_j = 0.62A_o$ is a good approximation.

REFERENCES

1. American Society of Mechanical Engineers. 2012. *ANSI/ASME Standard B31.1-2012: Power Piping.* New York: Author.

2. Menon, E. Shashi. 2005. *Piping Calculations Manual.* Clinton, NC: Construction Trades Press.

3. Nayyar, Mohinder. 2002. *Piping Databook.* Clinton, NC: Construction Trades Press.

4. Nayyar, Mohinder. 2000. *Piping Handbook,* 7th ed. Clinton, NC: Construction Trades Press.

5. Silowash, Brian. 2010. *Piping Systems Manual.* Clinton, NC: Construction Trades Press.

6. U.S. Army Corps of Engineers. 1999. *Liquid Process Piping* (Engineer Manual 1110-1-4008). Washington, DC: Author.

INTERNET RESOURCES

1. **The MacTutor History of Mathematics Archive:** An archive of over 1000 biographies and history topics, including biographies of Daniel Bernoulli and Evangelista Torricelli. From the home page, select *Biographies Index,* then the first letter of the last name, then scan the list for the specific person.

2. **TubeNet.org:** A listing of the dimensions, properties, and suppliers of steel pipe and tubing along with many other types of fluid-flow-related data. From the left side of the home page, select the region of interest: U.S., Europe, or Asia.

3. **Copper Development Association:** A professional association of the copper industry; the site offers a large amount of data on sizes, pressure ratings, and physical characteristics of copper tubing. A *Copper Tube Handbook* or parts of it can be downloaded from the site.

4. **Ductile Iron Pipe Research Association:** Technical information about ductile iron pipe including dimensions, flow performance data, and comparison with other types of pipe.

5. **Stainless Tubular Products:** A supplier of stainless steel pipe, tubing, fittings, flanges, and stock materials.

6. **Plastics Pipe Institute:** An association representing all segments of the plastics piping industry, promoting the effective use of plastics piping for water and gas distribution, sewer and wastewater, oil and gas production, industrial and mining uses, power and communications, ducts, and irrigation. Includes a list of members that manufacture plastic pipe from which much data for sizes and application information can be found.

7. **Charter Plastics:** A supplier of polyethylene plastic pipe and tubing for many applications including industrial and municipal uses such as water distribution, sewer applications, and chemical service.

8. **Expert Piping Supply:** A supplier of polyethylene, polypropylene, PVC, CPVC, copper, and steel pipe in a wide range of diameters and wall thicknesses.

9. **Independent Pipe Products, Inc.:** Listings of suppliers of high-density polyethylene pipe fittings in many size classifications that match the outside diameters of steel pipes, ductile iron pipes, and copper tubes. Also lists suppliers of other types and materials of pipe and tubing.

10. **The Piping Tool Box:** A site containing data and basic information for piping system design. It includes U.S. and metric data for piping dimensions, fluid flow and pressure loss in pipes, piping standards, piping design strategy, and many other related topics. It is part of The Engineering Toolbox site. Select *Piping Systems.*

11. **Hydraulic Supermarket.com:** From the home page, select *Technical Library* for access to an extensive set of articles and

technical data related to hydraulic systems and components, maintenance and troubleshooting, application guidelines, and formulas. From the home page, select *Product Library* for lists of suppliers of products for hydraulic systems such as pumps, valves, and actuators.

12. **Eaton Hydraulics:** Manufacturer of hydraulic systems and components, including hydraulic and industrial hose under the brand names Aeroquip and Weatherhead. From the home page, select *Products & Solutions*, then *Hydraulics.*

13. **Parker Steel Company:** Producer of metric round seamless and hydraulic tubing from carbon steel, stainless steel, alloy steel, aluminum, brass, copper, titanium, and nickel alloys.

14. **Epco Plastics:** Supplier of industrial plastics including piping, tubing, fittings, valves, and accessories in both U.S. and metric sizes. Also includes the Comer Spa product line of ABS, PVC, PE, and PP metric plastic pipe.

15. **Industrial Hydraulic Service, Inc.:** This site includes tables of data for sizing tubing for hydraulic systems in U.S. Customary and metric sizes considering flow rate and pressure ratings.

PRACTICE PROBLEMS

Conversion Factors

6.1 Convert a volume flow rate of 3.0 gal/min to m^3/s.

6.2 Convert 459 gal/min to m^3/s.

6.3 Convert 8720 gal/min to m^3/s.

6.4 Convert 84.3 gal/min to m^3/s.

6.5 Convert a volume flow rate of 125 L/min to m^3/s.

6.6 Convert 4500 L/min to m^3/s.

6.7 Convert 15 000 L/min to m^3/s.

6.8 Convert 459 gal/min to L/min.

6.9 Convert 8720 gal/min to L/min.

6.10 Convert 23.5 cm^3/s to m^3/s.

6.11 Convert 0.296 cm^3/s to m^3/s.

6.12 Convert 0.105 m^3/s to L/min.

6.13 Convert 3.58×10^{-3} m^3/s to L/min.

6.14 Convert 5.26×10^{-6} m^3/s to L/min.

6.15 Convert 459 gal/min to ft^3/s.

6.16 Convert 20 gal/min to ft^3/s.

6.17 Convert 2500 gal/min to ft^3/s.

6.18 Convert 2.50 gal/min to ft^3/s.

6.19 Convert 125 ft^3/s to gal/min.

6.20 Convert 0.060 ft^3/s to gal/min.

6.21 Convert 7.50 ft^3/s to gal/min.

6.22 Convert 0.008 ft^3/s to gal/min.

6.23 Table 6.2 lists the range of typical volume flow rates for centrifugal fire-fighting pumps to be 500 to 2500 gal/min. Express this range in the units of ft^3/s and m^3/s.

6.24 Table 6.2 lists the range of typical volume flow rates for pumps in industrial oil hydraulic systems to be 3 to 30 gal/min. Express this range in the units of ft^3/s and m^3/s.

6.25 A certain deep-well pump for a residence is rated to deliver 745 gal/h of water. Express this flow rate in ft^3/s.

6.26 A small pump delivers 0.85 gal/h of liquid fertilizer. Express this flow rate in ft^3/s.

6.27 A small metering pump delivers 11.4 gal of a water treatment chemical per 24 h. Express this flow rate in ft^3/s.

6.28 A small metering pump delivers 19.5 mL/min of water to dilute a waste stream. Express this flow rate in m^3/s.

Note: In the following problems you may be required to refer to an appendix for fluid properties, dimensions of pipe and tubing, or conversion factors. Assume that there are no energy losses in all problems. Unless otherwise stated, the pipe sizes given are actual inside diameters.

Fluid Flow Rates

6.29 Water at 10°C is flowing at 0.075 m^3/s. Calculate the weight flow rate and the mass flow rate.

6.30 Oil for a hydraulic system (sg = 0.90) is flowing at 2.35×10^{-3} m^3/s. Calculate the weight flow rate and mass flow rate.

6.31 A liquid refrigerant (sg = 1.08) is flowing at a weight flow rate of 28.5 N/h. Calculate the volume flow rate and the mass flow rate.

6.32 After the refrigerant from Problem 6.31 flashes into a vapor, its specific weight is 12.50 N/m^3. If the weight flow rate remains at 28.5 N/h, compute the volume flow rate.

6.33 A fan delivers 640 ft^3/min (CFM) of air. If the density of the air is 1.20 kg/m^3, compute the mass flow rate in slugs/s and the weight flow rate in lb/h.

6.34 A large blower for a furnace delivers 47 000 ft^3/min (CFM) of air having a specific weight of 0.075 lb/ft^3. Calculate the weight flow rate and mass flow rate.

6.35 A furnace requires 1200 lb/h of air for efficient combustion. If the air has a specific weight of 0.062 lb/ft^3, compute the required volume flow rate.

6.36 If a pump removes 1.65 gal/min of water from a tank, how long will it take to empty the tank if it contains 7425 lb of water?

Continuity Equation

6.37 Calculate the diameter of a pipe that would carry 75.0 ft^3/s of a liquid at an average velocity of 10.0 ft/s.

6.38 If the velocity of a liquid is 1.65 ft/s in a special pipe with an inside diameter of 12 in, what is the velocity in a 3-in-diameter jet exiting from a nozzle attached to the pipe?

6.39 When 2000 L/min of water flows through a circular section with an inside diameter of 300 mm that later reduces to a 150-mm diameter, calculate the average velocity of flow in each section.

6.40 Water flows at 1.20 m/s in a circular section with a 150 mm inside diameter. Calculate the velocity of flow in a 300-mm-diameter section connected to it.

6.41 Figure 6.16 shows a fabricated assembly made from three different sizes of standard steel tubing listed in Appendix G.2. The larger tube on the left carries 0.072 m^3/s of water. The tee branches into two smaller sections. If the velocity in the 50-mm tube is 12.0 m/s, what is the velocity in the 100-mm tube?

6.42 A standard Schedule 40 steel pipe is to be selected to carry 10 gal/min of water with a maximum velocity of 1.0 ft/s. What size pipe should be used?

6.43 If water at 180°F is flowing with a velocity of 4.50 ft/s in a standard 6-in Schedule 40 pipe, calculate the weight flow rate in lb/h.

6.44 A standard steel tube, 1.5 25-mm OD $\times$ 1.5-mm wall (Appendix G.2), is carrying 19.7 L/min of oil. Calculate the velocity of flow.

6.45 The recommended velocity of flow in the discharge line of an oil hydraulic system is in the range of 8.0 to 25.0 ft/s. If the pump delivers 30 gal/min of oil, specify

FIGURE 6.16   Problem 6.41.

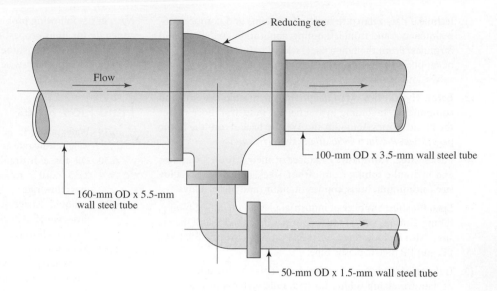

the smallest and largest suitable sizes of steel tubing from Appendix G.1.

6.46   Repeat Problem 6.45, except specify suitable sizes for the suction lines to maintain the velocity between 2.0 ft/s and 7.0 ft/s for 30 gal/min of flow.

6.47   Table 6.2 shows the typical volume flow rate for centrifugal fire-fighting pumps is in the range of 1800 L/min to 9500 L/min. Specify the smallest suitable DN size of Schedule 40 steel pipe for each flow rate that will maintain the maximum velocity of flow at 2.0 m/s.

6.48   Repeat Problem 6.47, but use Schedule 80 DN pipe.

6.49   Compute the resulting velocity of flow if 400 L/min of fluid flows through a DN 50 Schedule 40 pipe.

6.50   Repeat Problem 6.49 for a DN 50 Schedule 80 pipe.

6.51   Compute the resulting velocity of flow if 400 gal/min of fluid flows through a 4-in Schedule 40 pipe.

6.52   Repeat Problem 6.51 for a 4-in Schedule 80 pipe.

6.53   From the list of standard hydraulic steel tubing in Appendix G.2, select the smallest size that would carry 2.80 L/min of oil with a maximum velocity of 0.30 m/s.

6.54   A standard 6-in Schedule 40 steel pipe is carrying 95 gal/min of water. The pipe then branches into two standard 3-in pipes. If the flow divides evenly between the branches, calculate the velocity of flow in all three pipes.

For problems 6.55–6.57, use Fig. 6.3 to specify suitable Schedule 40 pipe sizes for carrying the given volume flow rate of water in the suction line and in the discharge line of a pumped distribution system. Select the pipe sizes both above and below the curve for the given flow rate and then calculate the actual velocity of flow in each.

6.55   Use $Q = 800$ gal/min.

6.56   Use $Q = 2000$ gal/min.

6.57   Use $Q = 60$ m^3/h.

6.58   A venturi meter is a device that uses a constriction in a flow system to measure the velocity of flow. Figure 6.17 illustrates one type of design. If the main pipe section is a standard hydraulic copper tube having a 100-mm outside diameter $\times$ 3.5-mm-wall thickness, compute the volume flow rate when the velocity there is 3.0 m/s. Then, for that volume flow rate, specify the required size of the throat section that would make the velocity there at least 15.0 m/s.

6.59   A flow nozzle, shown in Fig. 6.18, is used to measure the velocity of flow. If the nozzle is installed inside a 14-in Schedule 40 pipe and the nozzle diameter is 4.60 in, compute the velocity of flow at section 1 and the throat of the nozzle at section 2 when 7.50 ft^3/s of water flows through the system.

FIGURE 6.17   Venturi meter for Problem 6.58.

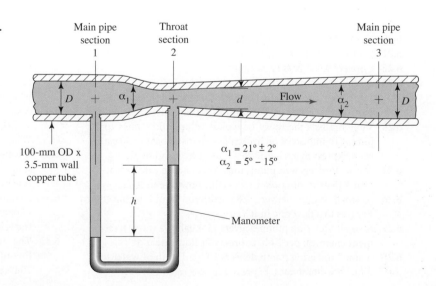

FIGURE 6.18 Nozzle meter for Problem 6.59.

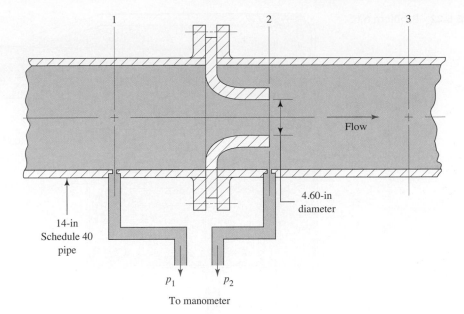

FIGURE 6.19 Problem 6.60.

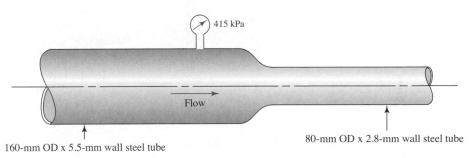

Note: For all remaining problems, assume that energy losses are zero. Systems with energy losses are covered in Chapters 7–13.

Bernoulli's Equation

6.60 Gasoline (sg = 0.67) is flowing at 0.11 m³/s in the fabricated tube shown in Fig. 6.19. If the pressure before the contraction is 415 kPa, calculate the pressure in the smaller tube.

6.61 Water at 10°C is flowing from point A to point B through the fabricated section shown in Fig. 6.20 at the rate of 0.37 m³/s. If the pressure at A is 66.2 kPa, calculate the pressure at B.

6.62 Calculate the volume flow rate of water at 5°C through the system shown in Fig. 6.21.

6.63 Calculate the pressure required in the larger section just ahead of the nozzle in Fig. 6.22 to produce a jet velocity of 75 ft/s. The fluid is water at 180°F.

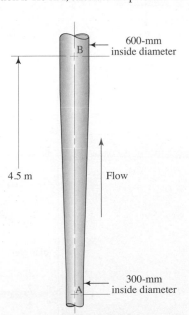

FIGURE 6.20 Problem 6.61.

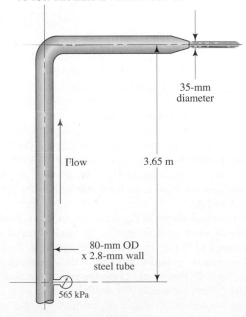

FIGURE 6.21 Problem 6.62.

FIGURE 6.22 Problem 6.63.

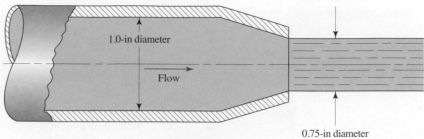

FIGURE 6.23 Problem 6.65.

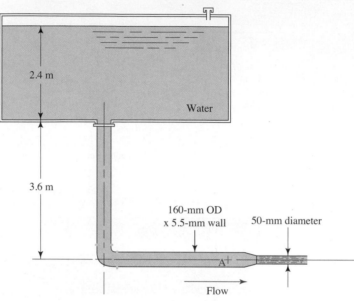

FIGURE 6.24 Problem 6.66.

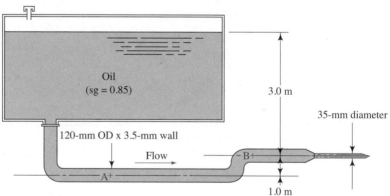

6.64 Kerosene with a specific weight of 50.0 lb/ft³ is flowing at 10 gal/min from a standard 1-in Schedule 40 steel pipe to a standard 2-in Schedule 40 steel pipe. Calculate the difference in pressure in the two pipes.

6.65 For the system shown in Fig. 6.23, calculate (a) the volume flow rate of water from the nozzle and (b) the pressure at point A.

6.66 For the system shown in Fig. 6.24, calculate (a) the volume flow rate of oil from the nozzle and (b) the pressures at A and B.

6.67 For the tank shown in Fig. 6.25, calculate the volume flow rate of water from the nozzle. The tank is sealed with a pressure of 20 psig above the water. The depth h is 8 ft.

6.68 Calculate the pressure of the air in the sealed tank shown in Fig. 6.25 that would cause the velocity of flow to be 20 ft/s from the nozzle. The depth h is 10 ft.

6.69 For the siphon in Fig. 6.26, calculate (a) the volume flow rate of water through the nozzle and (b) the pressure at points A and B. The distances X = 4.6 m and Y = 0.90 m.

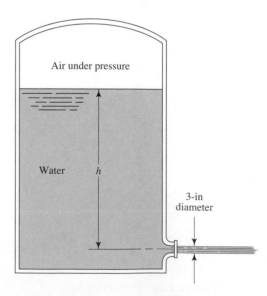

FIGURE 6.25 Problems 6.67 and 6.68.

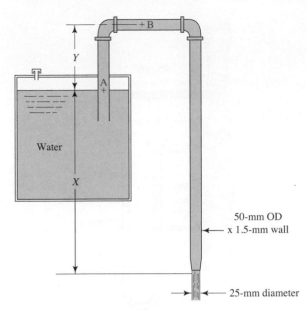

FIGURE 6.26 Problems 6.69, 6.70, and 6.71.

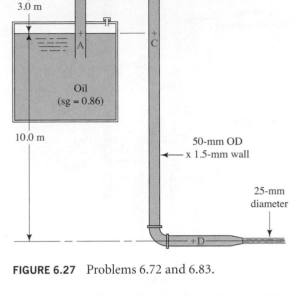

FIGURE 6.27 Problems 6.72 and 6.83.

6.70 For the siphon in Fig. 6.26, calculate the distance X required to obtain a volume flow rate of $7.1 \times 10^{-3} \, \mathrm{m^3/s}$.

6.71 For the siphon in Fig. 6.26, assume that the volume flow rate is $5.6 \times 10^{-3} \, \mathrm{m^3/s}$. Determine the maximum allowable distance Y if the minimum allowable pressure in the system is -18 kPa (gage).

6.72 For the siphon shown in Fig. 6.27, calculate (a) the volume flow rate of oil from the tank and (b) the pressures at points A, B, C, and D.

6.73 For the special fabricated reducer shown in Fig. 6.28, the pressure at A is 50.0 psig and the pressure at B is 42.0 psig. Calculate the velocity of flow of water at point B.

6.74 In the fabricated enlargement shown in Fig. 6.29, the pressure at A is 25.6 psig and the pressure at B is 28.2 psig. Calculate the volume flow rate of oil (sg = 0.90).

6.75 Figure 6.30 shows a manometer being used to indicate the pressure difference between two points in a fabricated system. Calculate the volume flow rate of water in the system if the manometer deflection h is 250 mm. (This arrangement is called a venturi meter, which is often used for flow measurement.)

6.76 For the venturi meter shown in Fig. 6.30, calculate the manometer deflection h if the velocity of flow of water in the 25-mm-diameter section is 10 m/s.

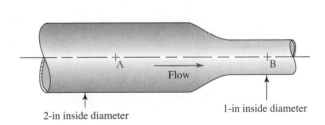

FIGURE 6.28 Problems 6.73 and 6.84.

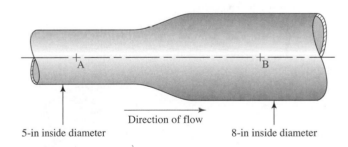

FIGURE 6.29 Problem 6.74.

FIGURE 6.30 Problems 6.75 and 6.76.

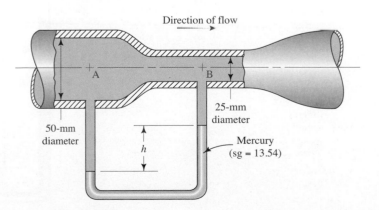

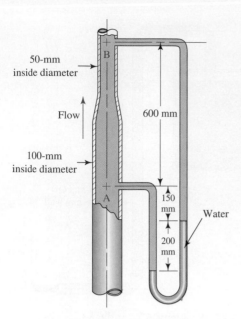

FIGURE 6.31 Problem 6.77.

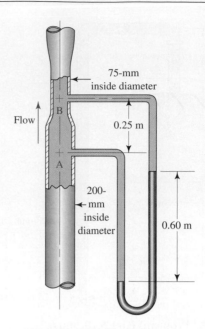

FIGURE 6.32 Problem 6.78.

6.77 Oil with a specific weight of 8.64 kN/m³ flows from A to B through the special fabricated system shown in Fig. 6.31. Calculate the volume flow rate of oil.

6.78 The venturi meter shown in Fig. 6.32 carries oil (sg = 0.90). The specific gravity of the gage fluid in the manometer is 1.40. Calculate the volume flow rate of oil.

6.79 Oil with a specific gravity of 0.90 is flowing downward through the venturi meter shown in Fig. 6.33. If the manometer deflection h is 28 in, calculate the volume flow rate of oil.

6.80 Oil with a specific gravity of 0.90 is flowing downward through the venturi meter shown in Fig. 6.33. If the velocity of flow in the 2-in-diameter section is 10.0 ft/s, calculate the deflection h of the manometer.

6.81 Gasoline (sg = 0.67) is flowing at 4.0 ft³/s in the fabricated reducer shown in Fig. 6.34. If the pressure before the reduction is 60 psig, calculate the pressure in the 3-in-diameter section.

6.82 Oil with a specific weight of 55.0 lb/ft³ flows from A to B through the system shown in Fig. 6.35. Calculate the volume flow rate of the oil.

FIGURE 6.33 Problems 6.79 and 6.80.

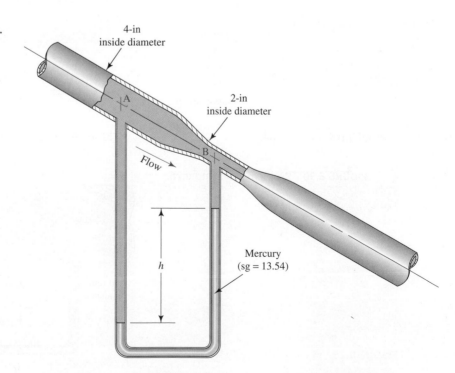

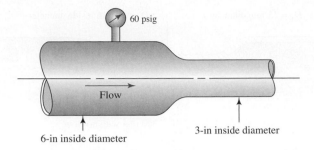

FIGURE 6.34 Problem 6.81.

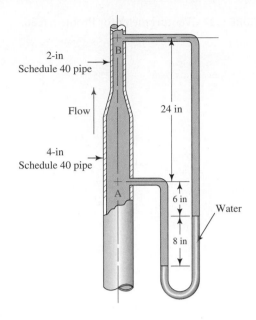

FIGURE 6.35 Problem 6.82.

6.83 Draw a plot of elevation head, pressure head, velocity head, and total head for the siphon system shown in Fig. 6.27 and analyzed in Problem 6.72.

6.84 Draw a plot of elevation head, pressure head, velocity head, and total head for the fabricated reducer shown in Fig. 6.28 and analyzed in Problem 6.73.

6.85 Figure 6.36 shows a system in which water flows from a tank through a pipe system having several sizes and elevations. For points A–G, compute the elevation head, the pressure head, the velocity head, and the total head. Plot these values on a sketch similar to that shown in Fig. 6.7.

6.86 Figure 6.37 shows a venturi meter with a U-tube manometer to measure the velocity of flow. When no flow occurs, the mercury column is balanced and its top is 300 mm below the throat. Compute the volume flow rate through the meter that will cause the mercury to flow into the throat. Note that for a given manometer deflection h, the left side will move down $h/2$ and the right side would rise $h/2$.

6.87 For the tank shown in Fig. 6.38, compute the velocity of flow from the outlet nozzle at varying depths from 10.0 ft

to 2.0 ft in 2.0-ft increments. Then, use increments of 0.5 ft to zero. Plot the velocity versus depth.

6.88 What depth of fluid above the outlet nozzle is required to deliver 200 gal/min of water from the tank shown in Fig. 6.37? The nozzle has a 3-in diameter.

Torricelli's Theorem

6.89 Derive Torricelli's theorem for the velocity of flow from a tank through an orifice opening into the atmosphere under a given depth of fluid.

6.90 Solve Problem 6.88 using the direct application of Torricelli's theorem.

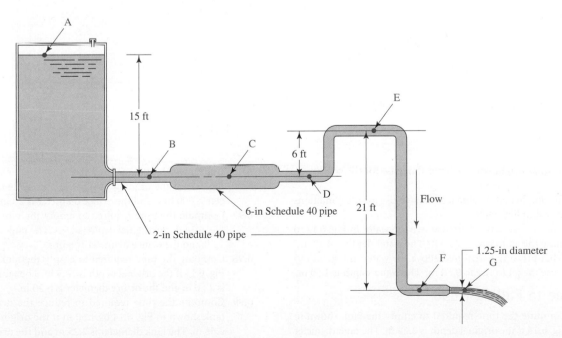

FIGURE 6.36 Flow system for Problem 6.85.

FIGURE 6.37   Venturi meter for Problem 6.86.

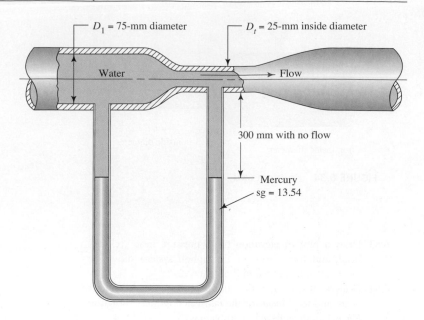

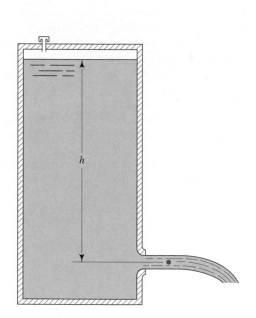

FIGURE 6.38   Tank for Problems 6.87–6.88.

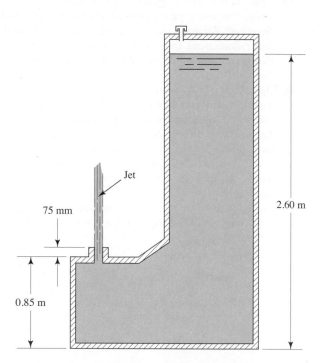

FIGURE 6.39   Problem 6.91.

6.91 To what height will the jet of fluid rise for the conditions shown in Fig. 6.39?

6.92 To what height will the jet of water rise for the conditions shown in Fig. 6.40?

6.93 What pressure is required above the water in Fig. 6.12 to cause the jet to rise to 28.0 ft? The water depth is 4.50 ft.

6.94 What pressure is required above the water in Fig. 6.13 to cause the jet to rise to 9.50 m? The water depth is 1.50 m.

Flow Due to Falling Head

6.95 Compute the time required to empty the tank shown in Fig. 6.14 if the original depth is 2.68 m. The tank diameter is 3.00 m and the orifice diameter is 150 mm.

6.96 Compute the time required to empty the tank shown in Fig. 6.14 if the original depth is 55 mm. The tank diameter is 300 mm and the orifice diameter is 20 mm.

6.97 Compute the time required to empty the tank shown in Fig. 6.14 if the original depth is 15 ft. The tank diameter is 12 ft and the orifice diameter is 6 in.

6.98 Compute the time required to empty the tank shown in Fig. 6.14 if the original depth is 18.5 in. The tank diameter is 22.0 in and the orifice diameter is 0.50 in.

6.99 Compute the time required to reduce the depth in the tank shown in Fig. 6.14 by 1.50 m if the original depth is 2.68 m. The tank diameter is 2.25 m and the orifice diameter is 50 mm.

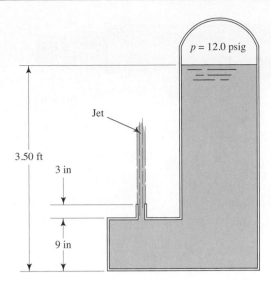

$p = 12.0$ psig

Jet

3.50 ft

3 in

9 in

FIGURE 6.40 Problem 6.92.

6.100 Compute the time required to reduce the depth in the tank shown in Fig. 6.14 by 225 mm if the original depth is 1.38 m. The tank diameter is 1.25 m and the orifice diameter is 25 mm.

6.101 Compute the time required to reduce the depth in the tank shown in Fig. 6.14 by 12.5 in if the original depth is 38 in. The tank diameter is 6.25 ft and the orifice diameter is 0.625 in.

6.102 Compute the time required to reduce the depth in the tank shown in Fig. 6.14 by 21.0 ft if the original depth is 23.0 ft. The tank diameter is 46.5 ft and the orifice diameter is 8.75 in.

6.103 Repeat Problem 6.97 if the tank is sealed and a pressure of 5.0 psig is above the water in the tank.

6.104 Repeat Problem 6.101 if the tank is sealed and a pressure of 2.8 psig is above the water in the tank.

6.105 Repeat Problem 6.96 if the tank is sealed and a pressure of 20 kPa(gage) is above the water in the tank.

6.106 Repeat Problem 6.100 if the tank is sealed and a pressure of 35 kPa(gage) is above the water in the tank.

Supplemental Problems

6.107 A village currently carries water by hand from a lake that is 1200 m from the village center. It is later determined that the surface of the lake is 3 m above the elevation of the village, so someone began to wonder if a simple plumbing line could deliver the water. If a flexible plastic line with a 20-mm inside diameter could be installed from the lake to the village, what theoretical flow rate is possible, ignoring all losses?

6.108 A "spa tub" is to be designed to replace bath tubs in renovations. There are to be 6 outlet nozzles, each with a diameter of 12 mm, and each should have an outlet velocity of 12 m/s. What is the required flow rate from the single pump that supplies all of these nozzles? If there is one suction line leading to the pump, what is the minimum diameter to limit the velocity at the inlet of the pump to 2.5 m/s?

6.109 A simple soft drink system relies on pressurized CO_2 to force the soft drink (sg = 1.08) from its tank sitting on the floor up to the outlet where cups are filled. Determine the required CO_2 pressure to allow a 16 oz cup to be filled in 6 s, when the beverage tank is nearly empty, given Fig. 6.41.

6.110 A concept team for a toy company is considering a new squirt gun. They have an idea for one that could shoot a vertical stream to a height 7 m from a 5-mm-diameter nozzle. People like squirt guns that shoot for a long time, but also do not like water tanks that are too big or heavy. If the tank of this squirt gun can hold 3 L, how long can the squirt gun shoot?

6.111 Bernoulli's principle applies to Venturi tubes that are used in many practical devices such as "air brush" painters, vacuum systems, carburetors, water bed drains and many other devices. One such system used to spray fertilizer is

FIGURE 6.41 Problem 6.109.

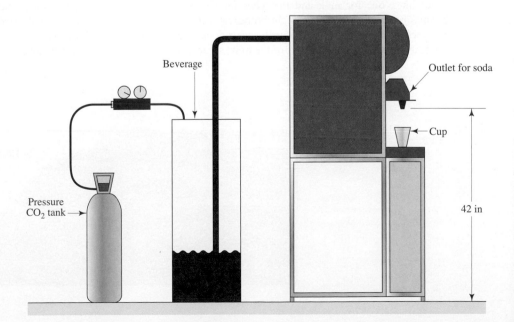

Beverage

Outlet for soda

Cup

Pressure
CO_2 tank

42 in

FIGURE 6.42 Problem 6.111.

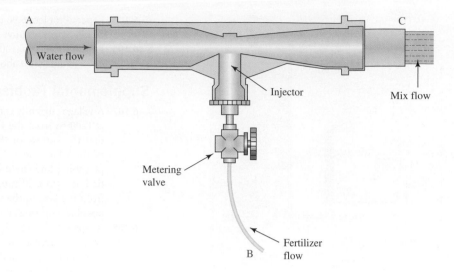

shown in Fig. 6.42. Port A is connected to a water supply that is directed through the venturi. At the throat of the Venturi, Port B is connected to the supply of fertilizer concentrate from a container below. Port C is the spray nozzle that directs the diluted fertilizer solution out to the plants. Port A, 10 mm in diameter, is connected to a water supply that reads 180 kPa while flowing at 12 L/min. Determine the vacuum pressure in the 3.5-mm-diameter throat if the metering valve is completely closed. Explain what will happen to the fertilizer concentrate in the container below as the metering valve is opened.

6.112 A decorative fountain for a corporate world headquarters is to be designed to shoot a stream of water straight up in the air. If the designers would like the fountain to reach at least 50 ft into the air, what pressure must exist at the nozzle inlet? The nozzle has an inlet diameter of 5.0 in and an outlet diameter of 2.0 in.

6.113 You are to develop a mixing valve for use in a dairy processing facility. The rated output of the valve is to be 10 gal/min of chocolate milk. There will be two separate input lines, one for milk and the other for chocolate syrup. Your valve is to ensure that the proper ratio of milk to chocolate syrup is 16:1. As a start for your design, determine the minimum diameters of the milk, syrup, and

chocolate milk fittings if they are to limit the velocity in each line to a maximum of 8.0 ft/s.

6.114 While maneuvering at the scene of a fire, a truck accidently backs over a fire hydrant and breaks it. The diameter of the water line to the hydrant is 6 in, but due to internal plumbing, the effective diameter at the water outlet is 4 in. If the flow rate of the water leaving the hydrant is 1000 gal/min, what height will the water reach?

6.115 You would like to empty the in-ground pool in the back yard but the drain at the bottom of the pool is no longer functional. Given the dimensions in Fig. 6.43, determine the flow rate from the pool at the instant shown if the hose has an inside diameter of 0.5 in. What had to happen to initiate flow from the drain hose? What will happen to the flow rate as the level of the pool drops?

6.116 A pressure washer available to home owners lists 1300 psi and 2 gpm among its specifications. We know, however, that the actual pressure of the water is atmospheric (0 gage) once it exits the nozzle. The key feature of the so-called pressure washer then is actually the velocity with which it exits the nozzle. Neglecting any losses, what would be the velocity of the stream from this machine if it achieves the specified flow rate through an outlet nozzle having a diameter of 0.062 in?

FIGURE 6.43 Problem 6.115.

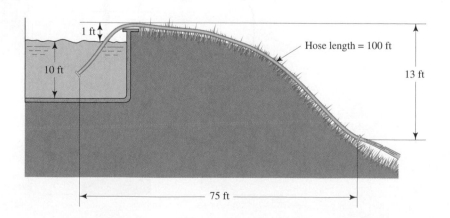

ANALYSIS PROJECTS USING BERNOULLI'S EQUATION AND TORRICELLI'S THEOREM

1. Create a spreadsheet for computing the values of the pressure head, the velocity head, the elevation head, and the total head for given values of pressure, velocity, and elevation.

2. Enhance the spreadsheet in Project 1 by causing it to list side by side in several combinations the various head components in order to compare one with another as done when using Bernoulli's equation.

3. In the spreadsheet in Project 1, include the ability to compute the velocity of flow from given data for volume flow rate and pipe size.

4. Create a spreadsheet for computing, using Eq. (6–26), the time required to decrease the fluid level in a tank between two values for any combination of tank size and nozzle diameter. Apply it to Problems 6.95–6.102.

5. Add the ability to pressurize the system to the spreadsheet in Project 4. Apply it to Problems 6.103–6.106.

6. Create a spreadsheet for computing the velocity of flow from an orifice, using Torricelli's theorem for any depth of fluid and any amount of pressure above the fluid. Apply it to Problems 6.90–6.94.

GENERAL ENERGY EQUATION

You will now expand your ability to analyze the energy in fluid flow systems by adding terms to Bernoulli's equation which was introduced in Chapter 6. You will account for a variety of forms of energy that were neglected before, such as:

- Energy lost from a system through friction as the fluid flows through pipes
- Energy lost as the fluid flows through valves, or fittings where the fluid must travel in complex paths, accelerate or decelerate, or change direction
- Energy added to the system by a pump as it provides the impetus for the fluid to move and increases the fluid pressure
- Energy removed from the system by fluid motors or turbines that use the energy to drive other mechanical systems.

Adding these terms to Bernoulli's equation eliminates many of the restrictions that were identified in Section 6.7 and transforms it into the *general energy equation* that you will apply as you study Chapters 7–13.

As an example of a system where energy losses and additions occur, refer now to Fig. 7.1 showing a portion of an industrial fluid distribution system. The fluid enters from the left, where the suction line draws fluid from a storage tank. The inline pump adds energy to the fluid and causes it to flow into the discharge line and then through the rest of the piping system. Note that the suction pipe is larger than the discharge pipe. If the sizes of the pump suction inlet and the discharge ports provided by the pump manufacturer are different from the pipe sizes, a gradual reducer or a gradual enlargement may be needed. This is a common occurrence. The fluid then passes straight through the run of a tee, where a valve in the branch line can be opened to draw some fluid off to another destination point. After leaving the tee, the fluid passes through a valve that can be used to shut off the discharge line. Just downstream from the valve is another tee where now the fluid takes the branch path, passes around a 90° elbow, and passes through another valve. The discharge pipe beyond the valve is insulated and the fluid travels through the long, straight pipe line to its ultimate destination.

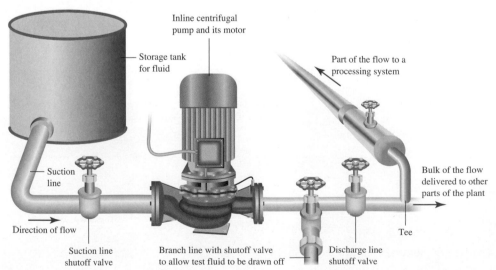

FIGURE 7.1 In systems like this typical industrial pipeline installation, showing a pump, valves, tees, and other fittings, you must use the general energy equation to analyze its performance.

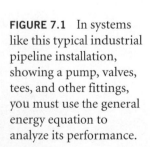

Each valve, tee, elbow, reducer, and enlargement causes energy to be lost from the fluid. In addition, as the fluid flows through straight lengths of pipe, energy is lost due to friction. Your task might be to design the system, specify the sizes of the pipes and the kinds of valves and fittings, analyze the pressure at various points within the system, determine the demands placed on the pump, and specify a suitable pump for the system. The information in Chapters 7–13 gives you the tools to accomplish these tasks. In this chapter you will learn how to analyze the changes in energy that occur throughout the system, the corresponding changes in pressure, the power delivered by a pump to the fluid, and the efficiency of the pump. For systems that employ a fluid motor or a turbine, you will learn how to analyze the energy removed from the fluid, the power delivered to the fluid motor or turbine and their efficiency.

Exploration

Think again about the fluid flow systems discussed in the Big Picture section of Chapter 6. Perhaps you considered the water distribution system in your home, a lawn sprinkling system, the piping for a fluid power system, or fluid distribution systems in a manufacturing industry. Try to answer the following questions about each system:

- In what ways do those systems include energy losses caused by valves or other flow control devices?

- How does the fluid make changes in direction as it travels through the system?

- Are there places where the size of the flow path changes, either getting smaller or larger?

- Do some of the systems include pumps to deliver the energy that causes flow and increases the pressure in the fluid?

- Is there a fluid motor or a turbine that extracts energy from the fluid to drive a shaft to do work?

Note also that there will always be energy losses as the fluid flows through the straight pipes and tubes that cause the pressure to drop.

Introductory Concepts

You should now have a basic understanding of how to analyze fluid flow systems from your work in Chapter 6.

You should be able to compute volume flow rate, weight flow rate, and mass flow rate. You should be comfortable with various uses of the principle of continuity, which states that the mass flow rate is the same throughout a steady flow system. We use the following form of the continuity equation involving volume flow rate most often when liquids are flowing in the system:

$$Q_1 = Q_2$$

Because $Q = Av$, we can write this as

$$A_1v_1 = A_2v_2$$

These relationships allow us to determine the velocity of flow at any point in a system if we know the volume flow rate and the areas of the pipes at the sections of interest.

You should also be familiar with the terms that express the energy possessed by a fluid per unit weight of the fluid flowing in the system:

p/γ is the pressure head.

z is the elevation head.

$v^2/2g$ is the velocity head.

The sum of these three terms is called the total head. All of this comes together in Bernoulli's equation,

$$\frac{p_1}{\gamma} + z_1 + \frac{v_1^2}{2g} = \frac{p_2}{\gamma} + z_2 + \frac{v_2^2}{2g}$$

where the subscripts 1 and 2 refer to two different points of interest in the fluid flow system. However, as you learned in Section 6.7, there are several restrictions on the use of Bernoulli's equation.

1. It is valid only for incompressible fluids.
2. There can be no mechanical devices such as pumps, fluid motors, or turbines between the two sections of interest.
3. There can be no energy lost due to friction or to the turbulence created by valves and fittings in the flow system.
4. There can be no heat transferred into or out of the fluid.

In reality, no system satisfies all these restrictions so now we develop the general energy equation, adding terms for energy losses of all kinds, energy additions due to pumps, and energy removals due to fluid motors or turbines.

7.1 OBJECTIVES

After completing this chapter, you should be able to:

1. Identify the conditions under which energy losses occur in fluid flow systems.

2. Identify the means by which energy can be added to a fluid flow system.

3. Identify the means by which energy can be removed from a fluid flow system.

4. Expand Bernoulli's equation to form the general energy equation by considering energy losses, energy additions, and energy removals and apply it to a variety of practical problems.

FIGURE 7.2 Gear pump.
(a) Cutaway view of pump
(*Source:* Danfoss Power Solutions,
Ames, IA); (b) Sketch of meshing
gears and flow path of the fluid.
(*Source:* Machine Design Magazine)

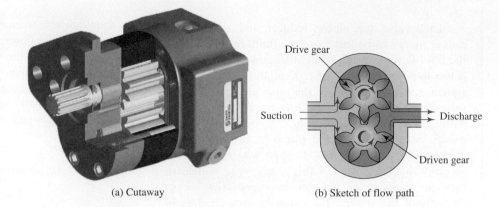

(a) Cutaway   (b) Sketch of flow path

5. Compute the power added to a fluid by pumps, the power required to drive the pumps, and the efficiency of the pumps.

6. Compute the power delivered by a fluid to a fluid motor, the power actually used by the motor to drive a mechanical system, and the efficiency of the fluid motor.

7.2 ENERGY LOSSES AND ADDITIONS

The objective of this section is to describe, in general terms, the various types of devices and components of fluid flow systems. They occur in most fluid flow systems and they either add energy to the fluid, remove energy from the fluid, or cause undesirable losses of energy from the fluid.

At this time we are only describing these devices in conceptual terms. We discuss pumps, fluid motors, friction losses as fluid flows in pipes and tubes, energy losses from changes in the size of the flow path, and energy losses from valves and fittings.

In later chapters, you will learn in more detail about how to compute the amount of energy losses in pipes and specific types of valves and fittings. You will learn the method of using performance curves for pumps to apply them properly.

7.2.1 Pumps

A pump is a common example of a mechanical device that adds energy to a fluid. An electric motor or some other prime power device drives a rotating shaft in the pump. The pump then takes this kinetic energy and delivers it to the fluid, resulting in fluid flow and increased fluid pressure.

Many configurations are used in pump designs. The system shown in Fig. 7.1 contains a centrifugal pump mounted inline with the process piping. Figures 7.2 and 7.3 show two types of fluid power pumps capable of producing very high pressures in the range of 1500 to 5000 psi (10.3–34.5 MPa). Chapter 13 extensively discusses these and several other styles of pumps along with their selection and application.

7.2.2 Fluid Motors

Fluid motors, turbines, rotary actuators, and linear actuators are examples of devices that take energy from a fluid and deliver it in the form of work, causing the rotation of a shaft or the linear movement of a piston.

Many fluid motors have the same basic configurations as the pumps shown in Figs. 7.2 and 7.3. The major difference between a pump and a fluid motor is that, when acting as a motor, the fluid drives the rotating elements of the device. The reverse is true for pumps. For some designs, such as the gear-on-gear type shown in Fig. 7.2, a pump could act as a motor by forcing a flow of fluid through the device. In other types, a change in the valve arrangement or in the configuration of the rotating elements would be required.

Series 90 Variable Pump

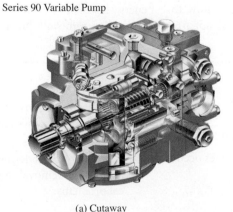

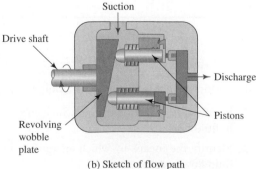

FIGURE 7.3 Piston pump.
(a) Cutaway view of pump
(*Source:* Danfoss Power Solutions,
Ames, IA); (b) Sketch of cross
section of pump and flow path
of the fluid. (*Source:* Machine
Design Magazine)

(a) Cutaway   (b) Sketch of flow path

FIGURE 7.4 Hydraulic motor. (a) Cutaway view of motor (*Source:* Danfoss Power Solutions, Ames, IA); (b) Rotor and internal gear. (*Source:* Machine Design Magazine)

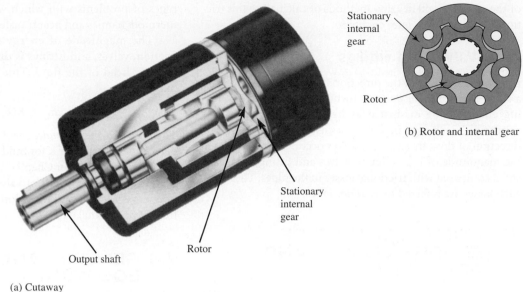

(b) Rotor and internal gear

(a) Cutaway

The hydraulic motor shown in Fig. 7.4 is often used as a drive for the wheels of construction equipment and trucks and for rotating components of material transfer systems, conveyors, agricultural equipment, special machines, and automation equipment. The design incorporates a stationary internal gear with a special shape. The rotating component is like an external gear, sometimes called a *gerotor* that has one fewer teeth than the internal gear. The external gear rotates in a circular orbit around the center of the internal gear. High-pressure fluid entering the cavity between the two gears acts on the rotor and develops a torque that rotates the output shaft. The magnitude of the output torque depends on the pressure difference between the input and output sides of the rotating gear. The speed of rotation is a function of the displacement of the motor (volume per revolution) and the volume flow rate of fluid through the motor.

Figure 7.5 is a photograph of a cutaway model of a fluid power cylinder or linear actuator.

7.2.3 Fluid Friction

A fluid in motion offers frictional resistance to flow. Part of the energy in the system is converted into *thermal energy* (heat), which is dissipated through the walls of the pipe in which the fluid is flowing. The magnitude of the energy loss is dependent on the properties of the fluid, the flow velocity, the pipe size, the smoothness of the pipe wall, and the length

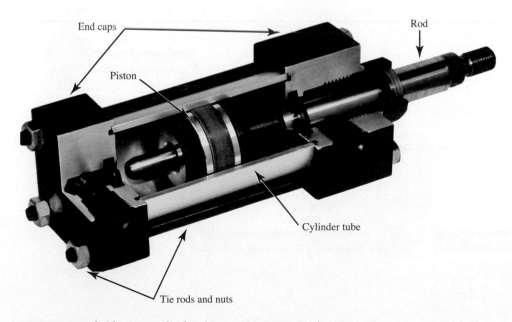

FIGURE 7.5 Fluid power cylinder. (*Source:* Norgren, Inc.)

of the pipe. We will develop methods of calculating this frictional energy loss in later chapters.

7.2.4 Valves and Fittings

Elements that control the direction or flow rate of a fluid in a system typically set up local turbulence in the fluid, causing energy to be dissipated as heat. Whenever there is a restriction, a change in flow velocity, or a change in the direction of flow, these energy losses occur. In a large system the magnitude of losses due to valves and fittings is usually small compared with frictional losses in the pipes. Therefore, such losses are referred to as *minor losses*.

7.3 NOMENCLATURE OF ENERGY LOSSES AND ADDITIONS

We will account for energy losses and additions in a system in terms of energy per unit weight of fluid flowing in the system. This is also known as "head," as described in Chapter 6. As an abbreviation for head we will use the symbol h for energy losses and additions. Specifically, we will use the following terms throughout the next several chapters:

h_A = *Energy added* to the fluid with a mechanical device such as a pump; this is often referred to as the *total head* on the pump

h_R = *Energy removed* from the fluid by a mechanical device such as a fluid motor

h_L = *Energy losses* from the system due to friction in pipes or minor losses due to valves and fittings

We will not consider the effects of heat transferred into or out of the fluid at this time because they are negligible in the types of problems with which we are dealing. Courses in thermodynamics and heat transfer cover heat energy.

The magnitude of energy losses produced by fluid friction, valves, and fittings is directly proportional to the velocity head of the fluid. This can be expressed mathematically as

$$h_L = K(v^2/2g)$$

The term K is the *resistance coefficient*. You will learn how to determine the value of K for fluid friction in Chapter 8 using the Darcy equation. In Chapter 10, you will see methods of finding K for many kinds of valves, fittings, and changes in flow cross-section and direction. Most of these are found from experimental data.

7.4 GENERAL ENERGY EQUATION

The general energy equation as used in this text is an expansion of Bernoulli's equation, which makes it possible to solve problems in which energy losses and additions occur. The logical interpretation of the energy equation can be seen in Fig. 7.6, which represents a flow system. The terms E_1' and E_2' denote the energy possessed by the fluid per unit weight at sections 1 and 2, respectively. The respective energy additions, removals, and losses h_A, h_R, and h_L are shown. For such a system the expression of the principle of conservation of energy is

$$E_1' + h_A - h_R - h_L = E_2' \tag{7–1}$$

The energy possessed by the fluid per unit weight is

$$E' = \frac{p}{\gamma} + z + \frac{v^2}{2g} \tag{7–2}$$

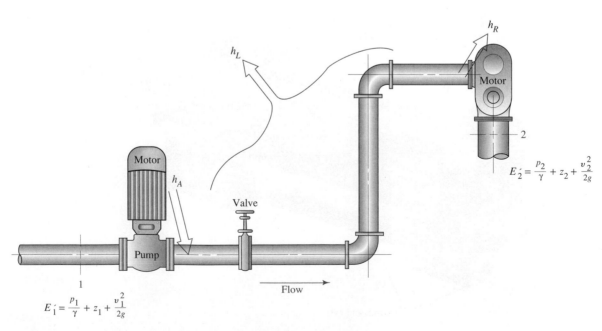

FIGURE 7.6 Fluid flow system illustrating the general energy equation.

Equation (7–1) then becomes

⮞ **General Energy Equation**

$$\frac{p_1}{\gamma} + z_1 + \frac{v_1^2}{2g} + h_A - h_R - h_L = \frac{p_2}{\gamma} + z_2 + \frac{v_2^2}{2g} \qquad (7\text{–}3)$$

This is the form of the energy equation that we will use most often in this book. As with Bernoulli's equation, each term in Eq. (7–3) represents a quantity of energy per unit weight of fluid flowing in the system. Typical SI units are N·m/N, or meters. U.S. Customary System units are lb-ft/lb, or feet.

It is essential that the general energy equation be written *in the direction of flow*, that is, *from* the reference point on the left side of the equation *to* that on the right side. Algebraic signs are critical because the left side of Eq. (7–3) states that an element of fluid having a certain amount of energy per unit weight at section 1 may have energy added $(+h_A)$, energy removed $(-h_R)$, or energy lost $(-h_L)$ from it before it reaches section 2. There it contains a different amount of energy per unit weight, as indicated by the terms on the right side of the equation.

For example, in Fig. 7.6, reference points are shown to be points 1 and 2 with the pressure head, elevation head, and velocity head indicated at each point. After the fluid leaves point 1 it enters the pump, where energy is added. A prime mover such as an electric motor drives the pump, and the impeller of the pump transfers the energy to the fluid $(+h_A)$. Then the fluid flows through a piping system composed of a valve, elbows, and the lengths of pipe, in which energy is dissipated from the fluid and is lost $(-h_L)$. Before reaching point 2, the fluid flows through a fluid motor, which removes some of the energy to drive an external device $(-h_R)$. The general energy equation accounts for all of these energies.

In a particular problem, it is possible that not all of the terms in the general energy equation will be required. For example, if there is no mechanical device between the sections of interest, the terms h_A and h_R will be zero and can be left out of the equation. If energy losses are so small that they can be neglected, the term h_L can be left out. If both of these conditions exist, it can be seen that Eq. (7–3) reduces to Bernoulli's equation.

PROGRAMMED EXAMPLE PROBLEMS

Example Problem 7.1

Water flows from a large reservoir at the rate of 1.20 ft³/s through a pipe system as shown in Fig. 7.7. Calculate the total amount of energy lost from the system because of the valve, the elbows, the pipe entrance, and fluid friction.

Using an approach similar to that used with Bernoulli's equation, select two sections of interest and write the general energy equation before looking at the next panel.

The sections at which we know the most information about pressure, velocity, and elevation are the surface of the reservoir and the free stream of fluid at the exit of the pipe. Call these section 1 and section 2, respectively. Then, the complete general energy equation [Eq. (7–3)] is

$$\frac{p_1}{\gamma} + z_1 + \frac{v_1^2}{2g} + h_A - h_R - h_L = \frac{p_2}{\gamma} + z_2 + \frac{v_2^2}{2g}$$

FIGURE 7.7 Pipe system for Example Problem 7.1.

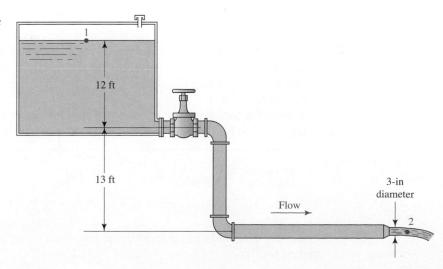

The value of some of these terms is zero. Determine which are zero and simplify the energy equation accordingly.

The value of the following terms is zero:

$$p_1 = 0 \quad \text{Surface of reservoir exposed to the atmosphere}$$
$$p_2 = 0 \quad \text{Free stream of fluid exposed to the atmosphere}$$
$$v_1 = 0 \quad \text{(Approximately) Surface area of reservoir is large}$$
$$h_A = h_R = 0 \quad \text{No mechanical device in the system}$$

Then, the energy equation becomes

$$\cancelto{0}{\frac{p_1}{\gamma}} + z_1 + \cancelto{0}{\frac{v_1^2}{2g}} + \cancelto{0}{h_A} - \cancelto{0}{h_R} - h_L = \cancelto{0}{\frac{p_2}{\gamma}} + z_2 + \frac{v_2^2}{2g}$$

$$z_1 - h_L = z_2 + v_2^2/2g$$

Because we are looking for the total energy lost from the system, solve this equation for h_L.

You should have

$$h_L = (z_1 - z_2) - v_2^2/2g$$

Now evaluate the terms on the right side of the equation to determine h_L in the units of lb-ft/lb.

The answer is $h_L = 15.75$ lb-ft/lb. Here is how it is found. First,

$$z_1 - z_2 = +25 \text{ ft}$$
$$v_2 = Q/A_2$$

Because Q was given as 1.20 ft³/s and the area of a 3-in-diameter jet is 0.0491 ft², we have

$$v_2 = \frac{Q}{A_2} = \frac{1.20 \text{ ft}^3}{s} \times \frac{1}{0.0491 \text{ ft}^2} = 24.4 \text{ ft/s}$$

$$\frac{v_2^2}{2g} = \frac{(24.4)^2 \text{ ft}^2}{s^2} \times \frac{s^2}{(2)(32.2) \text{ ft}} = 9.25 \text{ ft}$$

Then the total amount of energy lost from the system is

$$h_L = (z_1 - z_2) - v_2^2/2g = 25 \text{ ft} - 9.25 \text{ ft}$$
$$h_L = 15.75 \text{ ft, or } 15.75 \text{ lb-ft/lb}$$

Example Problem 7.2   The volume flow rate through the pump shown in Fig. 7.8 is 0.014 m³/s. The fluid being pumped is oil with a specific gravity of 0.86. Calculate the energy delivered by the pump to the oil per unit weight of oil flowing in the system. Energy losses in the system are caused by the check valve and friction losses as the fluid flows through the piping. The magnitude of such losses has been determined to be 1.86 N·m/N.

Using the sections where the pressure gages are located as the sections of interest, write the energy equation for the system, including only the necessary terms.

You should have

$$\frac{p_A}{\gamma} + z_A + \frac{v_A^2}{2g} + h_A - h_L = \frac{p_B}{\gamma} + z_B + \frac{v_B^2}{2g}$$

Notice that the term h_R has been left out of the general energy equation because no fluid motor is in the system.

FIGURE 7.8 Pump system for
Example Problem 7.2.

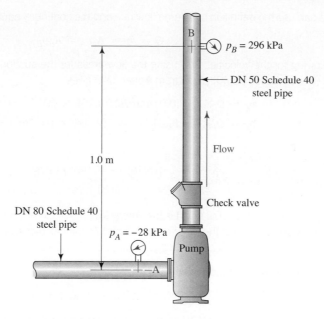

$p_B = 296$ kPa

DN 50 Schedule 40
steel pipe

Flow

1.0 m

DN 80 Schedule 40
steel pipe

$p_A = -28$ kPa

Check valve

Pump

The objective of the problem is to calculate the energy added to the oil by the pump. Solve for h_A before looking at the next panel.

One correct solution is

$$h_A = \frac{p_B - p_A}{\gamma} + (z_B - z_A) + \frac{v_B^2 - v_A^2}{2g} + h_L \qquad (7\text{--}4)$$

Notice that similar terms have been grouped. This will be convenient when performing the calculations.

Equation (7–4) should be studied well. It indicates that the total head on the pump h_A is a measure of all of the tasks the pump is required to do in a system. It must increase the pressure from that at point A at the inlet to the pump to the pressure at point B. It must raise the fluid by the amount of the elevation difference between points A and B. It must supply the energy to increase the velocity of the fluid from that in the larger pipe at the pump inlet (called the suction pipe) to the velocity in the smaller pipe at the pump outlet (called the discharge pipe). In addition, it must overcome any energy losses that occur in the system such as those due to the check valve and friction in the discharge pipe.

We recommend that you evaluate each of the terms in Eq. (7–4) separately and then combine them at the end. The first term is the difference between the pressure head at point A and that at point B. What is the value of γ?

Remember that the specific weight of the fluid being pumped must be used. In this case, the specific weight of the oil is

$$\gamma = (sg)(\gamma_w) = (0.86)(9.81 \text{ kN/m}^3) = 8.44 \text{ kN/m}^3$$

Now complete the evaluation of $(p_B - p_A)/\gamma$.

Because $p_B = 296$ kPa and $p_A = -28$ kPa, we have

$$\frac{p_B - p_A}{\gamma} = \frac{[296 - (-28)] \text{ kN}}{m^2} \times \frac{m^3}{8.44 \text{ kN}} = 38.4 \text{ m}$$

Now evaluate the elevation difference, $z_B - z_A$.

You should have $z_B - z_A = 1.0$ m. Notice that point B is at a higher elevation than point A and, therefore, $z_B > z_A$. The result is that $z_B - z_A$ is a positive number.
Now compute the velocity head difference term, $(v_B^2 - v_A^2)/2g$.

We can use the definition of volume flow rate and the continuity equation to determine each velocity:

$$Q = Av = A_A v_A = A_B v_B$$

Then, solving for the velocities and using the flow areas for the suction (DN 80 Schedule 40) and discharge (DN 50 Schedule 40) pipes from Appendix F gives

$$v_A = Q/A_A = (0.014 \text{ m}^3/\text{s})/(4.768 \times 10^{-3} \text{ m}^2) = 2.94 \text{ m/s}$$

$$v_B = Q/A_B = (0.014 \text{ m}^3/\text{s})/(2.168 \times 10^{-3} \text{ m}^2) = 6.46 \text{ m/s}$$

Finally,

$$\frac{v_B^2 - v_A^2}{2g} = \frac{[(6.46)^2 - (2.94)^2] \text{ m}^2/\text{s}^2}{2(9.81 \text{ m/s}^2)} = 1.69 \text{ m}$$

The only remaining term in Eq. (7–4) is the energy loss h_L, which is given to be 1.86 N·m/N, or 1.86 m. We can now combine all of these terms and complete the calculation of h_A.

The energy added to the system is

$$h_A = 38.4 \text{ m} + 1.0 \text{ m} + 1.69 \text{ m} + 1.86 \text{ m} = 42.9 \text{ m, or } 42.9 \text{ N·m/N}$$

That is, the pump delivers 42.9 N · m of energy to each newton of oil flowing through it. This completes the programmed instruction.

7.5 POWER REQUIRED BY PUMPS

Power is defined as the rate of doing work. In fluid mechanics we can modify this statement and consider that power is the rate at which energy is being transferred. We first develop the basic concept of power in SI units. Then we show the units for the U.S. Customary System.

The unit for energy in the SI system is joule (J) or N·m. The unit for power in the SI system is watt (W), which is equivalent to 1.0 N·m/s or 1.0 J/s.

In Example Problem 7.2 we found that the pump was delivering 42.9 N·m of energy to each newton of oil as it flowed through the pump. To calculate the power delivered to the oil, we must determine how many newtons of oil are flowing through the pump in a given amount of time. This is called the *weight flow rate W*, which we defined in Chapter 6, and is expressed in units of N/s. Power is calculated by multiplying the energy transferred per newton of fluid by the weight flow rate. This is

$$P_A = h_A W$$

Because $W = \gamma Q$, we can also write

▷ **Power Added to a Fluid by a Pump**

$$P_A = h_A \gamma Q \qquad (7\text{–}5)$$

where P_A denotes power added to the fluid, γ is the specific weight of the fluid flowing through the pump, and Q is the volume flow rate of the fluid.

By using the data of Example Problem 7.2, we can find the power delivered by the pump to the oil as follows:

$$P_A = h_A \gamma Q$$

We know from Example Problem 7.2 that

$$h_A = 42.9 \text{ N·m/N}$$
$$\gamma = 8.44 \text{ kN/m}^3 = 8.44 \times 10^3 \text{ N/m}^3$$
$$Q = 0.014 \text{ m}^3/\text{s}$$

Substituting these values into Eq. (7–5), we get

$$P_A = \frac{42.9 \text{ N·m}}{\text{N}} \times \frac{8.44 \times 10^3 \text{ N}}{\text{m}^3} \times \frac{0.014 \text{ m}^3}{\text{s}}$$

$$= 5069 \text{ N·m/s}$$

Because 1.0 W = 1.0 N·m/s, we can express the result in watts:

$$P_A = 5069 \text{ W} = 5.07 \text{ kW}$$

7.5.1 Power in the U.S. Customary System

The standard unit for energy in the U.S. Customary System is the lb-ft. The unit for power is lb-ft/s. Because it is common practice to refer to power in horsepower (hp), the conversion factor required is

$$1 \text{ hp} = 550 \text{ lb-ft/s}$$

In Eq. (7–5) the energy added h_A is expressed in feet of the fluid flowing in the system. Then, expressing the specific weight of the fluid in lb/ft³ and the volume flow rate in ft³/s would yield the weight flow rate γQ in lb/s. Finally, in the power equation $P_A = h_A \gamma Q$, power would be expressed in lb-ft/s.

To convert these units to the SI system we use the factors

$$1 \text{ lb-ft/s} = 1.356 \text{ W}$$
$$1 \text{ hp} = 745.7 \text{ W}$$

7.5.2 Mechanical Efficiency of Pumps

The term *efficiency* is used to denote the ratio of the power delivered by the pump to the fluid to the power supplied to the pump. Because of energy losses due to mechanical friction in pump components, fluid friction in the pump, and excessive fluid turbulence in the pump, not all of the input power is delivered to the fluid. Then, using the symbol e_M for mechanical efficiency, we have

▷ **Pump Efficiency**

$$e_M = \frac{\text{Power delivered to fluid}}{\text{Power put into pump}} = \frac{P_A}{P_I} \qquad (7\text{--}6)$$

The value of e_M will always be less than 1.0.

Continuing with the data of Example Problem 7.2, we could calculate the power input to the pump if e_M is known. For commercially available pumps the value of e_M is published as part of the performance data. If we assume that for the pump in this problem the efficiency is 82 percent, then

$$P_I = P_A/e_M = 5.07/0.82 = 6.18 \text{ kW}$$

The value of the mechanical efficiency of pumps depends not only on the design of the pump, but also on the conditions under which it is operating, particularly the total head and the flow rate. For pumps used in hydraulic systems, such as those shown in Figs. 7.2 and 7.3, efficiencies range from about 70 percent to 90 percent. For centrifugal pumps used primarily to transfer or circulate liquids, the efficiencies range from about 50 percent to 85 percent.

See Chapter 13 for more data and discussion of pump performance. Efficiency values for positive-displacement fluid power pumps are reported differently from those for centrifugal pumps. The values often used are: *overall efficiency e_o,* and *volumetric efficiency e_v.* More is said in Chapter 13 about the details of these efficiencies. In general, the overall efficiency is analogous to the mechanical efficiency discussed for other types of pumps in this section. Volumetric efficiency is a measure of the actual delivery from the pump compared with the ideal delivery found from the displacement per revolution times the rotational speed of the pump. A high volumetric efficiency is desired because the operation of the fluid power system depends on a nearly uniform flow rate of fluid through all operating conditions.

The following programmed example problem illustrates a possible setup for measuring pump efficiency.

PROGRAMMED EXAMPLE PROBLEM

Example Problem 7.3   For the pump test arrangement shown in Fig. 7.9, determine the mechanical efficiency of the pump if the power input is measured to be 3.85 hp when pumping 500 gal/min of oil ($\gamma = 56.0 \text{ lb/ft}^3$).

To begin, write the energy equation for this system.

Using the points identified as 1 and 2 in Fig. 7.9, we have

$$\frac{p_1}{\gamma} + z_1 + \frac{v_1^2}{2g} + h_A = \frac{p_2}{\gamma} + z_2 + \frac{v_2^2}{2g}$$

FIGURE 7.9 Pump test system for Example Problem 7.3.

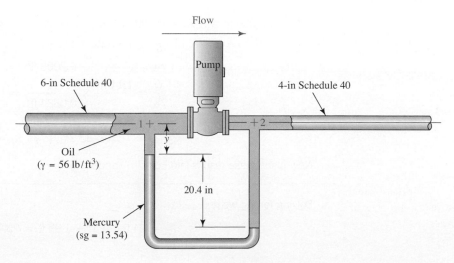

Because we must find the power delivered by the pump to the fluid, we should now solve for h_A.

We use the following equation:

$$h_A = \frac{p_2 - p_1}{\gamma} + (z_2 - z_1) + \frac{v_2^2 - v_1^2}{2g} \tag{7-7}$$

It is convenient to solve for each term individually and then combine the results. The manometer enables us to calculate $(p_2 - p_1)/\gamma$ because it measures the pressure difference. Using the procedure outlined in Chapter 3, write the manometer equation between points 1 and 2.

Starting at point 1, we have

$$p_1 + \gamma_o y + \gamma_m(20.4 \text{ in}) - \gamma_o(20.4 \text{ in}) - \gamma_o y = p_2$$

where y is the unknown distance from point 1 to the top of the mercury column in the left leg of the manometer. The terms involving y cancel out. Also, in this equation γ_o is the specific weight of the oil and γ_m is the specific weight of the mercury gage fluid.

The desired result for use in Eq. (7–7) is $(p_2 - p_1)/\gamma_o$. Solve for this now and compute the result.

The correct solution is $(p_2 - p_1)/\gamma_o = 24.0$ ft. Here is one way to find it:

$$\gamma_m = (13.54)(\gamma_w) = (13.54)(62.4 \text{ lb/ft}^3) = 844.9 \text{ lb/ft}^3$$

$$p_2 = p_1 + \gamma_m(20.4 \text{ in}) - \gamma_o(20.4 \text{ in})$$

$$p_2 - p_1 = \gamma_m(20.4 \text{ in}) - \gamma_o(20.4 \text{ in})$$

$$\frac{p_2 - p_1}{\gamma_o} = \frac{\gamma_m(20.4 \text{ in})}{\gamma_o} - 20.4 \text{ in} = \left(\frac{\gamma_m}{\gamma_o} - 1\right) 20.4 \text{ in}$$

$$= \left(\frac{844.9 \text{ lb/ft}^3}{56.0 \text{ lb/ft}^3} - 1\right) 20.4 \text{ in} = (15.1 - 1)(20.4 \text{ in})$$

$$\frac{p_1 - p_2}{\gamma_o} = (14.1)(20.4 \text{ in})\left(\frac{1 \text{ ft}}{12 \text{ in}}\right) = 24.0 \text{ ft}$$

The next term in Eq. (7–7) is $z_2 - z_1$. What is its value?

It is zero. Both points are at the same elevation. These terms could have been cancelled from the original equation. Now find $(v_2^2 - v_1^2)/2g$.

You should have $(v_2^2 - v_1^2)/2g = 1.99$ ft, obtained as follows. First, write

$$Q = (500 \text{ gal/min})\left(\frac{1 \text{ ft}^3/\text{s}}{449 \text{ gal/min}}\right) = 1.11 \text{ ft}^3/\text{s}$$

Using $A_1 = 0.2006 \text{ ft}^2$ and $A_2 = 0.0884 \text{ ft}^2$ from Appendix F, we get

$$v_1 = \frac{Q}{A_1} = \frac{1.11 \text{ ft}^3}{\text{s}} \times \frac{1}{0.2006 \text{ ft}^2} = 5.55 \text{ ft/s}$$

$$v_2 = \frac{Q}{A_2} = \frac{1.11 \text{ ft}^3}{\text{s}} \times \frac{1}{0.0884 \text{ ft}^2} = 12.6 \text{ ft/s}$$

$$\frac{v_2^2 - v_1^2}{2g} = \frac{(12.6)^2 - (5.55)^2}{(2)(32.2)} \frac{\text{ft}^2}{\text{s}^2} \frac{\text{s}^2}{\text{ft}} = 1.99 \text{ ft}$$

Now place these results into Eq. (7–7) and solve for h_A.

Solving for h_A, we get

$$h_A = 24.0 \text{ ft} + 0 + 1.99 \text{ ft} = 25.99 \text{ ft}$$

We can now calculate the power delivered to the oil, P_A.

The result is $P_A = 2.95$ hp, found as follows:

$$P_A = h_a \gamma Q = 25.99 \text{ ft} \left(\frac{56.0 \text{ lb}}{\text{ft}^3} \right) \left(\frac{1.11 \text{ ft}^3}{\text{s}} \right)$$

$$P_A = 1620 \text{ lb-ft/s} \left(\frac{1 \text{ hp}}{550 \text{ lb-ft/s}} \right) = 2.95 \text{ hp}$$

The final step is to calculate e_M, the mechanical efficiency of the pump.

From Eq. (7–6) we get

$$e_M = P_A / P_I = 2.95/3.85 = 0.77$$

Expressed as a percentage, the pump is 77 percent efficient at the stated conditions. This completes the programmed instruction.

7.6 POWER DELIVERED TO FLUID MOTORS

The energy delivered by the fluid to a mechanical device such as a fluid motor or a turbine is denoted in the general energy equation by the term h_R. This is a measure of the energy delivered by each unit weight of fluid as it passes through the device. We find the power delivered by multiplying h_R by the weight flow rate W:

▷ **Power Delivered by a Fluid to a Motor**

$$P_R = h_R W = h_R \gamma Q \qquad (7\text{–}8)$$

where P_R is the power delivered by the fluid to the fluid motor.

7.6.1 Mechanical Efficiency of Fluid Motors

As was described for pumps, energy losses in a fluid motor are produced by mechanical and fluid friction. Therefore, not all the power delivered to the motor is ultimately converted to power output from the device. Mechanical efficiency is then defined as

▷ **Motor Efficiency**

$$e_M = \frac{\text{Power output from motor}}{\text{Power delivered by fluid}} = \frac{P_O}{P_R} \qquad (7\text{–}9)$$

Here again, the value of e_M is always less than 1.0. Refer to Section 7.5 for power units.

PROGRAMMED EXAMPLE PROBLEM

Example Problem 7.4

Water at 10°C is flowing at a rate of 115 L/min through the fluid motor shown in Fig. 7.10. The pressure at A is 700 kPa and the pressure at B is 125 kPa. It is estimated that due to friction in the tubing there is an energy loss of 4.0 N·m/N of water flowing. At A the tubing entering the fluid motor is a standard steel hydraulic tube having an OD = 25 mm and a wall thickness of 2.0 mm. At B, the tube leaving the motor has OD = 80 mm and wall thickness of 2.8 mm. See Appendix G.2. (a) Calculate the power delivered to the fluid motor by the water. (b) If the mechanical efficiency of the fluid motor is 85 percent, calculate the power output.

Start the solution by writing the energy equation.

Choosing points A and B as our reference points, we get

$$\frac{p_A}{\gamma} + z_A + \frac{v_A^2}{2g} - h_R - h_L = \frac{p_B}{\gamma} + z_B + \frac{v_B^2}{2g}$$

The value of h_R is needed to determine the power output. Solve the energy equation for this term.

FIGURE 7.10 Fluid motor for Example Problem 7.4.

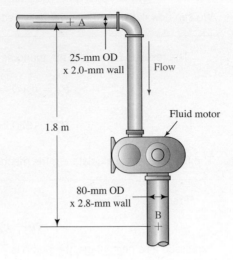

25-mm OD
x 2.0-mm wall

Flow

Fluid motor

1.8 m

80-mm OD
x 2.8-mm wall

Compare this equation with your result:

$$h_R = \frac{p_A - p_B}{\gamma} + (z_A - z_B) + \frac{v_A^2 - v_B^2}{2g} - h_L \qquad (7\text{–}10)$$

Before looking at the next panel, solve for the value of each term in this equation using the unit of N·m/N or m.

The correct results are as follows:

1. $\dfrac{p_A - p_B}{\gamma} = \dfrac{(700 - 125)(10^3)\text{N}}{\text{m}^2} \times \dfrac{\text{m}^3}{9.81 \times 10^3 \text{ N}} = 58.6 \text{ m}$

2. $z_A - z_B = 1.8 \text{ m}$

3. Solving for $(v_A^2 - v_B^2)/2g$, we obtain

$$Q = 115 \text{ L/min} \times \frac{1.0 \text{ m}^3/\text{s}}{60\,000 \text{ L/min}} = 1.92 \times 10^{-3} \text{ m}^3/\text{s}$$

$$v_A = \frac{Q}{A_A} = \frac{1.92 \times 10^{-3} \text{ m}^3}{\text{s}} \times \frac{1}{3.464 \times 10^{-4} \text{ m}^2} = 5.543 \text{ m/s}$$

$$v_B = \frac{Q}{A_B} = \frac{1.92 \times 10^{-3} \text{ m}^3}{\text{s}} \times \frac{1}{4.347 \times 10^{-3} \text{ m}^2} = 0.442 \text{ m/s}$$

$$\frac{v_A^2 - v_B^2}{2g} = \frac{(5.543)^2 - (0.442)^2}{(2)(9.81)} \frac{\text{m}^2}{\text{s}^2} \frac{\text{s}^2}{\text{m}} = 1.56 \text{ m}$$

4. $h_L = 4.0 \text{ m}$ (given)

Complete the solution of Eq. (7–10) for h_R now.

The energy delivered by the water to the turbine is

$$h_R = (58.6 + 1.8 + 1.56 - 4.0) \text{ m} = 57.96 \text{ m}$$

To complete part (a) of the problem, calculate P_R.

Substituting the known values into Eq. (7–8), we get

$$P_R = h_R \gamma Q$$

$$P_R = 57.96 \text{ m} \times \frac{9.81 \times 10^3 \text{ N}}{\text{m}^3} \times \frac{1.92 \times 10^{-3} \text{ m}^3}{\text{s}} = 1092 \text{ W}$$

$$P_R = 1.092 \text{ kW}$$

This is the power delivered to the fluid motor by the water. How much useful power can the motor put out?

Because the efficiency of the motor is 85 percent, we get 0.928 kW of power out. Using Eq. (7–09), $e_M = P_O/P_R$, we get

$$P_O = e_M P_R$$
$$= (0.85)(1.092 \text{ kW})$$
$$P_O = 0.928 \text{ kW}$$

This completes the programmed example problem.

PRACTICE PROBLEMS

It may be necessary to refer to the appendices for data concerning the dimensions of pipes or the properties of fluids. Assume there are no energy losses unless stated otherwise.

7.1 A horizontal pipe carries oil with a specific gravity of 0.83. If two pressure gages along the pipe read 74.6 psig and 62.2 psig, respectively, calculate the energy loss between the two gages.

7.2 Water at 40°F is flowing downward through the fabricated reducer shown in Fig. 7.11. At point A the velocity is 10 ft/s and the pressure is 60 psig. The energy loss between points A and B is 25 lb-ft/lb. Calculate the pressure at point B.

7.3 Find the volume flow rate of water exiting from the tank shown in Fig. 7.12. The tank is sealed with a pressure of 140 kPa above the water. There is an energy loss of 2.0 N·m/N as the water flows through the nozzle.

7.4 A long DN 150 Schedule 40 steel pipe discharges 0.085 m³/s of water from a reservoir into the atmosphere as shown in Fig. 7.13. Calculate the energy loss in the pipe.

7.5 Figure 7.14 shows a setup to determine the energy loss due to a certain piece of apparatus. The inlet is through a 2-in Schedule 40 pipe and the outlet is a 4-in Schedule 40 pipe. Calculate the energy loss between points A and B if water is flowing upward at 0.20 ft³/s. The gage fluid is mercury (sg = 13.54).

7.6 A test setup to determine the energy loss as water flows through a valve is shown in Fig. 7.15. Calculate the energy loss if 0.10 ft³/s of water at 40°F is flowing. Also calculate the resistance coefficient K if the energy loss is expressed as $K(v^2/2g)$.

7.7 The setup shown in Fig. 7.16 is being used to measure the energy loss across a valve. The velocity of flow of the oil is 1.2 m/s. Calculate the value of K if the energy loss is expressed as $K(v^2/2g)$.

7.8 A pump is being used to transfer water from an open tank to one that has air at 500 kPa above the water, as shown in Fig. 7.17. If 2250 L/min is being pumped, calculate the power delivered by the pump to the water. Assume that the level of the surface in each tank is the same.

7.9 In Problem 7.8 (Fig. 7.17), if the left-hand tank is also sealed and air pressure above the water is 68 kPa, calculate the pump power.

7.10 A commercially available sump pump is capable of delivering 2800 gal/h of water through a vertical lift of 20 ft. The inlet to the pump is just below the water surface and the discharge is to the atmosphere through a 1¼-in

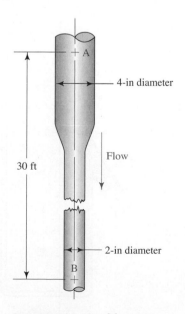

FIGURE 7.11 Problem 7.2.

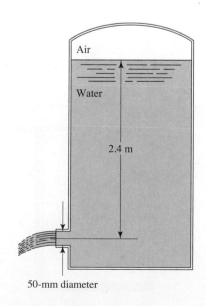

FIGURE 7.12 Problem 7.3.

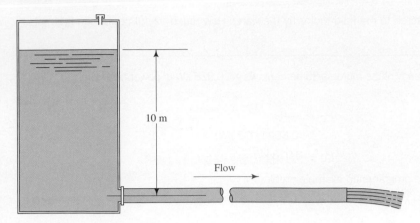

FIGURE 7.13 Problem 7.4.

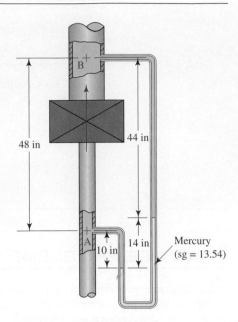

FIGURE 7.14 Problem 7.5.

Schedule 40 pipe. (a) Calculate the power delivered by the pump to the water. (b) If the pump draws 0.5 hp, calculate its efficiency.

7.11 A submersible deep-well pump delivers 745 gal/h of water through a 1-in Schedule 40 pipe when operating in the system sketched in Fig. 7.18. An energy loss of 10.5 lb-ft/lb occurs in the piping system. (a) Calculate the power delivered by the pump to the water. (b) If the pump draws 1 hp, calculate its efficiency.

7.12 In a pump test the suction pressure at the pump inlet is 30 kPa below atmospheric pressure. The discharge pressure at a point 750 mm above the inlet is 520 kPa. Hydraulic steel tubing is used for both the suction and discharge lines with 80 mm OD × 2.8 mm wall. If the volume flow rate of water is 75 L/min, calculate the power delivered by the pump to the water.

7.13 The pump shown in Fig. 7.19 is delivering hydraulic oil with a specific gravity of 0.85 at a rate of 75 L/min. The pressure at A is −275 kPa and the pressure at B is 275 kPa. The energy loss in the system is 2.5 times the velocity head in the discharge pipe. Calculate the power delivered by the pump to the oil.

7.14 The pump in Fig. 7.20 delivers water from the lower to the upper reservoir at the rate of 2.0 ft³/s. The energy loss between the suction pipe inlet and the pump is 6 lb-ft/lb and that between the pump outlet and the upper reservoir is 12 lb-ft/lb. Both pipes are 6-in Schedule 40 steel pipe. Calculate (a) the pressure at the pump inlet, (b) the pressure at the pump outlet, (c) the total head on the pump, and (d) the power delivered by the pump to the water.

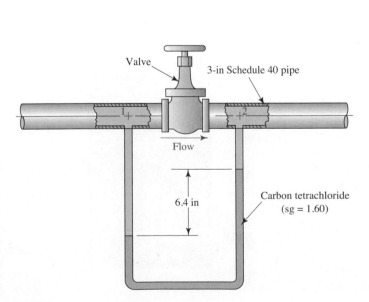

FIGURE 7.15 Problem 7.6.

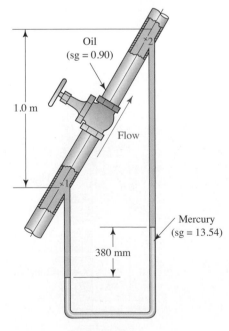

FIGURE 7.16 Problem 7.7.

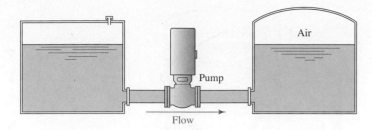

FIGURE 7.17 Problems 7.8 and 7.9.

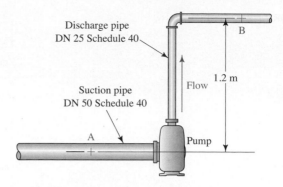

FIGURE 7.19 Problem 7.13.

7.15 Repeat Problem 7.14, but assume that the level of the lower reservoir is 10 ft above the pump instead of below it. All other data remain the same.

7.16 Figure 7.21 shows a pump delivering 840 L/min of crude oil (sg = 0.85) from an underground storage drum to the first stage of a processing system. (a) If the total energy loss in the system is 4.2 N·m/N of oil flowing, calculate the power delivered by the pump. (b) If the energy loss in the suction pipe is 1.4 N·m/N of oil flowing, calculate the pressure at the pump inlet.

7.17 Figure 7.22 shows a submersible pump being used to circulate 60 L/min of a water-based coolant (sg = 0.95) to the cutter of a milling machine. The outlet is through a DN 20 Schedule 40 steel pipe. Assuming a total energy loss due to the piping of 3.0 N·m/N, calculate the total head developed by the pump and the power delivered to the coolant.

7.18 Figure 7.23 shows a small pump in an automatic washer discharging into a laundry sink. The washer tub is 525 mm in diameter and 250 mm deep. The average head above the pump is 375 mm as shown. The discharge hose has an inside diameter of 18 mm. The energy loss in the hose system is 0.22 N·m/N. If the pump empties the tub in 90 s, calculate the average total head on the pump.

7.19 The water being pumped in the system shown in Fig. 7.24 discharges into a tank that is being weighed. It is found that 556 lb of water is collected in 10 s. If the pressure at A is 2.0 psi below atmospheric pressure, calculate the

horsepower delivered by the pump to the water. Neglect energy losses.

7.20 A manufacturer's rating for a gear pump states that 0.85 hp is required to drive the pump when it is pumping 9.1 gal/min of oil (sg = 0.90) with a total head of 257 ft. Calculate the mechanical efficiency of the pump.

7.21 The specifications for an automobile fuel pump state that it should pump 1.0 L of gasoline in 40 s with a suction pressure of 150 mm of mercury vacuum and a discharge pressure of 30 kPa. Assuming that the pump efficiency is 60 percent, calculate the power drawn from the engine. See Fig. 7.25. The suction and discharge lines are the same size. Elevation changes can be neglected.

7.22 Figure 7.26 shows the arrangement of a circuit for a hydraulic system. The pump draws oil with a specific gravity of 0.90 from a reservoir and delivers it to the hydraulic cylinder. The cylinder has an inside diameter of 5.0 in, and in 15 s the piston must travel 20 in while exerting a force of 11000 lb. It is estimated that there are energy losses of 11.5 lb-ft/lb in the suction pipe and 35.0 lb-ft/lb in the discharge pipe. Both pipes are ⅜-in Schedule 80 steel pipes. Calculate:

a. The volume flow rate through the pump.
b. The pressure at the cylinder.
c. The pressure at the outlet of the pump.
d. The pressure at the inlet to the pump.
e. The power delivered to the oil by the pump.

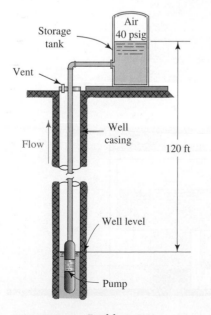

FIGURE 7.18 Problem 7.11.

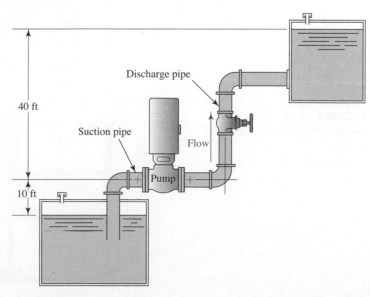

FIGURE 7.20 Problems 7.14 and 7.15.

FIGURE 7.21 Problem 7.16.

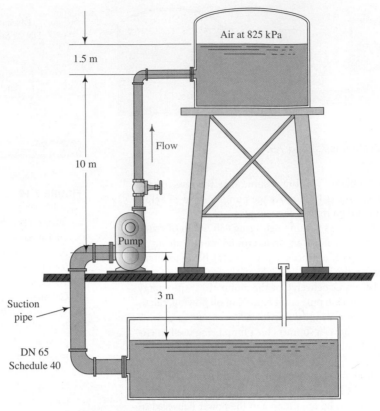

FIGURE 7.22 Problem 7.17.

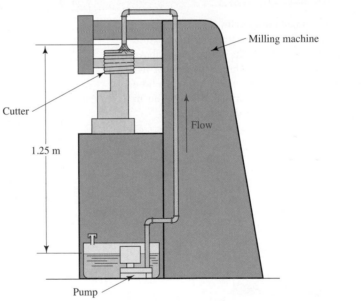

FIGURE 7.23 Problem 7.18.

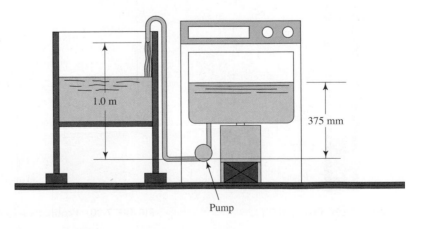

FIGURE 7.24 Problem 7.19.

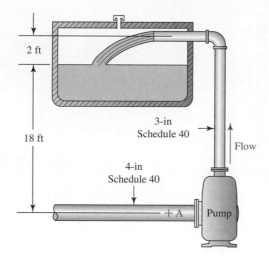

FIGURE 7.25 Automobile fuel pump for Problem 7.21.

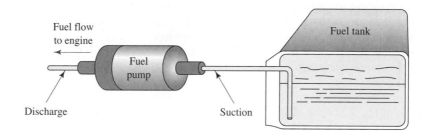

FIGURE 7.26 Problem 7.22.

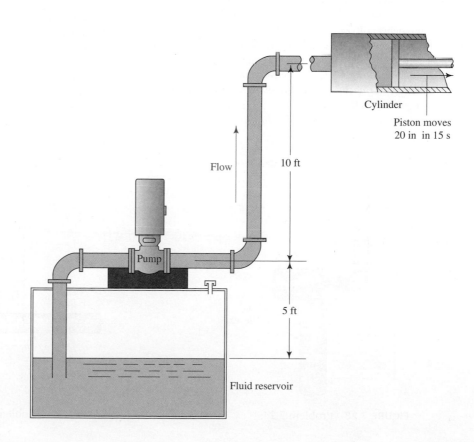

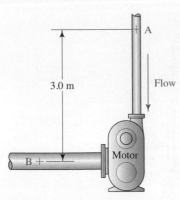

FIGURE 7.27 Problem 7.23.

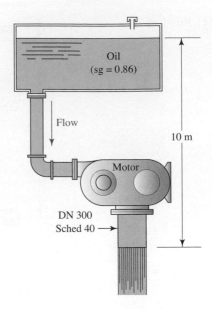

FIGURE 7.29 Problem 7.25.

7.23 Calculate the power delivered to the hydraulic motor in Fig. 7.27 if the pressure at A is 6.8 MPa and the pressure at B is 3.4 MPa. The motor inlet is a steel hydraulic tube with 25 mm OD × 1.5 mm wall and the outlet is a tube with 50 mm OD × 2.0 mm wall. The fluid is oil (sg = 0.90) and the velocity of flow is 1.5 m/s at point B.

7.24 Water flows through the turbine shown in Fig. 7.28, at a rate of 3400 gal/min when the pressure at A is 21.4 psig and the pressure at B is −5 psig. The friction energy loss between A and B is twice the velocity head in the 12-in pipe. Determine the power delivered by the water to the turbine.

7.25 Calculate the power delivered by the oil to the fluid motor shown in Fig. 7.29 if the volume flow rate is 0.25 m^3/s. There is an energy loss of 1.4 N·m/N in the piping system. If the motor has an efficiency of 75 percent, calculate the power output.

7.26 What hp must the pump shown in Fig. 7.30 deliver to a fluid having a specific weight of 60.0 lb/ft^3 if energy losses of 3.40 lb-ft/lb occur between points 1 and 2? The pump delivers 40 gal/min of fluid.

7.27 If the pump in Problem 7.26 operates with an efficiency of 75 percent, what is the power input to the pump?

7.28 The system shown in Fig. 7.31 delivers 600 L/min of water. The outlet is directly into the atmosphere. Determine the energy losses in the system.

7.29 Kerosene (sg = 0.823) flows at 0.060 m^3/s in the pipe shown in Fig. 7.32. Compute the pressure at B if the total energy loss in the system is 4.60 N·m/N.

7.30 Water at 60°F flows from a large reservoir through a fluid motor at the rate of 1000 gal/min in the system shown in Fig. 7.33. If the motor removes 37 hp from the fluid, calculate the energy losses in the system.

7.31 Figure 7.34 shows a portion of a fire protection system in which a pump draws 1500 gal/min of water at 50°F from a reservoir and delivers it to point B. The energy loss be-tween the reservoir and point A at the inlet to the pump is 0.65 lb-ft/lb. Specify the required depth h to maintain at least 5.0 psig pressure at point A.

7.32 For the conditions of Problem 7.31, and if we assume that the pressure at point A is 5.0 psig, calculate the power

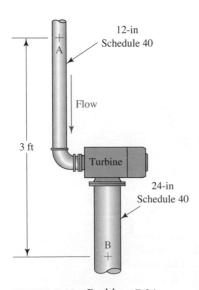

FIGURE 7.28 Problem 7.24.

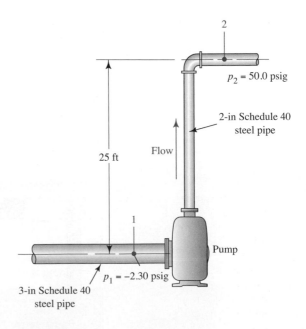

FIGURE 7.30 Problems 7.26 and 7.27.

FIGURE 7.31 Problem 7.28.

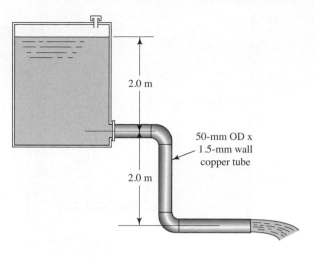

50-mm OD x
1.5-mm wall
copper tube

2.0 m

2.0 m

FIGURE 7.32 Problem 7.29.

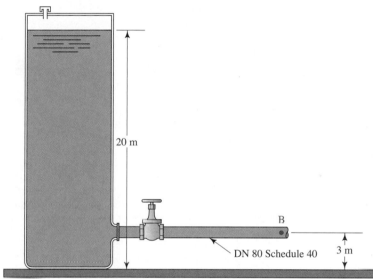

20 m

B

DN 80 Schedule 40

3 m

FIGURE 7.33 Problem 7.30.

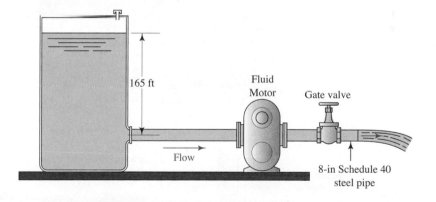

165 ft

Fluid
Motor

Gate valve

Flow

8-in Schedule 40
steel pipe

delivered by the pump to the water to maintain a pressure of 85 psig at point B. Energy losses between the pump and point B total 28.0 lb-ft/lb.

7.33 In Fig. 7.35, kerosene at 25°C is flowing at 500 L/min from the lower tank to the upper tank through hydraulic copper tubing (50 mm OD × 1.5 mm wall) and a valve. If the pressure above the fluid is 100 kPa gage, how much energy loss occurs in the system?

7.34 For the system shown in Fig. 7.35 and analyzed in Problem 7.33, assume that the energy loss is proportional to the velocity head in the tubing. Compute the pressure in the tank required to cause a flow of 1000 L/min.

General Data for Problems 7.35–7.40

Figure 7.36 shows a diagram of a fluid power system for a hydraulic press used to extrude rubber parts. The following data are known:

1. The fluid is oil (sg = 0.93).
2. Volume flow rate is 175 gal/min.
3. Power input to the pump is 28.4 hp.
4. Pump efficiency is 80 percent.
5. Energy loss from point 1 to 2 is 2.80 lb-ft/lb.
6. Energy loss from point 3 to 4 is 28.50 lb-ft/lb.
7. Energy loss from point 5 to 6 is 3.50 lb-ft/lb.

FIGURE 7.34 Problems 7.31
and 7.32.

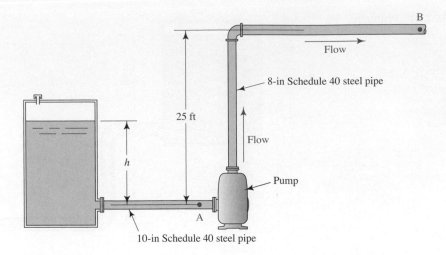

FIGURE 7.35 Problems 7.33
and 7.34.

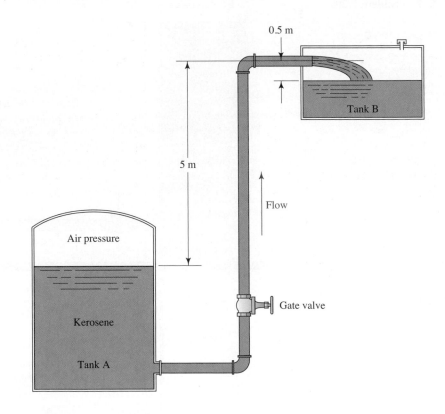

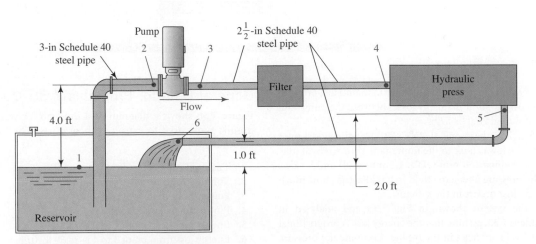

FIGURE 7.36 Problems 7.35–7.40.

FIGURE 7.37   Problem 7.41.

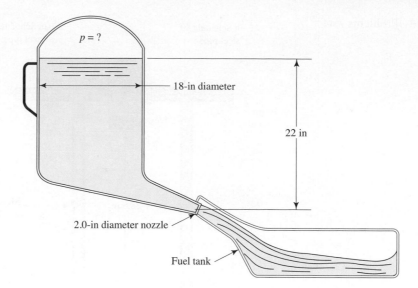

7.35 Compute the power removed from the fluid by the press.

7.36 Compute the pressure at point 2 at the pump inlet.

7.37 Compute the pressure at point 3 at the pump outlet.

7.38 Compute the pressure at point 4 at the press inlet.

7.39 Compute the pressure at point 5 at the press outlet.

7.40 Evaluate the suitability of the sizes for the suction and discharge lines of the system as compared with Fig. 6.3 in Chapter 6 and the results of Problems 7.35–7.39.

7.41 The portable, pressurized fuel can shown in Fig. 7.37 is used to deliver fuel to a race car during a pit stop. What

pressure must be above the fuel to deliver 40 gal in 8.0 s? The specific gravity of the fuel is 0.76. An energy loss of 4.75 lb-ft/lb occurs at the nozzle.

7.42 Professor Crocker is building a cabin on a hillside and has proposed the water system shown in Fig. 7.38. The distribution tank in the cabin maintains a pressure of 30.0 psig above the water. There is an energy loss of 15.5 lb-ft/lb in the piping. When the pump is delivering 40 gal/min of water, compute the horsepower delivered by the pump to the water.

FIGURE 7.38   Problems 7.42 and 7.43.

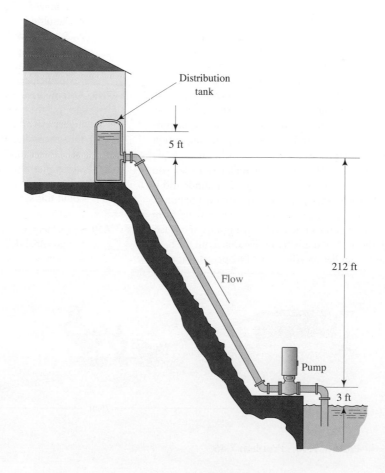

FIGURE 7.39 Problems 7.44 and 7.45.

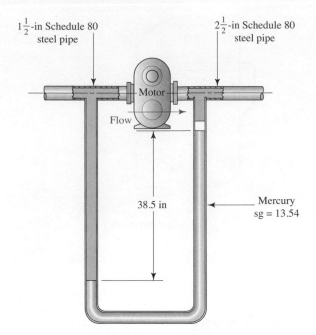

$1\frac{1}{2}$-in Schedule 80 steel pipe

$2\frac{1}{2}$-in Schedule 80 steel pipe

Motor

Flow

38.5 in

Mercury sg = 13.54

7.43 If Professor Crocker's pump, described in Problem 7.42, has an efficiency of 72 percent, what size motor is required to drive the pump?

7.44 The test setup in Fig. 7.39 measures the pressure difference between the inlet and the outlet of the fluid motor. The flow rate of hydraulic oil (sg = 0.90) is 135 gal/min. Compute the power removed from the fluid by the motor.

7.45 If the fluid motor in Problem 7.44 has an efficiency of 78 percent, how much power is delivered by the motor?

Supplemental Problems

7.46 A village with a need for a simple irrigation system proposes to have a collection pool for rain water and then a simple human-powered pump mounted to a stationary bicycle frame to deliver the water to crops during dry times. If the required lift is 2.5 m as shown in Fig. 7.40, and the total loss in the line is 1.8 m, determine what flow rate, in liters per minute, is possible with a person producing 125 W by pedaling assuming the pumping station to be 65 percent efficient.

7.47 As a member of a development team for a new jet ski, you are testing a new prototype. It weighs 320 pounds and is outfitted with a 70 horsepower engine driving the pump. In the stationary test stand, the pump takes in water from an open tank set at the same level as the pump, and ejects it through the 4.25-in diameter outlet aimed out the back end. If the system has an efficiency of 82 percent, at what

velocity, in mph, should you expect the water to exit when the engine runs at rated power?

7.48 A fire truck utilizes its engine to drive a pump that is 82 percent efficient. The water needs to reach an elevation of 15 m above the spray tip as shown in Fig. 7.41. *Note:* Only the vertical component of the exit velocity contributes to the height, and the vertical component is zero at the peak. Determine:
   a. Required exit velocity from the spray tip to reach the required height
   b. Resulting flow rate if the spray tip is 45 mm in diameter
   c. Power added to the water by the pump if the water is drawn from an open tank at ground level
   d. Power required by the pump from the engine given its inefficiency.

7.49 A home has a sump pump to handle ground water from around the foundation that needs to be removed to a higher elevation. To achieve a flow rate of 1600 gal/h in the system shown in Fig. 7.42, how much power is needed if the losses total 3.8 feet? Storms that bring a lot of water sometimes also cut off electrical power to the home, so this unit comes with a 12V battery backup. How long could this system run on battery backup if the battery is capable of storing 800 watt-hours of energy?

7.50 In problem 6.107, an initial calculation was made regarding the potential delivery of water to a village via a tube from a nearby water source. No losses were considered, and the

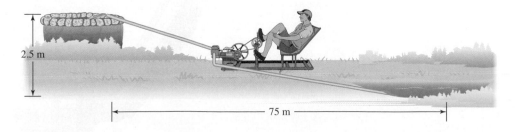

2.5 m

75 m

FIGURE 7.40   Problem 7.46.

FIGURE 7.41 Problem 7.48.

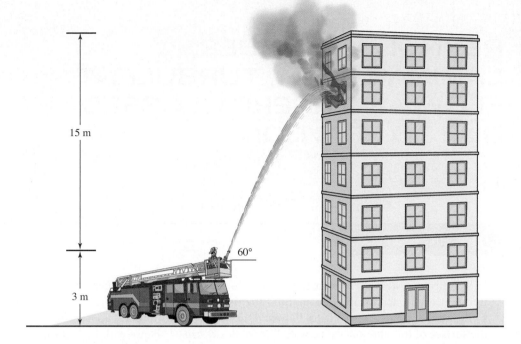

theoretical flow rate was determined to be 9.64×10^{-3} m³/s. Work the problem again but include estimated losses. The lake was at an elevation 3 m above the village, and the line was to be 20 mm in diameter. Determine the flow rate that would be carried from the lake to the village if the head loss is estimated to be 2.8 m. What conclusions can you draw?

7.51 A creek runs through a certain part of a campus where the water is falling about 2.5 m over a distance of just 8 m, and the creek before and after the fall is about 3 m wide. The sustainability club has asked you about the potential of harnessing this energy. It is difficult to measure exactly, but a rough estimate of the flow at this spot is 150 L/min. Determine the wattage that could be generated assuming an overall system efficiency of 60 percent. Sketch the arrangement that would permit this.

7.52 A hot tub is to have 40 outlets that are each 8 mm in diameter with water exiting at 7 m/s. Treating each of the outlets as if they are at the surface of the water and exit into atmospheric pressure, would a ½ HP pump be adequate? If so, what is the minimum efficiency that will still provide adequate power with this selection?

7.53 A large chipper/shredder is to be designed for use by commercial tree trimming companies. It would be mounted on a trailer to pull behind a large truck. The rotating blades of the unit protrude from a large flywheel that is driven by a fluid motor that runs on medium machine tool hydraulic oil from a hydraulic system. This is a great application for a fluid motor given the extreme and sudden variations in torque and speed. An average of 80 HP is required from this motor to drive the rotating blades. How much oil at 1200 psi is required given a motor with 87 percent efficiency? Assume the motor outlet to be at atmospheric pressure, and the same size as the inlet. Give your answer in gal/min.

FIGURE 7.42 Problem 7.49.

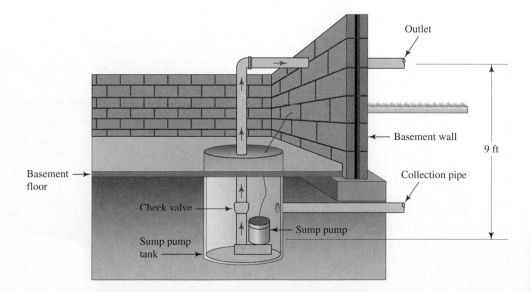

REYNOLDS NUMBER, LAMINAR FLOW, TURBULENT FLOW, AND ENERGY LOSSES DUE TO FRICTION

In any piping system there is energy loss due to the friction that occurs within the flowing fluid that is affected by the kind of fluid, the velocity of flow, and the nature of the surface of the stationary pipe wall. Friction losses can be quite significant, particularly in a case like the geothermal system for heating and cooling the home shown in Fig. 8.1, where the desire for heat transfer leads to a design with long- and small-diameter piping.

In this chapter you begin to develop your skills in analyzing the energy losses that occur as fluids flow in real pipeline systems. The only type of energy loss considered here is *energy loss due to friction in straight circular pipes or tubes*. In following chapters, you will develop skills to add other types of energy losses such as those created by valves, pipe fittings, changes in the areas of pipes, different shapes for the flow path, and others. Then, in Chapters 11, 12, and 13, you will learn how to analyze more comprehensive piping systems that combine these various kinds of energy losses along with pumps to move the fluid.

Characterizing the flow as laminar or turbulent is the first step in calculating friction losses. You must be able to determine the *Reynolds number,* introduced in this chapter, that is dependent on the velocity of flow, the size of the pipe, and the viscosity of the fluid. Flows with low Reynolds numbers appear slow and smooth and are called *laminar*. Flows with high Reynolds numbers appear fast, chaotic, and rough and are called *turbulent*. Because fluid viscosity is a critical component of Reynolds number, you should review Chapter 2.

Friction losses cause pressure to decrease along the pipe and they increase the amount of power that a pump must deliver to the fluid. You may have observed that the pressure drops as it flows from a faucet to the end of a long length of pipe, tubing, garden hose, or fire hose.

Exploration

By observing the flow of water from a simple faucet, you can see how the character of the flow changes as the velocity changes.

- Describe the appearance of the stream of water as you turn a faucet on with a very slow flow rate.

- Then slowly open the faucet fully and observe how the character of the flow stream changes.

FIGURE 8.1 In this geothermal heating and cooling system for a home, you must know the behavior of the fluid in the long lengths of tubing to accurately predict the energy losses and pressure drops in the system. (*Source:* Gunnar Assmy/Fotolia)

- Now close the faucet slowly and carefully while observing changes in the appearance of the flow stream as the flow velocity returns to the slow rate.

- Consider other kinds of fluid flow systems where you could observe the changing character of the flow from slow to fast.

- What happens when cool oil flows as compared with the flow of water? You know that cool oil has a much higher viscosity than water and you can observe that it flows more smoothly than water at comparable velocities.

- Check out Internet resource 1 to see a chart of pressure drop versus the flow rate and the length of a pipe.

Introductory Concepts

As the water flows from a faucet at a very low velocity, the flow appears to be smooth and steady. The stream has a fairly uniform diameter and there is little or no evidence of mixing of the various parts of the stream. This is called *laminar flow*, a term derived from the word *layer*, because the fluid appears to be flowing in continuous layers with little or no mixing from one layer to the adjacent layers. When the faucet is nearly fully open, the water has a rather high velocity. The elements of fluid appear to be mixing chaotically within the stream. This is a general description of *turbulent flow*.

Let's go back to when you observed laminar flow and then continued to open the faucet slowly. As you increased the velocity of flow, did you notice that the stream became less smooth with ripples developing along its length? The cross section of the flow stream might have appeared to oscillate in and out, even when the flow was generally smooth. This region of flow is called the *transition zone* in which the flow is changing from laminar to turbulent. Higher velocities produced more of these oscillations until the flow eventually became fully turbulent. The example of the flow of water from a faucet illustrates the importance of the flow velocity for the character of the flow.

Fluid viscosity is also important. In Chapter 2, both dynamic viscosity η (Greek eta) and kinematic viscosity ν (Greek nu) were defined. Recall that $\nu = \eta/\rho$, where ρ (rho) is the density of the fluid. One general observation you made is that fluids with low viscosity flow more easily than those with higher viscosity. To aid in your review, consider the following questions.

- What are some fluids that have a relatively low viscosity?
- What are some fluids that have a rather high viscosity?
- What happens with regard to the ease with which a high viscosity fluid flows when the temperature is increased?
- What happens when the temperature of a high viscosity fluid is decreased?

Heating a high viscosity fluid such as engine-lubricating oil lowers its viscosity and allows it to flow more easily. This happens as a car's engine warms up after initially starting. Conversely, reducing the oil's temperature increases the viscosity, and it flows more slowly. This happens after shutting off the engine and letting it sit overnight in a cold garage. These observations illustrate the concept that the character of the flow is also dependent on fluid viscosity. The flow of heavy viscous fluids like cold oil is more likely to be laminar. The flow of low viscosity fluids like water is more likely to be turbulent.

You will also see in this chapter that the size of the flow path affects the character of the flow. Much of our work will deal with fluid flow through circular pipes and tubes as discussed in Chapter 6. The inside flow diameter of the pipe plays an important role in characterizing the flow.

Figure 8.2 shows one way of visualizing laminar flow in a circular pipe. Concentric rings of fluid are flowing in a straight, smooth path. There is little or no mixing of the fluid across the "boundaries" of each layer as the fluid flows along in the pipe. Of course, in real fluids an infinite number of layers make up the flow.

Another way to visualize laminar flow is depicted in Fig. 8.3, which shows a transparent fluid such as water flowing in a clear glass tube. When a stream of a dark fluid such as a dye is injected into the flow, the stream remains intact as long as the flow remains laminar. The dye stream will not mix with the bulk of the fluid.

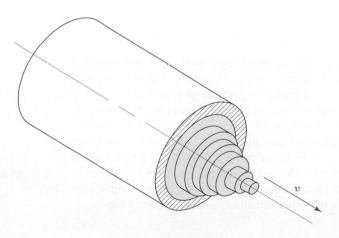

FIGURE 8.2 Illustration of laminar flow in a circular pipe.

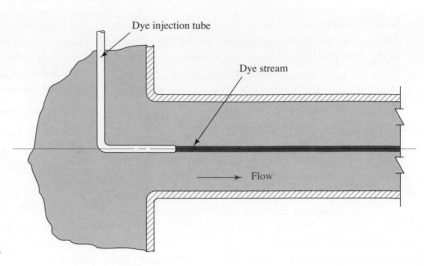

FIGURE 8.3 Dye stream in laminar flow.

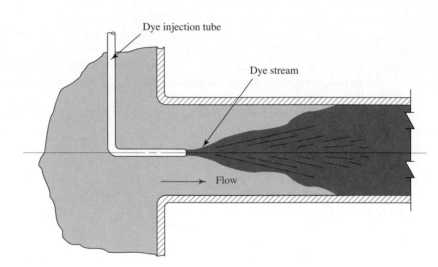

FIGURE 8.4 Dye stream mixing with turbulent flow.

In contrast to laminar flow, turbulent flow appears chaotic and rough with much intermixing of the fluid. Figure 8.4 shows that when a dye stream is introduced into turbulent flow, it immediately dissipates throughout the primary fluid. Indeed, an important reason for creating turbulent flow is to promote mixing in such applications as:

1. Blending two or more fluids.
2. Hastening chemical reactions.
3. Increasing heat transfer into or out of a fluid.

Open-channel flow is the type in which one surface of the fluid is exposed to the atmosphere. Figure 8.5 shows a reservoir discharging fluid into an open channel that eventually allows the stream to fall into a lower pool. Have you seen fountains that have this feature? Here, as with the flow in a circular pipe, laminar flow would appear to be smooth and layered. The discharge from the channel into the pool would be like a smooth sheet. Turbulent flow would appear to be chaotic. Have you seen Niagara Falls or some other fast-falling water? Open channel flow is covered in Chapter 14.

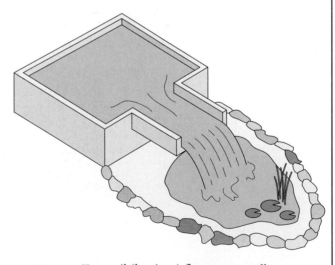

FIGURE 8.5 Tranquil (laminar) flow over a wall.

8.1 OBJECTIVES

After completing this chapter, you should be able to:

1. Describe the appearance of laminar flow and turbulent flow.

2. State the relationship used to compute the Reynolds number.

3. Identify the limiting values of the Reynolds number by which you can predict whether flow is laminar or turbulent.

4. Compute the Reynolds number for the flow of fluids in round pipes and tubes.

5. State *Darcy's equation* for computing the energy loss due to friction for either laminar or turbulent flow.

6. State the *Hagen–Poiseuille equation* for computing the energy loss due to friction in laminar flow.

7. Define the *friction factor* as used in Darcy's equation.

8. Determine the friction factor using Moody's diagram for specific values of Reynolds number and the relative roughness of the pipe.

9. Compute the friction factor using equations developed by Swamee and Jain.

10. Compute the energy loss due to friction for fluid flow in circular pipes, hoses, and tubes and use the energy loss in the general energy equation.

11. Use the *Hazen–Williams* formula to compute energy loss due to friction for the special case of the flow of water in circular pipes.

8.2 REYNOLDS NUMBER

The behavior of a fluid, particularly with regard to energy losses, is quite dependent on whether the flow is laminar or turbulent, as will be demonstrated later in this chapter. For this reason we need a means of predicting the type of flow without actually observing it. Indeed, direct observation is impossible for fluids in opaque pipes. It can be shown experimentally and verified analytically that the character of flow in a round pipe depends on four variables: fluid density ρ, fluid viscosity η, pipe diameter D, and average velocity of flow. Osborne Reynolds was the first to demonstrate that laminar or turbulent flow can be predicted if the magnitude

of a dimensionless number, now called the Reynolds number (N_R), is known. See Internet resource 1. The following equation shows the basic definition of the Reynolds number:

⇨ **Reynolds Number—Circular Sections**

$$N_R = \frac{vD\rho}{\eta} = \frac{vD}{\nu} \qquad (8-1)$$

These two forms of the equation are equivalent because $\nu = \eta/\rho$ as discussed in Chapter 2.

You must use a consistent set of units to ensure that the Reynolds number is dimensionless. Table 8.1 lists the required units in both the SI metric unit system and the U.S. Customary unit system. Converting to these standard units prior to entering data into the calculation for N_R is recommended. Of course, you could enter the given data with units into the calculation and perform the appropriate conversions as the calculation is being finalized. Review Sections 2.1 and 2.2 in Chapter 2 for the discussion of viscosity. Consult Appendix K for conversion factors.

We can demonstrate that the Reynolds number is dimensionless by substituting standard SI units into Eq. (8–1):

$$N_R = \frac{vD\rho}{\eta} = v \times D \times \rho \times \frac{1}{\eta}$$

$$N_R = \frac{m}{s} \times m \times \frac{kg}{m^3} \times \frac{m \cdot s}{kg}$$

Because all units can be cancelled, N_R is dimensionless.

The Reynolds number is one of several dimensionless numbers useful in the study of fluid mechanics and heat transfer. The process called *dimensional analysis* can be used to determine dimensionless numbers (see Reference 1).

The Reynolds number is the ratio of the inertia force on an element of fluid to the viscous force. The inertia force is developed from Newton's second law of motion, $F = ma$. As discussed in Chapter 2, the viscous force is related to the product of the shear stress times area.

Flows having large Reynolds numbers, typically because of high velocity and/or low viscosity, tend to be turbulent. Those fluids having high viscosity and/or moving at low velocities will have low Reynolds numbers and will tend to be laminar. The following section gives some quantitative data with which to predict whether a given flow system will be laminar or turbulent.

TABLE 8.1 Standard units for quantities used in the calculation of Reynolds number to ensure that it is dimensionless

Quantity	SI Units	U.S. Customary Units
Velocity	m/s	ft/s
Diameter	M	ft
Density	kg/m^3 or N·s^2/m^4	slugs/ft^3 or lb·s^2/ft^4
Dynamic viscosity	N·s/m^2 or Pa·s or kg/m·s	lb·s/ft^2 or slugs/ft·s
Kinematic viscosity	m^2/s	ft^2/s

The formula for Reynolds number takes a different form for noncircular cross sections, open channels, and the flow of fluid around immersed bodies. These situations are discussed elsewhere in this book.

8.3 CRITICAL REYNOLDS NUMBERS

For practical applications in pipe flow we find that if the Reynolds number for the flow is less than 2000, the flow will be laminar. If the Reynolds number is greater than 4000, the flow can be assumed to be turbulent. In the range of Reynolds numbers between 2000 and 4000, it is impossible to predict which type of flow exists; therefore, this range is called the *critical region*. Typical applications involve flows that are well within the laminar flow range or well within the turbulent flow range, so the existence of this region of uncertainty does not cause great difficulty. If the flow in a system is found to be in the critical region, the usual practice is to change the flow rate or pipe diameter to cause the flow to be definitely laminar or turbulent. More precise analysis is then possible.

By carefully minimizing external disturbances, it is possible to maintain laminar flow for Reynolds numbers as high as 50 000. However, when N_R is greater than about 4000, a minor disturbance of the flow stream will cause the flow to suddenly change from laminar to turbulent. For this reason, and because we are dealing with practical applications in this book, we will assume the following:

If $N_R < 2000$, the flow is laminar.

If $N_R > 4000$, the flow is turbulent.

Example Problem 8.1   Determine whether the flow is laminar or turbulent if glycerin at 25°C flows in a circular passage within a fabricated chemical processing device. The diameter of the passage is 150 mm. The average velocity of flow is 3.6 m/s.

Solution   We must first evaluate the Reynolds number using Eq. (8–1):

$$N_R = vD\rho/\eta$$
$$v = 3.6 \text{ m/s}$$
$$D = 0.15 \text{ m}$$
$$\rho = 1258 \text{ kg/m}^3 \quad \text{(from Appendix B)}$$
$$\eta = 9.60 \times 10^{-1} \text{ Pa·s} \quad \text{(from Appendix B)}$$

Then we have

$$N_R = \frac{(3.6)(0.15)(1258)}{9.60 \times 10^{-1}} = 708$$

Because $N_R = 708$, which is less than 2000, the flow is laminar. Notice that each term was expressed in consistent SI units before N_R was evaluated.

Example Problem 8.2   Determine whether the flow is laminar or turbulent if water at 70°C flows in a hydraulic copper tube with a 32 mm OD × 2.0 mm wall. The flow rate is 285 L/min.

Solution   Evaluate the Reynolds number, using Eq. (8–1):

$$N_R = \frac{vD\rho}{\eta} = \frac{vD}{\nu}$$

For the copper tube, $D = 28$ mm $= 0.028$ m, and $A = 6.158 \times 10^{-4}$ m^2 (from Appendix G.2). Then we have

$$v = \frac{Q}{A} = \frac{285 \text{ L/min}}{6.158 \times 10^{-4} \text{ m}^2} \times \frac{1 \text{ m}^3/\text{s}}{60\,000 \text{ L/min}} = 7.71 \text{ m/s}$$

$$\nu = 4.11 \times 10^{-7} \text{ m}^2/\text{s} \quad \text{(from Appendix A)}$$

$$N_R = \frac{(7.71)(0.028)}{4.11 \times 10^{-7}} = 5.25 \times 10^5$$

Because the Reynolds number is greater than 4000, the flow is turbulent.

Example Problem 8.3   Determine the range of average velocity of flow for which the flow would be in the critical region if SAE 10 oil at 60°F is flowing in a 2-in Schedule 40 steel pipe. The oil has a specific gravity of 0.89.

Solution   The flow would be in the critical region if $2000 < N_R < 4000$. First, we use the Reynolds number and solve for velocity:

$$N_R = \frac{vD\rho}{\eta}$$

$$v = \frac{N_R\eta}{D\rho} \tag{8–2}$$

Then we find the values for η, D, and ρ:

$$D = 0.1723 \text{ ft} \quad \text{(from Appendix F)}$$
$$\eta = 2.10 \times 10^{-3} \text{ lb-s/ft}^2 \quad \text{(from Appendix D)}$$
$$\rho = (sg)(1.94 \text{ slugs/ft}^3) = (0.89)(1.94 \text{ slugs/ft}^3) = 1.73 \text{ slugs/ft}^3$$

Substituting these values into Eq. (8–2), we get

$$v = \frac{N_R(2.10 \times 10^{-3})}{(0.1723)(1.73)} = (7.05 \times 10^{-3})N_R$$

For $N_R = 2000$, we have

$$v = (7.05 \times 10^{-3})(2 \times 10^3) = 14.1 \text{ ft/s}$$

For $N_R = 4000$, we have

$$v = (7.05 \times 10^{-3})(4 \times 10^3) = 28.2 \text{ ft/s}$$

Therefore, if $14.1 < v < 28.2$ ft/s, the flow will be in the critical region.

8.4 DARCY'S EQUATION

In the general energy equation

$$\frac{p_1}{\gamma} + z_1 + \frac{v_1^2}{2g} + h_A - h_R - h_L = \frac{p_2}{\gamma} + z_2 + \frac{v_2^2}{2g}$$

the term h_L is defined as the energy loss from the system. One component of the energy loss is due to friction in the flowing fluid. Friction is proportional to the velocity head of the flow and to the ratio of the length to the diameter of the flow stream, for the case of flow in pipes and tubes. This is expressed mathematically as Darcy's equation:

▷ **Darcy's Equation for Energy Loss**

$$h_L = f \times \frac{L}{D} \times \frac{v^2}{2g} \tag{8–3}$$

where

h_L = energy loss due to friction (N·m/N, m, lb-ft/lb, or ft)

L = length of flow stream (m or ft)

D = pipe diameter (m or ft)

v = average velocity of flow (m/s or ft/s)

f = friction factor (dimensionless)

Darcy's equation can be used to calculate the energy loss due to friction in long straight sections of round pipe for both laminar and turbulent flow. The difference between the two is in the evaluation of the dimensionless friction factor f as explained in the next two sections. Note that the calculation of velocity of flow for a given volume flow rate through a given pipe size requires the use of the equation, $Q = Av$, as introduced in Chapter 6. Now that you have mastered the use of this equation, you may find the website listed in Internet resource 2 to be a useful tool.

8.5 FRICTION LOSS IN LAMINAR FLOW

When laminar flow exists, the fluid seems to flow as several layers, one on another. Because of the viscosity of the fluid, a shear stress is created between the layers of fluid. Energy is lost from the fluid by the action of overcoming the frictional forces produced by the shear stress. Because laminar flow is so regular and orderly, we can derive a relationship between the energy loss and the measurable parameters of the flow system. This relationship is known as the *Hagen–Poiseuille equation*:

▷ **Hagen–Poiseuille Equation**

$$h_L = \frac{32\eta Lv}{\gamma D^2} \tag{8–4}$$

The parameters involved are the fluid properties of viscosity and specific weight, the geometrical features of length and pipe diameter, and the dynamics of the flow characterized by the average velocity. The Hagen–Poiseuille equation has been verified experimentally many times. You should observe from Eq. (8–4) that the energy loss in laminar flow is independent of the condition of the pipe surface. Viscous friction losses within the fluid govern the magnitude of the energy loss.

The Hagen–Poiseuille equation is valid only for laminar flow ($N_R < 2000$). However, we stated earlier that Darcy's equation, Eq. (8–3), could also be used to calculate the friction loss for laminar flow. If the two relationships for h_L are set equal to each other, we can solve for the value of the friction factor:

$$f \times \frac{L}{D} \times \frac{v^2}{2g} = \frac{32\eta L v}{\gamma D^2}$$

$$f = \frac{32\eta L v}{\gamma D^2} \times \frac{D2g}{L v^2} = \frac{64\eta g}{vD\gamma}$$

Because $\rho = \gamma/g$, we get

$$f = \frac{64\eta}{vD\rho}$$

The Reynolds number is defined as $N_R = vD\rho/\eta$. Then we have

⇨ **Friction Factor for Laminar Flow**

$$f = \frac{64}{N_R} \qquad (8\text{–}5)$$

In summary, the energy loss due to friction in *laminar flow* can be calculated either from the Hagen–Poiseuille equation,

$$h_L = \frac{32\eta L v}{\gamma D^2}$$

or from Darcy's equation,

$$h_L = f \times \frac{L}{D} \times \frac{v^2}{2g}$$

where $f = 64/N_R$.

Example Problem 8.4

Determine the energy loss if glycerin at 25°C flows 30 m through a standard DN 150-mm Schedule 80 pipe with an average velocity of 4.0 m/s.

Solution

First, we must determine whether the flow is laminar or turbulent by evaluating the Reynolds number:

$$N_R = \frac{vD\rho}{\eta}$$

From Appendix B, we find that for glycerin at 25°C

$$\rho = 1258 \text{ kg/m}^3$$
$$\eta = 9.60 \times 10^{-1} \text{ Pa·s}$$

Then, we have

$$N_R = \frac{(4.0)(0.1463)(1258)}{9.60 \times 10^{-1}} = 767$$

Because $N_R < 2000$, the flow is laminar.

Using Darcy's equation, we get

$$h_L = f \times \frac{L}{D} \times \frac{v^2}{2g}$$

$$f = \frac{64}{N_R} = \frac{64}{767} = 0.0835$$

$$h_L = 0.0835 \times \frac{30}{0.1463} \times \frac{(4.0)^2}{2(9.81)}\text{m} = 13.96 \text{ m}$$

Notice that each term in each equation is expressed in the units of the SI unit system. Therefore, the resulting units for h_L are m or N·m/N. This means that 13.96 N·m of energy is lost by each newton of the glycerin as it flows along the 30 m of pipe.

8.6 FRICTION LOSS IN TURBULENT FLOW

For turbulent flow of fluids in circular pipes it is most convenient to use Darcy's equation to calculate the energy loss due to friction. Turbulent flow is rather chaotic and is constantly varying. For these reasons we must rely on experimental data to determine the value of f.

Tests have shown that the dimensionless number f is dependent on two other dimensionless numbers, the Reynolds number and the relative roughness of the pipe. The relative roughness is the ratio of the pipe diameter D to the average

FIGURE 8.6 Pipe wall roughness (exaggerated).

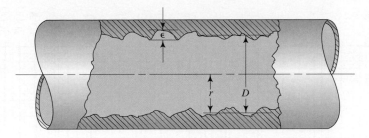

pipe wall roughness ε (Greek letter epsilon). Figure 8.6 illustrates pipe wall roughness (exaggerated) as the height of the peaks of the surface irregularities. The condition of the pipe surface is very much dependent on the pipe material and the method of manufacture. Because the roughness is somewhat irregular, averaging techniques are used to measure the overall roughness value.

For commercially available pipe and tubing, the design value of the average wall roughness ε has been determined as shown in Table 8.2. *These are only average values for new, clean pipe. Some variation should be expected. After a pipe has been in service for a time, the roughness could change due to the formation of deposits on the wall or due to corrosion.*

Glass tubing has an inside surface that is virtually hydraulically smooth, indicating a very small value of roughness. Therefore, the relative roughness, D/ε, approaches infinity. Plastic pipe and tubing are nearly as smooth as glass, and we use the listed roughness value in this book. Variations should be expected. Copper, brass, and some steel tubing are drawn to its final shape and size over an internal mandrel, leaving a fairly smooth surface. For standard steel pipe (such as Schedule 40 and Schedule 80) and welded steel tubing, we use the roughness value listed for commercial steel or welded steel. Galvanized iron has a metallurgically bonded zinc coating for corrosion resistance. Ductile iron pipe is typically coated on the inside with a cement mortar for corrosion protection and to improve surface roughness. In this book, we use the roughness values for coated ductile iron unless stated otherwise. Ductile iron pipe from some manufacturers has a smoother inside surface, approaching that of steel. Well-made concrete pipe can have roughness values similar to the values

for coated ductile iron as listed in the table. However, a wide range of values exist, and data should be obtained from the manufacturer. Riveted steel is used in some new large pipelines and in some existing installations.

8.6.1 The Moody Diagram

One of the most widely used methods for evaluating the friction factor employs the Moody diagram shown in Fig. 8.7. The diagram shows the friction factor f plotted versus the Reynolds number N_R, with a series of parametric curves related to the relative roughness D/ε. These curves were generated from experimental data by L. F. Moody (see Reference 2).

Both f and N_R are plotted on logarithmic scales because of the broad range of values encountered. At the left end of the chart, for Reynolds numbers less than 2000, the straight line shows the relationship $f = 64/N_R$ for laminar flow. For $2000 < N_R < 4000$, no curves are drawn because this is the critical zone between laminar and turbulent flow and it is not possible to predict the type of flow. The change from laminar to turbulent flow results in values for friction factors within the shaded band. Beyond $N_R = 4000$, the family of curves for different values of D/ε is plotted. Several important observations can be made from these curves:

1. For a given Reynolds number of flow, as the relative roughness D/ε is increased, the friction factor f decreases.

2. For a given relative roughness D/ε, the friction factor f decreases with increasing Reynolds number until the zone of complete turbulence is reached.

3. Within the zone of complete turbulence, the Reynolds number has no effect on the friction factor.

TABLE 8.2 Pipe roughness—design values		
Material	**Roughness ε (m)**	**Roughness ε (ft)**
Glass	Smooth	Smooth
Plastic	3.0×10^{-7}	1.0×10^{-6}
Drawn tubing; copper, brass, steel	1.5×10^{-6}	5.0×10^{-6}
Steel, commercial or welded	4.6×10^{-5}	1.5×10^{-4}
Galvanized iron	1.5×10^{-4}	5.0×10^{-4}
Ductile iron—coated	1.2×10^{-4}	4.0×10^{-4}
Ductile iron—uncoated	2.4×10^{-4}	8.0×10^{-4}
Concrete, well made	1.2×10^{-4}	4.0×10^{-4}
Riveted steel	1.8×10^{-3}	6.0×10^{-3}

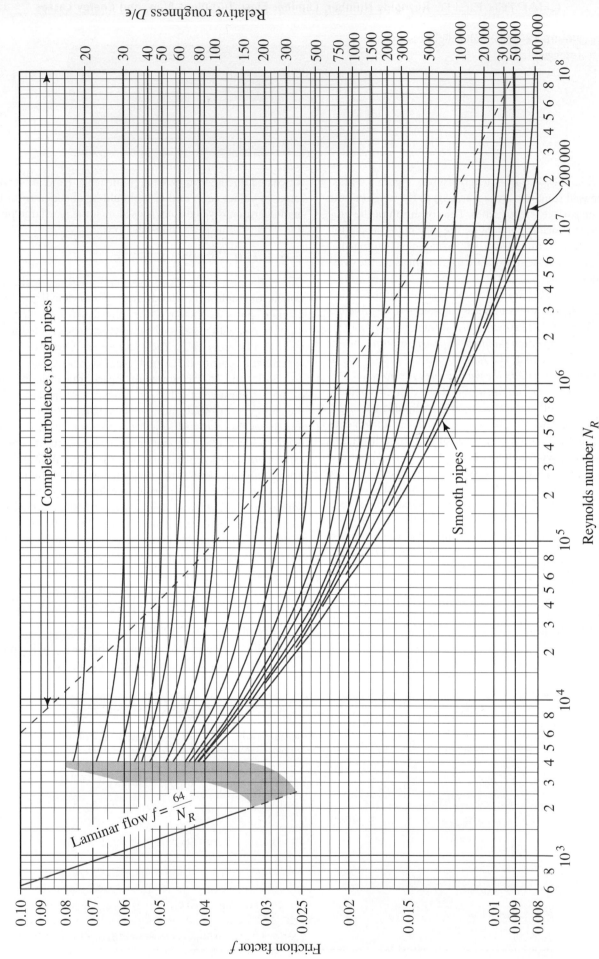

FIGURE 8.7 Mcody's diagram. (*Source:* Pao, R.H.E. *Fluid Mechanics*, p. 284. Copyright 9c. 1961. Reprinted by permission of the author.)

FIGURE 8.8 Explanation of parts of Moody's diagram.

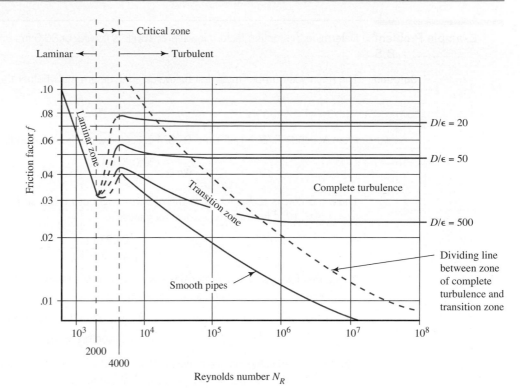

4. As the relative roughness D/ε increases, the value of the Reynolds number at which the zone of complete turbulence begins also increases.

Figure 8.8 is a simplified sketch of Moody's diagram in which the various zones are identified. The *laminar zone* at the left has already been discussed. At the right of the dashed line downward across the diagram is the *zone of complete turbulence*. The lowest possible friction factor for a given Reynolds number in turbulent flow is indicated by the *smooth pipes* line.

Between the smooth pipes line and the line marking the start of the complete turbulence zone is the *transition zone*. Here, the various D/ε lines are curved, and care must be exercised to evaluate the friction factor properly. You can see, for example, that the value of the friction factor for a relative roughness of 500 decreases from 0.0420 at $N_R = 4000$ to 0.0240 at $N_R = 6.0 \times 10^5$, where the zone of complete turbulence starts.

Check your ability to read the Moody diagram correctly by verifying the following values for friction factors for the given values of Reynolds number and relative roughness, using Fig. 8.7:

The critical zone between the Reynolds numbers of 2000 and 4000 is to be avoided if possible because within this range the type of flow cannot be predicted. The shaded band shows how the friction factor could change according to the value of the relative roughness. For low values of D/ε (indicating large pipe wall roughness), the increase in friction factor is great as the flow changes from laminar to turbulent. For example, for flow in a pipe with $D/\varepsilon = 20$, the friction factor would increase from 0.032 for $N_R = 2000$ at the end of the laminar range to approximately 0.077 at $N_R = 4000$ at the beginning of the turbulent range, an increase of 240 percent. Moreover, the value of the Reynolds number where this would occur cannot be predicted. Because the energy loss is directly proportional to the friction factor, changes of such magnitude are significant.

It should be noted that because relative roughness is defined as D/ε, a high relative roughness indicates a low value of ε, that is, a smoother pipe. In fact, the curve labeled *smooth pipes* is used for materials such as glass that have such a low roughness that D/ε would be an extremely large number, approaching infinity.

Some texts and references use other conventions for reporting relative roughness, such as ε/D, ε/r, or r/ε, where r is the pipe radius. We feel that the convention used in this book makes calculations and interpolations easier.

8.6.2 Use of the Moody Diagram

The Moody diagram is used to help determine the value of the friction factor f for turbulent flow. The value of the Reynolds number and the relative roughness must be known. Therefore, the basic data required are the pipe inside diameter (ID), the pipe material, the flow velocity, and the kind of fluid and its temperature, from which the viscosity can be found. The following example problems illustrate the procedure for finding f.

N_R	D/ε	f
6.7×10^3	150	0.0430
1.6×10^4	2000	0.0284
1.6×10^6	2000	0.0171
2.5×10^5	733	0.0223

Example Problem 8.5

Determine the friction factor f if water at 160°F is flowing at 30.0 ft/s in a 1-in Schedule 40 steel pipe.

Solution

The Reynolds number must first be evaluated to determine whether the flow is laminar or turbulent:

$$N_R = \frac{vD}{\nu}$$

From Appendix F: $D = 1.049$ in $= 0.0874$ ft. For the water, from Appendix A.2, $\nu = 4.38 \times 10^{-6}$ ft²/s, then

$$N_R = \frac{(30.0)(0.0874)}{4.38 \times 10^{-6}} = 5.98 \times 10^5$$

Thus, the flow is turbulent. Now the relative roughness must be evaluated. From Table 8.2 we find $\varepsilon = 1.5 \times 10^{-4}$ ft. Then the relative roughness is

$$\frac{D}{\varepsilon} = \frac{0.0874 \text{ ft}}{1.5 \times 10^{-4} \text{ ft}} = 583$$

Notice that for D/ε to be a dimensionless ratio, both D and ε must be in the same units.
  The final steps in the procedure are as follows:

1. Locate the Reynolds number on the abscissa of the Moody diagram:

$$N_R = 5.98 \times 10^5$$

2. Project vertically until the curve for $D/\varepsilon = 583$ is reached. You must interpolate between the curve for 500 and the one for 750 on the vertical line for $N_R = 5.98 \times 10^5$.

3. Project horizontally to the left, and read $f = 0.023$.

Example Problem 8.6

If the flow velocity of water in Problem 8.5 was 0.45 ft/s with all other conditions being the same, determine the friction factor f.

Solution

$$N_R = \frac{vD}{\nu} = \frac{(0.45)(0.0874)}{4.38 \times 10^{-6}} = 8.98 \times 10^3$$

$$\frac{D}{\varepsilon} = \frac{0.0874}{1.5 \times 10^{-4}} = 583$$

Then, from Fig. 8.7, $f = 0.0343$. Notice that this is on the curved portion of the D/ε curve and that there is a significant increase in the friction factor over that in Example Problem 8.5.

Example Problem 8.7

Determine the friction factor f if ethyl alcohol at 25°C is flowing at 5.3 m/s in a standard DN 40 Schedule 80 steel pipe.

Solution

Evaluating the Reynolds number, we use the equation

$$N_R = \frac{vD\rho}{\eta}$$

From Appendix B, $\rho = 787$ kg/m³ and $\eta = 1.00 \times 10^{-3}$ Pa·s. Also, for a DN 40 Schedule 80 pipe, $D = 0.0381$ m. Then we have

$$N_R = \frac{(5.3)(0.0381)(787)}{1.00 \times 10^{-3}} = 1.59 \times 10^5$$

Thus, the flow is turbulent. For a steel pipe, $\varepsilon = 4.6 \times 10^{-5}$ m, so the relative roughness is

$$\frac{D}{\varepsilon} = \frac{0.0381 \text{ m}}{4.6 \times 10^{-5} \text{ m}} = 828$$

From Fig. 8.7, $f = 0.0225$. You must interpolate on both N_R and D/ε to determine this value, and you should expect some variation. However, you should be able to read the value of the friction factor f within ± 0.0005 in this portion of the graph.

The following is a programmed example problem illustrating a typical fluid piping situation. The energy loss due to friction must be calculated as a part of the solution.

PROGRAMMED EXAMPLE PROBLEM

Example Problem 8.8

Figure 8.9 shows an industrial storage tank from which a horizontal 100 m long pipe carries water at 25°C to a process where a bulk food product is being prepared. The pipe is a DN 50 Schedule 40 steel pipe and the design delivery rate to the process is 520 L/min. Determine the amount of pressure drop that occurs in the pipe from the storage tank to the processing system.

First, lay out a plan for completing this problem.

Here is one approach:

1. Define point A in the pipe where it exits the storage tank and point B where the tank delivers the water to the processing system. The objective of the problem is to calculate $p_A - p_B$, the pressure drop between points A and B.

2. Use the energy equation to determine the pressure drop, considering the energy loss due to friction in the pipe.

3. Compute the energy loss using Darcy's equation.

Write the energy equation between points A and B now, and solve algebraically for the pressure drop.

The energy equation is:

$$p_A/\gamma + z_A + v_A^2/2g - h_L = p_B/\gamma + z_B + v_B^2/2g$$

Notice that $z_A = z_B$ and $v_A = v_B$ and therefore those terms can be cancelled from the equation. Now we can solve for the pressure drop.

$$p_A - p_B = \gamma h_L$$

How can you calculate the energy loss due to friction in the pipe?

Darcy's equation can be used:

$$h_L = f \times \frac{L}{D} \times \frac{v^2}{2g}$$

Determine the data needed to complete the calculation.

From the given data, we can show that $L = 100$ m, $Q = 520$ L/min, the pipe is a DN 50 Schedule 40 steel pipe, $g = 9.81$ m/s², and the fluid is water at 25°C for which Appendix A gives

$$\gamma = 9.78 \text{ kN/m}^3 \text{ and the kinematic viscosity is, } \nu = 8.94 \times 10^{-7} \text{ m}^2/\text{s}$$

For a DN 50 steel pipe, from Appendix F, $D = 52.5$ mm $= 0.0525$ m, and $A = 2.168 \times 10^{-3}$ m².

Now the velocity of flow can be computed.

$$v = \frac{Q}{A} = \frac{520 \text{ L/min}}{2.168 \times 10^{-3} \text{ m}^2} \times \frac{1.0 \text{ m}^3/\text{s}}{60\,000 \text{ L/min}} = 4.00 \text{ m/s}$$

FIGURE 8.9 Example Problem 8.8.

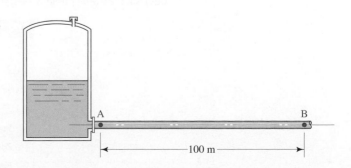

To determine the friction factor, f, we need to compute the Reynolds number.

$$N_R = \frac{vD}{\nu} = \frac{(4.00 \text{ m/s})(0.0525 \text{ m})}{8.94 \times 10^{-7} \text{ m}^2/\text{s}} = 2.349 \times 10^5$$

Then the flow is turbulent and we use the Moody diagram to find f. From Table 8.2 we find the roughness for a clean steel pipe to be $\varepsilon = 4.6 \times 10^{-5}$ m. Then,

$$D/\varepsilon = 0.0525 \text{ m}/4.6 \times 10^{-5} \text{ m} = 1141$$

From Moody's diagram, we can read $f = 0.0203$. Complete the calculation for h_L now.

Here is the result:

$$h_L = f \times \frac{L}{D} \times \frac{v^2}{2g} = 0.0203 \times \frac{100}{0.0525} \times \frac{(4.00)^2}{2(9.81)} = 31.53 \text{ m}$$

Now complete the calculation for the pressure drop in the pipe.

You should get $p_A - p_B = 308.4$ kPa. Here are the details:

$$p_A - p_B = \gamma h_L = (9.78 \text{ kN/m}^3)(31.53 \text{ m}) = 308.4 \text{ kN/m}^2 = 308.4 \text{ kPa}$$

This example problem is completed.

8.7 USE OF SOFTWARE FOR PIPE FLOW PROBLEMS

Calculations to achieve full understanding of fluid systems can be tedious and repetitive, making this process a good candidate for moving to software solutions. One program that automates the calculations presented in this text is called PIPE-FLO®, by Engineered Software Incorporated. Using powerful software always comes with a responsibility to fully understand the calculations that are being performed, and this application is no different. Many problems presented in this text can be effectively and efficiently solved with this software, and they will be presented in appropriate sections through Chapter 13 of this text. Using this software as a supplement to manual calculations while learning the principles not only aids in understanding, but also prepares one for the responsible use of such software throughout a career.

At this point in the course, students should go to:

http://www.eng-software.com/appliedfluidmechanics

Download the free "demo" version of the software.

Instructions for the download and other useful material for the course are available at the site. The demo version of the software limits the number of pipes and the types of fluids, but the limits are beyond most of the problems presented in this text, so the software will operate just like the full professional version throughout this course. Results that are printed for each solution are the same results that one would derive through manual calculation. The instructions, tutorials, and help functions available through the software minimize the need for written instructions in this text, but for each major new function this text provides an Example Problem to guide the process and give checkpoint answers to confirm correct usage. The demo version will also act as a "reader" for any pipe model, so you can open any system with the demo version, including large complex ones available at the website listed above.

Example Problem 8.9   Use the PIPE-FLO® software to determine the pressure drop in 100 m of horizontal DN 50 Schedule 40 pipe, carrying 25°C water at a velocity of 4 m/s. Report all applicable values related to the solution such as Reynolds number and friction factor.

Solution   1. Open a new project in PIPE-FLO® and select the "System" menu on the toolbar to initialize all key data such as units, fluid zones, and pipe specifications.

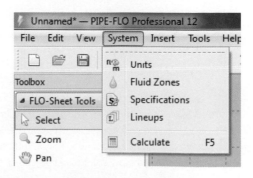

2. The "Units" menu allows us to choose how we want to define our units for the entire piping system. For this problem, select the typical SI units as shown below. Note that PIPE-FLO® defaults to English units at the start of a new project, but any units can be selected through this menu. The units for any individual parameter can be easily changed via drop down at any time when working with the model.

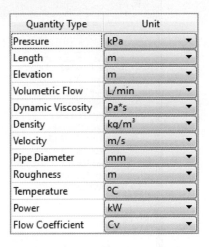

Quantity Type	Unit
Pressure	kPa ▼
Length	m ▼
Elevation	m ▼
Volumetric Flow	L/min ▼
Dynamic Viscosity	Pa*s ▼
Density	kg/m³ ▼
Velocity	m/s ▼
Pipe Diameter	mm ▼
Roughness	m ▼
Temperature	°C ▼
Power	kW ▼
Flow Coefficient	Cv ▼

3. The "Fluid Zone" menu allows us to define what fluid we wish to use in our system. All problems worked with PIPE-FLO® in this text will comply with the demo version constraint of a single fluid zone, meaning one fluid in the system. Under the "System" menu, select "Fluid Zone", then "New", then "Water." Upon selecting a fluid in the box on the left, and entering our initial temperature and pressure information, PIPE-FLO® will fill in the remaining information based on its internal database, reflecting the values we would manually look up in the appendix of this text or outside references. This database approach to properties is both convenient and powerful, also allowing us to edit a fluid's state, such as its temperature, and all associated properties are automatically updated. For this problem, enter 25°C and 101 kPa(abs), which designates atmospheric pressure. Note that in PIPE-FLO®, absolute pressure is written as "kPa a". We can also rename our fluid for clarity and convenience. Rename this fluid zone "Water @ 25C", as shown, which will be important later in this example.

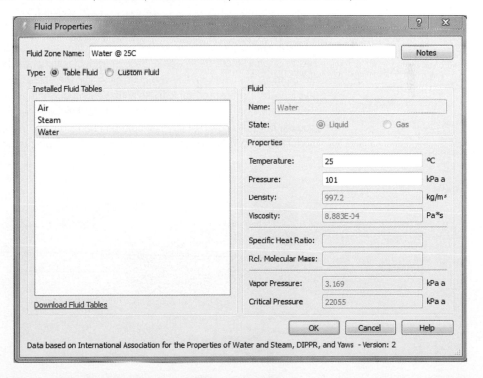

4. Now designate the type of pipe. For this problem, click "New" on the main pipe specification menu, scroll to "Steel A53-B36.10" to indicate commercial steel pipe, and then the number "40" to indicate the schedule. Be sure to use the small triangle to the left of the words to expose the pull-down menu that lists the various schedules available. The roughness factor shown corresponds to values shown in Table 8.2 in this chapter. Note that for special cases, the user can simply enter a roughness factor directly. As with the fluid zones, the pipe names can also be changed, shown in the example below as "Schedule 40". You will note that we haven't chosen a size for a particular pipe yet and there is no option to do so under this menu. We will complete the pipe size later in the problem. Here we simply establish the pipe type.

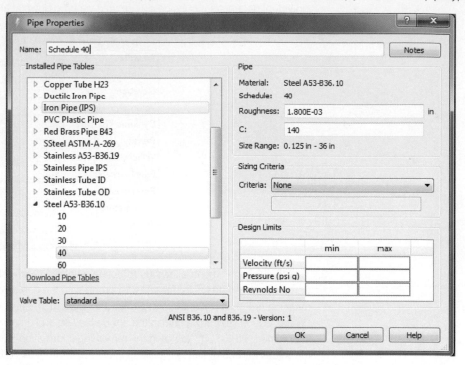

5. With all system data initialized, start building the system. There are many ways to model a horizontal pipe with flow to determine the pressure drop. Here let's do so with a supply tank at one end, and a "Flow Demand" at the other. Click and drag a "Tank" from the toolbox on the left, to the open background referred to as the FLO-Sheet®. We are using this tank as an arbitrary pressure source to model our horizontal run of pipe. Click on the tank on the FLO-Sheet® to display the "Property Grid" on the right-hand side of the page. Enter a tank elevation of 0 m, a surface pressure of 750 kPa, a liquid level of 10 m, and the fluid zone that we earlier named as "Water @ 25C". Note that these values are arbitrary for this problem because we are simply asked to compute the pressure loss over 100 m of pipe. More detail about these individual settings will be explained in future sections of the text as more complete models are built. Note that the default tank icon shows an open tank. Change that icon to a pressured tank by selecting "Symbol Settings" in the property grid and choosing a closed tank icon as shown below. Again, more detail will be given later with more complete system models, but note now that icons should be changed when necessary to more accurately reflect the model being built. An open tank icon, for example, should not be left to represent a pressurized tank.

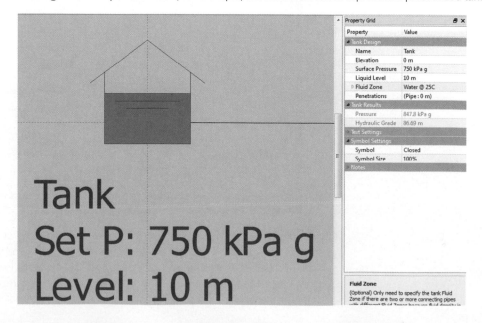

6. Next, add the pipe to the model. Choose "Pipe" from the toolbox menu on the left, click on the tank on the FLO-Sheet®, and drag the pipe to the right and click the FLO-Sheet® again. Don't worry about initial placement; after the pipe has been placed, we can easily modify its properties similar to the process used for the tank. Choose the same fluid zone. Choose the type of pipe established earlier, and specify size. This problem calls for DN 50 Schedule 40, and the actual size can be verified by dropping down the "Size" column in the property grid which shows the pipe I.D. as 52.5 mm. Enter a length of 100 m for this section of pipe. Recall from Chapter 6 that the DN 50 Schedule 40 pipe is identical to the 2-in Schedule 40 pipe.

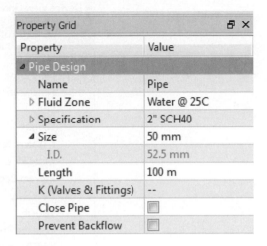

7. PIPE-FLO® can, of course, build complete systems, but it also provides ways to model segments of a system. In this case we're interested only in the pressure drop in one run of pipe. Rather than model components downstream, simply enter a "Flow Demand" at the end of the pipe indicating downstream flow without requiring any system details. Flow demand is found in the toolbox under the "Basic Devices" section. After placing it at the end of the pipe, we must enter values for the elevation, flow rate, and flow type of the demand. For this example, use an elevation of zero, assuming the pipe is horizontal. Calculate the flow rate corresponding to a velocity of 4 m/s. Enter that flow rate, 520.3 L/min. Since this demand represents flow leaving the system, choose the "Flow out" option under the flow type.

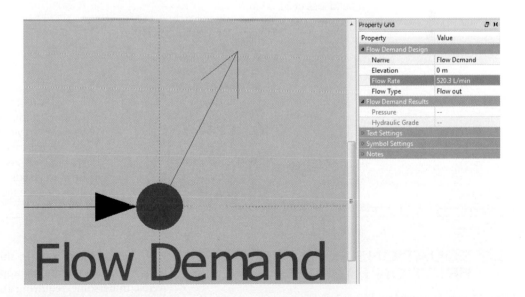

8. The final step in this problem is to calculate the results of the system. Click on the button that looks like a calculator on the toolbar.

To view the information that has been calculated about a particular item on the FLO-Sheet®, choose that item on the property grid under "Device View Options". Simply check the box of each item you want to display, and they will be shown on the FLO-Sheet® under that particular item. For this introductory problem select all data to be displayed. Note, in order to find the "Device View Options" box in the Property Grid, you must first click on the FLO-Sheet® background and make sure that no system component is selected.

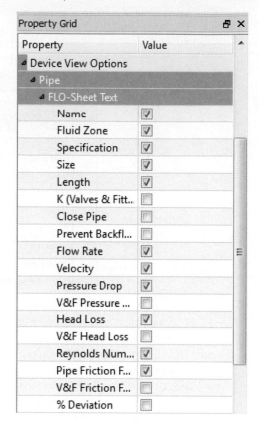

Final results are displayed under the pipe in text form. In this case, the pressure drop across this length of pipe is 309.8 kPa, with a Reynolds number of 236 068, and a friction factor of 0.02 033, resulting in a head loss of 31.68 m.

Pipe
Zone: Water @ 25C
Spec: 2" SCH40
Ø: 50 mm
L: 100 m
Flow: 520.3 L/min
Vel: 4.006 m/s
dP: 309.8 kPa
HL: 31.68 m
Re: 236068
ffp: 0.02033

L = Pipe length
Flow = Volume flow rate, Q
Vel = Velocity of flow, v
dP = Pressure drop, Δp
HL = Head loss, h_L
Re = Reynolds number, N_R
ffp = Friction factor, f

8.8 EQUATIONS FOR THE FRICTION FACTOR

The Moody diagram in Fig. 8.7 is a convenient and sufficiently accurate means of determining the value of the friction factor when solving problems by manual calculations. However, if the calculations are to be automated for solution on a computer or a programmable calculator, we need equations for the friction factor.

The equations used in the work by Moody form the basis of the computational approach.* But those equations were cumbersome, requiring an iterative approach. We show here two equations that allow the direct solution for the

*Earlier work in developing the equations was done by several researchers, most notably C. F. Colebrook, L. Prandtl, H. Rouse, T. van Karman, and J. Nikuradse, whose papers are listed in the bibliography of Reference 2.

friction factor. One covers laminar flow and the other is used for turbulent flow.

In the *laminar flow zone*, for values below 2000, f can be found from Eq. (8–5),

$$f = 64/N_R$$

This relationship, developed in Section 8.4, plots in the Moody diagram as a straight line on the left side of the chart.

Of course, for Reynolds numbers from 2000 to 4000, the flow is in the critical range and it is impossible to predict the value of f.

The following equation, which allows the direct calculation of the value of the friction factor for turbulent flow, was developed by P. K. Swamee and A. K. Jain and is reported in Reference 3:

⊳ **Friction Factor for Turbulent Flow**

$$f = \frac{0.25}{\left[\log\left(\dfrac{1}{3.7\,(D/\varepsilon)} + \dfrac{5.74}{N_R^{0.9}}\right)\right]^2} \tag{8–7}$$

Equation (8–7) produces values for f that are within ± 1.0 percent within the range of relative roughness D/ε from 100 to 1×10^6 and for Reynolds numbers from 5×10^3 to 1×10^8. This is virtually the entire turbulent zone of the Moody diagram. Other research has been published giving alternate equations for computing friction factors. Reference 5 includes a review of some of those research papers.

Summary   In this book, to calculate the value of the friction factor f when the Reynolds number and relative roughness are known, use Eq. (8–5) for laminar flow and Eq. (8–7) for turbulent flow.

Example Problem 8.10   Compute the value for the friction factor if the Reynolds number for the flow is 1×10^5 and the relative roughness is 2000.

Solution   Because this is in the turbulent zone, we use Eq. (8–7),

$$f = \frac{0.25}{\left[\log\left(\dfrac{1}{3.7(2000)} + \dfrac{5.74}{(1 \times 10^5)^{0.9}}\right)\right]^2}$$

$$f = 0.0204$$

This value compares closely with the value read from Fig. 8.7.

8.9 HAZEN–WILLIAMS FORMULA FOR WATER FLOW

The Darcy equation presented in this chapter for calculating energy loss due to friction is applicable for any Newtonian fluid. An alternate approach is convenient for the special case of the flow of water in pipeline systems.

The *Hazen–Williams formula* is one of the most popular formulas for the design and analysis of water systems. Its use is limited to the flow of water in pipes larger than 2.0 in and smaller than 6.0 ft in diameter. The velocity of flow should not exceed 10.0 ft/s. Also, it has been developed for water at 60°F. Use at temperatures much lower or higher would result in some error.

The Hazen–Williams formula is unit-specific. In the U.S. Customary unit system it takes the form

⊳ **Hazen–Williams Formula, U.S. Customary Units**

$$v = 1.32 C_h R^{0.63} s^{0.54} \tag{8–8}$$

where

$v = $ Average velocity of flow (ft/s)

$C_h = $ Hazen–Williams coefficient (dimensionless)

$R = $ Hydraulic radius of flow conduit (ft)

$s = $ Ratio of h_L/L: energy loss/length of conduit (ft/ft)

The use of the hydraulic radius in the formula allows its application to noncircular sections as well as circular pipes. Use $R = D/4$ for circular pipes. This is discussed in Chapter 9.

The coefficient C_h is dependent only on the condition of the surface of the pipe or conduit. Table 8.3 gives typical values. Note that some values are described for pipe in new, clean condition, whereas the design value accounts for the accumulation of deposits that develop on the inside surfaces of pipe after a time, even when clean water flows through them. Smoother pipes have higher values of C_h than rougher pipes.

The Hazen–Williams formula for SI units is

⊳ **Hazen–Williams Formula, SI Units**

$$v = 0.85 C_h R^{0.63} s^{0.54} \tag{8–9}$$

where

$v = $ Average velocity of flow (m/s)

$C_h = $ Hazen–Williams coefficient (dimensionless)

$R = $ Hydraulic radius of flow conduit (m)

$s = $ Ratio h_L/L: energy loss/length of conduit (m/m)

As before, the volume flow rate can be computed from $Q = Av$.

TABLE 8.3   Hazen–Williams coefficient C_h

Type of Pipe	C_h Average for New, Clean Pipe	C_h Design Value
Steel, ductile iron, or cast iron with centrifugally applied cement or bituminous lining	150	140
Plastic, copper, brass, glass	140	130
Steel, cast iron, uncoated	130	100
Concrete	120	100
Corrugated steel	60	60

Example Problem 8.11   For what velocity of flow of water in a new, clean, 6-in Schedule 40 steel pipe would an energy loss of 20 ft of head occur over a length of 1000 ft? Compute the volume flow rate at that velocity. Then refigure the velocity using the design value of C_h for steel pipe.

Solution   We can use Eq. (8–8). Write

$$s = h_L/L = (20 \text{ ft})/(1000 \text{ ft}) = 0.02$$
$$R = D/4 = (0.5054 \text{ ft})/4 = 0.126 \text{ ft}$$
$$C_h = 130$$

Then,

$$v = 1.32 \, C_h R^{0.63} s^{0.54}$$
$$v = (1.32)(130)(0.126)^{0.63}(0.02)^{0.54} = 5.64 \text{ ft/s}$$
$$Q = Av = (0.2006 \text{ ft}^2)(5.64 \text{ ft/s}) = 1.13 \text{ ft}^3/\text{s}$$

Now we can adjust the result for the design value of C_h.

Note that the velocity and volume flow rate are both directly proportional to the value of C_h. If the pipe degrades after use so the design value of $C_h = 100$, the allowable volume flow rate to limit the energy loss to the same value of 20 ft per 1000 ft of pipe length would be

$$v = (5.64 \text{ ft/s})(100/130) = 4.34 \text{ ft/s}$$
$$Q = (1.13 \text{ ft}^3/\text{s})(100/130) = 0.869 \text{ ft}^3/\text{s}$$

8.10 OTHER FORMS OF THE HAZEN–WILLIAMS FORMULA

Equations (8–8) and (8–9) allow the direct computation of the velocity of flow for a given type and size of flow conduit when the energy loss per unit length is known or specified. The volume flow rate can be simply calculated by using $Q = Av$. Other types of calculations that are often desired are:

1. To determine the required size of pipe to carry a given flow rate while limiting the energy loss to some specified value.

2. To determine the energy loss for a given flow rate through a given type and size of pipe of a known length.

Table 8.4 shows several forms of the Hazen–Williams formula that facilitate such calculations.

8.11 NOMOGRAPH FOR SOLVING THE HAZEN–WILLIAMS FORMULA

The nomograph shown in Fig. 8.10 allows the solution of the Hazen–Williams formula to be done by simply aligning known quantities with a straight edge and reading the desired unknowns at the intersection of the straight edge with the appropriate vertical axis. *Note that this nomograph is constructed for the value of the Hazen–Williams coefficient of* $C_h = 100$. If the actual pipe condition warrants the use of a different value of C_h, the following formulas can be used to adjust the results. The subscript "100" refers to the value read from the nomograph for $C_h = 100$. The subscript "c" refers to the value for the given C_h.

$$v_c = v_{100}(C_h/100) \qquad [\text{velocity}] \qquad (8\text{–}10)$$
$$Q_c = Q_{100}(C_h/100) \qquad [\text{volume flow rate}] \qquad (8\text{–}11)$$

TABLE 8.4 Alternate forms of the Hazen–Williams formula

U.S. Customary Units	SI Units
$v = 1.32C_hR^{0.63}s^{0.54}$	$v = 0.85C_hR^{0.63}s^{0.54}$
$Q = 1.32AC_hR^{0.63}s^{0.54}$	$Q = 0.85AC_hR^{0.63}s^{0.54}$
$h_L = L\left[\dfrac{Q}{1.32AC_hR^{0.63}}\right]^{1.852}$	$h_L = L\left[\dfrac{Q}{0.85AC_hR^{0.63}}\right]^{1.852}$
$D = \left[\dfrac{2.31Q}{C_hs^{0.54}}\right]^{0.380}$	$D = \left[\dfrac{3.59Q}{C_hs^{0.54}}\right]^{0.380}$
Note: Units must be consistent:	
v in ft/s	v in m/s
Q in ft³/s	Q in m³/s
A in ft²	A in m²
h_L, L, R, and D in ft	h_L, L, R, and D in m
s in ft/ft (dimensionless)	s in m/m (dimensionless)

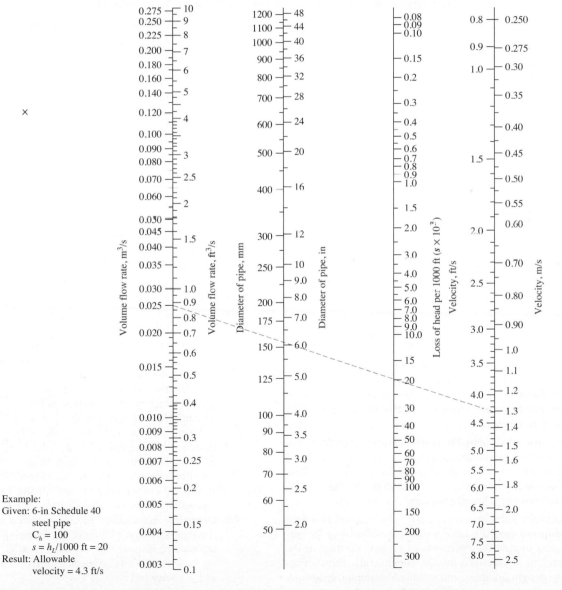

Example:
Given: 6-in Schedule 40
steel pipe
$C_h = 100$
$s = h_L/1000$ ft $= 20$
Result: Allowable
velocity = 4.3 ft/s

FIGURE 8.10 Nomograph for the solution of the Hazen–Williams formula for $C_h = 100$.

$$D_c = D_{100}(100/C_h)^{0.38} \quad [\text{pipe diameter}] \quad (8\text{--}12)$$

$$s_c = s_{100}(100/C_h)^{1.85} \quad [\text{head loss/length}] \quad (8\text{--}13)$$

The dashed line on the chart shows the use of the nomograph using data from Example Problem 8.11 for the case of $C_h = 100$.

One frequent use of a nomograph like that in Fig. 8.10 is to determine the required size of pipe to carry a given flow rate while limiting the energy loss to some specified value. Thus, it is a convenient design tool. See Reference 4.

**Example Problem
8.12**

Using Fig. 8.10, specify the required size of Schedule 40 steel pipe to carry 1.20 ft^3/s of water with no more than 4.0 ft of head loss over a 1000-ft length of pipe. Use the design value for C_h.

Solution

Table 8.3 suggests $C_h = 100$. Now, using Fig. 8.10, we can place a straight edge from $Q = 1.20$ ft^3/s on the volume flow rate line to the value of $s = (4.0 \text{ ft})/(1000 \text{ ft})$ on the energy loss line. The straight edge then intersects the pipe size line at approximately 9.7 in. The next larger standard Schedule 40 pipe size listed in Appendix F is the nominal 10-in pipe with an ID of 10.02 in.

Returning to the chart in Fig. 8.10 and slightly realigning $Q = 1.20$ ft^3/s with $D = 10.02$ in, we can read an average velocity of $v = 2.25$ ft/s. This is relatively low for a water distribution system, and the pipe is quite large. If the pipeline is long, the cost for piping would be excessively high.

If we allow the velocity of flow to increase to approximately 6.0 ft/s for the same volume flow rate, we can use the chart to show that a 6-in pipe could be used with a head loss of approximately 37 ft per 1000 ft of pipe. The lower cost of the pipe compared with the 10-in pipe would have to be compared with the higher energy cost required to overcome the additional head loss.

REFERENCES

1. Mory, Mathieu, 2011. *Fluid Mechanics for Chemical Engineering.* New York: John Wiley & Sons.

2. Moody, L. F. 1944. Friction Factors for Pipe Flow. *Transactions of the ASME* 66(8): 671–684. New York: American Society of Mechanical Engineers.

3. Swamee, P. K., and A. K. Jain. 1976. Explicit Equations for Pipe-flow Problems. *Journal of the Hydraulics Division* 102(HY5): 657–664. New York: American Society of Civil Engineers.

4. McGhee, T. J., T. McGhee, and E. W. Steel. 1990. *Water Supply and Sewerage*, 6th ed. New York: McGraw-Hill.

5. Avci, Atakan, and Irfan Karagoz. 2009. A Novel Explicit Equation for Friction Factor in Smooth and Rough Pipes, *ASME Journal of Fluids Engineering* 131(061203).

INTERNET RESOURCES

1. **The MacTutor History of Mathematics Archive:** An archive of over 1000 biographies and history topics, including the biography of Osborne Reynolds. From the home page, select *Biographies Index,* then the first letter of the last name, then scan the list for the specific person.

2. **1728 Software Systems-Flowrate:** An online calculator that solves the flow rate equation, $Q = Av$, for one unknown when any two of the three variables, pipe diameter, velocity of flow, or flow rate are entered. The site includes many other calculators and general technical resources.

3. **Engineered Software, Inc. (ESI):** www.eng-software.com Developer of the PIPE-FLO® software for calculating the performance of pipe line systems as an aid in understanding the interaction of pipelines, pumps, components, and controls and to design, optimize, and troubleshoot piping systems. A special demonstration version of PIPE-FLO® created for this book can be accessed by users of this book at http://www.eng-software.com/appliedfluidmechanics.

PRACTICE PROBLEMS

The following problems require the use of the reference data listed below:

- Appendices A–C: Properties of liquids
- Appendix D: Dynamic viscosity of fluids
- Appendices F–J: Dimensions of pipe and tubing
- Appendix K: Conversion factors
- Appendix L: Properties of areas

Reynolds Numbers

8.1 A 4-in-ductile iron pipe carries 0.20 ft^3/s of glycerin (sg = 1.26) at 100°F. Is the flow laminar or turbulent?

8.2 Calculate the minimum velocity of flow in ft/s of water at 160°F in a 2-in steel tube with a wall thickness of 0.065 in for which the flow is turbulent.

8.3 Calculate the maximum volume flow rate of fuel oil at 45°C at which the flow will remain laminar in a DN 100 Schedule 80 steel pipe. For the fuel oil, use sg = 0.895 and dynamic viscosity = 4.0×10^{-2} Pa·s.

8.4 Calculate the Reynolds number for the flow of each of the following fluids in a 2-in Schedule 40 steel pipe if the volume flow rate is 0.25 ft^3/s: (a) water at 60°F, (b) acetone at 77°F, (c) castor oil at 77°F, and (d) SAE 10 oil at 210°F (sg = 0.87).

8.5 Determine the smallest metric hydraulic copper tube size that will carry 4 L/min of the following fluids while maintaining laminar flow: (a) water at 40°C, (b) gasoline (sg = 0.68) 25°C, (c) ethyl alcohol (sg = 0.79) at 0°C, and (d) heavy fuel oil at 25°C.

8.6 In an existing installation, SAE 10 oil (sg = 0.89) must be carried in a DN 80 Schedule 40 steel pipe at the rate of 850 L/min. Efficient operation of a certain process requires that the Reynolds number of the flow be approximately 5×10^4. To what temperature must the oil be heated to accomplish this?

8.7 From the data in Appendix C, we can see that automotive hydraulic oil and the medium machine tool hydraulic oil have nearly the same kinematic viscosity at 212°F. However, because of their different viscosity index, their viscosities at 104°F are quite different. Calculate the Reynolds number for the flow of each oil at each temperature in a 5-in Schedule 80 steel pipe at 10 ft/s velocity. Are the flows laminar or turbulent?

8.8 Compute the Reynolds number for the flow of 325 L/min of water at 10°C in a standard hydraulic steel tube, 50 mm OD × 1.5 mm wall thickness. Is the flow laminar or turbulent?

8.9 Benzene (sg = 0.86) at 60°C is flowing at 25 L/min in a DN 25 Schedule 80 steel pipe. Is the flow laminar or turbulent?

8.10 Hot water at 80°C is flowing to a dishwasher at a rate of 15.0 L/min through a standard hydraulic copper tube, 15 mm OD × 1.2 mm wall. Is the flow laminar or turbulent?

8.11 A major water main is an 18-in ductile iron pipe. Compute the Reynolds number if the pipe carries 16.5 ft³/s of water at 50°F.

8.12 An engine crankcase contains SAE 10 motor oil (sg = 0.88). The oil is distributed to other parts of the engine by an oil pump through an $\frac{1}{8}$-in steel tube with a wall thickness of 0.032 in. The ease with which the oil is pumped is obviously affected by its viscosity. Compute the Reynolds number for the flow of 0.40 gal/h of the oil at 40°F.

8.13 Repeat Problem 8.12 for an oil temperature of 160°F.

8.14 At approximately what volume flow rate will propyl alcohol at 77°F become turbulent when flowing in a 3-in Type K copper tube?

8.15 SAE 30 oil (sg = 0.89) is flowing at 45 L/min through a 20 mm OD × 1.2 mm wall hydraulic steel tube. If the oil is at 110°C, is the flow laminar or turbulent?

8.16 Repeat Problem 8.15 for an oil temperature of 0°C.

8.17 Repeat Problem 8.15, except the tube is 50 mm OD × 1.5 mm wall thickness.

8.18 Repeat Problem 8.17 for an oil temperature of 0°C.

8.19 The lubrication system for a punch press delivers 1.65 gal/min of a light lubricating oil (see Appendix C) through 5⁄16-in steel tubes having a wall thickness of 0.049 in. Shortly after the press is started, the oil temperature is 104°F. Compute the Reynolds number for the oil flow.

8.20 After the press has run for some time, the lubricating oil described in Problem 8.19 heats to 212°F. Compute the Reynolds number for the oil flow at this temperature. Discuss the possible operating difficulty as the oil heats up.

8.21 A system is being designed to carry 500 gal/min of ethylene glycol at 77°F at a maximum velocity of 10.0 ft/s. Specify the smallest standard Schedule 40 steel pipe to meet this condition. Then, for the selected pipe, compute the Reynolds number for the flow.

8.22 The range of Reynolds numbers between 2000 and 4000 is described as the *critical region* because it is not possible to predict whether the flow is laminar or turbulent. One should avoid operation of fluid flow systems in this range. Compute the range of volume flow rates in gal/min of water at 60°F for which the flow would be in the critical region in a ¾-in Type K copper tube.

8.23 The water line described in Problem 8.22 was a cold water distribution line. At another point in the system, the same-size tube delivers water at 180°F. Compute the range of volume flow rates for which the flow would be in the critical region.

8.24 In a dairy, milk at 100°F is reported to have a kinematic viscosity of 1.30 centistokes. Compute the Reynolds number for the flow of the milk at 45 gal/min through a 1¼-in steel tube with a wall thickness of 0.065 in.

8.25 In a soft-drink bottling plant, the concentrated syrup used to make the drink has a kinematic viscosity of 17.0 centistokes at 80°F. Compute the Reynolds number for the flow of 215 L/min of the syrup through a 1-in Type K copper tube.

8.26 A certain jet fuel has a kinematic viscosity of 1.20 centistokes. If the fuel is being delivered to the engine at 200 L/min through a 1-in steel tube with a wall thickness of 0.065 in, compute the Reynolds number for the flow.

Energy Losses

8.27 Crude oil is flowing vertically downward through 60 m of DN 25 Schedule 80 steel pipe at a velocity of 0.64 m/s. The oil has a specific gravity of 0.86 and is at 0°C. Calculate the pressure difference between the top and bottom of the pipe.

8.28 Water at 75°C is flowing in a standard hydraulic copper tube, 15 mm OD × 1.2 mm wall, at a rate of 12.9 L/min. Calculate the pressure difference between two points 45 m apart if the tube is horizontal.

8.29 Fuel oil is flowing in a 4-in Schedule 40 steel pipe at the maximum rate for which the flow is laminar. If the oil has a specific gravity of 0.895 and a dynamic viscosity of 8.3×10^{-4} lb-s/ft², calculate the energy loss per 100 ft of pipe.

8.30 A 3-in Schedule 40 steel pipe is 5000 ft long and carries a lubricating oil between two points A and B such that the Reynolds number is 800. Point B is 20 ft higher than A. The oil has a specific gravity of 0.90 and a dynamic viscosity of 4×10^{-4} lb-s/ft². If the pressure at A is 50 psig, calculate the pressure at B.

8.31 Benzene at 60°C is flowing in a DN 25 Schedule 80 steel pipe at the rate of 20 L/min. The specific weight of the benzene is 8.62 kN/m³. Calculate the pressure difference between two points 100 m apart if the pipe is horizontal.

8.32 As a test to determine the effective wall roughness of an existing pipe installation, water at 10°C is pumped through it at the rate of 225 L/min. The pipe is standard steel tubing, 40 mm OD × 2.0 mm wall. Pressure gages located at 30 m apart in a horizontal run of the pipe read 1035 kPa and 669 kPa. Determine the effective pipe wall roughness.

8.33 Water at 80°F flows from a storage tank through 550 ft of 6-in Schedule 40 steel pipe, as shown in Fig. 8.11. Taking the energy loss due to friction into account, calculate the required head *h* above the pipe inlet to produce a volume flow rate of 2.50 ft³/s.

FIGURE 8.11 Problem 8.33.

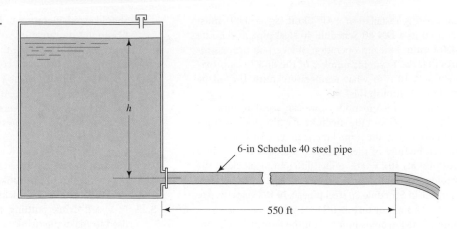

6-in Schedule 40 steel pipe

h

550 ft

8.34 A water main is an 18-in-diameter concrete pressure pipe. Calculate the pressure drop over a 1-mi length due to pipe friction if the pipe carries 15 ft³/s of water at 50°F.

8.35 Figure 8.12 shows a portion of a fire protection system in which a pump draws water at 60°F from a reservoir and delivers it to a point B at the flow rate of 1500 gal/min.

a. Calculate the required height *h* of the water level in the tank in order to maintain 5.0 psig pressure at point A.

b. Assuming that the pressure at point A is 5.0 psig, calculate the power delivered by the pump to the water in order to maintain the pressure at point B at 85 psig. Include energy losses due to friction, but neglect any other energy losses.

FIGURE 8.12 Problem 8.35.

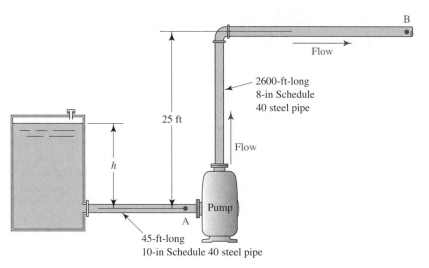

B

Flow

2600-ft-long
8-in Schedule
40 steel pipe

25 ft

Flow

h

Pump

A

45-ft-long
10-in Schedule 40 steel pipe

8.36 A submersible deep-well pump delivers 745 gal/h of water at 60°F through a 1-in Schedule 40 steel pipe when operating in the system shown in Fig. 8.13. If the total length of pipe is 140 ft, calculate the power delivered by the pump to the water.

8.37 On a farm, water at 60°F is delivered from a pressurized storage tank to an animal watering trough through 300 ft of 1½-in Schedule 40 steel pipe as shown in Fig. 8.14. Calculate the required air pressure above the water in the tank to produce 75 gal/min of flow.

8.38 Figure 8.15 shows a system for delivering lawn fertilizer in liquid form. The nozzle on the end of the hose requires 140 kPa of pressure to operate effectively. The hose is smooth plastic with an ID of 25 mm. The fertilizer solution has a specific gravity of 1.10 and a dynamic viscosity of 2.0 × 10⁻³ Pa·s. If the length of the hose is 85 m, determine (a) the power delivered by the pump to the solution and (b) the pressure at the outlet of the pump. Neglect the energy losses on the suction side of the pump. The flow rate is 95 L/min.

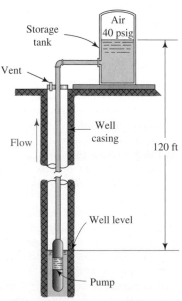

Air
40 psig

Storage
tank

Vent

Well
casing

Flow

120 ft

Well level

Pump

FIGURE 8.13 Problem 8.36.

FIGURE 8.14 Problem 8.37.

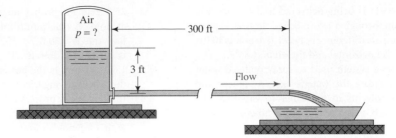

FIGURE 8.15 Problem 8.38.

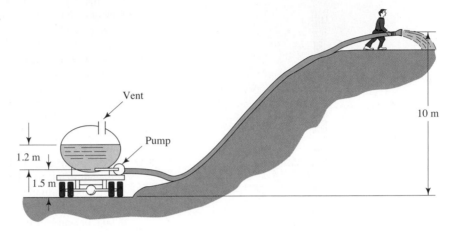

8.39 A pipeline transporting crude oil (sg = 0.93) at 1200 L/min is made of DN 150 Schedule 80 steel pipe. Pumping stations are spaced 3.2 km apart. If the oil is at 10°C, calculate (a) the pressure drop between stations and (b) the power required to maintain the same pressure at the inlet of each pump.

8.40 For the pipeline described in Problem 8.39, consider that the oil is to be heated to 100°C to decrease its viscosity.
  a. How does this affect the pump power requirement?
  b. At what distance apart could the pumps be placed with the same pressure drop as that from Problem 8.39?

8.41 Water at 10°C flows at the rate of 900 L/min from the reservoir and through the pipe shown in Fig. 8.16. Compute the pressure at point B, considering the energy loss due to friction, but neglecting other losses.

8.42 For the system shown in Fig. 8.17, compute the power delivered by the pump to the water to pump 50 gal/min of water at 60°F to the tank. The air in the tank is at 40 psig.

Consider the friction loss in the 225-ft-long discharge pipe, but neglect other losses. Then, redesign the system by using a larger pipe size to reduce the energy loss and reduce the power required to no more than 5.0 hp.

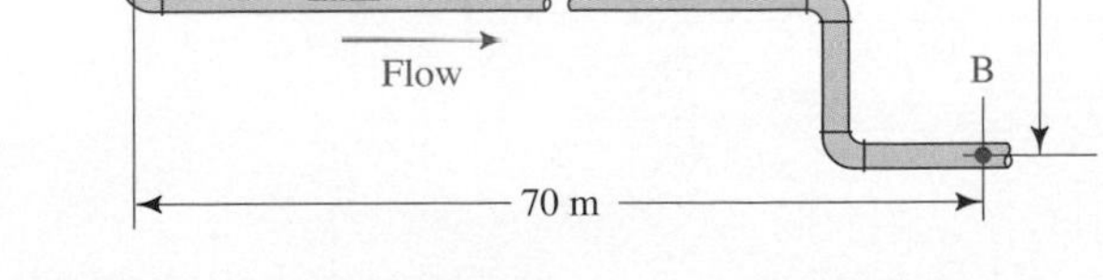

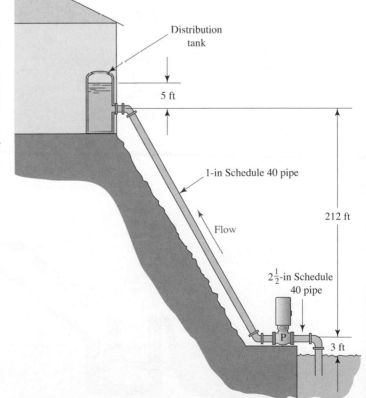

FIGURE 8.16 Problem 8.41.

FIGURE 8.17 Problem 8.42.

8.43 Fuel oil (sg = 0.94) is being delivered to a furnace at a rate of 60 gal/min through a 1½-in Schedule 40 steel pipe. Compute the pressure difference between two points 40.0 ft apart if the pipe is horizontal and the oil is at 85°F.

8.44 Figure 8.18 shows a system used to spray polluted water into the air to increase the water's oxygen content and to cause volatile solvents in the water to vaporize. The pressure at point B just ahead of the nozzle must be

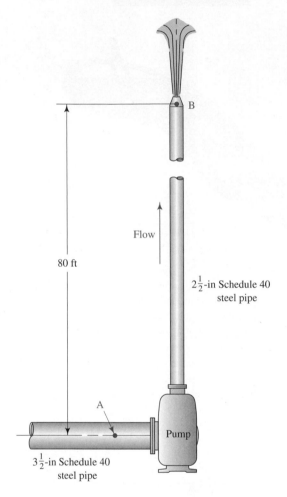

Flow

80 ft

$2\frac{1}{2}$-in Schedule 40 steel pipe

A

Pump

$3\frac{1}{2}$-in Schedule 40 steel pipe

B

FIGURE 8.18 Problem 8.44.

25.0 psig for proper nozzle performance. The pressure at point A (the pump inlet) is −3.50 psig. The volume flow rate is 0.50 ft³/s. The dynamic viscosity of the fluid is 4.0×10^{-5} lb·s/ft². The specific gravity of the fluid is 1.026. Compute the power delivered by the pump to the fluid, considering friction energy loss in the discharge line.

8.45 In a chemical processing system, the flow of glycerin at 60°F (sg = 1.24) in a copper tube must remain laminar with a Reynolds number approximately equal to but not exceeding 300. Specify the smallest standard Type K copper tube that will carry a flow rate of 0.90 ft³/s. Then, for a flow of 0.90 ft³/s in the tube you specified, compute the pressure drop between two points 55.0 ft apart if the pipe is horizontal.

8.46 Water at 60°F is being pumped from a stream to a reservoir whose surface is 210 ft above the pump. See Fig. 8.19. The pipe from the pump to the reservoir is an 8-in Schedule 40 steel pipe, 2500 ft long. If 4.00 ft³/s is being pumped, compute the pressure at the outlet of the pump. Consider the friction loss in the discharge line, but neglect other losses.

8.47 For the pump described in Problem 8.46, if the pressure at the pump inlet is −2.36 psig, compute the power delivered by the pump to the water.

8.48 Gasoline at 50°F flows from point A to point B through 3200 ft of standard 10-in Schedule 40 steel pipe at the rate of 4.25 ft³/s. Point B is 85 ft above point A and the pressure at B must be 40.0 psig. Considering the friction loss in the pipe, compute the required pressure at A.

8.49 Figure 8.20 shows a pump recirculating 300 gal/min of heavy machine tool lubricating oil at 104°F to test the oil's stability. The total length of 4-in pipe is 25.0 ft and the total length of 3-in pipe is 75.0 ft. Compute the power delivered by the pump to the oil.

8.50 Linseed oil at 25°C flows at 3.65 m/s in a standard hydraulic copper tube, 20 mm OD × 1.2 mm wall. Compute the pressure difference between two points in the tube 17.5 m apart if the first point is 1.88 m above the second point.

8.51 Glycerin at 25°C flows through a straight hydraulic copper tube, 80 mm OD × 2.8 mm wall, at a flow rate of 180 L/min. Compute the pressure difference between two points 25.8 m apart if the first point is 0.68 m below the second point.

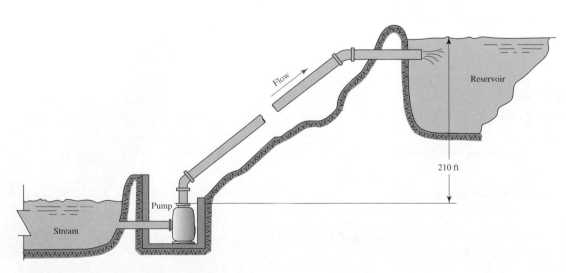

Flow

Reservoir

210 ft

Pump

Stream

FIGURE 8.19 Problems 8.46 and 8.47.

FIGURE 8.20 Problem 8.49.

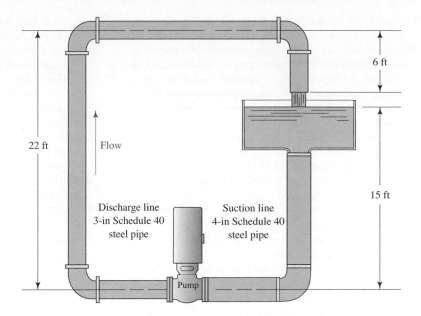

22 ft

Flow

6 ft

15 ft

Discharge line
3-in Schedule 40
steel pipe

Suction line
4-in Schedule 40
steel pipe

Pump

Note: For Problems 8.52–8.62, use the equations for friction factor from Section 8.7 to compute the friction factor.

8.52 Water at 75°C flows in a standard hydraulic copper tube, 15 mm OD $\times$ 1.2 mm wall, at a rate of 12.9 L/min.

8.53 Benzene (sg = 0.88) at 60°C flows in a DN 25 Schedule 80 steel pipe at the rate of 20 L/min.

8.54 Water at 80°F flows in a 6-in coated ductile iron pipe at a rate of 2.50 ft³/s.

8.55 Water at 50°F flows at 15.0 ft³/s in a concrete pipe with an ID of 18.0 in.

8.56 Water at 60°F flows at 1500 gal/min in a 10-in Schedule 40 steel pipe.

8.57 A liquid fertilizer solution (sg = 1.10) with a dynamic viscosity of 2.0×10^{-3} Pa·s flows at 95 L/min through a 25-mm-diameter smooth plastic hose.

8.58 Crude oil (sg = 0.93) at 100°C flows at a rate of 1200 L/min in a DN 150 Schedule 80 steel pipe.

8.59 Water at 65°C flows in a DN 40 Schedule 40 steel pipe at a rate of 10 m/s.

8.60 Propyl alcohol flows in a standard hydraulic copper tube, 80 mm OD $\times$ 2.8 mm wall, at 25°C at a rate of 0.026 m³/s.

8.61 Water at 70°F flows in a 12-in-diameter concrete pipe at 3.0 ft³/s.

8.62 Heavy fuel oil at 77°F flows in a 6-in Schedule 40 steel pipe at 12 ft/s.

Energy Loss Using the Hazen–Williams Formula

Use the design values for the coefficient C_h from Table 8.3 unless stated otherwise. Use either of the various forms of the formula or the nomograph in Fig. 8.10 as assigned.

8.63 Water flows at a rate of 1.50 ft³/s through 550 ft of 6-in cement-lined ductile iron pipe. Compute the energy loss.

8.64 Compute the energy loss as water flows in a standard hydraulic copper tube, 120 mm OD $\times$ 3.5 mm wall, at a rate of 1000 L/min over a length of 45 m.

8.65 A water main is an 18-in-diameter concrete pressure pipe. Calculate the energy loss over a 1-mile length if it carries 7.50 ft³/s of water.

8.66 A fire protection system includes 1500 ft of 10-in Schedule 40 steel pipe. Compute the energy loss in the pipe when it carries 1500 gal/min of water.

8.67 A standard hydraulic copper tube, 120 mm OD $\times$ 3.5 mm wall, carries 900 L/min of water over a length of 80 m. Compute the energy loss.

8.68 Compute the energy loss as 0.20 ft³/s of water flows through a length of 80 ft of 2½-in Schedule 40 steel pipe.

8.69 It is desired to flow 2.0 ft³/s of water through 2500 ft of 8-in nominal size pipe. Compute the head loss for both plain Schedule 40 steel pipe and ductile iron pipe coated with a centrifugally applied cement lining.

8.70 Specify a suitable size of new, clean Schedule 40 steel pipe that would carry 300 gal/min of water over a length of 1200 ft with no more than 10 ft of head loss. For the selected pipe, compute the actual expected head loss.

8.71 For the pipe selected in Problem 8.70, compute the head loss using the design value for C_h rather than that for new, clean pipe.

8.72 Compare the head loss that would result from the flow of 100 gal/min of water through 1000 ft of new, clean Schedule 40 steel pipe for 2-in and 3-in sizes.

Supplemental Problems

8.73 In problem 6.107, a theoretical flow rate of water to a village was calculated without any consideration of line losses. In problem 7.50, an assumed value of 2.8 m was included as an estimate for losses in the line and the resulting flow rate was only 6.22×10^{-4} m³/s. Now rework the problem and determine the actual losses that would occur. Use 25°C water running in a flexible smooth tube that is 1200 m long and has the same 20-mm diameter. Let's check the assumed value for losses now. Using a flow rate of 6.22×10^{-4} m³/s, what would be the actual losses? What conclusions can we draw about the original proposal to install this line? Why is the calculation of losses so important?

8.74 A pipeline is needed to transport medium fuel oil at 77°F. The pipeline needs to traverse 80 mi in total, and the initial proposal is to space pumping stations 2 mi apart. The line needs to carry 750 gal/min and would be made of 6-in Schedule 80 steel pipe. Calculate the pressure drop between stations and the power required to maintain the same pressure at the inlet of each pump. Comment on the design.

8.75 Medium fuel oil at 25°C is to be pumped at a flow rate of 200 m³/h through a DN 125 Schedule 40 pipe over a total horizontal distance of 15 km. The maximum working pressure of the piping is to be limited to 4800 kPa gage and the pumps being used require an inlet pressure of at least 70 kPa absolute. Determine the number of pumping stations needed to traverse the total distance. Sketch your design. What would be an advantage to changing the pipe to Schedule 80 or Schedule 160?

8.76 A tremendous amount of study has gone into the fluid effects of air over various spheres due to the impact on sports and recreation. Golf balls, for example, are dimpled due to the tremendous effect on flow characteristic and resulting drag force. Chapter 17 states that for a spherical body moving through a static fluid, the standard Reynolds number Eq. (8–1) can be used if the value of D is taken to be the diameter of the sphere. Calculate the Reynolds number through air at standard conditions at sea level (see Appendix E) for the following applications:

	Diameter	Velocity
Volleyball serve:	8.5 in	55 mph
Cricket pitch:	7 cm	135 km/h
Baseball pitch:	2.88 in	95 mph
Musket ball shot:	13 mm	440 m/s

8.77 In a given installation, it is determined that the pipe size used for the project was 1-in Schedule 40 pipe rather than the 2 in size that was specified. Some have said that it won't be a problem since a factor of two was built into the system anyway. Others say it must be changed. If the run is 100-ft of horizontal pipe carrying 150 gal/min of water at 60°F, find the head loss for each size pipe and comment on the difference that results from this mistake at the construction site.

8.78 "Laminar" fountains have become quite popular due to the desirable aesthetics that result from a smooth shaped fluid held together with its own surface tension during flight. Check out videos of "laminar fountain" on the web. To convert turbulent to laminar flow a conduit is often transitioned to a large diameter, and then subdivided into many smaller ones, sometimes called straighteners. Calculate the Reynolds number for a pipe that is originally 25 mm in diameter and carrying 8 m³/h of 20°C water. That flow is then directed to a 75 mm pipe and stuffed with plastic drinking straws, each with a diameter of 3 mm. Did the flow change from turbulent in the small pipe to laminar in the larger subdivided pipe?

8.79 Use PIPE-FLO® to model a straight horizontal run of 100 ft of 1-in Schedule 40 pipe carrying 20 gal/min of 75°F water from a tank with a water level of 25 ft. Display the calculated pressure drop in the pipe, Reynolds number, and friction factor on the FLO-Sheet®.

COMPUTER AIDED ENGINEERING ASSIGNMENTS

1. Write a program for computing the friction factor for the flow of any fluid through pipes and tube using Eqs. (8–5) and (8–7). The program must compute the Reynolds number and the relative roughness. Then, decisions must be made as follows:
   a. If $N_R < 2000$, use $f = 64/N_R$ [Eq. (8–5)].
   b. If $2000 < N_R < 4000$, the flow is in the critical range and no reliable value can be computed for f. Print a message to the user of the program.
   c. If $N_R > 4000$, the flow is turbulent. Use Eq. (8–7) to compute f.
   d. Print out N_R, D/ε, and f.

2. Incorporate Program 1 into an enhanced program for computing the pressure drop for the flow of any fluid through a pipe of any size. The two points of interest can be any distance apart, and one end can be any elevation relative to the other. The program should be able to complete such analyses as required for Problems 8.27, 8.28, and 8.31. The program also can be set up to determine the energy loss only in order to solve problems such as Problem 8.29.

3. Write a program for solving the Hazen–Williams formula in any of its forms listed in Table 8.4. Allow the program operator to specify the unit system to be used, which values are known, and which values are to be solved for.

4. Create a spreadsheet for solving the Hazen–Williams formula in any of its forms listed in Table 8.4. Different parts of the spreadsheet can compute different quantities: velocity, head loss, or pipe diameter. Provide for solutions in both U.S. Customary and SI units.

VELOCITY PROFILES FOR CIRCULAR SECTIONS AND FLOW IN NONCIRCULAR SECTIONS

THE BIG PICTURE

Chapters 6–8 considered fluid flow in pipes and tubes with circular cross sections, as will most of this book. When using the velocity of flow in analysis for energy losses, the average velocity was used, calculated from $v = Q/A$. This is very convenient and many ancillary factors such as Reynolds number and resistance coefficients (covered in Chapter 10) are also based on average velocity. Also, no attention was given to the velocity of flow at specific points within the pipe.

You will now consider two new topics that build on those in Chapters 6–8 and they treat situations that are less frequently encountered. However, they are important to help you gain better understanding of the nature of fluid flow.

When fluids are flowing in a pipe or any other shape of conduit, the velocity is *not* uniform across the cross section. You will learn the nature of the velocity profile and how to predict the velocity at any point in circular pipes or tubes for both laminar and turbulent flow.

What about flow paths that are *not* circular? Examples exist within the human body in the cardiovascular, circulatory, and respiratory systems. In these fluid systems, energy losses and pressure distributions must be considered in judging one's overall health. In the case of the cardio system, the heart, acting as a pump, is stressed if the losses in the system become excessive as shown in Fig. 9.1. The top shows the healthy artery with a substantially circular cross section. As cholesterol builds up, it tends to clump in local

Healthy artery

Build-up begins

Plaque forms

Plaque ruptures; blood clot forms

FIGURE 9.1 A common medical condition involving fluid flow is shown here where blood flow through an artery is restricted because of the effects of cholesterol buildup on the walls. (*Source:* Alila Medical Images/Fotolia)

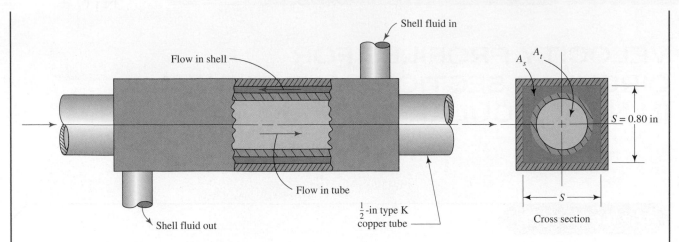

FIGURE 9.2   Shell-and-tube heat exchanger.

areas, not uniformly around the wall of the artery. There-fore, the flow path is decidedly *noncircular*. The build-up of cholesterol reduces the cross-sectional area of the artery, causing increased restriction and pressure losses, with the result that a higher pressure must be developed by the heart to deliver adequate blood flow to all parts of the body.

Many ducts in buildings, automobiles, and engines are square, rectangular, oval, or some very unique shape to fit the available space. Some heat exchangers are of the *tube within a tube* type with a smaller tube centered inside a larger tube that may be circular or square, as illustrated in Fig. 9.2. A hot process fluid may flow inside the smaller tube that does have a circular cross section, but cooling water flows in the space between the outside of the inner tube and the inside of the outer tube. The flow area for the cold water is shaded in dark blue in the figure. In this chapter, you will learn how to analyze the flow in noncircular cross sections flowing full, calculating velocity, Reynolds number, and energy losses due to friction.

Exploration

Look around for examples of flow conduits that do not have a circular cross section. Consider the HVAC system in your home or your college, the ducting under the hood or within the instrument panel in a car, and the ducts that carry moist air from a clothes dryer to the outside of the home. If you work in industry, or visit a manufacturing plant on a tour, seek examples of noncircular flow systems within automation equipment, furnaces, heat treatment equipment, or other processing systems.

Introductory Concepts

We demonstrate in this chapter that the velocity of flow in a circular pipe varies from point to point in the cross section. The velocity right next to the pipe wall is actually zero because it is in contact with the stationary pipe. At points away from the wall, the velocity increases, reaching a maximum at the centerline of the pipe.

Why would you want to know how the velocity varies? One important reason is in the study of heat transfer. For example, when hot water flows along a copper tube in your home, heat transfers from the water to the tube wall and then to the surrounding air. The amount of heat transferred depends on the velocity of the water in the thin layer closest to the wall, called the boundary layer.

Another example involves the measurement of the flow rate in a conduit. Some types of flow measurement devices you will study in Chapter 15 actually detect the local velocity at a small point within the fluid. To use these devices to determine the volume flow rate from $Q = Av$, you will need the average velocity, not one local velocity. You will learn that you must traverse across the diameter of the conduit, making several velocity measurements at specific locations and then compute the average.

Many of the calculations in previous chapters depended on the inside diameter D of a pipe. You will learn in this chapter that you can characterize the size of a noncircular cross section by computing the value of the *hydraulic radius, R,* as discussed in Section 9.5.

9.1 OBJECTIVES

After completing this chapter, you should be able to:

1. Describe the velocity profile for laminar and turbulent flow in circular pipes, tubes, or hose.

2. Describe the laminar boundary layer as it occurs in turbulent flow.

3. Compute the local velocity of flow at any given radial position in a circular cross section.

4. Compute the average velocity of flow in noncircular cross sections.

5. Compute the Reynolds number for flow in noncircular cross sections using the *hydraulic radius* to characterize the size of the cross section.

FIGURE 9.3 Velocity profiles for pipe flow.

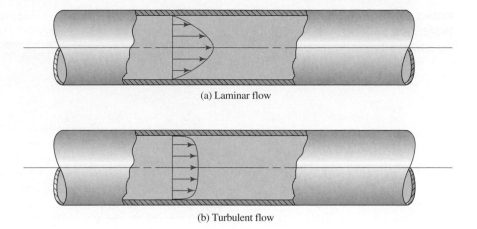

(a) Laminar flow

(b) Turbulent flow

6. Determine the energy loss for the flow of a fluid in a noncircular cross section, considering special forms for the relative roughness and Darcy's equation.

9.2 VELOCITY PROFILES

The magnitude of the local velocity of flow is very nonuniform across the cross section of a circular pipe, tube, or hose. Figure 9.3 shows the general shape of the velocity profiles for laminar and turbulent flow. We observed in Chapter 2 that the velocity of a fluid in contact with a stationary solid boundary is zero. This corresponds to the inside wall of any conduit. The velocity then increases at points away from the wall, reaching a maximum at the centerline of a circular pipe.

It was shown in Fig. 8.2 that laminar flow can be thought of as a series of concentric layers of the fluid sliding along each other. This smooth flow results in a parabolic shape for the velocity profile.

Conversely, we have described turbulent flow as chaotic with significant amounts of intermixing of the particles of fluid and a consequent transfer of momentum among the particles. The result is a more nearly uniform velocity across much of the cross section. Still, the velocity at the pipe wall is zero. The local velocity increases rapidly over a short distance from the wall and then more gradually to the maximum velocity at the center.

9.3 VELOCITY PROFILE FOR LAMINAR FLOW

Because of the regularity of the velocity profile in laminar flow, we can define an equation for the local velocity at any point within the flow path. If we call the local velocity U at a radius r, the maximum radius r_o, and the average velocity v, then

$$U = 2v\left[1 - (r/r_o)^2\right] \qquad (9\text{–}1)$$

Example Problem 9.1

In Example Problem 8.1 we found that the Reynolds number is 708 when glycerin at 25°C flows with an average flow velocity of 3.6 m/s in a circular passage through a chemical processing device having a 150-mm inside diameter. Thus, the flow is laminar. Compute points on the velocity profile from the wall to the centerline of the passage in increments of 15 mm. Plot the data for the local velocity U versus the radius r.

Solution

We can use Eq. (9–1) to compute U. First we compute the maximum radius r_o:

$$r_o = D/2 = 150/2 = 75 \text{ mm}$$

At $r = 75$ mm $= r_o$ at the pipe wall, $r/r_o = 1$ and $U = 0$ from Eq. (9–1). This is consistent with the observation that the velocity of a fluid at a solid boundary is equal to the velocity of that boundary.

At $r = 60$ mm,

$$U = 2(3.6 \text{ m/s})\left[1 - (60/75)^2\right] = 2.59 \text{ m/s}$$

Using a similar technique, we can compute the following values:

FIGURE 9.4 Results of Example Problems 9.1 and 9.2. Velocity profile for laminar flow.

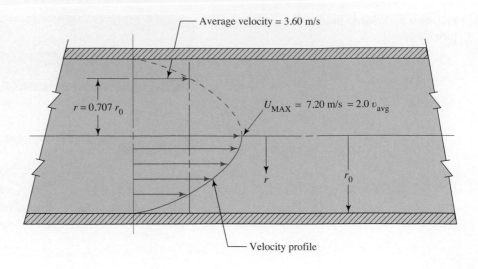

r (mm)	r/r₀	U (m/s)
75	1.0	0 (at the pipe wall)
60	0.8	2.59
45	0.6	4.61
30	0.4	6.05
15	0.2	6.91
0	0.0	7.20 (middle of the pipe)

Notice that the local velocity at the middle of the pipe is 2.0 times the average velocity. Figure 9.4 shows the plot of U versus r.

Example Problem 9.2   Compute the radius at which the local velocity U would equal the average velocity v for laminar flow and show its location on the velocity profile plot

Solution   In Eq. (9–1), for the condition that $U = v$, we can first divide by U to obtain

$$1 = 2[1 - (r/r_0)^2]$$

Now, solving for r gives

$$r = \sqrt{0.5}\, r_0 = 0.707 r_0 \qquad \text{(9–2)}$$

For the data from Example Problem 9.1, the local velocity is equal to the average velocity 3.6 m/s at

$$r = 0.707(75\,\text{mm}) = 53.0\,\text{mm}$$

The radial location of the average velocity is shown in Fig. 9.4.

FIGURE 9.5 General form of velocity profile for turbulent flow.

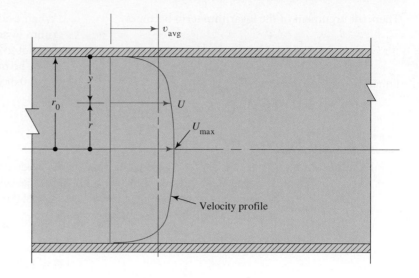

9.4 VELOCITY PROFILE FOR TURBULENT FLOW

The velocity profile for turbulent flow is far different from the parabolic distribution for laminar flow. As shown in Fig. 9.5, the fluid velocity near the wall of the pipe changes rapidly from zero at the wall to a nearly uniform velocity distribution throughout the bulk of the cross section. The actual shape of the velocity profile varies with the friction factor f, which in turn varies with the Reynolds number and the relative roughness of the pipe. The governing equation (from Reference 1) is

$$U = v[1 + 1.43\sqrt{f} + 2.15\sqrt{f}\log_{10}(1 - r/r_o)] \quad \textbf{(9–2)}$$

Figure 9.6 compares the velocity profiles for laminar flow and for turbulent flow at a variety of Reynolds numbers.

An alternate form of this equation can be developed by defining the distance from the wall of the pipe as $y = r_o - r$.

FIGURE 9.6 Velocity profiles in laminar and turbulent flow in a smooth pipe. (*Source:* From Miller, R.W. *Flow Measurement Engineering Handbook*, 3/e © 1983. Reprinted with permission of McGraw-Hill Companies, Inc.)

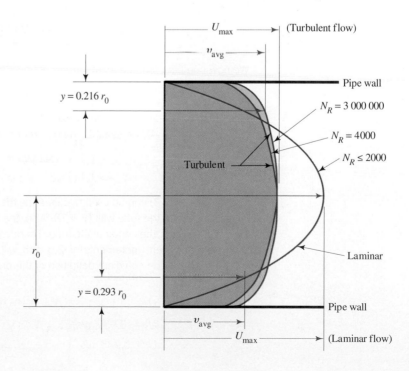

Then, the argument of the logarithm term becomes

$$1 - \frac{r}{r_o} = \frac{r_o}{r_o} - \frac{r}{r_o} = \frac{r_o - r}{r_o} = \frac{y}{r_o}$$

Equation (9–2) is then

$$U = v\left[1 + 1.43\sqrt{f} + 2.15\sqrt{f}\,\log_{10}(y/r_o)\right] \quad \textbf{(9–3)}$$

When evaluating Eq. (9–2) or (9–3), recall that the logarithm of zero is undefined. You may allow r to *approach* r_o, but not to equal it. Similarly, y can only approach zero.

The maximum velocity occurs at the center of the pipe ($r = 0$ or $y = r_o$), and its value can be computed from

$$U_{max} = v(1 + 1.43\sqrt{f}) \quad \textbf{(9–4)}$$

Example Problem 9.3

A specially fabricated plastic tube has an inside diameter of 50.0 mm and it carries 110 L/min of benzene at 50°C (sg = 0.86). Compute the average velocity of flow, the expected maximum velocity of flow, and several points on the velocity profile. Plot the velocity versus the distance from the tube wall and show where the average velocity occurs.

Solution

Given the following data:

$$Q = 110 \text{ L/min}$$
$$D = 50.0 \text{ mm} = 0.050 \text{ m}$$
$$\text{Benzene at } 50°C \text{ (sg} = 0.86)$$

In order to apply Eqs. (9–3) and (9–4), we need to compute the Reynolds number and then find the friction factor for the plastic tube.

For the benzene:

$$\rho = \text{sg} \times \rho_w = (0.86)(1000 \text{ kg/m}^3) = 860 \text{ kg/m}^3$$

From Appendix D, the dynamic viscosity is: $\eta = 4.2 \times 10^{-4}$ Pa·s

The average velocity of flow is:

$$v = Q/A$$

$$Q = 110 \text{ L/min}\left[\frac{1 \text{ m}^3/\text{s}}{60\ 000 \text{ L/min}}\right] = 1.83 \times 10^{-3} \text{ m/s}$$

$$A = \pi D^2/4 = \pi(0.050 \text{ m})^2/4 = 1.963 \times 10^{-3} \text{ m}^2$$

Then the average velocity is:

$$v = Q/A = (1.83 \times 10^{-3} \text{ m/s})/(1.963 \times 10^{-3} \text{ m}^2) = 0.932 \text{ m/s}$$

Now compute the Reynolds number, $N_R = vD\rho/\eta$

$$N_R = \frac{(0.932)(0.050)(860)}{4.2 \times 10^{-4}} = 9.54 \times 10^4 \text{ (turbulent)}$$

We now need to compute the relative roughness, D/ε. From Table 8.2, we find $\varepsilon = 3.0 \times 10^{-7}$ m. Then

$$D/\varepsilon = 0.050/3.0 \times 10^{-7} = 1.667 \times 10^5$$

From Moody's diagram, we find $f = 0.018$,
Now, from Eq. (9–4), we see that the maximum velocity of flow is

$$U_{max} = v(1 + 1.43\sqrt{f}) = (0.932 \text{ m/s})(1 + 1.43\sqrt{0.018})$$
$$U_{max} = 1.111 \text{ m/s} \quad \text{at the center of the tube}$$

Equation (9–3) can be used to determine the points on the velocity profile. We know that the velocity equals zero at the tube wall ($y = 0$). Also, the rate of change of velocity with position is greater near the wall than near the center of the tube. Therefore, increments of 0.5 mm will be used from $y = 0.5$ to $y = 2.5$ mm. Then, increments of 2.5 mm will be used up to $y = 10$ mm. Finally, increments of 5.0 mm will provide sufficient definition of the profile near the center of the tube. At $y = 1.0$ mm and $r_o = 25$ mm,

$$U = v\left[1 + 1.43\sqrt{f} + 2.15\sqrt{f}\,\log_{10}(y/r_o)\right]$$
$$U = (0.932 \text{ m/s})\left[1 + 1.43\sqrt{0.018} + 2.15\sqrt{0.018}\,\log_{10}(1/25)\right]$$
$$U = 0.735 \text{ m/s}$$

Using similar calculations, we can compute the following values:

y (mm)	y/r_o	U (m/s)
0.5	0.02	0.654
1.0	0.04	0.735
1.5	0.06	0.782
2.0	0.08	0.816
2.5	0.10	0.842
5.0	0.20	0.923
7.5	0.30	0.970
10.0	0.40	1.004
15.0	0.60	1.051
20.0	0.80	1.085
25.0	1.00	1.111 (U_{max} at center of tube)

Figure 9.7 is the plot of *y* versus velocity in the form in which the velocity profile is normally shown. Because the plot is symmetrical, only one-half of the profile is shown. Note that the position of the average velocity on this chart is at approximately *y* = 5.4 mm from the tube wall, about 22 percent of the radius.

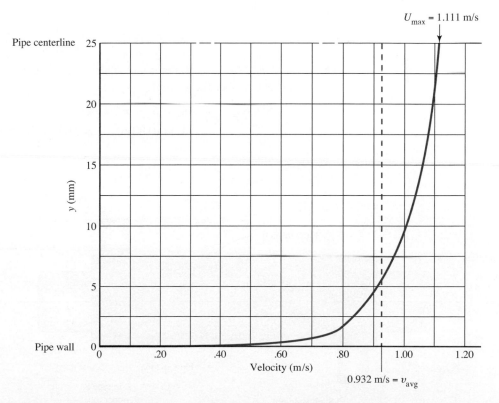

FIGURE 9.7   Velocity profile for turbulent flow for Example Problem 9.3.

9.5 FLOW IN NONCIRCULAR SECTIONS

Here we show how fluid flow calculations for flow in noncircular sections vary from those developed in Chapters 6–8. We discuss average velocity, hydraulic radius used as the characteristic size of the section, Reynolds number, and energy loss due to friction. All flow conduit sections considered here are full of liquid. Noncircular sections for open-channel flow or partially filled sections are discussed in Chapter 14.

9.5.1 Average Velocity

The definition of volume flow rate and the continuity equation first used in Chapter 6 are applicable to noncircular sections as well as for circular pipes, tubes, and hose:

$$Q = Av$$
$$v = Q/A$$
$$A_1 v_1 = A_2 v_2$$

Care must be exercised to compute the net cross-sectional area for flow from the specific geometry of the noncircular section.

Example Problem 9.4 Figure 9.8 shows a heat exchanger used to transfer heat from the fluid flowing inside the inner tube to that flowing in the space between the outside of the tube and the inside of the square shell that surrounds the tube. Such a device is often called a *shell-and-tube heat exchanger*. Compute the volume flow rate in gal/min that would produce a velocity of 8.0 ft/s both inside the tube and in the shell.

Solution We use the formula for volume flow rate, $Q = Av$, for each part.

a. Inside the $\frac{1}{2}$-in Type K copper tube: From Appendix H, we can read

$$OD = 0.625 \text{ in}$$
$$ID = 0.527 \text{ in}$$
$$\text{Wall thickness} = 0.049 \text{ in}$$
$$A_t = 1.515 \times 10^{-3} \text{ ft}^2 = \text{flow area in tube}$$

Then, the volume flow rate inside the tube is

$$Q_t = A_t v = (1.515 \times 10^{-3} \text{ ft}^2)(8.0 \text{ ft/s}) = 0.01212 \text{ ft}^3/\text{s}$$

Converting to gal/min gives

$$Q_t = 0.01212 \text{ ft}^3/\text{s} \frac{449 \text{ gal/min}}{1.0 \text{ ft}^3/\text{s}} = 5.44 \text{ gal/min}$$

b. In the shell: The net flow area is the difference between the area *inside* the square shell and the *outside* of the tube. Then,

$$A_s = S^2 - \pi OD^2/4$$
$$A_s = (0.80 \text{ in})^2 - \pi(0.625 \text{ in})^2/4 = 0.3332 \text{ in}^2$$

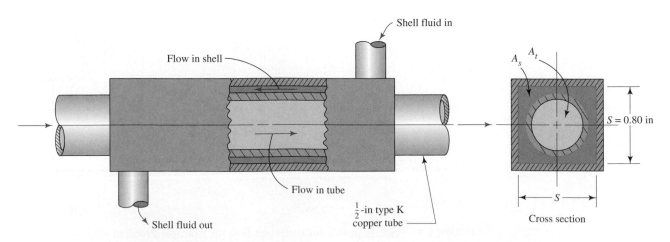

FIGURE 9.8 Shell-and-tube heat exchanger.

Converting to ft² gives

$$A_s = 0.3332 \text{ in}^2 \frac{1.0 \text{ ft}^2}{144 \text{ in}^2} = 2.314 \times 10^{-3} \text{ ft}^2$$

The required volume flow rate is then

$$Q_s = A_s \nu = (2.314 \times 10^{-3} \text{ ft}^2)(8.0 \text{ ft/s}) = 0.01851 \text{ ft}^3/\text{s}$$

$$Q_s = 0.01851 \text{ ft}^3/\text{s} \frac{449 \text{ gal/min}}{1.0 \text{ ft}^3/\text{s}} = 8.31 \text{ gal/min}$$

The ratio of the flow in the shell to the flow in the tube is

$$\text{Ratio} = Q_s/Q_t = 8.31/5.44 = 1.53$$

9.5.2 Hydraulic Radius for Noncircular Cross Sections

Examples of typical closed, noncircular cross sections are shown in Fig. 9.9. The sections shown could represent (a) a shell-and-tube heat exchanger, (b) and (c) air distribution ducts, and (d) a shell-and-tube heat exchanger, or a flow path inside a machine.

The characteristic dimension of noncircular cross sections is called the *hydraulic radius R*, defined as the ratio of the net cross-sectional area of a flow stream to the wetted perimeter of the section. That is,

↬ **Hydraulic Radius**

$$R = \frac{A}{WP} = \frac{\text{Area}}{\text{Wetted perimeter}} \qquad (9\text{--}5)$$

The unit for R is the meter in the SI unit system. In the U.S. Customary System, R is expressed in feet.

In the calculation of the hydraulic radius, the net cross-sectional area should be evident from the geometry of the section.

The wetted perimeter is defined as the sum of the length of the boundaries of the section actually in contact with (that is, wetted by) the fluid.

Expressions for the area A and the wetted perimeter WP are given in Fig. 9.9 for the sections illustrated. In each case, the fluid flows in the shaded portion of the section. A dashed line is shown adjacent to the boundaries that make up the wetted perimeter.

FIGURE 9.9 Examples of closed noncircular cross sections.

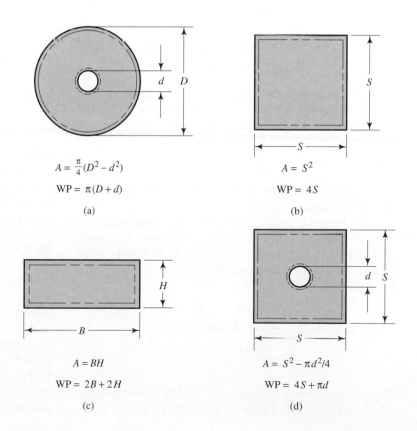

$$A = \frac{\pi}{4}(D^2 - d^2)$$

$$WP = \pi(D + d)$$

(a)

$$A = S^2$$

$$WP = 4S$$

(b)

$$A = BH$$

$$WP = 2B + 2H$$

(c)

$$A = S^2 - \pi d^2/4$$

$$WP = 4S + \pi d$$

(d)

FIGURE 9.10 Cross section for a duct for Example Problems 9.5–9.7.

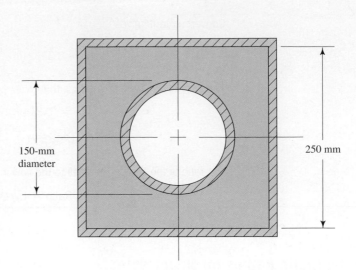

150-mm diameter

250 mm

Example Problem 9.5

Determine the hydraulic radius of the section shown in Fig. 9.10 if the inside dimension of each side of the square is 250 mm and the outside diameter of the tube is 150 mm.

Solution

The net flow area is the difference between the area of the square and the area of the circle:

$$A = S^2 - \pi d^2/4 = (250)^2 - \pi(150)^2/4 = 44\,829 \text{ mm}^2$$

The wetted perimeter is the sum of the four sides of the square and the circumference of the circle:

$$WP = 4S + \pi d = 4(250) + \pi(150) = 1471 \text{ mm}$$

Then, the hydraulic radius R is

$$R = \frac{A}{WP} = \frac{44\,829 \text{ mm}^2}{1471 \text{ mm}} = 30.5 \text{ mm} = 0.0305 \text{ m}$$

9.5.3 Reynolds Number for Closed Noncircular Cross Sections

When the fluid completely fills the available cross-sectional area and is under pressure, the average velocity of flow is determined by using the volume flow rate and the net flow area in the familiar equation,

$$v = Q/A$$

Note that the area is the same as that used to compute the hydraulic radius.

The Reynolds number for flow in noncircular sections is computed in a very similar manner to that used for circular pipes and tubes. The only alteration to Eq. (8–1) is the replacement of the diameter D with $4R$, four times the hydraulic radius. The result is

⇨ **Reynolds Number—Noncircular Sections**

$$N_R = \frac{v(4R)\rho}{\eta} = \frac{v(4R)}{\nu} \qquad (9\text{–}6)$$

The validity of this substitution can be demonstrated by calculating the hydraulic radius for a circular pipe:

$$R = \frac{A}{WP} = \frac{\pi D^2/4}{\pi D} = \frac{D}{4}$$

Then,

$$D = 4R$$

Therefore, $4R$ is equivalent to D for the circular pipe. Thus, by analogy, the use of $4R$ as the characteristic dimension for noncircular cross sections is appropriate. This approach will give reasonable results as long as the cross section has an *aspect ratio* not much different from that of the circular cross section. In this context, aspect ratio is the ratio of the width of the section to its height. So, for a circular section, the aspect ratio is 1.0. In Fig. 9.9, all the examples shown have reasonable aspect ratios.

An example of a shape that has an unacceptable aspect ratio is a rectangle for which the width is more than four times the height. For such shapes, the hydraulic radius is approximately one-half the height. Some annular shapes, similar to that shown in Fig. 9.9(a), would have high aspect ratios if the space between the two pipes was small. However, general data are not readily available for what constitutes a "small" space or for how to determine the hydraulic radius. Performance testing of such sections is recommended. More on flow in noncircular sections can be found in References 2 and 3.

Example Problem 9.6 Compute the Reynolds number for the flow of ethylene glycol at 25°C through the section shown in Fig. 9.10. The volume flow rate is 0.16 m³/s. Use the dimensions given in Example Problem 9.5.

Solution The Reynolds number can be computed from Eq. (9–6). The results for the flow area and the hydraulic radius for the section from Example Problem 9.5 can be used: $A = 44\,829$ mm² and $R = 0.0305$ m. We can use $\eta = 1.62 \times 10^{-2}$ Pa·s and $\rho = 1100$ kg/m³ (from Appendix B). The area must be converted to m². We have

$$A = (44\,829 \text{ mm}^2)(1 \text{ m}^2/10^6 \text{ mm}^2) = 0.0448 \text{ m}^2$$

The average velocity of flow is

$$v = \frac{Q}{A} = \frac{0.16 \text{ m}^3/\text{s}}{0.0448 \text{ m}^2} = 3.57 \text{ m/s}$$

The Reynolds number can now be calculated:

$$N_R = \frac{v(4R)\rho}{\eta} = \frac{(3.57)(4)(0.0305)(1100)}{1.62 \times 10^{-2}}$$

$$N_R = 2.96 \times 10^4$$

9.5.4 Friction Loss in Noncircular Cross Sections

Darcy's equation for friction loss can be used for noncircular cross sections if the geometry is represented by the hydraulic radius instead of the pipe diameter, as is used for circular sections. After computing the hydraulic radius, we can compute the Reynolds number from Eq. (9–6). In Darcy's equation, replacing D with $4R$ gives

▷ **Darcy's Equation for Noncircular Sections**

$$h_L = f\frac{L}{4R}\frac{v^2}{2g} \tag{9–7}$$

The relative roughness D/ε becomes $4R/\varepsilon$. The friction factor can be found from the Moody diagram.

Example Problem 9.7 Determine the pressure drop for a 50-m length of a duct with the cross section shown in Fig. 9.10. Ethylene glycol at 25°C is flowing at the rate of 0.16 m³/s. The inside dimension of the square is 250 mm and the outside diameter of the tube is 150 mm. Use $\varepsilon = 3 \times 10^{-5}$ m, somewhat smoother than commercial steel pipe.

Solution The area, velocity, hydraulic radius, and Reynolds number were computed in Example Problems 9.5 and 9.6. The results are

$$A = 0.0448 \text{ m}^2$$
$$v = 3.57 \text{ m/s}$$
$$R = 0.0305 \text{ m}$$
$$N_R = 2.96 \times 10^4$$

The flow is turbulent, and Darcy's equation can be used to calculate the energy loss between two points 50 m apart. To determine the friction factor, we must first find the relative roughness:

$$4R/\varepsilon = (4)(0.0305)/(3 \times 10^{-5}) = 4067$$

From the Moody diagram, $f = 0.0245$. Then, we have

$$h_L = f \times \frac{L}{4R} \times \frac{v^2}{2g} = 0.0245 \times \frac{50}{(4)(0.0305)} \times \frac{(3.57)^2}{(2)(9.81)}\text{m}$$

$$h_L = 6.52 \text{ m}$$

If the duct is horizontal,

$$h_L = \Delta p/\gamma$$

$$\Delta p = \gamma h_L$$

where Δp is the pressure drop caused by the energy loss. Use $\gamma = 10.79\,\text{kN/m}^3$ from Appendix B. Then, we have

$$\Delta p = \frac{10.79\,\text{kN}}{\text{m}^3} \times 6.52\,\text{m} = 70.4\,\text{kPa}$$

9.6 COMPUTATIONAL FLUID DYNAMICS

Fluid systems addressed with conventional processes in this text are well understood and the governing principles applied to them have been empirically tested over time. For systems that follow these basic principles, manual calculations are sufficient. For systems that have a large number of components, several segments, and varying pipe sizes, these calculations can become time-consuming and tedious, so using software such as PIPE-FLO® is helpful and saves time. (See Internet resource 3 in Chapter 8.) Keep in mind, however, that PIPE-FLO® and similar packages simply automate the process of calculations using the same basic principles of Darcy–Weisbach and the others presented in this text.

There are, however, many applications that are not conducive to such calculation methods. There are new, different, and untested fluid applications that must be understood using methods that are better suited for such a high degree of complexity. Such applications are better addressed

through the use of computational fluid dynamics, or CFD. Computational fluid dynamics uses the power of computers to perform a tremendous number of calculations for very small fluid elements in a very short amount of time. Rather than breaking a system into a component level as we do in this text and with software such as PIPE-FLO®, CFD analyzes fluid flow with very small, elemental, flow volumes, and could be used to help design the components we apply in this text. Those small elements are then combined into a grid or mesh for overall analysis. In some ways similar to finite element analysis (FEA) that is used for stress and deformation analysis of solid objects, CFD typically generates a graphical output showing gradients in various colors to indicate key flow parameters. See Figs. 9.11 and 9.12 for typical results generated by CFD for fluid flow within a globe valve. Such valves are described in Chapter 10. *Note that the two figures do not show the same identical valve.*

Figure 9.11 shows the total flow path from the inlet pipe, through the valve, and through the outlet pipe, with a cutaway drawing of the valve superimposed on the graphic

FIGURE 9.11 Flow through a globe valve as represented by computational fluid dynamics analysis (CFD). (*Source:* Autodesk screen shots reprinted with the permission of Autodesk, Inc.)

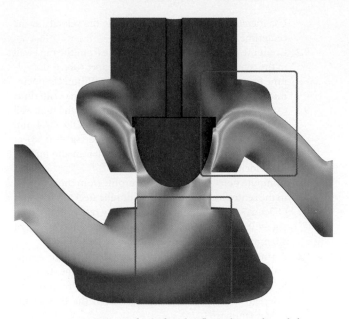

FIGURE 9.12 CFD analysis for the flow through a globe valve in the area of the seat. (*Source:* Image and model courtesy of DASSAULT SYSTEMES SOLIDWORKS CORPORATION)

representation of the CFD results. The varying degree of shading indicates variations in flow velocity and pressure in the fluid as it navigates the complex path through the valve.

Figure 9.12 isolates the port within a globe valve. The flow enters from the left, travels downward, then turns upward where it flows through an annular passage between the adjustable globe-shaped plug and the fixed seat in the valve body. The fluid then rejoins in the upper part of the valve body, turns downward, and flows into the outlet pipe. High velocity and a significant pressure drop occur around the plug and both velocity and pressure drop vary widely as the valve is opened and closed. The red squares in the figure highlight two areas where special attention to design details is needed.

The partial differential equations that govern fluid flow and heat transfer are not only complex, but they are intimately coupled and nonlinear, making a general analytic solution impossible in most cases. Computational fluid dynamics was developed many years ago to address these applications, but it required special computing capability, expensive software, and much advanced training. In recent years, however, CFD software has taken the form of affordable modules within products such as AutoDesk and SolidWorks, and can easily be run on conventional personal computers. See Internet resources 1–6 for a variety of vendors of CFD software. Reference 4 is an extensive treatment of CFD.

With CFD within the reach of so many designers now, application has become more common. Examples of CFD applications from the aerospace industry include the flow over airfoils and the flow through a jet engine over turbine blades. In the area of fluid-moving equipment, CFD modeling now aids in the design of valves, pumps, fans, blowers, and compressors. Automotive engine designers rely on CFD to simulate flow in intake and exhaust manifolds. The effectiveness of a plastic injection mold depends greatly on the way in which molten plastic will flow and transfer heat

though the unique passageways and part cavities. The use of CFD can simulate the mold flow in the design stage and ensure that the flow characteristics will result in desirable mold performance and part quality. Mold flow CFD programs also account for the decidedly non-Newtonian behavior of the molten plastic and the changes in its characteristics during the solidification process.

Adequate accuracy requires that the elements in the CFD model be very small, so that the complete finite-element model may contain literally millions of elements. High-speed computing and efficient computer codes make this analysis practical. The results include flow velocity profiles, pressure and temperature variations, and streamlines that can be displayed graphically, usually in color, to assist the user in interpreting the results.

The steps required to use CFD include the following:

1. Define the three-dimensional geometry of the object being analyzed using 3D CAD software.

2. Establish the boundary conditions that define known values of pressure, velocity, temperature, and heat transfer coefficients in the fluid.

3. Assign a mesh size to each element, with the nominal size being 0.10 mm.

4. Most commercially available CFD software will then automatically create the mesh and the complete finite-element model.

5. Specify material types for solid components (such as steel, aluminum, and plastic) and fluids (such as air, water, and oil). The software typically includes the necessary properties of such materials, for example, specific heats, thermal conductivities, and the coefficients of thermal expansion.

6. Initiate the computational process. This process may take a significant amount of time because of the huge number of calculations to be made. The total time depends on the complexity of the model.

7. When the analysis is completed, the user can select the type of display pertinent to the factors being investigated. It may be fluid trajectories, velocity profiles, isothermal temperature plots, pressure distributions, or others.

Internet resource 1 includes more detail about the CFD software called Autodesk Simulation that can run on typical personal computers. It can be integrated with many popular three-dimensional computer-aided design software packages, such as Inventor, Mechanical Desktop, SolidWorks, ProEngineer, and others to import the solid model directly into the simulation software. Mesh generation is automatic with optimized mesh geometry around small features. Laminar and turbulent flow regimes can be analyzed for compressible or incompressible fluids in subsonic, transonic, or supersonic velocity regions. The heat transfer modes of conduction, convection (natural or forced), or radiation are included.

The use of CFD software can provide a dramatic reduction in the time needed to develop new products. Modeling the flow and heat transfer characteristics of a proposed design while still just a solid model on the designer's desktop, allows for quick improvement and optimization through iterations, thus saving the time and expense of prototyping and testing actual hardware. Internet resources 2–5 identify several other CFD software packages, some of which are general purpose, whereas others specialize in such applications as thermal analysis of electronic system cooling, engine flows, aeroacoustics (combined flow and noise analysis in ducts), airfoil analysis, polymer processing, fire modeling, open-channel flow analysis, heating, ventilating and air conditioning, and marine systems.

REFERENCES

1. Miller, R. W. 1996. *Flow Measurement Engineering Handbook,* 3rd ed. New York: McGraw-Hill.

2. Basniev, Kaplan S., Nikolay M. Dmitriev, George V. Chilingar, Misha Gorfunkle, and Amir G. Mohammed Hejad. 2012. *Mechanics of Fluid Flow.* New York: Wiley Publishing Co.

3. Crane Company. 2011. *Flow of Fluids through Valves, Fittings, and Pipe* (Technical Paper No. 410). Stamford, CT: Crane Company.

4. Biringen, Sedat, and Chuen-Yen Chow. 2011. *An Introduction to Computational Fluid Mechanics by Example.* New York: Wiley Publishing Co.

INTERNET RESOURCES

1. **Autodesk Simulation CFD:** Producer of computational fluid dynamics (CFD) software for analyzing fluid flow and thermal behavior for complex flow paths such as valves, manifolds, pumps, fans, and heat exchangers. Formerly known as CFD Software, it is now integrated within the broad Autodesk product line.

2. **ANSYS Fluent Software:** Producer of the computational fluid dynamics software packages ANSYS Fluent, ANSYS CFX, ANSYS CFD, and ANSYS Workbench, that include model building, applying a mesh, and post-processing.

3. **Flow Science, Inc.:** Producer of FLOW-3D™ software, with special emphasis on free surface flows, also handling external flows and confined flows and providing assistance in creating the geometry, preprocessing, and post-processing.

4. **CFD-Online:** An online center for computational fluid dynamics, listing CFD resources, events, news, books, and discussion forums.

5. **Solidworks Flow Simulation:** Producer of flow simulation software integrated within the Solidworks computer aided design and computer aided engineering packages. Analysis of fluid flow, heat transfer, and fluid forces applied to HVAC systems, electronics cooling, valves, fittings, and thermal comfort systems.

PRACTICE PROBLEMS

Velocity Profile—Laminar Flow

9.1 Compute points on the velocity profile from the pipe wall to the centerline of a 2-in Schedule 40 steel pipe if the volume flow rate of castor oil at 77°F is 0.25 ft^3/s.

Use increments of 0.20 in and include the velocity at the centerline.

9.2 Compute points on the velocity profile from the pipe wall to the centerline of a 3/4-in Type K copper tube if the volume flow rate of water at 60°F is 0.50 gal/min. Use increments of 0.05 in and include the velocity at the centerline.

9.3 Compute points on the velocity profile from the tube wall to the centerline of a plastic pipe, 125 mm OD × 7.4 mm wall, if the volume flow rate of gasoline (sg = 0.68) at 25°C is 3.0 L/min. Use increments of 8.0 mm and include the velocity at the centerline.

9.4 Compute points on the velocity profile from the tube wall to the centerline of a standard hydraulic steel tube, 50 mm OD × 1.5 mm wall, if the volume flow rate of SAE 30 oil (sg = 0.89) at 110°C is 25 L/min. Use increments of 4.0 mm and include the velocity at the centerline.

9.5 A small velocity probe is to be inserted through a pipe wall. If we measure from the outside of the DN 150 Schedule 80 pipe, how far (in mm) should the probe be inserted to sense the average velocity if the flow in the pipe is laminar?

9.6 If the accuracy of positioning the probe described in Problem 9.5 is plus or minus 5.0 mm, compute the possible error in measuring the average velocity.

9.7 An alternative scheme for using the velocity probe described in Problem 9.5 is to place it in the middle of the pipe, where the velocity is expected to be 2.0 times the average velocity. Compute the amount of insertion required to center the probe. Then, if the accuracy of placement is again plus or minus 5.0 mm, compute the possible error in measuring the average velocity.

9.8 An existing fixture inserts the velocity probe described in Problem 9.5 exactly 60.0 mm from the outside surface of the pipe. If the probe reads 2.48 m/s, compute the actual average velocity of flow, assuming the flow is laminar. Then, check to see if the flow actually is laminar if the fluid is a heavy fuel oil with a kinematic viscosity of 850 centistokes.

Velocity Profile—Turbulent Flow

9.9 For the flow of 12.9 L/min of water at 75°C in a plastic pipe, 16 mm OD × 1.5 mm wall, compute the expected maximum velocity of flow from Eq. (9–4).

9.10 A large pipeline with a 1.200-m inside diameter carries oil similar to SAE 10 at 40°C (sg = 0.8). Compute the volume flow rate required to produce a Reynolds number of 3.60×10^4. Then, if the pipe is clean steel, compute several points of the velocity profile and plot the data in a manner similar to that shown in Fig. 9.7.

9.11 Repeat Problem 9.10 if the oil is at 110°C but with the same flow rate. Discuss the differences in the velocity profile.

9.12 Using Eq. (9–3), compute the distance y for which the local velocity U is equal to the average velocity v.

9.13 The result for Problem 9.12 predicts that the average velocity for turbulent flow will be found at a distance of $0.216r_o$ from the wall of the pipe. Compute this distance for a 24-in Schedule 40 steel pipe. Then, if the pipe carries water at 50°F at a flow rate of 16.75 ft^3/s, compute the velocity at points 0.50 in on either side of the average velocity point.

9.14 Using Eq. (9–4), compute the ratio of the average velocity to the maximum velocity of flow in smooth pipes with Reynolds numbers of 4000, 10^4, 10^5, and 10^6.

FIGURE 9.13 Shell-and-tube heat exchanger for Problems 9.19, 9.25, and 9.38.

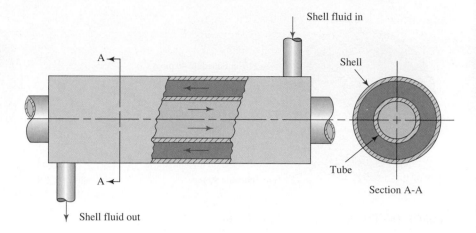

Shell fluid in

Shell

Tube

Section A-A

Shell fluid out

9.15 Using Eq. (9–4), compute the ratio of the average velocity to the maximum velocity of flow for the flow of a liquid through a concrete pipe with an inside diameter of 8.00 in with Reynolds numbers of 4000, 10^4, 10^5, and 10^6.

9.16 Using Eq. (9–3), compute several points on the velocity profile for the flow of 400 gal/min of water at 50°F in a new, clean, 4-in Schedule 40 steel pipe. Make a plot similar to Fig. 9.7 with a fairly large scale.

9.17 Repeat Problem 9.16 for the same conditions, except that the inside of the pipe is roughened by age so that $\varepsilon = 5.0 \times 10^{-3}$. Plot the results on the same graph as that used for the results of Problem 9.16.

9.18 For both situations described in Problems 9.16 and 9.17, compute the pressure drop that would occur over a distance of 250 ft of horizontal pipe.

Noncircular Sections—Average Velocity

9.19 A shell-and-tube heat exchanger is made of two standard steel tubes, as shown in Fig. 9.13. The outer tube has an OD of 7/8 in and the OD for the inner tube is ½ in. Each tube has a wall thickness of 0.049 in. Calculate the required

ratio of the volume flow rate in the shell to that in the tube if the average velocity of flow is to be the same in each.

9.20 Figure 9.14 shows a heat exchanger in which each of two DN 150 Schedule 40 pipes carries 450 L/min of water. The pipes are inside a rectangular duct whose inside dimensions are 200 mm by 400 mm. Compute the velocity of flow in the pipes. Then, compute the required volume flow rate of water in the duct to obtain the same average velocity.

9.21 Figure 9.15 shows the cross section of a shell-and-tube heat exchanger. Compute the volume flow rate required in each small pipe and in the shell to obtain an average velocity of flow of 25 ft/s in all parts.

Noncircular Cross Sections—Reynolds Number

9.22 Air with a specific weight of 12.5 N/m³ and a dynamic viscosity of 2.0×10^{-5} Pa·s flows through the shaded portion of the duct shown in Fig. 9.16 at the rate of 150 m³/h. Calculate the Reynolds number of the flow.

FIGURE 9.14 Problems 9.20, 9.26, and 9.39.

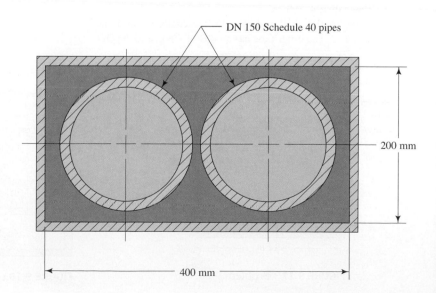

DN 150 Schedule 40 pipes

200 mm

400 mm

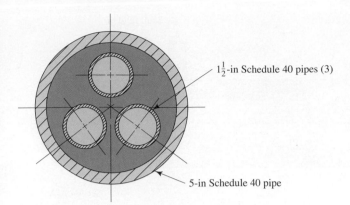

FIGURE 9.15 Problems 9.21, 9.27, and 9.40.

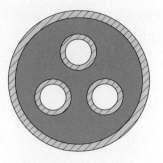

FIGURE 9.18 Problems 9.28 and 9.41.

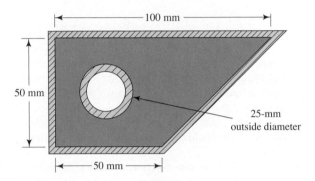

FIGURE 9.16 Problem 9.22.

of the inner tube is ½ in. Both tubes are standard steel tubes with 0.049-in wall thicknesses. The inside tube carries 4.75 gal/min of water at 200°F and the shell carries 30.0 gal/min of ethylene glycol at 77°F to carry heat away from the water. Compute the Reynolds number for the flow in both the tube and the shell.

9.26 Refer to Fig. 9.14, which shows two DN 150 Schedule 40 pipes inside a rectangular duct. Each pipe carries 450 L/min of water at 20°C. Compute the Reynolds number for the flow of water. Then, for benzene (sg = 0.862) at 70°C flowing inside the duct, compute the volume flow rate required to produce the same Reynolds number.

9.27 Refer to Fig. 9.15, which shows three pipes inside a larger pipe. The inside pipes carry water at 200°F and the large pipe carries water at 60°F. The average velocity of flow is 25.0 ft/s in each pipe; compute the Reynolds number for each.

9.28 Water at 10°C is flowing in the shell shown in Fig. 9.18 at the rate of 850 L/min. The shell is a 50 mm OD × 1.5 mm wall copper tube and the inside tubes are 15 mm OD × 1.2 mm wall copper tubes. Compute the Reynolds number for the flow.

9.29 Figure 9.19 shows the cross section of a heat exchanger used to cool a bank of electronic devices. Ethylene glycol at 77°F flows in the shaded area. Compute the volume flow rate required to produce a Reynolds number of 1500.

9.23 Carbon dioxide with a specific weight of $0.114\ \text{lb/ft}^3$ and a dynamic viscosity of $3.34 \times 10^{-7}\ \text{lb-s/ft}^2$ flows in the shaded portion of the duct shown in Fig. 9.17. If the volume flow rate is $200\ \text{ft}^3/\text{min}$, calculate the Reynolds number of the flow.

9.24 Water at 90°F flows in the space between 6-in Schedule 40 steel pipe and a square duct with inside dimensions of 10.0 in. The shape of the duct is similar to that shown in Fig. 9.10. Compute the Reynolds number if the volume flow rate is $4.00\ \text{ft}^3/\text{s}$.

9.25 Refer to the shell-and-tube heat exchanger shown in Fig. 9.13. The outer tube has an OD of 7/8 in and the OD

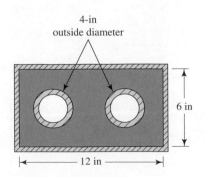

FIGURE 9.17 Problem 9.23.

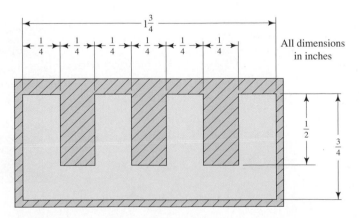

FIGURE 9.19 Problems 9.29 and 9.42.

FIGURE 9.20 Problem 9.30.

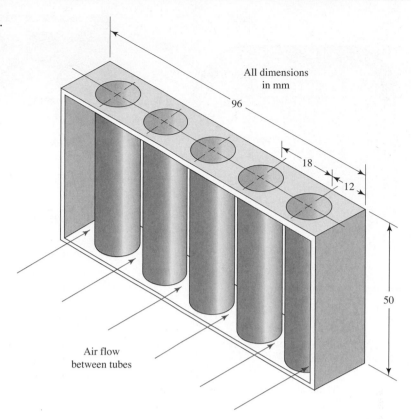

All dimensions
in mm

96

18

12

50

Air flow
between tubes

Both 150 mm square outside

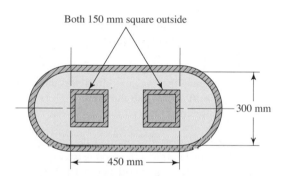

300 mm

450 mm

FIGURE 9.21 Problems 9.31, 9.32, 9.43, and 9.44.

Electronic
devices

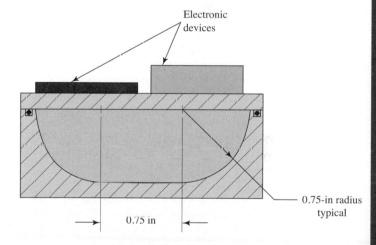

0.75-in radius
typical

0.75 in

FIGURE 9.22 Problems 9.33 and 9.45.

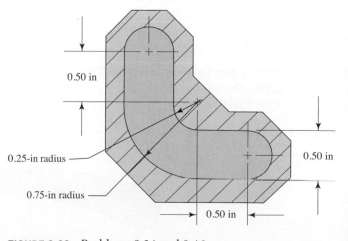

0.50 in

0.25-in radius

0.75-in radius

0.50 in

0.50 in

0.50 in

FIGURE 9.23 Problems 9.34 and 9.46.

9.30 Figure 9.20 shows a liquid-to-air heat exchanger in which air flows at 50 m³/h inside a rectangular passage and around a set of five vertical tubes. Each tube is a standard hydraulic steel tube, 15 mm OD × 1.2 mm wall. The air has a density of 1.15 kg/m³ and a dynamic viscosity of 1.63×10^{-5} Pa·s. Compute the Reynolds number for the air flow.

9.31 Glycerin (sg = 1.26) at 40°C flows in the portion of the duct outside the square tubes shown in Fig. 9.21. Calculate the Reynolds number for a flow rate of 0.10 m³/s.

9.32 Each of the square tubes shown in Fig. 9.21 carries 0.75 m³/s of water at 90°C. The thickness of the walls of the tubes is 2.77 mm. Compute the Reynolds number of the flow of water.

9.33 A heat sink for an electronic circuit is made by machining a pocket into a block of aluminum and then covering it with a flat plate to provide a passage for cooling water as shown in Fig. 9.22. Compute the Reynolds number if the flow of water at 50°F is 78.0 gal/min.

9.34 Figure 9.23 shows the cross section of a cooling passage for an odd-shaped device. Compute the volume flow rate

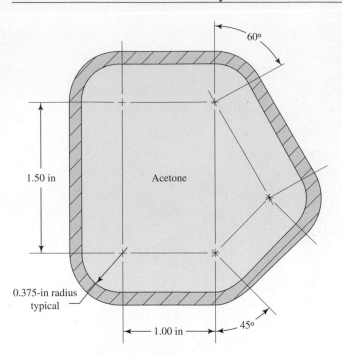

FIGURE 9.24   Problem 9.35.

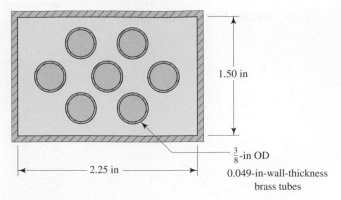

FIGURE 9.26   Problem 9.47.

of water at 50°F that would produce a Reynolds number of 1.5×10^5.

9.35   Figure 9.24 shows the cross section of a flow path machined from a casting using a ¾-in-diameter milling cutter. Considering all the fillets, compute the hydraulic radius for the passage, and then compute the volume flow rate of acetone at 77°F required to produce a Reynolds number for the flow of 2.6×10^4.

9.36   The blade of a gas turbine engine contains internal cooling passages, as shown in Fig. 9.25. Compute the volume flow rate of air required to produce an average velocity of flow in each passage of 25.0 m/s. The air flow distributes evenly to all six passages. Then, compute the Reynolds number if the air has a density of 1.20 kg/m^3 and a dynamic viscosity of 1.50×10^{-5} Pa·s.

Noncircular Cross Sections—Energy Losses

9.37   For the system described in Problem 9.24, compute the pressure difference between two points 30.0 ft apart if the duct is horizontal. Use $\varepsilon = 8.5 \times 10^{-5}$ ft.

9.38   For the shell-and-tube heat exchanger described in Problem 9.25, compute the pressure difference for both fluids between two points 5.25 m apart if the heat exchanger is horizontal.

9.39   For the system described in Problem 9.26, compute the pressure drop for both fluids between two points 3.80 m apart if the duct is horizontal. Use the roughness for steel pipe for all surfaces.

9.40   For the system described in Problem 9.27, compute the pressure difference in both the small pipes and the large pipe between two points 50.0 ft apart if the pipes are horizontal. Use the roughness for steel pipe for all surfaces.

9.41   For the shell-and-tube heat exchanger described in Problem 9.28, compute the pressure drop for the flow of water in the shell. Use the roughness for copper for all surfaces. The length is 3.60 m.

9.42   For the heat exchanger described in Problem 9.29, compute the pressure drop for a length of 57 in.

9.43   For the glycerin described in Problem 9.31, compute the pressure drop for a horizontal duct 22.6 m long. All surfaces are copper.

9.44   For the flow of water in the square tubes described in Problem 9.32, compute the pressure drop over a length of 22.6 m. All surfaces are copper and the duct is horizontal.

9.45   If the heat sink described in Problem 9.33 is 105 in long, compute the pressure drop for the water. Use $\varepsilon = 2.5 \times 10^{-5}$ ft for the aluminum.

9.46   Compute the energy loss for the flow of water in the cooling passage described in Problem 9.34 if its total length is 45 in. Use ε for steel. Also compute the pressure difference across the total length of the cooling passage.

9.47   In Fig. 9.26, ethylene glycol (sg = 1.10) at 77°F flows around the tubes and inside the rectangular passage. Calculate the volume flow rate of ethylene glycol in gal/min required for the flow to have a Reynolds number of 8000. Then, compute the energy loss over a length of 128 in. All surfaces are brass.

9.48   Figure 9.27 shows a duct in which methyl alcohol at 25°C flows at the rate of 3000 L/min. Compute the energy loss over a 2.25-m length of the duct. All surfaces are smooth plastic.

9.49   A furnace heat exchanger has a cross section like that shown in Fig. 9.28. The air flows around the three thin passages in which hot gases flow. The air is at 140°F and has a density of $2.06 \times 10^{-3} \text{ slugs/ft}^3$ and a dynamic

FIGURE 9.25   Problem 9.36.

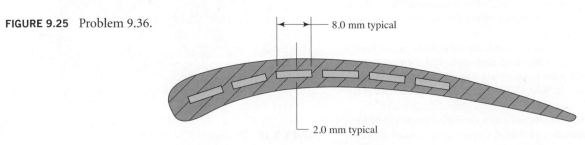

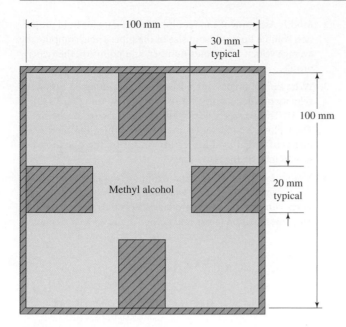

FIGURE 9.27 Problem 9.48.

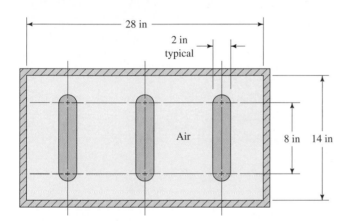

FIGURE 9.28 Problem 9.49.

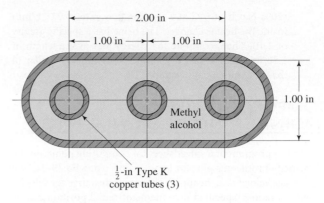

FIGURE 9.29 Problem 9.50.

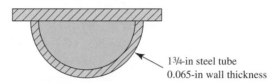

FIGURE 9.30 Problem 9.51.

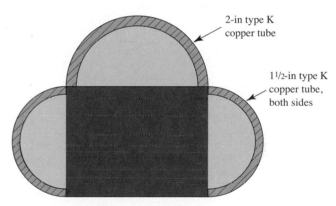

FIGURE 9.31 Problem 9.52.

viscosity of 4.14×10^{-7} lb·s/ft². Compute the Reynolds number for the flow if the velocity is 20 ft/s.

9.50 Figure 9.29 shows a system in which methyl alcohol at 77°F flows outside the three tubes while ethyl alcohol at 0°F flows inside the tubes. Compute the volume flow rate of each fluid required to produce a Reynolds number of 3.5×10^4 in all parts of the system. Then, compute the pressure difference for each fluid between two points 10.5 ft apart if the system is horizontal. All surfaces are copper.

9.51 A simple heat exchanger is made by welding one-half of a 1¾-in drawn steel tube to a flat plate as shown in Fig. 9.30. Water at 40°F flows in the enclosed space and cools the plate. Compute the volume flow rate required so that the Reynolds number of the flow is 3.5×10^4. Then, compute the energy loss over a length of 92 in.

9.52 Three surfaces of an instrument package are cooled by soldering half-sections of copper tubing to it as shown in Fig. 9.31. Compute the Reynolds number for each section if ethylene glycol at 77°F flows with an average velocity of 15 ft/s. Then compute the energy loss over a length of 54 in.

9.53 Figure 9.32 shows a heat exchanger with internal fins. Compute the Reynolds number for the flow of brine

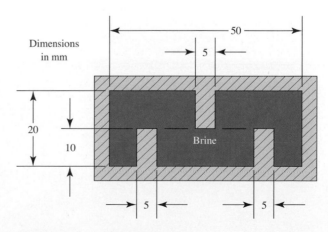

FIGURE 9.32 Problem 9.53.

(20% NaCl) at 0°C at a volume flow rate of 225 L/min inside the heat exchanger. The brine has a specific gravity of 1.10. Then, compute the energy loss over a length of 1.80 m. Assume that the surface roughness is similar to that of commercial steel pipe.

COMPUTER AIDED ENGINEERING ASSIGNMENTS

1. Write a program or a spreadsheet for computing points on the velocity profile in a pipe for laminar flow using Eq. (9–1). The average velocity can be input. Then, plot the curve for velocity versus radius. Specified increments of radial position can be input, but should include the centerline.

2. Modify Assignment 1 to require input of data for fluid properties, volume flow rate, and size of the pipe. Then, compute the average velocity, Reynolds number, and points on the velocity profile.

3. Write a program or a spreadsheet for computing points on the velocity profile in a pipe for turbulent flow using Eq. (9–2) or (9–3). The average velocity and friction factor can be input. Then, plot the curve for velocity versus radius. Specified increments of radial position can be input by the operator, but should include the centerline.

4. Modify Assignment 3 to require input of data for fluid properties, volume flow rate, pipe wall roughness, and size of the pipe. Then, compute the average velocity, Reynolds number, relative roughness, friction factor, and points on the velocity profile.

MINOR LOSSES

In Chapter 6, the importance of including all forms of energy in the analysis of fluid flow systems was introduced and you learned to apply *Bernoulli's equation*. In Chapter 7, you applied the *general energy equation*, which extended Bernoulli's equation to account for energy losses and additions that typically occur in real flow systems. In Chapter 8, you learned how to calculate the magnitude of energy losses due to friction as fluids flow through pipes and tubes. For long piping systems, friction losses can be quite large.

However, most piping systems also contain other elements that cause energy losses; valves, fittings (e.g., elbows, tees, expansions, contractions), entrances to the piping and exits from the piping, and special equipment such as gages, flow meters, heat exchangers, filters, and strainers. We generally refer to such losses as *minor losses*. However, the actual magnitude of these losses can be significant and, when considering that a large number of valves and fittings may exist, the cumulative amount of energy loss may be substantial and all minor losses should be accounted for. Figure 10.1 shows an industrial piping installation that illustrates numerous minor losses.

Exploration

Study Fig. 7.1 again from the Big Picture part of Chapter 7. The drawing shows an industrial piping system delivering fluid from storage tanks to processes that use the fluid. List all of the components in the drawing that are used to control the flow or to direct it to specific destinations. These are examples of devices that cause energy to be lost from the flowing fluid.

Also, describe other fluid flow systems that you can observe, and identify the path of the piping and the other components that cause energy losses. Discuss these systems with your colleagues and with the course instructor or facilitator.

Introductory Concepts

Here you continue to learn techniques for analyzing real pipeline problems in which several types of flow system components exist. You are close to the goal we set in Chapter 6, where Bernoulli's equation was introduced. We said that in Chapters 6–11 you would continue to develop concepts related to the flow of fluids in pipeline systems. The goal is to put them all together to analyze

FIGURE 10.1 This industrial piping system containing numerous valves, elbows, tees, gages, and flow meters is an example of the real systems you will learn how to analyze in this chapter and in Chapters 11–13. (*Source:* Aleksey Stemmer/Fotolia)

the performance of such systems. You will do this in Chapter 11.

From your study of the industrial piping system in Fig. 7.1, how does your list of fluid control components compare with this?

1. The fluid exits from the storage tank at the rear and flows through a pipe, called the *suction line*, to the left side of the pump. Note that the suction line is somewhat larger than the *discharge line* on the right side of the pump, a typical design feature of pumped fluid flow systems. It is also possible that the suction line size is larger than the size of the inlet port for the pump and that the discharge line size is larger than the size of the discharge port.

2. As it approaches the pump, the flow passes through the *suction line shutoff valve* that permits the piping system to be isolated from the pump during pump service or replacement.

3. At the suction port flange, the pipe size may be reduced through a *gradual reducer* that would be needed if the suction pipe size is larger than the standard connection provided by the pump manufacturer. As a result, the fluid velocity would increase somewhat as it moves from the pipe into the suction inlet of the pump.

4. The pump, driven by an electric motor, pulls the fluid from the suction line and adds energy to it as it moves the fluid into the *discharge line*. The fluid in the discharge line now has a higher energy level, resulting in a higher pressure head.

5. Because the discharge line may be larger than the pump outlet size, an *enlargement* may be used that increases the size to the full size of the discharge line. As the fluid moves through the enlargement, the flow velocity decreases.

6. Just to the right of the discharge flange there is a *tee* in the pipe with another line heading toward the front of the drawing. This allows the operator of the system to direct the flow in either of two ways. The normal direction is to continue through the main discharge line. This would happen if the valve to the front side of the tee is shut off. But if that valve is opened, all or part of the flow would turn into the branch line through the tee and flow through the adjacent *valve*. It would then continue on through the branch line.

7. Let's assume that the valve in the branch line is shut off. The fluid continues in the discharge line and encounters another *valve*. Normally, this valve is fully open, allowing the fluid to go on to its destination. The valve permits the system to be shut down after the pump is stopped, allowing for pump replacement or service without draining the piping system downstream from the pump.

8. After flowing through the valve in the discharge line, another *tee* allows some of the fluid to go into a branch to the long pipe that proceeds toward the rear of the drawing while the bulk of the flow is delivered to other parts of the plant. Let's assume that some flow does go into the branch line as described next.

9. After leaving the tee through the branch line, the fluid immediately encounters an *elbow* that redirects it from a vertical to a horizontal direction.

10. After moving through a short length of pipe, another *valve* is in the line to control the flow to the rest of the system.

11. Also in this section, there is a *flow meter* to permit the operator to measure how much fluid is flowing in the pipe.

12. After flowing through the meter, the fluid continues through the long pipe to the process that will use it.

Look at the numerous control devices (shown in italics) in this list. Energy is lost from the system through each of these devices. When you design such a system, you will need to account for these energy losses.

Now, study the list of other fluid flow systems you have seen and identify other kinds of elements that could cause energy losses to occur. Examples are listed next.

- Consider the plumbing system in your home. Track how the water gets from the main supply point to the kitchen sink. Write down each element that causes an obstruction to the flow (such as a valve), that changes the direction of the flow, or that changes the velocity of flow.

- Consider how the water gets to an outside faucet that can be used to water the lawn or garden. Track the flow all the way to the sprinkler head.

- How does the water get from the city supply wells or reservoir to your home?

- How does the cooling fluid in an automotive engine move from the radiator through the engine and back to the radiator?

- How does the windshield-washing fluid get from the reservoir to the windshield?

- How does the gasoline in your car or a truck get from the fuel tank to the engine intake ports?

- How does fuel on an airplane get from its fuel tanks in the wings to the engines?

- How does the refrigerant in your car's air conditioning system flow from the compressor attached to the engine through the system that makes the car cool?

- How does the refrigerant in your refrigerator move through its cooling system?

- How does the water in a clothes washer get from the house piping system into the wash tub?

- How does the wash water drain from the tub and get pumped into the sewer drain?

- How does the water flow through a squirt toy?
- Have you seen a high-pressure washing system that can be used to remove heavy dirt from a deck, a driveway, or a boat? Track the flow of fluid through that kind of system.
- How does water in an apartment building or a hotel get from the city supply line to each apartment or hotel room?
- How does the water flow from the city supply line through the sprinkler system in an office building or warehouse to protect the people, products, and equipment from a fire?
- How does the oil in a fluid power system flow from the pump through the control valves, cylinders, and other fluid power devices to actuate industrial automation systems, construction equipment, agricultural machinery, or aircraft landing gear?
- How does engine oil get pumped from the oil pan to lubricate the moving parts of the engine?

- How does the lubricating fluid in a complex piece of manufacturing equipment get distributed to critical moving parts?
- How do liquid components of chemical processing systems move through those systems?
- How does milk, juice, or soft-drink mix flow through the systems that finally deliver it to the bottling station?

What other fluid flow systems did you think of?

Now let's learn how to analyze the energy losses in these kinds of systems. In this chapter you will learn how to determine the magnitude of minor losses. Included here are descriptions of methods for analyzing energy losses for changes in the flow area, changes in direction, valves, and fittings. Several comprehensive references are included at the end of the chapter that present additional information. See References 2, 3, 5–7, 9, 11, and 13.

10.1 OBJECTIVES

After completing this chapter, you should be able to:

1. Recognize the sources of minor losses.
2. Define *resistance coefficient*.
3. Determine the energy loss for flow through the following types of minor losses:
   a. Sudden enlargement of the flow path.
   b. Exit loss when fluid leaves a pipe and enters a static reservoir.
   c. Gradual enlargement of the flow path.
   d. Sudden contraction of the flow path.
   e. Gradual contraction of the flow path.
   f. Entrance loss when fluid enters a pipe from a static reservoir.
4. Define the term *vena contracta*.
5. Define and use the *equivalent-length technique* for computing energy losses in valves, fittings, and pipe bends.
6. Describe the energy losses that occur in a typical fluid power system.
7. Demonstrate how the *flow coefficient* C_V is used to evaluate energy losses in some types of valves.
8. Use the PIPE-FLO® software to analyze fluid flow systems having minor losses.

10.2 RESISTANCE COEFFICIENT

Energy losses are proportional to the velocity head of the fluid as it flows around an elbow, through an enlargement or contraction of the flow section, or through a valve.

Experimental values for energy losses are usually reported in terms of a resistance coefficient K as follows:

⇨ **Minor Loss Using Resistance Coefficient**

$$h_L = K(v^2/2g) \qquad (10\text{–}1)$$

In Eq. (10–1), h_L is the minor loss, K is the resistance coefficient, and v is the average velocity of flow in the pipe in the vicinity where the minor loss occurs. In some cases, there may be more than one velocity of flow, as with enlargements or contractions. It is most important for you to know which velocity is to be used with each resistance coefficient.

The resistance coefficient is dimensionless because it represents a constant of proportionality between the energy loss and the velocity head. The magnitude of the resistance coefficient depends on the geometry of the device that causes the loss and sometimes on the velocity of flow. In the following sections, we will describe the process for determining the value of K and for calculating the energy loss for many types of minor loss conditions.

As in the energy equation, the velocity head $v^2/2g$ in Eq. (10–1) is typically in the SI units of meters (or, N·m/N of fluid flowing) or the U.S. Customary units of feet (or, ft-lb/lb of fluid flowing). Because K is dimensionless, the energy loss has the same units.

Reference 4 provides extensive discussion and tables of data for K-factors for energy losses due to changes in flow area and other minor losses.

FIGURE 10.2 Sudden enlargement.

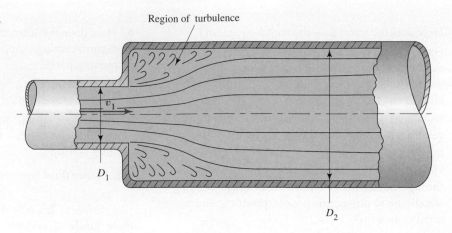

10.3 SUDDEN ENLARGEMENT

As a fluid flows from a smaller pipe into a larger pipe through a sudden enlargement, its velocity abruptly decreases, causing turbulence, which generates an energy loss. See Fig. 10.2. Tests have shown that the amount of turbulence, and therefore the amount of energy loss, is dependent on the ratio of the sizes of the two pipes and the magnitude of the flow velocity in the smaller pipe. Then we can adapt Equation 10-1 to form the following equation for this type of minor loss

$$h_L = K(v_1^2/2g) \tag{10–2}$$

where v_1 is the average velocity of flow in the smaller pipe ahead of the enlargement. The data for values of K are illustrated graphically in Figure 10.3 and in tabular form in Table 10.1A. Table 10.1B gives data in metric units.

FIGURE 10.3 Resistance coefficient—sudden enlargement.

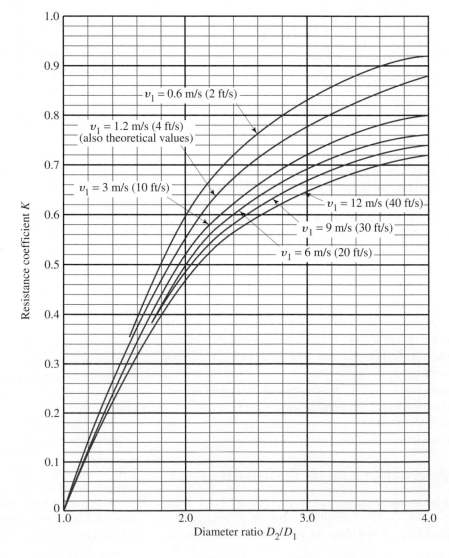

TABLE 10.1A   Resistance coefficient—sudden enlargement. Data for Figure 10.3

| | Velocity v_1 | | | | | | |
D_2/D_1	0.6 m/s 2 ft/s	1.2 m/s 4 ft/s	3 m/s 10 ft/s	4.5 m/s 15 ft/s	6 m/s 20 ft/s	9 m/s 30 ft/s	12 m/s 40 ft/s
1.0	0.0	0.0	0.0	0.0	0.0	0.0	0.0
1.2	0.11	0.10	0.09	0.09	0.09	0.09	0.08
1.4	0.26	0.25	0.23	0.22	0.22	0.21	0.20
1.6	0.40	0.38	0.35	0.34	0.33	0.32	0.32
1.8	0.51	0.48	0.45	0.43	0.42	0.41	0.40
2.0	0.60	0.56	0.52	0.51	0.50	0.48	0.47
2.5	0.74	0.70	0.65	0.63	0.62	0.60	0.58
3.0	0.83	0.78	0.73	0.70	0.69	0.67	0.65
4.0	0.92	0.87	0.80	0.78	0.76	0.74	0.72
5.0	0.96	0.91	0.84	0.82	0.80	0.77	0.75
10.0	1.00	0.96	0.89	0.86	0.84	0.82	0.80
∞	1.00	0.98	0.91	0.88	0.86	0.83	0.81

By making some simplifying assumptions about the character of the flow stream as it expands through the sudden enlargement, it is possible to analytically predict the value of K from the following equation:

$$K = [1 - (A_1/A_2)]^2 = [1 - (D_1/D_2)^2]^2 \quad \textbf{(10–3)}$$

The subscripts 1 and 2 refer to the smaller and larger sections, respectively, as shown in Fig. 10.2. Values for K from this equation agree well with experimental data when the velocity v_1 is approximately 1.2 m/s (4 ft/s). At higher velocities, the actual values of K are lower than the theoretical values, and at lower velocities K-values are higher. We recommend that experimental values from charts or tables be used if the velocity of flow is known.

TABLE 10.1B   Resistance coefficient—sudden enlargement—Metric data

| | Velocity v_1, m/s | | | | | | | | | |
D_2/D_1	0.5	1.0	2.0	3.0	4.0	5.0	6.0	7.0	8.0	9.0	10.0
1.2	0.11	0.10	0.09	0.09	0.09	0.09	0.09	0.09	0.09	0.09	0.09
1.4	0.26	0.26	0.24	0.23	0.23	0.22	0.22	0.22	0.21	0.21	0.21
1.6	0.40	0.39	0.36	0.35	0.35	0.34	0.33	0.33	0.32	0.32	0.32
1.8	0.51	0.49	0.46	0.45	0.44	0.43	0.42	0.42	0.41	0.41	0.41
2.0	0.60	0.58	0.54	0.52	0.52	0.51	0.50	0.50	0.49	0.48	0.48
2.5	0.74	0.72	0.67	0.65	0.64	0.63	0.62	0.62	0.61	0.60	0.59
3.0	0.84	0.80	0.75	0.73	0.71	0.70	0.69	0.68	0.67	0.67	0.66
4.0	0.93	0.89	0.83	0.80	0.79	0.77	0.76	0.75	0.74	0.74	0.73
5.0	0.97	0.93	0.87	0.84	0.83	0.81	0.80	0.79	0.78	0.77	0.76
10.0	1.00	0.98	0.92	0.89	0.87	0.85	0.84	0.83	0.82	0.82	0.81
∞	1.00	1.00	0.94	0.91	0.89	0.87	0.86	0.85	0.84	0.83	0.82

D_2/D_1—ratio of diameter of larger pipe to diameter of smaller pipe; v_1—velocity in smaller pipe.

Source: Brater, Ernest F, et al. © 1996. *Handbook of Hydraulics,* 7th ed. New York: McGraw-Hill, Table 6–5.

**Example Problem
10.1**

Determine the energy loss that will occur as 100 L/min of water flows through a sudden enlargement made from two sizes of copper hydraulic tubing. The small tube is 25 mm OD × 1.5 mm wall; the large tube is 80 mm OD × 2.8 mm wall. See Appendix G.2 for tube dimensions and areas.

Solution

Using the subscript 1 for the section just ahead of the enlargement and subscript 2 for the section downstream from the enlargement, we get

$$D_1 = 22.0 \text{ mm} = 0.022 \text{ m}$$

$$A_1 = 3.801 \times 10^{-4} \text{ m}^2$$

$$D_2 = 74.4 \text{ mm} = 0.0744 \text{ m}$$

$$A_2 = 4.347 \times 10^{-3} \text{ m}^2$$

$$v_1 = \frac{Q}{A_1} = \frac{100 \text{ L/min}}{3.801 \times 10^{-4} \text{ m}^2} \times \frac{1 \text{ m}^3/\text{s}}{60\,000 \text{ L/min}} = 4.385 \text{ m/s}$$

$$\frac{v_1^2}{2g} = \frac{(4.385)^2}{(2)(9.81)} \text{ m} = 0.980 \text{ m}$$

To find a value for K, the diameter ratio is needed. We find that

$$D_2/D_1 = 74.4/22.0 = 3.382$$

From Fig. 10.3, we read $K = 0.740$. Then we have

$$h_L = K(v_1^2/2g) = (0.740)(0.980 \text{ m}) = 0.725 \text{ m}$$

This result indicates that 0.725 N·m of energy is dissipated from each newton of water that flows through the sudden enlargement.

The following problem illustrates the calculation of the pressure difference between points 1 and 2.

**Example Problem
10.2**

For the data from Example Problem 10.1, determine the difference between the pressure ahead of the sudden enlargement and the pressure downstream from the enlargement.

Solution

First, we write the energy equation:

$$\frac{p_1}{\gamma} + z_1 + \frac{v_1^2}{2g} - h_L = \frac{p_2}{\gamma} + z_2 + \frac{v_2^2}{2g}$$

Solving for $p_1 - p_2$ gives

$$p_1 - p_2 = \gamma[(z_2 - z_1) + (v_2^2 - v_1^2)/2g + h_L]$$

If the enlargement is horizontal, $z_2 - z_1 = 0$. Even if it were vertical, the distance between points 1 and 2 is typically so small that it is considered negligible. Now, calculating the velocity in the larger pipe, we get

$$v_2 = \frac{Q}{A_2} = \frac{100 \text{ L/min}}{4.347 \times 10^{-3} \text{ m}^2} \times \frac{1 \text{ m}^3/\text{s}}{60\,000 \text{ L/min}} = 0.383 \text{ m/s}$$

Using $\gamma = 9.81 \text{ kN/m}^3$ for water and $h_L = 0.725 \text{ m}$ from Example Problem 10.1, we have

$$p_1 - p_2 = \frac{9.81 \text{ kN}}{\text{m}^3}\left[0 + \frac{(0.383)^2 - (4.385)^2}{(2)(9.81)} \text{ m} + 0.725 \text{ m}\right]$$

$$= -2.43 \text{ kN/m}^2 = -2.43 \text{ kPa}$$

Therefore, p_2 is 2.43 kPa greater than p_1.

10.4 EXIT LOSS

As a fluid flows from a pipe into a large reservoir or tank, as shown in Fig. 10.4, its velocity is decreased to very nearly zero. In the process, the kinetic energy that the fluid possessed in the pipe, indicated by the velocity head $v_1^2/2g$, is dissipated. Therefore, the energy loss for this condition is

$$h_L = 1.0(v_1^2/2g) \qquad (10\text{–}4)$$

This is called the *exit loss*. The value of $K = 1.0$ is used regardless of the form of the exit where the pipe connects to the tank wall.

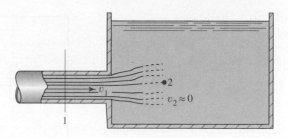

FIGURE 10.4 Exit loss as fluid flows from a pipe into a static reservoir.

Example Problem 10.3 Determine the energy loss that will occur as 100 L/min of water flows from a copper hydraulic tube, 25.0 mm OD × 1.5 mm wall, into a large tank.

Solution Using Eq. (10–4), we have

$$h_L = 1.0(v_1^2/2g)$$

From the calculations in Example Problem 10.1, we know that

$$v_1 = 4.385 \text{ m/s}$$
$$v_1^2/2g = 0.740 \text{ m}$$

Then the energy loss is

$$h_L = (1.0)(0.740 \text{ m}) = 0.740 \text{ m}$$

10.5 GRADUAL ENLARGEMENT

If the transition from a smaller to a larger pipe can be made less abrupt than the square-edged sudden enlargement, the energy loss is reduced. This is normally done by placing a conical section between the two pipes as shown in Fig. 10.5. The sloping walls of the cone tend to guide the fluid during the deceleration and expansion of the flow stream. Therefore, the size of the zone of separation and the amount of turbulence are reduced as the cone angle is reduced.

The energy loss for a gradual enlargement is calculated from

$$h_L = K(v_1^2/2g) \qquad (10\text{–}5)$$

where v_1 is the velocity in the smaller pipe ahead of the enlargement. The magnitude of K is dependent on both the diameter ratio D_2/D_1 and the cone angle θ. Data for various values of θ and D_2/D_1 are given in Fig. 10.6 and Table 10.2.

The energy loss calculated from Eq. (10–5) does not include the loss due to friction at the walls of the transition. For relatively steep cone angles, the length of the transition is short and, therefore, the wall friction loss is negligible. However, as the cone angle decreases, the length of the transition increases and wall friction becomes significant. Taking both wall friction loss and the loss due to the enlargement into account, we can obtain the minimum energy loss with a cone angle of about 7°.

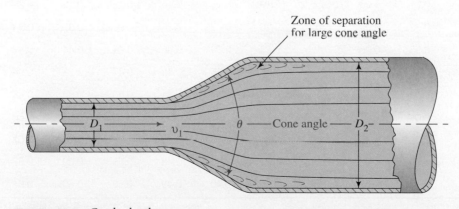

FIGURE 10.5 Gradual enlargement.

FIGURE 10.6 Resistance coefficient—gradual enlargement.

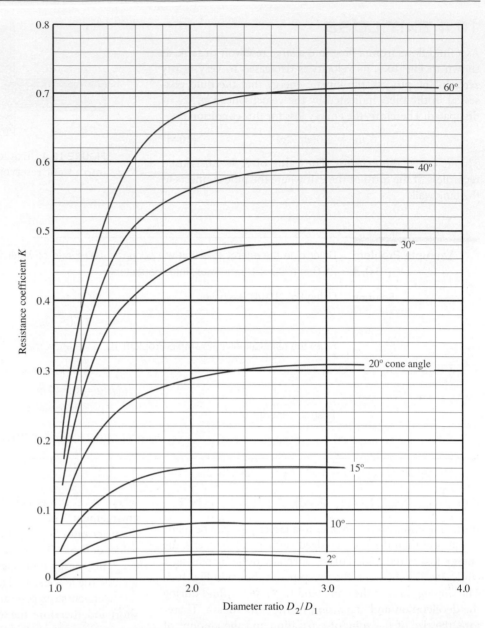

TABLE 10.2 Resistance coefficient—gradual enlargement

	Angle of Cone θ											
D_2/D_1	2°	6°	10°	15°	20°	25°	30°	35°	40°	45°	50°	60°
1.1	0.01	0.01	0.03	0.05	0.10	0.13	0.16	0.18	0.19	0.20	0.21	0.23
1.2	0.02	0.02	0.04	0.09	0.16	0.21	0.25	0.29	0.31	0.33	0.35	0.37
1.4	0.02	0.03	0.06	0.12	0.23	0.30	0.36	0.41	0.44	0.47	0.50	0.53
1.6	0.03	0.04	0.07	0.14	0.26	0.35	0.42	0.47	0.51	0.54	0.57	0.61
1.8	0.03	0.04	0.07	0.15	0.28	0.37	0.44	0.50	0.54	0.58	0.61	0.65
2.0	0.03	0.04	0.07	0.16	0.29	0.38	0.46	0.52	0.56	0.60	0.63	0.68
2.5	0.03	0.04	0.08	0.16	0.30	0.39	0.48	0.54	0.58	0.62	0.65	0.70
3.0	0.03	0.04	0.08	0.16	0.31	0.40	0.48	0.55	0.59	0.63	0.66	0.71
∞	0.03	0.05	0.08	0.16	0.31	0.40	0.49	0.56	0.60	0.64	0.67	0.72

Source: Brater, Ernest F, Horace W. King, James E. Lindell, and C. Y. Wei. 1996. *Handbook of Hydraulics,* 7th ed. New York: McGraw-Hill, Table 6–6.

Example Problem 10.4 Determine the energy loss that will occur as 100 L/min of water flows from a small copper hydraulic tube to a larger tube through a gradual enlargement having an included angle of 30°. The small tube has a 25 mm OD × 1.5 mm wall; the large tube has an 80 mm OD × 2.8 mm wall.

Solution Using data from Appendix G.2 and the results of some calculations in preceding example problems, we know that

$$v_1 = 4.385 \text{ m/s}$$
$$v_1^2/2g = 0.980 \text{ m}$$
$$D_2/D_1 = 74.4/22.0 = 3.382$$

From Fig. 10.6, we find that $K = 0.48$. Then, we have

$$h_L = K(v_1^2/2g) = (0.48)(0.980 \text{ m}) = 0.470 \text{ m}$$

Compared with the sudden enlargement described in Example Problem 10.1, the energy loss decreases by 35 percent when the 30° gradual enlargement is used.

Diffuser Another term for an enlargement is a *diffuser*. The function of a diffuser is to convert kinetic energy (represented by velocity head $v^2/2g$) to pressure energy (represented by the pressure head p/γ) by decelerating the fluid as it flows from the smaller to the larger pipe. The diffuser can be either sudden or gradual, but the term is most often used to describe a gradual enlargement.

An ideal diffuser is one in which no energy is lost as the flow decelerates. Of course, no diffuser performs in the ideal fashion. If it did, the theoretical maximum pressure after the expansion could be computed from Bernoulli's equation,

$$p_1/\gamma + z_1 + v_1^2/2g = p_2/\gamma + z_2 + v_2^2/2g$$

If the diffuser is in a horizontal plane, the elevation terms can be cancelled out. Then the pressure increase across the ideal diffuser is

▷ **Pressure Recovery—Ideal Diffuser**

$$\Delta p = p_2 - p_1 = \gamma(v_1^2 - v_2^2)/2g$$

This is often called *pressure recovery*.

In a *real diffuser*, energy losses do occur and the general energy equation must be used:

$$p_1/\gamma + z_1 + v_1^2/2g - h_L = p_2/\gamma + z_2 + v_2^2/2g$$

The pressure increase becomes

▷ **Pressure Recovery—Real Diffuser**

$$\Delta p = p_2 - p_1 = \gamma[(v_1^2 - v_2^2)/2g - h_L]$$

The energy loss is computed using the data and procedures in this section. The ratio of the pressure recovery from the

real diffuser to that of the ideal diffuser is a measure of the effectiveness of the diffuser.

10.6 SUDDEN CONTRACTION

The energy loss due to a sudden contraction, such as that sketched in Fig. 10.7, is calculated from

$$h_L = K(v_2^2/2g) \tag{10-6}$$

where v_2 is the velocity in the small pipe downstream from the contraction. The resistance coefficient K is dependent on the ratio of the sizes of the two pipes and on the velocity of flow, as Fig. 10.8 and Table 10.3 show.

The mechanism by which energy is lost due to a sudden contraction is quite complex. Figure 10.9 illustrates what happens as the flow stream converges. The lines in the figure represent the paths of various parts of the flow stream called *streamlines*. As the streamlines approach the contraction, they assume a curved path and the total stream continues to neck down for some distance beyond the contraction. Thus, the effective minimum cross section of the flow is smaller than that of the smaller pipe. The section where this minimum flow area occurs is called the *vena contracta*. Beyond the vena contracta, the flow stream must decelerate and expand again to fill the pipe. The turbulence caused by the contraction and the subsequent expansion generates the energy loss.

Comparing the values for the loss coefficients for sudden contraction (Fig. 10.8) with those for sudden enlargements (Fig. 10.3), we see that the energy loss from a sudden contraction is somewhat smaller. In general, accelerating a fluid causes less turbulence than decelerating it for a given ratio of diameter change.

FIGURE 10.7 Sudden contraction.

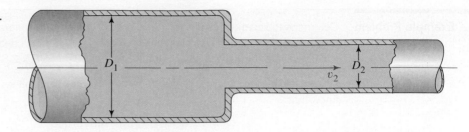

FIGURE 10.8 Resistance coefficient—sudden contraction.

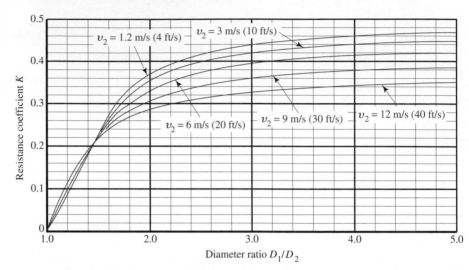

FIGURE 10.9 Vena contracta formed in a sudden contraction.

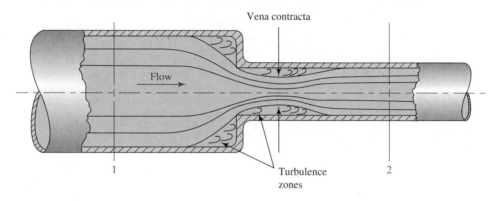

Example Problem 10.5

Determine the energy loss that will occur as 100 L/min of water flows from a large copper hydraulic tube to a smaller one with a sudden contraction. The large tube has an 80 mm OD × 2.8 mm wall. The small tube has a 25 mm × 2.5 mm wall. The tube sizes and the flow rate are the same as those used in previous example problems.

Solution

From Eq. (10–6), we have

$$h_L = K(v_2^2/2g)$$

For the copper tube we know that $D_1 = 74.4$ mm, $D_2 = 22.0$ mm, and $A_2 = 3.801 \times 10^{-4}$ m^2. Then we can find the following values

$$D_1/D_2 = 74.4/22.0 = 3.383$$

$$v_2 = \frac{Q}{A_2} = \frac{100 \text{ L/min}}{3.801 \times 10^{-4} \text{ m}^2} \times \frac{1 \text{ m}^3/\text{s}}{60\,000 \text{ L/min}} = 4.385 \text{ m/s}$$

$$v_2^2/2g = 0.980 \text{ m}$$

From Fig. 10.7 we can find $K = 0.415$. Then we have

$$h_L = K(v_2^2/2g) = (0.415)(0.980 \text{ m}) = 0.407 \text{ m}$$

TABLE 10.3A   Resistance coefficient—sudden contraction—Data for Figure 10.8

	Velocity v_2								
D_2/D_1	0.6 m/s 2 ft/s	1.2 m/s 4 ft/s	1.8 m/s 6 ft/s	2.4 m/s 8 ft/s	3 m/s 10 ft/s	4.5 m/s 15 ft/s	6 m/s 20 ft/s	9 m/s 30 ft/s	12 m/s 40 ft/s
1.0	0.0	0.0	0.0	0.0	0.0	0.0	0.0	0.0	0.0
1.1	0.03	0.04	0.04	0.04	0.04	0.04	0.05	0.05	0.06
1.2	0.07	0.07	0.07	0.07	0.08	0.08	0.09	0.10	0.11
1.4	0.17	0.17	0.17	0.17	0.18	0.18	0.18	0.19	0.20
1.6	0.26	0.26	0.26	0.26	0.26	0.25	0.25	0.25	0.24
1.8	0.34	0.34	0.34	0.33	0.33	0.32	0.31	0.29	0.27
2.0	0.38	0.37	0.37	0.36	0.36	0.34	0.33	0.31	0.29
2.2	0.40	0.40	0.39	0.39	0.38	0.37	0.35	0.33	0.30
2.5	0.42	0.42	0.41	0.40	0.40	0.38	0.37	0.34	0.31
3.0	0.44	0.44	0.43	0.42	0.42	0.40	0.39	0.36	0.33
4.0	0.47	0.46	0.45	0.45	0.44	0.42	0.41	0.37	0.34
5.0	0.48	0.47	0.47	0.46	0.45	0.44	0.42	0.38	0.35
10.0	0.49	0.48	0.48	0.47	0.46	0.45	0.43	0.40	0.36
∞	0.49	0.48	0.48	0.47	0.47	0.45	0.44	0.41	0.38

TABLE 10.3B   Resistance coefficient—sudden enlargement—Metric data

	Velocity v_2, m/s									
D_2/D_1	0.5	1.0	2.0	3.0	4.0	5.0	6.0	7.0	8.0	9.0	10.0
1.1	0.03	0.04	0.04	0.04	0.04	0.04	0.05	0.05	0.05	0.05	0.05
1.2	0.07	0.07	0.07	0.08	0.08	0.08	0.09	0.09	0.10	0.10	0.10
1.4	0.17	0.17	0.17	0.18	0.18	0.18	0.18	0.19	0.19	0.19	0.19
1.6	0.26	0.26	0.26	0.26	0.26	0.26	0.25	0.25	0.25	0.25	0.24
1.8	0.34	0.34	0.34	0.33	0.32	0.31	0.31	0.30	0.29	0.29	0.28
2.0	0.38	0.38	0.37	0.36	0.35	0.34	0.33	0.33	0.32	0.31	0.30
2.2	0.40	0.40	0.39	0.38	0.37	0.36	0.35	0.35	0.34	0.33	0.32
2.5	0.42	0.42	0.41	0.40	0.39	0.38	0.37	0.36	0.35	0.34	0.33
3.0	0.44	0.44	0.43	0.42	0.41	0.40	0.39	0.38	0.37	0.36	0.35
4.0	0.47	0.46	0.45	0.44	0.43	0.42	0.41	0.40	0.38	0.37	0.36
5.0	0.48	0.48	0.46	0.45	0.45	0.44	0.42	0.41	0.39	0.38	0.37
10.0	0.49	0.48	0.47	0.46	0.46	0.44	0.43	0.42	0.41	0.40	0.39
∞	0.49	0.49	0.47	0.47	0.46	0.45	0.44	0.43	0.42	0.41	0.40

D_2/D_1—ratio of diameter of larger pipe to diameter of smaller pipe; v_2—velocity in smaller pipe.

Source: Brater, Ernest F, Horace W. King, James E. Lindell, and C. Y. Wei. 1996. *Handbook of Hydraulics,* 7th ed. New York: McGraw-Hill, Table 6–7.

FIGURE 10.10 Gradual contraction.

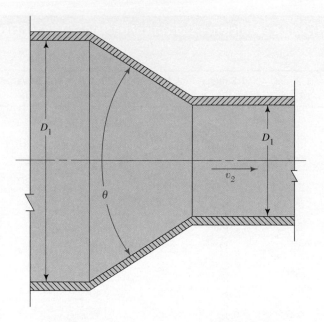

10.7 GRADUAL CONTRACTION

The energy loss in a contraction can be decreased substantially by making the contraction more gradual. Figure 10.10 shows such a gradual contraction, formed by a conical section between the two diameters with sharp breaks at the junctions. The angle θ is called the *cone angle*.

Figure 10.11 shows the data (from Reference 8) for the resistance coefficient versus the diameter ratio for several values of the cone angle. The energy loss is computed from Eq. (10–6), where the resistance coefficient is based on the velocity head in the smaller pipe after the contraction. These data are for Reynolds numbers greater than 1.0×10^5. Note that for angles over the wide range of 15° to 40°, $K = 0.05$ or less, a very low value. For angles as high as 60°, K is less than 0.08.

As the cone angle of the contraction decreases below 15°, the resistance coefficient actually increases, as shown in Fig. 10.12. The reason is that the data include the effects of both the local turbulence caused by flow separation and pipe friction. For the smaller cone angles, the transition between the two diameters is very long, which increases the friction losses.

Rounding the end of the conical transition to blend it with the smaller pipe can decrease the resistance coefficient to below the values shown in Fig. 10.11. For example, in Fig. 10.13, which shows a contraction with a 120° included angle and $D_1/D_2 = 2.0$, the value of K decreases from approximately 0.27 to 0.10 with a radius of only $0.05(D_2)$, where D_2 is the inside diameter of the smaller pipe.

FIGURE 10.11 Resistance coefficient—gradual contraction with $\theta \geq 15°$.

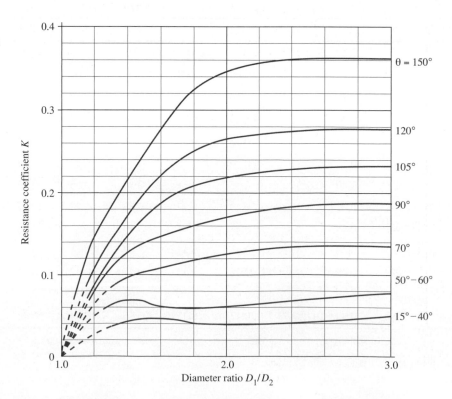

FIGURE 10.12 Resistance coefficient—gradual contraction with $\theta < 15°$.

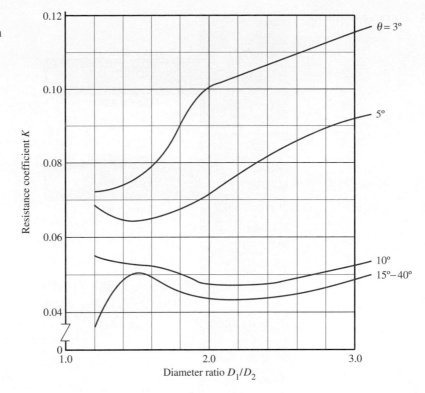

FIGURE 10.13 Gradual contraction with a rounded end at the small diameter.

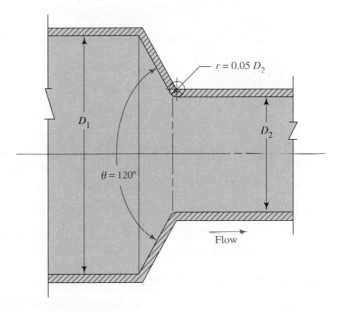

10.8 ENTRANCE LOSS

A special case of a contraction occurs when a fluid flows from a relatively large reservoir or tank into a pipe. The fluid must accelerate from a negligible velocity to the flow velocity in the pipe. The ease with which the acceleration is accomplished determines the amount of energy loss, and, therefore, the value of the entrance resistance coefficient is dependent on the geometry of the entrance.

Figure 10.14 shows four different configurations and the suggested value of K for each. The streamlines illustrate the flow of fluid into the pipe and show that the turbulence associated with the formation of a vena contracta in the tube is a major cause of the energy loss. This condition is most severe for the inward-projecting entrance, for which Reference 2 recommends a value of $K = 0.78$ that will be used for problems in this book. A more precise estimate of the resistance coefficient for an inward-projecting entrance is given in Reference 8. For a *well-rounded entrance with $r/D_2 > 0.15$*, no vena contracta is formed, the energy loss is quite small, and we use $K = 0.04$.

In summary, after selecting a value for the resistance coefficient from Fig. 10.14, we can calculate the energy loss at an entrance from

$$h_L = K(v_2^2/2g) \qquad \textbf{(10–7)}$$

where v_2 is the velocity of flow in the pipe.

FIGURE 10.14 Entrance resistance coefficients.

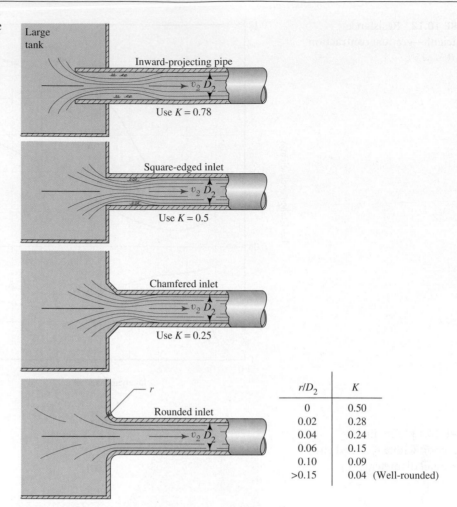

r/D_2	K
0	0.50
0.02	0.28
0.04	0.24
0.06	0.15
0.10	0.09
>0.15	0.04 (Well-rounded)

Example Problem 10.6

Determine the energy loss that will occur as 100 L/min of water flows from a reservoir into a copper hydraulic tube having a 25 mm OD × 1.5 mm wall, (a) through an inward-projecting tube and (b) through a well-rounded inlet.

Solution

Part (a): For the tube, $D_2 = 22.0$ mm and $A_2 = 3.801 \times 10^{-4}$ m^2. Then, we get

$$v_2 = Q/A_2 = 4.385 \text{ m/s} \qquad \text{(from Example Problem 10.1)}$$
$$v_2^2/2g = 0.980 \text{ m}$$

For an inward-projecting entrance, $K = 0.78$. Then, we have

$$h_L = (0.78)(0.980 \text{ m}) = 0.764 \text{ m}$$

Part (b): For a well-rounded inlet, $K = 0.04$. Then, we have

$$h_L = (0.04)(0.980 \text{ m}) = 0.039 \text{ m}$$

10.9 RESISTANCE COEFFICIENTS FOR VALVES AND FITTINGS

Many different kinds of valves and fittings are available from several manufacturers for specification and installation into fluid flow systems. Valves are used to control the amount of flow and may be globe valves, angle valves, gate valves, butterfly valves, any of several types of check valves, and many more. See Figs. 10.15–10.22 for some examples. Fittings direct the path of flow or cause a change in the size of the flow path. Included are elbows of several designs, tees, reducers, nozzles, and orifices. See Figs. 10.23 and 10.24.

FIGURE 10.15 Globe valve. (Reprinted with permission from "Flow of Fluids Through Valves, Fittings and Pipe, Technical Paper 410" 2009. Crane Co. All Rights Reserved)

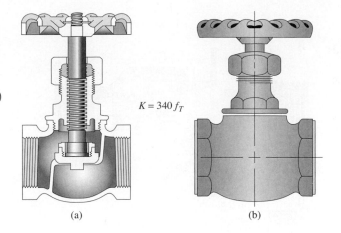

$K = 340 f_T$

(a)   (b)

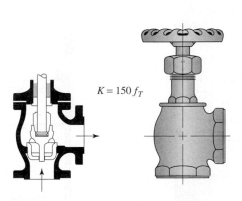

$K = 150 f_T$

FIGURE 10.16 Angle valve. (Reprinted with permission from "Flow of Fluids Through Valves, Fittings and Pipe, Technical Paper 410" 2009. Crane Co. All Rights Reserved)

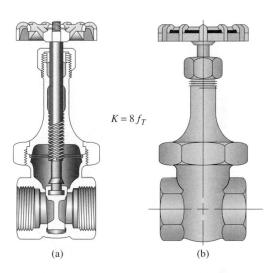

$K = 8 f_T$

(a)   (b)

FIGURE 10.17 Gate valve. (Reprinted with permission from "Flow of Fluids Through Valves, Fittings and Pipe, Technical Paper 410" 2009. Crane Co. All Rights Reserved)

FIGURE 10.18 Check valve—swing type. (Reprinted with permission from "Flow of Fluids Through Valves, Fittings and Pipe, Technical Paper 410" 2009. Crane Co. All Rights Reserved)

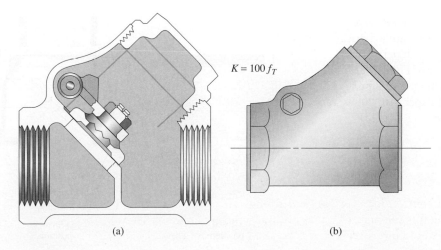

$K = 100 f_T$

(a)   (b)

FIGURE 10.19 Check valve—ball type. (Reprinted with permission from "Flow of Fluids Through Valves, Fittings and Pipe, Technical Paper 410" 2009. Crane Co. All Rights Reserved)

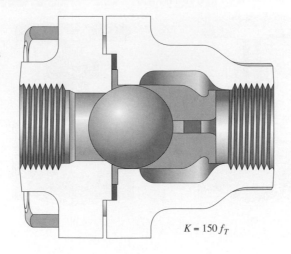

$$K = 150\,f_T$$

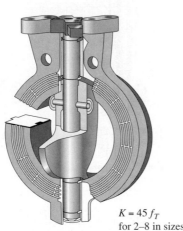

$$K = 45\,f_T$$
for 2–8 in sizes

FIGURE 10.20 Butterfly valve. (Reprinted with permission from "Flow of Fluids Through Valves, Fittings and Pipe, Technical Paper 410" 2009. Crane Co. All Rights Reserved)

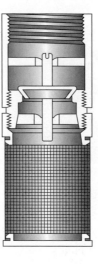

Closed position

$$K = 420\,f_T$$

Open position

FIGURE 10.21 Foot valve with strainer—poppet disc type. (Reprinted with permission from "Flow of Fluids Through Valves, Fittings and Pipe, Technical Paper 410" 2009. Crane Co. All Rights Reserved)

FIGURE 10.22 Foot valve with strainer—hinged disc. (Reprinted with permission from "Flow of Fluids Through Valves, Fittings and Pipe, Technical Paper 410" 2009 Crane Co. All Rights Reserved)

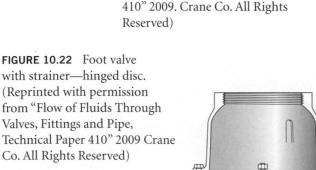

Closed position

Open position

$$K = 75\,f_T$$

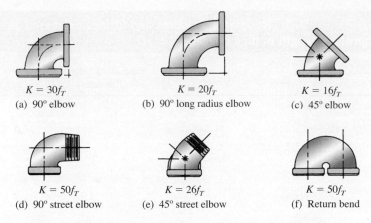

$K = 30f_T$
(a) 90° elbow

$K = 20f_T$
(b) 90° long radius elbow

$K = 16f_T$
(c) 45° elbow

$K = 50f_T$
(d) 90° street elbow

$K = 26f_T$
(e) 45° street elbow

$K = 50f_T$
(f) Return bend

FIGURE 10.23 Pipe elbows. (Reprinted with permission from "Flow of Fluids Through Valves, Fittings and Pipe, Technical Paper 410" 2009 Crane Co. All Rights Reserved)

FIGURE 10.24 Standard tees. (Reprinted with permission from "Flow of Fluids Through Valves, Fittings and Pipe, Technical Paper 410" 2009 Crane Co. All Rights Reserved)

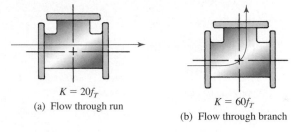

$K = 20f_T$
(a) Flow through run

$K = 60f_T$
(b) Flow through branch

It is important to determine the resistance data for the particular type and size chosen because the resistance is dependent on the geometry of the valve or fitting. Also, different manufacturers may report data in different forms.

Data reported here are taken from Reference 2, which includes a much more extensive list. See also Internet resource 1. References 2, 6, 10, 12, and 13 provide extensive discussion and information about valves.

Energy loss incurred as fluid flows through a valve or fitting is computed from Eq. (10–1) as used for the minor losses already discussed. However, the method of determining the resistance coefficient K is different. The value of K is reported in the form

$$K = (L_e/D)f_T \qquad \textbf{(10–8)}$$

The value of L_e/D, called the equivalent length ratio, is reported in Table 10.4, and it is considered to be constant for a given type of valve or fitting. The value of L_e is called the equivalent length and is the length of a straight pipe of the same nominal diameter as the valve that would have the same resistance as the valve. The term D is the actual inside diameter of the pipe.

The term f_T is the friction factor in the pipe to which the valve or fitting is connected, *taken to be in the zone of*

complete turbulence. Note in Fig. 8.7, the Moody diagram, that the zone of complete turbulence lies in the far right area where the friction factor is independent of Reynolds number. The dashed line running generally diagonally across the diagram divides the zone of complete turbulence from the transition zone to the left.

Values for f_T vary with the size of the pipe and the valve, causing the value of the resistance coefficient K to also vary. Table 10.5 lists the values of f_T for standard sizes of new, clean, commercial steel pipe.

Some system designers prefer to compute the equivalent length of pipe for a valve and combine that value with the actual length of pipe. Equation (10–8) can be solved for L_e:

$$L_e = KD/f_T \qquad \textbf{(10–9)}$$

We can also compute $L_e = (L_e/D)D$. Note, however, that this would be valid only if the flow in the pipe is in the zone of complete turbulence.

If the pipe is anything different from a new, clean, Schedule 40 commercial steel pipe, it is necessary to compute the relative roughness D/ε, and then use the Moody diagram to determine the friction factor in the zone of complete turbulence, f_T.

TABLE 10.4 Resistance in valves and fittings expressed as equivalent length in pipe diameters, L_e/D

Type	Equivalent Length in Pipe Diameters L_e/D
Globe valve—fully open	340
Angle valve—fully open	150
Gate valve—fully open	8
—¾ open	35
—½ open	160
—¼ open	900
Check valve—swing type	100
Check valve—ball type	150
Butterfly valve—fully open, 2–8 in	45
—10–14 in	35
—16–24 in	25
Foot valve—poppet disc type	420
Foot valve—hinged disc type	75
90° standard elbow	30
90° long radius elbow	20
90° street elbow	50
45° standard elbow	16
45° street elbow	26
Close return bend	50
Standard tee—with flow through run	20
—with flow through branch	60

(Reprinted with permission from "Flow of Fluids Through Valves, Fittings and Pipe, Technical Paper 410" 2011. Crane Co. All Rights Reserved.)

TABLE 10.5 Friction factor in zone of complete turbulence for new, clean, commercial Schedule 40 steel pipe

U.S. (in)	Metric (mm)	Friction factor, f_T	U.S. (in)	Metric (mm)	Friction factor, f_T
½	DN 15	0.026	3, 3½	DN 80, DN 90	0.017
¾	DN 20	0.024	4	DN 100	0.016
1	DN 25	0.022	5, 6	DN 125, DN 150	0.015
1¼	DN 32	0.021	8	DN 200	0.014
1½	DN 40	0.020	10–14	DN 250 to DN 350	0.013
2	DN 50	0.019	16–22	DN 400 to DN 550	0.012
2½	DN 65	0.018	24–36	DN 600 to DN 900	0.011

PROCEDURE FOR COMPUTING THE ENERGY LOSS CAUSED BY VALVES AND FITTINGS USING EQ. (10–8)

1. Find L_e/D for the valve or fitting from Table 10.4.
2a. *If the pipe is new clean Schedule 40 steel:*
   Find f_T from Table 10.5.
2b. *For other pipe materials or schedules:*
   Determine the pipe wall roughness ε from Table 8.2.
   Compute D/ε.
   Use the Moody diagram, Fig. 8.7, to determine f_T in the zone of complete turbulence.
3. Compute $K = f_T(L_e/D)$.
4. Compute $h_L = K(v_p^2/2g)$, where v_p is the velocity in the pipe.

Example Problem 10.7

Determine the resistance coefficient K for a fully open globe valve placed in a 6-in Schedule 40 steel pipe.

Solution

From Table 10.4 we find that the equivalent-length ratio L_e/D for a fully open globe valve is 340. From Table 10.5 we find $f_T = 0.015$ for a 6-in pipe. Then,

$$K = (L_e/D)f_T = (340)(0.015) = 5.10$$

Using $D = 0.5054$ ft for the pipe, we find the equivalent length

$$L_e = KD/f_T = (5.10)(0.5054 \text{ ft})/(0.015) = 172 \text{ ft}$$

Or, if the flow is in the zone of complete turbulence,

$$l_c = (L_e/D)D = (340)(0.5054 \text{ ft}) = 172 \text{ ft}$$

Example Problem 10.8

Calculate the pressure drop across a fully open globe valve placed in a 4-in Schedule 40 steel pipe carrying 400 gal/min of oil (sg = 0.87).

Solution

A sketch of the installation is shown in Fig. 10.25. To determine the pressure drop, the energy equation should be written for the flow between points 1 and 2:

$$\frac{p_1}{\gamma} + z_1 + \frac{v_1^2}{2g} - h_L = \frac{p_2}{\gamma} + z_2 + \frac{v_2^2}{2g}$$

The energy loss h_L is the minor loss due to the valve only. The pressure drop is the difference between p_1 and p_2. Solving the energy equation for this difference gives

$$p_1 - p_2 = \gamma\left[(z_2 - z_1) + \frac{v_2^2 - v_1^2}{2g} + h_L\right]$$

But $z_1 = z_2$ and $v_1 = v_2$. Then we have

$$p_1 - p_2 = \gamma h_L$$

FIGURE 10.25 Globe valve for Example Problem 10.8.

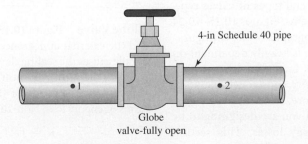

4-in Schedule 40 pipe

Globe valve-fully open

Equation (10–1) is used to determine h_L:

$$h_L = K \times \frac{v^2}{2g} = f_T \times \frac{L_e}{D} \times \frac{v^2}{2g}$$

The velocity v is the average velocity of flow in the 4-in pipe. For the pipe, $D = 0.3355$ ft and $A = 0.0884$ ft^2. Then, we have

$$v = \frac{Q}{A} = \frac{400 \text{ gal/min}}{0.0884 \text{ ft}^2} \times \frac{1 \text{ ft}^3/\text{s}}{449 \text{ gal/min}} = 10.08 \text{ ft/s}$$

From Table 10.5 we find $f_T = 0.016$ for a 4-in pipe. For the globe valve, $L_e/D = 340$. Then,

$$K = f_T \frac{L_e}{D} = (0.016)(340) = 5.44$$

$$h_L = K \times \frac{v^2}{2g} = (5.44)\frac{(10.08)^2}{(2)(32.2)} \text{ ft} = 8.58 \text{ ft}$$

For the oil, $\gamma = (0.870)(62.4 \text{ lb/ft}^3)$. Then, we have

$$p_1 - p_2 = \gamma h_L = \frac{(0.870)(62.4) \text{ lb}}{\text{ft}^3} \times 8.58 \text{ ft} \times \frac{1 \text{ ft}^2}{144 \text{ in}^2}$$

$$p_1 - p_2 = 3.24 \text{ psi}$$

Therefore, the pressure in the oil drops by 3.24 psi as it flows through the valve. Also, an energy loss of 8.58 lb-ft is dissipated as heat from each pound of oil that flows through the valve.

Example Problem 10.9

Calculate the energy loss for the flow of 500 m^3/h of water through a standard tee connected to a 6-in uncoated ductile iron pipe. The flow is through the branch.
Use the *Procedure for Computing the Energy Loss.*

Solution

1. From Table 10.4, $L_e/D = 60$.

2. For the ductile iron pipe, $\varepsilon = 2.4 \times 10^{-4}$ m (Table 8.2) and $D = 0.159$ m (Appendix I). The relative roughness is $D/\varepsilon = (0.159 \text{ m})/(2.4 \times 10^{-4} \text{ m}) = 663$.

From the Moody diagram, $f_T = 0.022$ in the zone of complete turbulence.

3. The resistance coefficient is $K = f_T(L_e/D) = (0.022)(60) = 1.32$.

4. The velocity in the pipe is

$$v_p = \frac{Q}{A} = \frac{500 \text{ m}^3}{\text{h}} \frac{1\text{h}}{3600 \text{ s}} \frac{1}{0.01993 \text{ m}^2} = 6.97 \text{ m/s}$$

Then the energy loss is

$$h_L = K(v_p^2/2g) = (1.32)(6.97 \text{ m/s})^2/[(2)(9.81 \text{ m/s}^2)] = 3.27 \text{ m}$$

10.10 APPLICATION OF STANDARD VALVES

The preceding section showed several types of valves typically used in fluid distribution systems. Figures 10.15–10.24 show drawings and cutaway photographs of the configuration of these valves. The resistance is heavily dependent on the path of the fluid as it travels into, through, and out from the valve. A valve with a more constricted path will cause more energy losses. Therefore, select the valve type with care if you desire the system you are designing to be efficient with relatively low energy losses. This section

describes the general characteristics of the valves shown. You should seek similar data for other types of valves. See Internet resources 1–6.

Globe Valve   Figure 10.15 shows the internal construction and the external appearance of the globe valve. Turning the handle causes the sealing device to lift vertically off the seat. It is one of the most common valves and is relatively inexpensive. However, it is one of the poorest performing valves in terms of energy loss. Note that the resistance factor K is

$$K = f_T(L_e/D) = 340f_T$$

This is among the highest of those listed in Table 10.4. It would be used where there is no real problem created by the energy loss. The energy loss occurs because the fluid must travel a complex path from input to output, first traveling upward, then down around the valve seat, then turning again to move to the outlet. Much turbulence is created.

Another use for the globe valve is to *throttle the flow* in a system. The term *throttle* refers to purposely adding resistance to the flow to control the amount of fluid delivered. An example is the simple faucet for a garden hose. You may choose to open the valve completely to get the maximum flow of water to your garden or lawn. By partially closing the valve, however, you could get a lesser flow rate for a more gentle spray or for washing the dog. Partially closing the valve provides more restriction, and the pressure drop from the inlet to the outlet increases. The result is less flow. Refer back to Figs. 9.11 and 9.12 for CFD illustrations of the flow through a globe valve. You can see the complex path that the fluid takes, causing large energy losses.

If the globe valve were used in a commercial pipeline system where throttling is not needed, it would be very wasteful of energy. More-efficient valves with lower L_e/D values should be considered.

Angle Valves   Figure 10.16 shows the external appearance of the angle valve and a sketch of its internal passages. The construction is very similar to that of the globe valve. However, the path is somewhat simpler because the fluid comes in through the lower port, moves around the valve seat, and turns to exit to the right. The resistance factor K is

$$K = f_T(L_e/D) = 150f_T$$

Gate Valves   The gate valve in Fig. 10.17 is shown in the closed position. Turning the handle lifts the gate vertically out of the flow path. When fully open, there is very little obstruction in the flow path to cause turbulence in the fluid flow stream. Therefore, this is one of the best types of valve for limiting the energy loss. The resistance factor K is

$$K = f_T(L_e/D) = 8f_T$$

In a given installation, the fully open gate valve would have only 2.4% ($8/340 \times 100\%$) of the amount of energy loss as a globe valve. The higher cost of the valve is usually justified by the saving of energy during the lifetime of the system.

The gate valve could be used for throttling by partially closing the valve, bringing the gate back into the flow stream to some degree. Sample data are given in Table 10.4 for the partially closed positions. Note that it is highly nonlinear and care must be taken to obtain the desired flow rate by throttling. Wear on guides and sealing surfaces must also be considered.

A modified version of a gate valve, called a *knife gate valve*, is a standard element that can be obtained from selected vendors. The design of this type of valve is similar to the gate valve shown in Fig. 10.17 except that the gate is a thin sheet instead of the thicker style shown. The operational characteristics and the K-factors of these two designs are similar. Some users prefer the knife gate valve, particularly

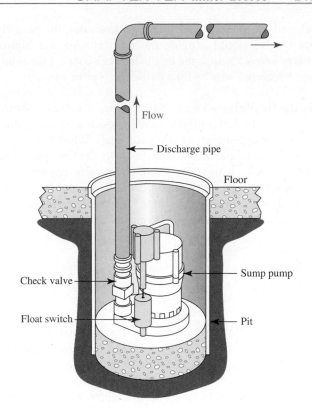

FIGURE 10.26   Sump pump system with check valve.

when handling heavier fluids or slurries that may contain significant quantities of solids.

Check Valves   The function of a check valve is to allow flow in one direction while stopping flow in the opposite direction. A typical use is shown in Fig. 10.26, in which a sump pump is moving fluid from a sump below grade to the outside of a home or commercial building to maintain a dry basement area. The pump draws water from the sump and forces it up through the discharge pipe. When the water level in the sump drops to an acceptable level, the pump shuts off. At that time, you would not want the water that was in the pipe to flow back down through the pump and partially refill the sump. The use of a check valve just outside the discharge port of the pump precludes this from happening. The check valve closes immediately when the pressure on the output side exceeds that on the input side.

Two kinds of check valve are shown in Figs. 10.18 and 10.19, the ball type and the swing type. There are several other designs available. When open, the swing check provides a modest restriction to the flow of fluid, resulting in the resistance factor of

$$K = f_T(L_e/D) = 100f_T$$

The ball check causes more restriction because the fluid must flow completely around the ball. However, the ball check is typically smaller and simpler than the swing check. Its resistance is

$$K = f_T(L_e/D) = 150f_T$$

An important application factor for check valves is that a certain minimum flow velocity is required to cause the valve to completely open. At lower flow rates, the partially open valve would provide more restriction and higher energy losses. Consult the manufacturer's data for the minimum required velocity for a particular type of valve.

Butterfly Valve Figure 10.20 shows a cutaway photograph of a typical butterfly valve in which a relatively thin, smooth disc pivots about a vertical shaft. When fully open, only the thin dimension of the disc faces the flow, providing only a small obstruction. Closing the valve requires only one-quarter turn of the handle, and this is often accomplished by a motorized operator for remote operation. The fully open butterfly valve has a resistance of

$$K = f_T(L_e/D) = 45f_T$$

This value is for the smaller valves from 2 in to 8 in. From 10 in to 14 in, the factor is $35f_T$. Larger valves from 16 in to 24 in have a resistance factor of $25f_T$.

Foot Valves with Strainers Foot valves perform a similar function to that of check valves. They are used at the inlet of suction pipes that deliver fluid from a source tank or reservoir to a pump as illustrated in Fig. 10.27. They are typically equipped with an integral strainer to keep foreign objects out of the piping system. This is especially necessary when drawing water from an open pit or a natural lake or stream. There may be fish in the lake!

The resistances for the two kinds of foot valves shown in Figs. 10.21 and 10.22 are

$$K = f_T(L_e/D) = 420f_T \qquad \text{Poppet disc type}$$
$$K = f_T(L_e/D) = 75f_T \qquad \text{Hinged disc type}$$

The poppet disc type is similar to the globe valve in internal construction, but it is even more constricted. The hinge type is similar to the swing-type check valve. Some extra resistance should be planned for if the strainer could become clogged during service. See Internet resources 1–6 for descriptions of more valves and fittings. See Reference 2 for much more information about resistance of valves and fittings.

10.11 PIPE BENDS

It is frequently more convenient to bend a pipe or tube than to install a commercially made elbow. The resistance to flow of a bend is dependent on the ratio of the bend radius r to the pipe inside diameter D. Figure 10.28 shows that the minimum resistance for a 90° bend occurs when the ratio r/D is approximately three. The resistance is given in terms of the equivalent length ratio L_e/D, and therefore Eq. (10–8) must be used to calculate the resistance coefficient. The resistance shown in Fig. 10.28 includes both the bend resistance and the resistance due to the length of the pipe in the bend. See Reference 1.

When we compute the r/D ratio, r is defined as the radius to the *centerline* of the pipe or tube, called the *mean radius* (see Fig. 10.29). That is, if R_o is the radius to the outside of the bend, R_i is the radius to the inside of the bend, and D_o is the *outside diameter* of the pipe or tube:

$$r = R_i + D_o/2$$
$$r = R_o - D_o/2$$
$$r = (R_o + R_i)/2$$

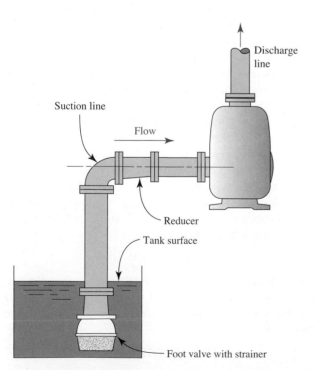

FIGURE 10.27 Pumping system with a foot valve in the suction line.

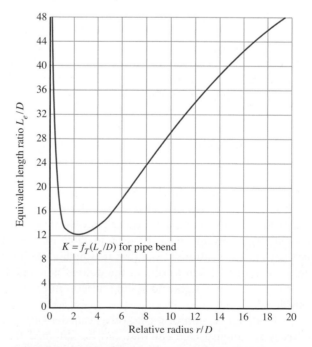

FIGURE 10.28 Resistance due to 90° pipe bends. (*Source:* Beij, K. H., Pressure Losses for Fluid Flow in 90 Degree Pipe Bends. 1938. *Journal of Research of the National Bureau of Standards* 21: 1–18) See Reference 1.

FIGURE 10.29 90° pipe bend.

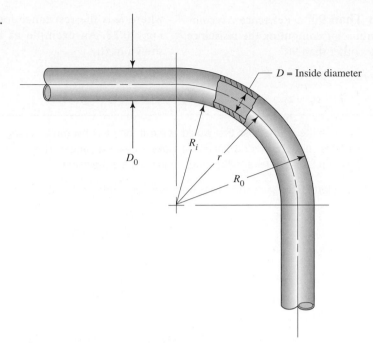

D = Inside diameter

R_i

r

D_0

R_0

Example Problem
10.10

A distribution system for liquid propane is made from steel hydraulic tube, 32 mm OD × 2.0 mm wall. Several 90° bends are required to fit the tubes to the other equipment in the system. The specifications call for the radius to the inside of each bend to be 200 mm. When the system carries 160 L/min of propane at 25°C, compute the energy loss due to each bend.

Solution

Darcy's equation should be used to compute the energy loss with the L_e/D ratio for the bends found from Fig. 10.28. First, let's determine r/D, recalling that D is the inside diameter of the tube and r is the radius to the centerline of the tube. From Appendix G.2 we find $D = 28.0\,\text{mm} = 0.028\,\text{m}$. The radius r must be computed from

$$r = R_i + D_o/2$$

where $D_o = 32.0\,\text{mm}$, the outside diameter of the tube, as found from Appendix G.2. Completion of the calculation gives

$$r = 200\,\text{mm} + (32.0\,\text{mm})/2 = 216\,\text{mm}$$

and

$$r/D = 216\,\text{mm}/28.0\,\text{mm} = 7.71$$

From Fig. 10.28 we find the equivalent-length ratio to be 23.

We must now compute the velocity to complete the evaluation of the energy loss from Darcy's equation:

$$v = \frac{Q}{A} = \frac{160\,\text{L/min}}{6.158 \times 10^{-4}\,\text{m}^2} \frac{1.0\,\text{m}^3/\text{s}}{60\,000\,\text{L/min}} = 4.33\,\text{m/s}$$

The relative roughness is

$$D/\varepsilon = (0.028\,\text{m})(1.5 \times 10^{-6}\,\text{m}) = 18\,667$$

Then, we can find $f_T = 0.0108$ from the Moody diagram (Fig. 8.7) in the zone of complete turbulence. Then,

$$K = f_T\left(\frac{L_e}{D}\right) = 0.0108(23) = 0.248$$

Now the energy loss can be computed:

$$h_L = K\frac{v^2}{2g} = 0.248\frac{(4.33)^2}{(2)(9.81)} = 0.237\,\text{m} = 0.237\,\text{N·m/N}$$

Bends at Angles Other Than 90° Reference 2 recommends the following formula for computing the resistance factor K for bends at angles other than 90°:

$$K_B = (n - 1)[0.25\pi f_T(r/D) + 0.5\,K] + K \qquad (10\text{--}10)$$

where K is the resistance for one 90° bend found from Fig. 10.28. An example of the use of this equation is shown next.

Example Problem 10.11

Evaluate the energy loss that would occur if the steel hydraulic tubing described in Example Problem 10.10 is coiled for 4½ revolutions to make a heat exchanger. The inside radius of the bend is the same 200 mm used earlier and the other conditions are the same.

Solution

Let's start by bringing some data from Example Problem 10.10.

$$r/D = 7.71$$
$$f_T = 0.0108$$
$$K = 0.248$$
$$v = 4.33\,\text{m/s}$$

Now we can compute the value of K_B for the complete coil using Eq. (10–10). Note that each revolution in the coil contains four 90° bends. Then,

$$n = 4.5\,\text{revolutions (4.0 90° bends/rev)} = 18$$

The total bend resistance K_B is

$$K_B = (n - 1)[0.25\pi f_T(r/D) + 0.5\,K] + K$$
$$K_B = (18 - 1)[0.25\pi(0.0108)(7.71) + 0.5(0.248)] + 0.248$$
$$K_B = 3.47$$

Then the energy loss is found from

$$h_L = K_B(v^2/2g) = 3.47(4.33)^2/[2(9.81)] = 3.32\,\text{N·m/N}$$

10.12 PRESSURE DROP IN FLUID POWER VALVES

The field of fluid power encompasses both the flow of liquid hydraulic fluids and air flow systems called *pneumatic systems*. Liquid hydraulic fluids are generally some form of petroleum oil, although many types of blended and synthetic materials can be used. We will refer to the liquid hydraulic fluids simply as oil.

You may be familiar with fluid power systems that operate automation equipment in a production system. They move products through an assembly and packaging system. They actuate forming presses that can exert huge forces. They raise components or products to different elevations, similar to an elevator. They actuate processes to perform a variety of functions such as cutting metal, clamping, slitting, compressing bulk materials, and driving fasteners such as screws, bolts, nuts, nails, and staples.

Another large use is for agricultural and construction equipment. Consider the classic bulldozer that shapes the land for a construction project. The level of the bulldozer's blade is adjusted by the operator using fluid power controls to ensure that the grade of the land meets the design goals. When excess dirt must be removed, a front-end loader is often used to pick it up and dump it into a truck. Numerous hydraulic actuators drive the interesting linkage system that allow the bucket to pick up the dirt and maintain it in a safe position throughout the motion to the truck and then to dump it. The truck is then emptied at another site by actuating hydraulic cylinders to raise the truck bed. In farm work, most modern tractors and harvesting equipment employ hydraulic systems to raise and lower components, to drive rotary motors, and sometimes to even drive the units themselves.

Common elements for a liquid hydraulic system include:

- A pump to provide fluid to a system at an adequate pressure and at the appropriate volume flow rate to accomplish the desired task.

- A tank or reservoir of hydraulic fluid from which the pump draws fluid and to which the fluid is returned after accomplishing the task. Most fluid power systems are closed circuits in which the fluid is continuously circulated.

- One or more directional control valves to manage the flow as it moves through the system.

- Linear actuators, often called hydraulic cylinders, that provide the forces and motion needed to perform the actuation tasks.

- Rotary actuators, called fluid motors, to operate rotating cutting tools, agitators, wheels, linkages, and other devices needing rotary motion.

- Pressure control valves to ensure that an adequate and safe level of pressure exists at all parts of the system.
- Flow control devices that ensure that the correct volume flow rate is delivered to the actuators to provide the proper linear velocity or rotational angular velocity.

See Internet resource 5 for manufacturers' data for fluid power devices. Reference 14 provides a comprehensive coverage of fluid power systems.

Fluid power systems consist of a very wide variety of components arranged in numerous ways to accomplish specific tasks. Also, the systems are inherently *not operating in steady flow* as was assumed in the examples in most of this book. Therefore, different methods of analysis are typically used for fluid power components than for the general-purpose fluid-handling devices discussed earlier in the chapter.

However, the principles of energy loss that we discussed still apply. You should be concerned with the energy loss due to any change in direction, change in the size of the flow path, restrictions such as within the valves, and friction as the fluids flow through pipes and tubing.

Example Fluid Power System Consider the fluid power system shown in Fig. 10.30. The basic purpose and operation of the system are described here.

Forward Actuation of the Load to the Right: Figure 10.30(a)

- The function of the system is to exert a force of 20 000 lb on a load while providing linear actuation motion to the load. A large part of the force is required to accomplish a forming operation near the end of the stroke.
- An oil hydraulic linear actuator provides the force.
- Fluid is delivered to the actuator by a positive-displacement pump, which draws the fluid from a tank.
- The fluid leaves the pump and flows to the directional control valve. When it is desired to actuate the load, the flow passes through the valve from the P port to the A delivery port (P — A).
- The flow control valve is placed between the directional control valve and the actuator to permit the system to be adjusted for optimum performance under load.
- The fluid flows into the piston end of the actuator.
- The fluid pressure acts on the face area of the piston exerting the force required to move the load and accomplish the forming operation.
- Simultaneously, the fluid in the rod end of the actuator flows out of the cylinder and proceeds to and through the directional control valve and back to the tank.
- A protection device called a pressure relief valve is placed in the line between the pump and the directional control valve to ensure that the pressure in the system never exceeds the level set by the relief valve. When the pressure rises above the set point, the valve opens and delivers part of the flow back to the tank. Flow can continue through the directional control valve, but its pressure will be less

than it would have been without the pressure relief valve in the system.

Return Actuation of the Piston Rod to the Left: Figure 10.30(b) For the return action much less force is required because the load is relatively light and no forming action takes place. The sequence proceeds like this:

- The directional control valve is shifted to the right, changing the direction of the flow. The fluid that comes from the pump to the P port is directed to the B port and thus to the rod end of the actuator.
- As the fluid flows into the cylinder, the piston is forced to the left toward its home position.
- Simultaneously, the fluid in the piston end is forced out from port A on the cylinder, passes to the A port of the valve, and is directed back to the tank.
- Because less pressure is required to accomplish this task, the pressure relief valve does not open.

Idle Position of the System: Figure 10.30(c)

- When the load is returned to its home position, it may be required to idle at that position until some other action has been completed and a signal is received to start a new cycle. To accomplish that, the valve is placed in its center position.
- The flow from the pump is directed immediately to the tank.
- The A port and the B port are blocked in the valve and thus no flow can come back from the actuator. This holds the actuator in position.
- When the conditions are right for another forming stroke, the directional control valve is switched back to the left and the cycle begins again.

Pressure Levels and Energy Losses and Additions in this Fluid Power System Let's now identify where energy additions and losses occur in this system and how the pressure levels will vary at critical points.

1. Let's start with the fluid in the tank. Assume that it is at rest and that the tank is vented with atmospheric pressure above the surface of the fluid.

2. As the pump draws fluid, we see that a suction line must accelerate the fluid from its resting condition in the tank to the velocity of flow in the suction line. Thus, there will be *an entrance loss* that depends on the configuration of the inlet. The pipe may simply be submerged in the hydraulic fluid or it may have a strainer at the inlet to keep foreign particles out of the pump and the valves.

3. There will be *friction losses in the pipe* as the fluid flows to the suction port of the pump.

4. Along the way, there may be *energy losses in any elbows or bends* in the pipe.

5. We must be concerned with the pressure at the inlet to the pump to ensure that cavitation does not happen and that there is an adequate supply of fluid.

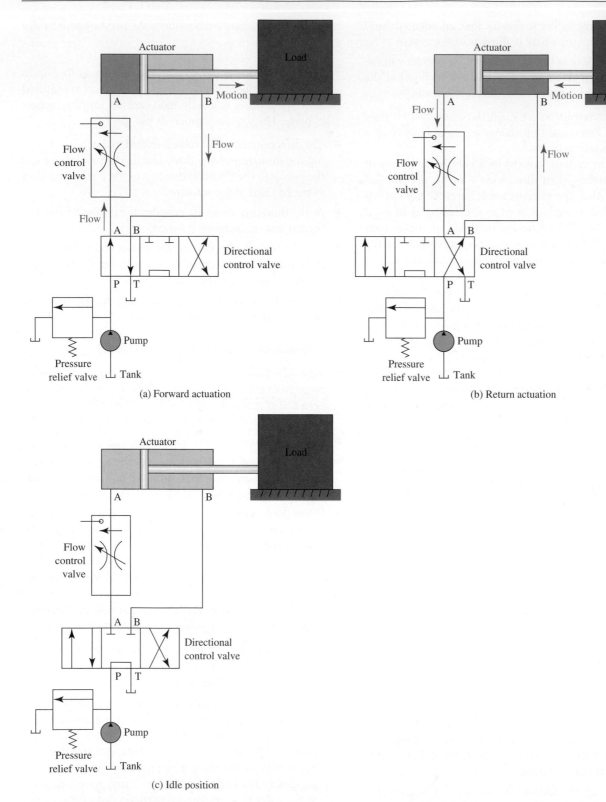

(a) Forward actuation

(b) Return actuation

(c) Idle position

FIGURE 10.30 Fluid power system.

6. The *pump adds energy to the fluid* to cause the flow and to increase the fluid pressure to that required to operate the system. The energy comes from the prime mover, typically an electric motor or an engine. Some of the *input energy is lost because of the volumetric efficiency and mechanical efficiency* of the pump. (See Chapters 7

and 13.) Together these combine to produce the overall efficiency defined as follows:

Overall efficiency e_o = (Volumetric efficiency e_v)
(Mechanical efficiency e_M)

Input power P_I = (Power delivered to the fluid)/e_o

7. As the fluid leaves the pump and travels to the directional control valve, *friction losses occur in the piping system, including any elbows, tees, or bends in the pipe.* These losses will cause the pressure that appears at the P port of the valve to be less than that at the outlet of the pump.

8. If the pressure relief valve has actuated because the pump discharge pressure exceeds the set point of the valve, there will be *pressure drop across this valve.* The pressure actually reduces from the discharge line pressure p_d to the atmospheric pressure in the tank p_T. Much energy is lost during this process. If we apply the energy equation to the inlet and outlet of the pressure relief valve, we can show that

$$h_L = (p_d - p_T)/\gamma$$

9. Back at the directional control valve, the fluid passes through the valve from the P port to the A port. *Energy losses occur in the valve* because the fluid must flow through several restrictions and changes in direction in the ports and around the movable spool in the valve that directs the fluid to the proper outlet port. These energy losses cause a pressure drop across the valve. The amount of the pressure drop is dependent on the design of the valve. Manufacturer's literature will typically include data from which you can estimate the magnitude of the pressure drop. Figure 10.31 shows a typical graph of pressure drop across the valve versus flow rate. In the fluid power industry, these graphs are used rather than reporting resistance factors as is done for the standard fluid distribution valves discussed earlier in this chapter.

10. As the fluid flows from the A port to the flow control valve, *energy losses occur in the piping* as before.

11. The flow control valve ensures that the flow of fluid into the cylinder at the left end of the actuator is proper to cause the load to be moved at the desired speed. Control is affected by adjustable internal restrictions that can be set during system operation. *The restrictions cause energy losses* and, therefore, there is a pressure drop across the valve.

12. *Energy is lost at the actuator* as the fluid flows into the left end of the cylinder at A and out from the right end at B.

13. On the return path, *energy losses occur in the piping system.*

14. More *energy losses occur in the directional control valve* as the fluid passes back through the B port and on to the tank. The reasons for these losses are similar to those described in item 9.

This summary identifies 14 ways in which energy is either added to or lost from the hydraulic fluid in this relatively simple fluid power system. Each energy loss results in a pressure drop that could affect the performance of the system.

However, designers of fluid power systems do not always analyze each pressure drop. The transient nature of the operation makes it critical that there is sufficient pressure and flow at the actuator under all reasonable conditions. It is not uncommon for designers to provide extra capacity in the basic system design to overcome unforeseen circumstances. In the circuit just described, the critical pressure drops occur at the pressure relief valve, through the directional control valve, and through the flow control valve. These elements will be analyzed carefully. Other losses will often be only estimated in the initial design. In many cases, the actual configuration of the piping system is not defined during the design process, leaving it to skilled technicians to properly fit the components to the machine. Then, when the system is in operation, some fine tuning will be done to ensure proper operation.

This scenario applies most to systems designed for a special purpose when one or only a few systems will be built. When a system is designed for a production application or for a very critical application, more time spent on analysis and optimization of system performance is justified. Examples are aircraft control systems and actuators for construction and agricultural equipment that are made in quantity.

10.13 FLOW COEFFICIENTS FOR VALVES USING C_V

A large number of manufacturers of valves used for controlling liquids, air, and other gases prefer to rate the performance of the valves using the flow coefficient C_V. One basis for this flow coefficient is that a valve having a flow coefficient of 1.0 will pass 1.0 gal/min of water at 1.0-psi pressure drop across the valve. The test is convenient to run and gives a reliable means of comparing the overall performance characteristics of different valves.

The basic liquid flow equation is

$$Q = \text{Flow in gal/min} = C_V\sqrt{\Delta p/sg} \quad \textbf{(10–10)}$$

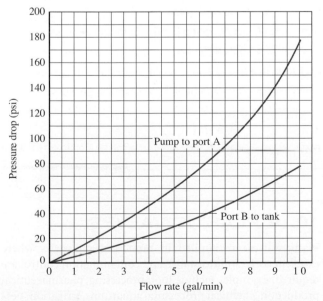

FIGURE 10.31 Pressure drop in a directional control valve.

where Δp is in lb/in². Δp is called the pressure drop, calculated from $p_U - p_D$, the difference in pressure between points upstream and downstream from the valve. The term sg is the unitless specific gravity of the fluid. *Be careful to note that C_V is not a unitless factor.* Let's examine the units by solving Eq. 10–10 for C_V.

$$C_V = \frac{Q}{\sqrt{\Delta p/sg}} \frac{\text{gal/min}}{(\text{psi})^{1/2}}$$

Note that the value for C_V is often reported without units, but you must understand that its definition in Eq. 10–10 is a unit-specific equation.

Equation 10–10 can be solved for the pressure drop, Δp:

$$\Delta p = sg(Q/C_V)^2$$

Data reported in a manufacturer's catalog typically list the value of C_V for the valve in the fully open condition. But the valve is often used to control flow rate by partially closing the valve manually or automatically. Therefore, many manufacturers will report the effective C_V as a function of the number of turns of the valve stem from full closed to full open. Alternatively, C_V may be reported as a percentage (e.g., 50%

open, 60% open, and so forth). Such data are highly dependent on the construction of the internal parts of the valve, particularly the closure device, sometimes called the *plug*. More is said about control valves in Chapters 11, 12, and 13.

When using a valve as a control valve, it is often selected to operate at the mid-point of its range, permitting control of flow rate upward and downward from this setting.

Some valves employ a pointed stem that is moved away from a seat as the valve is opened, progressively expanding the flow area around the stem. This type of valve is called *a needle valve* and it is often used in high-precision instruments, pharmaceutical production, or other critical applications.

Users of such valves, for controlling the flow of air or other gases, must account for the compressibility of the gas and the effect of the overall pressure difference across the valve. As discussed in a later chapter on the flow of gases, when the ratio of the upstream to the downstream pressure in a gas reaches the *critical pressure ratio*, no further increase of flow occurs as the downstream pressure is lowered. At the critical pressure ratio, the velocity of flow through the nozzle or valve is equal to the speed of sound in the gas at the local conditions.

| **Example Problem** | A particular design for a ½-in needle valve has a C_V rating of 1.5. Compute the pressure drop when |
| **10.12** | 5.0 gal/min of water at 60°F flows through the valve. |

Solution   We can use the form of Eq. (10–10) giving Δp. Note that this is a unit-specific equation with Δp in psi, Q in gal/min, and C_V has the unit described above. The specific gravity, sg, is unitless and the water has a specific gravity sg = 1.0. Then,

$$\Delta p = sg\left(\frac{Q}{C_V}\right)^2 = 1.0\left(\frac{5.0}{1.5}\right)^2 = 11.1 \text{ psi}$$

| **Example Problem** | A particular design for a 4-in plastic butterfly valve has a C_V rating of 550. Compute the pressure drop |
| **10.13** | when 875 gal/min of turpentine at 77°F flows through the valve. |

Solution   We can use Eq. (10–10) after solving for Δp. Note that this is a unit-specific equation with Δp in psi and Q in gal/min, and C_V has the unit described above. The turpentine has a specific gravity sg = 0.87 (Appendix B). Then,

$$\Delta p = sg\left(\frac{Q}{C_V}\right)^2 = (0.87)\left(\frac{875}{550}\right)^2 = 2.20 \text{ psi}$$

Metric Flow Coefficient, K_V   When working in metric units, an alternate form of the flow coefficient is used and it is called K_V instead of C_V. It is defined as the amount of water in m³/h at a pressure drop of one bar across the valve. Use the following equation for conversion between C_V and K_V:

$$C_V = 1.156 K_V$$

10.14 PLASTIC VALVES

Plastic valves are applied in numerous industries where excellent corrosion resistance and contamination control are required. Examples include food processing, pharmaceutical production, chemical processing, aquariums, irrigation, pesticide delivery, and water purification. Materials used are similar to those discussed in Section 6.3.5 in Chapter 6 for plastic pipe and tubing; polyvinyl chloride (PVC), chlorinated polyvinyl chloride (CPVC), polyvinylidene fluoride (PVDF), polyethylene (PE), and polypropylene (PP or PPL). Seats and seals are typically made from polytetrafluoroethylene (PTFE), ethylene propylene diene monomer (EPDM), Buna-N (NBR or Nitrile), or fluorocarbon elastomers (FKM) such as Viton (a registered trademark of DuPont Dow Elastomers) and Fluorel (a registered trademark of 3M Corporation).

Temperature and pressure limits are typically lower for plastic valves than for metal valves. For example, PVC is limited to approximately 140°F (60°C); CPVC to 190°F (88°C); PP to 250°F (121°C); EPDM to 300°F (149°C); FKM to 400°F (204°C); and PTFE to 500°F (260°C). Low-temperature limits range from approximately −20°F (−29°C) for most sealing materials to −80°F (−62°C) for PVDF. Pressure ratings of plastic valves range from 100 psi to 225 psi (690 kPa to 1550 kPa) at moderate temperatures, depending on design and size.

The following discussion highlights a sampling of the types of plastic valves available. In most cases the general design is similar to the metal types discussed previously in this chapter and shown in Figs. 10.15–10.20. More complete descriptions and performance data are available in Internet resources 6–8.

Ball Valves Used most often for on/off operation, ball valves require only one-quarter turn to actuate them from full closed to full open. The rotating spherical ball is typically bored with a hole of the same diameter as the pipe or tube to which it is connected to provide low energy loss and pressure drop. They can be directly connected to the pipe or tube with adhesive or connected by flanges, unions, or screwed ends. Some ball valves are designed especially for the proportional control of flow by tailoring the shape of the hole. See Internet resource 6.

Butterfly Valves Similar to the metal valve shown in Fig. 10.20, the plastic butterfly disc provides simple opening and closure with one-quarter turn of the handle. Actuation can be manual, electric, or pneumatic. The torque required to turn the valve varies dramatically from closed to open and large valves may require powered actuators or a gearing mechanism. All parts that contact the flowing fluid are made from noncorroding materials. The shaft for the disc is typically made from stainless steel and is isolated from fluid contact. Most valves are very thin and can be mounted between standard pipe flanges for easy installation and removal. Some designs can replace existing metal valves where appropriate.

Diaphragm Valves The diaphragm, typically made from EPDM, PTFE, or FKM fluorocarbon elastomers, is designed to rise from a seat when the hand wheel is turned. Opposite rotation then reseals the valve. The valve is suitable for both on/off and modulated flow operation. The diaphragm isolates the brass hand-wheel shaft and other parts from the flowing fluid. Materials for wetted parts are selected

for corrosion resistance to the particular fluid and temperatures to be encountered. The ends may be directly connected to the pipe or tube with adhesive or connected by flanges, unions, or screwed ends.

Swing Check Valves Designed similarly to the drawing in Fig. 10.18, this valve opens easily in the proper direction of flow, but closes rapidly to prevent backflow. All wetted parts are made from corrosion-resistant plastic, including the pin on which the disc pivots. External fasteners are typically made from stainless steel. The bonnet can be easily removed to clean the valve or to replace seals.

Sediment Strainers Strainers remove particulate impurities from the fluid stream to protect product quality or sensitive equipment. All fluid is directed to flow through perforated or screen-style filters as it passes through the body of the strainer. Plastic screens are made with perforations from $\frac{1}{32}$ in to $\frac{3}{16}$ in (0.8 mm to 4.8 mm) to remove coarse dirt and debris. Stainless steel screens can be made with large perforations or from fine mesh screening down to 325 mesh with openings only a few thousandths of an inch (approximately 0.05 mm or 50 μm). Screens must be removed periodically for cleaning. See Internet resource 6 for more detail and for C_V ratings.

Sample Data for C_V for Plastic Valves Table 10.6 gives representative sample data for plastic valves that can be used for problems in this book. Metric sizes are not equivalent to U.S. sizes. Final designs must be based on manufacturers' data for the specified valve. See Internet resources 6–8.

10.15 USING K-FACTORS IN PIPE-FLO® SOFTWARE

There are very few actual fluid flow systems that don't have elbows, valves, or other devices that change the rate or direction of flow in the system. PIPE-FLO® can handle not just straight pipe as shown in Chapter 8, but it can also incorporate various types of valves, fittings, and other commercially available components. In Example Problem 10.14 that follows, we show the general method for including such minor losses in a PIPE FLO® analysis problem when all elements are part of the program's library. Example Problem 10.15 then follows in which we demonstrate how to include devices that have custom K-values (e.g., heat exchangers, filters, and special processing equipment).

In these problems, the requested calculation is the pressure drop in a pipeline that contains fittings in addition to pipe friction losses. The method is similar to that used in Chapter 8 for which a pressurized tank causes the flow. Arbitrary values for the tank pressure and the depth of the fluid in the tank are selected. Note that the pressure at the pipe entrance is larger than the tank pressure because of the hydrostatic head created by the depth of fluid. After calculating the performance of the system, just the pressure drop in the pipeline can be selected for callout on the screen.

TABLE 10.6   Sample data for C_v for a variety of types and sizes of plastic valves

Ball valve				Butterfly valve			
Size (in)	C_V	Size (mm)	K_V	Size (in)	C_V	Size (mm)	K_V
1/2	12	20	21	1½	90	50	100
3/4	25	25	32	2	115	63	150
1	37	32	70	3	330	90	390
1½	120	50	150	4	550	110	540
2	170	63	390	6	1150	160	1120
3	450	90	510	8	2280	225	2840
4	640	110	590	10	4230	280	4350
6	1400	160	1510	12	5600	315	5190

Diaphragm valve				Swing check valve			
Size (in)	C_V	Size (mm)	K_V	Size (in)	C_V	Size (mm)	K_V
1/2	5	20	8	—			
3/4	9	25	13	¾	25	25	35
1	15	32	30	1	40	32	52
1½	34	50	56	1½	80	50	100
2	65	63	95	2	115	63	150
3	160	90	215	3	330	90	360
4	275	110	280	4	500	110	610
6	700	160	650	6	1240	160	1420
				8	2300	225	1560

Example Problem 10.14

Use PIPE-FLO® software to determine the pressure drop across 50 ft of 4-in Schedule 40 steel pipe, carrying 77°F kerosene at a velocity of 6 ft/s if along the pipe there are also (4) standard 90° elbows, (1) knife gate valve, and (1) sharp-edged pipe entrance at the tank. The bends where the elbows are installed are all in the same horizontal plane at the same elevation. Report all applicable values related to the solution such as Reynolds number and friction factor. Flow is caused by a pressurized tank with 50.0 psig above the kerosene and the depth of the tank is 7.0 ft.

Solution

1. Open a new project in PIPE-FLO® and select the "SYSTEM" menu on the toolbar to initialize all key data such as units, and pipe specifications the same way as the Example Problem in Chapter 8. For the "fluid zone" select "NEW" and follow the instructions to download the property data for kerosene and many other fluids from the ESI website. After downloading, initialize a fluid zone for kerosene at 77°F.

2. Place a closed pressurized tank on the FLO-Sheet® and draw in the pipe. Remember that we define the length of pipe in the property grid. The actual drawing of the pipe is not important, but the sketch below shows four turns to represent the elbows. Achieve this graphic by simply clicking the mouse in the desired corners and continuing with the pipe. The elevation of the pipe, since the entire problem is horizontal, is entered as 0 ft. Enter this value in the property grid after clicking on the pipe. Add a flow demand at the exit of the pipe to define velocity in the pipe. The flow rate, as calculated by using 6 ft/s velocity and 0.3355 ft² flow area, results in 238 gal/min. Use this value in the "Set Flow" category in the property grid for the flow demand.

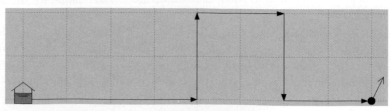

3. With the basic system in place, it is time to add components that will be installed along the pipe. Components such as these are simply considered a part of the pipe in PIPE-FLO® so they are not drawn separately on the FLO-Sheet.® Instead, they are entered as values in the properties grid for the associated pipe. Start by clicking on the pipe. Under the "K (Valves & Fittings)" box in the property grid, click the ". . ." button and the table below appears on the screen.

4. To choose a particular fitting or valve, drop down the category that would contain the desired component and select it. Also, select the quantity of that component, and then select "ADD" to place it in the system. Multiple fittings and valves can be entered at the same time in the same way for convenience. The 90° standard elbows are located under the "Fitting" heading. Select a quantity of (4) and then "Add". Similarly, choose a knife gate valve and its quantity, then "Add" to place it in the system. If you aren't sure whether the component is the one you need, verify that the "K" value specified in PIPE-FLO® matches the one you would use if you were doing manual calculations. The sharp-edged pipe entrance can be found under the "Fitting" heading; add it to the system the same way as the other components. When finished, the "Valves and Fittings" box should look like the box shown below.

5. After returning to the FLO-Sheet®, the symbols for the added components are visible. These symbols are both moveable and changeable so they can be placed in their "proper" positions and modified to the users liking. The pipe entrance symbol isn't shown, so be careful to make sure it is still included in the system. Also, only one symbol is shown for the elbows regardless of how many are placed in the system. In other words, the elbow icon indicates the presence of elbows, not the number of elbows. The final sketch of the system is shown below. Again, since the pipe can be drawn many ways, this is just one example of how it could be drawn. It is important to remember that calculations are done based on the data entered inside the worksheets, not the graphical image on the screen.

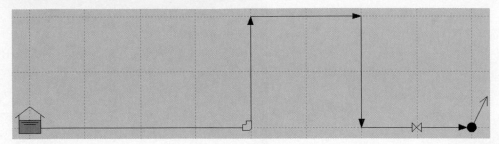

6. Since PIPE-FLO® treats the elbows, valves, and entrances as part of the pipe itself, results for individual components are not given, but rather the results for the pipe and all components taken together. To get results for an individual component, dedicate a single pipe to that component or do manual calculations. To find the total system pressure drop through the 50 ft of pipe, select "CALCULATE" as shown in the previous Example Problem in Chapter 8. The values given will reflect answers based on both minor losses and pipe friction losses. Remember that some values such as friction coefficient, Reynolds number, etc., aren't shown immediately and must be added by turning them on in the "Device View Options" in the Property Grid.

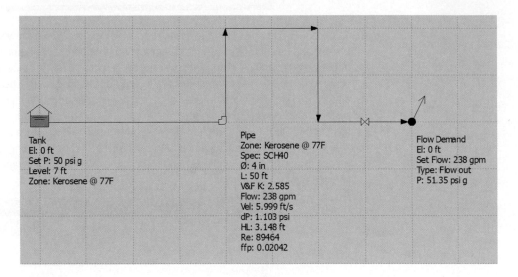

Tank
El: 0 ft
Set P: 50 psi g
Level: 7 ft
Zone: Kerosene @ 77F

Pipe
Zone: Kerosene @ 77F
Spec: SCH40
Ø: 4 in
L: 50 ft
V&F K: 2.585
Flow: 238 gpm
Vel: 5.999 ft/s
dP: 1.103 psi
HL: 3.148 ft
Re: 89464
ffp: 0.02042

Flow Demand
El: 0 ft
Set Flow: 238 gpm
Type: Flow out
P: 51.35 psi g

For this system then, the total head loss is 3.15 ft, while the pressure drop through the system is about 1.10 psi. Reynolds number and friction factor are 89 464 and 0.0204, respectively. Complete the problem by working it through hand calculations and compare your answers.

Using a Custom "K" Value in a System Example Problem 10.14 used elements such as standard elbows and a valve for which standard K values are listed in this text and other references, and that are a part of the database used in PIPE-FLO.® There are times, however, when a proprietary component is to be used and must be accounted for in the system. Generally, manufacturers provide the K value for such specialized components, and PIPE-FLO® allows for the entry of such a custom K value.

Example Problem 10.15

Calculate the pressure drop in a straight, horizontal, Schedule 40, DN 150 pipe due to friction and a custom filter with a "K" factor value of 0.75. The total pipe length is 15 m and water at 25°C flows through it at 3 m/s. The tank pressure is 500 kPa(g) and the depth of the water is 5.0 m.

Solution

1. Open a new project in PIPE-FLO® and select the "SYSTEM" menu on the toolbar to initialize all key data such as units, fluid zones, and pipe specifications the same way as the Example Problem in Chapter 8.

2. Put in the tank, pipe, and flow demand according to the variables as specified in the problem statement.

3. Go to the menu used to insert fittings and valves the same way as if putting in an elbow or valve as explained in the previous example problem, and click on the "OTHER" category. Under this drop down menu, find the "Fixed K" option and click on it. This gives the user the opportunity to enter any "K" value for a device and provide a description for it. Using this method in PIPE-FLO®, any device can be created and a "K" value assigned. We can also specify a name for the device to keep track of it in the "Valves and Fitting" menu. For this problem, enter "Custom Filter" as the description and "0.75" for the "K" value. Note that PIPE-FLO® does not add a graphic symbol to the FLO-Sheet® so it is very important to check back to the "Valves and Fitting" menu to verify what components are already placed in the system rather than simply depending on the graphic of the fluid circuit.

4. Return to the FLO-Sheet® and select "CALCULATE" to have PIPE-FLO® calculate the pressure difference across the pipe and other variables. Be sure to show the other values that PIPE-FLO® is able to give to verify your hand calculations.

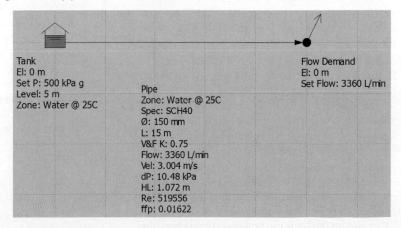

5. Calculations indicate a pressure drop of 10.48 kPa through the pipe, a total head loss of 1.072 m, and a friction factor of 0.016.

REFERENCES

1. Beij, K. H. 1938. Pressure Losses for Fluid Flow in 90 Degree Pipe Bends. *Journal of Research of the National Bureau of Standards* 21: 1–18.

2. Crane Co. 2011. *Flow of Fluids through Valves, Fittings and Pipe* (Technical Paper No. 410). Stamford, CT: Author.

3. The Hydraulic Institute. 1990. *Engineering Data Book*, 2nd ed. Parsippany, NJ: Author.

4. Brater, Ernest, C. Y. Wei, Horace W. King, and James E. Lindell. 1996. *Handbook of Hydraulics*, 7th ed. New York: McGraw-Hill.

5. Crocker, Sabin, and R. C. King. 1972. *Piping Handbook*, 6th ed. New York: McGraw-Hill.

6. Dickenson, T. C. 1999. *Valves, Piping, and Pipelines Handbook*, 3rd ed. New York: Elsevier Science.

7. Frankel, Michael. 2010. *Facility Piping Systems Handbook*, 3rd ed. New York: McGraw-Hill.

8. Idelchik, I. E. 2008. *Handbook of Hydraulic Resistance*, 3rd ed. Mumbai, India: Jaico Publishing House.

9. Nayyar, Mohinder L. 2000. *Piping Handbook*, 7th ed. New York: McGraw-Hill.

10. Skousen, Philip L. 2011. *Valve Handbook*, 3rd ed. New York: McGraw-Hill.

11. Willoughby, David A., Rick Sutherland, and R. Dodge Woodson. 2009. *Plastic Piping Handbook*. New York: McGraw-Hill.

12. Smith, Peter, and R. W. Zappe. 2004. *Valve Selection Handbook*, 5th ed. Houston, TX: Gulf.

13. Ruiz de, Burton A. 2013. *Handbook of Valves: Design, Selection, & Uses*. Nottingham, UK: Auris Reference.

14. Klette, Patrick J. 2010. *Fluid Power Systems*. Orland Park, IL: American Technical Publishers.

INTERNET RESOURCES

1. **Crane Energy Flow Solutions:** Manufacturer of numerous types of valves for piping applications in the refining, oil, gas, pulp, paper, wastewater treatment, and chemical process industries. Brands include Crane, Jenkins, Pacific, Krombach, and others. The site offers a valve selection guide. The useful reference *Crane Technical Paper 410* (Reference 2) can be ordered through this site. From the home page, select either *Products* or *Brands*.

2. **Flow of Fluids:** A special website for the Publication, Flow of Fluids Through Valves, Fittings and Pipe—Crane Technical Paper 410, the guide to understanding the flow of fluids through valves, pipes, and fittings. Other useful publications can be acquired from this site, operated by Engineered Software, Inc. (See also Internet resource 3.)

3. **Engineered Software, Inc. (ESI):** www.eng-software.com Developer of the PIPE-FLO® fluid flow analysis software to design, optimize, and troubleshoot fluid piping systems, as demonstrated in this book. A special demonstration version of PIPE-FLO® created for this book can be accessed by users of this book at http://www.eng-software.com/appliedfluidmechanics.

4. **Zurn Industries:** Manufacturer of control valves, faucets, strainers, pressure regulators, pressure relief valves, backflow preventers, and other devices for commercial and residential plumbing applications.

5. **Eaton Hydraulics:** Manufacturer of fluid power valves, pumps, actuators, and other components for fluid power systems for industrial, agricultural, construction, mining, marine, and lawn and garden care applications. Brands include Eaton, Vickers, Char-Lynn, Hydro-Line, Aeroquip, and others.

6. **Hayward Flow Control:** Manufacturer of plastic piping components for commercial and industrial applications. Products include ball valves, butterfly valves, diaphragm valves, check valves, control valves, pipeline strainers, pumps, and filters. Data sheets for each product include flow resistance data expressed as flow coefficients C_V. Sizes include ½ in to 24 in. Division of Hayward Industries, Inc.

7. **Kerotest Company:** From the home page, select *Products* to learn more about this manufacturer's fluid flow products such as alloy steel needle valves, Polyball plastic ball valves, gate valves, strainers and other products. Flow resistance data are given as flow coefficients, C_V.

8. **Thermoplastic Valves, Inc.:** Manufacturer of a diverse line of thermoplastic valves for industries such as water filtration, irrigation, chemical processing, pharmaceutical production, food processing, and others. Products include ball valves, butterfly valves, check valves, diaphragm valves, and strainers. Flow resistance data are given as flow coefficients C_V.

PRACTICE PROBLEMS

10.1 Determine the energy loss due to a sudden enlargement from a 50 mm OD × 2.4 mm wall plastic pipe to a 90 mm OD × 2.8 mm wall plastic pipe when the velocity of flow is 3 m/s in the smaller pipe.

10.2 Determine the energy loss due to a sudden enlargement from a standard DN 25 Schedule 80 steel pipe to a DN 90 mm Schedule 80 steel pipe when the rate of flow is 3×10^{-3} m^3/s.

10.3 Determine the energy loss due to a sudden enlargement from a standard 1-in Schedule 80 pipe to a $3\frac{1}{2}$-in Schedule 80 pipe when the rate of flow is 0.10 ft^3/s.

10.4 Determine the pressure difference between two points on either side of a sudden enlargement from a tube with a 2-in ID to one with a 6-in ID when the velocity of flow of water is 4 ft/s in the smaller tube.

10.5 Determine the pressure difference for the conditions in Problem 10.4 if the enlargement is gradual with a cone angle of 15°.

10.6 Determine the energy loss due to a gradual enlargement from a 25 mm OD × 2.0 mm wall copper hydraulic tube to a 80 mm OD × 2.8 mm wall tube when the velocity of flow is 3 m/s in the smaller tube and the cone angle of the enlargement is 20°.

10.7 Determine the energy loss for the conditions in Problem 10.6 if the cone angle is increased to 60°.

10.8 Compute the energy loss for gradual enlargements with cone angles from 2° to 60° in the increments shown in Fig. 10.5. For each case, water at 60°F is flowing at 85 gal/min in a 2-in Schedule 40 steel pipe that enlarges to a 6-in Schedule 40 pipe.

10.9 Plot a graph of energy loss versus cone angle for the results of Problem 10.8.

10.10 For the data in Problem 10.8, compute the length required to achieve the enlargement for each cone angle. Then compute the energy loss due to friction in that length using the velocity, diameter, and Reynolds number for the midpoint between the ends of the enlargement. Use water at 60°F.

10.11 Add the energy loss due to friction from Problem 10.10 to the energy loss for the enlargement from Problem 10.8 and plot the total versus the cone angle on the same graph used in Problem 10.9.

10.12 Another term for an enlargement is a diffuser. A diffuser is used to convert kinetic energy $(v^2/2g)$ to pressure energy (p/γ). An ideal diffuser is one in which no energy losses occur and Bernoulli's equation can be used to compute the pressure after the enlargement. Compute the pressure after the enlargement for an ideal diffuser for the flow of water at 20°C from a 25 mm OD × 2.0 mm wall copper tube to an 80 mm OD × 2.8 mm wall copper tube. The volume flow rate is 150 L/min and the pressure before the enlargement is 500 kPa.

10.13 Compute the resulting pressure after a "real" diffuser in which the energy loss due to the enlargement is considered for the data presented in Problem 10.12. The enlargement is sudden.

10.14 Compute the resulting pressure after a "real" diffuser in which the energy loss due to the enlargement is considered for the data presented in Problem 10.12. The enlargement is gradual with cone angles of (a) 60°, (b) 30°, and (c) 10°. Compare the results with those of Problems 10.12 and 10.13.

10.15 Determine the energy loss when 0.04 m³/s of water flows from a DN 150 standard Schedule 40 pipe into a large reservoir.

10.16 Determine the energy loss when 1.50 ft³/s of water flows from a 6-in standard Schedule 40 pipe into a large reservoir.

10.17 Determine the energy loss when oil with a specific gravity of 0.87 flows from a 4-in pipe to a 2-in pipe through a sudden contraction if the velocity of flow in the larger pipe is 4.0 ft/s.

10.18 For the conditions in Problem 10.17, if the pressure before the contraction was 80 psig, calculate the pressure in the smaller pipe.

10.19 True or false: For a sudden contraction with a diameter ratio of 3.0, the energy loss decreases as the velocity of flow increases.

10.20 Determine the energy loss for a sudden contraction from a DN 125 Schedule 80 steel pipe to a DN 50 Schedule 80 pipe for a flow rate of 500 L/min.

10.21 Determine the energy loss for a gradual contraction from a DN 125 Schedule 80 steel pipe to a DN 50 Schedule 80 pipe for a flow rate of 500 L/min. The cone angle for the contraction is 105°.

10.22 Determine the energy loss for a sudden contraction from a 4-in Schedule 80 steel pipe to a 1½-in Schedule 80 pipe for a flow rate of 250 gal/min.

10.23 Determine the energy loss for a gradual contraction from a 4-in Schedule 80 steel pipe to a 1½-in Schedule 80 pipe for a flow rate of 250 gal/min. The cone angle for the contraction is 76°.

10.24 For the data in Problem 10.22, compute the energy loss for gradual contractions with each of the cone angles listed in Figs. 10.10 and 10.11. Plot energy loss versus the cone angle.

10.25 For each contraction described in Problems 10.22 and 10.24, make a scale drawing of the device to observe its physical appearance.

10.26 Note in Figs. 10.10 and 10.11 that the minimum energy loss for a gradual contraction ($K = 0.04$ approximately) occurs when the cone angle is in the range of 15° to 40°. Make scale drawings of contractions at both of these extremes for a reduction from a 6-in to a 3-in ductile iron pipe.

10.27 If the contraction from a 6-in to a 3-in ductile iron pipe described in Problem 10.26 was made with a cone angle of 120°, what would the resulting resistance coefficient be? Make a scale drawing of this reducer.

10.28 Compute the energy loss that would occur as 50 gal/min of water flows from a tank into a steel tube with an OD of 2.0 in and a wall thickness of 0.065 in. The tube is installed flush with the inside of the tank wall with a square edge.

10.29 Determine the energy loss that will occur if water flows from a reservoir into a pipe with a velocity of 3 m/s if the configuration of the entrance is (a) an inward-projecting pipe, (b) a square-edged inlet, (c) a chamfered inlet, or (d) a well-rounded inlet.

10.30 Determine the equivalent length in meters of pipe of a fully open globe valve placed in a DN 250 Schedule 40 pipe.

10.31 Repeat Problem 10.30 for a fully open gate valve.

10.32 Calculate the resistance coefficient K for a ball-type check valve placed in a 2-in Schedule 40 steel pipe if water at 100°F is flowing with a velocity of 10 ft/s.

10.33 Calculate the pressure difference across a fully open angle valve placed in a 5-in Schedule 40 steel pipe carrying 650 gal/min of oil (sg − 0.90).

10.34 Determine the pressure drop across a 90° standard elbow in a DN 65 Schedule 40 steel pipe if water at 15°C is flowing at the rate of 750 L/min.

10.35 Repeat Problem 10.34 for a street elbow.

10.36 Repeat Problem 10.34 for a long radius elbow. Compare the results from Problems 10.34–10.36.

10.37 A simple heat exchanger is made by installing a close return bend on two ½-in Schedule 40 steel pipes as shown in Fig. 10.32. Compute the pressure difference between the inlet and the outlet for a flow rate of 12.5 gal/min of ethylene glycol at 77°F.

FIGURE 10.32 Problem 10.37.

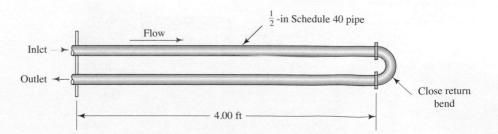

Flow

Inlet

Outlet

½-in Schedule 40 pipe

Close return bend

4.00 ft

FIGURE 10.33 Problem 10.38.

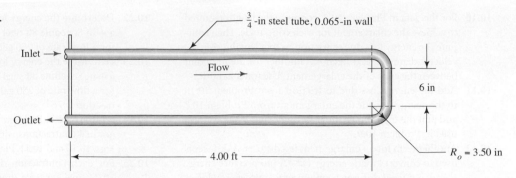

$\frac{3}{4}$-in steel tube, 0.065-in wall

Inlet →

Flow →

Outlet ←

6 in

4.00 ft

$R_o = 3.50$ in

10.38 A proposed alternate form for the heat exchanger described in Problem 10.37 is shown in Fig. 10.33. The entire flow conduit is a ¾-in steel tube with a wall thickness of 0.065 in. Note that the ID for this tube is 0.620 in, slightly smaller than that of the ½-in Schedule 40 pipe ($D = 0.622$ in). The return bend is formed by two 90° bends with a short length of straight tube between them. Compute the pressure difference between the inlet and the outlet of this design and compare it with the system from Problem 10.37.

10.39 A piping system for a pump contains a tee, as shown in Fig. 10.34, to permit the pressure at the outlet of the pump to be measured. However, there is no flow into

the line leading to the gage. Compute the energy loss as 0.40 ft³/s of water at 50°F flows through the tee.

10.40 A piping system for supplying heavy fuel oil at 25°C is arranged as shown in Fig. 10.35. The bottom leg of the tee is normally capped, but the cap can be removed to clean the pipe. Compute the energy loss as 0.08 m³/s flows through the tee.

10.41 A 25 mm OD × 2.0 mm wall copper tube supplies hot water (80°C) to a washing system in a factory at a flow rate of 250 L/min. At several points in the system, a 90° bend is required. Compute the energy loss in each bend if the radius to the outside of the bend is 300 mm.

FIGURE 10.34 Problem 10.39.

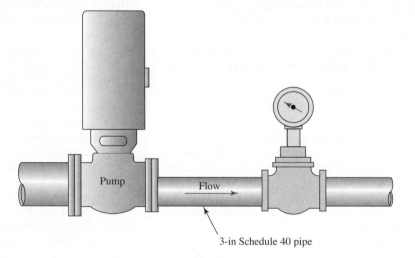

Pump

Flow

3-in Schedule 40 pipe

FIGURE 10.35 Problem 10.40.

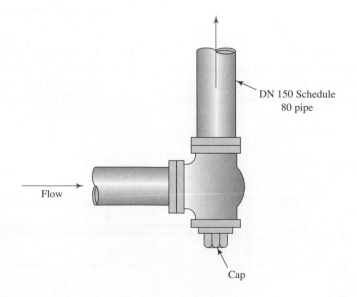

DN 150 Schedule 80 pipe

Flow

Cap

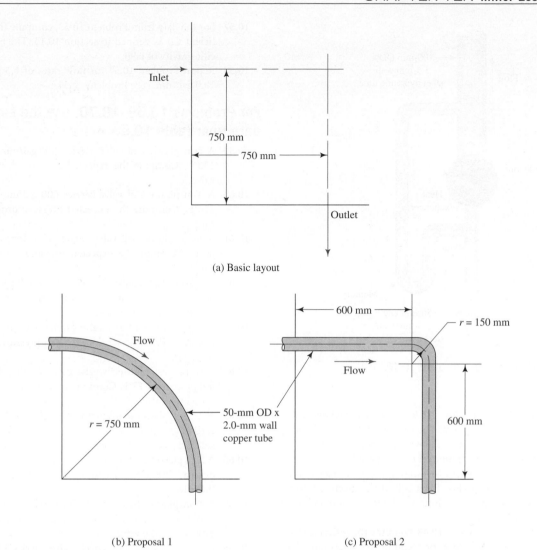

(a) Basic layout

(b) Proposal 1

(c) Proposal 2

FIGURE 10.36 Problem 10.43.

10.42 Specify the radius in mm to the centerline of a 90° bend in a 25 mm OD × 2.0 mm wall copper tube to achieve the minimum energy loss. For such a bend carrying 250 L/min of water at 80°C, compute the energy loss. Compare the results with those of Problem 10.41.

10.43 The inlet and the outlet shown in Fig. 10.36(a) are to be connected with a 50 mm OD × 2.0 mm wall copper tube to carry 750 L/min of propyl alcohol at 25°C. Evaluate the two schemes shown in parts (b) and (c) of the figure with regard to the energy loss. Include the losses due to both the bend and the friction in the straight tube.

10.44 Compare the energy losses for the two proposals from Problem 10.43 with the energy loss for the proposal in Fig. 10.37.

10.45 Determine the energy loss that occurs as 40 L/min of water at 10°C flows around a 90° bend in a commercial steel tube having an OD of 20 mm and a wall thickness of 1.5 mm. The radius of the bend to the centerline of the tube is 150 mm.

10.46 Figure 10.38 shows a test setup for determining the energy loss due to a heat exchanger. Water at 50°C is flowing vertically upward at 6.0×10^{-3} m³/s. Calculate the energy loss between points 1 and 2. Determine the resistance

coefficient for the heat exchanger based on the velocity in the inlet tube.

10.47 Compute the energy loss in a 90° bend in a steel tube used for a fluid power system. The tube has a ½-in OD and a wall thickness of 0.065 in. The mean bend radius is 2.00 in. The flow rate of hydraulic oil is 3.5 gal/min.

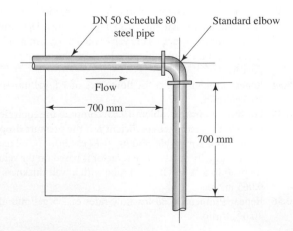

FIGURE 10.37 Problem 10.44.

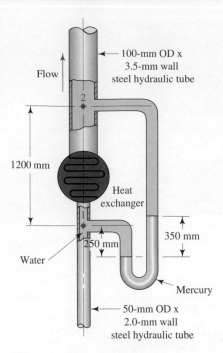

Flow ← 100-mm OD x
3.5-mm wall
steel hydraulic tube

2

1200 mm

Heat
exchanger

1

250 mm   350 mm

Water

Mercury

← 50-mm OD x
2.0-mm wall
steel hydraulic tube

FIGURE 10.38   Problem 10.46.

10.48 Compute the energy loss in a 90° bend in a steel tube used for a fluid power system. The tube has a 1¼-in OD and a wall thickness of 0.083 in. The mean bend radius is 3.25 in. The flow rate of hydraulic oil is 27.5 gal/min.

10.49 For the data in Problem 10.47, compute the resistance factor and the energy loss for a coil of the given tube that makes six complete revolutions. The mean bend radius is the same, 2.00 in.

10.50 For the data in Problem 10.48, compute the resistance factor and the energy loss for a coil of the given tube that makes 8.5 revolutions. The mean bend radius is the same, 3.50 in.

10.51 A tube similar to that in Problem 10.47 is being routed through a complex machine. At one point, the tube must be bent through an angle of 145°. Compute the energy loss in the bend.

10.52 A tube similar to that in Problem 10.48 is being routed through a complex machine. At one point, the tube must be bent through an angle of 60°. Compute the energy loss in the bend.

10.53 A fluid power system incorporates a directional control valve similar to that shown in Fig. 10.30(a). Determine the pressure drop across the valve when 5.0 gal/min of hydraulic oil flows through the valve from the pump port to port A.

10.54 Repeat Problem 10.53 for flow rates of 7.5 gal/min and 10.0 gal/min.

10.55 For the data from Problem 10.53, compute the equivalent value of the resistance coefficient K if the pressure drop is found from $\Delta p = \gamma h_L$ and $h_L = K(v^2/2g)$. The oil has a specific gravity of 0.90. The K factor is based on the velocity head in a ⅝-in-OD steel tube with a wall thickness of 0.065 in.

10.56 Repeat Problem 10.55 for flow rates of 7.5 gal/min and 10.0 gal/min.

10.57 For the data from Problem 10.53, compute the flow coefficient C_V as defined in section 10.13. The oil has a specific gravity of 0.90.

10.58 Repeat Problem 10.57 for flow rates of 7.5 gal/min and 10.0 gal/min. (See Problem 10.54.)

For Problems 10.59–10.70, use the sample data from Table 10.6.

10.59 A 2-in plastic ball valve carries 150 gal/min of water at 150°F. Compute the expected pressure drop across the valve.

10.60 A 4-in plastic ball valve carries 600 gal/min of water at 120°F. Compute the expected pressure drop across the valve.

10.61 A ¾-in plastic ball valve carries 15 gal/min of water at 80°F. Compute the expected pressure drop across the valve.

10.62 A 1½-in plastic butterfly valve carries 60 gal/min of carbon tetrachloride at 77°F. Compute the expected pressure drop across the valve.

10.63 A 3-in plastic butterfly valve carries 300 gal/min of gasoline at 77°F. Compute the expected pressure drop across the valve.

10.64 A 10-in plastic butterfly valve carries 5000 gal/min of liquid propane at 77°F. Compute the expected pressure drop across the valve.

10.65 A 1½-in plastic diaphragm valve carries 60 gal/min of carbon tetrachloride at 77°F. Compute the expected pressure drop across the valve.

10.66 A 3-in plastic diaphragm valve carries 300 gal/min of gasoline at 77°F. Compute the expected pressure drop across the valve.

10.67 A 6-in plastic diaphragm valve carries 1500 gal/min of liquid propane at 77°F. Compute the expected pressure drop across the valve.

10.68 A ¾-in plastic swing check valve carries 18 gal/min of seawater at 77°F. Compute the expected pressure drop across the valve.

10.69 A 3-in plastic swing check valve carries 300 gal/min of kerosene at 77°F. Compute the expected pressure drop across the valve.

10.70 An 8-in plastic swing check valve carries 3500 gal/min of glycerin at 77°F. Compute the expected pressure drop across the valve.

Supplemental Problems (PIPE-FLO® only)

10.71 Use PIPE-FLO® software to determine the pressure drop in a 20 m horizontal run of DN 100 Schedule 40 pipe, carrying 25°C kerosene at a velocity of 3 m/s. The pipe includes a sharp-edged pipe entrance at the tank, two standard 90° elbows, and a fully open globe valve. Report all applicable values related to the solution such as Reynolds number and friction factor.

10.72 Use PIPE-FLO® to calculate the head loss and pressure drop in a length of pipe that includes a filter. The pipe is a horizontal 6-in Schedule 40 pipe. The total length of pipe is 45 ft and the manufacturer of the filter specifies that it has a K-factor value of 0.82. Water at 60°F flows through the system at 9 ft/s. The pressure in the tank is 75 psig.

COMPUTER AIDED ANALYSIS AND DESIGN ASSIGNMENTS

The purpose of the following assignments is to prepare aids that a designer of fluid power systems can use to specify appropriate sizes of steel tubing for a system being designed. Some also help to evaluate energy losses and to ensure that losses due to bends in the tubing are as low as practical.

1. Your company designs special-purpose fluid power systems for the industrial automation market. The normal technique used to fabricate the systems is to route steel tubing among the pumps, control valves, and actuators for the system using straight tubing and 90° bends. Many different sizes of tubing are used in the systems depending on the flow rate of hydraulic oil required for the application. You are asked to create a chart of the recommended bend radii for each nominal size of steel tubing listed in Appendix G.1. The wall thickness for each size will always be the largest listed in the table because of the high pressures used in the hydraulic systems. According to Fig. 10.28, the minimum resistance will occur when the relative radius of the bend is approximately 3.0. Create the table of recommended bend radii, rounding the radii to the nearest ½ in, but ensure that the relative radius of any bend is never less than 2.0. A spreadsheet approach is suggested.

2. Section 6.4 includes the recommendation that the velocity of flow in discharge lines of fluid power systems be in the range 7–25 ft/s. The average of these values is 16 ft/s. Design a spreadsheet to determine the inside diameter of the discharge line to achieve this velocity for any design volume flow rate. Then, refer to Appendix G.1 to specify a suitable steel tube, using the largest of the given wall thicknesses for any size because of the high pressures used in fluid power systems. For the selected tube, compute the actual velocity of flow when carrying the design volume flow rate.

3. For each size of tubing used in Assignment 1, determine the value of f_T for use in the energy loss equation for any minor

loss calculation requiring that value for valves, fittings, and bends. See Example Problem 10.9 for an example. You will need to compute the ratio of D/ε for each tube size using the roughness for steel tubing. Then refer to the Moody diagram to determine the friction factor in the fully turbulent zone. List that value within the spreadsheet for Assignment 1 or make a separate spreadsheet for the list.

4. Combine Assignments 1–3 to include the computation of the energy loss for a given bend, using the following process:

   ■ Given a required volume flow rate for a fluid power system, determine an appropriate size for the discharge tubing to produce a velocity of flow in the recommended range.
   ■ For the selected tube size, recommend the bend radius for 90° bends.
   ■ For the selected tube size, determine the value of f_T, the friction factor in the fully turbulent range.
   ■ Compute the resistance factor K for the bend from $K = f_T(L_e/D)$.
   ■ Compute the actual velocity of flow for the given volume flow rate in the selected tube size.
   ■ Compute the energy loss in the bend from $h_L = K(v^2/2g)$.

5. Repeat Assignment 1 for each tube size but use the smallest wall thickness rather than the largest. Such tubes could be used for the suction lines that draw the oil from the tank and deliver it to the inlet of the pump. The pressure in suction tubing is very low.

6. Repeat Assignment 2 except recommend the size of suction line tubing to achieve the recommended velocity of flow in suction lines of 3.0 ft/s. Use the smallest wall thickness for any size of tube because of the low pressure in the suction lines.

7. Repeat any of Assignments 1–6 using SI metric data. Volume flow rates are to be in appropriate units assigned by the instructor, such as m^3/s, m^3/h, L/s, or L/min. Velocity calculations should be in m/s.

SERIES PIPELINE SYSTEMS

THE BIG PICTURE

This chapter is the capstone for the preceding Chapters 6–10, which considered specific aspects of the flow of fluids in pipes and tubes. Now we bring together the basic concepts of using the energy equation, identifying laminar and turbulent flows, evaluating friction losses in pipes, and examining minor losses to analyze series pipe line systems that may contain any combination of pumps, valves, fittings, and energy losses due to friction.

A series pipeline system is one in which the fluid follows a single flow path throughout the system.

You will develop the ability to identify three different classes of series pipeline systems and practice the techniques of analyzing them. Because most real systems include many different elements, the calculations can become highly involved. After mastering the basic principles of analyzing series systems, you should develop your ability to use computer-assisted analysis of fluid flow systems to perform most of the calculations.

Consider the industrial piping system shown in Fig. 11.1. The elements with hand-wheel actuators are valves used to start and stop flow or to direct the flow to various parts of the system. There are elbows, enlargements, and pipe sections all connected together. A pump delivers fluid into the system.

Exploration

Review Chapters 6–10 to remind you of the analytical tools presented there: the continuity equation, the general energy equation, energy losses due to friction, and minor losses. Study the various pipeline systems depicted in Chapter 7 and identify where energy losses occur. Review the Big Picture discussions from Chapters 8–10 where you identified laminar and turbulent flow types and energy losses in many kinds of piping elements.

Introductory Concepts

Recall the discussion in the Big Picture section of Chapter 10. There you examined real systems, following the path of the fluid flow and identifying the kinds of minor losses that occur in the systems. Each valve, fitting, or change in the size or direction of the flow path, causes energy loss from the system. The energy is lost in the form of heat dissipated from the fluid with resulting decreases in pressure throughout the system. The lost energy was first delivered into the system by pumps or because the source was at a higher elevation or in a main line at high pressure. Therefore, the loss of energy is wasteful, but the system elements are essential to the purpose of the system. Lower energy losses in those components generally mean that a smaller pump and motor could be used or a given system could produce a greater output.

FIGURE 11.1 Real industrial piping systems like this contain many types and sizes of pipes, valves, and fittings. (*Source:* Andrei Merkulov/Fotolia)

System analysis and design problems can be classified into three classes as follows:

Class I   The system is completely defined in terms of the size of pipes, the types of minor losses that are present, and the volume flow rate of fluid in the system. The typical objective is to compute the pressure at some point of interest, to compute the total head on a pump, or to compute the elevation of a source of fluid to produce a desired flow rate or pressure at selected points in the system.

Class II   The system is completely described in terms of its elevations, pipe sizes, valves and fittings, and allowable pressure drop at key points in the system. You desire to know the volume flow rate of the fluid that could be delivered by a given system.

Class III   The general layout of the system is known along with the desired volume flow rate. The size of the pipe required to carry a given volume flow rate of a given fluid is to be determined.

As you study the methods of analyzing and designing these three classes of systems, you should also learn what the desirable elements of a system are. What are the better types of valves to use in given applications? Where are the critical points in a system to evaluate pressures? Where should you place a pump in a system relative to the source of the fluid? How much total head must the pump be capable of delivering? What are reasonable velocities of flow in different parts of the systems? Some of these issues were brought up in earlier chapters. Now you will be using them together to evaluate the acceptability of a proposed system and to recommend improvements.

11.1 OBJECTIVES

After completing this chapter, you should be able to:

1. Identify series pipeline systems.

2. Determine whether a given system is Class I, Class II, or Class III.

3. Compute the total energy loss, elevation differences, or pressure differences for Class I systems with any combination of pipes, minor losses, pumps, or reservoirs when the system carries a given flow rate.

4. Determine for Class II systems the velocity or volume flow rate through the system with known pressure differences and elevation heads.

5. Determine for Class III systems the size of pipe required to carry a given fluid flow rate with a specified limiting pressure drop or for a given elevation difference.

6. Apply the PIPE-FLO® software to analyze the performance of series pipeline problems.

11.2 CLASS I SYSTEMS

This chapter deals only with series systems such as the one illustrated in Fig. 11.2. The energy equation for this system, using the surface of each reservoir as the reference points, is

$$\frac{p_1}{\gamma} + z_1 + \frac{v_1^2}{2g} + h_A - h_L = \frac{p_2}{\gamma} + z_2 + \frac{v_2^2}{2g} \quad (11\text{–}1)$$

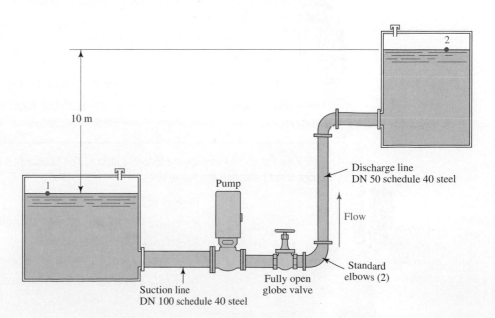

FIGURE 11.2 System for Example Problem 11.1.

The first three terms on the left-hand side of this equation represent the energy possessed by the fluid at point 1 in the form of pressure head, elevation head, and velocity head. The terms on the right-hand side of the equation represent the energy possessed by the fluid at point 2. The term h_A is the energy added to the fluid by a pump. A common name for this energy is *total head on the pump*, and it is used as one of the primary parameters in selecting a pump and in determining its performance. The term h_L denotes the total energy lost from the system anywhere between reference points 1 and 2. There are typically several factors that contribute to the total energy loss. Six different factors apply in this problem:

$$h_L = h_1 + h_2 + h_3 + h_4 + h_5 + h_6 \quad (11\text{--}2)$$

where

$h_L =$ Total energy loss per unit weight of fluid flowing

$h_1 =$ Entrance loss

$h_2 =$ Friction loss in the suction line

$h_3 =$ Energy loss in the valve

$h_4 =$ Energy loss in the two 90° elbows

$h_5 =$ Friction loss in the discharge line

$h_6 =$ Exit loss

In a series pipeline the total energy loss is the sum of the individual minor losses and all pipe friction losses. This statement is in agreement with the principle that the energy equation is a means of accounting for all of the energy in the system between the two reference points.

Our approach to the analysis of Class I systems is identical to that used throughout the previous chapters except that generally many types of energy losses will exist. The following programmed example problem will illustrate the solution of a Class I problem.

PROGRAMMED EXAMPLE PROBLEM

Example Problem 11.1 Calculate the power supplied to the pump shown in Fig. 11.2 if its efficiency is 76 percent. Methyl alcohol at 25°C is flowing at the rate of 54.0 m³/h. The suction line is a standard DN 100 Schedule 40 steel pipe, 15 m long. The total length of DN 50 Schedule 40 steel pipe in the discharge line is 200 m. Assume that the entrance from reservoir 1 is through a square-edged inlet and that the elbows are standard. The valve is a fully open globe valve.

To begin the solution, write the energy equation for the system.

Using the surfaces of the reservoirs as the reference points, you should have

$$\frac{p_1}{\gamma} + z_1 + \frac{v_1^2}{2g} + h_A - h_L = \frac{p_2}{\gamma} + z_2 + \frac{v_2^2}{2g}$$

Because $p_1 = p_2 = 0$ and v_1 and v_2 are approximately zero, the equation can be simplified to

$$z_1 + h_A - h_L = z_2$$

Because the objective of the problem is to calculate the power supplied to the pump, solve now for the total head on the pump, h_A.

The total head is

$$h_A = z_2 - z_1 + h_L$$

There are six components to the total energy loss. List them and write the formula for evaluating each one.

Your list should include the following items. The subscript s indicates the suction line and the subscript d indicates the discharge line:

$$h_1 = K(v_s^2/2g) \quad \text{(entrance loss)}$$
$$h_2 = f_s(L/D)(v_s^2/2g) \quad \text{(friction loss in suction line)}$$
$$h_3 = f_{dT}(L_e/D)(v_d^2/2g) \quad \text{(valve)}$$
$$h_4 = f_{dT}(L_e/D)(v_d^2/2g) \quad \text{(two 90° elbows)}$$
$$h_5 = f_d(L/D)(v_d^2/2g) \quad \text{(friction loss in discharge line)}$$
$$h_6 = 1.0(v_d^2/2g) \quad \text{(exit loss)}$$

Because the velocity head in the suction or discharge line is required for each energy loss, calculate these values now.

You should have $v_s^2/2g = 0.17$ m and $v_d^2/2g = 2.44$ m, found as follows:

$$Q = \frac{54.0 \text{ m}^3}{\text{hr}} \times \frac{1 \text{ hr}}{3600 \text{ s}} = 0.015 \text{ m}^3/\text{s}$$

$$v_s = \frac{Q}{A_s} = \frac{0.015 \text{ m}^3}{\text{s}} \times \frac{1}{8.213 \times 10^{-3} \text{m}^2} = 1.83 \text{ m/s}$$

$$\frac{v_s^2}{2g} = \frac{(1.83)^2}{2(9.81)}\text{m} = 0.17 \text{ m}$$

$$v_d = \frac{Q}{A_d} = \frac{0.015 \text{ m}^3}{\text{s}} \times \frac{1}{2.168 \times 10^{-3} \text{ m}^2} = 6.92 \text{ m/s}$$

$$\frac{v_d^2}{2g} = \frac{(6.92)^2}{2(9.81)}\text{m} = 2.44 \text{ m}$$

To determine the friction losses in the suction line and the discharge line and the minor losses in the discharge line, we need the Reynolds number, relative roughness, and friction factor for each pipe, and the friction factor in the zone of complete turbulence for the discharge line that contains a valve and pipe fittings. Find these values now.

For methyl alcohol at 25°C, $\rho = 789$ kg/m^3 and $\eta = 5.60 \times 10^{-4}$ Pa·s. Then, in the suction line, we have

$$N_R = \frac{vD\rho}{\eta} = \frac{(1.83)(0.1023)(789)}{5.60 \times 10^{-4}} = 2.64 \times 10^5$$

Because the flow is turbulent, the value of f_s must be evaluated from the Moody diagram, Fig. 8.7. For steel pipe, $\varepsilon = 4.6 \times 10^{-5}$ m. Write

$$D/\varepsilon = 0.1023/(4.6 \times 10^{-5}) = 2224$$
$$N_R = 2.64 \times 10^5$$

Then $f_s = 0.018$.

In the discharge line, we have

$$N_R = \frac{vD\rho}{\eta} = \frac{(6.92)(0.0525)(789)}{5.60 \times 10^{-4}} = 5.12 \times 10^5$$

This flow is also turbulent. Evaluating the friction factor f_d gives

$$D/\varepsilon = 0.0525/(4.6 \times 10^{-5}) = 1141$$
$$N_R = 5.12 \times 10^5$$
$$f_d = 0.020$$

We can find from Table 10.5 that $f_{dT} = 0.019$ for the DN 50 discharge pipe in the zone of complete turbulence.

Returning now to the energy loss calculations, evaluate h_1, the entrance loss, in N·m/N or m.

The result is $h_1 = 0.09$ m. For a square-edged inlet, $K = 0.5$ and

$$h_1 = 0.5(v_s^2/2g) = (0.5)(0.17 \text{ m}) = 0.09 \text{ m}$$

Now calculate h_2, the friction loss in the suction line.

The result is $h_2 = 0.45\,\text{m}$.

$$h_2 = f_s \times \frac{L}{D} \times \frac{v_s^2}{2g} = f_s\left(\frac{15}{0.1023}\right)(0.17)\,\text{m}$$

$$h_2 = (0.018)\left(\frac{15}{0.1023}\right)(0.17)\,\text{m} = 0.45\,\text{m}$$

Now calculate h_3, the energy loss in the valve in the discharge line.

From the data in Chapter 10, the equivalent-length ratio L_e/D for a fully open globe valve is 340. The friction factor is $f_{dT} = 0.019$. Then, we have

$$h_3 = f_{dT} \times \frac{L_e}{D} \times \frac{v_d^2}{2g} = (0.019)(340)(2.44)\,\text{m} = 15.76\,\text{m}$$

Now calculate h_4, the energy loss in the two 90° elbows.

For standard 90° elbows, $L_e/D = 30$. The value of f_{dT} is 0.019, the same as that used in the preceding panel. Then, we have

$$h_4 = 2f_{dT} \times \frac{L_e}{D} \times \frac{v_d^2}{2g} = (2)(0.019)(30)(2.44)\,\text{m} = 2.78\,\text{m}$$

Now calculate h_5, the friction loss in the discharge line.

The discharge-line friction loss is

$$h_5 = f_d \times \frac{L}{D} \times \frac{v_d^2}{2g} = (0.020)\left(\frac{200}{0.0525}\right)(2.44)\,\text{m} = 185.9\,\text{m}$$

Now calculate h_6, the exit loss.

The exit loss is

$$h_6 = 1.0(v_d^2/2g) = 2.44\,\text{m}$$

This concludes the calculation of the individual energy losses. The total loss h_L can now be determined.

$$h_L = h_1 + h_2 + h_3 + h_4 + h_5 + h_6$$
$$h_L = (0.09 + 0.45 + 15.76 + 2.78 + 185.9 + 2.44)\,\text{m}$$
$$h_L = 207.4\,\text{m}$$

From the energy equation the expression for the total head on the pump is

$$h_A = z_2 - z_1 + h_L$$

Then, we have

$$h_A = 10\,\text{m} + 207.4\,\text{m} = 217.4\,\text{m}$$

Now calculate the power supplied to the pump, P_A.

$$\text{Power} = \frac{h_A \gamma Q}{e_M} = \frac{(217.4\,\text{m})(7.74 \times 10^3\,\text{N/m}^3)(0.015\,\text{m}^3/\text{s})}{0.76}$$
$$P_A = 33.2 \times 10^3\,\text{N·m/s} = 33.2\,\text{kW}$$

This concludes the programmed example problem.

General Principles of Pipeline System Design Although the specific requirements of a given system may dictate some features of a pipeline system, the following guidelines should help you to design reasonably efficient systems.

1. Recall from Chapter 7 that the power required by the pump in a system is computed from

$$P_A = h_A \gamma Q$$

where h_A is the total head on the pump. Energy losses contribute much to this total head, making it desirable to minimize them.

2. Particular attention should be paid to the pressure at the inlet to a pump, keeping it as high as practical. The final design of the suction line must be checked to ensure that *cavitation* does not occur in the suction port of the pump by computing the *net positive suction head* (*NPSH*) as discussed in detail in Chapter 13.

3. System components should be selected to minimize energy losses while maintaining a reasonable physical size and cost for the components.

4. The selection of pipe sizes should be guided by the recommendations given in Section 6.4 in Chapter 6, considering the type of system being designed. Figure 6.3 should be used to determine the approximate sizes for suction and discharge lines of typical fluid transfer systems. Larger pipe sizes should be specified for very long pipes or when energy losses are to be minimized.

5. If the pipe sizes selected differ from the sizes of the suction and discharge connections of the pump, simple low-loss gradual reductions or enlargements can be used as discussed in Chapter 10. Standard components of this type are commercially available for many kinds of piping.

6. The length of suction lines should be as short as practical.

7. Low-loss shut-off and control valves, such as gate or butterfly valves, are recommended unless the system design calls for the valves to provide for throttling the flow to maintain control of the system to desired parameters. Then globe valves may be specified. In smaller systems, needle valves may be preferred for control.

8. It is often desirable to place a shut-off valve on either side of a pump to permit the repair or removal of the pump.

Critique of the System Shown in Fig. 11.2 and Analyzed in Example Problem 11.1 Problem solutions such as that just concluded can give you, the fluid flow system designer, much useful information on which you can evaluate the proposed design and make rational decisions about system improvement. Here we apply the principles just presented to the system analyzed in Example Problem 11.1. The goal is to propose several ways to redesign the system to reduce dramatically the power required by the pump

and to adjust the design of the suction line. The following are some observations:

1. The length of the suction line between the first reservoir and the pump, given to be 15 m, appears to be excessively long. We recommended that the pump be relocated closer to the reservoir so the suction line can be as short as practical.

2. It may be desirable to place a valve in the suction line before the inlet to the pump to allow the pump to be removed or serviced without draining the reservoir. A gate valve should be used so the energy loss is small during normal operation with the valve completely open.

3. Refer to Section 6.4 and Fig. 6.3 to determine an appropriate size for the suction line. For a volume flow rate of 54.0 m³/s, a pipe size of approximately DN 80 mm is suggested. The DN 100 size used in Example Problem 11.1 is acceptable, and the suction line velocity of 1.83 m/s produces a fairly low velocity head of 0.17 m and a correspondingly low friction loss, desirable for a suction line.

4. The energy loss in the 200-m-long discharge line is very high, due mostly to the high velocity of flow in the DN 50 pipe, 6.92 m/s. Figure 6.3 suggests a discharge-line size of approximately DN 65. However, because of the great length, let's specify a DN 80 Schedule 40 steel pipe that will produce a velocity of 3.15 m/s and a velocity head of 0.504 m. Compared to the original velocity head of 2.44 m for the DN 50 pipe, this is a *reduction of almost five times*. The energy loss will be reduced approximately proportionally.

5. The globe valve in the discharge line should be replaced by a type with less resistance. The equivalent-length ratio L_e/D of 340 is among the highest of any kind of valve. A fully open gate valve has $L_e/D = 8$, *a reduction of over 42 times!*

Summary of Design Changes The following changes are proposed:

1. Decrease the suction line length from 15 m to 1.5 m. Assuming that the two reservoirs must stay in the same position, the extra 13.5 m of length will be added to the discharge line, making it a total of 213.5 m long.

2. Add a fully open gate valve in the suction line.

3. Increase the discharge line size from DN 50 to DN 80 Schedule 40. Then, $v_d = 3.15$ m/s and the velocity head is 0.504 m.

4. Replace the globe valve in the discharge line with a fully open gate valve.

Making all of these changes would result in the reduction of the energy to be added by the pump from 217.4 m to 37.1 m. The power supplied to the pump would decrease from 33.2 kW to 5.66 kW, *a reduction of almost a factor of 6!*

11.3 SPREADSHEET AID FOR CLASS I PROBLEMS

The solution procedure for Class I series pipeline problems is direct in that the system is completely defined and the analysis leads to the final solution with no iteration or estimates of values. But it is a cumbersome procedure requiring many calculations. If several systems are to be designed or if the designer wants to try several modifications to a given design, it can take much time. The use of a spreadsheet can improve the procedure dramatically by doing most of the calculations for you after you enter the basic data.

Before implementing any computer-assisted approach to solving fluid flow problems, it is essential that you understand the fundamental principles and analysis techniques that underlie the program. Entering erroneous data or misinterpreting results can cause significant harm such as having the system perform poorly in service, damaging equipment, or even causing injury. The spreadsheets shown in this chapter are available for download by instructors for courses using this book. It is also quite instructive for you to reproduce the given spreadsheets and compare your results with those given in the book. You can also benefit from producing enhancements to the design of the spreadsheets. One important goal for using spreadsheets in this course is to alert you to the advantages of using computational tools to make your work easier and quicker so you can produce accurate, optimal solutions to complex design problems.

The use of commercially available analysis software, such as the PIPE-FLO® software used in this book in Chapters 8–13, provides much more capability than self-written spreadsheets or other computational aids you may develop. So you may choose to use both approaches during your career. In summary, you must become proficient in understanding the principles of fluid mechanics and also using computational aids to enhance productivity.

Figure 11.3 shows one approach to using a spreadsheet for Class I series pipeline problems. It is designed to model a system similar to that shown in Fig. 11.2, in which a pump draws fluid from some source and delivers it to a destination point. The data shown are from Example Problem 11.1, where the objective was to compute the power required to drive the pump. Compare the values in the spreadsheet with those found in the example problem. The minor differences are mostly due to rounding and the fact that the friction factors are computed by the spreadsheet, whereas they were read manually from the Moody diagram for the example problem.

The spreadsheet is somewhat more versatile, however. Its features are explained as follows.

Features of the Spreadsheet to Compute the Power Required by a Pump in a Class I Series Pipeline System (Si Metric Units Version)

1. Data that you must enter in appropriate cells are identified by the shaded areas.

2. At the top left of the sheet, you can enter the identification information for the system.

3. At the top right, you enter the description of the two reference points for use in the energy equation.

4. Then enter the system data. First enter the volume flow rate Q in the units of m³/s. Then enter the pressures and elevations at both reference points. In the example problem, the pressures are zero because both reference points are at the free surface of the reservoirs. The reference elevation is taken at the surface of reservoir 1. Therefore, the elevation of point 1 is 0.0 m and for point 2 it is 10.0 m.

5. Carefully study the required velocity data. In the example problem, the velocity at both reference points is zero because they are at the free, still surface of the reservoirs. The zero values were entered manually. But if either or both of the reference points are in a pipe instead of at the surface of a reservoir, actual pipe velocities are needed. The instruction to the right side of the spreadsheet calls for you to actually type a cell reference for the velocities. The cell reference "B20" refers to the cell where the velocity of flow in pipe 1 is computed below. The cell reference "E20" is for the cell where the velocity of flow for pipe 2 is computed. Then, after the proper data for the pipes are entered, the correct velocity and velocity head values will appear in the system data cells.

6. Enter the fluid properties data next. The specific weight γ and the kinematic viscosity ν are needed to compute the Reynolds number and the power required by the pump. Note that you must compute kinematic viscosity from $\nu = \eta/\rho$ if you originally know only the dynamic viscosity η and the density of the fluid ρ.

7. Pipe data are now entered. Provisions are made for systems with two different pipe sizes such as those in the example problem. It is typical for pumped systems to have a larger suction pipe and a smaller discharge pipe. For each, you must enter in the shaded areas the flow diameter, the wall roughness, and the total length of straight pipe. The system then computes the values in the unshaded areas. Note that the friction factors are computed using the Swamee–Jain equation from Chapter 8, Eq. (8–7).

8. The energy losses are addressed next in the spreadsheet. The energy loss is computed using the appropriate resistance factor K for each element. K for pipe friction is computed automatically. For minor losses you will have to obtain values from charts or compute them as described below. These are entered in the shaded areas and brief descriptions of each element can be listed. Room for eight losses in each of two pipes is provided. Values for cells not used should be entered as zero. Recall the following from Chapters 8 and 10:

   - For pipe friction, $K = f(L/D)$, where f is the friction factor, L is the length of straight pipe, and D is the flow diameter of the pipe. These data values were computed in the pipe data section, so this value is automatically computed by the spreadsheet.
   - For minor losses due to changes in the size of the flow path, refer to Sections 10.3–10.8 for values of K. It is essential that these values be entered for the

APPLIED FLUID MECHANICS		CLASS I SERIES SYSTEMS	
Objective: _Pump Power_		*Reference points for the energy equation:*	
Example Problem 11.1		Point 1: At surface of lower reservoir	
Figure 11.2		Point 2: At surface of upper reservoir	
System Data:	**SI Metric Units**		
Volume flow rate: Q =	*0.015 m³/s*	*Elevation at point 1 =* 0 m	
Pressure at point 1 =	*0 kPa*	*Elevation at point 2 =* 10 m	
Pressure at point 2 =	*0 kPa*	*If Ref. pt. is in pipe: Set v1 "= B20" OR Set v2 "= E20"*	
Velocity at point 1 =	*0 m/s →*	*Vel head at point 1 =* 0 m	
Velocity at point 2 =	*0 m/s →*	*Vel head at point 2 =* 0 m	
Fluid Properties:		*May need to compute $\nu = \eta/\rho$*	
Specific weight =	*7.74 kN/m³*	*Kinematic viscosity = 7.10E-07 m²/s*	
Pipe 1: DN 100 Schedule 40 steel pipe		**_Pipe 1: DN 80 Schedule 40 steel pipe_**	
Diameter: D = 0.1023 m		*Diameter: D =* 0.0525 m	
Wall roughness: ϵ = 4.60E-05 m		*Wall roughness: ϵ = 4.60E-05 m* [See Table 8.2]	
Length: L = 15 m		*Length: L =* 200 m	
Area: A = 8.22E-03 m²		Area: A = 2.16E-03 m² [A = πD²/4]	
D/ϵ = 2224		D/ϵ = 1141 Relative roughness	
L/D = 147		L/D = 3810	
Flow velocity = 1.82 m/s		Flow velocity = 6.93 m/s [v = Q/A]	
Velocity head = 0.170 m		Velocity head = 2.447 m [v²/2g]	
Reynolds No. = 2.63E+05		Reynolds No. = 5.13E+05 [$N_R = vD/\nu$]	
Friction factor: f = 0.0182		Friction factor: f = 0.0198 Using Eq. 8–7	

Energy Losses in Pipe 1:		Qty.		
Pipe: K_1 = f(L/D) =	2.67	1	Energy loss h_{L1} = 0.453 m Friction	
Entrance loss: K_2 =	0.50	1	Energy loss h_{L2} = 0.085 m	
Element 3: K_3 =	0.00	1	Energy loss h_{L3} = 0.000 m	
Element 4: K_4 =	0.00	1	Energy loss h_{L4} = 0.000 m	
Element 5: K_5 =	0.00	1	Energy loss h_{L5} = 0.000 m	
Element 6: K_6 =	0.00	1	Energy loss h_{L6} = 0.000 m	
Element 7: K_7 =	0.00	1	Energy loss h_{L7} = 0.000 m	
Element 8: K_8 =	0.00	1	Energy loss h_{L8} = 0.000 m	

Energy Losses in Pipe 2:		Qty.		
Pipe: K_1 = f(L/D) =	75.35	1	Energy loss h_{L1} = 184.40 m Friction	
Globe valve: K_2 =	6.46	1	Energy loss h_{L2} = 15.81 m	
2 std elbows: K_3 =	0.57	2	Energy loss h_{L3} = 2.79 m	
Exit loss: K_4 =	1.00	1	Energy loss h_{L4} = 2.45 m	
Element 5: K_5 =	0.00	1	Energy loss h_{L5} = 0.00 m	
Element 6: K_6 =	0.00	1	Energy loss h_{L6} = 0.00 m	
Element 7: K_7 =	0.00	1	Energy loss h_{L7} = 0.00 m	
Element 8: K_8 =	0.00	1	Energy loss h_{L8} = 0.00 m	

		Total energy loss $h_{L\,tot}$ =	205.98 m
	Results:	Total head on pump: h_A =	216.0 m
		Power added to fluid: P_A =	25.08 kW
		Pump efficiency =	*76.00 %*
		Power input to pump: P_I =	32.99 kW

FIGURE 11.3 Spreadsheet for Class I series pipeline systems. Data for Example Problem 11.1.

proper pipe. You must note which velocity is used as the reference velocity for the given type of minor loss. *K* factors for enlargements and contractions are based on the velocity head in the smaller pipe.

■ For minor losses due to valves, fittings, and bends, $K = f_T(L_e/D)$, where f_T is the friction factor in the fully turbulent zone for the size and type of pipe to which the element is connected. Table 10.5 is the source of such data for steel pipe. For other types of pipe or tubing, the method shown in Section 10.9 should be used. The relative roughness D/ε is used to find the value of *f* in the zone of complete turbulence from the Moody diagram. The values for the equivalent-length ratio L_e/D can be found in Table 10.4 or in Fig. 10.28.

9. The results are computed automatically at the bottom of the sheet. The total energy loss is the sum of all pipe friction and minor losses in both pipes.

10. The total head on the pump h_A is found by solving the general energy equation for that value:

$$h_A = \frac{p_2 - p_1}{\gamma} + (z_2 - z_1) + \frac{v_2^2 - v_1^2}{2g} + h_L$$

The spreadsheet makes the necessary calculations using data from appropriate cells in the upper part of the sheet.

11. The power added to the fluid is computed from

$$P_A = h_A \gamma Q$$

12. The pump efficiency e_M must be entered as a percentage.

13. The power input to the pump is computed from

$$P_I = P_A/e_M$$

Other types of Class I series pipeline problems can be analyzed in a similar manner by adjusting this form. Different sheets for different unit systems should be created because certain unit-specific constants, such as $g = 9.81$ m/s^2, are used in this version.

For example, if the objective of the problem is to compute the pressure at a particular upstream point A when the pressure is known at a downstream reference point B, the energy equation can be solved for the upstream pressure as

$$p_A = p_B + \gamma\left[(z_B - z_A) + \frac{v_B^2 - v_A^2}{2g} + h_L\right]$$

You must configure the spreadsheet to evaluate these terms as the final result. Note that it is assumed that no pump or fluid motor is in the system for this set of conditions.

11.4 CLASS II SYSTEMS

A Class II series pipeline system is one for which you desire to know the volume flow rate of the fluid that could be delivered by a given system. The system is completely described in terms of its elevations, pipe sizes, valves and fittings, and allowable pressure drop at key points in the system.

You know that pressure drop is directly related to the energy loss in the system and that the energy losses are typically proportional to the velocity head of the fluid as it flows through the system. Because velocity head is $v^2/2g$, the energy losses are proportional to the square of the velocity. Your task as the designer is to determine how high the velocity can be and still meet the goal of a limited pressure drop.

We will suggest three different approaches to designing Class II systems. They vary in their complexity and the degree of precision of the final result. The following list gives the type of system for which each method is used and a brief overview of the method. More details for each method are presented within Example Problems 11.2–11.4.

Method II-A   Used for a series system in which only pipe friction losses are considered, this direct solution process uses an equation, based on the work of Swamee and Jain (Reference 13), that includes the direct computation of the friction factor. This approach was introduced in Section 8.8. See Example Problem 11.2.

Method II-B   Used for a series system in which relatively small minor losses exist along with a relatively large pipe friction loss, this method adds steps to the process of Method II-A. Minor losses are initially neglected and the same equation used in Method II-A is used to estimate the allowable velocity and volume flow rate. Then a modestly lower volume flow rate is decided on, the minor losses are introduced, and the system is analyzed as a Class I system to determine the final performance at the specified flow rate. If the performance is satisfactory, the problem is finished. If not, different volume flow rates can be tried until satisfactory results are obtained. See the spreadsheet for Example Problem 11.3. This method requires some trial and error, but the process goes quickly once the data are entered into the spreadsheet.

Method II-C   Used for a series system in which minor losses are significant in comparison with the pipe friction losses and for which a high level of precision in the analysis is desired, this method is the most time-consuming, requiring an algebraic analysis of the behavior of the entire system and the expression of the velocity of flow in terms of the friction factor in the pipe. Both of these quantities are unknown because the friction factor also depends on velocity (Reynolds number). An *iteration* process is used to complete the analysis. Iteration is a controlled "trial-and-error" method in which each step of iteration yields a more accurate estimate of the limiting velocity of flow to meet the pressure drop limitation. The process typically converges in two to four iterations. See Example Problem 11.4.

Example Problem 11.2   A lubricating oil must be delivered through a horizontal DN 150 Schedule 40 steel pipe with a maximum pressure drop of 60 kPa per 100 m of pipe. The oil has a specific gravity of 0.88 and a dynamic viscosity of 9.5×10^{-3} Pa·s. Determine the maximum allowable volume flow rate of oil.

Solution   Figure 11.4 shows the system. This is a Class II series pipeline problem because the volume flow rate is unknown and, therefore, the velocity of flow is unknown. Method II-A is used here because only pipe friction losses exist in the system.

FIGURE 11.4 Reference points in the pipe for Example Problem 11.2.

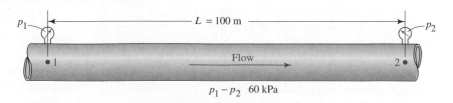

Step 1 Write the energy equation for the system.

Step 2 Solve for the limiting energy loss h_L.

Step 3 Determine the following values for the system:

Pipe flow diameter D

Relative roughness D/ε

Length of pipe L

Kinematic viscosity of the fluid ν; may require using $\nu = \eta/\rho$

Step 4 Use the following equation to compute the limiting volume flow rate, ensuring that all data are in the coherent units of the given system:

$$Q = -2.22D^2\sqrt{\frac{gDh_L}{L}}\log\left(\frac{1}{3.7D/\varepsilon} + \frac{1.784\nu}{D\sqrt{gDh_L/L}}\right) \tag{11–3}$$

Results We use points 1 and 2 shown in Fig. 11.3 to write the energy equation:

$$\frac{p_1}{\gamma} + z_1 + \frac{v_1^2}{2g} - h_L = \frac{p_2}{\gamma} + z_2 + \frac{v_2^2}{2g}$$

We can cancel some terms because in this problem, $z_1 = z_2$ and $v_1 = v_2$. The equation then becomes

$$\frac{p_1}{\gamma} - h_L = \frac{p_2}{\gamma}$$

Then we solve algebraically for h_L and evaluate the result:

$$h_L = \frac{p_1 - p_2}{\gamma} = \frac{60\text{ kN}}{\text{m}^2} \times \frac{\text{m}^3}{(0.88)(9.81\text{ kN})} = 6.95\text{ m}$$

Other data needed are:

Pipe flow diameter, $D = 0.1541$ m [Appendix F]

Pipe wall roughness, $\varepsilon = 4.6 \times 10^{-5}$m [Table 8.2]

Relative roughness, $D/\varepsilon = (0.1541\text{ m})/(4.6 \times 10^{-5}\text{ m}) = 3350$

Length of pipe, $L = 100$ m

Kinematic viscosity of the fluid; use

$$\rho = (0.88)(1000\text{ kg/m}^3) = 880\text{ kg/m}^3$$

Then

$$\nu = \eta/\rho = (9.5 \times 10^{-3}\text{ Pa·s})/(880\text{ kg/m}^3) - 1.08 \times 10^{-5}\text{ m}^2/\text{s}$$

We place these values into Eq. (11–3), ensuring that all data are in coherent SI units for this problem.

$$Q = -2.22(0.1541)^2\sqrt{\frac{(9.81)(0.1541)(6.95)}{100}}$$

$$\times \log\left[\frac{1}{(3.7)(3350)} + \frac{(1.784)(1.08 \times 10^{-5})}{(0.1541)\sqrt{(9.81)(0.1541)(6.95)/100}}\right]$$

$$Q = 0.057\text{ m}^3/\text{s}$$

Comment Thus, if the volume flow rate of oil through this pipe is no greater than 0.057 m³/s, the pressure drop over a 100-m length of the pipe will be no greater than 60 kPa.

Spreadsheet Solution for Method II-A Class II Series Pipeline Problems Figure 11.5 shows a simple spreadsheet to facilitate the calculations required for Method II-A. Its features are as follows.

1. The heading identifies the nature of the spreadsheet and allows the problem number or other description of the problem to be entered in the shaded area.

2. The system data consist of the pressures and elevations at two reference points. If a given problem gives the allowable difference in pressure Δp, you may assign the value for pressure at one point and then compute the pressure at the second point from $p_2 = p_1 + \Delta p$.

3. The energy loss is calculated in the spreadsheet using

$$h_L = (p_1 - p_2)/\gamma + z_1 - z_2$$

This is found from the energy equation, noting that the velocities are equal at the two reference points.

4. The fluid properties of specific weight and kinematic viscosity are entered.

FIGURE 11.5 Spreadsheet for Method II-A Class II series pipeline problems.

APPLIED FLUID MECHANICS		CLASS II SERIES SYSTEMS	
Objective: Volume Flow Rate		**Method II-A: No minor losses**	
Example Problem 11.2 Figure 11.4		Uses Eq. (11–3) to find maximum allowable volume flow rate to maintain desired pressure at point 2 for a given pressure at point 1	
System Data:	SI Metric Units		
Pressure at point 1 =	120 kPa	Elevation at point 1 =	0 m
Pressure at point 2 =	60 kPa	Elevation at point 2 =	0 m
Energy loss: h_L =	6.95 m		
Fluid Properties:		May need to compute $\nu = \eta/\rho$	
Specific weight =	8.63 kN/m^3	Kinematic viscosity = 1.08E-05 m^2/s	
Pipe Data: DN 150 Schedule 40 steel pipe			
Diameter: D =	0.1541 m		
Wall roughness: ϵ = 4.60E-05 m			
Length: L =	100 m	**Results: Maximum values**	
Area: A =	0.01865 m^2	Volume flow rate: Q = 0.0569 m^3/s	
D/ϵ =	3350	Velocity: v = 3.05 m/s	

5. Pipe data for flow diameter, roughness, and length are entered.

6. The spreadsheet completes the remaining calculations for area and relative roughness that are needed in Eq. (11–3).

7. The results are then computed using Eq. (11–3), and the maximum allowable volume flow rate and the corresponding velocity are shown at the bottom right of the spreadsheet. These values compare favorably with those found in Example Problem 11.2.

Spreadsheet for Solution Method II-B for Class II Series Pipeline Problems We use a new spreadsheet shown in Fig. 11.6 for solution Method II-B that is an extension of the spreadsheet for Method II-A. In fact, the first part of the spreadsheet is identical to Fig. 11.5 in which the allowable volume flow rate for a straight pipe with no minor losses is determined. Then a lower volume flow rate is assumed in the lower part of the spreadsheet that includes the effect of minor losses. Obviously, with minor losses

FIGURE 11.6 Spreadsheet for Method II-B Class II series pipeline problems.

APPLIED FLUID MECHANICS		CLASS II SERIES SYSTEMS	
Objective: Volume Flow Rate		**Method II–A: No minor losses**	
Example Problem 11.3 Figure. 11.7		Uses Eq. (11–3) to estimate the allowable volume flow rate to maintain desired pressure at point 2 for a given pressure at point 1	
System Data:	SI Metric Units		
Pressure at point 1 =	120 kPa	Elevation at point 1 =	0 m
Pressure at point 2 =	60 kPa	Elevation at point 2 =	0 m
Energy loss: h_L =	6.95 m		
Fluid Properties:		May need to compute $\nu = \eta/\rho$	
Specific weight =	8.63 kN/m^3	Kinematic viscosity = 1.08E-05 m^2/s	
Pipe Data: DN 150 Schedule 40 steel			
Diameter: D =	0.1541 m		
Wall roughness: ϵ = 4.60E-05 m			
Length: L =	100 m	**Results: Maximum values**	
Area: A =	0.01865 m^2	Volume flow rate: Q = 0.0569 m^3/s	
D/ϵ =	3350	Velocity: v = 3.05 m/s	

CLASS II SERIES SYSTEMS		Volume flow rate: Q =	0.0538 m^3/s	
Method II-B: Use results of Method II-A; include minor losses; then pressure at point 2 is computed		Given: Pressure p_1 =	120 kPa	
		Pressure p_2 =	**60.18 kPa**	
		NOTE: Should be >	**60 kPa**	
Additional Pipe Data:		**Adjust estimate for Q until p_2**		
L/D =	649	**is greater than desired pressure.**		
Flow velocity =	2.88 m/s	Velocity at point 1 =	2.88 m/s	→ If velocity is in pipe:
Velocity head =	0.424 m	Velocity at point 2 =	2.88 m/s	→ Enter "=B24"
Reynolds No. = 4.12E+04		Vel. head at point 1 =	0.424 m	
Friction factor: f =	0.0228	Vel. head at point 2 =	0.424 m	
Energy Losses in Pipe 1:	Qty.			
Pipe: K_1 = f(L/D) =	14.76	1	Energy loss h_{L1} =	6.26 m	Friction
2 std elbows: K_2 =	0.45	2	Energy loss h_{L2} =	0.38 m	
Butterfly valve: K_3 =	0.68	1	Energy loss h_{L3} =	0.29 m	
Element 4: K_4 =	0.00	1	Energy loss h_{L4} =	0.00 m	
Element 5: K_5 =	0.00	1	Energy loss h_{L5} =	0.00 m	
Element 6: K_6 =	0.00	1	Energy loss h_{L6} =	0.00 m	
Element 7: K_7 =	0.00	1	Energy loss h_{L7} =	0.00 m	
Element 8: K_8 =	0.00	1	Energy loss h_{L8} =	0.00 m	
			Total energy loss h_{Ltot} =	6.93 m	

added to the friction loss considered in Method II-A, a lower allowable volume flow rate will result. The method is inherently a two-step process, and more than one trial for the second step may be required.

To illustrate the use of Method II-B, we create the following new example problem. We take the same basic data from Example Problem 11.2 and add minor losses due to two standard elbows and a fully open butterfly valve.

Example Problem
11.3

A lubricating oil must be delivered through the piping system shown in Fig. 11.7 with a maximum pressure drop of 60 kPa between points 1 and 2. The oil has a specific gravity of 0.88 and a dynamic viscosity of 9.5×10^{-3} Pa·s. Determine the maximum allowable volume flow rate of oil.

FIGURE 11.7 Piping system for Example Problem 11.3

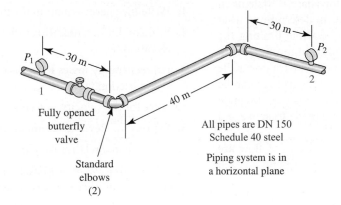

All pipes are DN 150
Schedule 40 steel

Piping system is in
a horizontal plane

Fully opened
butterfly
valve

Standard
elbows
(2)

Solution

The system is similar to that in Example Problem 11.2. The total length of DN 150 Schedule 40 steel pipe is 100 mm in a horizontal plane. The addition of the valve and the two elbows provide a moderate amount of energy loss that should be considered in addition to the friction losses in the pipes.

Initially, we ignore the minor losses and use Eq. (11–3) to compute a rough estimate of the allowable volume flow rate. This is accomplished in the upper part of the spreadsheet in Fig. 11.6, and it is identical to the solution shown in Fig. 11.5 for Example Problem 11.2. This is the starting point for Method II-B.

The features of the lower part of Fig. 11.6 are described next.

1. A revised estimate of the allowable volume flow rate Q is entered at the upper right, just under the computation of the initial estimate. The revised estimate must be lower than the initial estimate.

2. The spreadsheet then computes the "Additional Pipe Data" using the known pipe data from the upper part of the spreadsheet and the new estimated value for Q.

3. Note that at the middle right of the spreadsheet, the velocities at reference points 1 and 2 must be entered. If they are in the pipe, as they are in this problem, then the cell reference "=B24" can be entered because that is where the velocity in the pipe is computed. Other problems may have the reference points elsewhere, such as the surface of a reservoir where the velocity is zero. The appropriate value should then be entered in the shaded area.

4. Now the data for minor losses must be added in the section called "Energy Losses in Pipe 1." The K factor for the pipe friction loss is automatically computed from known data. The values for the other two K factors must be determined and entered in the shaded area in a manner similar to that used in the Class I spreadsheet. In this problem they are both dependent on the value of f_T for the DN 150 pipe. That value is 0.015 as found in Table 10.5.

   ■ Elbow (standard): $K = f_T(L_e/D) = (0.015)(30) = 0.45$
   ■ Butterfly valve: $K = f_T(L_e/D) = (0.015)(45) = 0.675$

5. The spreadsheet then computes the total energy loss and uses this value to compute the pressure at reference point 2. The equation is derived from the energy equation,

$$p_2 = p_1 + \gamma[z_1 - z_2 + v_1^2/2g - v_2^2/2g - h_L]$$

6. The computed value for p_2 must be larger than the desired value as entered in the upper part of the spreadsheet. This value is placed close to the assumed volume flow rate to give you a visual cue as to the acceptability of your current estimate for the limiting volume flow rate. Adjustments in the value of Q can then be quickly made until the pressure assumes an acceptable value.

Result   The spreadsheet in Fig. 11.6 shows that a volume flow rate of 0.0538 m³/s through the system in Fig. 11.7 will result in the pressure at point 2 being 60.18 kPa, slightly more than the minimum acceptable value.

Method II-C: Iteration Approach for Class II Series Pipeline Problems   Method II-C is presented here as a manual iteration process. It is used for Class II systems in which minor losses play a primary role in determining what the maximum volume flow rate can be while limiting the pressure drop in the system to a specified amount. As with all Class II systems except those for which pipe friction is the only significant loss, there are more unknowns than can be directly solved for. The process of iteration is used to guide you through the choices that need to be made to arrive at a satisfactory design or analysis.

Both the friction factor and the velocity of flow are unknown in a Class II system. Because they depend on each other, no direct solution is possible.

The iteration proceeds most efficiently if the problem is set up to facilitate the final cycle of estimating one unknown, the friction factor, to compute an approximate value of the other major unknown, the velocity of flow in the system. The procedure provides a means of checking the accuracy of the trial value of f and also indicates the new trial value to be used if an additional cycle is required. This is what distinguishes iteration from "trial and error," in which there are no discrete guidelines for subsequent trials.

The complete iteration process is illustrated within Example Problem 11.4. The following step-by-step procedure is used.

Solution Procedure for Class II Systems with One Pipe

1. Write the energy equation for the system.
2. Evaluate known quantities such as pressure heads and elevation heads.
3. Express energy losses in terms of the unknown velocity v and friction factor f.
4. Solve for the velocity in terms of f.
5. Express the Reynolds number in terms of the velocity.
6. Calculate the relative roughness D/ε.
7. Select a trial value of f based on the known D/ε and a Reynolds number in the turbulent range.
8. Calculate the velocity, using the equation from Step 4.
9. Calculate the Reynolds number from the equation in Step 5.
10. Evaluate the friction factor f for the Reynolds number from Step 9 and the known value of D/ε, using the Moody diagram, Fig. 8.7.
11. If the new value of f is different from the value used in Step 8, repeat Steps 8–11 using the new value of f.
12. If there is no significant change in f from the assumed value, then the velocity found in Step 8 is correct.

PROGRAMMED EXAMPLE PROBLEM

Example Problem 11.4   Water at 80°F is being supplied to an irrigation ditch from an elevated storage reservoir as shown in Fig. 11.8. Calculate the volume flow rate of water into the ditch.

Begin with Step 1 of the solution procedure by writing the energy equation. Use A and B as the reference points and simplify the equation as much as possible.

Compare this with your solution:

$$\frac{p_A}{\gamma} + z_A + \frac{v_A^2}{2g} - h_L = \frac{p_B}{\gamma} + z_B + \frac{v_B^2}{2g}$$

Because $p_A = p_B = 0$, and v_A is approximately zero, then

$$z_A - h_L = z_B + (v_B^2/2g)$$
$$z_A - z_B = (v_B^2/2g) + h_L \qquad\qquad (11\text{–}3)$$

Notice that the stream of water at point B has the same velocity as that inside the pipe.

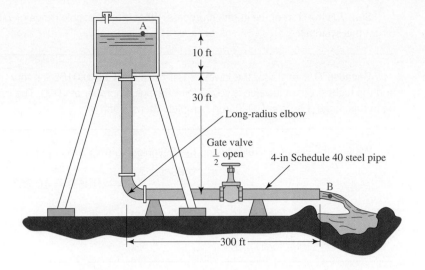

FIGURE 11.8 Pipeline system for Example Problem 11.4.

The elevation difference, $z_A - z_B$, is known to be 40 ft. However, the energy losses that make up h_L all depend on the unknown velocity, v_B. Thus, iteration is required. Do Step 3 of the solution procedure now.

There are four components of the total energy loss h_L:

$$h_L = h_1 + h_2 + h_3 + h_4$$

where

$$h_1 = 1.0(v_B^2/2g) \qquad \text{(entrance loss)}$$
$$h_2 = f(L/D)(v_B^2/2g) \qquad \text{(pipe friction loss)}$$
$$= f(330/0.3355)(v_B^2/2g)$$
$$= 985f(v_B^2/2g)$$
$$h_3 = f_T(L_e/D)(v_B^2/2g) \qquad \text{(long-radius elbow)}$$
$$= 20f_T(v_B^2/2g)$$
$$h_4 = f_T(L_e/D)(v_B^2/2g) \qquad \text{(half-open gate valve)}$$
$$= 160f_T(v_B^2/2g)$$

From Table 10.5, we find $f_T = 0.016$ for a 4-in steel pipe. Then we have

$$h_L = (1.0 + 985f + 20f_T + 160f_T)(v_B^2/2g)$$
$$= (3.88 + 985f)(v_B^2/2g) \tag{11-5}$$

Now substitute this expression for h_L into Eq. (11–4) and solve for v_B in terms of f.

You should have

$$v_B = \sqrt{2580/(4.88 + 985f)}$$

Now,

$$z_A - z_B = (v_B^2/2g) + h_L$$
$$40 \text{ ft} = (v_B^2/2g) + (3.88 + 985f)(v_B^2/2g)$$
$$= (4.88 + 985f)(v_B^2/2g)$$

Solving for v_B, we get

$$v_B = \sqrt{\frac{2g(40)}{4.88 + 985f}} = \sqrt{\frac{2580}{4.88 + 985f}} \tag{11-6}$$

Equation (11–6) represents the completion of Step 4 of the procedure. Now do Steps 5 and 6.

$$N_R = \frac{v_B D}{\nu} = \frac{v_B(0.3355)}{9.15 \times 10^{-6}} = (0.366 \times 10^5)v_B \tag{11-7}$$

$$D/\varepsilon = (0.3355/1.5 \times 10^{-4}) = 2235$$

Step 7 is the start of the iteration process. What is the possible range of values for the friction factor for this system?

Because $D/\varepsilon = 2235$, the lowest possible value of f is 0.0155 for very high Reynolds numbers and the highest possible value is 0.039 for a Reynolds number of 4000. The initial trial value of f must be in this range. Use $f = 0.020$ and complete Steps 8 and 9.

We find the values for velocity and the Reynolds number by using Eqs. (11–6) and (11–7):

$$v_B = \sqrt{\frac{2580}{4.88 + (985)(0.02)}} = \sqrt{105.0} = 10.25 \text{ ft/s}$$

$$N_R = (0.366 \times 10^5)(10.25) = 3.75 \times 10^5$$

Now do Step 10.

You should have $f = 0.0175$. Because this is different from the initial trial value of f, Steps 8–11 must be repeated now.

Using $f = 0.0175$, we get

$$v_B = \sqrt{\frac{2580}{4.88 + (985)(0.0175)}} = \sqrt{116.6} = 10.8 \text{ ft/s}$$

$$N_R = (0.366 \times 10^5)(10.8) = 3.94 \times 10^5$$

The new value of f is 0.0175, which is unchanged, and the computed value for v_B is correct. Therefore, we have

$$v_B = 10.8 \text{ ft/s}$$

$$Q = A_B v_B = (0.0884 \text{ ft}^2)(10.8 \text{ ft/s}) = 0.955 \text{ ft}^3/\text{s}$$

This programmed example problem is concluded.

11.5 CLASS III SYSTEMS

A Class III series pipeline system is one for which you desire to know the size of pipe that will carry a given volume flow rate of a given fluid with a specified maximum pressure drop due to energy losses.

You can use a similar logic to that used to discuss Class II series pipeline systems to plan an approach to designing Class III systems. You know that pressure drop is directly related to the energy loss in the system and that the energy losses are typically proportional to the velocity head of the fluid as it flows through the system. Because velocity head is $v^2/2g$, the energy losses are proportional to the square of the velocity. Velocity is, in turn, inversely proportional to the flow area found from

$$A = \pi D^2/4$$

Therefore, the energy loss is inversely proportional to the flow diameter *to the fourth power*. The size of the pipe is a major factor in how much energy loss occurs in a pipeline system. Your task as the designer is to determine how small the pipe can be and still meet the goal of a limited pressure drop. You don't want to use an unreasonably large pipe

because the cost of piping increases with increasing size. If the size of the pipe is too small, however, the energy wasted by excessive energy losses would generate a high operating cost for the life of the system. You should consider the total life-cycle cost.

We suggest two different approaches to designing Class III systems.

Method III-A This simplified approach considers only energy loss due to friction in the pipe. We assume that the reference points for the energy equation are in the pipe to be designed and at a set distance apart. There may be an elevation difference between the two points. Because the flow diameter is the same at the two reference points, however, there is no difference in the velocities or the velocity heads. We can write the energy equation and then solve for the energy loss,

$$\frac{p_1}{\gamma} + z_1 + \frac{v_1^2}{2g} - h_L = \frac{p_2}{\gamma} + z_2 + \frac{v_2^2}{2g}$$

But $v_1 = v_2$. Then, we have

$$h_L = \frac{p_1 - p_2}{\gamma} + z_1 - z_2$$

This value, along with other system data, can be entered into the following design equation (see References 12 and 13):

$$D = 0.66\left[\varepsilon^{1.25}\left(\frac{LQ^2}{gh_L}\right)^{4.75} + \nu Q^{9.4}\left(\frac{L}{gh_L}\right)^{5.2}\right]^{0.04} \quad (11\text{–}8)$$

The result is the smallest flow diameter that can be used for a pipe to limit the pressure drop to the desired value. Normally, you will specify a standard pipe or tube that has an inside diameter (ID) just larger than this limiting value.

Example Problem 11.5   Compute the required size of new clean Schedule 40 pipe that will carry 0.50 ft³/s of water at 60°F and limit the pressure drop to 2.00 psi over a length of 100 ft of horizontal pipe.

Solution   We first calculate the limiting energy loss. Note that the elevation difference is zero. Write

$$h_L = (p_1 - p_2)/\gamma + (z_1 - z_2) = (2.00\ \text{lb/in}^2)(144\ \text{in}^2/\text{ft}^2)/(62.4\ \text{lb/ft}^3) + 0 = 4.62\ \text{ft}$$

The following data are needed in Eq. (11–8):

$$Q = 0.50\ \text{ft}^3/\text{s} \quad L = 100\ \text{ft} \quad g = 32.2\ \text{ft/s}^2$$
$$h_L = 4.62\ \text{ft} \quad \varepsilon = 1.5 \times 10^{-4}\ \text{ft} \quad \nu = 1.21 \times 10^{-5}\ \text{ft}^2/\text{s}$$

Now we can enter these data into Eq. (11–8):

$$D = 0.66\left[(1.5 \times 10^{-4})^{1.25}\left[\frac{(100)(0.50)^2}{(32.2)(4.62)}\right]^{4.75} + (1.21 \times 10^{-5})(0.50)^{9.4}\left[\frac{100}{(32.2)(4.62)}\right]^{5.2}\right]^{0.04}$$

$$D = 0.309\ \text{ft}$$

The result shows that the pipe should be larger than $D = 0.309$ ft. The next larger standard pipe size is a 4-in Schedule 40 steel pipe having an ID of $D = 0.3355$ ft.

Spreadsheet for Completing Method III-A for Class III Series Pipeline Problems   Obviously, Eq. (11–8) is cumbersome to evaluate, and the opportunity for calculation error is great. The use of a spreadsheet to perform the calculation alleviates this problem.

Figure 11.9 shows an example of such a spreadsheet. Its features are as follows.

- The problem identification and given data are listed to the left side. Where the allowable pressure drop Δp is given, as it is in Example Problem 11.5, we specify an arbitrary value for the pressure at point 2 and then set the pressure at point 2 as

$$p_2 = p_1 + \Delta p$$

- Note that the spreadsheet computes the allowable energy loss h_L using the method shown in the solution of Example Problem 11.5.

- The fluid properties data are entered at the upper right side of the spreadsheet.

- The intermediate results are reported simply for reference. They represent factors from Eq. (11–8) and can be used by those solving the equation manually as a check on their calculation procedure. If you prepare a spreadsheet yourself, you should carefully verify the form of the equation that solves Eq. (11–8) because the programming is complex. Breaking it up into parts can simplify the final equation.

APPLIED FLUID MECHANICS		CLASS III SERIES SYSTEMS		
Objective: Minimum Pipe Diameter		**Method III-A:** *Uses Eq. (11–8) to compute the*		
Example Problem 11.5		*minimum size of pipe of a given length*		
		that will flow a given volume flow rate of fluid		
System Data:	**SI Metric Units**	*with a limited pressure drop (no minor losses)*		
Pressure at point 1 =	*102* psig	**Fluid Properties:**		
Pressure at point 2 =	*100* psig	*Specific weight =*	*62.4* lb/ft³	
Elevation at point 1 =	*0* ft	*Kinematic viscosity = 1.21E-05* ft²/s		
Elevation at point 2 =	*0* ft	**Intermediate Results in Eq. (11–8):**		
Allowable Energy Loss: h_L =	*4.62* ft	*L/gh_L = 0.672878*		
Volume flow rate: Q =	*0.5* ft³/s	Argument in bracket: 5.77E-09		
Length of pipe: L =	*100* ft	**Final Minimum Diameter:**		
Pipe wall roughness: ϵ =	*1.50E-04* ft	Minimum diameter: D = 0.3090 ft		

FIGURE 11.9   Spreadsheet for Method III-A for Class III series pipeline problems.

APPLIED FLUID MECHANICS		CLASS III SERIES PIPE LINE SYSTEMS
Objective: Minimum Pipe Diameter		**Method III-A:** *Uses Eq. (11–8) to compute the*
Example Problem 11.6		*minimum size of pipe of a given length*
		that will flow a given volume flow rate of fluid
System Data:	**SI Metric Units**	*with a limited pressure drop (no minor losses)*
Pressure at point 1 =	102 psig	**Fluid Properties:**
Pressure at point 2 =	100 psig	Specific weight = 62.4 lb/ft^3
Elevation at point 1 =	0 ft	Kinematic Viscosity = 1.21E-05 ft^2/s
Elevation at point 2 =	0 ft	**Intermediate Results in Eq. (11–8):**
Allowable Energy Loss: h_L =	4.62 ft	L/gh_L = 0.672878
Volume flow rate: Q =	0.5 ft^3/s	Argument in bracket: 5.77E-09
Length of pipe: L =	100 ft	**Final Minimum Diameter:**
Pipe wall roughness: ϵ = 1.50E-04 ft		Minimum diameter: D = 0.3090 ft

CLASS III SERIES SYSTEMS			Specified pipe diameter: D =	0.3355 ft
Method III-B: Use results of Method III-A;			4-in schedule 40 steel pipe	
specify actual diameter; include minor losses;			**If velocity is in the pipe, enter "=B23" for value**	
then pressure at point 2 is computed			Velocity at point 1 =	5.66 ft/s
Additional Pipe Data:			Velocity at point 2 =	5.66 ft/s
Flow area: A =	0.08840 ft^2		Vel. head at point 1 =	0.497 ft
Relative roughness: D/ϵ =	2237		Vel. head at point 2 =	0.497 ft
L/D =	298		**Results:**	
Flow velocity =	5.66 ft/s		Given pressure at point 1 =	102 psig
Velocity head =	0.497 ft		Desired pressure at point 2 =	100 psig
Reynolds No. =	1.57E+05		**Actual pressure at point 2 =**	100.48 psig
Friction factor: f =	0.0191		**(Compare actual with desired pressure at point 2)**	
Energy Losses in Pipe:		Qty.		
Pipe friction: K_1 = f(L/D) =	5.70	1	Energy loss h_{L1} =	2.83 ft
Two long rad. elbows: K_2 =	0.32	2	Energy loss h_{L2} =	0.32 ft
Butterfly valve: K_3 =	0.72	1	Energy loss h_{L3} =	0.36 ft
Element 4: K_4 =	0.00	1	Energy loss h_{L4} =	0.00 ft
Element 5: K_5 =	0.00	1	Energy loss h_{L5} =	0.00 ft
Element 6: K_6 =	0.00	1	Energy loss h_{L6} =	0.00 ft
Element 7: K_7 =	0.00	1	Energy loss h_{L7} =	0.00 ft
Element 8: K_8 =	0.00	1	Energy loss h_{L8} =	0.00 ft
			Total energy loss h_{Ltot} =	3.51 ft

FIGURE 11.10 Spreadsheet for Method III-B for Class III series pipeline problems.

- The Final Minimum Diameter is the result of the calculation from Eq. (11–8) and it represents the minimum acceptable size of pipe to carry the given volume flow rate with the stated limitation on pressure drop.

Method III-B When minor losses are to be considered, a modest extension of Method III-A can be used. The standard pipe size selected as a result of Method III-A is normally somewhat larger than the minimum allowable diameter. Therefore, modest additional energy losses due to a few minor losses will likely not produce a total pressure drop greater than that allowed. The selected pipe size will probably still be acceptable.

After making a tentative specification of pipe size, we can add the minor losses to the analysis and examine the resulting pressure at the end of the system to ensure that it is within the desired limits. If not, a simple adjustment to the next larger size pipe will almost surely produce an acceptable design. Implementing this procedure using a spreadsheet makes the calculations very rapid.

Figure 11.10 shows a spreadsheet that implements this design philosophy. It is actually a combination of two spreadsheets already described in this chapter. The upper

part is identical to Fig. 11.9, which was used for solving Example Problem 11.5 using Method III-A. From that we gain an estimate of the size of pipe that will carry the desired amount of fluid without any minor losses.

The lower part of the spreadsheet uses a technique similar to that in Fig. 11.3 for Class I series pipeline problems. It is simplified to include only one pipe size. Its objective is to compute the pressure at point 2 in a system when the pressure at point 1 is given. Minor losses are included.

The following procedure illustrates the use of this spreadsheet.

Spreadsheet for Method III-B Class III Series Pipeline Problems with Minor Losses

- Initially ignore the minor losses and use the upper part of the spreadsheet to estimate the size of pipe required to carry the given flow rate with less than the allowable pressure drop. This is identical to Method III-A described in the preceding example problem.

- Enter the next larger standard pipe size at the upper right part of the lower spreadsheet in the cell called "Specified pipe diameter: D."

- The spreadsheet automatically computes the values under **Additional Pipe Data.**

- The velocities listed in the right column are usually in the pipe being analyzed and are usually equal. The reference to cell **B23** will automatically enter the computed velocity from the pipe data. However, if the system being analyzed has a reference point outside the pipe, the actual velocity there must be entered. Then the velocity heads at the reference points are calculated.

- The section headed **Energy Losses in Pipe** requires you to enter the resistance factors K for each minor loss, as was done in earlier spreadsheet solution procedures. The K factor for pipe friction loss is computed automatically from the pipe data.

- The **Results** section lists the given pressure at point 1 and the desired pressure at point 2 taken from the initial data at the top of the spreadsheet. The **Actual pressure at point 2** is computed from an equation derived from the energy equation

$$p_2 = p_1 - \gamma(z_1 - z_2 + v_1^2/2g - v_2^2/2g - h_L)$$

- You as the designer of the system must compare the actual pressure at point 2 with the listed desired pressure.

- If the actual pressure is greater than the desired pressure, you have a satisfactory result, and the pipe size specified is acceptable.

- If the actual pressure is less than the desired pressure, simply pick the next larger standard pipe size and repeat the spreadsheet calculations. This step is virtually immediate because all calculations are automatic once the new pipe flow diameter is entered.

- Unless there are many high-loss minor losses, this size pipe should be acceptable. If not, continue to specify larger pipes until a satisfactory solution is achieved. Also, examine the magnitude of the energy losses contributed by the minor losses. You may be able to use a smaller pipe size if you change valves and fittings to more efficient, lower-loss designs.

The example problem that follows illustrates the use of this spreadsheet.

Example Problem 11.6   Extend the situation described in Example Problem 11.5 by adding a fully open butterfly valve and two long-radius elbows to the 100 ft of straight pipe. Will the 4-inch Schedule 40 steel pipe size selected limit the pressure drop to 2.00 psi with these minor losses added?

Solution   To simulate the desired pressure drop of 2.00 psi, we have set the pressure at point 1 to be 102 psig. Then we examine the resulting value of the pressure at point 2 to see that it is at or greater than 100 psig.

The spreadsheet in Fig. 11.10 shows the calculations. For each minor loss, a resistance factor K is computed as defined in Chapters 8 and 10. For the pipe friction loss,

$$K_1 = f(L/D)$$

and the friction factor f is computed by the spreadsheet using Eq. (8–7).

For the elbows and the butterfly valve, the method of Chapter 10 is applied. Write

$$K = f_T(L_e/D)$$

The values of (L_e/D) and f_T are found from Tables 10.4 and 10.5, respectively.

Result   The result shows that the pressure at point 2 at the end of the system is 100.48 psig. Thus, the design is satisfactory. Note that the energy loss due to pipe friction is 2.83 ft out of the total energy loss of 3.51 ft. The elbows and the valve contribute truly minor losses.

11.6 PIPE-FLO® EXAMPLES FOR SERIES PIPELINE SYSTEMS

The PIPE-FLO® problems presented in earlier chapters have relied on fluid pressure alone to produce the flow in the system and included only energy losses in steel piping and through single valves or fittings. In this section, two new applications of PIPE-FLO® are addressed. One uses gravity to drive the flow through a copper tube and the other one uses a pump to deliver a desired flow rate of fluid between two tanks. In each case, several minor losses are included in the flow path.

The following example problem is a guide to the use of copper tubing in PIPE-FLO® and the system uses strictly elevation head as the input energy. The problem is an analysis of just one section of a total system, a procedure that is commonly needed in larger systems.

Example Problem 11.7

Gravity flow through a copper tube

Water, at 10°C, flows from a large reservoir at the rate of 1.5×10^{-2} m³/s through the system shown in Fig. 11.11. Use PIPE-FLO® to calculate the pressure in the tube at point B.

FIGURE 11.11 Gravity flow through a copper tube for Example Problem 11.7.

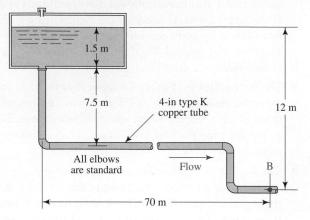

Solution

1. Begin by using the "system" menu in PIPE-FLO® to establish units and fluid zones as defined by the problem.

2. For the pipe specification, choose the "Copper Tube H23" drop down menu (SYSTEM/SPECIFICATIONS/NEW). Double click and select "K" for type K copper tubing similar to the way that "Schedule 40" was selected for steel pipe. After choosing Type K, verify the roughness value that appears on the right side of the pipe properties box, understanding that these values are good checkpoints, and also that they can be manually adjusted.

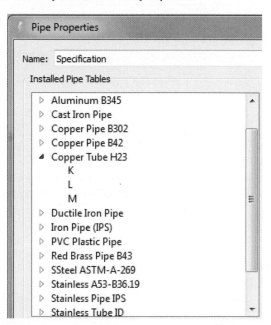

3. Next place a tank on the FLO-Sheet® and set up its variables in the properties grid. Also, draw the pipe in the relative shape as shown in Fig. 11.11. Be sure to include the three elbows and the pipe entrance under the "Valves and Fittings" category in the properties grid for the pipe.

4. Under "basic devices" select "flow demand" and insert it at point B to represent the point at which this section connects to the rest of the system. Specify the flow rate past this point, which is simply the total flow rate given for the problem. Independent analysis of just this section is possible with this approach.

5. After all components have been entered, choose "CALCULATE" to display all the answers required in the problem statement. Remember, you may have to turn on the values shown for each component as done in previous examples through the "Device View Options" menu on the right.

6. The results and the overall set-up of the system are given below. The pressure in the pipe at point B is 86.76 kPa gage. It is a positive gage pressure because the pipe is 12 m below the water surface in the open tank. However, the energy losses as the water flows out of the tank and through

the piping and fittings causes the pressure to be somewhat lower. Note that the pressure drop through the pipe is 72.05 kPa.

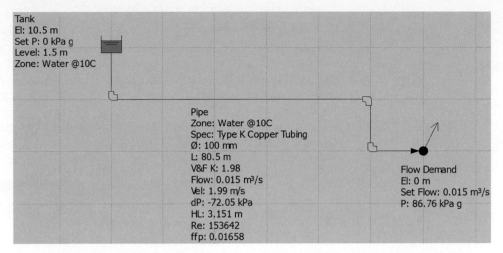

Tank
El: 10.5 m
Set P: 0 kPa g
Level: 1.5 m
Zone: Water @10C

Pipe
Zone: Water @10C
Spec: Type K Copper Tubing
Ø: 100 mm
L: 80.5 m
V&F K: 1.98
Flow: 0.015 m³/s
Vel: 1.99 m/s
dP: -72.05 kPa
HL: 3.151 m
Re: 153642
ffp: 0.01658

Flow Demand
El: 0 m
Set Flow: 0.015 m³/s
P: 86.76 kPa g

It is recommended that you also work this problem by hand and verify your answers with the PIPE-FLO® results shown here. (*Note*: This is the same problem as Practice Problem 11.1 at the end of the chapter.)

Most industrial systems use a pump to drive the flow. PIPE-FLO® is capable of advanced pump calculations using specific commercially available pumps. The software can also do simpler analysis, though that allows for sizing a theoretical pump for quick analysis and design. This *sizing pump* feature allows the user to input a volumetric flow rate for the pump along with the elevation of the suction and discharge ports and the nature of the piping in the system.

The following example problem offers a guide for analyzing flow through a series piping system that uses a pump, reporting just its most general parameters of head added, flow rate, suction pressure, and discharge pressure. Chapter 13 presents problems that illustrate the more detailed capability of the PIPE-FLO® software with more advanced terms than are introduced there.

Example Problem 11.8

Using a sizing-pump in PIPE-FLO®

The pump illustrated in Fig. 11.12 delivers water from the lower reservoir to the upper reservoir at a rate of 2.0 ft³/s. Both the suction and discharge pipes are 6-in Schedule 40 steel pipe. The length of the suction pipe leading to the pump is 12 ft, and 24 ft of discharge pipe extend from the pump outlet to the upper tank. There are three standard 90° elbows and a fully open gate valve. The depth of the fluid level inside the lower tank is 10 ft. Use PIPE-FLO® to calculate (a) the pressure at the pump inlet, (b) the pressure at the pump outlet, and (c) the total head on the pump.

FIGURE 11.12 Pumped fluid flow system for Example Problem 11.8.

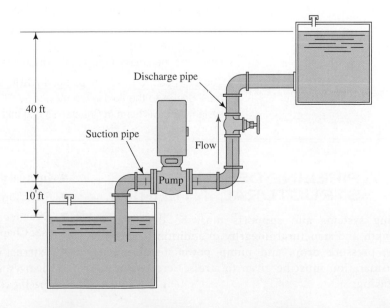

Solution

1. Open a new project in PIPE-FLO® and select the "SYSTEM" menu on the toolbar to initialize all key data such as units, fluid zones, and pipe specifications the same way as in the previous example problems using PIPE-FLO® in previous chapters.

2. Start by placing the tanks in the problem, and filling in their initial data in the property grid.

3. To insert a sizing pump, choose "sizing pump" from the "pumps" menu in the tool box and place it on the FLO-Sheet®. As stated earlier, the only data requested from the user are the suction and discharge elevation to define the relative position of the pump, and the desired volumetric flow rate. This flow rate can be specified in any form of a volumetric flow rate by double clicking the ". . ." button that appears on the right after selecting the value box for the flow rate. For this specific example, the volume flow rate was given as 2.0 ft³/s. Since the bottom of the lower tank was set at an elevation of 0 ft and the fluid level in the tank was given to be 10 ft, the elevation of the pump suction and discharge would be 20 ft since the pump is shown as being horizontal.

Sizing Pump Design	
Name	Sizing Pump
Suction Elevation	20 ft
Discharge Elevation	20 ft
Flow Rate	2 ft³/s

4. Continue the problem by adding in the pipe sections, elbows, pipe entrance, pipe exit, and fully open gate valve in the same manner as presented in past example problems.

5. After all components and information have been entered, be sure to turn on the information to be displayed for each component under the "Device View Options" in the property grid.

6. For the pump in this problem, it is requested that we show total head, suction pressure, and discharge pressure. In Chapter 13, other factors will be called for. The results from the problem are shown below.

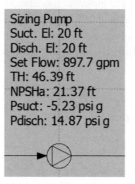

Sizing Pump
Suct. El: 20 ft
Disch. El: 20 ft
Set Flow: 897.7 gpm
TH: 46.39 ft
NPSHa: 21.37 ft
Psuct: -5.23 psi g
Pdisch: 14.87 psi g

7. Note the following output values:

   a. There is a negative gage pressure at the inlet of the pump, −5.23 psig, because the pump must raise the fluid 10 ft from the surface of the lower tank to the suction inlet port of the pump and overcome energy losses in the suction pipe.

   b. The pressure at the pump outlet is positive, 14.87 psig. The result is that the pump increased the pressure of the fluid by 20.1 psi which is necessary to elevate fluid to the upper tank and to overcome the energy losses in the discharge pipe, the elbows, and the valve.

   c. The total head value for the pump, 46.39 ft, is a measure of the amount of energy that the pump must deliver to the fluid when moving 2.0 ft³/s of water between the tanks for the given piping system. You will learn in Chapter 13 that this is an essential data point for pump selection.

11.7 PIPELINE DESIGN FOR STRUCTURAL INTEGRITY

Piping systems and supports must be designed for strength and structural integrity in addition to meeting flow, pressure drop, and pump power requirements. Consideration must be given to stresses created by the following:

- Internal pressure
- Static forces due to the weight of the piping and the fluid
- Dynamic forces created by moving fluids inside the pipe (see Chapter 16)
- External loads caused by seismic activity, temperature changes, installation procedures, or other application-specific conditions

The American Society of Mechanical Engineers (ASME), the American Water Works Association (AWWA), the National Fire Protection Association (NFPA), and others develop standards for such considerations. See References 1–17 and Internet resources 2–10.

Other details and practical considerations of piping system design are discussed in References 3 and 6–11 and in the various Internet resources listed at the end of the chapter.

Structural integrity evaluation should consider pipe stress due to internal pressure, static loads due to the weight of the pipe and its contents, wind loads, installation processes, thermal expansion and contraction, hydraulic transients such as water hammer caused by rapid valve actuation, long-term degradation of piping due to corrosion or erosion, pressure cycling, external loads and reactions at connections to other equipment, impact loads, dynamic performance in response to seismic events, flow-induced vibration, and vibration caused by other structures or equipment.

Careful selection of piping materials must consider strength at operating temperatures, ductility, toughness, impact resistance, resistance to ultraviolet radiation from sunlight, compatibility with the flowing fluid, atmospheric environment around the installation, paint or other corrosion-protection coatings, insulation, fabrication of pipe connections, and installation of valves, fittings, pressure gages, and flow measurement devices.

The nominal size of the pipe or tubing is typically determined from flow considerations as outlined in this chapter. The pressure class (a function of wall thickness) is based on calculations considering internal pressure, allowable stress of the pipe material at operating temperature, the actual wall thickness of the pipe, tolerances on wall thickness, method of fabrication of the pipe, allowance for long-term corrosion, and a wall thickness correction factor. The following equations are taken from Reference 1, and you are advised to consult that document for details and pertinent data. Reference 14 gives some discussion of the use of these equations along with example problems. These equations are based on the classic tangential (hoop) stress analysis for thin-walled cylinders with internal pressure.

Basic Wall Thickness Calculation:

$$t = \frac{pD}{2(SE + pY)} \tag{11–9}$$

where

- t = Basic wall thickness (in or mm)
- p = Design pressure [psig or Pa(gage)]
- D = Pipe outside diameter (in or mm)
- S = Allowable stress in tension (psi or MPa)
- E = Longitudinal joint quality factor
- Y = Correction factor based on material type and temperature

Careful attention to unit consistency must be exercised.

Values for allowable stresses for a variety of metals at temperatures from 100°F to 1500°F (38°C to 816°C) are listed in Reference 1. For example, for carbon steel pipe

(ASTM A106), $S = 20.0$ ksi (138 MPa) for temperatures up to 400°F (204°C).

The value of E depends on how the pipe is made. For example, for seamless steel and nickel alloy pipe, $E = 1.00$. For electric resistance welded steel pipe, $E = 0.85$. For welded nickel alloy pipe, $E = 0.80$.

The value of Y is 0.40 for steel, nickel alloys, and nonferrous metals at temperatures of 900°F and lower. It ranges as high as 0.70 for higher temperatures.

The basic wall thickness must be adjusted as follows:

$$t_{min} = t + A \tag{11–10}$$

where A is a corrosion allowance based on the chemical properties of the pipe and the fluid and the design life of the piping. One value sometimes used is 2 mm or 0.08 in.

Commercial piping is typically produced to a tolerance of $+0/-12.5\%$ on the wall thickness. Therefore, the nominal minimum wall thickness is computed from

$$t_{nom} = t_{min}/(1 - 0.125) = t_{min}/(0.875)$$
$$= 1.143 t_{min} \tag{11–11}$$

Combining Eqs. (11–9)–(11–11) gives

$$t_{nom} = 1.143\left[\frac{pD}{2(SE + pY)} + A\right] \tag{11–12}$$

Stresses Due to Piping Installation and Operation

External stresses on piping combine with the hoop and longitudinal stresses created by the internal fluid pressure. Horizontal spans of piping between supports are subjected to tensile and compressive bending stresses due to the weight of the pipe and the fluid. Vertical lengths of pipe experience tensile or compressive stresses depending on the manner of support. Torsional shear stresses in one pipe can be created by offset branches of the piping layout that exert twisting moments about the axis of the pipe. Most of these stresses are static or mildly varying for only a moderate number of cycles. However, frequent pressure or temperature cycling, machine vibration, or flow-induced vibration can create cyclical stresses that may cause fatigue failures.

You should carefully design the supports for the piping system to minimize external stresses and to obtain a balance between constraining the pipe and allowing for expansion and contraction due to pressure and temperature changes. Pumps, large valves, and other critical equipment are typically supported directly under the body or at their inlet and outlet connections. Piping can be supported on saddle-type supports that carry loads to the ground or to firm structural members. Some supports are fixed to the pipe, whereas others contain rollers to allow the pipe to move during expansion and contraction. Supports should be placed at regular intervals so that the spans are of moderate length, limiting bending stresses and deflections. Some designers limit bending deflection to no more than 0.10 in (2.5 mm) between support points. Elevated piping can be supported by hangers attached to overhead beams or roof structures. Some hangers contain springs to permit movement of the piping due to transient conditions while maintaining fairly equal forces in the pipe. In some installations, electrical isolation of the piping may be required. Internet resources 7 and 8 show a variety of commercially available clamps, hangers, and supports.

Finally, after the piping is installed it must be cleaned and pressure tested, typically using hydrostatic pressure at approximately 1.5 times the design pressure. Periodic testing should be done to ensure that no critical leaks or pipe failures occur over time.

REFERENCES

1. American Society of Mechanical Engineers. 2012. *ASME B31.3, Process Piping Code.* New York: Author.

2. Becht, Charles, IV. 2009. *Process Piping: The Complete Guide to ASME B31.3,* 3rd ed. New York: ASME Press.

3. Willoughby, David. 2009. *Plastic Piping Handbook.* New York: McGraw-Hill.

4. Crane Co. 2011. *Flow of Fluids through Valves, Fittings, and Pipe* (Technical Paper No. 410). Stamford, CT: Author.

5. Hardy, Ray T., and Jeffrey L. Sines. 2012. *Piping Systems Fundamentals,* 2nd ed. Lacey, WA: ESI Press, Engineered Software, Inc.

6. Heald, C. C., Ed. 2002. *Cameron Hydraulic Data,* 19th ed. Irving, TX: Flowserve, Inc.

7. Lin, Shun Dar, and C. C. Lee. 2007. *Water and Wastewater Calculations Manual,* 2nd ed. New York: McGraw-Hill.

8. Mohitpour, M., H. Golshan, and A. Murray. 2007. *Pipeline Design and Construction: A Practical Approach,* 3rd ed. New York: ASME Press.

9. Nayyar, Mohinder. 2002. *Piping Databook.* New York: McGraw-Hill.

10. Silowash, Brian. 2010. *Piping Systems Manual.* New York: McGraw-Hill.

11. Nayyar, Mohinder. 2000. *Piping Handbook,* 7th ed. New York: McGraw-Hill.

12. Pritchard, Philip J. 2011. *Fox and McDonald's Introduction to Fluid Mechanics,* 8th ed. New York: John Wiley & Sons.

13. Swamee, P. K., and A. K. Jain. 1976. Explicit Equations for Pipe-flow Problems. *Journal of the Hydraulics Division* 102(HY5): 657–664. New York: American Society of Civil Engineers.

14. U.S. Army Corps of Engineers. 1999. *Liquid Process Piping* (Engineer Manual 1110-1-4008). Washington, DC: Author. (See Internet resource 2.)

15. Frankel, Michael. 2009. *Facility Piping Systems Handbook,* 3rd ed. New York: McGraw-Hill.

16. Smith, Peter. 2007. *Fundamentals of Piping Design,* Vol. I. Houston, TX: Gulf Publishing Co.

17. Boterman, Rutger, and Peter Smith. 2008. *Advanced Piping Design,* Vol. II. Houston, TX: Gulf Publishing Co.

INTERNET RESOURCES

1. **Engineered Software, Inc. (ESI):** www.eng-software.com Developer of the PIPE-FLO® fluid flow analysis software to design, optimize, and troubleshoot fluid piping systems, as demonstrated in this book. A special demonstration version of PIPE-FLO® created for this book can be accessed by users of this book at http://www.eng-software.com/appliedfluidmechanics

2. **The Piping Designers.com:** A site containing data and basic information for piping system design. It includes data for piping dimensions, piping fittings, CAD tools, flanges, piping standards, and many other related topics. From the *Pipes &Piping Tools* page, the entire document listed as Reference 14 can be read or downloaded. Numerous links to commercial suppliers of piping, fittings, pumps, and valves are listed on the site.

3. **PipingDesigners.com:** This site contains the document, "Overview of Process Plant Piping System Design," an excellent set of 133 presentation slides created by Vincent A. Carucci, of Carmagen Engineering, Inc. Use the search box on the home page and search on *Process Plant Piping.* The presentation is made available by ASME International as part of its ASME Career Development Series.

4. **National Fire Protection Association:** Developer and publisher of codes and standards for fire protection including NFPA 13, *Standard for the Installation of Sprinkler Systems.* Also, publishes other references pertinent to the study of fluid mechanics, such as *The Fire Pump Handbook.*

5. **Ultimate Fire Sprinkler Guide:** A huge set of links to sources for components for fire sprinkler systems and similar industrial pumped piping systems. Included are pipe, fittings, pumps, valves, and many others.

6. **Anvil International:** Manufacturer of pipe fittings, pipe hangers, and supports. Site includes an extensive amount of design information on pipe hangers, sizes and weights of pipe, seismic effects, and thermal considerations.

7. **Cooper B-Line:** Manufacturer of pipe hangers, anchor systems, and supports for piping and electrical cables.

8. **eCompressedair:** From the bottom of the home page, select *Main Library,* then select *Piping Systems* for an extensive set of documents giving guidelines for the design and installation of piping for compressed air systems for industrial applications.

9. **American Water Works Association:** An international non-profit scientific and educational society dedicated to the improvement of drinking water quality and supply. It is the authoritative resource for knowledge, information, and advocacy for improving the quality and supply of drinking water in North America and beyond.

PRACTICE PROBLEMS

Class I Systems

11.1 Water at 10°C flows from a large reservoir at the rate of 1.5×10^{-2} m^3/s through the system shown in Fig. 11.13. Calculate the pressure at B.

11.2 For the system shown in Fig. 11.14, kerosene (sg = 0.82) at 20°C is to be forced from tank A to reservoir B by increasing the pressure in the sealed tank A above the kerosene. The total length of DN 50 Schedule 40 steel pipe is 38 m. The elbow is standard. Calculate the required pressure in tank A to cause a flow rate of 435 L/min.

11.3 Figure 11.15 shows a portion of a hydraulic circuit. The pressure at point B must be 200 psig when the volume flow rate is 60 gal/min. The hydraulic fluid has a specific gravity of 0.90 and a dynamic viscosity of 6.0×10^{-5} lb·s/ft^2. The total length of pipe between A and B is 50 ft. The

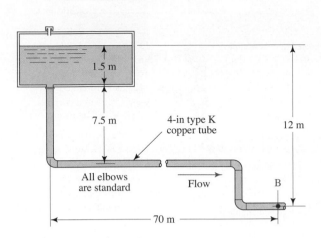

FIGURE 11.13   Problem 11.1.

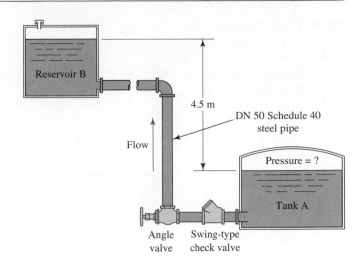

FIGURE 11.14   Problem 11.2.

elbows are standard. Calculate the pressure at the outlet of the pump at A.

11.4 Figure 11.16 shows part of a large hydraulic system in which the pressure at B must be 500 psig while the flow rate is 750 gal/min. The fluid is a medium machine tool hydraulic oil. The total length of the 4-in pipe is 40 ft. The elbows are standard. Neglect the energy loss due to friction in the 6-in pipe. Calculate the required pressure at A if the oil is (a) at 104°F and (b) at 212°F.

11.5 Oil is flowing at the rate of 0.015 m³/s in the system shown in Fig. 11.17. Data for the system are as follows:

- Oil specific weight = 8.80 kN/m³
- Oil kinematic viscosity = 2.12×10^{-5} m²/s
- Length of DN 150 pipe = 180 m
- Length of DN 50 pipe = 8 m

FIGURE 11.16   Problem 11.4.

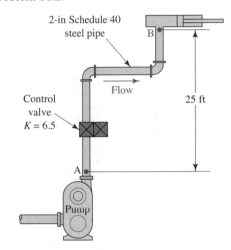

FIGURE 11.15   Problem 11.3.

FIGURE 11.17   Problem 11.5.

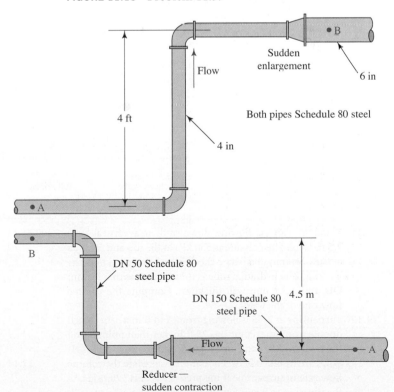

FIGURE 11.18   Problem 11.6.

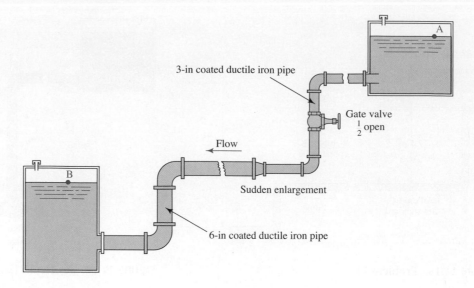

- Elbows are long-radius type
- Pressure at $B = 12.5$ MPa

  Considering all pipe friction and minor losses, calculate the pressure at A.

11.6 For the system shown in Fig. 11.18, calculate the vertical distance between the surfaces of the two reservoirs when water at 10°C flows from A to B at the rate of 0.03 m³/s. The elbows are standard. The total length of the 3-in pipe is 100 m. For the 6-in pipe it is 300 m.

11.7 A liquid refrigerant flows through the system, shown in Fig. 11.19, at the rate of 1.70 L/min. The refrigerant has a specific gravity of 1.25 and a dynamic viscosity of 3×10^{-4} Pa·s. Calculate the pressure difference between points A and B. The hydraulic tube is drawn steel, with an outside diameter (OD) of 15 mm, a wall thickness of 1.5 mm, and a total length of 30 m.

Class II Systems

11.8 Water at 100°F is flowing in a 4-in Schedule 80 steel pipe that is 25 ft long. Calculate the maximum allowable volume flow rate if the energy loss due to pipe friction is to be limited to 30 ft-lb/lb.

11.9 A hydraulic oil is flowing in a drawn steel hydraulic tube with an OD of 50 mm and a wall thickness of 1.5 mm. A pressure drop of 68 kPa is observed between two points in the tube 30 m apart. The oil has a specific gravity of 0.90 and a dynamic viscosity of 3.0×10^{-3} Pa·s. Calculate the velocity of flow of the oil.

11.10 In a processing plant, ethylene glycol at 77°F is flowing in a 6-in coated ductile iron pipe having a length of 5000 ft. Over this distance, the pipe falls 55 ft and the pressure drops from 250 psig to 180 psig. Calculate the velocity of flow in the pipe.

FIGURE 11.19   Problem 11.7.

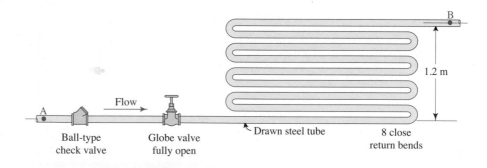

11.11 Water at 15°C is flowing downward in a vertical tube 7.5 m long. The pressure is 550 kPa at the top and 585 kPa at the bottom. A ball-type check valve is installed near the bottom. The hydraulic tube is drawn steel, with a 32 mm OD and a 2.0 mm wall thickness. Compute the volume flow rate of the water.

11.12 Turpentine at 77°F is flowing from A to B in a 3-in coated ductile iron pipe. Point B is 20 ft higher than point A and the total length of the pipe is 60 ft. Two 90° long-radius elbows are installed between A and B. Calculate the volume flow rate of turpentine if the pressure at A is 120 psig and the pressure at B is 105 psig.

11.13 A device designed to allow cleaning of walls and windows on the second floor of homes is similar to the system shown in Fig. 11.20. Determine the velocity of flow from the nozzle if the pressure at the bottom is (a) 20 psig and (b) 80 psig. The nozzle has a loss coefficient K of 0.15 based on the outlet velocity head. The tube is smooth drawn aluminum and has an ID of 0.50 in. The 90° bend has a radius of 6.0 in. The total length of straight tube is 20.0 ft. The fluid is water at 100°F.

11.14 Kerosene at 25°C is flowing in the system shown in Fig. 11.21. The total length of 50 mm OD × 1.5 mm wall hydraulic copper tubing is 30 m. The two 90° bends have

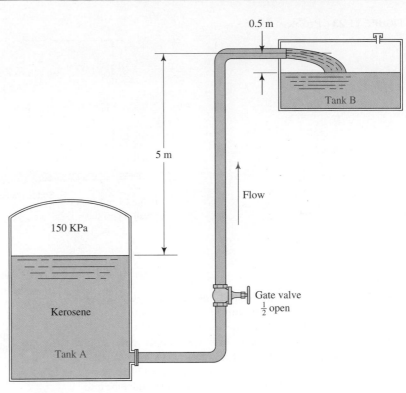

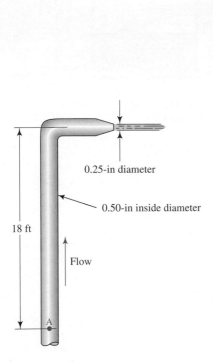

FIGURE 11.20 Problem 11.13.

FIGURE 11.21 Problem 11.14.

a radius of 300 mm. Calculate the volume flow rate into tank B if a pressure of 150 kPa is maintained above the kerosene in tank A.

11.15 Water at 40°C is flowing from A to B through the system shown in Fig. 11.22. Determine the volume flow rate of water if the vertical distance between the surfaces of the two reservoirs is 10 m. The elbows are standard.

11.16 Oil with a specific gravity of 0.93 and a dynamic viscosity of 9.5×10^{-3} Pa·s is flowing into the open tank shown in Fig. 11.23. The total length of 50 mm tubing is 30 m. For the 100 mm tubing the total length is 100 m. The elbows are standard. Determine the volume flow rate into the tank if the pressure at point A is 175 kPa.

FIGURE 11.22 Problem 11.15.

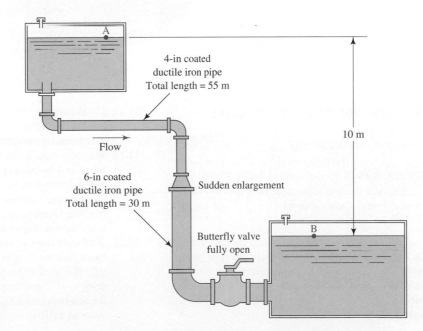

FIGURE 11.23   Problem 11.16.

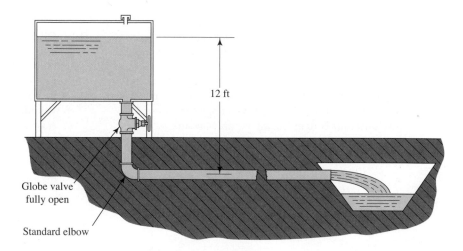

4.5 m

Flow

0.6 m

100-mm OD x 3.5-mm
wall copper tube

A

50-mm OD x 1.5-mm
wall copper tube

Sudden enlargement

Class III Systems

11.17 Determine the required size of new Schedule 80 steel pipe to carry water at 160°F with a maximum pressure drop of 10 psi per 1000 ft when the flow rate is 0.5 ft³/s.

11.18 What size of standard hydraulic copper tube from Appendix G.2 is required to transfer 0.06 m³/s of water at 80°C from a heater where the pressure is 150 kPa to an open tank? The water flows from the end of the tube into the atmosphere. The tube is horizontal and 30 m long.

11.19 Water at 60°F is to flow by gravity between two points, 2 mi apart, at the rate of 13 500 gal/min. The upper end is 130 ft higher than the lower end. What size concrete pipe is required? Assume that the pressure at both ends of the pipe is negligible.

11.20 The tank shown in Fig. 11.24 is to be drained to a sewer. Determine the size of new Schedule 40 steel pipe that will carry at least 400 gal/min of water at 80°F through the system shown. The total length of pipe is 75 ft.

FIGURE 11.24   Problem 11.20.

12 ft

Globe valve
fully open

Standard elbow

Practice Problems for Any System Class

11.21 Figure 11.25 depicts gasoline flowing from a storage tank into a truck for transport. The gasoline has a specific gravity of 0.68 and the temperature is 25°C. Determine the required depth h in the tank to produce a flow of 1500 L/min into the truck. Because the pipes are short, neglect the energy losses due to pipe friction, but do consider minor losses.

Note: Figure 11.26 shows a system used to pump coolant from a collector tank to an elevated tank, where it is cooled. The pump delivers 30 gal/min. The coolant then flows back to the machines as needed, by gravity. The coolant has a specific gravity of 0.92 and a dynamic viscosity of 3.6×10^{-5} lb·s/ft². This system is used in Problems 11.22–11.24.

11.22 For the system in Fig. 11.26, compute the pressure at the inlet to the pump. The filter has a resistance coefficient of 1.85 based on the velocity head in the suction line.

11.23 For the system in Fig. 11.26, compute the total head on the pump and the power delivered by the pump to the coolant.

11.24 For the system in Fig. 11.26, specify the size of Schedule 40 steel pipe required to return the fluid to the machines. Machine 1 requires 20 gal/min and Machine 2 requires 10 gal/min. The fluid leaves the pipes at the machines at 0 psig.

11.25 A manufacturer of spray nozzles specifies that the maximum pressure drop in the pipe feeding a nozzle be 10.0 psi per 100 ft of pipe. Compute the maximum allowable velocity of flow through a 1-in Schedule 80 steel pipe feeding the nozzle. The pipe is horizontal and the fluid is water at 60°F.

FIGURE 11.25 Problem 11.21.

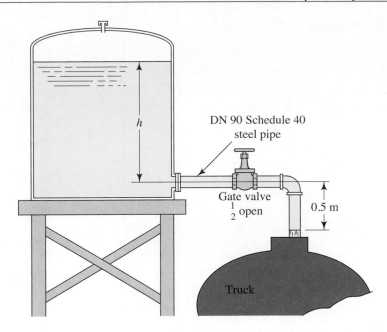

FIGURE 11.26 Problems 11.22–11.24.

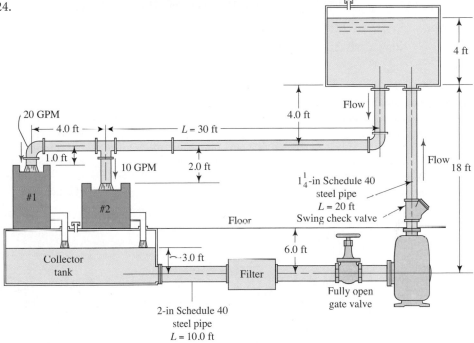

11.26 Specify the size of new Schedule 40 steel pipe required to carry gasoline at 77°F through 120 ft of horizontal pipe with no more than 8.0 psi of pressure drop at a volume flow rate of 100 gal/min.

11.27 Refer to Fig. 11.27. Water at 80°C is being pumped from the tank at the rate of 475 L/min. Compute the pressure at the inlet of the pump.

11.28 For the system shown in Fig. 11.27 and analyzed in Problem 11.27, it is desirable to change the system to increase the pressure at the inlet to the pump. The volume flow rate must stay at 475 L/min, but anything else can be changed. Redesign the system and recompute the pressure at the inlet to the pump for comparison with the result of Problem 11.27.

11.29 In a water pollution control project, the polluted water is pumped vertically upward 80 ft and then sprayed into the air to increase the oxygen content in the water and to evaporate volatile materials. The system is sketched in Fig. 11.28. The polluted water has a specific weight of 64.0 lb/ft^3 and a dynamic viscosity of 4.0×10^{-5} lb·s/ft^2. The flow rate is 0.50 ft^3/s. The pressure at the inlet to the pump is 3.50 psi below atmospheric pressure. The total length of discharge pipe is 82 ft. The nozzle has a resistance coefficient of 32.6 based on the velocity head in the discharge pipe. Compute the power delivered by the pump to the fluid. If the pump efficiency is 76 percent, compute the power input to the pump.

FIGURE 11.27 Problems 11.27 and 11.28.

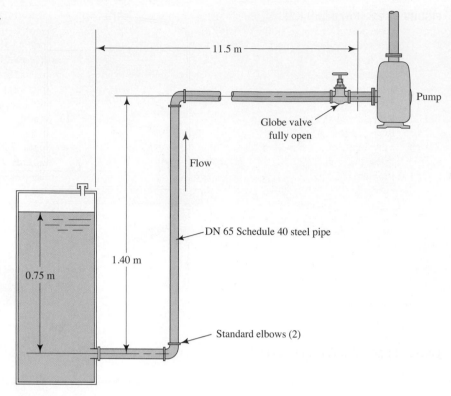

11.30 Repeat Problem 11.29, but use a 3-in Schedule 40 steel pipe for the discharge line instead of the 2½-in pipe. Compare the power delivered by the pump for the two designs.

11.31 Water at 10°C is being delivered to a tank on the roof of a building, as shown in Fig. 11.29. The elbow is standard. What pressure must exist at point A for 200 L/min to be delivered?

11.32 If the pressure at point A in Fig. 11.29 is 300 kPa, compute the volume flow rate of water at 10°C delivered to the tank.

11.33 Change the design of the system in Fig. 11.29 to replace the globe valve with a fully open gate valve. Then, if the pressure at point A is 300 kPa, compute the volume flow rate of water at 10°C delivered to the tank. Compare the result with that of Problem 11.32 to demonstrate the effect of the valve change.

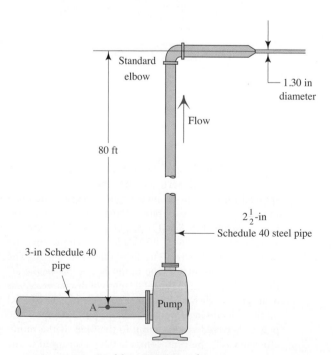

FIGURE 11.28 Problems 11.29 and 11.30.

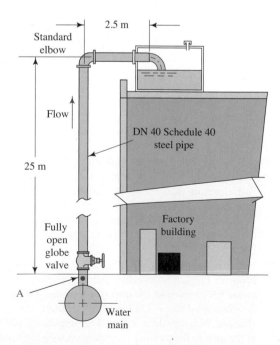

FIGURE 11.29 Problems 11.31–11.33.

11.34 It is desired to deliver 250 gal/min of ethyl alcohol at 77°F from tank A to tank B in the system shown in Fig. 11.30. The total length of pipe is 110 ft. Compute the required pressure in tank A.

11.35 For the system shown in Fig. 11.30, compute the volume flow rate of ethyl alcohol at 77°F that would occur if the pressure in tank A is 125 psig. The total length of pipe is 110 ft.

11.36 Repeat Problem 11.35, but consider the valve to be fully open.

11.37 Repeat Problem 11.35, but consider the valve to be fully open and the elbows to be the long-radius type instead of standard. Compare the results with those from Problems 11.35 and 11.36.

11.38 Figure 11.31 depicts a DN 100 Schedule 40 steel pipe delivering water at 15°C from a main line to a factory. The pressure at the main is 415 kPa. Compute the maximum allowable flow rate if the pressure at the factory must be no less than 200 kPa.

11.39 Repeat Problem 11.38, but replace the globe valve with a fully open butterfly valve.

11.40 Repeat Problem 11.38, but use a DN 125 Schedule 40 pipe.

11.41 Repeat Problem 11.38, but replace the globe valve with a butterfly valve and use a DN 125 Schedule 40 steel pipe. Compare the results of Problems 11.38–11.41.

FIGURE 11.30 Problems 11.34–11.37.

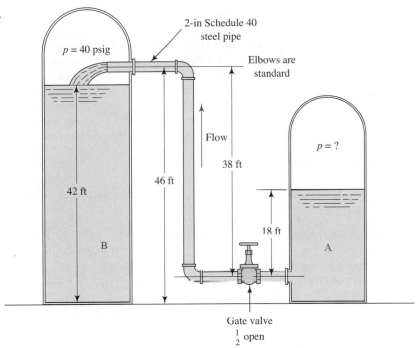

FIGURE 11.31 Problems 11.38–11.41.

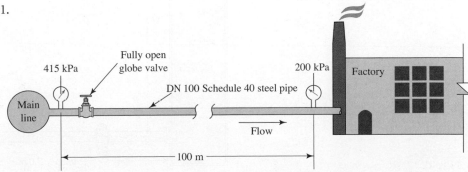

FIGURE 11.32   Problem 11.43

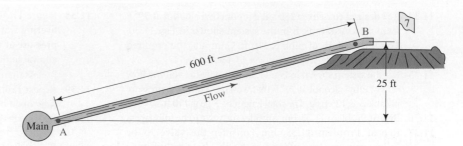

11.42 It is desired to drive a small, positive-displacement pump by chucking a household electric drill to the drive shaft of the pump. The pump delivers 1.0 in³ of water at 60°F per revolution, and the pump rotates at 2100 rpm. The outlet of the pump flows through a 100-ft smooth plastic hose with an ID of 0.75 in. How far above the source can the outlet of the hose be if the maximum power available from the drill motor is 0.20 hp? The pump efficiency is 75 percent. Consider the friction loss in the hose, but neglect other losses.

11.43 Figure 11.32 shows a pipe delivering water to the putting green on a golf course. The pressure in the main is at 80 psig and it is necessary to maintain a minimum of 60 psig at point B to adequately supply a sprinkler system. Specify the required size of Schedule 40 steel pipe to supply 0.50 ft³/s of water at 60°F.

11.44 Repeat Problem 11.43, except consider that there will be the following elements added to the system:
- A fully open gate value near the water main
- A fully open butterfly valve near the green (but before point B)
- Three standard 90° elbows
- Two standard 45° elbows
- One swing-type check valve

11.45 A sump pump in a commercial building sits in a sump at an elevation of 150.4 ft. The pump delivers 40 gal/min of water through a piping system that discharges the water at an elevation of 172.8 ft. The pressure at the pump discharge is 15.0 psig. The fluid is water at 60°F. Specify the

required size of plastic pipe if the system contains the following elements.
- A ball-type check valve
- Eight standard elbows
- A total length of pipe of 55.3 ft

The pipe is available in the same dimensions as Schedule 40 steel pipe.

11.46 For the system designed in Problem 11.45, compute the total head on the pump.

11.47 Figure 11.33 shows a part of a chemical processing system in which propyl alcohol at 25°C is taken from the bottom of a large tank and transferred by gravity to another part of the system. The length between the two tanks is 7.0 m. A filter is installed in the line and is known to have a resistance coefficient K of 8.5 based on the pipe velocity head. Drawn stainless steel tubing is to be used. Specify the standard size of tubing from Appendix G.2 that would allow a volume flow rate of 150 L/min.

11.48 For the system described in Problem 11.47, and using the tube size found in that problem, compute the expected volume flow rate through the tube if the elevation in the large tank drops to 12.8 m.

11.49 For the system described in Problem 11.47, and using the tube size found in that problem, compute the expected volume flow rate through the tube if the pressure above the fluid in the large tank at A is −32.5 kPa gage.

FIGURE 11.33   Problems 11.47–11.50

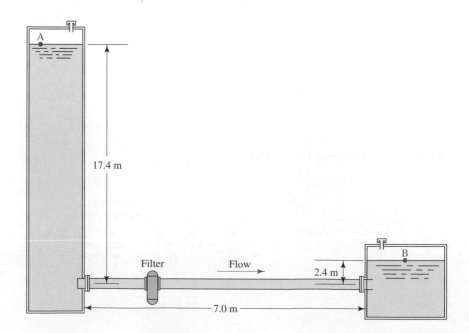

11.50 For the system described in Problem 11.47, and using the tube size found in that problem, compute the expected volume flow rate through the tube if a half-open gate valve is placed in the line ahead of the filter.

Supplemental Problems (PIPE-FLO® only)

11.51 Analyze the system shown in Fig. 11.11 with kerosene at 20°C as the working fluid. Use PIPE-FLO® software to determine the pressure at point B that results in a flow rate of 800 L/min. Report all key values such as Reynolds number and friction factor.

11.52 The pump pictured in Fig. 11.12 delivers water from the lower reservoir to the upper reservoir at a rate of 220 gal/min. There are 10 ft of 3-in Schedule 40 steel pipe before the pump, and 32 ft after. There are three standard 90° elbows and a fully open gate valve. The depth of the fluid inside the lower reservoir is 3 ft. Use PIPE-FLO® to calculate (a) the pressure at the pump inlet, (b) the pressure at the pump outlet, (c) the total head on the pump, and (d) the power delivered by the pump to the water.

COMPUTER AIDED ANALYSIS AND DESIGN ASSIGNMENTS

1. Create a program or a spreadsheet for analyzing Class I pipeline systems, including energy losses due to friction and minor losses due to valves and fittings.

2. Create a program or a spreadsheet for determining the velocity of flow and the volume flow rate in a given pipe with a limited pressure drop, considering energy loss due to friction only. Use the computational approach described in Section 11.4 and illustrated in Example Problem 11.2.

3. Create a program or a spreadsheet for determining the size of pipe required to carry a specified flow rate with a limited pressure drop, using the Class III solution procedure described in Example Problem 11.5.

4. Create a program or a spreadsheet for determining the size of pipe required to carry a specified flow rate with a limited pressure drop. Consider both energy loss due to friction and minor losses. Use a method similar to that described in Example Problem 11.6.

PARALLEL AND BRANCHING PIPELINE SYSTEMS

Parallel and branching pipeline systems are those having more than one path for the fluid to take as it flows from a source to a destination point. Figure 12.1 shows an example. This is part of a sprinkler system for fire protection in a building such as a commercial store, an office building, a school, a multiunit housing facility, or an industrial plant. Flow of water from a main source for the building enters the sprinkler system through the larger diameter pipe along the bottom. Called a *header*, this pipe presents water at a particular pressure to all of the branching pipes. While only the starts of the branches are shown, you can visualize that each branch would be routed to sprinkler heads via straight lengths of pipe having valves, elbows, and tees as needed to reach its destination. Each of the branches may have different paths requiring different lengths of pipe and different fittings. After installation, all valves leading into the branches would be opened, the pipes would fill and be fully pressurized.

Now consider what happens when two or more sprinkler heads are triggered on by fire or smoke detectors. Water flows out through the heads and is sprayed over a broad area to extinguish the fire. Different heads may have different capacities depending on the area needed to be controlled. Therefore, the flow rates through each pipe may be different. However, the principle of continuity indicates that all of the outflow from all of the heads must enter through the main header. Balancing the

flow at all parts of the system so that each receives the desired amount of water, requires control valves that can be adjusted, either manually or automatically.

What about the pressure distribution within the system? At the start of each branch, the pressure is essentially the same—the pressure in the header. Then, as the flow proceeds through the piping circuit, energy is lost due to friction and the minor losses in valves and fittings, causing the pressure to drop. At the ends of the branches, there will be a sprinkler head that is designed to require a certain minimum pressure in order to deliver the design value for flow rate when it is opened. As water leaves the sprinkler head, the pressure in the fluid is equal to atmospheric pressure so the total pressure drop in each line is from the header pressure to atmospheric pressure.

Try to visualize what would happen if all of the sprinkler heads at the ends of all of the branches opened at the same time. As noted above, each branch would have the same Δp from the header to the discharge stream from the heads. Using the relation between pressure and head, $\Delta p = \gamma h_L$, or, $h_L = \Delta p/\gamma$, we can say that the water in each branch experiences the same loss in head. However, because of differences in the design of various branches described before, different total resistance is present in each branch. The only way that the system can operate with different resistances in each branch is to adjust the amount of flow through each branch, with more flow

FIGURE 12.1 In parallel and branching pipeline systems, like this one, the fluid has alternative paths to progress through the circuit. They are quite common and require special analysis. (*Source:* mathisa/Fotolia)

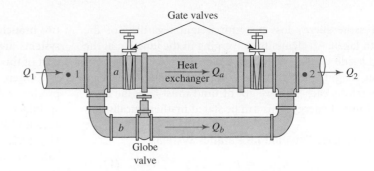

FIGURE 12.2 Parallel system with two branches.

traveling in the low resistance branches and lower flow rates in the branches with more resistance.

Now, it is likely that the flow rates through some branches may be lower than necessary to control a fire in the space served by the sprinkler head, while the flow in other branches may be greater than needed. This situation is typically rectified by placing control valves in each line that can be adjusted after the system is installed. Of course, the system design would have been done originally to produce reasonable flow rates through each branch. Notice in Fig. 12.1, for example, that some branch pipes are larger than others. Also, each control valve will be specified to have the ability to control the range of possible pressure drop across the valve so that the system can be balanced by a qualified technician after installation.

The sprinkler system just described is called a *branching system*, because the branches do not reconnect somewhere downstream from the header. Figure 12.2 shows a different approach, called a *parallel system*, in which two or more branch lines do reconnect. Both branching and parallel systems are covered in this chapter.

Exploration

- Find examples of parallel or branching flow systems around your home, in your car, or at your place of work.

- Sketch any system you find, showing the main pipe, all of the branches, the sizes of pipe or tube used, and any valves or fittings.

- Do the branches reconnect at some point or do they remain separate?

Introductory Concepts The analysis of parallel piping systems is based on the energy equation as you have done in Chapters 7–11, but with some additional observations and considerations. Look at Fig. 12.2, for example. Imagine that you are a small part of the fluid stream entering the system from the left and you find yourself at point 1. The total volume flow rate here is called Q_1 and you are a part of it. Then, as you enter the junction point, you have a decision to make. Which way do you go as you proceed to the destination? All the other parts of the flow must make the same decision.

Of course, some of the flow goes into each of the two branches that lead away from the junction, called a and b in the figure. These flow rates are called Q_a and Q_b respectively. You will learn in this chapter that the important thing for you to determine is how much fluid flows into each branch and how much pressure drop occurs as the fluid completes the circuit and ends up at the destination. In this case, the two paths rejoin at the right of the system and flow on through an outlet pipe to point 2, the destination. Here the volume flow rate is called Q_2.

When we apply the principle of steady flow to a parallel system, we reach the following conclusion:

▷ **Continuity Equation for Parallel Systems**

$$Q_1 = Q_2 = Q_a + Q_b \qquad (12\text{–}1)$$

The first part, $Q_1 = Q_2$, says that the volume flow rate is the same at any particular cross section when the total flow is considered. No fluid has been added to or taken away from the system between points 1 and 2. The second part defines that the branch flows, $Q_a + Q_b$, must sum to the total volume flow rate entering the first tee. This should seem logical because all the fluid that flows into the left junction must go somewhere and it splits into two parts. Finally, you should see that all of the flows from the branches must come together at the right junction for the total flow to continue as Q_2.

Now let's consider the pressure drop across the system. At point 1 there is a pressure p_1. At point 2 there is a different pressure p_2. The pressure drop is then $p_1 - p_2$. To help analyze the pressures, use the energy equation between points 1 and 2:

$$\frac{p_1}{\gamma} + z_1 + \frac{v_1^2}{2g} - h_L = \frac{p_2}{\gamma} + z_2 + \frac{v_2^2}{2g}$$

Solving for the pressure drop $p_1 - p_2$ gives

$$p_1 - p_2 = \gamma\left[(z_2 - z_1) + (v_2^2 - v_1^2)/2g + h_L\right]$$

This form of the energy equation says that the difference in pressure between points 1 and 2 depends on the elevation difference, the difference in the velocity heads, and the energy loss per unit weight of fluid flowing in the system. When any element of fluid reaches point 2 in the system shown in Fig. 12.2, each will have experienced the same elevation change, the same velocity change, and the same energy loss per unit weight regardless of the path taken. All elements converging in the junction at the right side of the system have the same total energy per unit weight. That is, they all have the same total head. Therefore, each unit weight of fluid must have the same amount of energy. This can be stated mathematically as

⇨ **Head Loss Equation for Parallel Systems**

$$h_{L_{1-2}} = h_a = h_b \qquad (12\text{--}2)$$

Equations (12–1) and (12–2) are the governing relationships for parallel pipeline systems. For a given configuration of the system, the flow in each branch automatically adjusts until the total system flow satisfies these equations. In general, more fluid will pass through the branch having the lower resistance and less flow will go through the higher resistance branch. For systems with more than two branches, these equations can be extended and such systems are often called *networks*, analyzed in the latter part of this chapter.

Often in industrial applications, the goal is to produce a desired flow rate in each branch, much like the sprinkler system described above. Control valves accomplish that objective. For example, in the system in Fig. 12.2, normal operation would have the two gate valves in branch *a* fully open to allow fluid to flow with low resistance through the heat exchanger. The globe valve in branch *b* may be closed. If a lower flow through the heat exchanger is desired, the globe valve can be opened to allow some flow to follow branch *b*. The valve can be adjusted to achieve different flow rates. The two gate valves in branch *a* allow the heat exchanger to be removed for cleaning or replacement and can also be used for additional flow control.

12.1 OBJECTIVES

After completing this chapter, you should be able to:

1. Discuss the difference between series pipeline systems and parallel or branching pipeline systems.

2. State the general relationships for flow rates and head losses for parallel or branching pipeline systems.

3. Compute the amount of flow that occurs in each branch of a two-branch parallel pipeline system and the head loss that occurs across the system when the total flow rate and the description of the system are known.

4. Determine the amount of flow that occurs in each branch of a two-branch parallel pipeline system and the total flow if the pressure drop across the system is known.

5. Apply the PIPE-FLO® software to analyze parallel and branching pipeline systems.

6. Use the Hardy Cross technique to compute the flow rates in all branches of a network having three or more branches.

12.2 SYSTEMS WITH TWO BRANCHES

A common parallel piping system includes two branches arranged as shown in Fig. 12.2. The lower branch is added to allow some fluid to bypass the heat exchanger. The branch could also be used to isolate the heat exchanger, allowing continuous flow while the equipment is serviced. The analysis of this type of system is relatively simple and straightforward, although some iteration is typically required. Because velocities are unknown, friction factors are also unknown.

Parallel systems having more than two branches are more complex because there are many more unknown quantities than there are equations relating the unknowns. A solution procedure is described in Section 12.4.

The following example problems are presented in the programmed form. You should pay careful attention to the logic of the solution procedure as well as to the details performed.

Method A—solution method for systems with two branches when the total flow rate and the description of the branches are known Example Problem 12.1 is of this type. The method of solution is as follows:

1. Equate the total flow rate to the sum of the flow rates in the two branches, as stated in Eq. (12–1). Then express the branch flows as the product of the flow area and the average velocity; that is,

$$Q_a = A_a v_a \quad \text{and} \quad Q_b = A_b v_b$$

2. Express the head loss in each branch in terms of the velocity of flow in that branch and the friction factor. Include all significant losses due to friction and minor losses.

3. Compute the relative roughness D/ε for each branch, estimate the value of the friction factor for each branch, and complete the calculation of head loss in each branch in terms of the unknown velocities.

4. Equate the expression for the head losses in the two branches to each other as stated in Eq. (12–2).

5. Solve for one velocity in terms of the other from the equation in Step 4.

6. Substitute the result from Step 5 into the flow rate equation developed in Step 1 and solve for one of the unknown velocities.

7. Solve for the second unknown velocity from the relationship developed in Step 5.

8. If there is doubt about the accuracy of the value of the friction factor used in Step 2, compute the Reynolds number for each branch and re-evaluate the friction factor from the Moody diagram or compute the values for the friction factors from Eq. (8–7) in Chapter 8.

9. If the values for the friction factor have changed significantly, repeat Steps 3–8, using the new values for friction factor.

10. When satisfactory precision has been achieved, use the now-known velocity in each branch to compute the vol-ume flow rate for that branch. Check the sum of the vol-ume flow rates to ensure that it is equal to the total flow in the system.

11. Use the velocity in either branch to compute the head loss across that branch, employing the appropriate rela-tionship from Step 3. This head loss is also equal to the head loss across the entire branched system. You can compute the pressure drop across the system, if desired, by using the relationship $\Delta p = \gamma h_L$.

PROGRAMMED EXAMPLE PROBLEM

Example Problem 12.1

In Fig. 12.2, 100 gal/min of water at 60°F is flowing in a 2-in Schedule 40 steel pipe at section 1. The heat exchanger in branch a has a loss coefficient of $K = 7.5$ based on the velocity head in the pipe. All three valves are wide open. Branch b is a bypass line composed of 1¼-in Schedule 40 steel pipe. The elbows are standard. The length of pipe between points 1 and 2 in branch b is 20 ft. Because of the size of the heat exchanger, the length of pipe in branch a is very short and friction losses can be neglected. For this arrangement, determine (a) the volume flow rate of water in each branch and (b) the pressure drop between points 1 and 2.

Solution

If we apply Step 1 of the solution method, Eq. (12–1) relates the two volume flow rates. How many quantities are unknown in this equation?

The two velocities v_a and v_b are unknown. Because $Q = Av$, Eq. (12–1) can be expressed as

$$Q_1 = A_a v_a + A_b v_b \qquad \textbf{(12–3)}$$

From the given data, $A_a = 0.02333 \ \text{ft}^2$, $A_b = 0.01039 \ \text{ft}^2$, and $Q_1 = 100 \ \text{gal/min}$. Expressing Q_1 in the units of ft^3/s gives

$$Q_1 = 100 \ \text{gal/min} \times \frac{1 \ \text{ft}^3/\text{s}}{449 \ \text{gal/min}} = 0.223 \ \text{ft}^3/\text{s}$$

Generate another equation that also relates v_a and v_b, using Step 2.

Equation (12–2) states that the head losses in the two branches are equal. Because the head losses h_a and h_b are dependent on the velocities v_a and v_b, respectively, this equation can be used in conjunction with Eq. (12–3) to solve for the velocities. Now, express the head losses in terms of the velocities for each branch.

You should have something similar to this for branch a:

$$h_a = 2K_1(v_a^2/2g) + K_2(v_a^2/2g)$$

where

$$K_1 = f_{aT}(L_e/D) = \text{Resistance coefficient for each gate valve}$$
$$K_2 = \text{Resistance coefficient for the heat exchanger} = 7.5 \ \text{(given in problem statement)}$$

The following data are known:

$$f_{aT} = 0.019 \ \text{for a 2-in Schedule 40 pipe (Table 10.5)}$$
$$L_e/D = 8 \ \text{for a fully open gate valve (Table 10.4)}$$

Then,

$$K_1 = (0.019)(8) = 0.152$$

Then,

$$h_a = (2)(0.152)(v_a^2/2g) + 7.5(v_a^2/2g) = 7.80(v_a^2/2g) \tag{12–4}$$

For branch b:

$$h_b = 2K_3(v_b^2/2g) + K_4(v_b^2/2g) + K_5(v_b^2/2g)$$

where $K_3 = f_{bT}(L_e/D) = $ Resistance coefficient for each elbow

$K_4 = f_{bT}(L_e/D) = $ Resistance coefficient for the globe valve

$K_5 = f_b(L_b/D) = $ Friction loss in the pipe of branch b for a pipe length of $L_b = 20$ ft.

The value of f_b is not known and will be determined through iteration. The known data are

$f_{bT} = 0.021$ for a 1¼-in Schedule 40 pipe (Table 10.5)

$L_e/D = 30$ for each elbow (Table 10.4)

$L_e/D = 340$ for a fully open globe valve (Table 10.4)

Then,

$$K_3 = (0.021)(30) = 0.63$$
$$K_4 = (0.021)(340) = 7.14$$
$$K_5 = f_b(20/0.1150) = 173.9f_b$$

Then,

$$h_b = (2)(0.63)(v_b^2/2g) + (7.14)(v_b^2/2g) + f_b(173.9)(v_b^2/2g)$$
$$h_b = (8.40 + 173.9f_b)(v_b^2/2g)$$

This equation introduces the additional unknown, f_b. We can use an iteration procedure similar to that used for Class II series pipeline systems in Chapter 11. The relative roughness for branch b will aid in the estimation of the first trial value for f_b:

$$D/\varepsilon = (0.1150/1.5 \times 10^{-4}) = 767$$

From the Moody diagram in Fig. 8.7, a logical estimate for the friction factor is $f_b = 0.023$. Substituting this into the equation for h_b gives

$$h_b = [8.40 + 173.9(0.023)](v_b^2/2g) = 12.40(v_b^2/2g) \tag{12–5}$$

We now have completed Step 3 of the solution procedure. Steps 4 and 5 can be done now to obtain an expression for v_a in terms of v_b.

You should have $v_a = 1.261v_b$, obtained as follows:

$$h_a = h_b$$
$$7.80(v_a^2/2g) = 12.40(v_b^2/2g)$$

Solving for v_a gives

$$v_a = 1.261v_b \tag{12–6}$$

At this time, you can combine Eqs. (12–3) and (12–6) to calculate the velocities (Steps 6 and 7).

The solutions are $v_a = 5.60$ ft/s and $v_b = 7.06$ ft/s. Here are the details:

$$Q_1 = A_a v_a + A_b v_t \tag{12–3}$$
$$v_a = 1.261v_b \tag{12–6}$$

Then, we have

$$Q_1 = A_a(1.261v_b) + A_b v_b = v_b(1.261A_a + A_b)$$

Solving for v_b, we get

$$v_b = \frac{Q_1}{1.261A_a + A_b} = \frac{0.223 \text{ ft}^3/\text{s}}{[(1.261)(0.02333) + 0.01039] \text{ ft}^2}$$

$$v_b = 5.60 \text{ ft/s}$$

$$v_a = (1.261)(5.60) \text{ ft/s} = 7.06 \text{ ft/s}$$

Because we made these calculations using an assumed value for f_b, we should check the accuracy of the assumption.

We can evaluate the Reynolds number for branch b:

$$N_{Rb} = v_b D_b / \nu$$

From Appendix A, Table A.2, we find $\nu = 1.21 \times 10^{-5} \text{ ft}^2/\text{s}$. Then,

$$N_{Rb} = (5.60)(0.1150)/(1.21 \times 10^{-5}) = 5.32 \times 10^4$$

Using this value and the relative roughness of 767 from before, in the Moody diagram, yields a new value, $f_b = 0.0248$. Because this is significantly different from the assumed value of 0.023, we can repeat the calculations for Steps 3–8. The results are summarized as follows:

$$h_b = [8.40 + 173.9(0.0248)](v_b^2/2g) = 12.71(v_b^2/2g) \qquad \textbf{(12–5)}$$

$$h_a = 7.80(v_a^2/2g) \quad \text{(same as for first trial)}$$

Equating the head losses in the two branches gives

$$h_a = h_b$$

$$7.80(v_a^2/2g) = 12.71(v_b^2/2g)$$

Solving for the velocities gives

$$v_a = 1.277 v_b$$

Substituting this into the equation for v_b used before gives

$$v_b = \frac{0.223 \text{ ft}^3/\text{s}}{[(1.277)(0.02333) + 0.01039] \text{ ft}^2} = 5.55 \text{ ft/s}$$

$$v_a = 1.277 v_b = 1.277(5.55) = 7.09 \text{ ft/s}$$

Recomputing the Reynolds number for branch b gives

$$N_{Rb} = v_b D_b / \nu$$

$$N_{Rb} = (5.55)(0.1150)/(1.21 \times 10^{-5}) = 5.27 \times 10^4$$

There is no significant change in the value of f_b. Therefore, the values of the two velocities computed above are correct. We can now complete Steps 10 and 11 of the procedure to find the volume flow rate in each branch and the head loss and the pressure drop across the entire system.

Now calculate the volume flow rates Q_a and Q_b (Step 10).

You should have

$$Q_a = A_a v_a = (0.02333 \text{ ft}^2)(7.09 \text{ ft/s}) = 0.165 \text{ ft}^3/\text{s}$$

$$Q_b = A_b v_b = (0.01039 \text{ ft}^2)(5.55 \text{ ft/s}) = 0.0577 \text{ ft}^3/\text{s}$$

Converting these values to the units of gal/min gives $Q_a = 74.1$ gal/min and $Q_b = 25.9$ gal/min.

We are also asked to calculate the pressure drop. How can this be done?

We can write the energy equation using points 1 and 2 as reference points. Because the velocities and elevations are the same at these points, the energy equation is simply

$$\frac{p_1}{\gamma} - h_L = \frac{p_2}{\gamma}$$

Solving for the pressure drop, we get

$$p_1 - p_2 = \gamma h_L \qquad (12\text{--}7)$$

What can be used to calculate h_L?

Because $h_{L_{1-2}} = h_a = h_b$, we can use either Eq. (12–4) or (12–5). Using Eq. (12–4), we get

$$h_a = 7.80(v_a^2/2g) = (7.80)(7.09)^2/64.4 \text{ ft} = 6.09 \text{ ft}$$

Note that this neglects the minor losses in the two tees. Then, we have

$$p_1 - p_2 = \gamma h_L = \frac{62.4 \text{ lb}}{\text{ft}^3} \times 6.09 \text{ ft} \times \frac{1 \text{ ft}^2}{144 \text{ in}^2} = 2.64 \text{ psi}$$

This example problem is concluded.

Note that in the system shown in Fig. 12.2, if the globe valve in pipe *b* is closed, all flow passes through the heat exchanger, and the pressure drop can be computed using the Class I series pipeline system analysis as discussed in Chapter 11. Similarly, if the gate valves in pipe *a* are closed, all flow passes through the bypass line.

Method B—solution method for systems with two branches when the pressure drop across the system is known and the volume flow rate in each branch and the total volume flow rate are to be computed   Example Problem 12.2 is of this type. The method of solution is as follows:

1. Compute the total head loss across the system using the known pressure drop Δp in the relation $h_L = \Delta p/\gamma$.

2. Write expressions for the head loss in each branch in terms of the velocity in that branch and the friction factor.

3. Compute the relative roughness D/ε for each branch, assume a reasonable estimate for the friction factor, and

complete the calculation for the head loss in terms of the velocity in each branch.

4. Letting the magnitude of the head loss in each branch equal the total head loss as found in Step 1, solve for the velocity in each branch by using the expression found in Step 3.

5. If there is doubt about the accuracy of the value of the friction factor used in Step 3, compute the Reynolds number for each branch and reevaluate the friction factor from the Moody diagram in Fig. 8.7 or compute the value of the friction factor from Eq. (8–7).

6. If the values for the friction factor have changed significantly, repeat Steps 3 and 4, using the new values for friction factor.

7. When satisfactory precision has been achieved, use the now-known velocity in each branch to compute the volume flow rate for that branch. Then, compute the sum of the volume flow rates, which is equal to the total flow in the system.

PROGRAMMED EXAMPLE PROBLEM

Example Problem 12.2   The arrangement shown in Fig. 12.3 is used to supply lubricating oil to the bearings of a large machine. The bearings act as restrictions to the flow. The resistance coefficients are 11.0 and 4.0 for the two bearings. The lines in each branch are 15 mm OD × 1.2 mm wall, drawn steel tubing. Each of the four bends in the tubing has a mean radius of 100 mm. Include the effect of these bends, but exclude the friction losses because the lines are short. Determine (a) the flow rate of oil in each bearing and (b) the total flow rate in L/min. The oil has a specific gravity of 0.881 and a kinematic viscosity of $2.50 \times 10^{-6} \text{ m}^2/\text{s}$. The system lies in one plane, so all elevations are equal.

Solution   Write the equation that relates the head loss h_L across the parallel system to the head losses in each line h_a and h_b.

You should have

$$h_L = h_a = h_b \qquad (12\text{--}8)$$

They are all equal. Determine the magnitude of these head losses by using Step 1.

FIGURE 12.3 Parallel system for Example Problem 12.2.

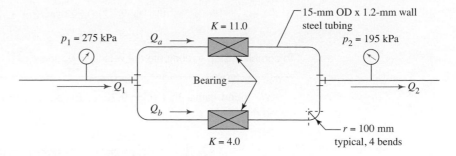

We can find h_L from the energy equation

$$\frac{p_1}{\gamma} + z_1 + \frac{v_1^2}{2g} - h_L = \frac{p_2}{\gamma} + z_2 + \frac{v_2^2}{2g}$$

Because $z_1 = z_2$ and $v_1 = v_2$,

$$\frac{p_1}{\gamma} - h_L = \frac{p_2}{\gamma}$$

$$h_L = (p_1 - p_2)/\gamma \qquad \text{(12–9)}$$

Using the given data, we get

$$h_L = \frac{(275 - 195)\ \text{kN}}{\text{m}^2} \times \frac{\text{m}^3}{(0.881)(9.81)\ \text{kN}}$$

$$h_L = 9.26\ \text{m}$$

Now write the expressions for h_a and h_b, Step 2.

Considering the losses in the bends and in the bearings, you should have

$$h_a = 2K_1(v_a^2/2g) + K_2(v_a^2/2g) \qquad \text{(12–10)}$$
$$h_b = 2K_1(v_b^2/2g) + K_3(v_b^2/2g) \qquad \text{(12–11)}$$

where

$K_1 = f_T(L_eD)$ = Resistance coefficient for each bend

K_2 = Resistance coefficient for the bearing in branch a = 11.0 (given in problem statement)

K_3 = Resistance coefficient for the bearing in branch b = 4.0 (given in problem statement)

f_T = Friction factor in the zone of complete turbulence in the steel tube

(L_e/D) = Equivalent length ratio for each bend Chapter 10, Fig. 10.28

We need the relative radius for the bends,

$$r/D = (100\ \text{mm})/(12.6\ \text{mm}) = 7.94$$

From Fig. 10.28, we find $L_e/D = 23.6$.

The friction factor in the zone of complete turbulence can be determined by using the relative roughness D/ε and the Moody diagram, reading at the right end of the relative roughness curve where it approaches a horizontal line:

$$D/\varepsilon = 0.0126\ \text{m}/1.5 \times 10^{-6}\ \text{m} = 8400$$

We can read $f_T = 0.0124$ from the Moody diagram. Now we can complete Step 3 by evaluating all of the resistance factors and expressing the energy loss in each branch in terms of the velocity head in the branch:

$$K_1 = f_T(L_e/D) = (0.0124)(23.6) = 0.293$$
$$K_2 = 11.0$$
$$K_3 = 4.0$$
$$h_a = (2)(0.293)(v_a^2/2g) + 11.0(v_a^2/2g)$$
$$h_a = 11.59 v_a^2/2g \qquad \text{(12–12)}$$

$$h_b = (2)(0.293)(v_b^2/2g) + 4.0(v_b^2/2g)$$
$$h_b = 4.59 v_b^2/2g \qquad\qquad \textbf{(12–13)}$$

To complete Step 4, compute the velocities v_a and v_b.

We found earlier that $h_L = 9.26$ m. Because $h_L = h_a = h_b$, Eqs. (12–12) and (12–13) can be solved directly for v_a and v_b:

$$h_a = 11.59 v_a^2/2g$$
$$v_a = \sqrt{\frac{2gh_a}{11.59}} = \sqrt{\frac{(2)(9.81)(9.26)}{11.59}} \text{ m/s} = 3.96 \text{ m/s}$$
$$h_b = 4.59 v_b^2/2g$$
$$v_b = \sqrt{\frac{2gh_b}{4.59}} = \sqrt{\frac{(2)(9.81)(9.26)}{4.59}} \text{ m/s} = 6.29 \text{ m/s}$$

Now find the volume flow rates, as called for in Step 7.

You should have $Q_a = 29.6$ L/min, $Q_b = 47.1$ L/min, and the total volume flow rate = 76.7 L/min. The area of each tube is 1.247×10^{-4} m². Then, we have

$$Q_a = A_a v_a = 1.247 \times 10^{-4} \text{ m}^2 \times 3.96 \text{ m/s} \times \frac{60\,000 \text{ L/min}}{\text{m}^3/\text{s}}$$

$$Q_a = 29.6 \text{ L/min}$$

Similarly,

$$Q_b = A_b v_b = 47.1 \text{ L/min}$$

Then the total flow rate is

$$Q_1 = Q_a + Q_b = (29.6 + 47.1)\text{L/min} = 76.7 \text{ L/min}$$

This example problem is concluded.

12.3 PARALLEL PIPELINE SYSTEMS AND PRESSURE BOUNDARIES IN PIPE-FLO®

The use of software in the analysis and design of parallel pipeline systems is particularly beneficial given the tedious nature of the required calculations. This example problem provides a guide to modeling and determining results for the parallel system shown below using PIPE-FLO® and it is repeated for manual solution as Practice Problem 12.3 at the end of the chapter.

PIPE-FLO®
Example Problem
12.3

In the branched pipe system shown in Fig. 12.4, 850 L/min of water at 10°C is flowing in a DN 100 Schedule 40 pipe at A where the pressure is 1000 kPa. The flow splits into two DN 50 Schedule 40 pipes as shown and then rejoins at B. Calculate (a) the flow rate in each of the branches and (b) the pressure at B. Include the effect of the minor losses in the lower branch of the system. The total length of pipe in the lower branch is 60 m. The elbows are standard. Assume all components lie in the same horizontal plane.

FIGURE 12.4 Parallel pipeline system for Example Problem 12.4

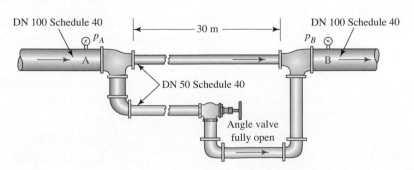

Solution

1. Begin by using the "SYSTEM" menu to establish the units, fluid zones, and pipe specifications for the system.

2. This problem statement indicates that this zone of analysis is a subset of a larger system. In cases like these, it is helpful to utilize another option available within PIPE-FLO® called the "Pressure Boundary." This option essentially puts a user-entered value for the pressure at a given point in the system. While there is no tank or explicit fluid supply shown in the schematic, the "pressure boundary" assures that fluid is still available and flowing at this start point. A flow demand will also be incorporated into the problem to define the fluid flow rate. For this example, a pressure boundary will be used to represent point A in the system shown above. The amount of pressure at this point is set at 1000 kPa.

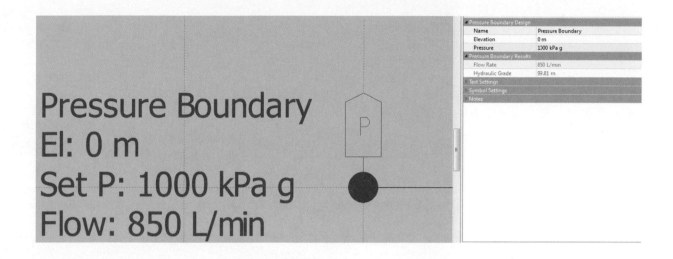

Pressure Boundary
El: 0 m
Set P: 1000 kPa g
Flow: 850 L/min

Pressure Boundary Design	
Name	Pressure Boundary
Elevation	0 m
Pressure	1000 kPa g
Pressure Boundary Results	
Flow Rate	850 L/min
Hydraulic Grade	99.81 m
Text Settings	
Symbol Settings	
Notes	

3. The main new feature of this example problem is the parallel nature of the piping system. To make a split or convergence of pipes, a "Node" must be used. A node acts as a point in the system where two pipes either split or come together. In this case, the node will represent the tee shown in the problem figure. To create a node at the required point in the problem, draw a pipe from the pressure boundary to an arbitrary point on the FLO-Sheet® and double click until the "dot" appears. Once the dot has appeared, the node has been created. The only input in the properties grid for a node is its elevation. Enter "0" in this case.

4. Ideally, the pressure boundary should be placed on the node so that a pipe isn't required to connect the pressure boundary to the node. However, PIPE-FLO® won't allow a pressure boundary to be located at a node. This is an example of the importance of fully understanding the principles before using any engineering software. To make this problem fit the software structure, insert a length of pipe just 1 mm in length. In reality, there will be no such pipe in the system, but it meets the goal of allowing the creation of the model, and keeping any associated loss to be negligible. The figure below represents the pressure boundary connected to the node with the short section of pipe.

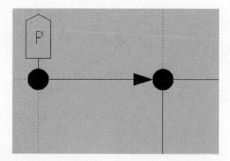

5. From this point, the rest of the pipe system can be drawn as done in previous problems. Another node will have to be used where the two pipes converge. To do this, simply double click the mouse after drawing the straight pipe, and connect the lower section of pipe to the same node.

6. As mentioned earlier, a flow demand will be used to represent the flow of the established fluid through the system. The flow demand will represent point B in Fig. 12.4. Enter the 850 L/min value in the properties grid for the flow demand. Although there isn't a tank to represent the fluid in the system, PIPE-FLO® assumes that the fluid established in the fluid zones menu is actually flowing through the system because of the pressure boundary.

7. Be sure to include the proper number of elbows (3) and the valve as shown in Fig. 12.4. The valve is located under the "Valves" section of the "Valves and Fittings" menu. It is called a "Globe—Angle 90°" valve. Enter that item just like the fittings and valves from previous problems.

8. After the system has been drawn and all components "defined" by entering all the required data, run the calculations as done in previous example problems. The results are shown below.

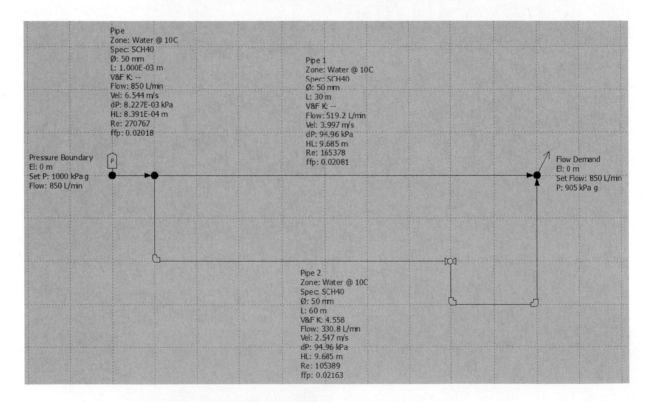

9. As previously mentioned, the values shown for the short section of pipe between the pressure boundary and the node can be considered negligible because of its short length. Summary of results:

a. The total pressure drop across each branch is 95 kPa.

b. The pressure at B is 1000 kPa − 95 kPa = 905 kPa as shown.

c. The flow rates through the upper and lower sections are 519 and 331 L/min respectively.

Be sure to work Example Problem 12.3 by hand to verify the answers and to demonstrate that the software model duplicates the manual analytical solution.

Use of PIPE-FLO® For More Advanced System Analysis And Design The example problem above has demonstrated the value of using PIPE-FLO® or other software to analyze parallel pipeline problems that typically require many calculations and careful analysis. Also, the analysis becomes much more difficult and time-consuming when multiple operating conditions are considered and for more complex systems, typical of industrial applications. More complex systems can be analyzed using the educational version or the full version of such software and you are encouraged to explore their capabilities. The inclusion of pump sizing is covered in Chapter 13 with an example of selecting a suitable pump for a given system design.

Another important use of software is for designing and analyzing piping systems that employ control valves. For example, reconsider the system shown in Fig. 12.2 that uses a bypass parallel pipeline to moderate the fluid temperature after a heat exchanger. Example Problem 12.1 demonstrated how to compute the flow rate through the two branches; one through the heat exchanger itself, and one through the bypass line, for one configuration of the system. It was noted there that by adjusting the globe valve in the lower branch, an operator can change the fluid temperature downstream. Partially closing the globe valve causes it to create higher resistance to flow with correspondingly higher pressure drop across the valve. The higher resistance causes a decrease in the flow through the line containing the valve and a consequent increase in the flow through the heat exchanger. The result is a moderated temperature of the fluid flowing to a downstream process.

In an industrial production operation, it is typically desirable to replace the globe valve with an automatic *flow control valve*, sometimes called an FCV. Temperature sensors can be used to monitor the condition of the fluid at key points in the system and compare them with desired conditions. As conditions move out of an acceptable range, a control signal is fed back to the FCV, causing it to automatically adjust its position to bring the fluid properties back into the desired range. The FCV includes a drive system that automatically moves the plug of the valve with respect to the seat inside the valve, in a manner similar to how the operator may do that task manually.

Manufacturers of flow control valves develop data for the flow coefficient, C_V, at various positions of the plug and provide that data to the user. Recall the discussion of C_V in Chapter 10. The system designer will select an FCV that has a range of C_V values that span the expected range of conditions that the system will experience. The nominal design value for C_V will be near the mid-point of that range, allowing adjustment above and below that point.

Another application of flow control valves is for open systems, similar to the sprinkler system for fire protection shown in Fig. 12.1. Consider the use of such a system delivering different feedstocks to a blending system in food production or for chemical processes. As the nature of the feedstocks or the desired product properties change, flow control valves in each delivery line can be automatically adjusted.

Reference 1 includes extensive coverage of the design and analysis of piping systems employing flow control valves.

12.4 SYSTEMS WITH THREE OR MORE BRANCHES—NETWORKS

When three or more branches occur in a pipe flow system, it is called a *network*. Networks are indeterminate because there are more unknown factors than there are independent equations relating the factors. For example, in Fig. 12.5 there are three unknown velocities, one in each pipe. The equations available to describe the system are

$$Q_1 = Q_2 = Q_a + Q_b + Q_c \qquad \text{(12–14)}$$

$$h_{L_{1-2}} = h_a = h_b = h_c \qquad \text{(12–15)}$$

A third independent equation is required to solve explicitly for the three velocities, and none is available.

A rational approach to complete the analysis of a system such as that shown in Fig. 12.5 employing an iteration procedure has been developed by Hardy Cross (see Reference 2). This procedure converges on the correct flow rates quite rapidly. Many calculations are still required, but they can be set up in an orderly fashion for use on a calculator or digital computer.

The Cross technique requires that the head loss terms for each pipe in the system be expressed in the form

$$h = kQ^n \qquad \text{(12–16)}$$

where k is an equivalent resistance to flow for the entire pipe and Q is the flow rate in the pipe. We will illustrate the creation of such an expression in the example problem that follows this general discussion of the Cross technique.

Recall that both friction losses and minor losses are proportional to the velocity head, $v^2/2g$. Then, using the continuity equation, we can express the velocity in terms of the volume flow rate. That is,

$$v = Q/A$$

and

$$v^2 = Q^2/A^2$$

This will allow the development of an equation of the form shown in Eq. (12–16).

The Cross iteration technique requires that initial estimates for the volume flow rate in each branch of the system be made. Two factors that help in making these estimates are as follows:

1. At each junction in the network, the sum of the flow into the junction must equal the flow out.

2. The fluid tends to follow the path of least resistance through the network. Therefore, a pipe having a lower value of k will carry a higher flow rate than those having higher values.

The network should be divided into a set of closed-loop circuits prior to the beginning of the iteration process.

FIGURE 12.5   Network with three branches.

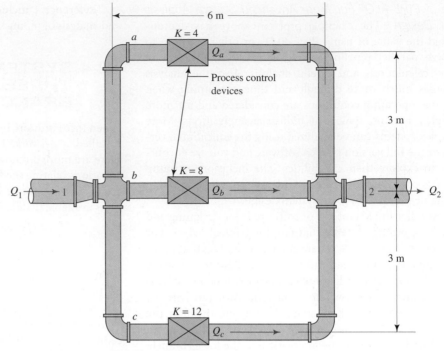

Note: Inlet and outlet pipes: DN 50 Sch. 40
Branch pipes a, b, and c: DN 25 Sch. 40
Elbows are standard

Figure 12.6 shows a schematic representation of a three-pipe system such as that shown in Fig. 12.5. The dashed arrows drawn in a clockwise direction assist in defining the signs for the flow rates Q and the head losses h in the various pipes of each loop according to the following convention:

> ***If the flow in a given pipe of a circuit is clockwise, Q and h are positive.***
>
> ***If the flow is counterclockwise, Q and h are negative.***

Then, for circuit 1 in Fig. 12.6, h_a and Q_a are positive and h_b and Q_b are negative. The signs are critical to the correct calculation of adjustments to the volume flow rates, indicated by ΔQ, that are produced at the end of each iteration cycle. Notice that pipe b is common to both circuits. Therefore, the adjustments ΔQ for each circuit must be applied to the flow rate in this pipe.

The Cross technique for analyzing the flow in pipe networks is presented in step-by-step form below.

A programmed example problem follows to illustrate the application of the procedure.

Cross Technique for Analysis of Pipe Networks

1. Express the energy loss in each pipe in the form $h = kQ^2$.
2. Assume a value for the flow rate in each pipe such that the flow into each junction equals the flow out of the junction.
3. Divide the network into a series of closed-loop circuits.
4. For each pipe, calculate the head loss $h = kQ^2$, using the assumed value of Q.
5. Proceeding around each circuit, algebraically sum all values for h using the following sign convention:

    If the flow is clockwise, h and Q are positive.

    If the flow is counterclockwise, h and Q are negative.

    The resulting summation is referred to as Σh.

6. For each pipe, calculate $2kQ$.

FIGURE 12.6   Closed-loop circuits used for the Cross technique for the analysis of pipe networks.

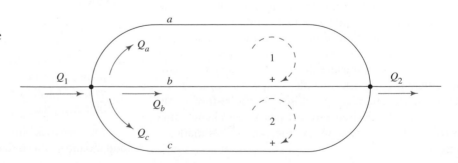

7. Sum all values of $2kQ$ for each circuit, assuming all are positive. This summation is referred to as $\Sigma(2kQ)$.

8. For each circuit, calculate the value of ΔQ from

$$\Delta Q = \frac{\Sigma h}{\Sigma(2kQ)} \qquad (12\text{--}17)$$

9. For each pipe, calculate a new estimate for Q from

$$Q' = Q - \Delta Q$$

10. Repeat Steps 4–8 until ΔQ from Step 8 becomes negligibly small. The Q' value is used for the next cycle of iteration.

PROGRAMMED EXAMPLE PROBLEM

Example Problem 12.4

For the system shown in Fig. 12.5, determine the volume flow rate of water at 15°C through each branch if 600 L/min (0.01 m³/s) is flowing into and out of the system through the DN 50 pipes.

Solution

The head loss in each pipe should now be expressed in the form of $h = kQ^2$ as Step 1 of the procedure. Consider branch a first and write an expression for the head loss h_a.

The total head loss for the branch is due to the two elbows (each with $L_e/D = 30$), the restriction (with $K = 4.0$ based on the velocity head in the pipe), and friction in the pipe. Then,

$$h_a = \underset{\text{(elbows)}}{2(f_{aT})(30)(v_a^2/2g)} + \underset{\text{(restriction)}}{4.0(v_a^2/2g)} + \underset{\text{(friction)}}{f_a(L_a/D_a)(v_a^2/2g)}$$

The friction factor f_a for flow in the pipe is dependent on the Reynolds number and, therefore, on the volume flow rate. Because that is the objective of the network analysis, we cannot determine that value explicitly at this time. Furthermore, the flow rate will, in general, be different in each segment of the flow system, resulting in different values for the friction factor. We will take that into account in the present analysis by computing the value of the friction factor after assuming the magnitude of the volume flow rate in each pipe, a step that is inherent in the Cross technique. We will use the Swamee–Jain method to compute the friction factor from Eq. (8–7). Then, we will recompute the values of the friction factors for each trial as the value of the volume flow rate is refined.

First, let's simplify the equation for h_a by completing as many calculations as we can.

What values can be determined?

The total length of pipe in branch a is 12 m, and for the DN 25 Schedule 40 pipe $D = 0.0266$ m and $A = 5.574 \times 10^{-4}$ m². From Table 10.5 we can find that the value of $f_{aT} = 0.022$ for a DN 25 Schedule 40 steel pipe with flow in the completely turbulent zone. The water at 15°C has a kinematic viscosity $\nu = 1.15 \times 10^{-6}$ m²/s. We can introduce the volume flow rate Q into the equation by noting that, as shown before,

$$v_a^2 = Q_a^2/A_a^2$$

Now substitute these values into the equation for h_a and simplify it as much as possible.

You should have something like this:

$$h_a = [60(f_{aT}) + 4.0 + (f_a)(12/0.0266)](v_a^2/2g)$$
$$h_a = [60(f_{aT}) + 4.0 + 451(f_a)](Q_a^2/2gA^2)$$
$$h_a = [60(0.0022) + 4.0 + 451(f_a)]\left[\frac{Q_a^2}{2(9.81)(5.574 \times 10^{-4})^2}\right]$$
$$h_a = [5.32 + 451(f_a)](1.64 \times 10^5)Q_a^2 \qquad (12\text{--}18)$$

It is also convenient to express the Reynolds number in terms of the volume flow rate Q and to compute the value for the relative roughness D/ε. Do that now.

Because all three branches have the same size and type of pipe, these calculations apply to each branch. If different pipes are used throughout the network, these calculations must be redone for each pipe. For the DN 25 steel pipe,

$$D/\varepsilon = (0.0266 \text{ m})/(4.6 \times 10^{-5} \text{ m}) = 578$$

We should modify the Reynolds number formula as

$$N_{Ra} = \frac{v_a D_a}{\nu} = \frac{Q_a D_a}{A_a \nu} = \frac{Q_a(0.0266)}{(5.574 \times 10^{-4})(1.15 \times 10^{-6})}$$

$$N_{Ra} = (4.15 \times 10^7)Q_a \qquad\qquad \textbf{(12–19)}$$

Now create expressions for the head losses in the other two pipes, h_b and h_c, using similar procedures.

Compare your results with these. Note that the pipe size in branches b and c is the same as that in branch a. For branch b:

$$h_b = 8.0(v_b^2/2g) + f_b(L_b/D_b)(v_b^2/2g)$$
$$\text{(restriction)} \qquad\qquad \text{(friction)}$$

$$h_b = [8.0 + f_b(6/0.0266)](Q_b^2/2gA^2)$$

$$h_b = [8.0 + 225.6(f_b)](1.64 \times 10^5)Q_b^2 \qquad\qquad \textbf{(12–20)}$$

For branch c:

$$h_c = 2(f_{cT})(30)(v_c^2/2g) + 12.0(v_c^2/2g) + f_c(L_c/D_c)(v_c^2/2g)$$
$$\text{(elbows)} \qquad\qquad \text{(restriction)} \qquad\qquad \text{(friction)}$$

$$h_c = [60(f_{cT}) + 12.0 + f_c(12/0.0266)(v_c^2/2g)$$

$$h_c = [60(0.022) + 12.0 + 451f_c](Q_c^2/2gA^2)$$

$$h_c = [13.32 + 451(f_c)](1.64 \times 10^5)Q_c^2 \qquad\qquad \textbf{(12–21)}$$

Equations (12–18) to (12–21) will be used in the calculations of head losses as the Cross iteration process continues. When the values for the friction factors are known or assumed, the head loss equations can be reduced to the form of Eq. (12–16). Often it is satisfactory to assume reasonable values for the various friction factors because minor changes have little effect on the flow distribution and the total head loss. However, we will demonstrate the more complete solution procedure in which new friction factors are calculated for each pipe for each trial.

Step 2 of the procedure calls for estimating the volume flow rate in each branch. Which pipe should have the greatest flow rate and which should have the least?

Although the final values for the friction factors could affect the magnitudes of the resistances, it appears that pipe b has the least resistance and, therefore, it should carry the greatest flow. Pipe c has the most resistance and it should carry the least flow. Many different first estimates are possible for the flow rates, but we know that

$$Q_a + Q_b + Q_c = Q_1 = 0.01 \text{ m}^3/\text{s}$$

Let's use the initial assumptions

$$Q_a = 0.0033 \text{ m}^3/\text{s} \qquad Q_b = 0.0036 \text{ m}^3/\text{s} \qquad Q_c = 0.0031 \text{ m}^3/\text{s}$$

Step 3 of the procedure is already shown in Fig. 12.6. To complete Step 4 we need values for the friction factor in each pipe. With the assumed values for the volume flow rates we can compute the Reynolds numbers and then the friction factors. Do that now.

You should have, using Eq. (12–21) and $D/\varepsilon = 578$,

$$N_{Ra} = (4.15 \times 10^7)Q_a = (4.15 \times 10^7)(0.0033 \text{ m}^3/\text{s}) = 1.37 \times 10^5$$
$$N_{Rb} = (4.15 \times 10^7)Q_b = (4.15 \times 10^7)(0.0036 \text{ m}^3/\text{s}) = 1.49 \times 10^5$$
$$N_{Rc} = (4.15 \times 10^7)Q_c = (4.15 \times 10^7)(0.0031 \text{ m}^3/\text{s}) = 1.29 \times 10^5$$

We now use Eq. (9–5) to compute the friction factor for each pipe:

$$f_a = \frac{0.25}{\left[\log_{10}\left(\frac{1}{3.7(D/\varepsilon)} + \frac{5.74}{N_{Ra}^{0.9}} \right) \right]^2}$$

$$f_a = \frac{0.25}{\left[\log_{10}\left(\frac{1}{3.7(578)} + \frac{5.74}{(1.37 \times 10^5)^{0.9}} \right) \right]^2} = 0.0241$$

In a similar manner we compute $f_b = 0.0240$ and $f_c = 0.0242$. These values are quite close in magnitude and such precision may not be justified. However, with greater disparity among the pipes in the network, more sizable differences would occur and the accuracy of the iteration technique would depend on the accuracy of evaluating the friction factors.

Now, insert the friction factors and the assumed values for Q into Eqs. (12–18), (12–20), and (12–23) to compute k_a, k_b, and k_c:

$$h_a = [5.32 + 451(f_a)](1.64 \times 10^5)Q_a^2 = k_a Q_a^2$$
$$h_a = [5.32 + 451(0.0241)](1.64 \times 10^5) Q_a^2 = 2.655 \times 10^6 Q_a^2$$

Then, $k_a = 2.655 \times 10^6$. Completing the calculation gives

$$h_a = 2.655 \times 10^6(0.0033)^2 = 28.91$$

Similarly, for branch b:

$$h_b = [8.0 + 225.6(f_b)](1.64 \times 10^5)Q_b^2 = k_b Q_b^2$$
$$h_b = [8.0 + 225.6(0.0240)](1.64 \times 10^5)Q_b^2 = 2.20 \times 10^6 Q_b^2$$
$$h_b = 2.20 \times 10^6(0.0036)^2 = 28.53$$

For branch c:

$$h_c = [13.32 + 451(f_c)](1.64 \times 10^5)Q_c^2 = k_c Q_c^2$$
$$h_c = [13.32 + 451(0.0242)](1.64 \times 10^5) Q_c^2 = 3.974 \times 10^6 Q_c^2$$
$$h_c = 3.974 \times 10^6(0.0031)^2 = 38.19$$

This completes Step 4. Now do Step 5.

For circuit 1,

$$\Sigma h_1 = h_a - h_b = 28.91 - 28.53 = 0.38$$

For circuit 2,

$$\Sigma h_2 = h_b - h_c = 28.53 - 38.19 = -9.66$$

Now do Step 6.

Here are the correct values for the three pipes.

$$2k_a Q_a = (2)(2.655 \times 10^6)(0.0033) = 17\,523$$
$$2k_b Q_b = (2)(2.20 \times 10^6)(0.0036) = 15\,850$$
$$2k_c Q_c = (2)(3.974 \times 10^6)(0.0031) = 24\,639$$

Round-off differences may occur. Now do Step 7.

For circuit 1,

$$\Sigma(2kQ)_1 = 17\,523 + 15\,850 = 33\,373$$

For circuit 2,

$$\Sigma(2kQ)_2 = 15\,850 + 24\,639 = 40\,489$$

Now you can calculate the adjustment for the flow rates ΔQ for each circuit, using Step 8.

For circuit 1,

$$\Delta Q_1 = \frac{\Sigma h_1}{\Sigma(2kQ)_1} = \frac{0.38}{33\,373} = 1.14 \times 10^{-5}$$

For circuit 2,

$$\Delta Q_2 = \frac{\Sigma h_2}{\Sigma(2kQ)_2} = \frac{-9.66}{40\,489} = -2.39 \times 10^{-4}$$

The values for ΔQ are estimates of the error in the originally assumed values for Q. We recommended that the process be repeated until the magnitude of ΔQ is less than 1 percent of the assumed value of Q. Special circumstances may warrant using a different criterion for judging ΔQ.

Step 9 can now be completed. Calculate the new value for Q_a before looking at the next panel.

The calculation is as follows:

$$Q_a' = Q_a - \Delta Q_1 = 0.0033 - 1.14 \times 10^{-5}$$
$$= 0.003\,29 \text{ m}^3/\text{s}$$

Calculate the new value for Q_c before Q_b. Pay careful attention to algebraic signs.

You should have

$$Q_c' = Q_c - \Delta Q_2 = -0.0031 - (-2.39 \times 10^{-4})$$
$$= -0.002\,86 \text{ m}^3/\text{s}$$

Notice that Q_c is negative because it flows in a counterclockwise direction in circuit 2. We can interpret the calculation for Q_c' as indicating that the magnitude of Q_c must be decreased in absolute value.

Now calculate the new value for Q_b. Remember, pipe *b* is in each circuit.

Both ΔQ_1 and ΔQ_2 must be applied to Q_b. For circuit 1,

$$Q_b' = Q_b - \Delta Q_1 = -0.0036 - 1.14 \times 10^{-5}$$

This would result in an increase in the absolute value of Q_b. For circuit 2,

$$Q_b' = Q_b - \Delta Q_2 = +0.0036 - (-2.39 \times 10^{-4})$$

This also results in increasing Q_b. Then Q_b is actually increased in absolute value by the sum of ΔQ_1 and ΔQ_2. That is,

$$Q_b' = 0.0036 + 1.14 \times 10^{-5} + 2.39 \times 10^{-4}$$
$$= 0.003\,85 \text{ m}^3/\text{s}$$

Remember that the sum of the absolute values of the flow rates in the three pipes must equal 0.01 m³/s, the total Q.

We can continue the iteration by using Q_a', Q_b', and Q_c' as the new estimates for the flow rates and repeating Steps 4–8. The results for four iteration cycles are summarized in Table 12.1. You should carry out the calculations before looking at the table.

Notice that in Trial 4, the values of ΔQ are below 1 percent of the respective values of Q. This is an adequate degree of precision. The results show that:

$$Q_a = 3.402 \times 10^{-3} \text{ m}^3/\text{s} = 0.003\,402 \text{ m}^3/\text{s} = 204.1 \text{ L/min}$$
$$Q_b = 3.785 \times 10^{-3} \text{ m}^3/\text{s} = 0.003\,785 \text{ m}^3/\text{s} = 227.1 \text{ L/min}$$
$$Q_c = 2.813 \times 10^{-3} \text{ m}^3/\text{s} = 0.002\,813 \text{ m}^3/\text{s} = 168.8 \text{ L/min}$$

The total $Q = 600$ L/min. Once again, observe that the branches having the lower resistances carry the greater flow rates.

The results of the iteration process for the Cross technique for the data of Example Problem 12.4 as shown in Table 12.1 were found using a spreadsheet on a computer. This facilitated the sequential, repetitive calculations typically required in such problems. Other computer-based computational software packages can also be used to advantage, especially if a large number of pipes and circuits exist in the network to be analyzed.

Many network analysis computer programs are commercially available. See Internet resources 1–8. Some of

TABLE 12.1

Trial 1

Circuit	Pipe	Q (m³/s)	N_R	f	k	$h = kQ^2$	$2kQ$	ΔQ	%Chg
1	a	3.300E − 03	1.37E + 05	0.0241	2.66E + 06	28.933	17535		0.38
	b	−3.600E − 03	1.50E + 05	0.0240	2.20E + 06	−28.518	15843		−0.35
					Summations:	0.415	33379	1.244E − 05	
2	b	3.600E − 03	1.50E + 05	0.0241	2.20E + 06	28.518	15843		−6.64
	c	−3.100E − 03	1.29E + 05	0.0242	3.98E + 06	−38.201	24646		7.71
					Summations:	−9.6830	40489	−2.391E − 04	

Trial 2

Circuit	Pipe	Q (m³/s)	N_R	f	k	$h = kQ^2$	$2kQ$	ΔQ	%Chg
1	a	3.288E − 03	1.37E + 05	0.0241	2.66E + 06	28.7196	17471.66		−3.43
	b	−3.852E − 03	1.60E + 05	0.0239	2.20E + 06	−32.5987	16927.42		2.93
					Summations:	−3.8792	34399.08	−1.128E − 04	
2	b	3.852E − 03	1.603E + 05	0.0239	2.20E + 06	32.5987	16927.42		−0.003
	c	−2.861E − 03	1.191E + 05	0.0243	3.98E + 06	−32.6040	22793.25		0.005
					Summations:	−0.0053	39720.67	−1.334E − 07	

Trial 3

Circuit	Pipe	Q (m³/s)	N_R	f	k	$h = kQ^2$	$2kQ$	ΔQ	%Chg
1	a	3.400E − 03	1.42E + 05	0.0241	2.65E + 06	30.6858	18048.74		−0.04
	b	−3.739E − 03	1.56E + 05	0.0240	2.20E + 06	−30.7382	16442.14		0.04
					Summations	−0.0523	34490.88	−1.518E − 06	
2	b	3.739E − 03	1.556E + 05	0.0240	2.20E + 06	30.7382	16442.14		−1.27
	c	−2.861E − 03	1.191E + 05	0.0243	3.98E + 06	−32.6010	22792.21		1.66
					Summations:	−1.8628	39234.35	−4.748E − 05	

Trial 4

Circuit	Pipe	Q (m³/s)	N_R	f	k	$h = kQ^2$	$2kQ$	ΔQ	%Chg
1	a	3.402E − 03	1.42E + 05	0.0241	2.65E + 06	30.7127	18056.51		−0.66
	b	−3.785E − 03	1.58E + 05	0.0240	2.20E + 06	−31.4908	16640.17		0.59
					Summations:	−0.7781	34696.68	−2.242E − 05	
2	b	3.785E − 03	1.58E + 05	0.0240	2.20E + 06	31.4908	16640.17		−0.03
	c	−2.813E − 03	1.17E + 05	0.0243	3.99E + 06	31.5424	22424.29		0.05
					Summations:	−0.0516	39064.46	−1.321E − 06	

Final flows in L/min

Circuit	Pipe	Q		
1	a	204.1 L/min		
	b	−227.1 L/min	Total Q = 600.0 L/min	
2	a	227.1 L/min		
	b	−168.8 L/min		

these are general purpose while others are focused on specific industrial applications such as oil and gas production and processing or chemical processing systems. The software listed in Internet resource 1 is highlighted in this book, using a special version tailored to the scope of problems encountered while learning the basic principles of fluid mechanics. The full, industrial scale version of the software has significantly higher capacity and it can be applied to virtually any major piping system design and analysis project.

REFERENCES

1. Hardee, Ray T. and Jeffrey I. Sines. 2012. *Piping System Fundamentals,* 2nd ed. Lacey WA: ESI Press, Engineered Software, Inc.

2. Cross, Hardy. 1936 (November). *Analysis of Flow in Networks of Conduits or Conductors* (University of Illinois Engineering Experiment Station Bulletin No. 286). Urbana: University of Illinois.

INTERNET RESOURCES

1. **Engineered Software, Inc.:** *www.eng-software.com* It is the producer of the PIPE-FLO® fluid flow analysis software for liquids, compressible fluids, and pulp and paper stock. PUMP-FLO® software aids in the selection of centrifugal pumps using manufacturers' electronic pump catalogs. A large database of physical property data for process chemicals and industrial fluids is available. A special demonstration version of PIPE-FLO® created for this book can be accessed by users of this book at http://www.eng-software.com/appliedfluidmechanics.

2. **Tahoe Design Software:** A producer of HYDROFLO™, HYDRONET™, and PumpBase™ software for analyzing series, parallel, and network piping systems. PumpBase™ aids in the selection of centrifugal pumps from a large database of manufacturers' performance curves.

2. **ABZ, Inc.:** It is the producer of Design Flow Solutions® software for solving a variety of fluid flow problems, including series, parallel, and network systems. A provider of engineering and consulting services to the power industry.

3. **SimSci-Esscor—Invensys Operations Management:** A provider of software for fluid flow system design and analysis, including the Process Engineering Suite (PES) that includes the INPLANT simulator for designing and analyzing plant piping systems. The PIPEPHASE software models single- and multiphase flow in oil and gas networks and pipeline systems.

4. **EPCON Software:** The producer of System 7 Process Explorer, Engineer's Aide SINET fluid flow simulation software, and CHEMPRO Engineering Suite for pipeline network analysis and process engineering in liquid, gas, and multiphase systems. It includes a large physical property database.

5. **KORF Technology:** A UK-based producer of KORF *Hydraulics* software for calculating flow rates and pressures in pipes and piping networks for liquids and isothermal, compressible, and two-phase fluids.

6. **Applied Flow Technology:** The producer of AFT Titan, AFT Arrow, AFT Fathom, and AFT Mercury software packages that can analyze piping and ducting systems for liquids, air, and other compressible fluids.

7. **Autodesk Plant Design:** The Plant Design Suite of software is used for analyzing and designing piping systems, based on industry-standard piping formats.

PRACTICE PROBLEMS

Systems with Two Branches

12.1 Figure 12.7 shows a branched system in which the pressure at A is 700 kPa and the pressure at B is 550 kPa. Each branch is 60 m long. Neglect losses at the junctions, but consider all elbows. If the system carries oil with a specific weight of 8.80 kN/m³, calculate the total volume flow rate. The oil has a kinematic viscosity of 4.8×10^{-6} m²/s.

12.2 Using the system shown in Fig. 12.2 and the data from Example Problem 12.1, determine (a) the volume flow rate of water in each branch and (b) the pressure drop between points 1 and 2 if the first gate valve is one-half closed and the other valves are wide open.

12.3 In the branched pipe system shown in Fig. 12.8, 850 L/min of water at 10°C is flowing in a DN 100 Schedule

FIGURE 12.7 Problem 12.1.

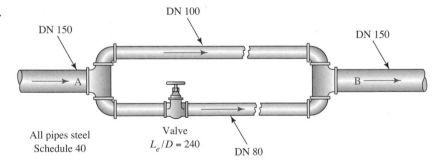

All pipes steel
Schedule 40

Valve
$L_e/D = 240$

FIGURE 12.8 Problems 12.3 and 12.8.

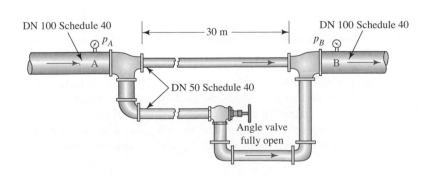

40 pipe at A. The flow splits into two DN 50 Schedule 40 pipes as shown and then rejoins at B. Calculate (a) the flow rate in each of the branches and (b) the pressure difference $p_A - p_B$. Include the effect of the minor losses in the lower branch of the system. The total length of pipe in the lower branch is 60 m. The elbows are standard.

12.4 In the branched-pipe system shown in Fig. 12.9, 1350 gal/min of benzene (sg = 0.87) at 140°F is flowing in the 8-in pipe. Calculate the volume flow rate in the 6-in and the 2-in pipes. All pipes are standard Schedule 40 steel pipes.

12.5 A 160-mm pipe branches into a 100-mm and a 50-mm pipe as shown in Fig. 12.10. Both pipes are hydraulic copper tubing and 30 m long. (The fluid is water at 10°C.) Determine what the resistance coefficient K of the valve

must be to obtain equal volume flow rates of 500 L/min in each branch.

12.6 For the system shown in Fig. 12.11, the pressure at A is maintained constant at 20 psig. The total volume flow rate exiting from the pipe at B depends on which valves are open or closed. Use $K = 0.9$ for each elbow, but neglect the energy losses in the tees. Also, because the length of each branch is short, neglect pipe friction losses. The steel pipe in branch 1 is 2-in Schedule 40, and branch 2 is 4-in Schedule 40. Calculate the volume flow rate of water for each of the following conditions:

  a. Both valves open.
  b. Valve in branch 2 only open.
  c. Valve in branch 1 only open.

12.7 Solve Problem 12.4, using the Cross technique.
12.8 Solve Problem 12.3, using the Cross technique.

FIGURE 12.9 Problems 12.4 and 12.7.

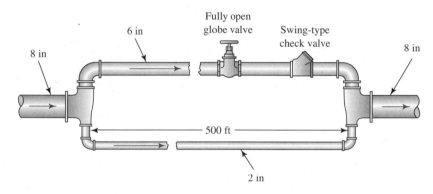

FIGURE 12.10 Problem 12.5.

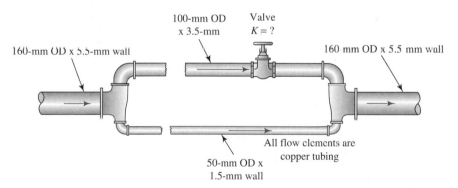

FIGURE 12.11 Problem 12.6

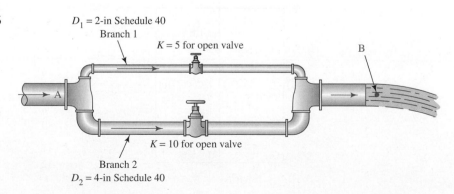

Networks

Note: Neglect minor losses.

12.9 Find the flow rate of water at 60°F in each pipe of Fig. 12.12.

12.10 Figure 12.13 represents a spray rinse system in which water at 15°C is flowing. All pipes are 3-in Type K copper tubing. Determine the flow rate in each pipe.

12.11 Figure 12.14 represents the water distribution network in a small industrial park. The supply of 15.5 ft³/s of water at 60°F enters the system at A. Manufacturing plants draw off the indicated flows at points C, E, F, G, H, and I. Determine the flow in each pipe in the system.

12.12 Figure 12.15 represents the network for delivering coolant to five different machine tools in an automated machining system. The grid is a rectangle 7.5 m by 15 m. All pipes are drawn steel tubing with a 0.065-in wall thickness. Pipes 1 and 3 are 2-in diameter, pipe 2 is 1½-in diameter, and all others are 1-in diameter. The coolant has a specific gravity of 0.92 and a dynamic viscosity of 2.00×10^{-3} Pa·s. Determine the flow in each pipe.

Supplemental Problem (PIPE-FLO® only)

12.13 Work Problem 12.4 using PIPE-FLO® software. Display the volume flow rate in each branch and all other relevant values on the FLO-Sheet®.

FIGURE 12.12   Problem 12.9.

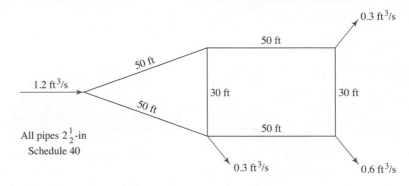

FIGURE 12.13   Problem 12.10.

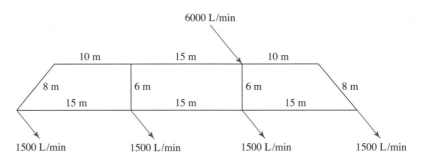

FIGURE 12.14   Problem 12.11.

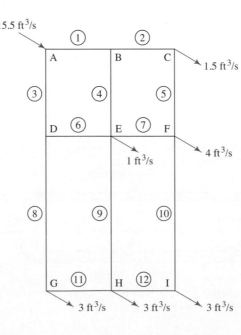

Pipe no.	Length (ft)	Size (in)
1	1500	16
2	1500	16
3	2000	18
4	2000	12
5	2000	16
6	1500	16
7	1500	12
8	4000	14
9	4000	12
10	4000	8
11	1500	12
12	1500	8

Pipe Data
All pipes Schedule 40

FIGURE 12.15 Problem 12.12.

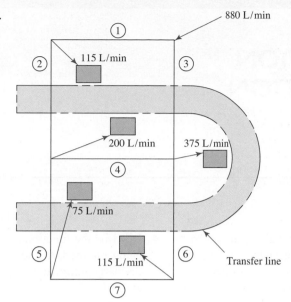

880 L/min

115 L/min

200 L/min   375 L/min

75 L/min

115 L/min

Transfer line

Pipe data
All pipes 7.5 m long
All pipes steel tubing
Wall thickness = 0.065 in

Pipe no.	Outside diameter (in)
1	2
2	$1\frac{1}{2}$
3	2
4	1
5	1
6	1
7	1

COMPUTER AIDED ENGINEERING ASSIGNMENTS

1. Write a program or a spreadsheet for analyzing parallel pipe-line systems with two branches of the type demonstrated in Example Problem 12.1. Part of the preliminary analysis, such as writing the expressions for head losses in the branches in terms of the velocities and the friction factors, may be done prior to entering data into the program.

2. Enhance the program from Assignment 1 so that it uses Eq. (8–7) from Chapter 8 to calculate the friction factor.

3. Write a program or a spreadsheet for analyzing parallel pipe-line systems with two branches of the type demonstrated in Example Problem 12.2. Use an approach similar to that described for Assignment 1.

4. Enhance the program from Assignment 3 so that it uses Eq. (8–7) from Chapter 8 to calculate the friction factor.

5. Write a program or a spreadsheet that uses the Cross technique, as described in Section 12.4 and illustrated in Example Problem 12.4, to perform the analysis of pipe flow networks. The following optional approaches may be taken:

   a. Consider single circuit networks with two branches as an alternative to the program from Assignment 1 or 2.

   b. Consider networks of two or more circuits similar to those described in Problems 12.9–12.12.

PUMP SELECTION AND APPLICATION

Pumps are used to deliver liquids through piping systems as shown in Fig. 13.1. They must deliver the desired volume flow rate of fluid while developing the required total dynamic head h_a created by elevation changes, differences in the pressure heads and velocity heads, and all energy losses in the system. You need to develop the ability to specify suitable pumps to satisfy system requirements. You also need to learn how to design efficient piping systems for the inlet to a pump (the suction line) and for the discharge side of the pump. The pressure at the inlet to the pump and its outlet must be measured to enable analysis that ensures proper operation of the pump. Most of the applications featured in this chapter are for industrial environments, fluid power systems, water supply, appliances, or other similar situations.

In this chapter you will learn how to analyze the performance of pumps and to select an appropriate pump for given application. You will also learn how to design an efficient system that minimizes the amount of energy required to drive the pump.

Exploration

- You probably encounter many different types of pumps performing many different jobs in the course of a given week. List some of them.

- For each pump, write down as much as you can about it and the system in which it operates.

- Describe the function of the pump, the kind of fluid being pumped, the source of the fluid, the ultimate discharge point, and the piping system with its valves and fittings.

Introductory Concepts   We saw the general application of pumps in earlier chapters. In Chapter 7, when the general energy equation was introduced you learned how to determine the energy added by a pump to the fluid, which we called h_a. Solving for h_a from the general energy equation yields

⇨ **Total Head on a Pump**

$$h_a = \frac{p_2 - p_1}{\gamma} + z_2 - z_1 + \frac{v_2^2 - v_1^2}{2g} + h_L \quad \textbf{(13–1)}$$

We will call this value of h_a the *total head on the pump*. Some pump manufacturers refer to this as the *total dynamic head* (*TDH*).

You should be able to interpret this equation as an expression for the total set of tasks the pump is asked to do in a given system:

- It must increase the fluid pressure from the source, p_1, to the fluid pressure at the destination point, p_2.

- It must raise the level of the fluid from the source, z_1, to the level at the destination, z_2.

FIGURE 13.1 Pumps serve as the prime movers in most fluid systems, and an understanding of their operation and how it relates to the rest of the system is critical, as in this multi-pump industrial system. (*Source:* ekipa/Fotolia)

- It must increase the velocity head from that at point 1 to that at point 2.

- It must overcome any energy losses that occur in the system due to friction in the pipes or energy losses in valves, fittings, process components, or changes in the flow area or direction of the flow.

It is your task to do the appropriate analysis to determine the value of h_a using the techniques discussed in Chapters 11 and 12.

You also learned how to compute the power delivered to the fluid by the pump, which we called P_A:

▷ **Power Delivered by a Pump to the Fluid**

$$P_A = h_a \gamma Q \qquad \textbf{(13–2)}$$

There are inevitable energy losses in the pump because of mechanical friction and the turbulence created in the fluid as it passes through it. Therefore, there is more power required to drive the pump than the amount that eventually gets delivered to the fluid. You also learned in Chapter 7 to use the efficiency of the pump e_M to determine the power input to the pump P_I:

▷ **Pump Efficiency**

$$e_M = P_A/P_I \qquad \textbf{(13–3)}$$

▷ **Power Input to a Pump**

$$P_I = P_A/e_M \qquad \textbf{(13–4)}$$

For the list of pumps you developed earlier, answer the following questions. Refer to Eq. (13–1) as you do this:

- Where does the fluid come from as it approaches the inlet of the pump?

- What is the elevation, pressure, and velocity of the fluid at the source?

- What kind of fluid is in the system?

- What is the temperature of the fluid?

- Would you consider the fluid to have a low viscosity similar to water or a high viscosity like heavy oil?

- Can you name the type of pump?

- How is the pump driven? By an electric motor? By a belt drive? Directly by an engine?

- What elements make up the *suction line* that brings fluid to the pump inlet? Describe the pipe, valves, elbows, or other elements.

- Where is the fluid delivered? Consider its elevation, the pressure at the destination, and the velocity of flow there.

- What elements make up the *discharge line* that takes fluid from the pump and delivers it to the destination? Describe the pipe, valves, elbows, or other elements.

There are many types of pumps described in this chapter; *centrifugal pumps* for general transfer of fluids from a source to a destination, *positive displacement pumps* for fluid power systems that may require very high pressures, *diaphragm pumps* that may be used to pump unwanted water from a construction site, *jet pumps* that provide drinking water to a farm home from a well, a *progressive cavity pump* used to deliver heavy, viscous fluids to a materials processing system and others. Look through the chapter to get a feel of the scope of the topics covered here. Some topics were introduced previously in Chapter 7 and you should review them now.

In this chapter you will learn how to analyze the performance of pumps and to select an appropriate pump for a given application. You will also see how the design of the fluid flow system affects the performance of the pump. This should help you to design an efficient system that minimizes the work required by the pump and, therefore, the amount of energy required to drive the pump.

13.1 OBJECTIVES

A wide variety of pumps is available to transport liquids in fluid flow systems. The proper selection and application of pumps requires an understanding of their performance characteristics and typical uses.

After completing this chapter, you should be able to:

1. List the parameters involved in pump selection.

2. List the types of information that must be specified for a given pump.

3. Describe the basic pump classifications.

4. List six types of *rotary positive-displacement pumps*.

5. List three types of *reciprocating positive-displacement pumps*.

6. List three types of *kinetic pumps*.

7. Describe the main features of *centrifugal pumps*.

8. Describe *deep-well jet pumps* and *shallow-well jet pumps*.

9. Describe the typical performance curve for rotary positive-displacement pumps.

10. Describe the typical performance curve for centrifugal pumps.

11. State the *affinity laws* for centrifugal pumps as they relate to the relationships among speed, impeller diameter, capacity, total head capability, and power required to drive the pump.

12. Describe how the *operating point* of a pump is related to the *system resistance curve (SRC)*.

13. Define the *net positive suction head required (NPSH$_R$)* for a pump and discuss its significance in pump performance.

14. Describe the importance of the vapor pressure of the fluid in relation to the *NPSH*.

15. Compute the *NPSH* available ($NPSH_A$) for a given suction line design and a given fluid.

16. Define the *specific speed* for a centrifugal pump and discuss its relationship to pump selection.

17. Describe the effect of increased viscosity on the performance of centrifugal pumps.

18. Describe the performance of parallel pumps and pumps connected in series.

19. Describe the features of a desirable suction line design.

20. Describe the features of a desirable discharge line design.

21. Consider the life cycle cost (LCC) for the pump, the entire system cost, and the operating cost over time, not just the acquisition price of the pump itself.

13.2 PARAMETERS INVOLVED IN PUMP SELECTION

When selecting a pump for a particular application, the following factors must be considered:

1. The nature of the liquid to be pumped

2. The required capacity (volume flow rate)

3. The conditions on the suction (inlet) side of the pump

4. The conditions on the discharge (outlet) side of the pump

5. The total head on the pump (the term h_a from the energy equation)

6. The type of system to which the pump is delivering the fluid

7. The type of power source (electric motor, diesel engine, steam turbine, etc.)

8. Space, weight, and position limitations

9. Environmental conditions, governing codes, and standards

10. Cost of pump purchase and installation

11. Cost of pump operation

12. The total LCC for the pumping system

The nature of the fluid is characterized by its temperature at the pumped condition, its specific gravity, its viscosity, its tendency to corrode or erode the pump parts, and its vapor pressure at the pumping temperature. The term *vapor pressure* is used to define the pressure at the free surface of a fluid due to the formation of a vapor. The vapor pressure gets higher as the temperature of the liquid increases, and it is essential that the pressure at the pump inlet stay above the vapor pressure of the fluid. We will learn more about vapor pressure in Section 13.11.

After pump selection, the following items must be specified:

1. Type of pump and manufacturer

2. Size of pump

3. Size of suction connection and type (flanged, screwed, etc.)

4. Size and type of discharge connection

5. Speed of operation

6. Specifications for driver (e.g., for an electric motor—power required, speed, voltage, phase, frequency, frame size, enclosure type)

7. Coupling type, manufacturer, and model number

8. Mounting details

9. Special materials and accessories required, if any

10. Shaft seal design and seal materials

Pump catalogs and manufacturers' representatives supply the necessary information to assist in the selection and specification of pumps and accessory equipment.

13.3 TYPES OF PUMPS

Pumps are typically classified as either positive-displacement or kinetic pumps. Table 13.1 lists several kinds of each. Positive displacement pumps deliver a specific volume of fluid for each revolution of the pump shaft or each cycle of motion of the active pumping elements. They often produce very high pressures at moderate volume flow rates. Kinetic pumps operate by transferring kinetic energy from a rotating element, called an impeller, to the fluid as it moves into and through the pump. Some of this energy is then converted to pressure energy at the pump outlet. The most frequently used type of kinetic pump is the centrifugal pump. The jet or ejector type of pump is a special version of a centrifugal kinetic pump and will be described later. A more extensive classification structure is available from Internet resource 1, with many of the variations dealing with the orientation of the pump (horizontal, vertical, in-line), the type of drive for the pump (close coupled, separately coupled, magnetic drive), or the mechanical design of certain features such as bearing supports and mountings.

13.4 POSITIVE-DISPLACEMENT PUMPS

Positive-displacement pumps ideally deliver a fixed quantity of fluid with each revolution of the pump rotor or drive shaft. The capacity of the pump is only moderately affected by pressure changes because of minor slippage caused by clearances between the housing and the rotor, pistons, vanes, or other active elements. Most positive-displacement pumps can handle liquids over a wide range of viscosities and can deliver fluids at high pressures.

13.4.1 Gear Pumps

Figure 7.2 in Chapter 7 shows the typical configuration of a *gear pump* that is used for fluid power applications and for delivering lubricants to specific machinery components. It is composed of two counter-rotating, tightly meshing gears rotating within a housing. The outer periphery of the gear teeth fit closely with the inside surface of the housing. Fluid is drawn in from the supply reservoir at the suction port and carried by the spaces between teeth to the discharge port,

TABLE 13.1 Classification of types of pumps

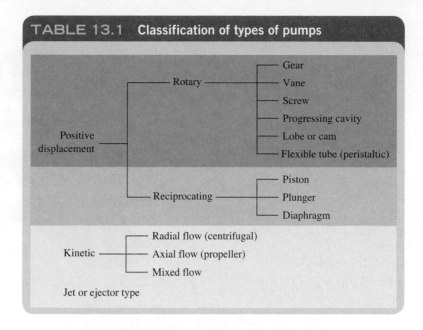

Positive displacement
- Rotary
  - Gear
  - Vane
  - Screw
  - Progressing cavity
  - Lobe or cam
  - Flexible tube (peristaltic)
- Reciprocating
  - Piston
  - Plunger
  - Diaphragm

Kinetic
- Radial flow (centrifugal)
- Axial flow (propeller)
- Mixed flow

Jet or ejector type

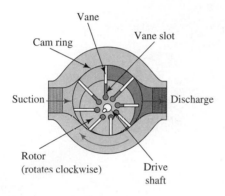

FIGURE 13.2 Vane pump.
(*Source:* Machine Design Magazine)

where it is delivered at high pressure to the system. The delivery pressure is dependent on the resistance of the system. A cutaway of a commercially available gear pump is shown in part (a) of the figure. Gear pumps develop system pressures in the range of 1500 psi to 4000 psi (10.3 MPa to 27.6 MPa). Delivery varies with the size of the gears and the rotational speed, which can be up to 4000 rpm. Deliveries from 1 to 50 gal/min (4–190 L/min) are possible with different size units. See Internet resources 8 and 9.

Advantages of gear pumps include low pulsation of the flow, good capability for handling high viscosity fluids, and it can be operated in either direction. Limiting factors include the capability of operating at only moderate pressures and that it is not recommended for handling fluids containing solids.

13.4.2 Piston Pumps for Fluid Power

Figure 7.3 shows an *axial piston pump*, which uses a rotating swash plate that acts like a cam to reciprocate the pistons. The pistons alternately draw fluid into their cylinders through suction valves and then force it out the discharge

valves against system pressure. Delivery can be varied from zero to maximum by changing the angle of the swash plate and, thus, changing the stroke of the pistons. Varying the speed of rotation of the pump also can be used to change the flow rate. Pressure capacity ranges up to 5000 psi (34.5 MPa). See Internet resources 8 and 9.

The ability to produce very high pressures is a major advantage, although only a moderate flow rate is typically available. Pressure pulsations of the output flow, being generally able to handle only low viscosity fluids, and potentially high wear of moving parts can be disadvantages.

13.4.3 Vane Pumps

Also used for fluid power, the *vane pump* (Fig. 13.3) consists of an eccentric rotor containing a set of sliding vanes that ride inside a housing. A cam ring in the housing controls the radial position of the vanes. Fluid enters the suction port at the left, is then captured in a space between two successive vanes, and is, thus, carried to the discharge port at the system pressure. The vanes then are retracted into their slots in the rotor as they travel back to the inlet, or suction, side of the pump. Variable-displacement vane pumps can deliver from zero to the maximum flow rate by varying the position of the rotor with respect to the cam ring and the housing. The setting of the variable delivery can be manual, electric, hydraulic, or pneumatically actuated to tailor the performance of the fluid power unit to the needs of the system being driven. The speed of rotation can also be varied to directly affect the delivery rate. Typical pressure capacities are from 2000 to 4000 psi (13.8 to 27.6 MPa). See Internet resources 8–9.

13.4.4 Screw Pumps

One disadvantage of the gear, piston, and vane pumps is that they deliver a pulsating flow to the output because each functional element moves a set, captured volume of

FIGURE 13.3 Screw pump.
(*Source:* Imo Pump three-screw technology, courtesy of Colfax Fluid Handling)

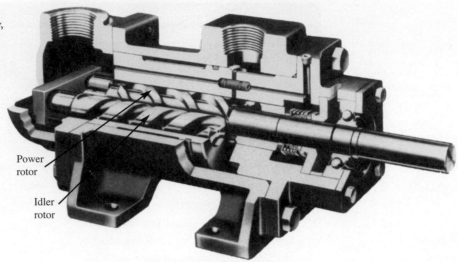

(a) Cutaway of pump assembly

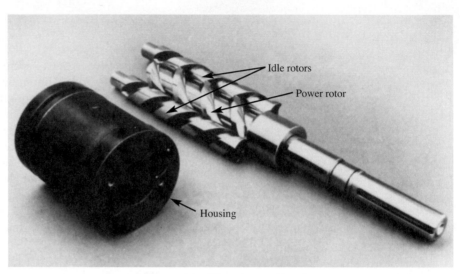

(b) Power rotor, idler rotors, and housing

fluid from suction to discharge. *Screw pumps* do not have this problem. Figure 13.3 shows a screw pump in which the central, thread-like power rotor meshes closely with the two idler rotors, creating an enclosure inside the housing that moves axially from suction to discharge, providing a continuous uniform flow. This style is called an *untimed multiple screw* pump. A *timed multiple* screw pump employs precision synchronized timing gears to maintain accurate location resulting in no contact with the casing. Screw pumps operate at nominally 3000 psi (20.7 MPa), can be run at high speeds, and run more quietly than most other types of hydraulic pumps. See Internet resource 11.

Other advantages of screw pumps are high pressure capability, quiet operation, ability to handle wide ranges of viscosities, and the availability of many different materials to ensure compatibility with the fluids. They are generally not used for fluids containing abrasives or solids.

13.4.5 Progressing Cavity Pumps

The *progressing cavity pump,* shown in Fig. 13.4, also produces a smooth, non-pulsating flow and is used mostly for the delivery of process fluids rather than hydraulic applications. As the long central rotor turns within the stator, cavities are formed which progress toward the discharge end of the pump carrying the material being handled. The rotor is typically made from steel plated with heavy layers of hard chrome to increase resistance to abrasion. For most applications, stators are made from natural rubber or any of several types and formulations of synthetic rubbers. A compression fit exists between the metal rotor and the rubber stator to reduce slippage and improve efficiency. Delivery for a given pump is dependent on the dimensions of the rotor/stator combination and is proportional to the speed of rotation. Flow capacities range up to 1860 gal/min (7040 L/min) and pressure capability is up to 900 psi (6.2 MPa).

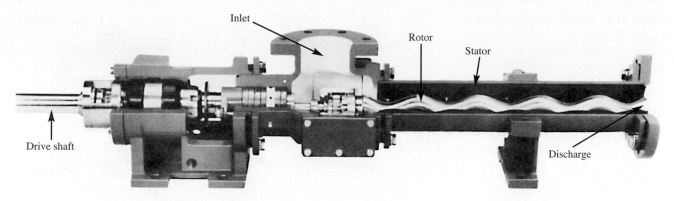

FIGURE 13.4   Progressing cavity pump. (*Source:* Robbins & Myers, Inc.)

This type of pump can handle a wide variety of fluids including clear water, slurries with heavy solids content, highly viscous liquids like adhesives and cement grout, abrasive fluids such as slurries of silicon carbide or ground limestone, pharmaceuticals such as shampoo and skin cream, corrosive chemicals such as cleaning solutions and fertilizers, and foods such as applesauce and even bread dough. They typically operate at relatively low speeds and may require high torque at startup. See Internet resources 13 and 14.

13.4.6 Lobe Pumps

The *lobe pump* (Fig. 13.5), sometimes called a cam pump, operates in a similar fashion to the gear pump. The two counter-rotating rotors may have two, three, or more lobes that mesh with each other and fit closely with the housing. Fluid is conducted around by the cavity formed between successive lobes. Advantages include very low pulsation of the flow, capability of handling large solids content and slurries, and that it is self-priming. Potential wear of the timing gears required to synchronize the rotors is a disadvantage.

13.4.7 Piston Pumps for Fluid Transfer

Piston pumps used for fluid transfer are classified as either single-acting simplex or double-acting duplex types as shown in Fig. 13.6. In principle, these are similar to the fluid power piston pumps, but they typically have a larger flow capacity and operate at lower pressures. In addition, they are usually driven through a crank-type drive rather than the swash plate described before.

13.4.8 Diaphragm Pumps

In the *diaphragm pump* shown in Fig. 13.7, a reciprocating rod moves a flexible diaphragm within a cavity, alternately discharging fluid as the rod moves to the left and drawing fluid in as it moves to the right. One advantage of this type of pump is that only the diaphragm contacts the fluid, eliminating contamination from the drive elements. The suction and discharge valves alternately open and close. See Internet resource 15.

Large-diaphragm pumps are used in construction, mining, oil and gas, food processing, chemical processing,

FIGURE 13.5   Lobe pump.

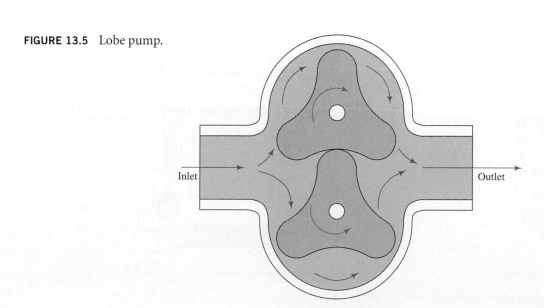

FIGURE 13.6 Piston pumps for fluid transfer.

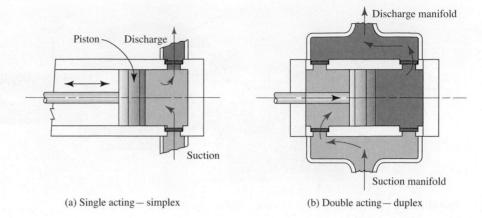

(a) Single acting— simplex

(b) Double acting— duplex

wastewater processing, and other industrial applications. Most are double acting with two diaphragms at opposite ends of the pump. Parallel suction and discharge ports and check valves provide a relatively smooth delivery while handling heavy solids content. The diaphragm can be made from many different rubber-like materials such as buna-N, neoprene, nylon, PTFE, polypropylene, and many special elastomeric polymers. Selection should be based on compatibility with the pumped fluid. Many such pumps are driven by compressed air controlled by a directional control valve.

Small-diaphragm pumps are also available that deliver very low fluid flow rates for applications like metering chemicals into a process, microelectronics manufacturing, and medical treatment. Most use electromagnetism to produce reciprocating motion of a rod that drives the diaphragm.

13.4.9 Peristaltic Pumps

Peristaltic pumps (Fig. 13.8) are unique in that the fluid is completely captured within a flexible tube throughout the pumping cycle. The tube is routed between a set of rotating rollers and a fixed housing. The rollers squeeze the tube, trapping a given volume between adjacent rollers. The design effectively eliminates the possibility of contaminating the product, making it attractive for chemical, medical, food processing, printing, water treatment, industrial, and scientific applications.

The tubing material is selected to be compatible with the fluid being pumped, whether it is alkaline, acid, or solvent. Typical materials are neoprene, PVC, PTFE, silicone, polyphenylene sulfide (PPS), and various formulations of proprietary thermoplastic elastomers. See Internet resource 16.

13.4.10 Performance Data for Positive-Displacement Pumps

In this section we will discuss the general characteristics of direct-acting reciprocating pumps and rotary pumps.

The operating characteristics of positive-displacement pumps make them useful for handling such fluids as water, hydraulic oils in fluid power systems, chemicals, paint, gasoline, greases, adhesives, and some food products. Because delivery is proportional to the rotational speed of the rotor, these pumps can be used for metering. In general, they are used for high-pressure applications requiring a relatively constant delivery. Some disadvantages of some designs include pulsating output, susceptibility to damage by solids and abrasives, and need for a relief valve.

13.4.11 Reciprocating Pump Performance

In its simplest form, the reciprocating pump (Fig. 13.6) employs a piston that draws fluid into a cylinder through an

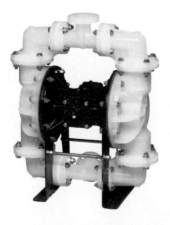

(a) Diaphragm pump with nonmetallic housing.

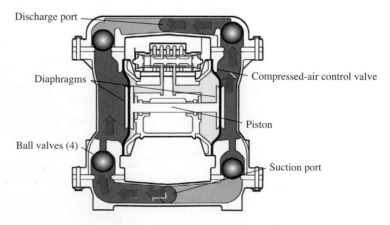

(b) Diagram of the flow through a double-piston diaphragm pump.

FIGURE 13.7 Diaphragm pump. (*Source:* Warren Rupp, Inc.)

(a) Peristaltic pump with variable-speed drive system.

(b) Peristaltic pump with case open to show tubing and rotating drive rollers.

FIGURE 13.8 Peristaltic pump. (*Source:* Watson-Marlow Pumps Group)

intake valve as the piston draws away from the valve. Then, as the piston moves forward, the intake valve closes and the fluid is pushed out through the discharge valve. Such a pump is called *simplex*, and its curve of discharge versus time looks like that shown in Fig. 13.9(a). The resulting intermittent delivery is often undesirable. If the piston is *double acting* or *duplex*, one side of the piston delivers fluid while the other takes fluid in, resulting in the performance curve shown in Fig. 13.9(b). The delivery can be smoothed even more by having three or more pistons. Piston pumps for hydraulic systems often have five or six pistons.

13.4.12 Rotary Pump Performance

Figure 13.10 shows a typical set of performance curves for rotary pumps such as gear, vane, screw, and lobe pumps. It is a plot of capacity, efficiency, and power versus discharge pressure. As pressure is increased, a slight decrease in capacity

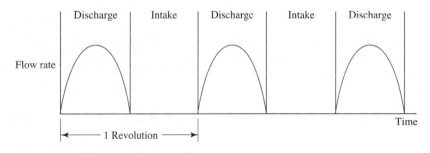

(a) Single-acting pump—simplex

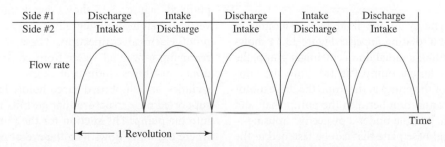

(b) Double-acting pump—duplex

FIGURE 13.9 Simplex and duplex pump delivery.

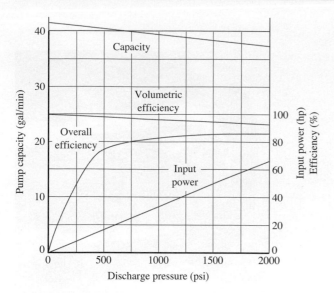

FIGURE 13.10 Performance curves for a positive-displacement rotary pump.

occurs due to internal leakage from the high-pressure side to the low-pressure side. This is often insignificant. The power required to drive the pump varies almost linearly with pressure. Also, because of the positive-displacement designs for rotary pumps, capacity varies almost linearly with the rotational speed, provided the suction conditions allow free flow into the pump.

The efficiency for positive-displacement pumps is typically reported in two ways, as shown in Fig. 13.10. *Volumetric efficiency* is a measure of the ratio of the volume flow rate delivered by the pump to the theoretical delivery, based on the displacement per revolution of the pump, times the speed of rotation. This efficiency is usually in the range from 90 percent to 100 percent, decreasing with increasing pressure in proportion to the decrease in capacity. *Overall efficiency* is a measure of the ratio of the power delivered to the fluid to the power input to the pump. Included in the overall efficiency is the volumetric efficiency, the mechanical friction from moving parts, and energy losses from the fluid as it passes through the pump. When operating at design conditions, rotary positive-displacement pumps exhibit an overall efficiency ranging from 80 percent to 90 percent.

13.5 KINETIC PUMPS

Kinetic pumps add energy to the fluid by accelerating it through the action of a rotating *impeller*. Figure 13.11 shows the basic configuration of a radial flow centrifugal pump, the most common type of kinetic pump. Part (a) shows the complete unit consisting of the pump at the front, the drive motor at the rear, and the connection between the pump shaft and the motor shaft in the middle under a protective housing—all mounted on a rigid base plate that can be fastened to the floor or another part of a machine where it is to be used. Part (b) shows a cutaway of the pump with the suction inlet on the right and the discharge port at the top. The fluid is drawn

into the center of the impeller and then thrown outward by the vanes. Leaving the impeller, the fluid passes through a spiral-shaped volute, where it is gradually slowed, and causing part of the kinetic energy to be converted to fluid pressure. The pump shaft, bearings, seal, and the housing are critical to efficient, reliable pump operation and long life. Part (c) shows an open radial-type impeller mounted in the pump case, oriented so that the discharge port is to the left. See Internet resource 7 for information and photographs for a wide variety of styles of centrifugal pumps.

Figure 13.12 shows the basic design of radial, axial, and mixed-flow impellers. The propeller type of pump (axial flow) depends on the hydrodynamic action of the propeller blades to lift and accelerate the fluid axially, along a path parallel to the axis of the propeller. The mixed-flow pump incorporates some actions from both the radial centrifugal and propeller types. See Section 13.15 for a comparison of the modes of operation of the three types of impellers.

13.5.1 Jet Pumps

Jet pumps, frequently used for household water systems, are composed of a centrifugal pump along with a jet or ejector assembly. Figure 13.13 shows a typical deep-well jet pump configuration where the main pump and motor are located above ground at the top of the well and the jet assembly is down near the water level. The pump delivers water under pressure down into the well through the pressure pipe to a nozzle. The jet issuing from the nozzle creates a vacuum behind it, which causes well water to be drawn up along with the jet. The combined stream passes through a diffuser, where the flow is slowed, thus, converting some of the kinetic energy of the water to pressure. Because the diffuser is inside the suction pipe, the water is carried to the inlet of the pump, where it is acted on by the impeller. Part of the output is discharged to the system being supplied and the remainder is recirculated to the jet to continue the operation.

If the well is shallow, with less than about 6.0 m (20 ft) from the pump to the water level, the jet assembly can be built into the pump body. Then the water is lifted through a single suction pipe, as shown in Fig. 13.14.

13.5.2 Submersible Pumps

Submersible pumps are designed so the entire assembly of the centrifugal pump, the drive motor, and the suction and discharge apparatus can be submerged in the fluid to be pumped. Figure 13.15 shows one design that has the sealed, vertical-shaft motor integrally mounted on top with a waterproof electrical connection. These pumps are useful for removing unwanted water from construction sites, mines, utility manholes, industrial tanks, waste water treatment facilities, and shipboard cargo holds. The pump is typically supported on a structure that permits free flow of the fluid into the pump. The suction for the pump is at the bottom, where the water flows into the eye of the abrasion resistant impeller, specially designed to handle large solids mixed with the water. The discharge flows out through the discharge port at the left.

(a) Centrifugal pump with drive motor on a mounting base.

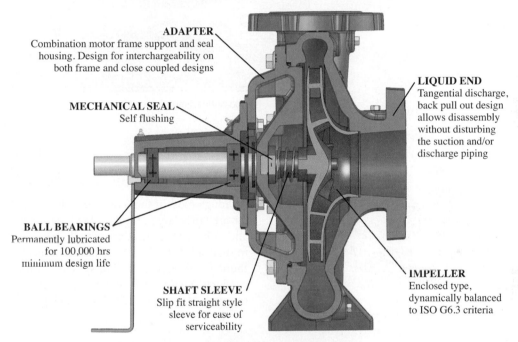

ADAPTER
Combination motor frame support and seal
housing. Design for interchargeability on
both frame and close coupled designs

MECHANICAL SEAL
Self flushing

BALL BEARINGS
Permanently lubricated
for 100,000 hrs
minimum design life

SHAFT SLEEVE
Slip fit straight style
sleeve for ease of
serviceability

LIQUID END
Tangential discharge,
back pull out design
allows disassembly
without disturbing
the suction and/or
discharge piping

IMPELLER
Enclosed type,
dynamically balanced
to ISO G6.3 criteria

(b) Cutaway view of a centrifugal pump with an enclosed type impeller.

(c) Radial, open-type impeller in the rear part of its pump case. Fluid enters
at the center of the impeller (called the eye), is thrown radially outward
by the vanes, travels around the volute, and exits through the discharge
port at the left. Rotation is counterclockwise. The front part of the case
contains the suction port and completes the volute.

FIGURE 13.11 Centrifugal pump and its components. (*Source:* Crane Pumps and Systems, Inc.)

FIGURE 13.12 Three styles of impellers for kinetic pumps.

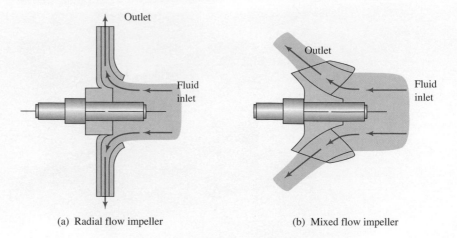

(a) Radial flow impeller

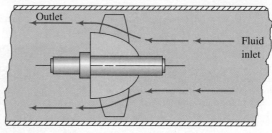

(b) Mixed flow impeller

(c) Axial flow impeller (propeller)

13.5.3 Small Centrifugal Pumps

Although most of the centrifugal pump styles discussed thus far are fairly large and have been designed for commercial and industrial applications, small units are available for use in small appliances such as clothes washers and dishwashers, fountains, machine cooling systems, and other small-scale products. Figure 13.16 shows one such design. See Internet resource 7 for more examples of such pumps.

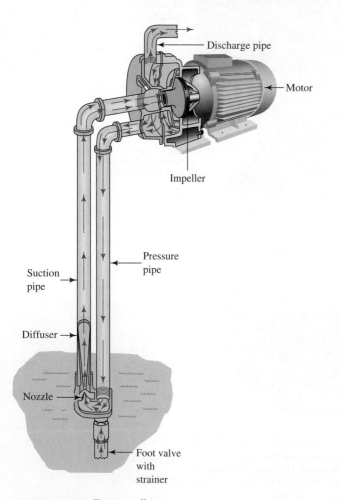

FIGURE 13.13 Deep-well jet pump.

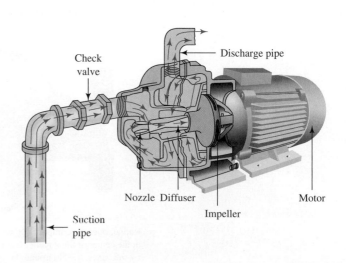

FIGURE 13.14 Shallow-well jet pump.

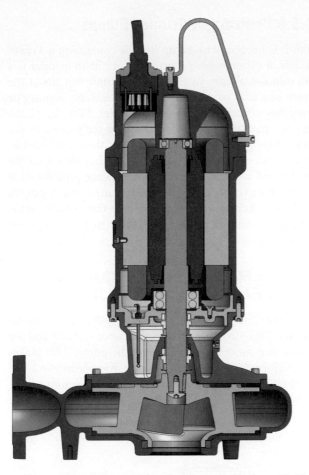

FIGURE 13.15 Submersible solids handling pump. Fluid enters through the eye at the bottom and exits to the left through the discharge port. (*Source:* Crane Pumps and Systems)

Suction port

Discharge port

FIGURE 13.16 Small centrifugal pump with integral motor for use in appliances and similar applications. (*Source:* Crane Pumps & Systems, Piqua, OH)

13.5.4 Self-Priming Pumps

It is essential that proper conditions exist at a pump's suction port when the pump is started to ensure that fluid will flow into the impeller and establish a steady flow of liquid. The term *priming* describes this process. The preferred method of priming a pump is to place the fluid source above the centerline of the impeller, relying on the effect of gravity to flood the suction port. However, it is often necessary to draw the fluid from a source below the pump, requiring the pump to create a partial vacuum to lift the fluid while simultaneously expelling any air that collects in the suction piping. See Internet resource 7.

Figure 13.17 shows one of several styles of self-priming pumps. The enlarged inlet chamber retains some of the

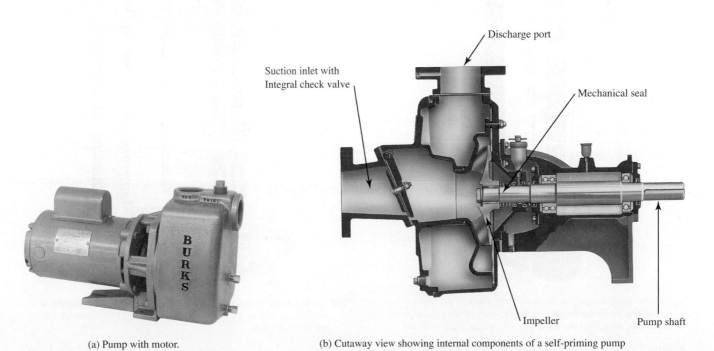

Suction inlet with Integral check valve

Discharge port

Mechanical seal

Impeller

Pump shaft

(a) Pump with motor.

(b) Cutaway view showing internal components of a self-priming pump

FIGURE 13.17 Self-priming pump. (*Source:* Crane Pumps & Systems)

liquid inside the housing during periods of shutdown with the action of the check valve in the suction port. When started, the impeller begins pulling air and water from the suction pipe into the housing and then opening the check valve. Some of the pumped water recirculates to maintain the pumping action. Simultaneously, the air flows out the discharge port, and the process continues until full liquid flow is established. Such pumps can lift fluid as much as 25 ft, although lower lifts are more common.

13.5.5 Column Pumps

When drawing fluid from a tank, sump, or other source with moderate depth, the column pump like that shown in Fig. 13.8 is a useful design to consider. Part (b) of the figure shows the internal arrangement of the pump assembly consisting of the suction port, the impeller, the case, and the discharge port that sits on the bottom of the tank, delivering the fluid through the vertical discharge line on the right side of the assembly. A support structure or short legs may be provided if needed to permit free flow of the fluid into the suction port. The long column on the left houses the drive shaft that extends from the top of the pump case to the top where a vertical motor is mounted and connected to the pump's drive shaft. See Internet resource 7.

13.5.6 Centrifugal Grinder Pumps

When it is necessary to pump liquids containing a variety of solids, a submersible pump with a built-in grinder is a good solution. Figure 13.19 shows a design that sits at the bottom of a tank or sump and handles sewage, laundry or dishwasher effluent, or other wastewater. Part (b) of the figure is a partial cutaway view showing the grinder at the bottom attached to the impeller shaft at the pump inlet so that it reduces the size of solids before they flow into the impeller and are delivered up the discharge pipe for final disposal. Part (c) shows the grinder cutters. Such pumps are often equipped with float switches that actuate automatically to control the level of fluid in the sump. See Internet resource 7.

13.6 PERFORMANCE DATA FOR CENTRIFUGAL PUMPS

Because centrifugal pumps are not positive-displacement types, there is a strong dependency between capacity and the pressure that must be developed by the pump. This makes their performance ratings somewhat more complex. The typical rating curve plots the total head on the pump h_a versus the capacity or discharge Q, as shown in Fig. 13.20. The

(a) Column pump designed to be installed in a tank or pit. Vertical drive motor mounts on the structure at the top and connects to the pump shaft. (Source: Crane Pumps & Systems, Piqua, OH)

FIGURE 13.18 Column pump. (*Source:* Crane Pumps & Systems)

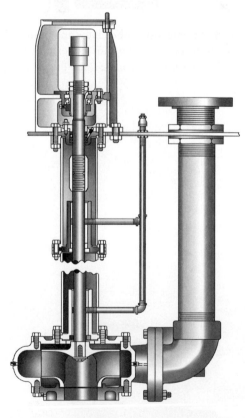

(b) Column pump—section view. Fluid enters the eye of the impeller at the bottom, exits to the right from the case and flows upward through the discharge pipe.

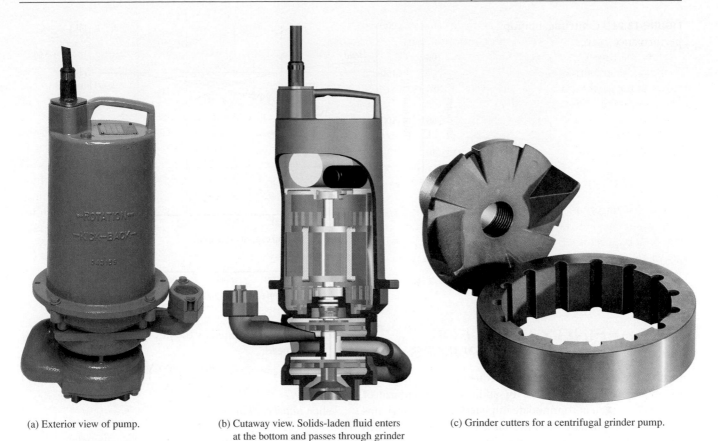

(a) Exterior view of pump.

(b) Cutaway view. Solids-laden fluid enters at the bottom and passes through grinder cutters before entering the impeller and outward through the discharge port.

(c) Grinder cutters for a centrifugal grinder pump.

FIGURE 13.19 Centrifugal grinder pump. (*Source:* Crane Pumps & Systems)

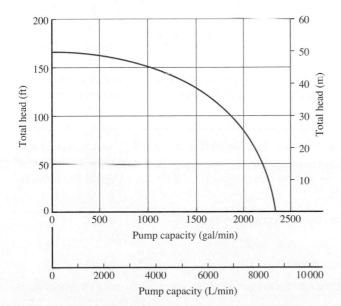

FIGURE 13.20 Performance curve for a centrifugal pump—total head versus capacity.

total head h_a is calculated from the general energy equation, as described in Chapter 7. It represents the amount of energy added to a unit weight of the fluid as it passes through the pump. See also Eq. (13–1).

As shown in Fig. 13.11, there are large clearances between the rotating impeller and the casing of the pump. This accounts for the decrease in capacity as the total head increases. Indeed, at a cut-off head, the flow is stopped completely when all of the energy input from the pump goes to maintain the head. Of course, the typical operating head is well below the cut-off head so that high capacity can be achieved.

The efficiency and power required are also important to the successful operation of a pump. Figure 13.21 shows a more complete performance rating of a pump, superimposing head, efficiency, and power curves and plotting all three versus capacity. Normal operation should be in the vicinity of the peak of the efficiency curve, with peak efficiencies in the range of 60–80 percent being typical for centrifugal pumps.

FIGURE 13.21 Centrifugal pump performance curves.

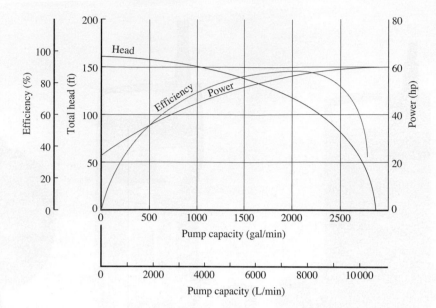

13.7 AFFINITY LAWS FOR CENTRIFUGAL PUMPS

Most centrifugal pumps can be operated at different speeds to obtain varying capacities. In addition, a given size of pump casing can accommodate impellers of differing diameters. It is important to understand the manner in which capacity, head, and power vary when either speed or impeller diameter is varied. These relationships, called *affinity laws*, are listed here. The symbol N refers to the rotational speed of the impeller, usually in revolutions per minute (r/min, or rpm).

When *speed varies*:

a. Capacity varies directly with speed:

$$\frac{Q_1}{Q_2} = \frac{N_1}{N_2} \qquad (13\text{--}5)$$

b. The total head capability varies with the square of the speed:

$$\frac{h_{a_1}}{h_{a_2}} = \left(\frac{N_1}{N_2}\right)^2 \qquad (13\text{--}6)$$

c. The power required by the pump varies with the cube of the speed:

$$\frac{P_1}{P_2} = \left(\frac{N_1}{N_2}\right)^3 \qquad (13\text{--}7)$$

When *impeller diameter varies*:

a. Capacity varies directly with impeller diameter:

$$\frac{Q_1}{Q_2} = \frac{D_1}{D_2} \qquad (13\text{--}8)$$

b. The total head varies with the square of the impeller diameter:

$$\frac{h_{a_1}}{h_{a_2}} = \left(\frac{D_1}{D_2}\right)^2 \qquad (13\text{--}9)$$

c. The power required by the pump varies with the cube of the impeller diameter:

$$\frac{P_1}{P_2} = \left(\frac{D_1}{D_2}\right)^3 \qquad (13\text{--}10)$$

Efficiency remains nearly constant for speed changes and for small changes in impeller diameter. (See Internet resource 10.)

Example Problem 13.1 Assume that the pump for which the performance data are plotted in Fig. 13.21 was operating at a rotational speed of 1750 rpm and that the impeller diameter was 13 in. First determine the head that would result in a capacity of 1500 gal/min and the power required to drive the pump. Then, compute the performance at a speed of 1250 rpm.

Solution From Fig. 13.21, projecting upward from $Q_1 = 1500$ gal/min gives

$$\text{Total head} = 130\,\text{ft} = h_{a_1}$$
$$\text{Power required} = 50\,\text{hp} = P_1$$

When the speed is changed to 1250 rpm, the new performance can be computed by using the affinity laws:

$$\text{Capacity: } Q_2 = Q_1(N_2/N_1) = 1500(1250/1750) = 1071\,\text{gal/min}$$
$$\text{Head: } h_{a_2} = h_{a_1}(N_2/N_1)^2 = 130(1250/1750)^2 = 66.3\,\text{ft}$$
$$\text{Power: } P_2 = P_1(N_2/N_1)^3 = 50(1250/1750)^3 = 18.2\,\text{hp}$$

Note the significant decrease in the power required to run the pump. If the capacity and the available head are adequate, large savings in energy costs can be obtained by varying the speed of operation of a pump. See also Section 13.14.1 on variable speed drives.

It should be noted here that these calculations apply only to the performance of the pump. Consequent changes occur in the behavior of the fluid flow system resulting in changes to the system head curve so the new operating point of the system must also be determined. This subject is discussed further in Sections 13.13 and 13.15.

13.8 MANUFACTURERS' DATA FOR CENTRIFUGAL PUMPS

It has been stated before that the basic data needed to specify a suitable pump for a given system is the required volume flow rate, called *capacity,* and the total head, h_a, for the system in which the pump is to operate. Because pump manufacturers are able to use different impeller diameters and speeds, they can cover a wide range of requirements for capacity and head with a few basic pump sizes. Many manufacturers of centrifugal pumps for industrial applications use a designation system that provides useful data for the size of the pump. For example, a pump may carry the designation, $2 \times 3 - 10$, and each of these numbers describes an important feature of the pump. The $2 \times 3 - 10$ centrifugal pump is one with a 2-in discharge connection, a 3-in suction connection, and a casing that can accommodate an impeller with a diameter of 10 in or smaller.

Figure 13.22 shows an example of a composite rating chart for one line of pumps operating at a speed of 3500 rpm, which allows the quick determination of the pump size. Then, for each pump size, the manufacturer prepares more complete performance charts as shown next. Note that Fig. 13.22 does

not represent a particular pump manufacturer's data, but it is typical of how the data are displayed in catalogs that can be found at the Internet resources 5–10 and 12.

13.8.1 Effect of Impeller Size

Figure 13.23 shows how the performance of the sample $2 \times 3 - 10$ centrifugal pump varies as the size of the impeller varies. Shown are the capacity-versus-head curves for five different sizes of impellers from 6 in to 10 in within the same casing. The operating speed is 3500 rpm, which corresponds to the full load speed of a common two-pole AC electric motor. Users will often modify an existing pump by trimming its impeller diameter to more nearly match the pump output to the needs of a particular system.

13.8.2 Effect of Speed

Figure 13.24 shows the performance of the same $2 \times 3 - 10$ pump operating at 1750 rpm (a typical operating speed for a four-pole AC motor) instead of 3500 rpm. If we compare the maximum total heads for each impeller size, we illustrate the affinity law; that is, doubling the speed increases the total

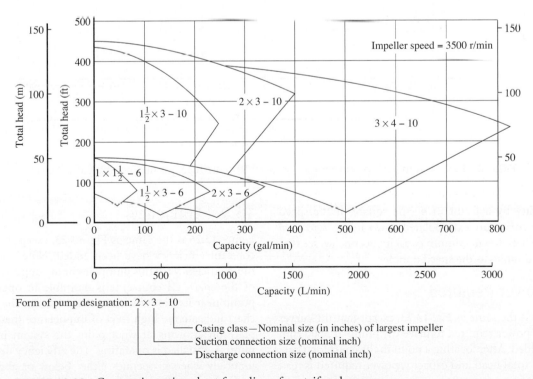

FIGURE 13.22 Composite rating chart for a line of centrifugal pumps.

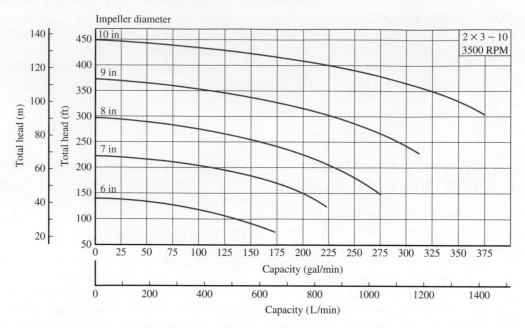

FIGURE 13.23 Illustration of pump performance for different impeller diameters. Performance chart for a $2 \times 3 - 10$ centrifugal pump at 3500 rpm.

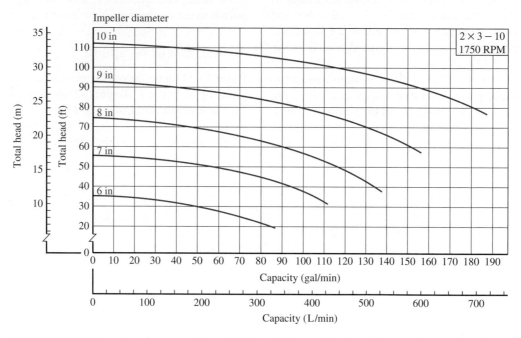

FIGURE 13.24 Pump performance for a $2 \times 3 - 10$ centrifugal pump operating at 1750 rpm.

head capability by a factor of 4 (the square of the speed ratio). If the curves are extrapolated down to the zero total head point where the maximum capacity occurs, we see that the capacity doubles as the speed doubles.

13.8.3 Power Required

Figure 13.25 is the same as Fig. 13.23, except that the curves showing the power required to drive the pump at 3500 rpm have been added. After locating a point in the chart for a particular set of total head and capacity, power required is read from the set of power curves.

13.8.4 Efficiency

Figure 13.26 is the same as Fig. 13.23, except that curves of constant efficiency have been added. Note that the maximum efficiency for this pump lies in the upper-right portion of the chart. Of course, it is desirable to operate a given pump near its best efficiency point (BEP). The data in this chart indicate the high level of importance that you should place on knowing at what point the system in which the pump is installed is operating. The efficiency decreases dramatically when it operates either below or above the BEP and other performance issues are likely to arise that can

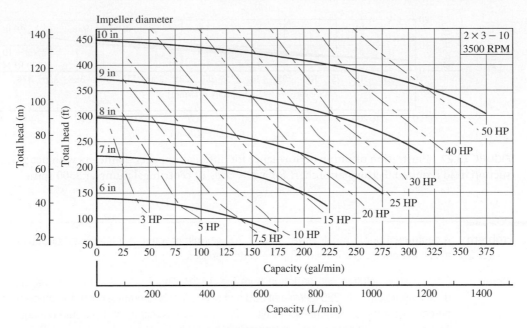

FIGURE 13.25 Illustration of pump performance for different impeller diameters with power required. Performance chart for a 2 × 3 − 10 centrifugal pump at 3500 rpm.

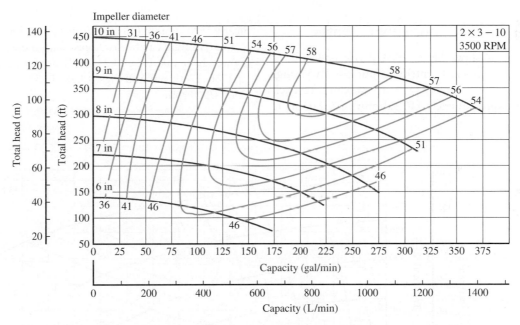

FIGURE 13.26 Illustration of pump performance for different impeller diameters with efficiency. Performance chart for a 2 × 3 − 10 centrifugal pump at 3500 rpm.

affect the life of the pump or its bearings. More is said about these issues in later sections of this chapter.

13.8.5 Net Positive Suction Head Required

Net positive suction head required ($NPSH_R$) is an important factor to consider in applying a pump, as will be discussed in Section 13.9. $NPSH_R$ is related to the pressure at the inlet to the pump. For this discussion, it is sufficient to say that a low $NPSH_R$ is desirable. Again, after locating a point in the chart for a particular set of total head and capacity, $NPSH_R$ is read from the given set of curves.

Figure 13.27 shows curves for $NPSH_R$ in relation to the range of capacity for the same pump as used for Figure 13.23. These curves are placed below the pump curves as shown next.

13.8.6 Complete Performance Chart

Figure 13.28 puts all these data together on one chart so the user can see all important parameters at the same time. The chart seems complicated at first, but having considered each individual part separately should help you to interpret it correctly. The example problem below illustrates the interpretation of this chart.

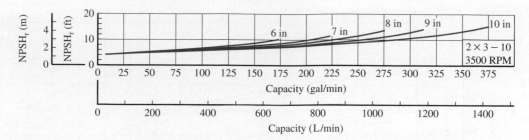

FIGURE 13.27 Illustration of pump performance for different impeller diameters with net positive suction head required. Performance chart for a 2 × 3 − 10 centrifugal pump at 3500 rpm.

Example Problem 13.2

A centrifugal pump must deliver at least 200 gal/min of water at a total head of 300 ft of water. Specify a suitable pump. List its performance characteristics.

Solution

One possible solution can be found from Fig. 13.28. The 2 × 3 − 10 pump with a 9-in impeller will deliver approximately 229 gal/min at 300 ft of head. At this operating point, the efficiency would be 58.0 percent, near the maximum for this type of pump. Approximately 30 hp would be required. The $NPSH_R$ at the suction inlet to the pump is approximately 8.8 ft of water.

13.8.7 Additional Performance Charts

Figures 13.29–13.34 show the complete performance charts for six other medium-sized centrifugal pumps. They range in size from 1½ × 3 − 6 to 6 × 8 − 17. Maximum capacities range from approximately 110 gal/min (416 L/min) to nearly 3500 gal/min (13 250 L/min). A total head up to 700 ft (213 m) of fluid can be developed within the pumps in these figures. Note that Figs. 13.29–13.32 are for pumps operating at 1750–1780 rpm and Figs. 13.28, 13.33, and 13.34 are for 3500–3560 rpm.

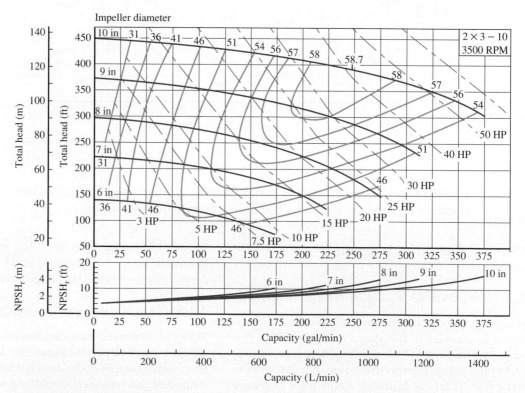

FIGURE 13.28 Complete pump performance chart for a 2 × 3 − 10 centrifugal pump at 3500 rpm. (*Source:* Reprinted by permission of ITT Corporation.)

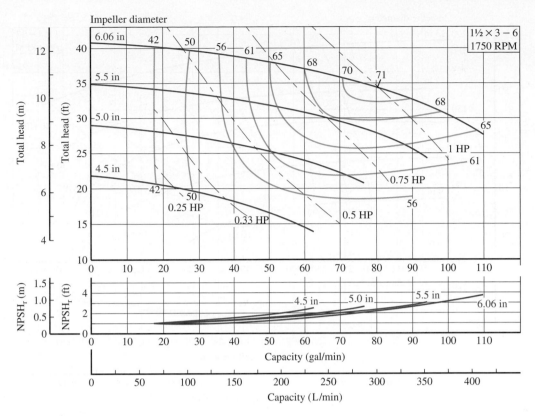

FIGURE 13.29   Performance for a 1½ × 3 − 6 centrifugal pump at 1750 rpm.
(*Source:* Reprinted by permission of ITT Corporation.)

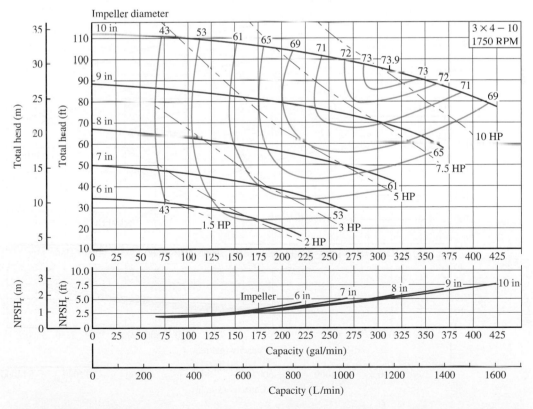

FIGURE 13.30   Performance for a 3 × 4 − 10 centrifugal pump at 1750 rpm. (*Source:* Reprinted by permission of ITT Corporation.)

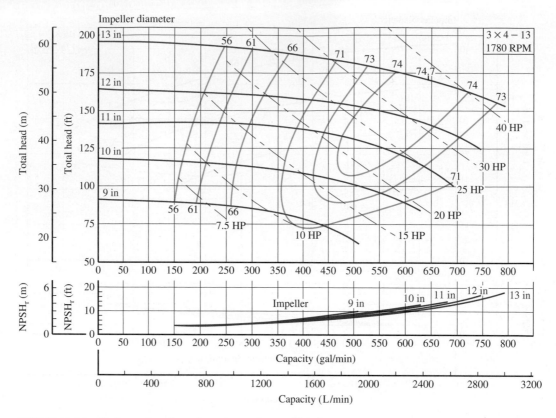

FIGURE 13.31 Performance for a 3 × 4 − 13 centrifugal pump at 1780 rpm. (*Source:* Reprinted by permission of ITT Corporation.)

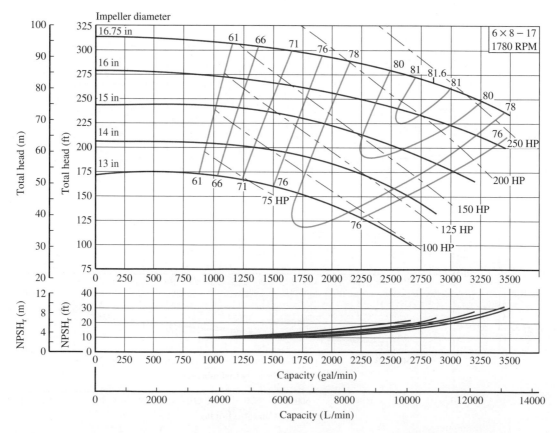

FIGURE 13.32 Performance for a 6 × 8 − 17 centrifugal pump at 1780 rpm. (*Source:* Reprinted by permission of ITT Corporation.)

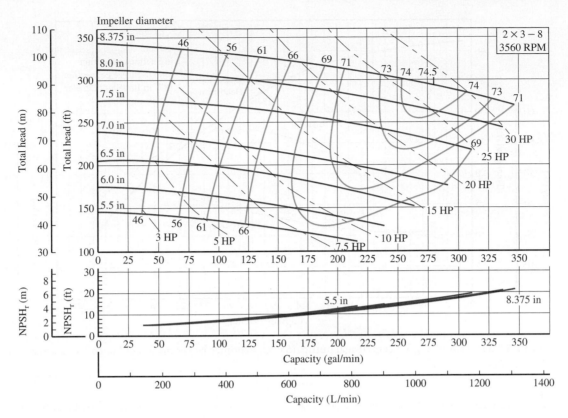

FIGURE 13.33 Performance for a 2 × 3 − 8 centrifugal pump at 3560 rpm. (*Source:* Reprinted by permission of ITT Corporation.)

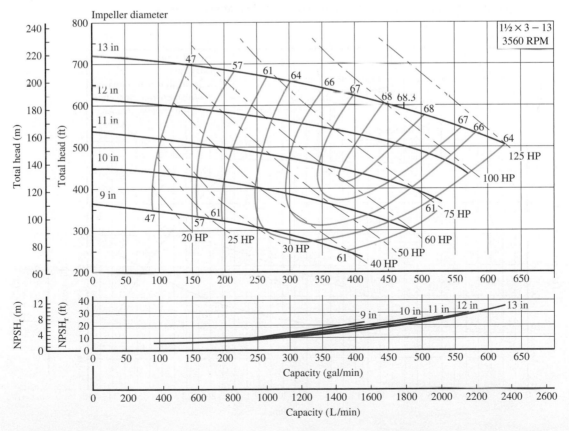

FIGURE 13.34 Performance for a 1½ × 3 − 13 centrifugal pump at 3560 rpm. (*Source:* Reprinted by permission of ITT Corporation.)

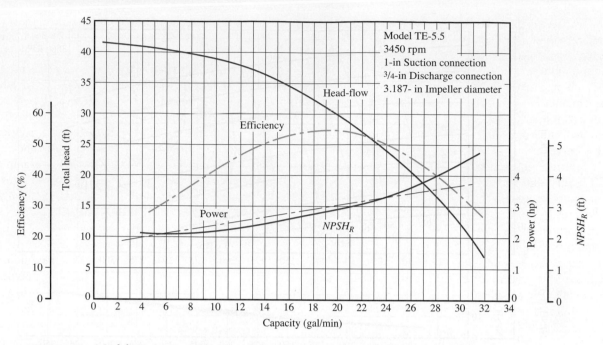

FIGURE 13.35 Model TE-5.5 centrifugal pump. (*Source:* Reprinted by permission of March Manufacturing, Inc., Glenview, IL)

Figures 13.35 and 13.36 show two additional performance curves for smaller centrifugal pumps. Because these pumps are generally offered with only one impeller size, the manner of displaying the performance parameters uses the format shown earlier in Fig. 13.21. Complete curves for total head, efficiency, input power required, and *NPSH* required are given versus the pump capacity. Each pump will deliver approximately 19 gal/min at the peak efficiency point, but the pump in Fig. 13.35 has a smaller impeller diameter giving a total head capability of 32 ft at 19 gal/min, whereas the larger pump in Fig. 13.36 has a total head capability of 43 ft at the same capacity.

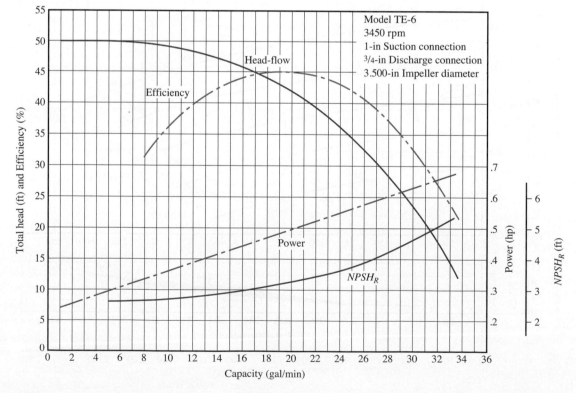

FIGURE 13.36 Model TE-6 centrifugal pump. (*Source:* Reprinted by permission of March Manufacturing, Inc., Glenview, IL)

You need to develop the ability to interpret performance data from these charts so you can specify a suitable pump for a given application. Several design projects are described at the end of this chapter that require you to specify a pump to deliver a particular volume flow rate at a given head to meet the demands of a particular system. After locating a point in the chart for particular set of total head and capacity, power required, efficiency, and $NPSH_R$ are read from the complete set of curves.

Internet resources 5–10 and 12 offer on-line performance curves for many types and sizes of centrifugal pumps. Some allow searching for suitable pumps when the total head and desired capacity are input. The PIPE-FLO® software featured in this book includes pump performance data from numerous commercial suppliers of centrifugal pumps, with the ability to automatically search for suitable pumps required for a given system being analyzed, as illustrated in Section 13.14 of this chapter. Other software packages listed in the Internet resources for Chapter 12 also include pump selection capability.

13.9 NET POSITIVE SUCTION HEAD

The descriptions of the several aspects of the performance of centrifugal pumps in the preceding sections emphasized the importance of the *net positive suction head, NPSH*. This concept is defined more completely here. The basic issues include:

1. Preventing a condition called *cavitation,* because of its extreme detrimental effects on the pump.

2. The effect of the vapor pressure of the fluid being pumped on the onset of cavitation.

3. The piping system design considerations that affect *NPSH.*

4. The $NPSH_R$ for the selected pump must be satisfied.

It is essential that the design of the piping system leading into the pump, called the *suction line,* permits full liquid flow all the way through the pump and into the discharge line. The primary factor is the fluid pressure at the pump suction inlet. The design of the suction piping system must provide a sufficiently high pressure that will avoid the development of cavitation in which vapor bubbles form within the flowing fluid. It is your responsibility to ensure that cavitation does not occur. The tendency for vapor bubbles to form depends on the nature of the fluid, its temperature, and the suction pressure. These factors are discussed in this section.

13.9.1 Cavitation

When the suction pressure at the pump inlet is too low, vapor bubbles form in the fluid in a manner similar to boiling. As an aid in understanding the formation of vapor bubbles, place an open pan of water on a cooking unit and observe its behavior as the temperature increases. At some point, a few small bubbles of water vapor will form at the bottom of the pan. Continued heating causes more bubbles to form; they rise to the surface, escape the liquid surface, and diffuse into the surrounding air. Finally, the water comes to full boiling with continuous and rapid vaporization.

To illustrate the effect of pressure on the formation of vapor bubbles, first consider that you perform the boiling experiment at a low altitude where the water in the open pan is at an atmospheric pressure of approximately 101 kPa or 14.7 psi. The temperature of the water at boiling is approximately 100°C or 212°F. However, at high altitudes, the atmospheric pressure is noticeably lower and the boiling temperature is correspondingly lower. For example, Appendix Table E.3 (Properties of the atmosphere) shows that the atmospheric pressure at 5000 ft (1524 m) is only 12.2 psi (84.3 kPa). This is the approximate elevation of Denver, Colorado, often called the "Mile-High City", where water boils at approximately 94°C or 201°F.

Relate this simple experiment with the conditions at the suction inlet of a pump. If the pump must pull fluid from below or if there are excessive energy losses in the suction line, the pressure at the pump may be sufficiently low to cause vapor bubbles to form in the fluid. Now consider what happens to the fluid as it begins its journey through the pump. Refer back to Fig. 13.11, which shows the design of a radial centrifugal pump.

The fluid enters the pump through the suction port at the central eye of the impeller and this where the lowest pressure occurs. The rotation of the impeller then accelerates the fluid outward along the vanes toward the casing, called a *volute.* Fluid pressure continues to rise throughout this process. If vapor bubbles had formed in the suction port because of excessively low pressure there, they would collapse as they flowed into the higher-pressure zones. Collapsing bubbles release large amounts of energy, which effectively exerts impact forces on the impeller vanes and cause rapid surface erosion.

When cavitation occurs, the performance of the pump is severely degraded as the volume flow rate delivered drops. The pump vibrates and becomes noisy, giving off a loud, rattling sound as if gravel was flowing with the fluid. If this was allowed to continue, the pump would be destroyed in a short time. The pump should be promptly shut down and the cause of the cavitation should be identified and corrected before resuming operation. Obviously, it is preferred to ensure that cavitation does not occur under expected operating conditions as will be demonstrated in Sections 13.10 and 13.13.

13.9.2 Vapor Pressure

The fluid property that determines the conditions under which vapor bubbles form in a fluid is its *vapor pressure* p_{vp}, typically reported as an absolute pressure in the units of kPa absolute or psia. When both vapor and liquid forms of a substance exist in equilibrium, there is a balance of vapor being driven off from the liquid by thermal energy and condensation of vapor to the liquid because of the attractive forces between molecules. The pressure of the liquid at this condition is called the vapor pressure. A liquid is called *volatile* if it has a relatively high vapor pressure and vaporizes rapidly at ambient conditions. Following is a list of six familiar

TABLE 13.2 Vapor pressure and vapor pressure head of water

Temperature °C	Vapor Pressure kPa (abs)	Specific Weight (kN/m³)	Vapor Pressure Head (m)	Temperature °F	Vapor Pressure (psia)	Specific Weight (lb/ft³)	Vapor Pressure Head (ft)
0	0.6105	9.806	0.06226	32	0.08854	62.42	0.2043
5	0.8722	9.807	0.08894	40	0.1217	62.43	0.2807
10	1.228	9.804	0.1253	50	0.1781	62.41	0.4109
20	2.338	9.789	0.2388	60	0.2563	62.37	0.5917
30	4.243	9.765	0.4345	70	0.3631	62.30	0.8393
40	7.376	9.731	0.7580	80	0.5069	62.22	1.173
50	12.33	9.690	1.272	90	0.6979	62.11	1.618
60	19.92	9.642	2.066	100	0.9493	62.00	2.205
70	31.16	9.589	3.250	120	1.692	61.71	3.948
80	47.34	9.530	4.967	140	2.888	61.38	6.775
90	70.10	9.467	7.405	160	4.736	61.00	11.18
100	101.3	9.399	10.78	180	7.507	61.58	17.55
				200	11.52	60.12	27.59
				212	14.69	59.83	35.36

liquids, ranked by increasing volatility: water, carbon tetrachloride, acetone, gasoline, ammonia, and propane. Several standards have been established by ASTM International to measure vapor pressure for different kinds of fluids. See References 1 and 2 for examples.

In the discussion of *net positive suction head* that follows, it is pertinent to use the *vapor pressure head* h_{vp} rather than the basic vapor pressure p_{vp}, where

$h_{vp} = p_{vp}/\gamma$ = Vapor pressure head of the liquid in meters or feet

The vapor pressure at any temperature must be divided by the specific weight of the liquid at that temperature.

The vapor pressure head of any liquid rises rapidly with increasing temperature. Table 13.2 lists the values of vapor pressure and vapor pressure head for water. Figure 13.37 shows graphs of vapor pressure head versus temperature, in both SI metric and U.S. Customary System units, for four different fluids: water, carbon tetrachloride, gasoline, and propane. Pumping any of these fluids at the higher temperatures requires careful consideration of the *NPSH*.

13.9.3 NPSH

Pump manufacturers test each pump design to determine the level of suction pressure required to avoid cavitation, reporting the result as the *net positive suction head required*, $NPSH_R$, for the pump at each operating condition of capacity (volume flow rate) and total head on the pump. It is the responsibility of the pump system designer to ensure that the *available net positive suction head, $NPSH_A$*, is significantly above $NPSH_R$.

Standards have been set jointly by the American National Standards Institute (ANSI) and the Hydraulic Institute (HI) calling for a minimum of a 10 percent margin for $NPSH_A$ over $NPSH_R$. We can define the *NPSH margin M* to be

⮑ **NPSH Margin**

$$M = NPSH_A - NPSH_R \qquad (13-11)$$

Higher margins, up to 100 percent, are expected for critical applications such as flood control, oil pipelines, and power generation service. Some designers call for a margin of 5.0 ft for large pumping systems. See ANSI/HI 9.6.1, *Standard for Centrifugal and Vertical Pumps for NPSH Margin*.

In design problems in this book, we call for a minimum of 10 percent margin. That is,

$$NPSH_A > 1.10 \, NPSH_R \qquad (13-12)$$

Computing $NPSH_A$ The value of $NPSH_A$ is dependent on the vapor pressure of the fluid being pumped, energy losses in the suction piping, the elevation of the fluid reservoir, and the pressure applied to the fluid in the reservoir. This can be expressed as

⮑ **NPSH Available**

$$NPSH_A = h_{sp} \pm h_s - h_f - h_{vp} \qquad (13-13)$$

These terms are illustrated in Fig. 13.38 and defined below. Figure 13.38(a) includes a pressurized reservoir placed above the pump. Part (b) shows the pump drawing fluid from an open reservoir below the pump.

p_{sp} = Static pressure (absolute) above the fluid in the reservoir

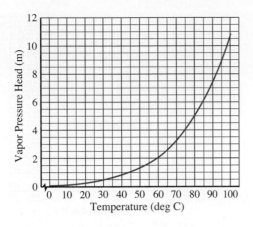

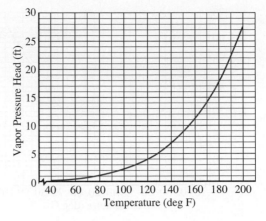

(a) Water

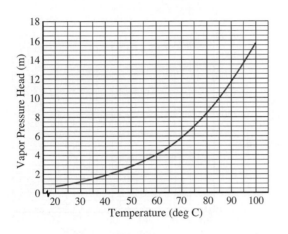

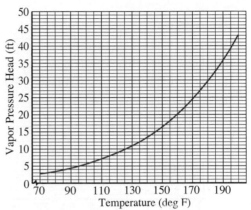

(b) Carbon Tetrachloride

FIGURE 13.37 Vapor pressure versus temperature for common liquids. The data for gasoline are approximate because there are many different formulations that have widely varying volatility for vehicle operation in different climates and altitudes.

h_{sp} = Static pressure head (absolute) above the fluid in the reservoir, expressed in meters or feet of the liquid; $h_{sp} = p_{sp}/\gamma$

h_s = Elevation difference from the level of fluid in the reservoir to the centerline of the pump suction inlet, expressed in meters or feet

If the pump is below the reservoir, h_s is positive [preferred; Fig. 13.38(a)]

If the pump is above the reservoir, h_s is negative [Fig. 13.38(b)]

h_f = Head loss in the suction piping due to friction and minor losses, expressed in meters or feet

p_{vp} = Vapor pressure (absolute) of the liquid at the pumping temperature

h_{vp} = Vapor pressure head of the liquid at the pumping temperature, expressed in meters or feet of the liquid; $h_{vp} = p_{vp}/\gamma$

Note that Eq. (13–13) does not include terms representing the velocity heads in the system. It is assumed that the velocity at the source reservoir is very nearly zero because it is very large relative to the pipe. The velocity head in the suction pipe was included in the derivation of the equation, but it cancelled out.

Effect of Pump Speed on $NPSH_R$ The data given in pump catalogs for $NPSH_R$ are for water and apply only to the listed operating speed. If the pump is operated at a different speed, the $NPSH$ required at the new speed can be calculated from

$$(NPSH_R)_2 = \left(\frac{N_2}{N_1}\right)^2 (NPSH_R)_1 \qquad \textbf{(13–14)}$$

where the subscript 1 refers to catalog data and the subscript 2 refers to conditions at the new operating speed. The pump speed in rpm is N. This is an important observation when designing variable speed pump drives.

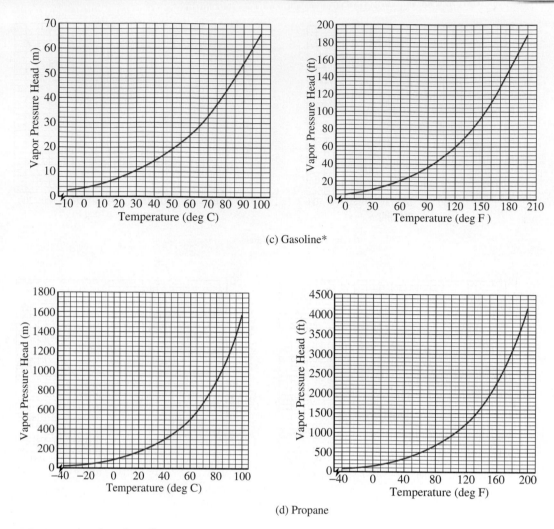

(c) Gasoline*

(d) Propane

FIGURE 13.37 (*continued*)

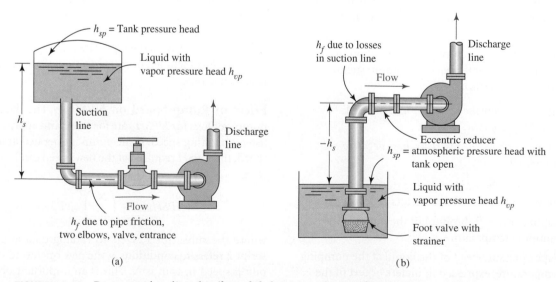

(a)

(b)

FIGURE 13.38 Pump suction-line details and definitions of terms for computing *NPSH*.

Example Problem
13.3

Determine the available *NPSH* for the system shown in Fig. 13.38(a). The fluid reservoir is a closed tank with a pressure of −20 kPa above water at 70°C. The atmospheric pressure is 100.5 kPa. The water level in the tank is 2.5 m above the pump inlet. The pipe is a DN 40 Schedule 40 steel pipe with a total length of 12.0 m. The elbow is standard and the valve is a fully open globe valve. The flow rate is 95 L/min. Then calculate the maximum allowable *NPSH_R* for the pump in this system.

Solution

Use Eq. (13–13). First, find h_{sp}:

$$\text{Absolute pressure} = \text{Atmospheric pressure} + \text{Tank gage pressure}$$

$$p_{abs} = 100.5 \text{ kPa} - 20 \text{ kPa} = 80.5 \text{ kPa}$$

But we know that

$$h_{sp} = p_{abs}/\gamma$$
$$= \frac{80.5 \times 10^3 \text{ N/m}^2}{9.59 \times 10^3 \text{ N/m}^3} = 8.39 \text{ m}$$

Now, based on the elevation of the tank, we have

$$h_s = +2.5 \text{ m}$$

To find the friction loss h_f, we must find the velocity, Reynolds number, and friction factor:

$$v = \frac{Q}{A} = \frac{95 \text{ L/min}}{1.314 \times 10^{-3} \text{ m}^2} \times \frac{1.0 \text{ m}^3/\text{s}}{60\,000 \text{ L/min}} = 1.21 \text{ m/s}$$

$$N_R = \frac{vD}{\nu} = \frac{(1.21)(0.0409)}{4.11 \times 10^{-7}} = 1.20 \times 10^5 \quad \text{(turbulent)}$$

$$\frac{D}{\varepsilon} = \frac{0.0409 \text{ m}}{4.6 \times 10^{-5} \text{ m}} = 889$$

Thus, from Fig. 8.7, $f = 0.0225$. From Table 10.5, $f_T = 0.020$. Now we have

$$h_f = f(L/D)(v^2/2g) + 2f_T(30)(v^2/2g) + f_T(340)(v^2/2g) + 1.0(v^2/2g)$$
$$\quad\text{(pipe)} \qquad\qquad \text{(elbows)} \qquad\qquad \text{(valve)} \qquad \text{(entrance)}$$

The velocity head is

$$\frac{v^2}{2g} = \frac{(1.21 \text{ m/s})^2}{2(9.81 \text{ m/s}^2)} = 0.0746 \text{ m}$$

Then, the friction loss is

$$h_f = (0.0225)(12/0.0409)(0.0746) + (0.020)(60)(0.0746)$$
$$\quad + (0.020)(340)(0.0746) + 0.0746$$
$$= (0.0746 \text{ m})[(0.0225)(12/0.0409) + (0.020)(60) + (0.020)(340) + 1.0]$$
$$= 1.16 \text{ m}$$

Finally, from Table 13.2 for the vapor pressure head of water, we get

$$h_{vp} = 3.25 \text{ m at } 70°C$$

Combining these terms gives

$$NPSH_A = 8.39 \text{ m} + 2.5 \text{ m} - 1.16 \text{ m} - 3.25 \text{ m} = 6.48 \text{ m}$$

We can compute the maximum allowable *NPSH_R* for the pump from Eq. (13–12),

$$NPSH_A > 1.10\, NPSH_R$$

Rearranging, we get

$$NPSH_R < NPSH_A/1.10 \qquad\qquad\qquad \textbf{(13–15)}$$

Then,

$$NPSH_R < 6.48 \text{ m}/1.10 = 5.89 \text{ m}$$

The result indicates that any pump that requires 5.89 m or less for *NPSH* is acceptable.

13.10 SUCTION LINE DETAILS

The *suction line* refers to all parts of the flow system from the source of the fluid to the inlet of the pump. Great care should be exercised in designing the suction line to ensure an adequate net positive suction head, as we discussed in Section 13.9. In addition, special conditions may require auxiliary devices. It is highly recommended to install a pressure gauge in the suction line near the pump to monitor the condition of the fluid and to detect the tendency for cavitation to develop.

Figure 13.38 shows two methods of providing fluid to a pump. In part (a), a positive head is created by placing the pump below the supply reservoir. This is an aid in ensuring a satisfactory *NPSH*. In addition, the pump will always be primed with a column of liquid at start-up.

In Fig. 13.38(b), a *suction lift* condition occurs because the pump must draw liquid from below. Most positive-displacement pumps can lift fluids about 8 m (26 ft). For most centrifugal pumps, however, the pump must be artificially primed by filling the suction line with fluid. This can be done by providing an auxiliary supply of liquid during start-up or by drawing a vacuum on the pump casing, causing the fluid to be sucked up from the source. Then, with the pump running, it will maintain the flow. See also Section 13.5.4 for self-priming centrifugal pumps.

Unless the fluid is known to be very clean, a strainer should be installed either at the inlet or elsewhere in the suction piping to keep debris out of the pump and out of the process to which the fluid is to be delivered. A foot valve at the inlet (Figures 10.21 and 10.22), acting as a check valve to maintain a column of liquid up to the pump, allows free flow to the pump, but shuts when the pump stops. This precludes the need to prime the pump each time it is started. If a valve is used near the pump, a gate valve, offering very little flow resistance when fully open, is preferred. The valve stem should be horizontal to avoid air pockets.

Although the pipe size for the suction line should never be smaller than the inlet connection on the pump, it can be somewhat larger to reduce flow velocity and friction losses. Pipe alignment should eliminate the possibility of forming air bubbles or air pockets in the suction line because this will cause the pump to lose capacity and possibly to lose prime. Long pipes should slope upward toward the pump. Elbows in a horizontal plane should be avoided. If a reducer is required, it should be of the eccentric type, as shown in Fig. 13.38(b). Concentric reducers place part of the supply pipe above the pump inlet where an air pocket could form.

The discussion in Section 6.4 and Fig. 6.3 in Chapter 6 includes recommendations for the ranges of desirable pipe sizes to carry a given volume flow rate. In general, the larger sizes and lower velocities are recommended based on the ideal of minimizing the energy losses in the lines leading into pumps. Practical installation considerations and cost, however, may lead to the selection of smaller pipes with the resulting higher velocities.

Some of these practical considerations include the cost of pipe, valves, and fittings; the physical space available to accommodate these elements; and the attachment of the suction pipe to the suction connection of the pump. See References 3–7 and 11–18 for additional details on piping systems. Reference 15 in particular includes extensive discussion on the details of suction line design.

13.11 DISCHARGE LINE DETAILS

In general, the discharge line should be as short and direct as possible to minimize the head on the pump. Elbows should be of the standard or long-radius type if possible. Pipe size should be chosen according to velocity or allowable friction losses. Figure 6.3 in Chapter 6 includes recommendations for the ranges of desirable pipe sizes to carry a given volume flow rate. In general, the larger sizes and lower velocities are recommended based on the ideal of minimizing the energy losses. Practical installation considerations and cost, however, may lead to the selection of smaller pipes with the resulting higher velocities.

The discharge line should contain a valve close to the pump to allow service or pump replacement. This valve acts with the valve in the suction line to isolate the pump. For low resistance, a gate or butterfly valve is preferred. If flow must be regulated during service, a globe valve is better because it allows a smooth throttling of the discharge. This, in effect, increases the system head and causes the pump delivery to decrease to the desired value.

FIGURE 13.39 Discharge line details.

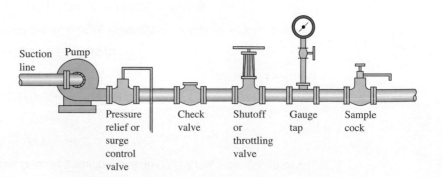

As shown in Fig. 13.39, other elements may be added to the discharge line as required. A pressure relief valve will protect the pump and other equipment in case of a blockage of the flow or accidental shut-off of a valve. A check valve prevents flow back through the pump when it is not running and it should be placed between the shut-off valve and the pump. If an enlarger is used from the pump discharge port it should be placed between the check valve and the pump. A tap into the discharge line for a gauge with its shut-off valve is highly recommended. Combined with the pressure gauge in the suction line, the operator can determine the total head on the pump and compare that to design requirements. A sample cock will allow a small flow of the fluid to be drawn off for testing without disrupting operation. Figure 13.1 and Figure 7.1 in Chapter 7 show illustrations of actual installations.

In many industrial piping installations, other process elements related to manufacturing are frequently included in the discharge line of the system. Examples are heat exchangers, filters, strainers, fluid power actuators, spray heads, and lubrication systems for machinery. Each of these elements provides additional resistance to the system. Furthermore, such systems are continuously subject to changes in demand for fluid flow rate due to changes in the needs of the processes being supplied by the system, changes in the depths of fluids in source and destination tanks, and changes in the nature of the product being pumped. In oil, gas, chemical, and food processing, and machinery lubrication systems, pressures, flow rates, fluid temperatures, and viscosities may be monitored and adjusted continuously. As a result, most systems of these types require flow control valves that adjust the flow rate in response to changing system needs. The control valves may be manually operated or automatically controlled by valve actuators in response to operator inputs or sensors placed in the system to monitor tank levels, pressures, temperatures, or other process needs. See also Section 13.15 for more discussion of control valves.

13.12 THE SYSTEM RESISTANCE CURVE

The operating point of a pump is defined as the volume flow rate it will deliver when installed in a given system and working against a particular total head. The piping system typically includes several elements described in previous sections on the design of suction and discharge lines; valves, elbows, process elements, and connecting straight lengths of pipe. The pump must accomplish the following tasks:

1. Elevate the fluid from a lower tank or other source to an upper tank or destination point.

2. Increase the pressure of the fluid from the source point to the destination point.

3. Overcome the resistance caused by pipe friction, valves, and fittings.

4. Overcome the resistance caused by processing elements as described in Section 13.11.

5. Supply energy related to the operation of flow control valves that inherently cause changes to the system head to achieve the desired flow rates.

The first two items in this list are components of the *static head*, h_0, for the system, where the name refers to the fact that the pump must overcome these resistances before any fluid begins to move, that is, the fluid is static. The *static head* h_0, is defined as,

⊃ **Total Static Head**

$$h_0 = (p_2 - p_1)/\gamma + (z_2 - z_1) \qquad (13\text{–}16)$$

But the pump is expected to work against a higher head and, in fact, to deliver fluid to the system at a specified rate. As soon as fluid starts to flow through the pipes, valves, fittings, and processing elements of the system, more head is developed because of the energy losses that occur. Recall that the energy losses are proportional to the velocity head in the pipes ($v^2/2g$) and, therefore, they increase according to the square of the volume flow rate. This causes the characteristic shape of a *system resistance curve* (SRC), sometimes called a *second degree curve*, as shown in Fig. 13.40.

As an initial example for developing a system curve, consider the system shown in Fig. 13.41 in which a pump is required to raise fluid from a vented lower tank to a pressurized elevated tank. Here we analyze the system at a steady flow condition without using a control valve. Systems with varying demands and control valves are discussed later. The data for this curve are determined next within Example Problem 13.4.

FIGURE 13.40 General shape of the system resistance curve (SRC) for a pumped fluid flow system.

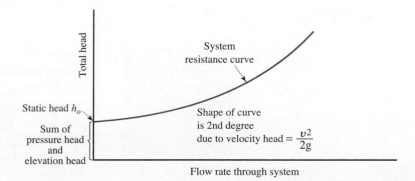

Example Problem
13.4

Figure 13.41 shows a pumped fluid flow system that is being designed to transfer 225 gal/min of water at 60°F from a lower vented reservoir to an elevated tank at a pressure of 35.0 psig. The suction line has 8.0 ft of steel pipe and the discharge line has 360 ft of pipe. Pipe sizes are shown in the figure. Prepare the SRC for this application, considering flow rates from zero to 250 gal/min, computing head values in 50 gal/min increments.

Solution

Objective: Develop the SRC for the pumped system shown in Fig. 13.41.

Given: Fluid: Water at 60°F: $\gamma = 62.4\ \text{lb/ft}^3$; $\nu = 1.21 \times 10^{-5}\ \text{ft}^2/\text{s}$; $h_{vp} = 0.5917\ \text{ft}$.

(Appendix A.2 and Table 13.2)

Range of volume flow rates: $Q = 0$ to 250 gal/min at increments of 50 gal/min

Select reference point 1 for the energy equation at the surface of the lower reservoir and point 2 at the surface of the upper tank.

Source: Lower reservoir; $p_1 = 0$ psig; Elevation $= 8.0$ ft above the pump inlet

Destination: Upper reservoir; $p_2 = 35.0$ psig; Elevation $= 88$ ft above pump inlet

Suction pipe is 3½-in Schedule 40 steel pipe; $D = 0.2957\ \text{ft}$, $A = 0.06868\ \text{ft}^2$; $L = 8.0$ ft

Discharge line is 2½-in Schedule 40 steel pipe; $D = 0.2058\ \text{ft}$, $A = 0.03326\ \text{ft}^2$; $L = 360$ ft

Step 1 Figure 13.41 shows the proposed layout.

Step 2 For the static head: $z_2 - z_1 = 88.0\ \text{ft} - 8.0\ \text{ft} = 80.0\ \text{ft}$; $p_2 = 35.0$ psig; $p_1 = 0$

Then,

$$\frac{P_2}{\gamma} = \frac{35.0\ \text{lb}}{\text{in}^2} \times \frac{\text{ft}^3}{62.4\ \text{lb}} \times \frac{144\ \text{in}^2}{\text{ft}^2} = 80.77\ \text{ft}$$

The total static head $h_o = (p_2 - p_1)/\gamma + (z_2 - z_1) = 80.77\ \text{ft} + 80\ \text{ft} = 160.77\ \text{ft}$.

Step 3 To compute the total head at many different flow rates, it is convenient to use the spreadsheet shown in Fig. 13.42, adapted from Chapter 11. The data shown in the figure are for the total dynamic head h_a at the design flow rate of 225 gal/min, given by

$$h_a = (z_2 - z_1) + p_2/\gamma + h_L = 80.0\ \text{ft} + 80.8\ \text{ft} + 139.0\ \text{ft} = 299.8\ \text{ft}$$

FIGURE 13.41 System for Example Problem 13.4.

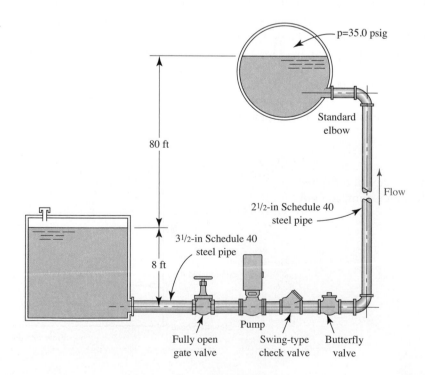

FIGURE 13.42 Spreadsheet calculation for the total head on the pump at the desired operating point for Example Problem 13.4.

APPLIED FLUID MECHANICS			CLASS I SERIES SYSTEMS		
Objective: *System Curve*			*Reference points for the energy equation:*		
Example Problem 13.4			Point 1: Surface of lower reservoir		
Figure 13.41			Point 2: Surface of upper reservoir		
System Data: **U.S. Customary Units**					
Volume flow rate: Q =	0.5011	*ft³/s*	*Elevation at point 1 =*	0 ft	
Pressure at point 1 =	0	*psig*	*Elevation at point 2 =*	80 ft	
Pressure at point 2 =	35	*psig*	*If Ref. pt. is in pipe: Set v₁" = B20" OR Set v₂" = E20"*		
Velocity at point 1 =	0.00	*ft/s →*	*Vel. head at point 1 =*	0.00 ft	
Velocity at point 2 =	0.00	*ft/s →*	*Vel. head at point 2 =*	0.00 ft	
Fluid Properties:			*May need to compute ν = η/ρ*		
Specific weight =	62.40	*lb/ft³*	*Kinematic viscosity =* 1.21E-05 ft²/s		
Pipe 1: 3½-in Schedule 40 steel pipe			**Pipe 2: 2½-in Schedule 40 steel pipe**		
Diameter: D =	0.2957	*ft*	*Diameter: D =*	0.2058	*ft*
Wall roughness: ε = 1.50E-04		*ft*	*Wall roughness: ε =* 1.50E-04		*ft*
Length: L =	8	*ft*	*Length: L =*	360	*ft*
Area: A =	0.06867	ft²	Area: A =	0.03326 ft²	[A = πD²/4]
D/ε =	1971		D/ε =	1372	Rel. roughness
L/D =	27		L/D =	1749	
Flow velocity =	7.30	ft/s	Flow velocity =	15.06 ft/s	[v = Q/A]
Velocity head =	0.827	ft	Velocity head =	3.524 ft	[v²/2g]
Reynolds No. = 1.78E+05			Reynolds No. = 2.56E+05		[N_R = vD/ν]
Friction factor: f =	0.0192		Friction factor: f =	0.0197	Using Eq. 8–7

Energy Losses in Pipe 1:		Qty.	Total K		
Pipe: K₁ =	0.519	1	0.519	Energy loss h_{L1} =	0.43 ft
Entrance: K₂ =	0.500	1	0.500	Energy loss h_{L2} =	0.41 ft
Gate valve: K₃ =	0.136	1	0.136	Energy loss h_{L3} =	0.11 ft
Element 4: K₄ =	0.000	1	0.000	Energy loss h_{L4} =	0.00 ft
Element 5: K₅ =	0.000	1	0.000	Energy loss h_{L5} =	0.00 ft
Element 6: K₆ =	0.000	1	0.000	Energy loss h_{L6} =	0.00 ft
Element 7: K₇ =	0.000	1	0.000	Energy loss h_{L7} =	0.00 ft
Element 8: K₈ =	0.000	1	0.000	Energy loss h_{L8} =	0.00 ft

Energy Losses in Pipe 2:		Qty.	Total K		
Pipe: K₁ =	34.488	1	34.488	Energy loss h_{L1} =	121.53 ft
Check valve: K₂ =	1.800	1	1.800	Energy loss h_{L2} =	6.34 ft
Butterfly valve: K₃ =	0.810	1	0.810	Energy loss h_{L3} =	2.85 ft
Standard elbow: K₄ =	0.540	2	1.080	Energy loss h_{L4} =	3.81 ft
Exit loss: K₅ =	1.000	1	1.000	Energy loss h_{L5} =	3.52 ft
Element 6: K₆ =	0.000	1	0.000	Energy loss h_{L6} =	0.00 ft
Element 7: K₇ =	0.000	1	0.000	Energy loss h_{L7} =	0.00 ft
Element 8: K₈ =	0.000	1	0.000	Energy loss h_{L8} =	0.00 ft
			Total energy loss h_{Ltot} =		139.01 ft
	Results:		Total head on pump: h_A =		299.8 ft

Step 4 Points on the SRC are shown in Table 13.3, computed using the spreadsheet and varying only the volume flow rate from zero to 250 gal/min in 50 gal/min increments (zero to 0.557 ft³/s in 0.111 ft³/s increments).

TABLE 13.3 Data points on the system resistance curve (SRC)

Q (gal/min)	Q (cfs)	h_a (ft)
0	0	160.8
50	0.111	168.6
100	0.223	189.9
150	0.334	224.1
200	0.445	271.3
225	0.501	299.8 (Design point)
250	0.557	331.4

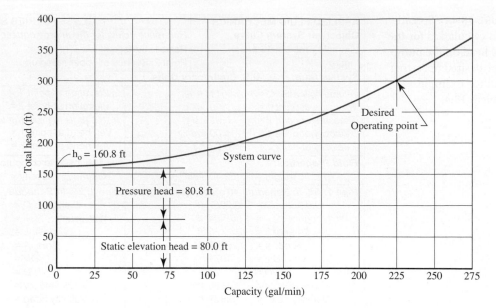

FIGURE 13.43 System curve for Example Problem 13.4. The system is shown in Figure 13.41.

Step 5 Figure 13.43 shows the system resistance curve.

Note the static head of 166.77 ft at the left axis for zero flow, the 2nd order shape of the curve, and the total head of 299.8 ft at the design flow rate of 225 gal/min.

13.13 PUMP SELECTION AND THE OPERATING POINT FOR THE SYSTEM

The basic data required for specifying a pump are the design flow rate and the total head required at that flow rate. This requires that the system in which the pump will operate has been designed and analyzed to determine its system resistance curve, using a method like that used in the preceding section and illustrated in Example Problem 13.4. Following is an example problem that extends the design of the system shown in Fig. 13.41 to include the pump selection process and the documentation of the operating parameters for the selected pump. For this process, we will use the pump curves shown in Figs. 13.28 to 13.36 that cover a wide range of flow rates (capacities) and total head capabilities.

Another concept introduced here is the identification of the *operating point* for the selected pump used in the system analyzed in Example Problem 13.4. To find the operating point, we will superimpose the pump performance curve for head versus flow onto the system resistance curve. The intersection of the two curves is the operating point. Figure 13.44 shows a generic example of the diagram from which the operating point can be seen. Following is a set of guidelines for selecting a suitable pump.

Guidelines for Pump Selection

Given the desired operating point for the system with the desired flow rate and the expected total head on the pump:

1. Seek a pump with high efficiency at the design point and one for which the operating point is near the *best efficiency point (BEP)* for the pump.

2. Standards set jointly by the American National Standards Institute (ANSI) and the Hydraulic Institute (HI) call for a *preferred operating region (POR)* for centrifugal pumps to be between 70 percent and 120 percent of the BEP. See Standard ANSI/HI 9.6.3-2012, *Standard for Centrifugal and Vertical Pumps for Allowable Operating Region.*

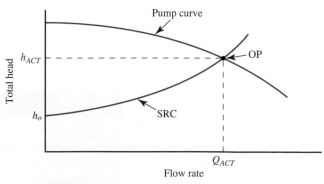

SRC = System resistance curve
OP = Operating point
Q_{ACT} = Actual flow rate in system
h_{ACT} = Actual total head on pump
h_o = Static head for system

FIGURE 13.44 Generic illustration of the operating point for a pump in a fluid flow system.

3. For the selected pump, specify the model designation, speed, impeller size, and the sizes for the suction and discharge ports.

4. At the actual operating point, determine the power required, the actual volume flow rate delivered, efficiency, and the $NPSH_R$. Also, check the type of pump, mounting requirements, and types and sizes for the suction and discharge lines to ensure that they are compatible with the intended installation.

5. Compute the $NPSH_A$ for the system, using Eq. (13–13).

6. Ensure that $NPSH_A > 1.10\, NPSH_R$ for all expected operating conditions.

7. If necessary, provide a means of connecting the specified pipe sizes to the connections for the pump if they are of different sizes. Use a gradual reducer or a gradual expander to minimize energy losses added to the system by these elements. See Figures 7.1, 13.38, and 13.39 for examples.

Example Problem 13.5

For the pumped fluid flow system shown in Fig. 13.41, select a suitable pump to enable the system to deliver at least 225 gal/min of water at 60°F at the total head of 299.8 ft as found in Example Problem 13.4. Then show the operating point for the pump in that system. Also, list the efficiency of the pump at the operating point, the power required to drive the pump, and the $NPSH_R$. Analyze the suction line part of the system to determine the $NPSH_A$ and ensure that it is adequate for the chosen pump. Recommend desirable adjustments to the piping system to accommodate the selected pump.

Solution

Given From Example Problem 13.4, we need to select a pump that can deliver 225 gal/min at 299.8 ft of total head.

Solution **Step 1** Figure 13.28 shows the performance curves for a suitable centrifugal pump, the 2 × 3 − 10 operating at 3500 rpm. The desired operating point lies between the curves for the 8-in and 9-in impellers. We specify the 9-in impeller that has a capacity slightly greater than the minimum of 225 gal/min. We find the following data near the desired operating point:

- The pump efficiency is approximately 58 percent, near the BEP for this pump.
- The suction port is 3 in and the discharge port is 2 in.

Step 2 Figure 13.45 shows the pump curve for the selected pump on the graph of the SRC that is shown in Fig. 13.43 as the result of Example Problem 13.4. The intersection of the pump curve and the SRC defines the operating point for this pump in this system.

Step 3 From Fig. 13.45, at the operating point, the following data are found:

$$\text{Capacity} = Q = 229\, \text{gal/min}$$

$$\text{Total head } h_a = 303\, \text{ft}$$

PROBLEM 13.45 Operating point for Example Problem 13.5.

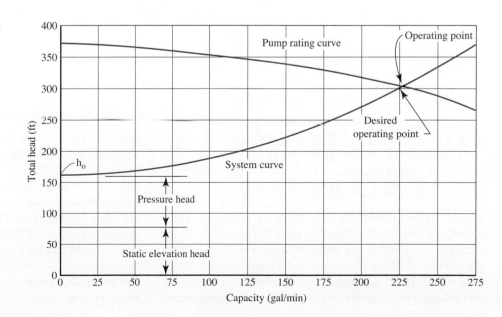

$$\text{Input power} = P = 30.0\,\text{hp}$$

$$NPSH_R = 8.8\,\text{ft}$$

Step 4 We can now compute the $NPSH_A$ for the system at the suction port of the pump, using the equation,

$$NPSH_A = h_{sp} \pm h_s - h_f - h_{vp}.$$

Assume $p_{sp} = 14.7$ psia (atmospheric) above the water in the source reservoir. Then,

$$h_{sp} = \frac{p_{sp}}{\gamma} = \frac{14.7\,\text{lb}}{\text{in}^2}\,\frac{144\,\text{in}^2}{\text{ft}^2}\,\frac{\text{ft}^3}{62.4\,\text{lb}} = 33.9\,\text{ft}$$

$h_s = +8.0\,\text{ft}$ (positive because the pump is below source level)

$h_f = $ Total energy loss in suction line $=$ Entrance loss $+$ Loss in valve $+$ Loss in pipe

$h_f = 0.421\,\text{ft} + 0.114\,\text{ft} + 0.436\,\text{ft} = 0.971\,\text{ft}$ (values found in Fig. 13.42)

$h_{vp} = 0.5917\,\text{ft}$ (from Table 13.2)

The values for head loss, h_f, used above were determined using the spreadsheet shown in Figure 13.42, but with the flow rate of 229 gal/min, instead of 225 gal/min. Then,

$$NPSH_A = 33.9\,\text{ft} + 8.0\,\text{ft} - 0.971\,\text{ft} - 0.5917\,\text{ft} = 40.34\,\text{ft}$$

Step 5 Now the value for the $NPSH_A$ can be compared with the $NPSH_R$ for the selected pump. It is recommended that the $NPSH_A$ be at least 10 percent greater than the $NPSH_R$ given by the manufacturer. Compute

$$1.10\,NPSH_R = 1.10(8.8\,\text{ft}) = 9.68\,\text{ft}$$

Then, because $NPSH_A = 40.34$ ft, the pump is acceptable.

Step 6 Now we can see if any issues exist for placing the selected pump in the system shown in Fig. 13.41. The sizes of the suction and discharge pipes are different from the sizes of the pump ports. A gradual reducer must be used from the 3½-in Schedule 40 suction line pipe to the 3-in suction port. A gradual enlargement must be used from the 2-in discharge port to the 2½-in Schedule 40 discharge pipe. These elements add energy losses that should be evaluated. The diameter ratio for each is approximately 1.2. Referring to Fig. 10.6 for a gradual enlargement and Fig. 10.11 for a gradual reducer, and specifying a 15° included angle, we find the K value will be 0.09 for the enlargement and 0.03 for the reducer. The additional energy losses are

$$h_{Ls} = 0.03(v^2/2g) = 0.03(0.842\,\text{ft}) = 0.025\,\text{ft}$$

$$h_{Ld} = 0.09(v^2/2g) = 0.09(3.587\,\text{ft}) = 0.323\,\text{ft}$$

These values are negligible compared to the other energy losses in the suction and discharge lines and, therefore, should not significantly affect the pump selection or its performance. Also, the $NPSH_A$ will still be acceptable.

13.14 USING PIPE-FLO® FOR SELECTION OF COMMERCIALLY AVAILABLE PUMPS

Earlier, we used PIPE-FLO® to simulate a generic sizing pump in a given system. Another powerful capability of the software is its ability to integrate commercially available pumps into a system to see real performance results. To do this, PIPE-FLO® has collected pump performance curves and other data from numerous manufacturers and integrated

them into the software for the user to simulate their system. The demo version of PIPE-FLO® shown throughout this text uses a "sample catalogue" with a limited number of typical pumps. The full version of the software, available upon request for academic use, includes many more pumps and identifies manufacturers and model numbers. Another convenient feature of PIPE-FLO® is its ability to make pipe size recommendations based on certain given inputs. This example problem outlines the use of a commercially available pump in a given system and shows the use of the pipe size calculator.

Example Problem 13.5
Pump Selection Using PIPE-FLO®

Water at 80°F is flowing in a river 130 ft below a water storage tank. The tank holds water drawn from the river with the use of a pump delivering 100 gal/min. The inlet and discharge elevations of the pump are both 15 ft. Set the river depth at 13 ft and the fluid level in the tank at 15 ft. The pipe to be used is Schedule 40 plastic PVC, and the total pipe length is 155 ft. Design the pipe size for a velocity of 8 ft/s. There is a pressure of 30 psi in the storage tank so it can be distributed for various uses. There are (2) 90° standard elbows and a foot valve on the suction line. There are (2) 45° standard elbows on the discharge line. Use PIPE-FLO® to calculate the total head on the pump, select a commercially available pump, show the pump curve, and summarize the major system design decisions.

Solution

1. Begin by using the "system" menu to establish the units for the problem, fluid zones, and pipe specifications. Since the problem statement says to design the piping at a nominal velocity of 8 ft/s, this must be accounted for when establishing the pipe specifications. In the pipe specifications menu, under "Sizing Criteria," choose "Design Velocity" then enter the 8 ft/s value. This will allow for later calculation of a pipe size for a given flow rate. See the figure below for reference.

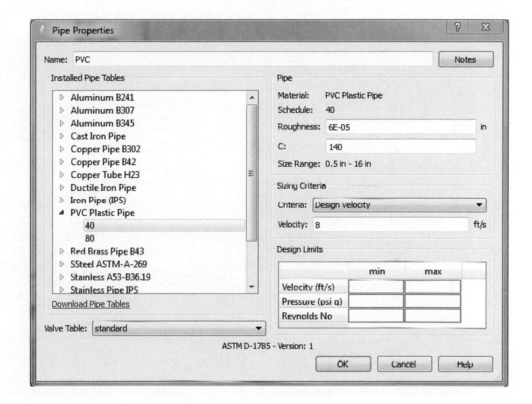

2. Next place the two "tanks" into the system. Right click on the symbols and select "CHANGE SYMBOL." Change one tank to make it appear as a river, and the other as a closed tank. Be sure to include the values from the problem statement for these two items in the properties grid.

3. Add in the sizing pump and enter its data into the properties grid.

4. Draw in the pipe to connect from the river to the pump. Assume that the length of this piece of pipe is 15 ft. Since a pipe size hasn't been specified, the "Design for Velocity" feature of PIPE-FLO® can be used to calculate an appropriate size. To do this, click on the pipe and begin entering the appropriate values in the property grid. To calculate an appropriate pipe size, click the "..." button that appears to the right of the pipe size box in the properties grid. A sizing calculator appears on the screen; fill in the required data and PIPE-FLO® will show the calculated size. For this example problem, a specified flow rate of 100 gal/min and velocity of 8 ft/s gives a recommended pipe size of 2.5-in Schedule 40. Continue entering the data in the property grid, including the foot valve and elbows, until the pipe is fully defined. See the figure below as a reference for the pipe size calculator.

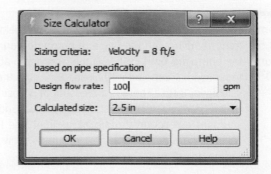

5. Repeat the same process for the discharge pipe from the pump to the storage tank. Assume the total length of this pipe is 140 ft. Be sure to include all fittings and valves when drawing in the pipes.

6. Once all the data have been entered run the PIPE-FLO® calculations to find the total head on the pump, suction pressure, discharge pressure, etc. Again, these calculated values will be used to help determine an appropriate commercially available pump to simulate the actual results. The solution to the first half of the problem is shown below. The total head calculated is 218 ft as seen in the system drawing and in the enlarged image of the sizing pump data.

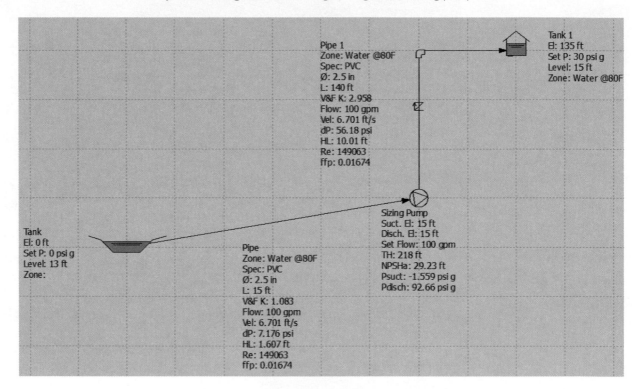

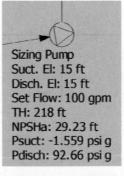

7. Now that the initial data have been established for the system, the results can be used to select a commercially available pump. To do this, right-click on the sizing pump and choose "Select Catalog Pump. . ." from the menu. A pump selection menu appears on the screen. In the demo version of the software, these are not actual pumps, but are very typical of commercially available pumps. The full version connects to actual pump curves that you could then specify and procure for the application.

8. From the "Select Catalog" drop-down box, choose "Sample Catalog.60." This is the sample catalog for pumps with 60 Hz motors; the common frequency for power supply in the United States. 50 Hz options are available for users in countries other than the United States. From the "Types" menu, check all of the boxes so that all pump manufacturers are shown. In the "Speeds" menu, check only the 3600 and 1800 options. These values represent the nominal speeds for the most common AC motors used to drive pumps. The actual speeds will be slightly lower and they are listed with the pump data. Pump curves now appear below in the same "Pump Selection" menu for all the applicable pumps; see the figure below.

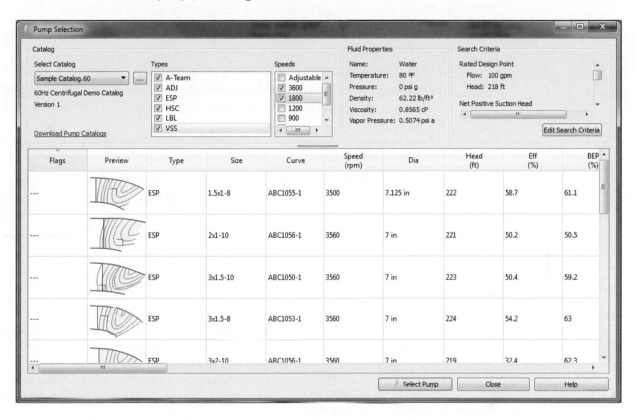

9. PIPE-FLO® automatically sorts the available pumps such that the optimum pump is first on the list, as indicated by the efficiency at the operating point for the pump. A sample pump curve is displayed, and a marker on the pump curve indicates where the particular system you are using falls on the curve for that specific pump. For this example, we will choose the first pump on the list and use the PIPE-FLO® recommendation.

10. Highlight the pump that is to be used in the system, and click the "Select Pump" button at the bottom of the page. This automatically replaces the sizing pump in the problem with the commercially available pump that was just chosen. If the calculate button is pressed again after the pump has been replaced, other values such as power and efficiency for the commercially available pump are displayed on the FLO-Sheet®. The results for this example problem are shown below.

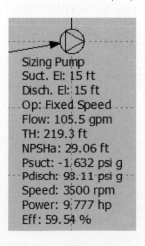

11. Note that PIPE-FLO® doesn't rename the pump after a commercially available one has been chosen since it is assuming the user is entering names for all the components. The way to tell if the pump has actually changed from the sizing pump to the commercially available one is by the calculated results shown. If power and efficiency are listed, the user knows that the commercially available pump is being listed in the problem since these values aren't given for a sizing pump.

12. Summary: The pump designation and its major performance parameters are:

   a. Operating point: 100 gal/min flow rate at 218 ft of total head

   b. Pump: 1 ½ × 1 × 8 centrifugal pump; ESP-type; Pump curve ABC1055-1

   c. Pump speed: 3500 rpm   Impeller diameter: 7.125 in

   d. BEP: 61.1 percent

   e. Efficiency at operating point: 58.7 percent (within 96 percent of BEP and to the left of BEP)

   f. The software chose a plastic pipe size for the system to be 2½-in Schedule 40 for both the suction and discharge pipes based on the desired flow velocity of 8.0 ft/s. The actual flow velocity is 6.70 ft/s.

   g. A gradual reducer is required at the pump inlet from 2½-in to 1½-in sizes, and a gradual enlarger is required at the pump outlet from 1-in to 2½-in sizes.

   h. The $NPSH_A$ for the suction inlet to the pump is computed to be 29.06 ft.

   i. The value for $NPSH_R$ for the pump and additional operational data for the pump can be accessed from the software.

13.15 ALTERNATE SYSTEM OPERATING MODES

In Sections 13.9 to 13.14, the focus was on the design and analysis of pumped fluid delivery systems that employed a single flow path and that operated at one fixed condition of flow rate, pressures, and elevations. Important principles of system operation were discussed such as the performance of centrifugal pumps, system resistance curves, the operating point of a pump in a given system, *NPSH*, efficiency, and power required to operate the pump. These fundamentals form the basis for understanding how a fluid flow system works.

Many alternate modes of system operation are in frequent use in a wide variety of industrial applications that build on those fundamentals, but that include additional features and that require different methods of analysis. This section will describe the following:

- Use of control valves to enable system operators to adjust the system's behavior to meet varying needs, either manually or automatically

- Variable speed drives that permit continuous variation of flow rates to fine tune system operation and to match levels of delivery to product or process needs

- Effect of fluid viscosity on pump performance

- Operating pumps in parallel

- Operating pumps in series

- Multistage pumps

Reference 7 and Internet resource 2 offer more extensive treatment of these topics beyond what is practical for inclusion in this book, and in a manner that is highly compatible with the terminology and analysis methods presented here. Several other references offer additional coverage as well, particularly References 3–6 and 9–20. Reference 8 is a source for an extensive set of fluid property data (viscosity, density, specific gravity, vapor pressure), steam data, friction losses in valves and fittings, steel and cast iron pipe data, electrical wiring, motors, and controls. Other references relevant to these topics appear in the References section for Chapter 11.

13.15.1 Use of Control Valves

It was stated in Section 13.13 that the operating point of a pump is defined as the volume flow rate it will deliver when installed in a given system and working against a particular total head. The piping system typically includes several elements described in previous sections on the design of suction and discharge lines; valves, elbows, process elements, and connecting straight lengths of pipe. Valves were placed in the system to allow the lines to be shut off when performing service or when the system is shut down; thus, they are often called *shut-off valves*. They were typically low-resistance types such as gate valves or butterfly valves and modeled in their fully open position as part of the SRC.

However, when there is a need for varying flow rates to meet different needs, control valves are used that can be adjusted either manually or automatically. Initial sizing of a control valve is often based on the mid-point between the high and low flow rate limits expected in the application. Then the valve can be adjusted to a more open position (less resistance) or more closed position (more resistance) to produce higher or lower flow rates, respectively.

It is important to obtain data from the supplier for control valve performance across its entire range, typically in

PROBLEM 13.46 Operating points for a system containing a control valve at varying control valve settings

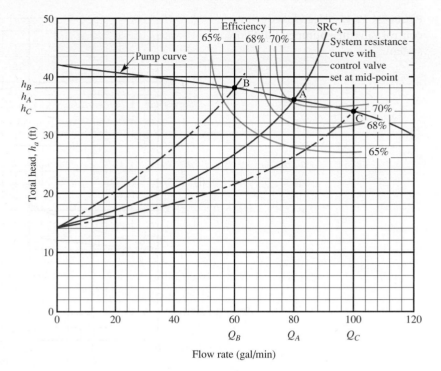

terms of the flow coefficient, C_v, as defined in Chapter 10. In U.S. Customary System units with Q in gal/min and pressure in psi, the definition of flow coefficient is:

$$C_v = \frac{Q}{\sqrt{\Delta p/sg}}$$

The basis for the flow coefficient is that a valve having a flow coefficient of 1.0 will pass 1.0 gal/min of water at 1.0-psi pressure drop across the valve. Alternate forms of this equation are useful:

$$Q = \text{Flow in gal/min} = C_v\sqrt{\Delta p/sg}$$
$$\Delta p = sg(Q/C_v)^2$$

When working in metric units, an alternate form of the flow coefficient is used and it is called K_v instead of C_v. It is defined as the amount of water in m³/h at a pressure drop of one bar across the valve. Use the following equation for conversion between C_v and K_v:

$$C_v = 1.156\,K_v$$

Now, with a control valve (set at its mid-point) in the system along with all other elements, the modeling of the SRC can be done, and a suitable pump can be selected for the operating point A as shown in Fig. 13.46. The flow rate at the operating point is the desired nominal flow rate for the system and the resulting total head on the pump can be read from the chart. For the sample data in Fig. 13.46, we read $Q = 80$ gal/min and $h_a = 36.0$ ft.

Let's explore what would need to be done if the system operator desired a flow rate of 60 gal/min instead of 80 gal/min. The control valve would be turned to a more restrictive position, placing more resistance to the flow through the system. Then the pressure drop across the control valve would have increased, with a corresponding

decrease in the flow and an increase in the total head on the pump. The result is that the SRC would pivot toward the left, reaching a new operating point B. At that point, the total head on the pump is 38.2 ft and an additional 2.2 ft of head will be dissipated from the control valve.

If the production system requires a greater flow rate, say 100 gal/min, the control valve will be opened to provide less resistance and the system curve pivots to the right to operating point C. At this point, the total head on the pump is 33.5 ft or 2.5 ft less than at point A.

It is important to note that other aspects of the pump operation are affected by changing the control valve setting. Figure 13.46 shows pump efficiency curves in the vicinity of the operating points discussed above. The initial operating point A results in the pump operating at about 70 percent efficiency, very near the BEP for this pump. When operating at C, the efficiency drops to about 68 percent, and at B it is about 66 percent. The range of flow rates from 60 gal/min to 100 gal/min is approximately the limit of the range recommended in Hydraulic Institute standards, between 70 percent and 120 percent of the flow at the BEP.

The operation of the control valve inherently involves the dissipation of energy from the system—energy that must be provided by the pump. Therefore, a cost is incurred to perform the control function. Figure 13.47 illustrates the nature of the energy used by the control valve. Curve A is the same as that shown in Fig. 13.46 for the system that includes a control valve set at its mid-point. Curve D is for the same system, but without the control valve. The difference in total head between these two curves represents the additional energy required to perform the control function and the cost for that energy can be calculated. Reference 7 contains extensive discussion about the types of control valves, sizing them to the needs of a particular system, and the costs incurred in their operation.

PROBLEM 13.47 Head loss due to a control valve in a pumped fluid flow system.

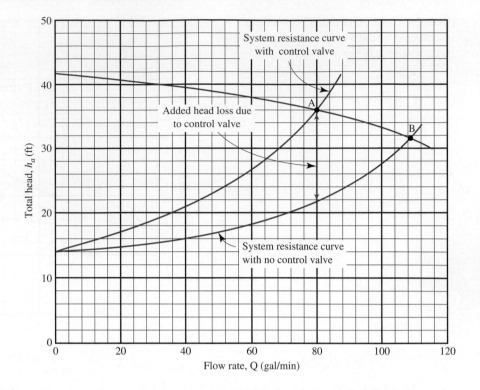

13.15.2 Variable-Speed Drives

Variable-speed drives offer an attractive alternative to use of a control valve. Several types of mechanical variable-speed drives and a variable-frequency electronic control for a standard AC electric motor are available. The standard frequency for AC power in the United States and many other countries is 60 hertz (Hz), or 60 cycles per second. In Europe and some other countries, 50 Hz is standard. Because the speed of an AC motor is directly proportional to the frequency of the AC current, varying the frequency causes the motor speed to vary. Because of the affinity laws, as the motor speed decreases, the capacity of the pump decreases; this allows the pump to operate at the desired delivery without use of a control valve and the attendant energy loss across the valve. Further benefit is obtained because the power required by the pump decreases in proportion to the speed reduction ratio cubed. Of course, the variable-speed drive is more expensive than a standard motor alone, and the overall economics of the system with time should be evaluated. See References 7 and 10.

The effect of implementing a variable-speed drive for a system with a centrifugal pump depends on the nature of the system curve as shown in Fig. 13.48. Part (a) shows a system curve that includes only friction losses. The system curve in part (b) includes a substantial static head comprising an elevation change and pressure change from the source to the destination. When only friction losses occur, the variation in pump performance tends to follow the constant efficiency curves, indicating that the affinity laws discussed in Section 13.7 closely apply. Flow rate changes in proportion to speed change; head changes as the square of the speed change; and power changes as the cube of the speed.

For the system curve having a high static head [Fig. 13.48(b)], the pump performance curve will move into different efficiency zones of operation, so the affinity law on power required will not strictly apply. However, the use of variable-speed drives for centrifugal pumps will always provide the lowest-energy method of varying pump delivery.

In addition to energy savings, other benefits result from using variable-speed drives:

- **Improved process control** Pump delivery can be matched closely to requirements, resulting in improved product quality.

- **Control of rate of change** Variable-speed drives control not only the final speed, but also the rate of change of speed, reducing pressure surges.

- **Reduced wear** Lower speeds dramatically reduce forces on seals and bearings, resulting in longer life and greater reliability of the pumping system.

Operating pumps over a wide range of speeds can produce undesirable effects as well. Moving fluids set up flow-induced vibrations that change with fluid velocity. Resonances can occur in the pump itself, the pump mounting structure, the piping support system, and in connected equipment. Monitoring of system operation over the complete expected range of speeds is required to identify such conditions. Often the resonances can be overcome by using vibration dampers, isolators, or different pipe supports.

The effects of lower or higher flow on fluid system components should also be checked. Check valves require a certain minimum flow to ensure full opening and secure closing of the internal valve components. Solids in slurries may tend to settle out and collect in undesirable regions of the system

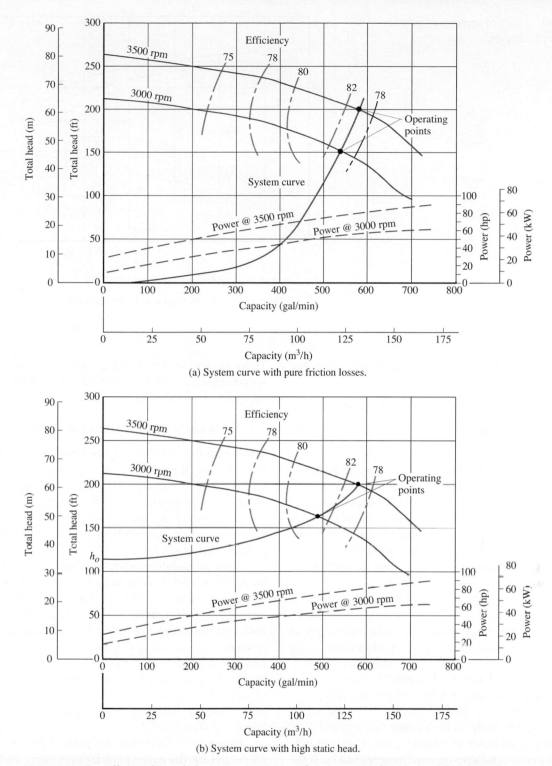

(a) System curve with pure friction losses.

(b) System curve with high static head.

FIGURE 13.48 Effects of speed changes on pump performance as a function of the type of system curve.

at low velocities. Operating pumps and drives at lower speeds may impair their lubrication or cooling, requiring supplemental systems. Speeds higher than normal may require a higher power than the prime mover is capable of delivering, and greater loads are placed on couplings and other drive components.

13.15.3 Effect of Fluid Viscosity

The performance rating curves for centrifugal pumps, such as those shown in Figs. 13.28–13.36, are generated from test data using water as the fluid. These curves are reasonably accurate for any fluid with a viscosity similar to that of

FIGURE 13.49 Effect of increased viscosity on pump performance.

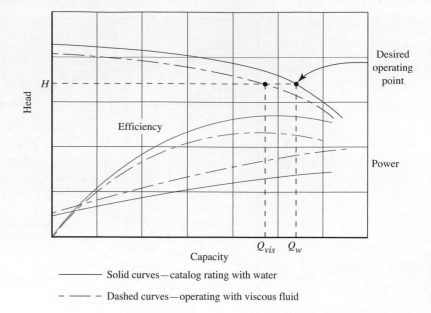

water. However, pumping more-viscous fluids causes the following effects:

- The power required to drive the pump increases
- The flow delivered against a given head decreases
- The efficiency decreases

Figure 13.49 illustrates the effect of pumping a viscous fluid if the pump was selected for the desired operating point without applying corrections. The symbol Q_w denotes the rated capacity of the pump with cool water (typically water at 60°F or 15.6°C) against a given head H. Against the same head, the pump would deliver the viscous fluid at the lower flow rate Q_{vis}; the efficiency would be lower and the power required to drive the pump would increase.

Reference 8 gives data for correction factors that can be used to compute expected performance with fluids of different viscosity. Some pump selection software automatically applies these correction factors to adjust pump performance curves after the user inputs the viscosity of the fluid being pumped. The PIPE-FLO® suite of software does provide viscosity corrections for pump performance. See Internet resource 2.

As an example of the effect of viscosity on performance, one set of data were analyzed for a pump that would deliver 750 gal/min of cool water at a head of 100 ft with an efficiency of 82 percent and a power requirement of 23 hp. If the fluid being pumped had a kinematic viscosity of approximately 2.33×10^{-3} ft²/s (2.16×10^{-4} m²/s; 1000 SUS), the following performance would be predicted:

1. At 100 ft of head, the delivery of the pump would be reduced to 600 gal/min.

2. To obtain 750 gal/min of flow, the head capability of the pump would be reduced to 88 ft.

3. At 88 ft of head and 750 gal/min of flow, the pump efficiency would be 51 percent and the power required would be 30 hp.

These are significant changes. The given viscosity corresponds approximately to that of a heavy machine lubricating oil, a heavy hydraulic fluid, or glycerin.

13.15.4 Operating Pumps in Parallel

Many fluid flow systems require largely varying flow rates that are difficult to provide with one pump without calling for the pump to operate far off its best efficiency point. An example is a multistory hotel where required water delivery varies with occupancy and time of day. Industry applications calling for varying amounts of process fluids or coolants present other examples.

A popular solution to this problem is to use two or more pumps in parallel, each drawing from the same inlet source and delivering to a common manifold and on to the total system. Figure 13.1 shows such a system with three pumps operating in parallel. Predicting the performance of parallel systems requires an understanding of the relationship between the pump curves and the system curve for the application. Adding a second pump theoretically doubles the capacity of the system. However, as greater flow rate occurs in the piping system, a greater head is created, causing each pump to deliver less flow.

Figure 13.50 illustrates this point. Observe that pump 1 operates on the lower performance curve and that at a head of H_1 it delivers a flow rate Q_1, which is near its maximum practical capacity at operating point 1. If greater flow is needed, a second, identical pump is activated and the flow increases. But the energy losses due to friction and minor losses continue to increase as well, as indicated by the system curve, eventually reaching operating point 2 and delivering the total flow Q_2 against the head H_2. However, pump 1 experiences the higher head and its delivery falls back to Q_1'. After the new equilibrium condition is reached, pump 1 and pump 2 deliver equal flows, each one half of the total flow. The pumps should be selected so

FIGURE 13.50 Performance of two pumps in parallel.

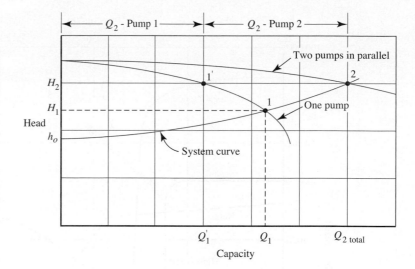

they have a reasonable efficiency at all expected capacities and heads.

Similar analyses can be applied to systems with three or more pumps, but careful consideration of each pump's operation at all possible combinations of head and flow are needed because other difficulties may arise. In addition, some designers use two identical pumps, operating one at a constant speed and the second with a variable-speed drive to more continuously match delivery with demand. Such systems also require special analysis, and the pump manufacturer should be consulted.

13.15.5 Operating Pumps in Series

Directing the output of one pump to the inlet of a second pump allows the same capacity to be obtained at a total head equal to the sum of the ratings of the two pumps. This method permits operation against unusually high heads.

Figure 13.51 illustrates the operation of two pumps in series. Obviously, each pump carries the same flow rate Q_{total}. Pump 1 brings the fluid from the source, increases the pressure somewhat, and delivers the fluid at this higher pressure to pump 2. Pump 1 is operating against the head H_1

produced by the suction line losses and the initial increase in pressure. Pump 2 then takes the output from pump 1, further increases the pressure, and delivers the fluid to the final destination. The head on pump 2, H_2, is the difference between the total dynamic head *TDH* at the operating point for the combined pumps and H_1.

13.15.6 Multistage Pumps

A performance similar to that achieved by using pumps in series can be obtained by using multistage pumps. Two or more impellers are arranged in the same housing in such a way that the fluid flows successively from one to the next. Each stage increases the fluid pressure so that a high total head can be developed.

13.16 PUMP TYPE SELECTION AND SPECIFIC SPEED

Figure 13.52 shows one method for deciding what type of pump is suitable for a given service. Some general conclusions can be drawn from such a chart, but it should be emphasized that boundaries between zones are approximate.

FIGURE 13.51 Performance of two pumps operating in series.

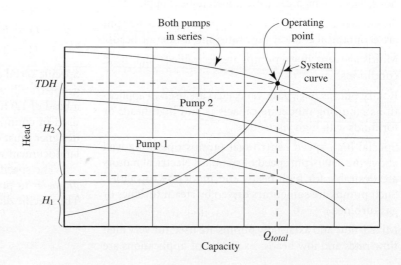

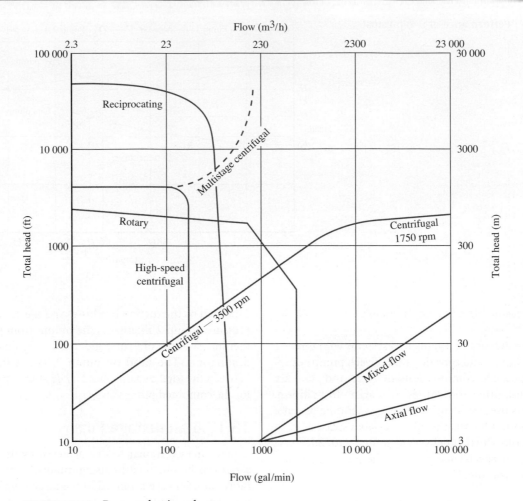

FIGURE 13.52 Pump selection chart.

Two or more types of pumps may give satisfactory service under the same conditions. Such factors as cost, physical size, suction conditions, and the type of fluid may dictate a particular choice. In general:

1. Reciprocating pumps are used for flow rates up to about 500 gal/min and from very low heads to as high as 50 000 ft.

2. Centrifugal pumps are used over a wide range of conditions, mostly in high-capacity, moderate-head applications.

3. Single-stage centrifugal pumps operating at 3500 rpm are economical at lower flow rates and moderate heads.

4. Multistage centrifugal pumps are desirable at high-head conditions.

5. Rotary pumps (e.g., gear, vane, etc.) are used in applications requiring moderate capacities and high heads or for fluids with high viscosities.

6. Special high-speed centrifugal pumps operating well above the 3500-rpm speed of standard electrical motors are desirable for high heads and moderate capacities. Such pumps are sometimes driven by steam turbines or gas turbines.

7. Mixed-flow and axial-flow pumps are used for very high flow rates and low heads. Examples of applications are

flood control and removal of ground water from construction sites.

Another parameter that is useful in selecting the type of pump for a given application is the *specific speed*, defined as

$$N_s = \frac{N\sqrt{Q}}{H^{3/4}} \qquad \textbf{(13–17)}$$

where

N = Rotational speed of the impeller (rpm)

Q = Flow rate through the pump (gal/min)

H = Total head on the pump (ft)

Specific speed can be thought of as the rotational speed of a geometrically similar impeller pumping 1.0 gal/min against a head of 1.0 ft (Reference 8). Different units are sometimes used in countries outside the United States, so the pump designer must determine what units were used in a particular document when making comparisons.

The specific speed is often combined with the *specific diameter* to produce a chart like that shown in Fig. 13.53. The specific diameter is

$$D_s = \frac{DH^{1/4}}{\sqrt{Q}} \qquad \textbf{(13–18)}$$

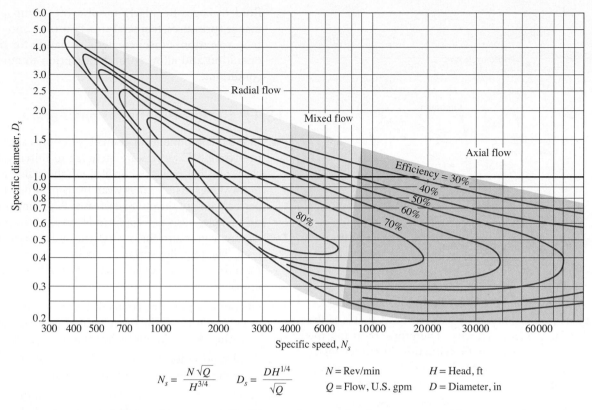

$$N_s = \frac{N\sqrt{Q}}{H^{3/4}} \qquad D_s = \frac{DH^{1/4}}{\sqrt{Q}} \qquad \begin{matrix} N = \text{Rev/min} \\ Q = \text{Flow, U.S. gpm} \end{matrix} \qquad \begin{matrix} H = \text{Head, ft} \\ D = \text{Diameter, in} \end{matrix}$$

FIGURE 13.53 Specific speed versus specific diameter for centrifugal pumps—An aid to pump selection. (Excerpted by special permission from *Chemical Engineering*, April 3, 1978. Copyright © 1978 by McGraw-Hill, Inc., New York)

where D is the impeller diameter in inches. The other terms were defined earlier for Equation 13-17.

From Fig. 13.53 we can see that radial-flow centrifugal pumps are recommended for specific speeds from about 400 to 4000. Mixed-flow pumps are used from 4000 to about 7000. Axial flow pumps are used from 7000 to over 60 000. See Fig. 13.12 for the shapes of the impeller types.

13.17 LIFE CYCLE COSTS FOR PUMPED FLUID SYSTEMS

The term *life cycle cost* (LCC) refers here to the consideration of all factors that make up the cost of acquiring, maintaining, and operating a pumped fluid system. Good design practice seeks to minimize LCC by quantifying and computing the sum of the following factors:

1. Initial cost to purchase the pump, piping, valves and other accessories, and controls.

2. Cost to install the system and put it into service.

3. Energy cost required to drive the pump and auxiliary components of the system over the expected life.

4. Operating costs related to managing the system including labor and supervision.

5. Maintenance and repair costs over the life of the system to keep the pump operating at design conditions.

6. Cost of lost production of a product during pump failures or when the pump is shut down for maintenance.

7. Environmental costs created by spilled fluids from the pump or related equipment.

8. Decommissioning costs at the end of the useful life of the pump including disposal of the pump and cleanup of the site.

More detail on each of these items and the larger context of LCC can be found in Reference 9.

Minimizing Energy Costs For pumps that operate continuously for long periods of time, the energy cost is the highest component of the total LCC. Even for a pump operating for just 8 hours a day, 5 days a week, the cumulative operating time is over 2000 h/yr. Pumps feeding continuous processes such as electric power generation may operate over 8000 h/yr. Therefore, minimizing the energy required to operate the pump is a major goal of good fluid system design. The following list summarizes the approaches to system design that can reduce energy cost and help to ensure reliable operation. Some of these items have already been discussed elsewhere in this chapter:

1. Perform a careful, complete analysis of the proposed design for the piping system to understand where energy losses occur and to predict accurately the design operating point for the pump.

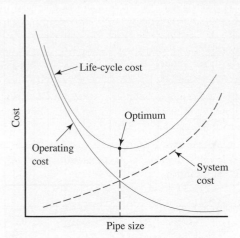

FIGURE 13.54 Life cycle cost principle for pumped fluid distribution systems.

2. Recognize that energy losses in piping, valves, and fittings are proportional to the velocity head, that is, to the square of the flow velocity. Therefore, reducing the velocity makes a dramatic reduction in energy losses and the total dynamic head required for the pump. A smaller, less expensive pump can then be used.

3. Use the largest practical size of pipe for both the suction and discharge lines of the system to keep the flow velocity at a minimum. Understand that larger pipes are more expensive than smaller pipes and they require larger, more expensive valves and fittings as well. However, the energy saving accumulated over the operating life of the system typically overcomes these higher costs. Figure 13.54 illustrates this point conceptually by comparing costs for the system with operating costs as a function of pipe size. Another practical consideration is the relationship between the pipe sizes and the sizes for the suction and discharge ports of the pump. Some designers recommend that the pipes be one size larger than the ports.

4. Carefully match the pump to the head and capacity requirements of the system to ensure that the pump operates at or near its BEP and to avoid using an oversized pump that will cause it to operate at a lower efficiency.

5. Use the most efficient pump for the application and operate the pump as close as possible to its BEP.

6. Use high-efficiency electric motors or other prime movers to drive the pump.

7. Consider the use of variable-speed drives (VSD) for the pumps to permit matching the pump delivery to the requirements of the process. See Section 13.15.2.

8. Consider two or more pumps operating in parallel for systems requiring widely varying flow rates. See Section 13.15.4.

9. Provide diligent maintenance of the pump and piping system to minimize the decrease in performance due to wear, build-up of corrosion on pipe surfaces, and

fluid leakage. Regularly monitoring pump performance (pressures, temperatures, flow rates, motor current, vibration, noise) creates a signature for normal operation and allows prompt attention to abnormal conditions.

13.17.1 Other Practical Considerations

1. Internal components of centrifugal pumps suffer wear over time. Wear rings are included to initially set the clearances between the impeller and the casing to optimum values. As the rings wear, the gaps widen and the performance of the pump decreases. Replacing the rings on a regular basis as recommended by the pump manufacturer can return the pump to its design performance level. The impeller surfaces may wear due to abrasion from the fluid. This could lead to the need to replace the impeller.

2. Operating a pump at points distant from the best efficiency point causes higher loads on bearings, seals, and wear rings and will reduce the life of the pump.

3. Rigid support for the pump and its drive system is critical to satisfactory operation and long life.

4. Careful alignment of the drive motor with the pump is essential to avoid excessive deflection of the pump's shaft that can cause early failure. Follow the pump manufacturer's recommendations and check alignment periodically.

5. Assure that the flow from the suction line to the pump inlet is smooth with no vortices or swirling. Some designers recommend a minimum of 10 diameters of straight pipe ($10 \times D$) between any valve or fitting and the pump inlet. However, if a reducer is required, it should be installed right at the pump.

6. Support piping and valves independently from the pump and do not allow significant pipe loads to be transferred to the pump casing. High loads tend to cause extra loads on bearings and shaft deflection that may change the clearance between the impeller and the casing.

7. Use clean oil, grease, or other lubricants for the pump bearings.

8. Do not allow the pump to operate dry or with entrained air in the fluid being pumped. This requires careful design of the intake to the suction line and the tank, sump, or reservoir from which the fluid is drawn.

REFERENCES

Note: See also the extensive list of references for Chapter 11, many of which treat the subject of pumps and pumped systems.

1. ASTM International. 2008. *ASTM D323-08 Standard Test Method for Vapor Pressure of Petroleum Products (Reid Method)*. DOI: 10.1520/D0323-08. West Conshohocken, PA: Author.

2. ———. 2012. *ASTM D4953-06(2012). Standard Test Method for Vapor Pressure of Gasoline and Gasoline-Oxygenate Blends*

(Dry Method). DOI: 10.1520/D4953-06(2012). West Conshohocken, PA: Author.

3. Bachus, Larry. 2003. *Know and Understand Centrifugal Pumps.* New York: Elsevier Science.

4. Bloch, Heinz P. 2011. *Pump Wisdom: Problem Solving for Operators and Specialists.* New York: Wiley.

5. Bloch, Heinz P., and Allan R. Budris. 2010. *Pump User's Handbook: Life Extension.* Boca Raton, FL: CRC Press.

6. Gülich, Johann F. 2010. *Centrifugal Pumps,* 2nd ed. New York: Springer Science+Business Media.

7. Hardee, Ray T., and Jeffrey L. Sines. 2012. *Piping Systems Fundamentals,* 2nd ed. Lacey, WA: ESI Press – Engineered Software, Inc.

8. Heald, C. C., ed. 2002. *Cameron Hydraulic Data*, 19th ed. Irving, TX: Flowserve, Inc.

9. Hydraulic Institute and Europump. 2001. *Pump Life Cycle Costs: A Guide to LCC Analysis for Pumping Systems.* Parsippany, NJ: Hydraulic Institute.

10. Hydraulic Institute and Europump. 2004. *Variable Speed Pumping: A Guide to Successful Applications.* Parsippany, NJ: Hydraulic Institute.

11. Hydraulic Institute. 1990. *Engineering Data Book,* 2nd ed. Parsippany, NJ: Author.

12. Hydraulic Institute. 2012. *Pump Standards.* Parsippany, NJ: Author. [Individual standards or complete sets for centrifugal pumps, reciprocating pumps, rotary pumps, vertical pumps, and air-operated pumps.]

13. Hydraulic Institute and Pump Systems Matter™. 2008. *Optimizing Pumping Systems.* Parsippany, NJ: Author.

14. Jones, Garr M., and Sanks, Robert L. 2008. *Pumping Station Design*, 3rd ed. New York: Elsevier.

15. Karassik, Igor J., Joseph P. Messina, Paul Cooper, and Charles C. Heald. 2008. *Pump Handbook*, 4th ed. New York: McGraw-Hill.

16. Manring, Noah D. 2013. *Fluid Power Pumps and Motors: Analysis, Design, and Control.* New York: McGraw-Hill.

17. Menon, E., and Shashi. 2009. *Working Guide to Pumps and Pumping Stations.* New York: Elsevier.

18. Michael, A. M., S. D. Khepar, and S. K. Sondhi. 2008. *Water Wells and Pumps.* New York: McGraw-Hill.

19. Rishel, J. B. 2010. *HVAC Pump Handbook.*, 2nd ed. New York: McGraw-Hill.

20. Volk, Michael. 2005. *Pump Characteristics and Applications,* 2nd ed. Boca Raton, FL: CRC Press.

INTERNET RESOURCES

Note: See the set of Internet Resources at the end of Chapter 12, which includes several commercially available pipeline system design software packages, many of which include pump selection features.

1. **Hydraulic Institute:** An association of pump manufacturers and users that provides product standards and a forum for the exchange of industry information in the engineering, manufacture, and application of pumping equipment.

2. **Engineered Software, Inc. (ESI):** www.eng-software.com Producer of the PIPE-FLO® and Pump Base® software for analyzing fluid flow piping systems and selecting optimum pumps for such systems, as featured in this book. An extensive list of pump suppliers' catalogs is included from which selections can be made. A special demonstration version of PIPE-FLO® can be accessed at http://www.eng-software.com/appliedfluidmechanics

3. **Pumps & Systems Magazine:** A publication serving pump users and manufacturers, with special emphasis on operation and maintenance of pumps and systems.

4. **Animated Software Company—Pumps:** Producer of *All About Pumps*, a set of graphical images of over 75 different types of pumps with animation showing fluid flow and mechanical action.

5. **Armstrong Pumps, Inc.:** Manufacturer of pumps for residential and commercial applications including HVAC, hydronic systems, and fire protection. Performance curves are available on the website.

6. **Bell & Gossett:** Manufacturer of centrifugal pumps for HVAC, hydronic, water systems, and industrial applications.

7. **Crane Pumps and Systems:** Manufacturer of a wide variety of centrifugal pump designs and configurations marketed under the brand names of Barnes, Burks, Prosser, Deming, Weinman, and Crown.

8. **Eaton Hydraulics:** Manufacturer of hydraulic pumps, motors, and valves under the brands Eaton, Vickers, Char-Lynn, HydroKraft, and Hydro-Line. Division of Eaton Corporation.

9. **Flowserve Corporation:** Manufacturer of centrifugal and rotary pumps under several brand names such as Flowserve, Durco, Pacific, Worthington, and others. Applications in the power generation, oil and gas, chemical processing, water resources, marine, pulp and paper, mining, primary metals, and general industry markets. A leader in the chemical processing and corrosion-resistant pump field.

10. **Goulds Pumps:** Manufacturer of a wide range of centrifugal pumps for water and wastewater, agricultural, irrigation, boiler feed, HVAC, and general industry applications. A subsidiary of ITT Industries, Inc.

11. **IMO Pump Company:** Manufacturer of screw and gear pumps for industries such as oil transport, hydraulic machinery, refinery, marine, aircraft fuel handling, and fluid power. A subsidiary of Colfax Corporation.

12. **March Pumps:** Manufacturer of small- and medium-capacity centrifugal pumps.

13. **Moyno, Inc.:** Manufacturer of the Moyno brand of progressing cavity pump used in environmental, specialty chemical, pulp and paper, building materials, food and beverage, mining, and many other applications. A part of Robbins & Myers, Inc., Process and Flow Control Group.

14. **Seepex Pumps:** Manufacturer of progressive cavity pumps for industrial applications.

15. **Warren Rupp, Inc.:** Manufacturer of diaphragm pumps under the SandPIPER and Marathon brands for the chemical, paint, food processing, construction, mining, and general industry markets. A unit of Idex Corporation.

16. **Watson-Marlow Pumps Group:** Manufacturer of peristaltic pumps for the chemical, printing, water treatment, mining, science, and general industry markets. Also produces small-diaphragm pumps for gases and liquids. A unit of Spirax-Sarco Engineering Company.

PRACTICE PROBLEMS

13.1 List 12 factors that should be considered when selecting a pump.

13.2 List 10 items that must be specified for pumps.

13.3 Describe a positive-displacement pump.

13.4 Name four examples of rotary positive-displacement pumps.

13.5 Name three types of reciprocating positive-displacement pumps.

13.6 Describe a kinetic pump.

13.7 Name three classifications of kinetic pumps.

13.8 Describe the action of the impellers and the general path of flow in the three types of kinetic pumps.

13.9 Describe a jet pump.

13.10 Distinguish between a shallow-well jet pump and a deep-well jet pump.

13.11 Describe the difference between a simplex reciprocating pump and a duplex type.

13.12 Describe the general shape of the plot of pump capacity versus discharge pressure for a positive-displacement rotary pump.

13.13 Describe the general shape of the plot of total head versus pump capacity for centrifugal pumps.

13.14 To the head-versus-capacity plot of Problem 13.13, add plots for efficiency and power required.

13.15 To what do the *affinity laws* refer in regard to pumps?

13.16 For a given centrifugal pump, if the speed of rotation of the impeller is cut in half, how does the capacity change?

13.17 For a given centrifugal pump, if the speed of rotation of the impeller is cut in half, how does the total head capability change?

13.18 For a given centrifugal pump, if the speed of rotation of the impeller is cut in half, how does the power required to drive the pump change?

13.19 For a given size of centrifugal pump casing, if the diameter of the impeller is reduced by 25 percent, how much does the capacity change?

13.20 For a given size of centrifugal pump casing, if the diameter of the impeller is reduced by 25 percent, how much does the total head capability change?

13.21 For a given size of centrifugal pump casing, if the diameter of the impeller is reduced by 25 percent, how much does the power required to drive the pump change?

13.22 Describe each part of this centrifugal pump designation: $1\frac{1}{2} \times 3 - 6$.

13.23 For the line of pumps shown in Fig. 13.22, specify a suitable size for delivering 100 gal/min of water at a total head of 300 ft.

13.24 For the line of pumps shown in Fig. 13.22, specify a suitable size for delivering 600 L/min of water at a total head of 25 m.

13.25 For the $2 \times 3 - 10$ centrifugal pump performance curve shown in Fig. 13.28, describe the performance that can be expected from a pump with an 8-in impeller operating against a system head of 200 ft. Give the expected capacity, the power required, the efficiency, and the required *NPSH*.

13.26 For the $2 \times 3 - 10$ centrifugal pump performance curve shown in Fig. 13.28, at what head will the pump having an 8-in impeller operate at its highest efficiency? List the pump's capacity, power required, efficiency, and the required *NPSH* at that head.

13.27 Using the result from Problem 13.26, describe how the performance of the pump changes if the system head increases by 15 percent.

13.28 For the $2 \times 3 - 10$ centrifugal pump performance curve shown in Fig. 13.28, list the total head and capacity at which the pump will operate at maximum efficiency for each of the impeller sizes shown.

13.29 For a given centrifugal pump and impeller size, describe how the *NPSH* required varies as the capacity increases.

13.30 State some advantages of using a variable-speed drive for a centrifugal pump that supplies fluid to a process requiring varying flow rates of a fluid as compared with adjusting throttling valves.

13.31 Describe how the capacity, efficiency, and power required for a centrifugal pump vary as the viscosity of the fluid pumped increases.

13.32 If two identical centrifugal pumps are connected in parallel and operated against a certain head, how would the total capacity compare with that of a single pump operating against the same head?

13.33 Describe the effect of operating two pumps in series.

13.34 For each of the following sets of operating conditions, list at least one appropriate type of pump. See Fig. 13.52.
   a. 500 gal/min of water at 80 ft of total head
   b. 500 gal/min of water at 800 ft of head
   c. 500 gal/min of a viscous adhesive at 80 ft of head
   d. 80 gal/min of water at 8000 ft of head
   e. 80 gal/min of water at 800 ft of head
   f. 8000 gal/min of water at 200 ft of head
   g. 8000 gal/min of water at 60 ft of head
   h. 8000 gal/min of water at 12 ft of head

13.35 For the $1\frac{1}{2} \times 3 - 13$ centrifugal pump performance curve shown in Fig. 13.34, determine the capacity that can be expected from a pump with a 12-in impeller operating against a system head of 550 ft. Then, compute the specific speed and specific diameter and locate the corresponding point on Fig. 13.53.

13.36 For the $6 \times 8 - 17$ centrifugal pump performance curve shown in Fig. 13.32, determine the capacity that can be expected from a pump with a 15-in impeller operating against a system head of 200 ft. Then, compute the specific speed and specific diameter and locate the corresponding point on Fig. 13.53.

13.37 Figure 13.52 shows that a mixed-flow pump is recommended for delivering 10 000 gal/min of water at a head of 40 ft. If such a pump operates with a specific speed of 5000, compute the appropriate operating speed of the pump.

13.38 Compute the specific speed for a pump operating at 1750 rpm delivering 5000 gal/min of water at a total head of 100 ft.

13.39 Compute the specific speed for a pump operating at 1750 rpm delivering 12 000 gal/min of water at a total head of 300 ft.

13.40 Compute the specific speed for a pump operating at 1750 rpm delivering 500 gal/min of water at a total head of 100 ft.

13.41 Compute the specific speed for a pump operating at 3500 rpm delivering 500 gal/min of water at a total head of 100 ft. Compare the result with that of Problem 13.40 and with Fig. 13.52.

13.42 It is desired to operate a pump at 1750 rpm by driving it with a four-pole electric motor. For each of the following conditions, compute the specific speed using Eq. (13–17). Then, recommend whether to use an axial pump, a

mixed-flow pump, a radial-flow pump, or none of these, based on the discussion related to Fig. 13.53.
   a. 500 gal/min of water at 80 ft of total head
   b. 500 gal/min of water at 800 ft of head
   c. 3500 gal/min of water at 80 ft of head
   d. 80 gal/min of water at 8000 ft of head
   e. 80 gal/min of water at 800 ft of head
   f. 8000 gal/min of water at 200 ft of head
   g. 8000 gal/min of water at 60 ft of head
   h. 8000 gal/min of water at 12 ft of head

13.43 Define *net positive suction head (NPSH)*.

13.44 Distinguish between *NPSH available* and *NPSH required*.

13.45 Describe what happens to the vapor pressure of water as the temperature increases.

13.46 Describe why it is important to consider *NPSH* when designing and operating a pumping system.

13.47 For what point in a pumping system is the *NPSH* computed? Why?

13.48 Discuss why it is desirable to elevate the reservoir from which a pump draws liquid.

13.49 Discuss why it is desirable to use relatively large pipe sizes for the suction lines in pumping systems.

13.50 Discuss why an *eccentric reducer* should be used when it is necessary to decrease the size of a suction line as it approaches a pump.

13.51 If we assume that a given pump requires 7.50 ft of *NPSH* when operating at 3500 rpm, what would be the *NPSH* required at 2850 rpm?

13.52 Determine the available *NPSH* for the pump in Practice Problem 7.14 if the water is at 80°F and the atmospheric pressure is 14.4 psia. Repeat the calculations for water at 180°F.

13.53 Find the available *NPSH* when a pump draws water at 140°F from a tank whose level is 4.8 ft below the pump inlet. The suction line losses are 2.2 lb-ft/lb and the atmospheric pressure is 14.7 psia.

13.54 A pump draws benzene at 25°C from a tank whose level is 2.6 m above the pump inlet. The suction line has a head loss of 0.8 N·m/N. The atmospheric pressure is measured to be 98.5 kPa(abs). Find the available *NPSH*. The vapor pressure of benzene is 13.3 kPa.

13.55 Determine the available *NPSH* for the system shown in Fig. 13.38(h). The fluid is water at 80°C and the atmospheric pressure is 101.8 kPa. The water level in the tank is 2.0 m below the pump inlet. The vertical leg of the suction line is a DN 80 Schedule 40 steel pipe, whereas the horizontal leg is a DN 50 Schedule 40 pipe, 1.5 m long. The elbow is of the long-radius type. Neglect the loss in the reducer. The foot valve and strainer are of the hinged-disk type. The flow rate is 300 L/min.

13.56 Determine the *NPSH* available when a pump draws carbon tetrachloride at 150°F (sg = 1.48) from a tank whose level is 3.6 ft below the pump inlet. The energy losses in the suction line total 1.84 ft and the atmospheric pressure is 14.55 psia.

13.57 Determine the *NPSH* available when a pump draws carbon tetrachloride at 65°C (sg = 1.48) from a tank whose level is 1.2 m below the pump inlet. The energy losses in the suction line total 0.72 m and the atmospheric pressure is 100.2 kPa absolute.

13.58 Determine the *NPSH* available when a pump draws gasoline at 40°C (sg = 0.65) from an underground tank whose level is 2.7 m below the pump inlet. The energy losses in the suction line total 1.18 m and the atmospheric pressure is 99.2 kPa absolute.

13.59 Determine the *NPSH* available when a pump draws gasoline at 110°F (sg = 0.65) from an outside storage tank whose level is 4.8 ft above the pump inlet. The energy losses in the suction line total 0.87 ft and the atmospheric pressure is 14.28 psia.

13.60 Repeat Problem 13.56 if the pump is 44 in below the fluid surface.

13.61 Repeat Problem 13.59 if the pump is 27 in above the fluid surface.

13.62 Repeat Problem 13.57 if the pump is 1.2 m below the fluid surface.

13.63 Repeat Problem 13.58 if the pump is installed under the tank, 0.65 m below the fluid surface.

13.64 A pump draws propane at 110°F (sg = 0.48) from a tank whose level is 30 in above the pump inlet. The energy losses in the suction line total 0.73 ft and the atmospheric pressure is 14.32 psia. Determine the required pressure above the propane in the tank to ensure that the *NPSH* available is at least 4.0 ft.

13.65 A pump draws propane at 45°C (sg = 0.48) from a tank whose level is 1.84 m below the pump inlet. The energy losses in the suction line total 0.92 m and the atmospheric pressure is 98.4 kPa absolute. Determine the required pressure above the propane in the tank to ensure that the $NPSH_A$ is at least 1.50 m.

SUPPLEMENTAL PROBLEM (PIPE-FLO® ONLY)

13.66 Select a pump from the sample catalogue within the PIPE-FLO® demo software to run a system that pumps water at 30°C from a reservoir up into a storage tower, 20 m higher, at a rate of 1800 L/min. Position the pump at the same level as the reservoir. On the suction side, use 8 m of DN 100 Schedule 40 pipe and install an angle foot valve and two standard elbows. On the discharge line use 30 m of DN 80 pipe, a fully open globe valve, and four standard 90° elbows. In addition to identifying a pump for the job, also determine $NPSH_R$, $NPSH_A$, pump efficiency, pressure at the pump outlet, and total head on the pump as reported on the FLO-Sheet® in PUMP-FLO®.

DESIGN PROBLEMS

Several situations are presented here in which a system is being designed to pump a fluid from some source to a given destination. In each case, the objective is to completely define the configuration of the system, including:

- Pipe sizes and types
- Location of the pump
- Length of pipe for all parts of the system
- Valves and fittings
- Specification of a pump with adequate performance
- Carefully rendered drawing of the piping layout
- List of materials required for the system
- Analysis of the pressure at pertinent points

FIGURE 13.55 Design Problem 1.

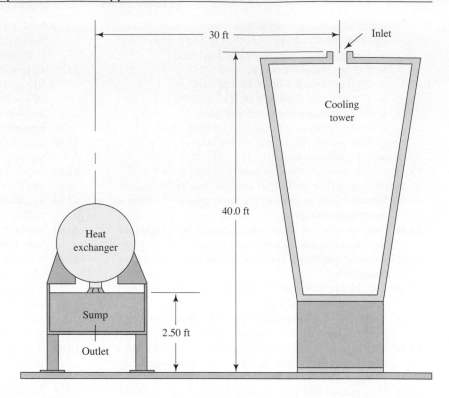

See Section 13.12 and Example Problem 13.4 for the procedure. Present the results in a written report using good technical report writing practice.

DESIGN PROBLEM STATEMENTS

1. Design a system to pump water at 140°F from a sump below a heat exchanger to the top of a cooling tower, as shown in Fig. 13.55. The desired minimum flow rate is 200 gal/min.

2. Design a system to pump water at 80°C from a water heater to a washing system, as shown in Fig. 13.56. The desired minimum flow rate is 750 L/min (198 gal/min).

3. Design a system to pump water at 90°F from a river to a tank elevated 55 ft above the surface of the river. The desired minimum flow rate is 1500 gal/min. The tank is to be set back 125 ft from the river bank.

4. Design the water system for Professor Crocker's cabin, as described in Fig. 7.38. The desired minimum flow rate is 40 gal/min, and the distribution tank is to be maintained at a pressure of 30 psig above the water. The cabin sits 150 ft from

FIGURE 13.56 Design Problem 2.

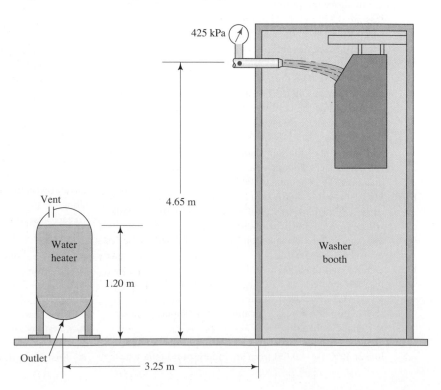

the side of the stream from which the water is to be drawn. The slope of the hillside is approximately 30° from the vertical. The water is at 80°F.

5. Design a system similar to that shown in Fig. 7.35, in which air pressure at 400 kPa above the kerosene at 25°C is used to cause the flow. The horizontal distance between the two tanks is 32 m. The desired minimum flow rate is 500 L/min.

6. Design a system similar to that shown in Fig. 8.12, which must supply at least 1500 gal/min of water at 60°F for a fire protection system. The pressure at point B must be at least 85 psig. The depth of water in the tank is 5.0 ft. Ignore the specified sizes for the pipes and make your own decisions. Add appropriate valves and redesign the suction line.

7. Design a system similar to that shown in Fig. 8.18 and described in Practice Problem 8.44. Ignore the given pipe sizes and the given pressure at the pump inlet. Add appropriate valves. The pump draws the polluted water from a still pond whose surface is 30 in below the centerline of the pump inlet. Use the vapor pressure for water at 100°F.

8. Design a system similar to that shown in Fig. 7.22 to deliver 60 L/min of a water-based cutting fluid (sg = 0.95) to the cutter of a milling machine. Assume that the viscosity and vapor pressure

are 10 percent greater than that for water at 40°C. Assume that the pump is submerged and that the minimum depth above the suction inlet is 75 mm. The total length of the path required for the discharge line is 1.75 m.

9. Design a system similar to that shown in Fig. 7.21 to deliver 840 L/min of water at 100°F from an underground storage tank to a pressurized storage tank. Ignore original pipe sizes and make your own decision. Add appropriate valves. The upper tank pressure is 500 kPa.

10. Specify a suitable pump for the system shown in Fig. 13.57. It is a combination series/parallel system that operates as follows.

- Water at 160°F is drawn at the rate of 275 gal/min from a tank into the 4-in suction line of the pump. The suction line has a total length of 10 ft.
- The 3-in discharge line elevates the water 15 ft to the level of a large heat exchanger. The discharge line has a total length of 40 ft.
- The flow splits into two branches with the primary 3-in line feeding a large heat exchanger that has a K-factor of 12 based on the velocity head in the pipe. The total length of pipe in this branch is 8 ft.

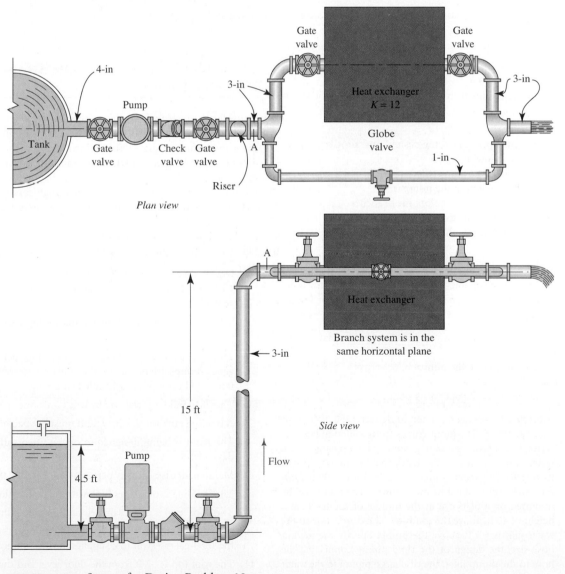

FIGURE 13.57   System for Design Problem 10.

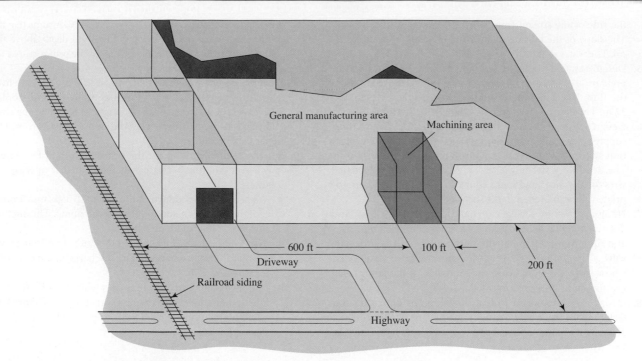

FIGURE 13.58   Plot plan for a factory building for a comprehensive design problem.

- The 1-in line is a bypass around the heat exchanger with a total length of 30 ft.
- The two lines join at the right and discharge to the atmosphere through a short 3-in pipe.
- All pipes are Schedule 40 steel.

For this system, operating at the desired operating conditions, determine the following:

a. The pressure at the pump inlet
b. The *NPSH* available at the pump inlet
c. The pressure at point A before the branches
d. The volume flow rate through the heat exchanger line
e. The volume flow rate through the bypass line
f. The total head on the pump
g. The power delivered to the water by the pump.

Then specify a suitable pump for this system that will deliver at least the desired 275 gal/min of flow. For the selected pump, determine the following:

h. The actual expected flow rate produced by the pump at the operating point
i. The power input to the pump
j. The *NPSH* required
k. The efficiency at the operating point.

11. A fire truck is being designed to deliver 1250 gal/min of water at 100°F. The input comes from a suction hose inserted into a lake, a river, or a pond. The discharge is to a water gun mounted on the truck that requires 150 psi at its nozzle. The water source could be up to 200 ft from the truck and 10 ft below the roadway. The pump will be mounted on a platform at the middle of the truck, at a height of 40 in above the roadway. The connection to the water gun is 6.5 ft above the pump. Specify the suction hose size, the design of the rigid piping connecting the hose to the pump inlet, the discharge piping to the water cannon, valves, and other fittings.

COMPREHENSIVE DESIGN PROBLEM

Consider yourself to be a plant engineer for a company that is planning a new manufacturing facility. As a part of the new plant, there will be an automated machining line in which five machines will be supplied with coolant from the same reservoir. You are responsible for the design of the system to handle the coolant from the time it reaches the plant in railroad tank cars until the dirty coolant is removed from the premises by a contract firm for reclaim.

The layout of the planned facility is shown in Fig. 13.58. The following data, design requirements, and limitations apply.

1. New coolant is delivered to the plant by tank cars carrying 15 000 gal each. A holding tank for new coolant must be specified.

2. The reservoir for the automated machining system must have a capacity of 1000 gal.

3. The 1000-gal tank is normally emptied once per week. Emergency dumps are possible if the coolant becomes overly contaminated prior to the scheduled emptying.

4. The dirty fluid is picked up by truck only once per month.

5. A holding tank for the dirty fluid must be specified.

6. The plant is being designed to operate two shifts per day, 7 days a week.

7. Maintenance is normally performed during the third shift.

8. The building is one-story high with a concrete floor.

9. The floor level is at the same elevation as the railroad track.

10. No storage tank can be inside the plant or under the floor except the 1000-gal reservoir that supplies the machining system.

11. The roof top is 32 ft from the floor level and the roof can be designed to support a storage tank.

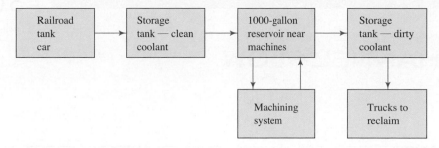

FIGURE 13.59 Block diagram of coolant system.

12. The building is to be located in Dayton, Ohio, where the outside temperature may range from −20°F to +105°F.

13. The frost line is 30 in below the surface.

14. The coolant is a solution of water and a soluble oil with a specific gravity of 0.94 and a freezing point of 0°F. Its corrosiveness is approximately the same as that of water.

15. Assume that the viscosity and vapor pressure of the coolant are 1.50 times that of water at any temperature.

16. You are not asked to design the system to supply the machines.

17. The basic coolant storage and delivery system is to have the functional design sketched in the block diagram in Fig. 13.59.

The following tasks are to be completed by you, the system designer:

a. Specify the location and size of all storage tanks.

b. Specify the layout of the piping system, the types and sizes of all pipes, and the lengths required.

c. Specify the number, type, and size of all valves, elbows, and fittings.

d. Specify the number of pumps, their types, capacities, head requirements, and power required.

e. Specify the installation requirements for the pumps, including the complete suction line system. Evaluate the $NPSH_A$ for your design, and demonstrate that your pump has an acceptable $NPSH$ required.

f. Determine the time required to fill and empty all tanks.

g. Sketch the layout of your design in both a plan view (top) and an elevation view (side). An isometric sketch may also be used.

h. Include the analysis of all parts of the system, including energy losses due to friction and minor losses.

i. Submit the results of your design in a neat and complete report, including a narrative description of the system, the sketches, a list of materials, and the analysis to show that your design meets the specifications.

OPEN-CHANNEL FLOW

An open channel is a flow system in which the top surface of the fluid is exposed to the atmosphere. Examples are rain gutters on buildings, storm sewers, natural rivers and streams, and channels constructed to drain fluids in a controlled manner. The analysis of open channels requires special techniques somewhat different from those you have used to analyze flow in closed pipe and tubing.

Consider the irrigation stream shown in Fig. 14.1 that carries water to fields. The sides and bottom of the stream are excavated from the earth and the only way water can flow is by gravity with the stream sloped downward from some source toward the destination for the water.

Introductory Concepts

In contrast to the closed conduits that have been discussed in preceding chapters, an *open channel* is a flow system in which the top surface of the fluid is exposed to the atmosphere. Many examples of open channels occur in nature and in systems designed to supply water to communities or to carry storm drainage and sewage safely away.

Rivers and streams are obvious examples of natural channels. Rain gutters on buildings and at the sides of streets carry rainwater. Storm sewers, usually beneath the streets, collect the runoff from the streets and conduct it

to a stream or to a larger, man-made ditch or canal. In industry, open channels are often used to convey cooling water away from heat exchangers or coolants away from machining systems.

The cross-sectional shape of the open channel is critical to its ability to deliver a particular volume flow rate of fluid and you will learn how to characterize the shape using the term *hydraulic radius, R,* that is dependent on the net cross-sectional area of the flow stream, *A,* and the *wetted perimeter, WP.* These terms were used also in Chapter 9 to describe noncircular conduits that were running full of fluid under pressure. Adjustment of the methods for computing *A, WP,* and *R* for open channels is described in this chapter. Figure 14.2 shows some common cross-sectional shapes for open channels along with the formulas for *A* and *WP.*

Exploration

- Observe where open channels exist in areas with which you are familiar.

- Look for rain gutters, natural streams, and other drainage structures.

- What new ones can you find?

Make sketches of the shapes of the cross sections of the open channels you find and describe how the channel is sloped to enable the water or other fluid to flow in

FIGURE 14.1 This irrigation stream is an example of an open channel because its top surface is exposed to the atmosphere. Also shown is a *weir*, an obstruction in the channel designed to make measurements of the flow rate in the stream. (*Source:* Ruud Morijn/Fotolia)

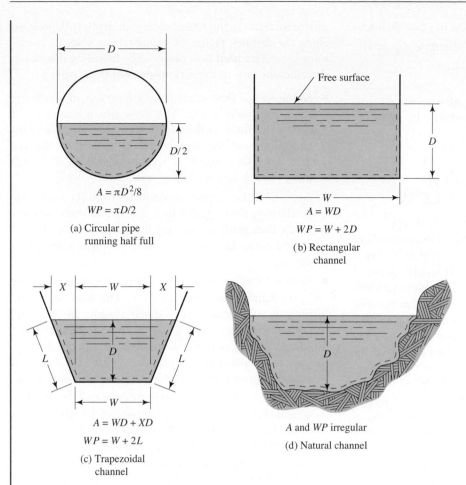

FIGURE 14.2 Examples of cross sections of open channels.

$A = \pi D^2/8$
$WP = \pi D/2$

(a) Circular pipe running half full

Free surface

$A = WD$
$WP = W + 2D$

(b) Rectangular channel

$A = WD + XD$
$WP = W + 2L$

(c) Trapezoidal channel

A and WP irregular

(d) Natural channel

a preferred direction. Answer such questions as the following:

- What is the channel used for?
- What fluid is flowing in the channel?
- Does the flow appear to be smooth and tranquil or chaotic and turbulent?
- What is the shape of the cross section of the channel and what are its dimensions?
- Is the cross-section uniform along its length or does it vary?
- How deep was the fluid when you saw it? How deep can the fluid be under very heavy flow conditions before overflowing?

- How does the shape of the flow stream change, if at all, as the depth increases?
- Could you detect whether the channel is installed on a slope?

This chapter will present some of the methods of analyzing open-channel flow. Complete coverage of the subject is an extensive undertaking and is treated in entire books such as References 3–5, 7–11, and 13–15 listed at the end of this chapter.

Measurement of the rate of flow in an open channel is also important to monitor natural streams and aid in predicting impending floods, to measure how much flow is being withdrawn from a reservoir, and as part of water pollution control plans. Open-channel flow measurement is presented in Section 14.10.

14.1 OBJECTIVES

After completing this chapter, you should be able to:

1. Compute the hydraulic radius for open channels.
2. Describe *uniform flow* and *varied flow*.
3. Use Manning's equation to analyze uniform flow.
4. Define the slope of an open channel and compute its value.

5. Compute the normal discharge for an open channel.
6. Compute the normal depth of flow for an open channel.
7. Design an open channel to transmit a given discharge with uniform flow.
8. Define the *hydraulic depth* and the *Froude number*.
9. Describe *critical flow*, *subcritical flow*, and *supercritical flow*.

10. Define the specific energy of the flow in open channels.

11. Define the terms *critical depth*, *alternate depth*, and *sequent depth*.

12. Describe the term *hydraulic jump*.

13. Describe *weirs* and *flumes* as they are used for measuring flow in open channels, and perform the necessary computations.

14.2 CLASSIFICATION OF OPEN-CHANNEL FLOW

Open-channel flow can be classified into several types.

Uniform steady flow occurs when the volume flow rate (typically called *discharge* in open-channel flow analysis) remains constant in the section of interest and the depth of the fluid in the channel does not vary. To achieve uniform steady flow, the cross section of the channel may not change along its length. Such a channel is *prismatic*. Figure 14.3 shows uniform flow in a side view. Note that the free, top surface of the fluid is parallel to the bottom surface of the channel.

Varied steady flow occurs when the discharge remains constant, but the depth of the fluid varies along the section of interest. This will occur if the channel is not prismatic.

Unsteady varied flow occurs when the discharge varies with time, resulting in changes in the depth of the fluid along the section of interest whether the channel is prismatic or not.

Varied flow can be further classified into *rapidly varying flow* or *gradually varying flow*. As the names imply, the difference refers to the rate of change in depth with position along the channel. Figure 14.4 illustrates a series of conditions in which varied flow occurs. The following discussion describes the flow in the various parts of this figure.

- *Section 1* The flow starts from a reservoir in which the fluid is virtually at rest. The *sluice gate* is a device that allows the fluid to flow from the reservoir at a point beneath the surface. Rapidly varying flow occurs near the gate as the fluid accelerates, and the velocity of flow is likely to be quite large in this area.

- *Section 2* If the channel downstream from the sluice gate is relatively short, and if its cross section does not vary much, then gradually varying flow occurs. If the channel is prismatic and long enough, uniform flow could develop.

- *Section 3* The formation of a *hydraulic jump* is a curious open-channel flow phenomenon. The flow before the jump is quite rapid and relatively shallow. In the jump, the flow becomes very turbulent, and a large amount of energy is dissipated. Then, following the jump, the flow velocity is much lower, and the depth of the fluid is greater. More will be said later about the hydraulic jump.

- *Section 4* A *weir* is an obstruction placed in the flow stream that causes an abrupt change in the cross section of the channel. Weirs can be used as control devices or to measure the volume flow rate. Typically the flow is rapidly varying as it travels over the weir, with a "waterfall" (called the *nappe*) formed downstream.

- *Section 5* As with Section 2, the flow downstream from a weir is usually gradually varying if the channel is prismatic.

FIGURE 14.3 Uniform steady open-channel flow—side view.

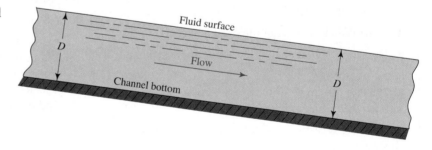

FIGURE 14.4 Conditions causing varied flow.

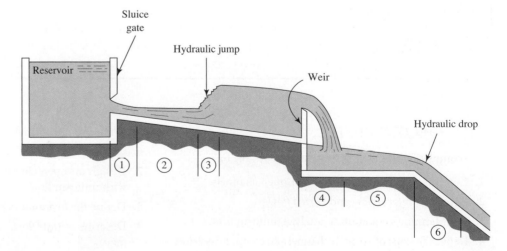

■ *Section 6* The *hydraulic drop* occurs when the slope of the channel suddenly increases to a steep angle. The flow accelerates because of gravity, and rapidly varying flow occurs.

14.3 HYDRAULIC RADIUS AND REYNOLDS NUMBER IN OPEN-CHANNEL FLOW

The characteristic dimension of open channels is the *hydraulic radius*, defined as the ratio of the net cross-sectional area of a flow stream to the wetted perimeter of the section. That is,

⇨ **Hydraulic Radius**

$$R = \frac{A}{WP} = \frac{\text{Area}}{\text{Wetted perimeter}} \quad (14\text{--}1)$$

The unit for R is the meter in the SI unit system and feet in the English system.

In the calculation of the hydraulic radius, the net cross-sectional area should be evident from the geometry of the section, particularly for commonly used manufactured channels as shown in parts (a), (b), and (c) of Fig. 14.2. For natural streams as sketched in part (d) of the figure, approximations are typically used. The *wetted perimeter* is defined as the sum of the length of the solid boundaries of the section actually in contact with (i.e., wetted by) the fluid. Expressions for the area A and the wetted perimeter WP are given in Fig. 14.2 for those sections illustrated. In each case, the fluid flows in the shaded portion of the section. A dashed line is shown adjacent to the boundaries that make up the wetted perimeter. Notice that the length of the free surface of an open channel is *NOT* included in WP.

The following example problem illustrates the calculations required to compute R.

Example Problem 14.1

Determine the hydraulic radius of the trapezoidal section shown in Fig. 14.2(c) if $W = 4$ ft, $X = 1$ ft, and $D = 2$ ft.

Solution

The net flow area is the sum of a rectangle and two triangles:

$$A = WD + 2(XD/2) = WD + XD$$
$$= (4)(2) + (1)(2) = 10 \text{ ft}^2$$

To find the wetted perimeter, we must determine the value of L using the Pythagorean Theorem:

$$WP = W + 2L$$
$$L = \sqrt{X^2 + D^2} = \sqrt{(1)^2 + (2)^2} = 2.24 \text{ ft}$$
$$WP = 4 + 2(2.24) = 8.48 \text{ ft}$$

Then, we have

$$R = A/WP = 10 \text{ ft}^2/8.48 \text{ ft} = 1.18 \text{ ft}$$

Recall that the Reynolds number for closed circular cross-sections running full is

$$N_R = \frac{vD}{\nu} \quad (14\text{--}2)$$

where v = Average velocity of flow, D = Pipe diameter, and ν = Kinematic viscosity of the fluid. We have seen that laminar flow occurs when $N_R < 2000$ and turbulent flow occurs when $N_R > 4000$ for most practical pipe flow situations. The Reynolds number represents the effects of fluid viscosity relative to the inertia of the fluid.

In open-channel flow, the characteristic dimension is the hydraulic radius R. It was shown in Chapter 9 that for a full circular cross section, $D = 4R$. For closed, noncircular cross sections, it was convenient to substitute $4R$ for D so that the Reynolds number would have the same order of magnitude as that for circular pipes and tubes. However, this is not usually done in open-channel flow

analysis. The Reynolds number for open-channel flow is then

⇨ **Reynolds Number for Open Channels**

$$N_R = \frac{vR}{\nu} \quad (14\text{--}3)$$

Experimental evidence (Reference 4) shows that, in open channels, laminar flow occurs when $N_R < 500$. The range from 500 to 2000 is the transition region. Turbulent flow normally occurs when $N_R > 2000$.

14.4 KINDS OF OPEN-CHANNEL FLOW

The Reynolds number and the terms *laminar* and *turbulent*, while helpful, are not sufficient to characterize all kinds of open-channel flow. In addition to the viscosity-versus-inertial effects, the ratio of inertial forces to gravity forces is also important, given by the *Froude number* N_F, defined as

▷ **Froude Number**

$$N_F = \frac{v}{\sqrt{gy_h}} \qquad (14\text{–}4)$$

where y_h, called the hydraulic depth, is given by

▷ **Hydraulic Depth**

$$y_h = A/T \qquad (14\text{–}5)$$

and T is the width of the free surface of the fluid at the top of the channel.

When the Froude number is equal to 1.0, that is, when $v = \sqrt{gy_h}$, the flow is called *critical flow*. When $N_F < 1.0$, the flow is *subcritical*, and when $N_F > 1.0$, the flow is *supercritical*. See also Section 14.8.

Then, the following kinds of flow are possible:

1. Subcritical-laminar: $N_R < 500$ and $N_F < 1.0$
2. Subcritical-turbulent: $N_R > 2000$ and $N_F < 1.0$
3. Supercritical-turbulent: $N_R > 2000$ and $N_F > 1.0$
4. Supercritical-laminar: $N_R < 500$ and $N_F > 1.0$

In addition, flows can be in the transition region. However, such flows are unstable and very difficult to characterize. See References 12 and 17.

In this discussion, the terms *laminar* and *turbulent* have the same significance as they did for pipe flow. Laminar flow is quite tranquil and there is little or no mixing of the fluid, so that a stream of dye injected into such a flow remains virtually intact. In turbulent flow, however, chaotic intermixing occurs, and the dye stream rapidly dissipates throughout the fluid.

14.5 UNIFORM STEADY FLOW IN OPEN CHANNELS

Figure 14.3 is a schematic illustration of uniform steady flow in an open channel. The distinguishing feature of uniform flow is that the fluid surface is parallel to the slope of the channel bottom. We will use the symbol S to indicate the slope of the channel bottom and S_w for the slope of the water surface. Then, for uniform flow, $S = S_w$. Theoretically, uniform flow can exist only if the channel is prismatic, that is, if

its sides are of uniform geometry aligned to an axis in the direction of flow. Examples of prismatic channels are rectangular, trapezoidal, triangular, and circular sections running partially full. In addition, the channel slope S must be constant. If the cross section or slope of the channel is changing, then the flow stream would be either converging or diverging and varied flow would occur.

The slope S of a channel can be reported in several ways. It is ideally defined as the ratio of the vertical drop h to the horizontal distance over which the drop occurs. For small slopes, which are typical in open-channel flow, it is more practical to use h/L where L is the length of the channel as shown in Fig. 14.5. Normally, the magnitude of the slope for natural streams and drainage structures is very small, a typical value being 0.001. This number can also be expressed as a percentage, where $0.01 = 1$ percent. Then, $0.001 = 0.1$ percent. Because $\sin \theta = h/L$, the angle that the channel bottom makes with the horizontal could also be used. In summary, the slope of 0.001 could be reported as:

1. The channel falls 1 m per 1000 m of channel.
2. The slope is 0.1 percent.
3. $\sin \theta = 0.001$. Then $\theta = \sin^{-1}(0.001) = 0.057°$.

Because the angle is so small, it is rarely used as a measure of the slope.

In uniform flow, the driving force for the flow is provided by the component of the weight of the fluid that acts along the channel, as shown in Fig. 14.5. This force is $w \sin \theta$, where w is the weight of a given element of fluid and θ is the angle of the slope of the channel bottom. If the flow is uniform, it cannot be accelerating. Therefore, there must be an equal opposing force acting along the channel surface. This is a friction force that depends on the roughness of the channel surfaces and on the cross-sectional size and shape.

By equating the expressions for the driving force and the opposing force, we can derive an expression for the average velocity of uniform flow. A commonly used form of the resulting equation was developed by Robert Manning. In SI units, Manning's equation is written as

▷ **Manning's Equation—SI Units**

$$v = \frac{1.00}{n} R^{2/3} S^{1/2} \qquad (14\text{–}6)$$

Units must be consistent in Eq. (14–6). The average velocity of flow v will be in m/s when the hydraulic radius R

FIGURE 14.5 Uniform open-channel flow: forces and slope

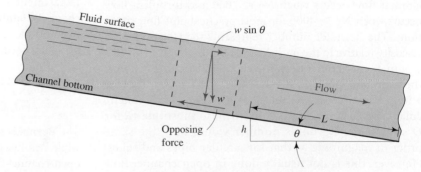

TABLE 14.1 Values for Manning's n

Channel Description	n
Glass, copper, plastic, or other smooth surfaces	0.010
Smooth, unpainted steel, planed wood	0.012
Painted steel or coated cast iron	0.013
Smooth asphalt, common clay drainage tile, trowel-finished concrete, glazed brick	0.013
Uncoated cast iron, black wrought iron pipe, vitrified clay sewer tile	0.014
Brick in cement mortar, float-finished concrete, concrete pipe	0.015
Formed, unfinished concrete, spiral steel pipe	0.017
Smooth earth	0.018
Clean excavated earth	0.022
Corrugated metal storm drain	0.024
Natural channel with stones and weeds	0.030
Natural channel with light brush	0.050
Natural channel with tall grasses and reeds	0.060
Natural channel with heavy brush	0.100

is in meters. The channel slope S, defined as $S = h/L$, is dimensionless. The final term n is a resistance factor sometimes called *Manning's n*. The value of n depends on the condition of the channel surface and is, therefore, somewhat analogous to the pipe wall roughness ε used previously. The form of Manning's equation in U.S. Customary System units is given later in this section.

Typical design values of n are listed in Table 14.1 for materials commonly used for artificial channels and natural streams. A very extensive discussion of the determination of Manning's n and a more complete table of values are given by V. T. Chow (Reference 4). The values listed in Table 14.1 are average values that will give good estimates for use in design or for rough analysis of existing channels. Variations from the average should be expected.

We can calculate the volume flow rate in the channel from the continuity equation, which is the same as that used for pipe flow:

$$Q = Av \qquad (14–7)$$

In open-channel flow analysis, Q is typically called the *discharge*. Substituting Eq. (14–6) into (14–7) gives an equation that directly relates the discharge to the physical parameters of the channel:

⮡ **Normal Discharge—SI Units**

$$Q = \left(\frac{1.00}{n}\right)AR^{2/3}S^{1/2} \qquad (14–8)$$

This is the only value of discharge for which uniform flow will occur for the given channel depth, and it is called the *normal discharge*. The units of Q are m^3/s when the area A is expressed in square meters (m^2) and the hydraulic radius in meters (m).

Another useful form of this equation is

$$AR^{2/3} = \frac{nQ}{S^{1/2}} \qquad (14–9)$$

The term on the left side of Eq. (14–9) is solely dependent on the geometry of the section. Therefore, for a given discharge, slope, and channel surface type, we can determine the geometrical features of a channel. Alternatively, for a given size and shape of channel, we can calculate the depth at which the normal discharge Q would occur. This depth is called the *normal depth*.

In analyzing uniform flow, typical problems encountered are the calculations of the normal discharge, the normal depth, the geometry of the channel section, the slope, and the value of Manning's n. We can make these calculations by using Eqs. (14–6)–(14–9).

14.5.1 Manning's Equation in U.S. Customary System

Though not strictly true, it is conventional to take the values of Manning's n to be dimensionless so that the same data can be used in either the SI form of the equation [Eq. (14–6)] or the U.S. Customary System form. A careful conversion of units (see Reference 4) allows the use of the same values of n in the following equation:

⮡ **Manning's Equation—U.S. Customary Units**

$$v = \frac{1.49}{n}R^{2/3}S^{1/2} \qquad (14–10)$$

The velocity will then be expressed in feet per second (ft/s) when R is in feet. This is the form of Manning's equation for the U.S. Customary System.

We can also create forms of this equation analogous to Eqs. (14–8) and (14–9). That is,

⮡ **Normal Discharge—U.S. Customary Units**

$$Q = Av = \left(\frac{1.49}{n}\right)AR^{2/3}S^{1/2} \qquad (14–11)$$

and

$$AR^{2/3} = \frac{nQ}{1.49S^{1/2}} \qquad (14–12)$$

In these equations, Q is the *normal discharge* in cubic feet per second (ft^3/s) when A is the flow area in square feet (ft^2) and R is expressed in feet (ft).

Example Problem 14.2   Determine the normal discharge for a 200-mm inside diameter common clay drainage tile running half full if it is laid on a slope that drops 1 m over a run of 1000 m.

Solution   Equation (14–8) will be used:

$$Q = \left(\frac{1.00}{n}\right)AR^{2/3}S^{1/2}$$

The slope $S = 1/1000 = 0.001$. From Table 14.1 we find $n = 0.013$. Figure 14.6 shows the cross section of the tile half full. Write

$$A = \frac{1}{2}\left(\frac{\pi D^2}{4}\right) = \frac{\pi D^2}{8} = \frac{\pi(200)^2}{8}\,mm^2 = 5000\pi\ mm^2$$

$$A = 15\,708\ mm^2 = 0.0157\ m^2$$

$$WP = \pi D/2 = 100\pi\ mm$$

Then, we have

$$R = A/WP = 5000\pi\ mm^2/100\pi\ mm = 50\ mm = 0.05\ m$$

Then in Eq. (14–8),

$$Q = \frac{(0.0157)(0.05)^{2/3}(0.001)^{1/2}}{0.013}$$

$$Q = 5.18 \times 10^{-3}\ m^3/s$$

FIGURE 14.6   Circular drain tile running half full for Example Problem 14.2 .

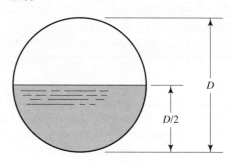

Example Problem 14.3   Calculate the minimum slope on which the channel shown in Fig. 14.7 must be laid if it is to carry 50 ft^3/s of water with a depth of 2 ft. The sides and bottom of the channel are made of formed, unfinished concrete.

Solution   Equation (14–11) can be solved for the slope S:

$$Q = \left(\frac{1.49}{n}\right)AR^{2/3}S^{1/2}$$

$$S = \left(\frac{Qn}{1.49AR^{2/3}}\right)^2 \qquad\qquad \textbf{(14–13)}$$

From Table 14.1 we find $n = 0.017$. The values of A and R can be calculated from the geometry of the section as illustrated in Fig. 14.2:

$$A = (4)(2) + (2)(2)(2)/2 = 12\ ft^2$$

$$WP = 4 + 2\sqrt{4 + 4} = 9.66\ ft$$

$$R = A/WP = 12/9.66 = 1.24\ ft$$

FIGURE 14.7   Trapezoidal channel for Example Problem 14.3.

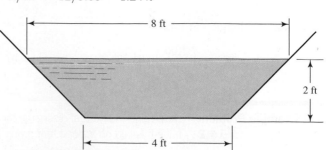

Then from Eq. (14–13) we have

$$S = \left[\frac{(50)(0.017)}{(1.49)(12)(1.24)^{2/3}} \right]^2 = 0.00169$$

Therefore, the channel must drop at least 1.69 ft per 1000 ft of length.

Example Problem 14.4

Design a rectangular channel to be made of formed, unfinished concrete to carry 5.75 m³/s of water when laid on a 1.2-percent slope. The normal depth should be one-half the width of the channel bottom.

Solution

Because the geometry of the channel is to be determined, Eq. (14–9) is most convenient:

$$AR^{2/3} = \frac{nQ}{S^{1/2}} = \frac{(0.017)(5.75)}{(0.012)^{1/2}} = 0.892$$

Figure 14.8 shows the cross section. Because $y = b/2$, both A and R can be expressed in terms of b and only b must be determined.

$$A = by = \frac{b^2}{2}$$

$$WP = b + 2y = 2b$$

$$R = A/WP = \frac{b^2}{(2)(2b)} = \frac{b}{4}$$

Then, we have

$$AR^{2/3} = 0.892$$

$$\frac{b^2}{2}\left(\frac{b}{4}\right)^{2/3} = 0.892$$

$$\frac{b^{8/3}}{5.04} = 0.892$$

$$b = (4.50)^{3/8} = 1.76 \text{ m}$$

The width of the channel must be 1.76 m.

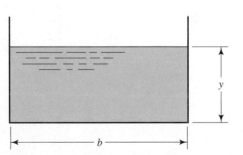

FIGURE 14.8 Rectangular channel for Example Problem 14.4.

Example Problem 14.5

In the final design of the channel described in Example Problem 14.4, the width was made 2 m. The maximum expected discharge for the channel is 12 m³/s. Determine the normal depth for this discharge.

Solution

Equation (14–9) will be used again:

$$AR^{2/3} = \frac{nQ}{S^{1/2}} = \frac{(0.017)(12)}{(0.012)^{1/2}} = 1.86$$

Both A and R must be expressed in terms of the dimension y in Fig. 14.8, with $b = 2.0$ m:

$$A = 2y$$

$$WP = 2 + 2y$$

$$R = A/WP = 2y/(2 + 2y)$$

Then, we have

$$1.86 = AR^{2/3} = 2y \left(\frac{2y}{2 + 2y} \right)^{2/3}$$

Algebraic solution for *y* is not simply done. A trial-and-error approach can be used. The results are as follows:

y (m)	A (m²)	WP (m)	R (m)	R²/³	AR²/³	Required Change in y
2.0	4.0	6.0	0.667	0.763	3.05	Make y lower
1.5	3.0	5.0	0.600	0.711	2.13	Make y lower
1.35	2.7	4.7	0.574	0.691	1.86	y is OK

Therefore, the channel depth would be 1.35 m when the discharge is 12 m³/s.

14.6 THE GEOMETRY OF TYPICAL OPEN CHANNELS

Frequently used shapes for open channels include circular, rectangular, trapezoidal, and triangular. Table 14.2 gives the formulas for computing the geometric features pertinent to open-channel flow calculations.

The trapezoid is popular for several reasons. It is an efficient shape because it gives a large flow area relative to the wetted perimeter. The sloped sides are convenient for channels made in the earth because the slopes can be set at an angle at which the construction materials are stable.

The slope of the sides can be defined by the angle with respect to the horizontal or by means of the *pitch*, the ratio of the horizontal distance to the vertical distance. The pitch in Table 14.2 is indicated by the value of *z*, which is the horizontal distance corresponding to one unit of vertical distance. Practical earth channels made in the trapezoidal shape use values of *z* from 1.0 to 3.0.

The rectangle is a special case of the trapezoid with a side slope of 90° or *z* = 0. Formed concrete channels are often made in this shape. The triangular channel is also a special case of the trapezoid with a bottom width of zero. Simple ditches in earth are often made in this shape.

The computation of the data for circular sections at various depths can be facilitated by the graph in Fig. 14.9. At the left side of the figure is shown half of a circular section

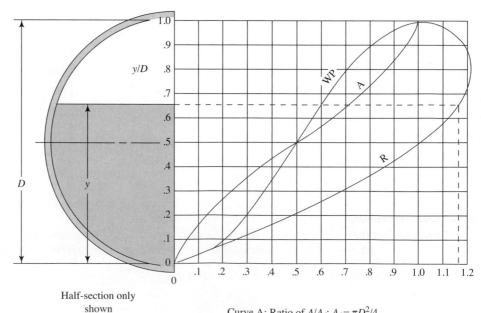

Half-section only shown

Curve A: Ratio of A/A_f; $A_f = \pi D^2/4$
Curve WP: Ratio of WP/WP_f; $WP_f = \pi D$
Curve R: Ratio of R/R_f; $R_f = D/4$

Example: $D = 2.0$ ft; $y = 1.30$ ft; $y/D = 0.65$

$A_f = 3.14$ ft²; $A/A_f = .7$; $A = 0.7(3.14) = 2.20$ ft²
$WP_f = 6.28$ ft; $WP/WP_f = 0.6$; $WP = 0.6(6.28) = 3.77$ ft
$R_f = 0.50$ ft; $R/R_f = 1.16$; $R = 1.16(0.50) = 0.580$ ft

FIGURE 14.9 Geometry for partially full circular section.

TABLE 14.2 Geometry of open-channel sections

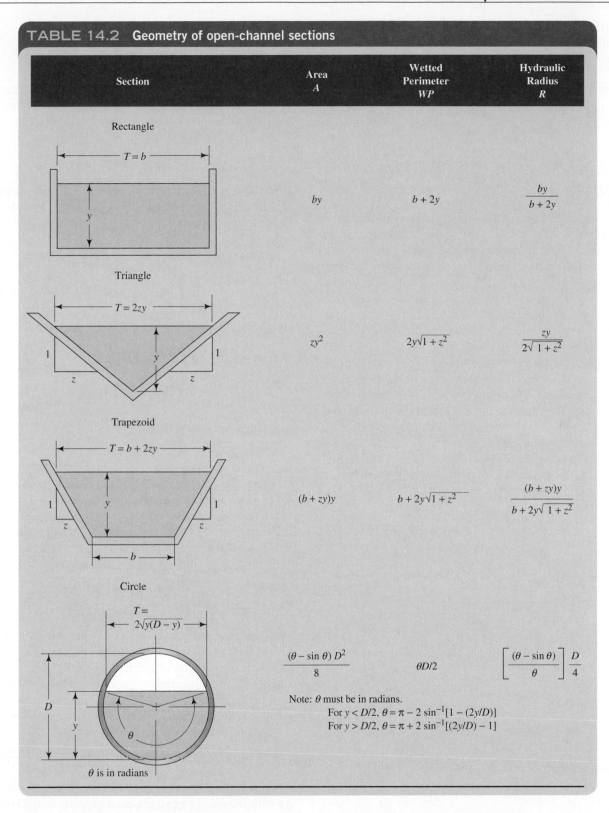

Section	Area A	Wetted Perimeter WP	Hydraulic Radius R
Rectangle	by	$b + 2y$	$\dfrac{by}{b + 2y}$
Triangle	zy^2	$2y\sqrt{1 + z^2}$	$\dfrac{zy}{2\sqrt{1 + z^2}}$
Trapezoid	$(b + zy)y$	$b + 2y\sqrt{1 + z^2}$	$\dfrac{(b + zy)y}{b + 2y\sqrt{1 + z^2}}$
Circle	$\dfrac{(\theta - \sin\theta)\,D^2}{8}$	$\theta D/2$	$\left[\dfrac{(\theta - \sin\theta)}{\theta}\right]\dfrac{D}{4}$

Note: θ must be in radians.
For $y < D/2$, $\theta = \pi - 2\sin^{-1}[1 - (2y/D)]$
For $y > D/2$, $\theta = \pi + 2\sin^{-1}[(2y/D) - 1]$

θ is in radians

running partially full with the depth of the fluid called y. The vertical scale for the graph is the ratio y/D. Curve A gives the ratio of A/A_f, in which A is the actual fluid flow area and A_f is the full cross-sectional area of the circle, easily calculated from $A_f = \pi D^2/4$. The use of curve A is demonstrated by noting that the figure is drawn for the case $y/D = 0.65$. Follow the dashed horizontal line from this value on the vertical scale over to curve A and then project down to the horizontal scale and read the value of 0.70. This means $A/A_f = 0.70$ for $y/D = 0.65$. As an example, assume $D = 2.00$ ft. Then,

$$A_f = \pi D^2/4 = \pi(2.00\text{ ft})^2/4 = 3.14\text{ ft}^2$$
$$A = (0.70)A_f = (0.70)(3.14\text{ ft}^2) = 2.20\text{ ft}^2$$

In a similar manner, you should be able to read the wetted perimeter ratio to be $WP/WP_f = 0.60$ and the hydraulic radius ratio to be $R/R_f = 1.16$. Then,

$$WP_f = \pi D \text{ for a full circle} = \pi(2.00 \text{ ft}) = 6.28 \text{ ft}$$

$$WP = (0.60)WP_f = (0.60)(6.28 \text{ ft}) = 3.77 \text{ ft}$$

and

$$R_f = D/4 \text{ for a full circle} = (2.00 \text{ ft})/4 = 0.50 \text{ ft}$$

$$R = (1.16)R_f = (1.16)(0.50 \text{ ft}) = 0.580 \text{ ft}$$

Thus, the curves in Fig. 14.9 will enable you to compute the values of A, WP, and R for partially full circular sections with simple formulas using values of the three ratios read from the chart. Otherwise, the equations for direct calculation of A, WP, and R are quite complex. Internet resource 1 includes an online calculator for determining area, wetted perimeter, and hydraulic radius for partially full circular pipes or culverts when the diameter and depth are input.

Figure 14.10 shows three other shapes used for open channels. Natural streams frequently can be approximated as shallow parabolas. The triangle with a rounded bottom is more practical to make in the earth than the sharp-V triangle. The round-cornered rectangle performs somewhat better than the square-cornered rectangle and is easier to maintain. However, it is more difficult to form. Reference 4 gives formulas for the geometric features of these types of cross sections.

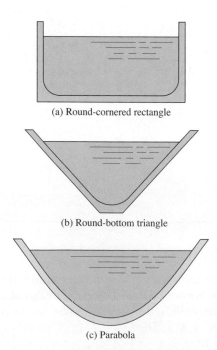

(a) Round-cornered rectangle

(b) Round-bottom triangle

(c) Parabola

FIGURE 14.10 Other shapes for open channels.

14.7 THE MOST EFFICIENT SHAPES FOR OPEN CHANNELS

The term *conveyance* is used to indicate the carrying capacity of open channels. Its value can be deduced from Manning's equation. In SI metric units, we have Eq. (14–8),

$$Q = \left(\frac{1.00}{n}\right)AR^{2/3}S^{1/2}$$

Everything on the right side of this equation is dependent on the design of the channel except the slope. We can then define the conveyance K to be

▷ **Conveyance—SI Units**

$$K = \left(\frac{1.00}{n}\right)AR^{2/3} \qquad (14\text{–}14)$$

In U.S. Customary units,

▷ **Conveyance—U.S. Customary Units**

$$K = \left(\frac{1.49}{n}\right)AR^{2/3} \qquad (14\text{–}15)$$

Manning's equation is then

$$Q = KS^{1/2} \qquad (14\text{–}16)$$

The conveyance of a channel would be maximum when the wetted perimeter is the least for a given area. Using this criterion, we find that the most efficient shape is the semicircle, that is, the circular section running half full. Table 14.3 shows the most efficient designs of other shapes.

14.8 CRITICAL FLOW AND SPECIFIC ENERGY

Consideration of energy in open-channel flow usually involves a determination of the energy possessed by the fluid at a particular section of interest. The total energy is measured relative to the channel bottom and is composed of potential energy due to the depth of the fluid plus kinetic energy due to its velocity.

Letting E denote the total energy, we get

$$E = y + v^2/2g \qquad (14\text{–}17)$$

where y is the depth and v is the average velocity of flow. As with the energy equation used previously, the terms in Eq. (14–17) have the units of energy per unit weight of fluid flowing. In open-channel flow analysis, E is usually referred to as the *specific energy*. For a given discharge Q, the velocity is Q/A. Then,

$$E = y + Q^2/2gA^2 \qquad (14\text{–}18)$$

Because the area can be expressed in terms of the depth of flow, Eq. (14–18) relates the specific energy to the depth of

TABLE 14.3 Most efficient sections for open channels

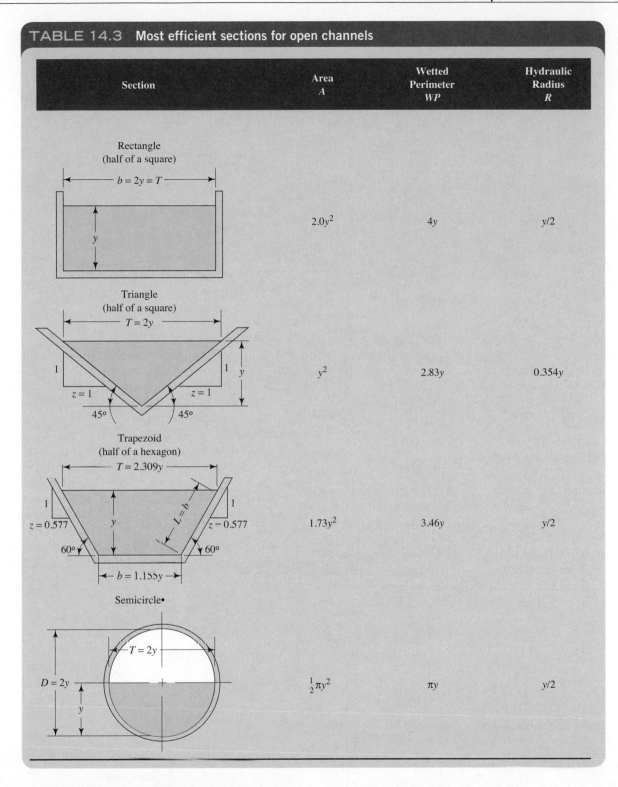

Section	Area A	Wetted Perimeter WP	Hydraulic Radius R
Rectangle (half of a square) $b = 2y = T$	$2.0y^2$	$4y$	$y/2$
Triangle (half of a square) $T = 2y$, $z = 1$, $45°$	y^2	$2.83y$	$0.354y$
Trapezoid (half of a hexagon) $T = 2.309y$, $z = 0.577$, $60°$, $b = 1.155y$	$1.73y^2$	$3.46y$	$y/2$
Semicircle $D = 2y$, $T = 2y$	$\frac{1}{2}\pi y^2$	πy	$y/2$

flow. A graph of the depth y versus the specific energy E is useful in visualizing the possible regimes of flow in a channel. For a particular channel section and discharge, the specific energy curve appears as shown in Fig. 14.11.

Several features of this curve are important. The 45° line on the graph represents the plot of $E = y$. Then, for any point on the curve, the horizontal distance to this line from the y axis represents the potential energy y. The remaining distance to the specific energy curve is the kinetic energy $v^2/2g$. A definite minimum value of E appears, and we can

show that this occurs when the flow is at the critical state, that is, when $N_F = 1$. See Section 14.4, Eq. (14–4), for the definition of the Froude number N_F.

The depth corresponding to the minimum specific energy is, therefore, called the *critical depth* y_c. For any depth greater than y_c, the flow is subcritical. Conversely, for any depth lower than y_c, the flow is supercritical. Notice that for any energy level greater than the minimum, there can exist two different depths. In Fig. 14.12, both y_1 below the critical depth y_c, and y_2 above y_c, have the same energy. In the case

FIGURE 14.11 Variation of specific energy with depth.

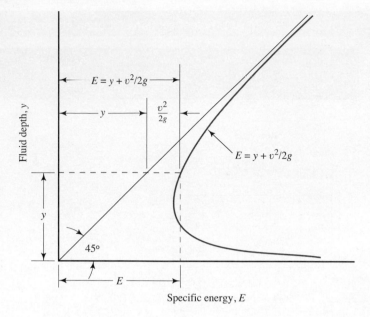

of y_1, the flow is supercritical and much of the energy is kinetic energy due to the high velocity of flow. At the greater depth y_2, the flow is slower and only a small portion of the energy is kinetic energy. The two depths y_1 and y_2 are called the *alternate depths* for the specific energy E.

14.9 HYDRAULIC JUMP

To understand the significance of the phenomenon known as *hydraulic jump*, consider one of its most practical uses, illustrated in Fig. 14.13. The water flowing over the spillway normally has a high velocity in the supercritical range when it reaches the bottom of the relatively steep slope at section 1. If this velocity were to be maintained into the natural stream bed beyond the paved spillway structure, the sides and bottom of the stream would be severely eroded. Instead, good design would cause a hydraulic jump to occur as shown, where the depth of flow abruptly changes from y_1 to y_2. Two beneficial effects result from the hydraulic jump. First, the velocity of flow is decreased substantially, decreasing the tendency for the flow to erode the stream bed. Second, much of the excess energy contained in the high-velocity flow is dissipated in the jump. Energy dissipation occurs because the flow in the jump is extremely turbulent.

For a hydraulic jump to occur, the flow before the jump must be the supercritical range. That is, at section 1 in Fig. 14.13, y_1 is less than the critical depth for the channel and the Froude number N_{F_1} is greater than 1.0. The depth at section 2 after the jump can be calculated from the equation

$$y_2 = (y_1/2)(\sqrt{1 + 8N_{F_1}^2} - 1) \qquad \textbf{(14–19)}$$

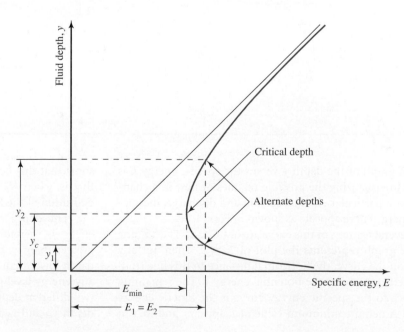

FIGURE 14.12 Critical depth and alternate depths.

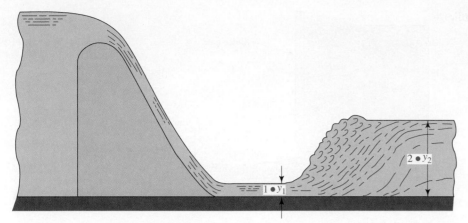

FIGURE 14.13 Hydraulic jump at the bottom of a spillway.

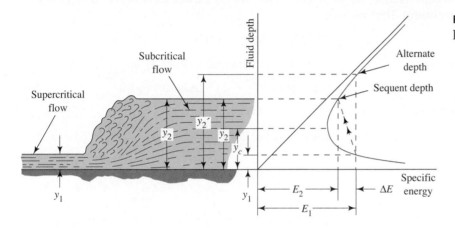

FIGURE 14.14 Energy and depths in a hydraulic jump.

The energy loss in the jump is dependent on the two depths y_2 and y_1:

$$E_1 - E_2 = \Delta E = (y_2 - y_1)^3/4y_1y_2 \qquad \textbf{(14–20)}$$

Figure 14.14 illustrates what happens in a hydraulic jump by using a specific energy curve. The flow enters the jump with an energy E_1 corresponding to a supercritical depth y_1. In the jump, the depth abruptly increases. If no energy were lost, the new depth would be y_2', which is the alternate depth for y_1. However, because there was some energy dissipated ΔE, the actual new depth y_2 corresponds to the energy level E_2. Still, y_2 is in the subcritical range, and tranquil flow would be maintained downstream from the jump. The name given to the actual depth y_2 after the jump is the *sequent depth*.

The following example problem illustrates another practical case in which hydraulic jump might occur.

Example Problem 14.6

As shown in Fig. 14.15, water is being discharged from a reservoir under a sluice gate at the rate of $18 \text{ m}^3/\text{s}$ into a horizontal rectangular channel, 3 m wide, made of unfinished formed concrete. At a point where the depth is 1 m, a hydraulic jump is observed to occur. Determine the following:

a. The velocity before the jump
b. The depth after the jump
c. The velocity after the jump
d. The energy dissipated in the jump

Solution

a. The velocity before the jump is

$$v_1 = Q/A_1$$
$$A_1 = (3)(1) = 3 \text{ m}^2$$
$$v_1 = (18 \text{ m}^3/\text{s})/3 \text{ m}^2 = 6.0 \text{ m/s}$$

b. Equation (14–19) can be used to determine the depth after the jump y_2:

$$y_2 = (y_1/2)(\sqrt{1 + 8N_{F_1}^2} - 1)$$
$$N_{F_1} = v_1/\sqrt{gy_h}$$

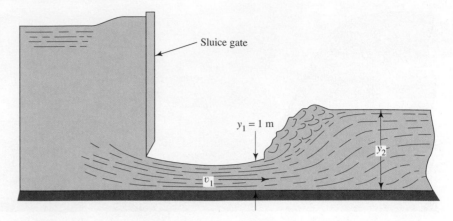

FIGURE 14.15 Hydraulic jump for Example Problem 14.6.

The hydraulic depth is equal to A/T, where T is the width of the free surface. Note that $y_h = y$ for a rectangular channel. Then we have

$$N_{F_1} = 6.0/\sqrt{(9.81)(1)} = 1.92$$

The flow is in the supercritical range. The value for the sequent depth, y_2, is

$$y_2 = (1/2)(\sqrt{1 + (8)(1.92)^2} - 1) = 2.26 \text{ m}$$

c. Because of continuity,

$$v_2 = Q/A_2 = (18 \text{ m}^3/\text{s})/(3)(2.26) \text{ m}^2 = 2.65 \text{ m/s}$$

d. From Eq. (14–20), we get

$$\Delta E = (y_2 - y_1)^3/4y_1y_2$$

$$= \frac{(2.26 - 1.0)^3}{(4)(1.0)(2.26)}\text{m} = 0.221 \text{ m}$$

This means that 0.221 N·m of energy is dissipated from each newton of water as it flows through the jump.

14.10 OPEN-CHANNEL FLOW MEASUREMENT

As stated at the beginning of this chapter, an open channel is one that has its top surface open to the prevailing atmosphere. Familiar examples are natural streams, sewers running partially full, wastewater management systems, and storm drainage structures. Industries often use open channels to conduct coolants away from machinery and to collect excess process fluids and return them to holding tanks. Flow measurement is important for such systems.

Two widely used devices for open-channel flow measurement are *weirs* and *flumes*. Each causes the area of the stream to change, which in turn changes the level of the fluid surface. The resulting level of the surface relative to some feature of the device is related to the quantity of flow. Large volume flow rates of liquids can be measured with weirs and flumes. See References 1, 2, 4, 6, and 16.

14.10.1 Weirs

A *weir* is a specially shaped barrier installed in an open channel over which the fluid flows as a free jet into a stream beyond the barrier. Figure 14.16 shows a side view of the typical design of a weir. The crest should be sharp and is often made from thin sheet metal attached to a substantial

FIGURE 14.16 Flow over a weir.

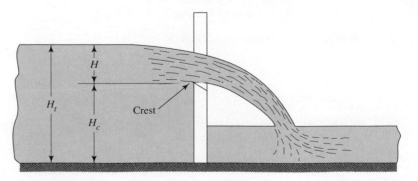

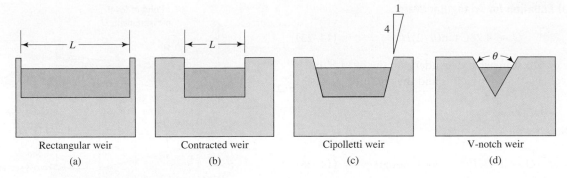

FIGURE 14.17 Notch geometry for weirs.

base. The top surface of the base is cut away at a steep angle on the downstream side to ensure that the fluid springs away as a free jet, called the nappe, with good aeration below the nappe.

Figure 14.17 shows four common shapes for weirs for which rating equations have been developed to enable the calculation of discharge Q as a function of the dimensions of the weir and the head of fluid above the crest H. Figure 14.1 shows a contracted rectangular weir in an irrigation stream. For all of these designs, the head H must be measured upstream from the face of the weir at a distance at least $4H_{max}$. The reason for this requirement is that the top surface of the stream slopes down as it approaches the crest because of the acceleration of the fluid as it contracts to fall over the crest, as shown in Fig. 14.16. See References 1, 2, 4, 6, and 16 for more details.

Measurement of the head can be made by a fixed gage, called a *staff gage*, mounted at the side of the stream for which the zero reading is at the level of the crest of the weir. Float-type devices are also used that can generate a signal that can be transmitted to a control panel or recorded to show a continuous record of flow. Electronic devices may be used that sense the top surface of the flowing fluid. See Internet resources 2 and 6 for commercially available units.

A *rectangular weir*, also called a *suppressed weir*, has a crest length L that extends the full width of the channel into which it is installed. The standard design requires:

1. The crest height above the bottom of the channel $H_c \geq 3H_{max}$
2. The minimum head above the crest $H_{min} > 0.2$ ft
3. The maximum head above the crest $H_{max} < L/3$

The rating equation is

▷ **Rectangular Weir**

$$Q = 3.33LH^{3/2} \qquad \textbf{(14–21)}$$

where L and H are in ft and Q is in ft³/s.

A *contracted weir* is a rectangular weir having sides extended inward from the sides of the channel by a distance of at least $2H_{max}$. The fluid stream must then contract as it flows around the sides of the weir, decreasing slightly the effective length of the weir. The standard design requires:

1. The crest height above the bottom of the channel $H_c \geq 2H_{max}$
2. The minimum head above the crest $H_{min} > 0.2$ ft
3. The maximum head above the crest $H_{max} < L/3$

The rating equation is

▷ **Contracted Weir**

$$Q = 3.33(L - 0.2H)H^{3/2} \qquad \textbf{(14–22)}$$

where L and H are in ft and Q is in ft³/s.

A *Cipolletti weir* is also contracted from the sides of the stream by a distance at least $2H_{max}$ and has sides that are sloped outward as shown in Fig. 14.17(c). The same requirements listed for the contracted rectangular weir apply. The rating equation is

▷ **Cipolletti Weir**

$$Q = 3.367LH^{3/2} \qquad \textbf{(14–23)}$$

The adjustment of the length included for the contracted rectangular weir is not applied here because the sloped sides tend to compensate. Internet resource 10 shows a commercially available Cipolletti weir.

The *triangular weir* is used primarily for low flow rates because the V-notch produces a larger head H than can be obtained with a rectangular notch. The angle of the V-notch is a factor in the discharge equation. Angles from 35° to 120° are satisfactory, but angles of 60° and 90° are quite commonly used. Internet resources 9 and 13 show commercially available V-notch weirs. The theoretical equation for a triangular weir is

$$Q = {}^{8}\!/_{15}C\sqrt{2g}\tan(\theta/2)H^{5/2} \qquad \textbf{(14–24)}$$

where θ is the total included angle between the sides of the notch. An additional reduction of this equation gives

▷ **General Equation for Triangular Weir**

$$Q = 4.28C \tan(\theta/2)H^{5/2} \qquad (14\text{--}25)$$

The value of C is somewhat dependent on the head H, but a nominal value is 0.58. Using this and the common values of 60° and 90° for θ, we get

▷ **60°V-Notch Weir**

$$Q = 1.43H^{5/2} \quad (60° \text{ notch}) \qquad (14\text{--}26)$$

▷ **90°V-Notch Weir**

$$Q = 2.48H^{5/2} \quad (90° \text{ notch}) \qquad (14\text{--}27)$$

14.10.2 Flumes

Critical flow flumes are contractions in the stream that cause the flow to achieve its critical depth within the structure. There is a definite relationship between depth and discharge when critical flow exists. See Internet resources 8–10 for a sample of commercially available flumes. A widely used type of critical flow flume is the *Parshall flume*, the geometry of which is shown in Fig. 14.18. The discharge is dependent on the width of the throat section L and the head H, where H is measured at a specific location along the converging section of the flume.

The discharge equations for the Parshall flume were developed empirically for flumes designed and constructed in dimensions in the U.S. Customary System. Table 14.4 lists the discharge equations for several sizes of flumes. The

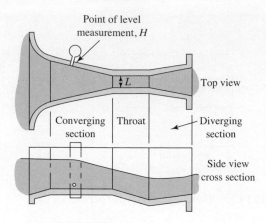

FIGURE 14.18 Parshall flume.

resulting value of Q can be converted to SI units by use of the factor

$$1.0 \text{ ft}^3/\text{s} = 0.028 \ 32 \text{ m}^3/\text{s}$$

Long-throated flumes, rather than Parshall flumes, are recommended for new construction because they are simpler and less expensive to build and can more easily be adapted to channels with a variety of shapes. They can be installed in rectangular, trapezoidal, or circular channels. Table 14.5 shows the general shape consisting of a straight ramp up from the bottom of the channel, a flat throat section, and a sudden drop. Also shown are the basic rating equations and a few sample dimensions for each shape. The dimension Y is the maximum depth of the channel. References 2, 6, and 16 give extensive discussions and data for design of long-throated flumes.

TABLE 14.4	Discharge equations for Parshall flumes		
Throat Width L	**Flow Range (ft^3/s)**		**Equation** (H and L in ft, Q in (ft^3/s)
	Min.	**Max.**	
3 in	0.03	1.9	$Q = 0.992H^{1.547}$
6 in	0.05	3.9	$Q = 2.06H^{1.58}$
9 in	0.09	8.9	$Q = 3.07H^{1.53}$
1 ft	0.11	16.1	$n=1.55$
2 ft	0.42	33.1	$n=1.55$
4 ft	1.3	67.9	$Q = 4.00LH^n$ $n=1.58$
6 ft	2.6	103.5	$n=1.59$
8 ft	3.5	139.5	$n=1.61$
10 ft	6	200	
20 ft	10	1000	
30 ft	15	1500	$Q = (3.6875L + 2.5)H^{1.6}$
40 ft	20	2000	
50 ft	25	3000	

TABLE 14.5 Long-throated flume data

Design A		Design B		Design C	
Rectangular channels: $Q = b_c K_1 (H + K_2)^n$					
b_c	0.500 ft	b_c	1.000 ft	b_c	1.500 ft
L	0.750 ft	L	1.000 ft	L	2.250 ft
p_1	0.125 ft	p_1	0.250 ft	p_1	0.500 ft
K_1	3.996	K_1	3.696	K_1	3.375
K_2	0.000	K_2	0.004	K_2	0.011
n	1.612	n	1.617	n	1.625
H_{min}	0.057 ft	H_{min}	0.082 ft	H_{min}	0.148 ft
H_{max}	0.462 ft	H_{max}	0.701 ft	H_{max}	1.500 ft
Q_{min}	0.020 ft³/s	Q_{min}	0.070 ft³/s	Q_{min}	0.255 ft³/s
Q_{max}	0.575 ft³/s	Q_{max}	2.100 ft³/s	Q_{max}	9.900 ft³/s
Trapezoidal channels: $Q = K_1 (H + K_2)^n$					
b_1	1.000 ft	b_1	1.000 ft	b_1	2.000 ft
b_c	2.000 ft	b_c	4.000 ft	b_c	5.000 ft
L	0.750 ft	L	1.000 ft	L	1.000 ft
p_1	0.500 ft	p_1	1.500 ft	p_1	1.500 ft
K_1	9.290	K_1	14.510	K_1	16.180
K_2	0.030	K_2	0.053	K_2	0.035
n	1.878	n	1.855	n	1.784
H_{min}	0.400 ft	H_{min}	0.579 ft	H_{min}	0.580 ft
H_{max}	0.893 ft	H_{max}	0.808 ft	H_{max}	1.456 ft
Q_{min}	1.900 ft³/s	Q_{min}	6.200 ft³/s	Q_{min}	6.800 ft³/s
Q_{max}	8.000 ft³/s	Q_{max}	11.000 ft³/s	Q_{max}	33.000 ft³/s
Circular channels: $Q = D^{2.5} K_1 (H/D + K_2)^n$					
D	1.000 ft	D	2.000 ft	D	3.000 ft
b_c	0.866 ft	b_c	1.834 ft	b_c	2.940 ft
L_a	0.600 ft	L_a	1.100 ft	L_a	1.350 ft
L_b	0.750 ft	L_b	1.800 ft	L_b	3.600 ft
L	1.125 ft	L	2.100 ft	L	2.700 ft
p_1	0.250 ft	p_1	0.600 ft	p_1	1.200 ft
K_1	3.970	K_1	3.780	K_1	3.507
K_2	0.004	K_2	0.000	K_2	0.000
n	1.689	n	1.625	n	1.573
H_{min}	0.069 ft	H_{min}	0.140 ft	H_{min}	0.180 ft
H_{max}	0.599 ft	H_{max}	1.102 ft	H_{max}	1.343 ft
Q_{min}	0.048 ft³/s	Q_{min}	0.283 ft³/s	Q_{min}	0.655 ft³/s
Q_{max}	1.689 ft³/s	Q_{max}	8.112 ft³/s	Q_{max}	15.448 ft³/s

Example Problem 14.7

Select a design from Table 14.5 for a long-throated flume for measuring a flow rate within the range of 2.5 to 6.0 ft³/s of water. Then compute the discharge Q for several values of head H.

Solution

Either design C for a rectangular channel, design A for a trapezoidal channel, or design B for a circular channel is appropriate for the given desired flow range. The trapezoidal channel will be illustrated here. The rating equation and the values for its variables are found in Table 14.5. We have

$$Q = K_1(H + K_2)^n$$
$$Q = 9.29(H + 0.03)^{1.878}$$

Evaluating this equation for $H = 0.50$ ft to 0.80 ft gives the following results:

Head H (ft)	Flow Q (ft³/s)
0.50	2.820
0.60	3.901
0.70	5.114
0.80	6.547

The flow equation can also be solved for the value of H that will give a desired Q,

$$H = \left(\frac{Q}{K_1}\right)^{1/n} - K_2$$

Now, we can determine what values of head correspond to the ends of the desired range of flow:

For $Q = 2.50$ ft³/s, $H = 0.467$ ft

For $Q = 6.00$ ft³/s, $H = 0.762$ ft

REFERENCES

1. Baker, R. C. 2004. *Introductory Guide to Flow Measurement.* New York: ASME Press.

2. Bos, Marinus. G. 1991. *Flow Measuring Flumes for Open Channel Systems.* St. Joseph, MI: American Society of Agricultural Engineers.

3. Chanson, Hubert. 2004. *Hydraulics of Open Channel Flow,* 2nd ed. New York: Elsevier Science & Technology.

4. Chow, Ven. T. 2009. *Open Channel Hydraulics.* Caldwell, NJ: Blackburn Press. [A classic reference for open-channel flow, initially published in 1959 by McGraw-Hill.]

5. Chow, Ven T., D. R. Maidment, and L. W. Mays. 1988. *Applied Hydrology.* New York: McGraw-Hill.

6. Clemmens, A. J., T. L. Wahl, M. G. Bos, and J. A. Replogle. 2010. *Water Measurement with Flumes and Weirs,* 3rd ed. Littleton, CO: Water Resources Publications.

7. Houghtalen Robert J., A. OsmanAkan, and Ned H. C. Hwang. 2009. *Fundamentals of Hydraulic Engineering Systems,* 4th ed. Upper Saddle River, NJ: Pearson/Prentice-Hall.

8. Jain, C. Subhash. 2000. *Open-Channel Flow.* New York: Wiley.

9. Mays, Larry W. 1999. *Hydraulic Design Handbook.* New York: McGraw-Hill.

10. Mays, Larry W. 2010. *Water Resources Engineering,* 2nd ed. New York: Wiley.

11. Montes, S. 1998. *Hydraulics of Open Channel Flow.* Reston, VA: American Society of Civil Engineers.

12. Munson, Bruce R., Alric P. Rothmayer, Theodore H. Okiishi, and Wade W. Huebsch. 2012. *Fundamentals of Fluid Mechanics,* 7th ed. New York: Wiley.

13. Prakash, Anand. 2004. *Water Resources Engineering.* Reston, VA: American Society of Civil Engineers.

14. Simon, A. L., and S. F. Korom. 2002. *Hydraulics,* 5th ed. San Diego, CA: Simon Publications.

15. Sturm, Terry. 2009. *Open Channel Hydraulics,* 2nd ed. New York: McGraw-Hill.

16. U.S. Bureau of Reclamation and the U.S. Department of Agriculture. 2001. *Water Measurement Manual,* 3rd ed. Washington, DC: U.S. Department of the Interior.

17. White, Frank. M. 2010. *Fluid Mechanics,* 7th ed. New York: McGraw-Hill.

DIGITAL PUBLICATIONS

1. Bengston, Harlan. 2012. *The Manning Equation for Open Channel Flow Calculations* (Kindle Edition). Amazon Digital Services, Inc.

2. Bengston, Harlan. 2012. *Partially Full Pipe Flow Calculations with Spreadsheets.* Amazon Digital Services, Inc.

INTERNET RESOURCES

1. **LMNO Engineering, Research, and Software, Ltd.:** LMNO Engineering is a software development and consulting firm.

The site lists numerous software products for open-channel flow, pipe flow, flow measurement, hydrology, and groundwater calculations. Some programs are free, including those for flow in rectangular and triangular channels, calculations for the Manning equation, circular culvert geometry, the V-notch weir, and the Cipolletti weir.

2. **Teledyne ISCO:** Manufacturers of flow meters, velocity meters, and depth measurement devices for open channel flow.

3. **U.S. Bureau of Reclamation—Water Resources Research Laboratory:** The Water Resources Research Laboratory provides hydraulic testing, analysis, and research services and applies hydraulic modeling expertise to the solution of water resources, hydraulics, and fluid mechanics problems.

4. **U.S. Bureau of Reclamation—WinFlume:** The U.S. Bureau of Reclamation, in cooperation with the U.S. Water Conservation Laboratory and the International Institute for Land Reclamation and Improvement, developed a computer program called WinFlume to design and calibrate long-throated flume and broad-crested weir flow measurement structures. The software can be downloaded from this site.

5. **U.S. Bureau of Reclamation—Water Measurement Manual:** The U.S. Bureau of Reclamation, in cooperation with the U.S. Department of Agriculture, has published the *Water Measurement Manual* as a guide to effective water measurement practices for better water management. This document contains extensive information about the design, installation, and operation of flumes and weirs and can be viewed from this site.

6. **Marsh-McBirney—A Hach Company Brand:** Manufacturer of a variety of flow meters for open-channel flow using electromagnetic, radar, ultrasonic, and pressure measurement techniques. Velocity, level, and depth measurements are combined to indicate volume flow rate. Some devices are portable and can be used in streams, canals, drainage structures, and rivers.

7. **HawsEDC:** The website describes engineering services, online calculators, and specialized tools for AutoCAD, mostly for applications in the civil engineering field. Calculators are given for flow in open channels such as trapezoidal and partially full circular pipes. Basic weir calculations can also be made.

8. **Plasti-Fab, Inc.:** Manufacturer of a variety of corrosion-resistant, fiberglass-reinforced plastic flumes including Parshall, Palmer-Bowlus, Trapezoidal, Cutthroat, RBC H/HS/HL, and Step Flumes that can be installed in an existing channel.

9. **Tracom Fiberglass Products:** Manufacturer of seven types of flumes for open channel flow monitoring: H/HS/HL-type, Parshall, Cutthroat, Palmer-Bowlus, Trapezoidal, RBC, and Montana flumes with a large range of sizes to measure open-channel flows from 0.07 gal/min to over 50 000 gal/min. Also makes a variety of fiberglass weir boxes with 22½, 30, 45, 60, 90, and 120 degree V-notches.

10. **Openchannelflow.com (OCF):** Manufacturer of flumes and weirs for the measurement, conditioning, and control of the flow of water. Included are flume types of H/HS/HL, Parshall, Cutthroat, Trapezoidal, RBC, and Montana. Weir plates of 30, 60, 90, and 120 degree V-notches, and Cipolletti standard types are offered along with a variety of weir boxes.

PRACTICE PROBLEMS

14.1 Compute the hydraulic radius for a circular drain pipe running half full if its inside diameter is 300 mm.

14.2 A rectangular channel has a bottom width of 2.75 m. Compute the hydraulic radius when the fluid depth is 0.50 m.

14.3 A drainage structure for an industrial park has a trapezoidal cross-section similar to that shown in Fig. 14.2(c). The bottom width is 3.50 ft and the sides are inclined at an angle of 60° from the horizontal. Compute the hydraulic radius for this channel when the fluid depth is 1.50 ft.

14.4 Repeat Problem 14.3 if the side slope is 45°.

14.5 Compute the hydraulic radius for a trapezoidal channel with a bottom width of 150 mm and with sides that pitch 15 mm horizontally for a vertical change of 10 mm. That is, the ratio of X/D in Fig. 14.2(c) is 1.50. The depth of the fluid in the channel is 62 mm.

14.6 Compute the hydraulic radius for the section shown in Fig. 14.19 if water flows at a depth of 2.0 in. The section is that of a rain gutter for a house.

14.7 Repeat Problem 14.6 for a depth of 3.50 in.

14.8 Compute the hydraulic radius for the channel shown in Fig. 14.20 if the water depth is 0.50 m.

14.9 Compute the hydraulic radius for the channel shown in Fig. 14.20 if the water depth is 2.50 m.

14.10 Water is flowing in a formed, unfinished concrete rectangular channel 3.5 m wide. For a depth of 2.0 m, calculate

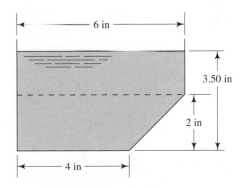

FIGURE 14.19 Problems 14.6, 14.7, and 14.11.

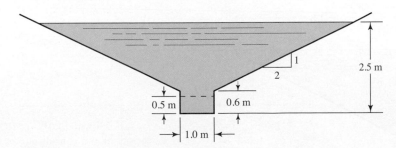

FIGURE 14.20 Problems 14.8, 14.9, and 14.14.

FIGURE 14.21   Problem 14.15.

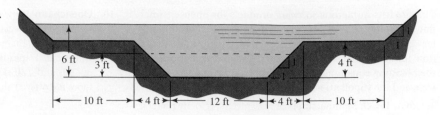

the normal discharge and the Froude number of the flow. The channel slope is 0.1 percent.

14.11 Determine the normal discharge for an aluminum rain spout with the shape shown in Fig. 14.19 that runs at the depth of 3.50 in. Use $n = 0.013$. The spout falls 4 in over a length of 60 ft.

14.12 A circular culvert under a highway is 6 ft in diameter and is made of corrugated metal. It drops 1 ft over a length of 500 ft. Calculate the normal discharge when the culvert runs half full.

14.13 A wooden flume is being built to temporarily carry 5000 L/min of water until a permanent drain can be installed. The flume is rectangular, with a 205-mm bottom width and a maximum depth of 250 mm. Calculate the slope required to handle the expected discharge.

14.14 A storm drainage channel in a city where heavy sudden rains occur has the shape shown in Fig. 14.20. It is made of unfinished concrete and has a slope of 0.5 percent. During normal times, the water remains in the small rectangular section. The upper section allows large volumes to be carried by the channel. Determine the normal discharge for depths of 0.5 m and 2.5 m.

14.15 Figure 14.21 represents the approximate shape of a natural stream channel with levees built on either side. The channel is earth with grass cover. Use $n = 0.04$. If the average slope is 0.000 15, determine the normal discharge for depths of 3 ft and 6 ft.

14.16 Calculate the depth of flow of water in a rectangular channel 10 ft wide, made of brick in cement mortar, for a discharge of 150 ft³/s. The slope is 0.1 percent.

14.17 Calculate the depth of flow in a trapezoidal channel with a bottom width of 3 m and whose walls slope 40° with the horizontal. The channel is made of unfinished concrete and is laid on a 0.1-percent slope. The discharge is 15 m³/s.

14.18 A rectangular channel must carry 2.0 m³/s of water from a water-cooled refrigeration condenser to a cooling pond. The available slope is 75 mm over a distance of 50 m. The maximum depth of flow is 0.40 m. Determine the width of the channel if its surface is trowel-finished concrete.

14.19 The channel shown in Fig. 14.22 has a surface of float-finished concrete and is laid on a slope that falls 0.1 m per 100 m of length. Calculate the normal discharge and the Froude number for a depth of 1.5 m. For that discharge, calculate the critical depth.

14.20 A square storage room is equipped with automatic sprinklers for fire protection that spray 1000 gal/min of water. The floor is designed to drain this flow evenly to troughs near each outside wall. The troughs are shaped as shown in Fig. 14.23. Each trough carries 250 gal/min, is laid on a 1-percent slope, and is formed of unfinished concrete. Determine the minimum depth h.

14.21 The flow from two of the troughs described in Problem 14.20 passes into a sump, from which a round common clay drainage tile carries it to a storm sewer. Determine the size of tile required to carry the flow (500 gal/min) when running half full. The slope is 0.1 percent.

14.22 For a rectangular channel with a bottom width of 1.00 m, compute the flow area and hydraulic radius for depths ranging from 0.10 m to 2.0 m. Plot a graph of area and hydraulic radius versus depth.

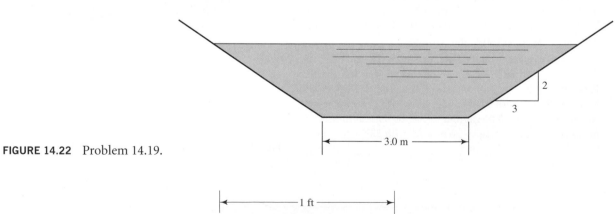

FIGURE 14.22   Problem 14.19.

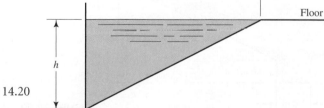

FIGURE 14.23   Problems 14.20 and 14.21.

14.23 It is desired to carry 2.00 m³/s of water at a velocity of 3.0 m/s in a rectangular open channel. The bottom width is 0.80 m. Compute the depth of the flow and the hydraulic radius.

14.24 For the channel designed in Problem 14.23, compute the required slope if the channel is float-finished concrete.

14.25 It is desired to carry 2.00 m³/s of water at a velocity of 3.0 m/s in a rectangular open channel. Compute the depth and hydraulic radius for a range of designs for the channel, with bottom widths of 0.50 m to 2.00 m. Plot depth and hydraulic radius versus bottom width.

14.26 For each of the channels designed in Problem 14.25, compute the required slope if the channel is float-finished concrete. Plot slope versus bottom width.

14.27 A trapezoidal channel has a bottom width of 2.00 ft and a pitch of its sides of $z = 1.50$. Compute the flow area and hydraulic radius for a depth of 20 in.

14.28 For the channel described in Problem 14.27, compute the normal discharge that would be expected for a slope of 0.005 if the channel is made from formed unfinished concrete.

14.29 Repeat Problem 14.28, except that the channel is lined with smooth plastic sheets.

14.30 A trapezoidal channel has a bottom width of 2.00 ft and a pitch of its sides of $z = 1.50$. Compute the flow area and hydraulic radius for depths ranging from 6.00 in to 24.00 in. Plot flow area and hydraulic radius versus depth.

14.31 For each channel designed in Problem 14.30, compute the normal discharge that would be expected for a slope of 0.005 if the channel is made from formed unfinished concrete.

14.32 Compute the flow area and hydraulic radius for a circular drain pipe 375 mm in diameter for a depth of 225 mm.

14.33 Repeat Problem 14.32 for a depth of 135 mm.

14.34 For the channel designed in Problem 14.32, compute the normal discharge that is expected for a slope of 0.12 percent if the channel is made from painted steel.

14.35 For the channel designed in Problem 14.33, compute the normal discharge that is expected for a slope of 0.12 percent if the channel is made from painted steel. Compare the result with that from Problem 14.34.

14.36 It is desired to carry 1.25 ft³/s of water at a velocity of 2.75 ft/s. Design the channel cross-section for each of the shapes shown in Table 14.3, which lists the most efficient sections for open channels.

14.37 For each section designed in Problem 14.36, compute the required slope if the channel is made of float-finished concrete. Compare the results.

14.38 For each section designed in Problem 14.36, compute the Froude number and tell whether the flow is subcritical or supercritical.

Complete the following list of tasks for each of Problems 14.39–14.42:
a. Calculate the critical depth.
b. Calculate the minimum specific energy.
c. Plot the specific energy curve.
d. Determine the specific energy for the given depth and the alternate depth for this energy.
e. Determine the velocity of flow and the Froude number for each depth in (d).
f. Calculate the required slopes of the channel if the depths from (d) are to be normal depths for the given flow rate.

14.39 A rectangular channel 2.00 m wide carries 5.5 m³/s of water and is made of formed unfinished concrete. Use $y = 0.50$ m in (d).

14.40 A circular finished concrete drainage pipe with a diameter of 1.20 m carries 1.45 m³/s. Use y = 0.50 m in (d).

14.41 A triangular channel with side slopes having a ratio of 1:1.5 carries 0.68 ft³/s of water and is made from clean, excavated earth. Use $y = 0.25$ ft in (d).

14.42 A trapezoidal channel with a bottom width of 3.0 ft and side slopes having a ratio of 1:0.75 carries 0.80 ft³/s of water and is made from trowel-finished concrete. Use $y = 0.05$ ft in (d).

Weirs and Flumes

14.43 Determine the maximum possible flow rate over a 60° V-notch weir if the width of the notch at the top is 12 in.

14.44 Determine the required length of a contracted weir similar to that shown in Fig. 14.17(b) to pass 15 ft³/s of water. The height of the crest is to be 3 ft from the channel bottom, and the maximum head above the crest is to be 18 in.

14.45 Plot a graph of Q versus H for a full-width weir with a crest length of 6 ft and whose crest is 2 ft from the channel bottom. Consider values of the head H from 0 to 12 inches in 2-in steps.

14.46 Repeat the calculations of Q versus H for a weir with the same dimensions as used in Problem 14.45 except that it is placed in a channel wider than 6 ft. It, thus, becomes a contracted weir.

14.47 Compare the discharges over the following weirs when the head H is 18 in:
a. Full-width rectangular: $L = 3$ ft, $H_c = 4$ ft
b. Contracted rectangular: $L = 3$ ft, $H_c = 4$ ft
c. 90° V notch (top width also 3 ft)

14.48 Plot a graph of Q versus H for a 90° V-notch weir for values of the head from 0 to 12 in. in 2-in steps.

14.49 For a Parshall flume with a throat width of 9 in, calculate the head H corresponding to the minimum and maximum flows.

14.50 For a Parshall flume with a throat width of 8 ft, calculate the head H corresponding to the minimum and maximum flows. Plot a graph of Q versus H, using five values of H spaced approximately equally between the minimum and the maximum.

14.51 A flow rate of 50 ft³/s falls within the range of both the 4-ft- and the 10-ft-wide Parshall flume. Compare the head H for this flow rate in each size.

14.52 A long-throated flume is installed in a trapezoidal channel using design C from Table 14.5. Compute the discharge for a head of 0.84 ft.

14.53 A long-throated flume is installed in a trapezoidal channel using design B from Table 14.5. Compute the discharge for a head of 0.65 ft.

14.54 A long-throated flume is installed in a rectangular channel using design A from Table 14.5. Compute the discharge for a head of 0.35 ft.

14.55 A long-throated flume is installed in a rectangular channel using design C from Table 14.5. Compute the discharge for a head of 0.40 ft.

14.56 A long-throated flume is installed in a circular pipe using design B from Table 14.5. Compute the discharge for a head of 0.25 ft.

14.57 A long-throated flume is installed in a circular channel using design A from Table 14.5. Compute the discharge for a head of 0.09 ft.

14.58 For a long-throated flume of design B in a rectangular channel, compute the head corresponding to a volume flow rate of 1.25 ft³/s.

14.59 For a long-throated flume of design C in a circular channel, compute the head corresponding to a volume flow rate of 6.80 ft³/s.

14.60 Select a long-throated flume from Table 14.5 that will carry a range of flow from 30 gal/min to 500 gal/min. Compute the head for each of these flows and then compute the flow that would result from four additional heads spaced approximately equally between them.

14.61 Select a long-throated flume from Table 14.5 that will carry a range of flow from 50 m³/h to 180 m³/h. Compute the head for each of these flows and then compute the flow that would result from four additional heads spaced approximately equally between them.

COMPUTER AIDED ENGINEERING ASSIGNMENTS

1. Develop a spreadsheet or a program for computing the geometry features for each section shown in Table 14.2. Include the area, wetted perimeter, and hydraulic radius.

2. Develop a spreadsheet or a program for computing the geometry features for each section shown in Table 14.3. Include the area, wetted perimeter, and hydraulic radius.

3. Develop a spreadsheet or a program for computing the normal discharge for the shapes of open channel shown in Table 14.2 for a given slope. Include the ability to compute the geometry features of the channel and a list of values for Manning's n from which the user can select a design value. Verify your work using data from Example Problem 14.2.

4. Develop a spreadsheet or a program for computing the slope required for a channel of any shape shown in Table 14.2 with given dimensions and desired normal discharge. Verify your work using data from Example Problem 14.3.

5. Develop a spreadsheet or a program for computing the normal depth for a rectangular channel of a given width carrying a given normal discharge at a given slope. A trial-and-error solution method is required. Verify your work using data from Example Problem 14.5.

6. Develop a spreadsheet or a program for computing the discharge through a full-width rectangular weir by using Eq. (14–21); through a contracted weir by using Eq. (14–22); through a Cipolletti weir by using Eq. (14–23); and through a triangular (V-notch) weir by using Eqs. (14–26) and (14–27).

7. Develop a spreadsheet or a program for computing the discharge through any of the standard Parshall flumes listed in Table 14.4.

8. Use Assignment 6 to solve Practice Problems 14.45–14.48.

9. Use Assignment 7 to solve Practice Problems 14.49–14.51.

10. Develop a spreadsheet or a program for computing flow rate Q for any of the long-throated flumes for rectangular channels shown in Table 14.5 for any input value of head H.

11. Include in Assignment 10 the computation of the head H that corresponds to any input value of flow rate Q.

12. Develop a spreadsheet or a program for computing flow rate Q for any of the long-throated flumes for trapezoidal channels shown in Table 14.5 for any input value of head H.

13. Include in Assignment 12 the computation of the head H that corresponds to any input value of flow rate Q.

14. Develop a spreadsheet or program for computing flow rate Q for any of the long-throated flumes for circular channels shown in Table 14.5 for any input value of head H.

15. Include in Assignment 14 the computation of the head H that corresponds to any input value of flow rate Q.

FLOW MEASUREMENT

Flow measurement refers to the ability to measure the velocity, volume flow rate, or mass flow rate of any liquid or gas. Accurate measurement of flow is essential to the control of industrial processes, the transfer of custody of fluids, and the evaluation of the performance of engines, refrigeration systems, and other systems employing moving fluids. There are many types of commercially available flowmeters with which you should be familiar.

Exploration

Think and talk with colleagues about the ways in which your life has been affected by flow measurement recently.

Look around your home, your school, your place of work, the shopping mall, an amusement park, or in your car.

List as many kinds of flowmeters as you can think of and describe them.

What fluid is involved? What is being measured? How is the measured quantity indicated?

Introductory Concepts

Flow measurement is an important function within any organization that employs fluids to carry on its regular operations. It refers to the ability to measure the velocity, volume flow rate, or mass flow rate of any liquid or gas. As you discuss flow measurement with your colleagues, compare the list of situations that you are aware of with the ones listed here.

- You buy gasoline from a service station and the pump system, perhaps like the one in Figure 15.1, includes a flowmeter to indicate to you and to the station operator how many gallons or liters you pumped so you can pay for just the amount you put into your car. The amount is also typically recorded by the station operator and accumulated to indicate the total number of gallons or liters sold. This can help in station management and indicate when a resupply is needed.

- The weather report indicates that showers are expected with winds of 30 mph.

- In the chemistry laboratory, you may monitor the heat input to a reaction by measuring the mass flow rate of fuel gas into a burner.

How many can you add to this list? Consider the following general reasons for measuring the flow of fluids:

- ***Custody transfer and accounting*** Any time a person acquires a fluid product from a supplier, an accurate

FIGURE 15.1 Consider the importance of accurate flow measurement in your own life. For the meters at your home for water or natural gas or a gasoline pump like this one, there are immediate financial ramifications. (*Source:* il-fede/Fotolia)

accounting is needed of the amount of fluid transferred. Have you noticed that the gasoline pump meter is checked periodically by a public agency responsible for enforcing standards for accuracy of weights and measures in general commerce?

■ *Performance evaluation* An engine requires fuel that provides the basic energy needed to run it. One indication of the performance of the engine is to measure the power output (energy per unit time) in relation to the rate of fuel used by the engine (gallons per hour). This is directly related to the fuel efficiency measure you typically use for your car, miles per gallon or kilometers/liter.

■ *Process control* Any industry that uses fluids in its processes must monitor the mass flow rate of key fluids into those processes. For example, beverages are blends of several constituents that must be precisely controlled to maintain the taste that the customer expects. Continuous monitoring and control of the flow rate of each constituent into the blending system is critical to producing consistently a quality product.

■ *Research and development* Numerous examples can be given. Consider the move from fluorocarbon refrigerants (freons) to more environmentally acceptable refrigerants. It is essential to test many candidate formulations to determine the refrigerating effect produced as a function of the mass flow rate of the refrigerant through the air conditioner or freezer.

■ *Medical treatment and research* The flow rate of oxygen, blood transfusions, or liquid medications must be carefully monitored to ensure that the patient receives the proper dosages. Inaccurate measurement could be dangerous!

This chapter will increase your awareness of the many types of flow measurement equipment available and help you to develop your skill in making the appropriate calculations to interpret the results obtained from them. You should also be able to recommend suitable types of flowmeters for a given application. It is most likely that you will use commercially available meters rather than designing and making your own. To do that efficiently and effectively, you must understand the physical principles on which the meters are based. This chapter describes many different types of flow measurement devices and identifies references from which much more detail can be learned. To insure consistency and accuracy, national and international standards for flow measurement are established. References 1–13 and 16 identify some of the standards developed by ASME (formerly known as the American Society of Mechanical Engineers) and ISO (International Standards Organization). Internet resources are listed at the end of the chapter that allow you to connect with numerous vendors of flow measurement equipment.

15.1 OBJECTIVES

After completing this chapter, you should be able to:

1. Describe six factors that should be considered when specifying a flow measurement system.

2. Describe four types of variable-head meters: the *venturi tube*, the *flow nozzle*, the *orifice*, and the *flow tube*.

3. Compute the velocity of flow and the volume flow rate for variable-head meters, including the determination of the discharge coefficient.

4. Describe the *rotameter variable-area meter*, *turbine flowmeter*, *magnetic flowmeter*, *vortex flowmeter*, and *ultrasonic flowmeter*.

5. Describe two methods of measuring mass flow rate.

6. Describe the *pitot-static tube* and compute the velocity of flow using data acquired from such a device.

7. Define the term *anemometer* and describe two kinds.

8. Describe seven types of level measurement devices.

15.2 FLOWMETER SELECTION FACTORS

Many devices are available for measuring flow. Some measure volume flow rate directly, whereas others measure an average velocity of flow that can then be converted to volume flow rate by using $Q = Av$. Some provide direct primary measurements, whereas others require calibration or the application of a discharge coefficient to the observed output of the device. The form of the flowmeter output also varies considerably from one type to another. The indication can be a pressure, a liquid level, a mechanical counter, the position of an indicator in the fluid stream, a continuous electrical signal, or a series of electrical pulses. The physical size of the meter, its cost, the system pressure, and the operator's skill should always be considered. References 1–5 and 14–21 give comprehensive standards, handbooks and reference texts covering the broad range of types of flow measurement devices. References 6–13 give standards for specific types of flow measurement devices that are discussed later in this chapter. Internet resources 1–19 provide links to a few of the many manufacturers and suppliers of commercially available flow meters of the types described in this chapter. Note that flow measurement for open channel flow was included in Chapter 14 with several references and Internet resources listed there for additional details and commercial vendors.

The choice of the basic type of fluid meter and its indication system depends on several factors, some of which we will discuss here.

15.2.1 Range

Commercially available meters can measure flows from a few milliliters per second (mL/s) for precise laboratory experiments to several thousand cubic meters per second (m³/s) for irrigation water or municipal water and sewage systems. Then, for a particular meter installation, the general order of magnitude of the flow rate must be known as well as the range of the expected variations.

A term often used in flow measurement literature is *turndown*, the ratio of the maximum flow rate the meter can measure to the minimum flow rate that it can measure within the stated accuracy. It is a measure of the meter's ability to function under all flow conditions expected in the application.

15.2.2 Accuracy Required

Virtually any flow-measuring device properly installed and operated can produce an accuracy within 5 percent of the actual flow. Most commercial meters are capable of 2-percent accuracy, and several claim accuracy better than 0.5 percent. Cost usually becomes an important factor when great accuracy is desired.

15.2.3 Pressure Loss

Because the construction details of the various meters are quite different, they produce differing amounts of energy loss or pressure loss as the fluid flows through them. Except for a few types, fluid meters accomplish the measurement by placing a restriction or a mechanical device in the flow stream, thus, causing the energy loss.

15.2.4 Type of Indication

Factors to consider when choosing the type of flow indication include whether remote sensing or recording is required, whether automatic control is to be actuated by the output, whether an operator needs to monitor the output, and whether severe environmental conditions exist.

15.2.5 Type of Fluid

The performance of some fluid meters is affected by the properties and condition of the fluid. A basic consideration is whether the fluid is a liquid or a gas. Other factors that may be important are viscosity, temperature, corrosiveness, electrical conductivity, optical clarity, lubricating properties, and homogeneity. Slurries and multiphase fluids require special meters.

15.2.6 Calibration

Calibration is required for some types of flowmeters. Some manufacturers provide a calibration in the form of a graph or a chart of actual flow versus indicator reading. Some are equipped for direct reading, with scales calibrated in the desired units of flow. In the case of the more fundamental types of meters, such as the variable-head types, standard geometrical forms and dimensions have been determined for which empirical data are available. These data relate flow to an easily measured variable such as a pressure difference or a fluid level. References 1–13 at the end of this chapter give many of these calibration factors.

If calibration is required by the user of the device, he or she may use another precision meter as a standard against which the reading of the test device can be compared. Alternatively, primary calibration can be performed by adjusting the flow to a constant rate through the meter and then collecting the output during a fixed time interval. The fluid thus collected can either be weighed for a weight-per-unit-time calibration, or its volume can be measured for a volume-flow-rate calibration. One type of commercially available flow calibrator employs a precision piston moving at a controlled rate to move the test fluid through the flowmeter being calibrated. The meter output is compared with the known flow rate by a computer data acquisition and analysis system to prepare calibration charts and graphs.

15.3 VARIABLE-HEAD METERS

The basic principle on which variable-head meters are based is that when a fluid stream is restricted, its pressure decreases by an amount that is dependent on the rate of flow through the restriction. Therefore, the pressure difference between points before and after the restriction can be used to indicate flow rate. The most common types of variable-head meters, sometimes called *differential pressure meters*, or simply *dp meters*, are the venturi tube, the flow nozzle, the orifice, and the flow tube. The derivation of the relationship between the pressure difference and the volume flow rate is the same regardless of which type of device is used. See References 1–5 for extensive technical data and information and Internet resources 4, 7, 8, 10, and 15 for commercially available designs.

15.3.1 Venturi Tube

Figure 15.2 shows the basic appearance of a venturi tube. The flow from the main pipe at section 1 is caused to accelerate through a narrow section called the *throat*, where the fluid pressure is decreased. The flow then expands through the diverging portion to the same diameter as the main pipe. Pressure taps are located in the pipe wall at section 1 and in the wall of the throat, which we will call section 2. These

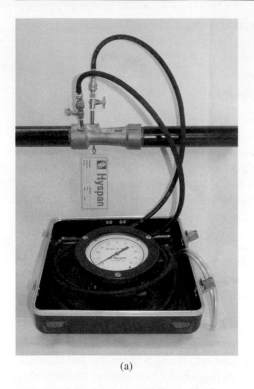

(a)

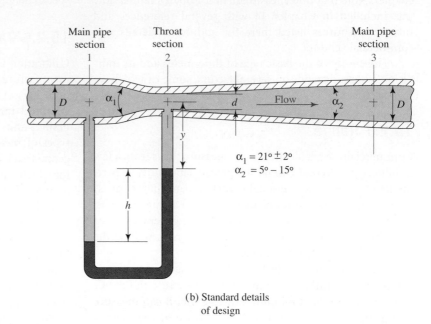

(b) Standard details
of design

FIGURE 15.2 A commercially available venturi meter installed in a pipe with a differential pressure measuring device to indicate flow rate. (*Source:* Hyspan Precision Products, Inc.)

pressure taps are attached to the two sides of a differential manometer so that the deflection h is an indication of the pressure difference $p_1 - p_2$. Of course, other types of differential pressure gages could be used. See Internet resource 20 for an example of a supplier of industrial venturi flow measurement systems for a large range of flow capacities.

The energy equation and the continuity equation can be used to derive the relationship from which we can calculate the flow rate. Using sections 1 and 2 in Fig. 15.2(b) as the reference points, we can write the following equations:

$$\frac{p_1}{\gamma} + z_1 + \frac{v_1^2}{2g} - h_L = \frac{p_2}{\gamma} + z_2 + \frac{v_2^2}{2g} \quad \text{(15–1)}$$

$$Q = A_1 v_1 = A_2 v_2 \quad \text{(15–2)}$$

These equations are valid only for incompressible fluids, that is, liquids. For the flow of gases, we must give special consideration to the variation of the specific weight γ with pressure. See Reference 5. The algebraic reduction of Eqs. (15–1) and (15–2) proceeds as follows:

$$\frac{v_2^2 - v_1^2}{2g} = \frac{p_1 - p_2}{\gamma} + (z_1 - z_2) - h_L$$

$$v_2^2 - v_1^2 = 2g[(p_1 - p_2)/\gamma + (z_1 - z_2) - h_L]$$

But $v_2^2 = v_1^2 (A_1/A_2)^2$. Then, we have

$$v_1^2[(A_1/A_2)^2 - 1] = 2g[(p_1 - p_2)/\gamma + (z_1 - z_2) - h_L]$$

$$v_1 = \sqrt{\frac{2g[(p_1 - p_2)/\gamma + (z_1 - z_2) - h_L]}{(A_1/A_2)^2 - 1}} \quad \text{(15–3)}$$

We can make two simplifications at this time. First, it is typical for the venturi to be installed in the horizontal position, so the elevation difference $z_1 - z_2$ is zero. If there is a significant elevation difference when installing the meter at an angle to the vertical, the elevation difference should be included in the calculations. Second, the term h_L is the energy loss from the fluid as it flows from section 1 to section 2. The value of h_L must be determined experimentally. But it is more convenient to modify Eq. (15–3) by dropping h_L and introducing a discharge coefficient C:

$$v_1 = C\sqrt{\frac{2g(p_1 - p_2)/\gamma}{(A_1/A_2)^2 - 1}} \quad \text{(15–4)}$$

Equation (15–4) can be used to calculate the velocity of flow in the throat of the meter. Note that the velocity depends on the difference in the pressure head between points 1 and 2. This is the reason these meters are called variable-head meters.

Normally we want to calculate the volume flow rate. Because $Q = A_1 v_1$, we have

$$Q = CA_1\sqrt{\frac{2g(p_1 - p_2)/\gamma}{(A_1/A_2)^2 - 1}} \quad \text{(15–5)}$$

The discharge coefficient C represents the ratio of the actual velocity through the venturi to the ideal velocity for a venturi with no energy loss at all. Therefore, the value of C will always be less than 1.0. The venturi of the Herschel type shown in Fig. 15.2(b) is designed to minimize energy losses by employing a smooth, gradual contraction to the throat and a smooth, long enlargement following the

FIGURE 15.3 Discharge coefficient for a rough-cast venturi tube of the Herschel type. (*Source:* From ASME Research Committee on Fluid Mechanics (1959). *Fluid Mechanics: Their Theory and Application,* 5/e © American Society of Mechanical Engineers. Reprinted with permission.)

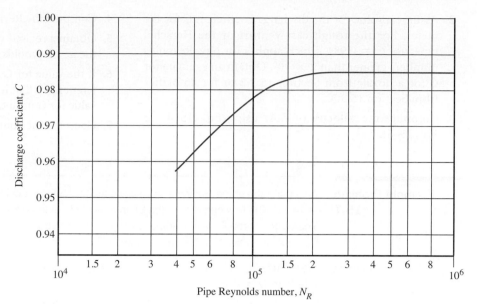

Pipe Reynolds number, N_R

throat. Therefore, the discharge coefficient is typically close to 1.0.

Figure 15.3 indicates that the actual value of C depends on the Reynolds number for the flow in the main pipe. Above a Reynolds number of 2×10^5 the value of C is taken to be 0.984. This value applies to a venturi of the Herschel type that is rough cast with a pipe diameter in the range from 4.0 in to 48.0 in (100 mm to 1200 mm). The throat diameter can vary over a fairly wide range, but the ratio of d/D, called the beta ratio or β, should be between 0.30 and 0.75.

For Reynolds numbers below 2×10^5 you must read the value of C from Fig. 15.3.

Smaller venturi meters, for pipe diameters in the range from 2 in to 10 in (50–250 mm), are typically machined, resulting in a better surface finish than the rough casting. The value of C is taken to be 0.995 for this type when $N_R > 2 \times 10^5$. Data are not available for C for lower Reynolds numbers for the machined venturi.

References 5, 6, and 14–21 give more information about the application of venturi meters, including extensive discussions of the corrections that must be made when using them for measuring the flow of air and other gases. We will limit our use of Eqs. (15–4) and (15–5) to liquid flow.

Flow Equation When a Manometer Is Used to Measure Pressure Difference A manometer is a popular method of measuring the pressure difference between the pipe and the throat of the venturi because it gives the required difference in terms of the manometer reading and the properties of the flowing fluid and the manometer fluid. For example, using the arrangement shown in Fig. 15.2(b), we use the following notation:

γ_f = Specific weight of the fluid flowing in the pipe

γ_m = Specific weight of the manometer fluid

y = Vertical distance from the centerline of the pipe to the top of the manometer fluid

Then we can write the manometer equation as

$$p_1 + \gamma_f y + \gamma_f h - \gamma_m h - \gamma_f y = p_2$$

Here we see that the term $\gamma_f y$ appears with both a positive and a negative sign. Thus, these terms can be cancelled.

Let's now solve the equation for the pressure difference we need in Eq. (15–4):

$$p_1 - p_2 = -\gamma_f h + \gamma_m h = \gamma_m h - \gamma_f h = h(\gamma_m - \gamma_f)$$

To get the pressure head difference, divide both sides of the equation by γ_f:

$$(p_1 - p_2)/\gamma_f = h(\gamma_m - \gamma_f)/\gamma_f = h(\gamma_m/\gamma_f - 1)$$

Substituting this into Eq. (15–4) gives

$$v_1 = C\sqrt{\frac{2gh[(\gamma_m/\gamma_f) - 1]}{(A_1/A_2)^2 - 1}} \qquad (15\text{–}6)$$

The principles developed above and the procedure outlined below are equally applicable to other types of differential pressure flow meters such as nozzles, orifice meters, and flow tubes that are described in following sections.

Procedure for Computing the Flow Rate of a Liquid through a Venturi, Nozzle, Orifice Meter, or Flow Tube

1. Obtain data for:
    a. Pipe inside diameter at the inlet to the meter, D_1.
    b. Diameter of throat of the meter, d_2.
    c. Specific weight γ_f and kinematic viscosity ν of the flowing fluid at the prevailing conditions in the pipe.
    d. Measurement of the differential pressure or pressure head difference between the pipe and the throat.
        i. Δp in pressure units.
        ii. Pressure head difference indicated by the deflection of a manometer.

2. Assume a value for the discharge coefficient C for the meter. For the rough-cast venturi of the Herschel type, use $C = 0.984$, which applies for pipe Reynolds numbers greater than 2×10^5. Obtain an estimate for C for a nozzle from Section 15.3.2 or for an orifice from Section 15.3.3.

3. Compute the velocity of flow using Eq. (15–4) or Eq. (15–6).

4. Compute the Reynolds number for the flow in the pipe.

5. Obtain a revised value for the discharge coefficient C at the new Reynolds number.

6. If the value for C assumed in Step 5 is significantly different from that in Step 2, repeat Steps 3–5 with the new value for C until there is agreement.

7. Compute the volume flow rate from $Q = A_1 v_1$.

Example Problem 15.1

A venturi tube of the Herschel type shown in Fig. 15.2(b) is being used to measure the flow rate of water at 140°F. The flow enters from the left in a 5-in Schedule 40 steel pipe. The throat diameter d is 2.200 in. The venturi is rough cast. The manometer fluid is mercury (sg = 13.54) and the deflection h is 7.40 in. Compute the velocity of flow in the pipe and the volume flow rate in gal/min.

Solution

We will use Eq. (15–6) to compute the velocity of flow in the pipe, v_1. Then, we will find the volume flow rate from $Q = A_1 v_1$.

First let's document pertinent data and compute some of the basic parameters in Eq. (15–6).

Fluid flowing in the pipe: water at 140°F; $\gamma_w = 61.4$ lb/ft^3, $\nu = 5.03 \times 10^{-6}$ ft^2/s (from Appendix A)

Manometer fluid: mercury (sg = 13.54); $\gamma_m = (13.54)(62.4$ lb/ft$^3) = 844.9$ lb/ft^3.

Pipe dimensions: $D = 0.4206$ ft, $A_1 = 0.1390$ ft^2 (from Appendix F)

Throat dimensions: $d = (2.20$ in$)(1.0$ ft/12 in$) = 0.1833$ ft, $A_2 = \pi d^2/4 = 0.02640$ ft^2

Then,

$$A_1/A_2 = (0.1390 \text{ ft}^2/(0.0264 \text{ ft}^2) = 5.265$$

$$\beta = d/D = (0.1833 \text{ ft})/(0.4206 \text{ ft}) = 0.436.$$

Note that $0.30 < \beta < 0.75$, within the recommended range

Figure 15.3 applies giving the value of the discharge coefficient C for the rough-cast venturi. Let's assume that the Reynolds number for the flow of water in the pipe is greater than 2.0×10^5 and use the value of $C = 0.984$ as the first estimate. This must be checked later when the Reynolds number is known and adjusted according to Fig. 15.3 if $N_R < 2.0 \times 10^5$.

Let's evaluate the term $[(\gamma_m/\gamma_f) - 1]$ first:

$$[(\gamma_m/\gamma_f) - 1] = [(844.9 \text{ lb/ft}^3/61.4 \text{ lb/ft}^3) - 1] = 12.76$$

Also, let's convert the h value to feet:

$$h = (7.40 \text{ in})(1 \text{ ft}/12 \text{ in}) = 0.6167 \text{ ft}$$

Now we can compute v, from Eq. (15–6):

$$v_1 = C\sqrt{\frac{2gh[(\gamma_m/\gamma_f) - 1]}{(A_1/A_2)^2 - 1}} = 0.984\sqrt{\frac{2(32.2 \text{ ft/s}^2)(0.6167 \text{ ft})(12.76)}{(5.265)^2 - 1}}$$

$$v_1 = 4.285 \text{ ft/s}$$

Now we must check the Reynolds number for the flow in the pipe using this value:

$$N_R = \frac{v_1 D}{\nu} = \frac{(4.285 \text{ ft/s})(0.4206 \text{ ft})}{5.03 \times 10^{-6}} = 3.58 \times 10^5$$

We note that this value is greater than 2×10^5 as we initially assumed. Then, the value for the discharge coefficient, $C = 0.984$, is correct and the calculation for v_1 is also correct. If the Reynolds number was less than 2×10^5, we would read a new value of C from Fig. 15.3 and recompute the velocity.

Result

Now we complete the problem by computing the volume flow rate Q:

$$Q = A_1 v_1 = (0.1390 \text{ ft}^2)(4.285 \text{ ft/s}) = 0.596 \text{ ft}^3/\text{s}$$

Converting this to gal/min, we obtain

$$Q = (0.596 \text{ ft}^3/\text{s})[(449 \text{ gal/min})/1.0 \text{ ft}^3/\text{s}] = 267 \text{ gal/min}$$

FIGURE 15.4 Flow nozzle.

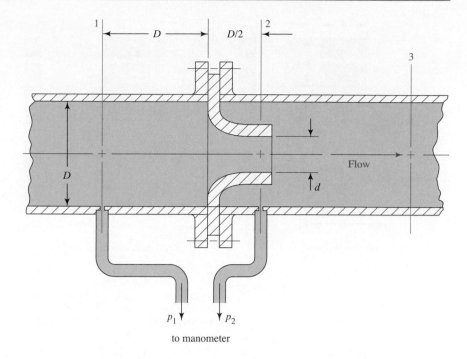

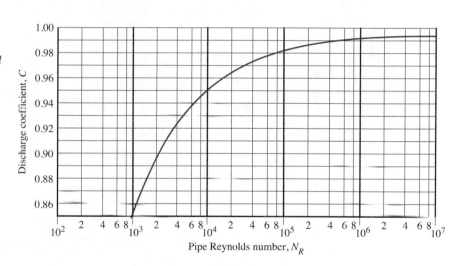

FIGURE 15.5 Flow nozzle discharge coefficient. (*Source:* From ASME Research Committee on Fluid Mechanics (1959). *Fluid Mechanics: Their Theory and Application,* 5/e © American Society of Mechanical Engineers. Reprinted with permission.)

15.3.2 Flow Nozzle

The *flow nozzle* is a gradual contraction of the flow stream followed by a short, straight cylindrical section as illustrated in Fig. 15.4. Several standard geometries for flow nozzles have been presented and adopted by organizations such as the ASME and the ISO. See References 5 and 16.

Equations (15–4)–(15–6) are used for the flow nozzle and the orifice as well as for the venturi tube. Because of the smooth, gradual contraction, there is very little energy loss between points 1 and 2 for a flow nozzle. A typical curve of C versus Reynolds number is shown in Fig. 15.5. At high Reynolds numbers C is above 0.99. At lower Reynolds numbers the sudden expansion outside the nozzle throat causes greater energy loss and a lower value for C.

Reference 13 recommends using the following equation for C:

$$C = 0.9975 - 6.53\sqrt{\beta/N_R} \qquad (15–7)$$

where $\beta - d/D$. Figure 15.5 is a plot of Eq. (15–7) for the value of $\beta = 0.50$.

References 1, 5, and 18 give extensive information on the proper selection and application of flow nozzles including corrections for gas flow that account for compressibility effects.

15.3.3 Orifice

A flat plate with an accurately machined, sharp-edged hole is referred to as an *orifice*. Figure 15.6 shows six styles for the bore through the orifice plate. A tab typically extends above the circular plate by 100 to 125 mm (4.0 to 5.0 in) to facilitate handling, insertion, and withdrawing the plates. When concentric orifices like those in parts *a* to *d* are placed between flanges on a pipe as shown in Fig. 15.6(g), it causes the flow to suddenly contract as it approaches the orifice and then suddenly expand back to the full pipe diameter. The

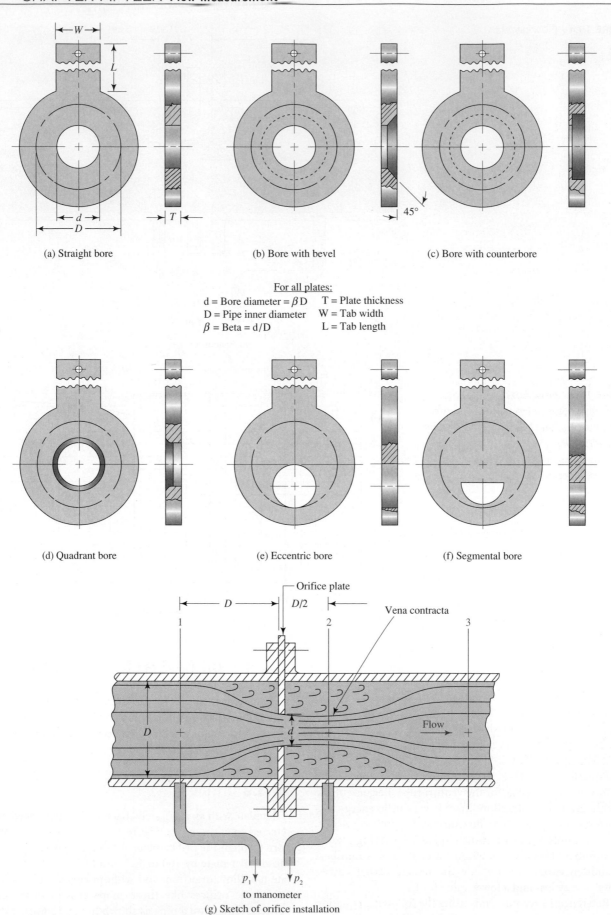

(a) Straight bore   (b) Bore with bevel   (c) Bore with counterbore

For all plates:
d = Bore diameter = β D   T = Plate thickness
D = Pipe inner diameter   W = Tab width
β = Beta = d/D   L = Tab length

(d) Quadrant bore   (e) Eccentric bore   (f) Segmental bore

(g) Sketch of orifice installation

FIGURE 15.6 Orifice plate styles and a typical installation between pipe flanges with pressure taps at distances D and $D/2$ from the plate.

stream flowing through the orifice forms a vena contracta and the rapid flow velocity results in a decreased pressure downstream from the orifice. See Sections 6.10 and 10.6 for more discussion of flows in which a vena contracta is formed. Pressure taps before and after the orifice (sections 1 and 2) allow the measurement of the differential pressure across the meter, which is related to the volume flow rate by Eq. (15–5).

When the orifice plates are thin, say < 3.0 mm (< 1/8 in), the bored hole can be machined straight and square clear through the plate as shown in Figure 15.6(a). Sometimes such a thin plate is susceptible to damage during installation or handling. Often, when thicker plates are used, about 3.0 to 6.0 mm (1/8 in to ¼ in), they may have a short, straight bore hole in the upstream face with a tapered relief downstream of the hole as shown in part (b). An alternative is a counter-bored design [part (c)] with relief on the back side. The quadrant bore style [part (d)] has a radius starting on the front face of the plate then blends with a short straight bore hole, a design recommended for viscous fluids operating in the laminar flow regime with $N_R < 4000$; typical fluids are heavy crude oil, slurries, glycerin, and syrup. The eccentric bore and the segmental bore shown in parts (e) and (f) are used when the fluid contains solids that may settle behind a concentric orifice plate. Using these designs for paper pulp, some petrochemical products, or sewage treatment sludge allows particles to flow through. Special flow coefficients for these styles must be obtained from vendors. Standard orifice plate sizes are commercially available for pipe sizes from DN 15 to DN 600 (1/2-in to 24 in) as featured in Internet resources 1, 7, 8, 10, and 13.

Commercially available units employing orifice plates incorporate all of the major subsystems needed to measure the differential pressure and the corresponding flow rate. The orifice plate is part of an integrated assembly that also includes:

- Pressure taps accurately located on both sides of the plate
- A manifold that facilitates the mounting of the differential producing (dp) cell
- A dp cell and transmitter to transmit the signal to a remote location

- A set of valves that allow fluid to bypass the dp cell for service
- Straight lengths of pipe into and out of the orifice to ensure predictable flow conditions at the orifice
- End flanges to connect the unit to process piping and flanges that accommodate the orifice plate
- A built-in microprocessor in the dp cell that linearizes the output signal across the entire range of the meter, giving a signal that is directly proportional to the flow; it performs the square root operation called for in Eq. (15–5)

The value of the discharge coefficient C is affected by small variations in the geometry of the edge of the orifice. Typical curves for sharp-edged concentric orifices are shown in Fig. 15.7, where D is the pipe diameter and d is the orifice diameter. The ratio of the diameters, d/D, is called the *beta ratio* and its effect is shown in the figure. The value of C is much lower than that for the venturi tube or the flow nozzle because the fluid is forced to make a sudden contraction followed by a sudden expansion. In addition, because measurements are based on the orifice diameter, the decrease in the diameter of the flow stream at the vena contracta tends to reduce the value of C.

The actual value of the discharge coefficient C depends also on the location of the pressure taps. Three possible locations are listed in Table 15.1.

TABLE 15.1 Location of pressure taps for orifice meters

	Inlet Pressure Tap, p_1	Output Pressure Tap, p_2
1.	One pipe diameter upstream from plate	One-half pipe diameter downstream from inlet face of plate
2.	One pipe diameter upstream from plate	At vena contracta (see Reference 5)
3.	In flange, 1-in upstream from plate	In flange, 1-in downstream from outlet face of plate

FIGURE 15.7 Orifice discharge coefficient. (*Source:* From ASME Research Committee on Fluid Mechanics (1959). *Fluid Mechanics: Their Theory and Application,* 5/e © American Society of Mechanical Engineers. Reprinted with permission.)

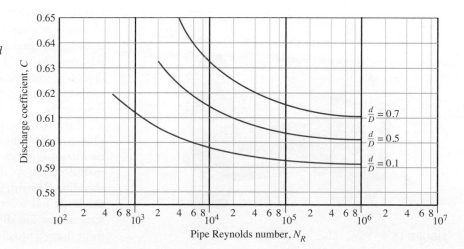

References 1, 5, and 14–16 give extensive information on the proper selection and application of orifices including adjustments for gas flow.

15.3.4 Flow Tubes

Several proprietary designs for modified variable-head flow meters called flow tubes are available. These can be used for applications similar to those for which the venturi, nozzle, or orifice meters are used, but flow tubes have somewhat lower pressure loss (higher pressure recovery). Figure 15.8 is drawing of a commercially available flow tube.

15.3.5 Overall Pressure Loss

In each of the four types of variable-head meters just described, the flow stream expands back to the main pipe diameter after passing the restriction. This is indicated as section 3 in Figs. 15.2, 15.4, and 15.6(g). Then, the difference between the pressure p_1 and p_3 is due to the meter. The difference can be evaluated by considering the energy equation:

$$\frac{p_1}{\gamma} + z_1 + \frac{v_1^2}{2g} - h_L = \frac{p_3}{\gamma} + z_3 + \frac{v_3^2}{2g}$$

Because the pipe sizes are the same at both sections, $v_1 = v_3$. We may also assume $z_1 = z_3$. Then,

$$p_1 - p_3 = \gamma h_L$$

The pressure drop is proportional to the energy loss. The careful streamlining of the venturi tube and the long gradual expansion after the throat cause very little excess turbulence in the flow stream. Therefore, energy loss is low and the pressure loss is low. The lack of a gradual expansion causes the nozzle to have a higher pressure loss, whereas that for an orifice is still higher. The lowest pressure loss is obtained from the flow tube. Figure 15.9 shows the comparison among the several types of variable-head meters with regard to pressure loss.

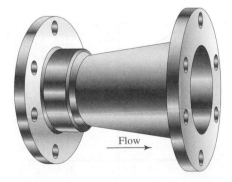

FIGURE 15.8 Flow tube.

15.4 VARIABLE-AREA METERS

The *rotameter* is a common type of variable area meter. Figure 15.10 shows a typical geometry. The fluid flows upward through a clear tube that has an accurate taper on the inside. A float is suspended in the flowing fluid at a position proportional to the flow rate. The upward forces due to the fluid dynamic drag on the float and buoyancy just balance the weight of the float. A different flow rate causes the float to move to a new position, changing the clearance area between the float and the tube until equilibrium is achieved again. The position of the float is measured against a calibrated scale graduated in convenient units of volume flow rate or weight flow rate.

Use of the type of rotameter shown in Fig. 15.10 requires that the fluid be transparent because the operator must visually see the position of the float. In addition, the transparent tube has a somewhat limited pressure capability. Some rotameters are made from opaque tubing to withstand higher pressures. The position of the float is sensed from outside the tube by an electromagnetic means and the flow rate is indicated on a gage. See Internet resources 11–15 for examples. Reference 6 is a standard for the use of variable area meters.

15.5 TURBINE FLOWMETER

Figure 15.11 shows a turbine flowmeter in which the fluid causes the turbine rotor to rotate at a speed dependent on the flow rate. As each blade of the rotor passes the magnetic coil, a voltage pulse is generated that can be input to a frequency meter, an electronic counter, or a similar device whose readings can be converted to flow rate. Flow rates from as low as 0.02 L/min (0.005 gal/min) to several thousand L/min or gal/min can be measured with turbine flowmeters of various sizes. See Internet resources 2, 3, 9, 11, and 13–15. Reference 7 is a standard for using turbine flowmeters.

15.6 VORTEX FLOWMETER

Figure 15.12 show a *vortex flowmeter*, in which a blunt obstruction placed in the flow stream causes vortices to be created and shed from the body at a frequency that is proportional to the flow velocity. A sensor in the flowmeter detects the vortexes and creates an indication for the meter readout device. Reference 8 is a standard for using vortex flowmeters.

Part (b) of Fig. 15.12 shows a sketch of the vortex-shedding phenomenon. The shape of the blunt body, also called the *vortex-shedding element*, may vary from manufacturer to manufacturer. As the flow approaches the front face of the shedding element, it divides into two streams. Fluid close to the body has a low velocity relative to that in the main streamlines. The difference in velocity causes shear layers to form that eventually break down into vortices alternately on

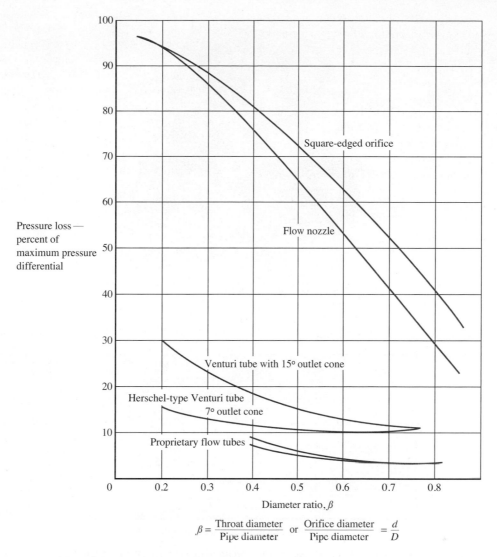

$$\beta = \frac{\text{Throat diameter}}{\text{Pipe diameter}} \quad \text{or} \quad \frac{\text{Orifice diameter}}{\text{Pipe diameter}} = \frac{d}{D}$$

FIGURE 15.9 Comparison of pressure loss for various variable head flowmeters. (*Source:* From H. S. Bean, ed. *Fluid Mechanics: Their Theory and Application*, 6/e Copyright © 1971 American Society of Mechanical Engineers. Reprinted with permission.)

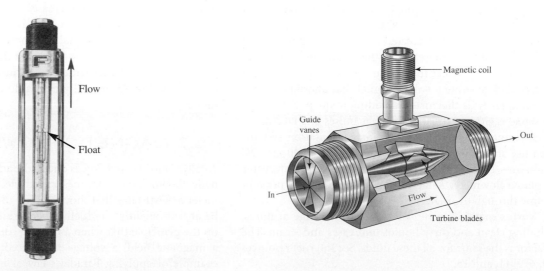

FIGURE 15.10 Rotameter. (*Source:* ABB, Inc., Automation Technology Products, Warminster, PA)

FIGURE 15.11 Turbine flowmeter.

(a) Photograph of a vortex flowmeter

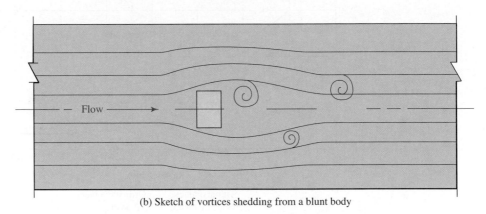

(b) Sketch of vortices shedding from a blunt body

FIGURE 15.12   Vortex flowmeter. (*Source:* ABB, Inc., Automation Technology Products, Warminster, PA)

the two sides of the shedding element. The trail of vortices is called the Karman street, named after Theodore von Karman who discovered and characterized the phenomenon. The frequency of the vortices created is directly proportional to the flow velocity and, therefore, to the volume flow rate. Sensors in the meter detect the pressure variations around the vortices and generate a voltage signal that alternates at the same frequency as the vortex-shedding frequency. The output signal is either a stream of voltage pulses or a DC (direct current) analog signal. Standard instrumentation systems often use an analog signal that varies from 4 to 20 mA DC (milliamps DC). For the pulse output, the manufacturer supplies a flowmeter *K*-factor that indicates pulses per unit volume through the meter.

Vortex meters can be used for a wide range of fluids, including clean and dirty liquids and gases and steam. The *K*-factor is the same for all these fluids. See Internet resources 1, 4, 6, 9, 11, and 15.

A related design useful for some fluids is called the *swirl meter*, comprised an inlet configuration that creates

a swirling motion in the fluid. At the core of the swirling fluid is a vortex system that creates a secondary pattern of vortex motion and its frequency is detected by a piezo sensor. The resulting signal is processed further in the pressure transmitter into a form that the meter can interpret as velocity of flow. Internet resource 4 is a source of data for one commercially available design for a swirl meter.

15.7 MAGNETIC FLOWMETER

Totally unobstructed flow is one of the advantages of a magnetic flowmeter (sometimes called electromagnetic flowmeter or EMF) like that shown in Fig. 15.13. The fluid must be at least slightly conducting because the meter operates on the principle that when a moving conductor cuts across a magnetic field, a voltage is induced. It is an excellent example of applying Faraday's law of induction. The primary components of the magnetic flowmeter include a tube lined with a non-conducting material, two electromagnetic

(a) Magnetic flowmeter

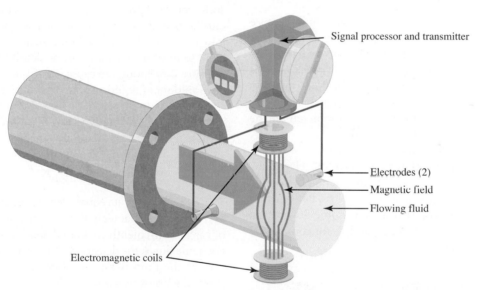

Signal processor and transmitter

Electrodes (2)
Magnetic field
Flowing fluid

Electromagnetic coils

(b) Schematic diagram of a magnetic flowmeter

FIGURE 15.13 Magnetic flowmeter. (*Source:* Endress+Hauser Flowtec AG, Reinach BL, Switzerland)

coils (one above the tube and one below), and two electrodes mounted 180° apart horizontally in the tube wall. The electrodes detect the voltage generated in the fluid. Because the generated voltage is directly proportional to the fluid velocity, a greater flow rate generates a greater voltage. An important feature of this type of meter is that its output is completely independent of temperature, viscosity, specific gravity, and turbulence. Tube sizes from 2.5 mm to 2.4 m (0.1 in to 8.0 ft) in diameter are available.

Flow measurement of many types of conductive fluids is accomplished with magnetic meters, including pulp and paper, mining solutions, two-phase fluids, foods, beverages, oil, pharmaceuticals, water and wastewater, and throughout the chemical process industries. Flow rates of fluids with high solids content are easily measured. Reference 9 is a standard for use of magnetic flowmeters. See Internet resources 1–4, 6, 11, 13, and 15 for commercial producers and suppliers.

15.8 ULTRASONIC FLOWMETERS

A major advantage of an ultrasonic flowmeter is that it is not necessary to penetrate the pipe in any way. An ultrasonic generator is strapped to the outside of the pipe and a high-frequency signal is transmitted through the wall of the pipe and across the flow stream, typically at an acute angle with respect to the axis of the pipe. The time for the signal to traverse the pipe depends on the velocity of flow of the fluid in the pipe. Some commercially available meters use detectors on the side opposite the transmitter, whereas others employ reflectors to return the signal to a receiver built into the transmitter.

Another approach is to use two transmitter/receiver units aligned with the axis of the pipe. Each delivers a signal at an angle to the flow that is reflected from the opposite side of the pipe and received by the other. The signal that is directed in the same direction as the flow takes a different time to reach the receiver than the signal that opposes the flow. The difference between these two times is proportional to the velocity of flow. A variety of orientations can be used for the transmitters, reflectors, and receivers of the signal. Most use two sets to reduce the sensitivity of the meter to the velocity profile of the fluid flow stream.

Transit-time meters work best with clean fluids because entrained particles in dirty fluids can affect the time readings and the strength of the signal that reaches the detectors.

A second type of meter, called the *Doppler-type meter*, is preferred for dirty fluids, slurries, and other fluids that may inhibit the transmission of the ultrasonic signal. The ultrasonic pressure wave does not traverse completely to the opposite wall of the pipe. Rather, it is reflected from the particles in the fluid itself and back to the receiver.

Because the ultrasonic flowmeter is completely noninvasive, the pressure loss is due only to the friction in the pipe itself. The meter contributes no additional loss. Reference 10 is a standard for use of transit-time ultrasonic flowmeters. See Internet resources 1–3, 9, 11, 13, and 15 for commercial suppliers of such meters.

15.9 POSITIVE-DISPLACEMENT METERS

Fluid entering a positive-displacement meter fills up a chamber that is moved from the input to the output side of the meter. The meter records or indicates the cumulative volume of fluid that has passed through the meter. The chambers can take many forms and are often proprietary to a given manufacturer.

Typical uses for positive-displacement meters are water delivered from the municipal system to a home or business, natural gas delivered to a customer, and gasoline delivered at a service station. They are also used in certain industrial applications in which blended materials are required to have a set volume of different constituents.

Gas meters like those used in homes employ flexible diaphragms that continuously capture and then deliver known volumes of the low-pressure natural gas. Other designs include meshing circular gears, meshing oval gears, lobed rotors, linear motion reciprocating pistons, and nutating disks. The nutating disk design incorporates a thin disk mounted at an angle on a shaft. A close fit seals the disk against the housing. As the fluid flows through the housing it induces rotation of the shaft, and a known volume passes through with each revolution. A counter or a recorder accumulates the number of revolutions over time, which can be reported in any convenient flow units. Internet resources 2, 3, 12, and 15 show a variety of types of positive-displacement meters along with capacity data.

15.10 MASS FLOW MEASUREMENT

The flowmeters discussed thus far in this chapter are designed to produce an output signal that is proportional to the average velocity of flow or the volume flow rate. This is satisfactory when only the *volume* delivered through the meter is needed. However, some processes require a measurement of the *mass* of fluid delivered. For example, in food-processing plants the production is often indicated as the amount delivered in kilograms, pounds-mass, or slugs. Some chemical processes are sensitive to the mass of the various constituents that are blended or that are introduced into a reaction. Two-phase fluids, such as steam, may be difficult to measure accurately if the density, temperature, and pressure vary enough to cause significant changes in the relative amount of liquid and vapor in the steam.

One way to obtain mass flow rate measurements is to use a flowmeter of the type just discussed, which indicates volume flow rate, and then simultaneously measure the density of the fluid. Then, the mass flow rate would be

$$M = \rho Q$$

That is, mass flow rate equals density times volume flow rate as discussed in Chapter 6. If the density of the fluid is known or can be conveniently measured, this is a simple calculation. For some fluids, the density can be calculated if the temperature of the fluid is known. Sometimes, particularly with gases, the pressure is also needed. Temperature probes and pressure transducers are readily available to provide the necessary data. Specific gravity can be measured with a device called a *gravitometer*. Density can be measured directly for some fluids with a *densitometer*. The signals related to volume flow rate, temperature, pressure, specific gravity, and density can all be input to special electronic devices that effectively perform the calculation of $M = \rho Q$. This is shown schematically in Fig. 15.14. This process, although straightforward, requires several separate measurements to be made, each of which is subject to small errors. Then the errors are compounded in the final calculation.

True mass flowmeters avoid the problems discussed above by generating a signal proportional to the mass flow rate directly. One such mass flowmeter is called the *Coriolis mass flowmeter*, shown in Fig. 15.15. The fluid enters the flowmeter from the process piping and is directed through

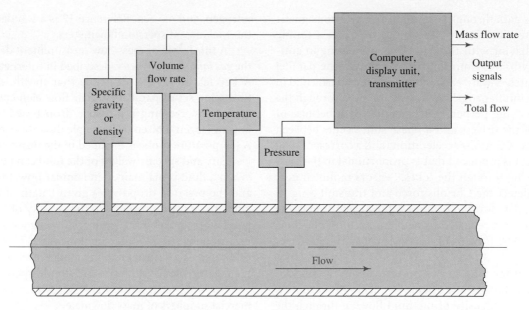

FIGURE 15.14 Schematic representation of mass flow measurement using multiple sensors.

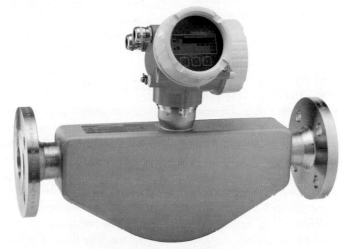

(a) Coriolis flowmeter-External view

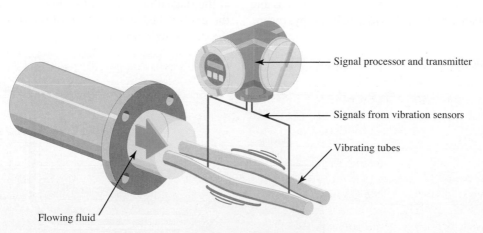

(b) Coriolis flowmeter-Schematic view

FIGURE 15.15 Coriolis mass flow meter. (*Source:* Endress+Hauser Flowtec AG, Reinach BL, Switzerland)

a continuous path through two tubes. Two electromagnetic drivers excite vibrations in the tubes to create a tuning fork-like behavior with the two tubes oscillating in antiphase. The vibratory motion created moves the parallel paths alternately toward each other and then apart. Fluid in the tubes is simultaneously following the path through the tubes and moving perpendicular to that path because of the action of the drivers and a phase shift occurs between the two tubes. A Coriolis acceleration (and a corresponding Coriolis force) is produced that is proportional to the mass of fluid flowing through the tubes. Sensors mounted near the drivers detect the Coriolis force and transmit a signal that can be related to the true mass flow rate through the meter. Accuracy is reported to be 0.2 percent of the indicated rate or 0.02 percent of full-scale capacity, whichever is greater.

Density of the fluid can be measured with the Coriolis mass flowmeter because the driving frequency of the tubes is dependent on the density of the fluid flowing through the tubes. A temperature probe is also included in the system, completing a comprehensive set of fluid properties and mass flow rate data. Non-conductive media, highly viscous liquids, or liquids with high solids content can be measured. Reference 11 is a standard for use of the Coriolis mass flowmeter. See Internet resources 1, 4, 6, 12, and 15 for commercially available Coriolis flowmeters.

Another form of mass flowmeter uses a thermal technique in which two probes, called *resistance temperature detectors* (RTDs), are inserted into the flow. One probe senses the temperature of the fluid stream as a reference. The other is heated to a set temperature above the reference temperature, and electronic circuitry (a form of Wheatstone bridge) continuously adjusts the power to this probe to maintain the set temperature difference. A greater mass flow rate around the probe causes more heat to be dissipated away from the heated probe, requiring a higher power. Therefore, there is a predictable relationship between the mass flow rate and the power input to the probes. A signal processing system in the control linearizes the output voltage signal with respect to mass flow rate. These devices can measure the mass flow of many kinds of gases such as air, natural gas, propane, carbon dioxide, helium, hydrogen,

nitrogen, and oxygen. Reference 12 is a standard for use of thermal mass dispersion flowmeters.

A third style of mass flow measurement device is called the *gas laminar flowmeter*, described in Internet resource 19. A key feature of the device is that the fluid is directed through a set of parallel laminar flow elements that act as restrictors, creating a pressure drop based on the well-known Hagen–Poiseuille principle described in Section 8.5. A temperature probe is included in the unit from which the viscosity and specific weight of the fluid can be determined. When a fluid is maintained in laminar flow, the energy loss and the pressure drop over a given length of conduit of a known size is proportional to the velocity of flow. Pressure taps before and after the laminar flow elements facilitate the measurement of the differential pressure. The entire unit is calibrated as a system enabling a high level of accuracy for the measured mass flow rate and simple operation.

See Internet resources 1, 4, 6, 11–15, and 19 for commercial suppliers of mass flowmeters.

15.11 VELOCITY PROBES

Several devices are available that measure the velocity of flow at a specific location rather than an average velocity. These are referred to as *velocity probes*. Some of the more common types will be described in this section.

15.11.1 Pitot Tube

When a moving fluid is caused to stop because it encounters a stationary object, a pressure is created that is greater than the pressure of the fluid stream. The magnitude of this increased pressure is related to the velocity of the moving fluid. The *pitot tube* uses this principle to indicate velocity, as illustrated in Fig. 15.16. The pitot tube is a hollow tube positioned so that the open end points directly into the fluid stream. The pressure at the tip causes a column of fluid to be supported. The fluid at or just inside the tip is then stationary or stagnant, and this point is referred to as the *stagnation point*. We can use the energy equation to relate the pressure at the stagnation point to the fluid velocity. If point 1 is in the undisturbed stream ahead of the tube and point s is at the stagnation point, then

FIGURE 15.16 Pitot tube.

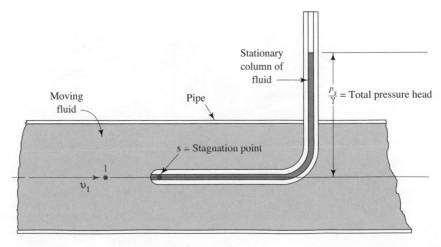

$$\frac{p_1}{\gamma} + z_1 + \frac{v_1^2}{2g} - h_L = \frac{p_s}{\gamma} + z_s + \frac{v_s^2}{2g} \quad \text{(15–8)}$$

Observe that $v_s = 0$, $z_1 = z_2$ or very nearly so, and $h_L = 0$ or very nearly so. Then, we have

$$\frac{p_1}{\gamma} + \frac{v_1^2}{2g} = \frac{p_s}{\gamma} \quad \text{(15–9)}$$

The names given to the terms in Eq. (15–9) are as follows:

$p_1 = $ Static pressure in the main fluid stream

$p_1/\gamma = $ Static pressure head

$p_s = $ Stagnation pressure or total pressure

$p_s/\gamma = $ Total pressure head

$v_1^2/2g = $ Velocity pressure head

The total pressure head is equal to the sum of the static pressure head and the velocity pressure head. Solving Eq. (15–9) for the velocity gives

$$v_1 = \sqrt{2g(p_s - p_1)/\gamma} \quad \text{(15–10)}$$

Notice that only the difference between p_s and p_1 is required to calculate the velocity. For this reason, most pitot tubes are made as shown in Fig. 15.17, providing for the measurement of both pressures with the same device.

The device shown in Fig. 15.17 facilitates the measurement of both the static pressure and the stagnation pressure simultaneously and so it is sometimes called a *pitot-static* tube. Its construction shown in part (b) is actually a tube within a tube. The small central tube is open at the end and functions in the same manner as the single pitot tube shown in Fig. 15.16. Thus, the stagnation pressure, also called the *total pressure*, is sensed through this tube. The *total pressure tap* at the end of this tube allows connection to a pressure-measuring device.

The larger outer tube is sealed around the central tube at its end, thus, creating a closed annular cavity between the central and the outer tube. Section A-A shows a series of small radial holes drilled through the outer tube, but not through the central tube. When the tube is aligned with the direction of flow, these radial holes are perpendicular to the flow and thus they sense the local static pressure, which we have called p_1. Notice that a static pressure tap is affixed at the end of the tube to allow connection to a measuring instrument.

The measuring instrument need not measure either p_s or p_1 because it is the *difference* $(p_s - p_1)$ that is needed in Eq. (15–10). Several manufacturers make differential pressure-measuring devices for such applications.

If a differential manometer is used, as shown in Fig. 15.18, the manometer deflection h can be related directly to the velocity. We can write the equation describing the difference between p_s and p_1 by starting at the static pressure holes in the side of the tube, proceeding through the manometer, and ending at the open tip of the tube at point s:

$$p_1 - \gamma x + \gamma y + \gamma_g h - \gamma h - \gamma y + \gamma x = p_s$$

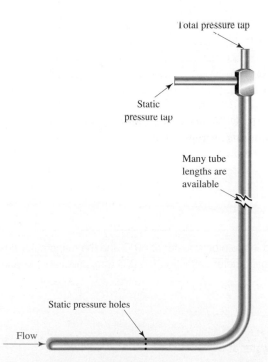

(a) Drawing of the external view of a pitot-static tube

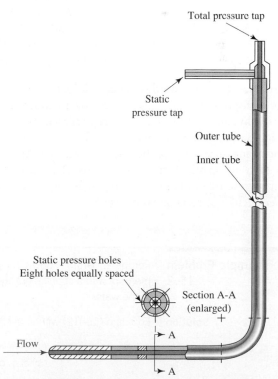

(b) Drawing of the construction of a pitot-static tube

FIGURE 15.17 Pitot-static tube.

FIGURE 15.18 Differential manometer used with a pitot-static tube.

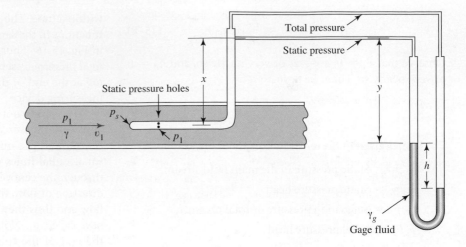

The terms involving the unknown distances x and y drop out. Then, solving for the pressure difference, we get

$$p_s - p_1 = \gamma_g h - \gamma h = h(\gamma_g - \gamma) \qquad \textbf{(15–11)}$$

Substituting this into Eq. (15–10) gives

$$v_1 = \sqrt{2gh(\gamma_g - \gamma)/\gamma} \qquad \textbf{(15–12)}$$

Pipe Traverse to Obtain Average Velocity   The velocity calculated by either Eq. (15–10) or Eq. (15–12) is the local velocity at the particular location of the tip of the tube. In Chapters 8 and 9 we found that the velocity of flow varies from point to point across a pipe. Therefore, if the average velocity of flow is desired, a traverse of the pipe should be made with the tip of the tube placed at the specific ten points indicated in Fig. 15.19. The dashed circles define concentric annular rings that have equal areas. The velocity at each point can be calculated, using Eq. (15–12). Then, the average velocity of flow is the average of these ten values. We can find the volume flow rate from $Q = Av$, using the average velocity.

Traverse of a Rectangular Duct   To obtain the average velocity for a rectangular duct, it is recommended that the area be divided into 16–64 equal rectangular areas (depending on the size of the duct and the desired precision), taking velocity measurements at the center of each of these areas, and then averaging all the readings. See Reference 13 and Internet resource 13 for additional details about traverses of either circular or rectangular ducts.

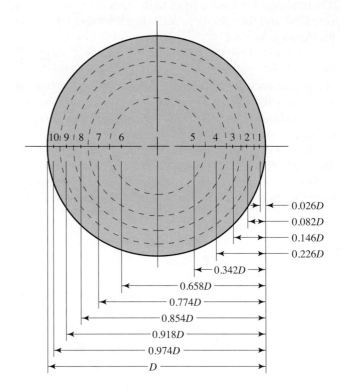

FIGURE 15.19   Velocity measurement points within a pipe for computing average velocity.

Example Problem 15.2   For the apparatus shown in Fig. 15.18, the fluid in the pipe is water at 60°C and the manometer fluid is mercury with a specific gravity of 13.54. If the manometer deflection h is 264 mm, calculate the velocity of the water.

Solution   Equation (15–12) will be used:

$$v_1 = \sqrt{2gh(\gamma_g - \gamma)/\gamma}$$

$\gamma = 9.65 \text{ kN/m}^3$ (water at 60°C)

$\gamma_g = (13.54)(9.81 \text{ kN/m}^3) = 132.8 \text{ kN/m}^3$ (mercury)

$h = 264 \text{ mm} = 0.264 \text{ m}$

Because all terms are in SI units, the velocity is in m/s:

$$v_1 = \sqrt{\frac{(2)(9.81)(0.264)(132.8 - 9.65)}{9.65}}$$

$$= 8.13 \text{ m/s}$$

The pressure differential created by a pitot tube can also be read by an electronic device such as that shown in Fig. 15.20. The individual readings taken during a traverse of a pipe or duct can be saved. Then the average is automatically computed in either SI units or U.S. Customary System units and it can be downloaded.

15.11.2 Cup Anemometer

Air velocity is often measured with a *cup anemometer* such as that shown in Fig. 15.21. The moving air strikes the open cups, causing rotation of the shaft on which they are mounted. The rotational speed of the shaft is proportional to the air velocity, which is indicated on a meter or transmitted electrically. See Internet resource 18.

15.11.3 Hot-Wire Anemometer

This type of velocity probe employs a very thin wire, about 12 μm in diameter, through which an electrical current is passed. The wire is suspended on two supports as shown in Fig. 15.22 and inserted into the fluid stream. The wire tends to heat because of the current flowing in it, but it is cooled by convection heat transfer to the moving fluid stream. The amount of cooling depends on the velocity of the fluid. In one type of hot-wire anemometer, a constant current is applied to the wire. A variation in the flow velocity causes a change in the wire temperature and, therefore, its resistance changes. The electronic measurement of the resistance change can be related to flow velocity. Another type senses a change in the resistance of the wire, but then varies the current flow to maintain a set wire temperature regardless of fluid velocity. The magnitude of the current flow is then related to fluid velocity. See Internet resources 17 and 18.

15.11.4 Flow Imaging

Various techniques are available for creating visual images of flow patterns that represent velocity distribution and flow direction for complex fluid flow systems. Internet resource 17 describes flow-imaging systems using constant-temperature anemometry (CTA) probes, particle image velocimetry (PIV), laser Doppler anemometry (LDA), computational fluid dynamics (CFD), and laser-induced fluorescence (LIF) techniques.

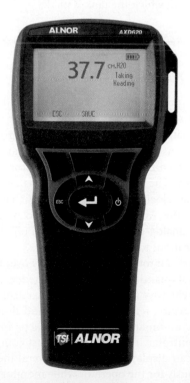

FIGURE 15.20 Electronic readout device for pitot tubes. (*Source:* TSI Incorporated)

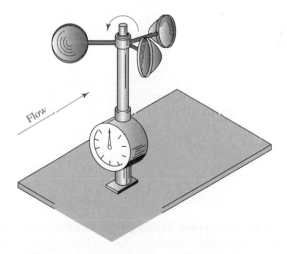

FIGURE 15.21 Rotating-cup anemometer.

FIGURE 15.22 Hot-wire anemometer tip.

15.12 LEVEL MEASUREMENT

Bulk storage tanks are integral parts of many fluid flow systems, and it is often necessary to monitor the level of fluid in such tanks. Level measurements are typically transmitted to remote monitors or central control stations and can trigger automatic level control. Several types of level measurement devices are available for tanks holding liquids or solids. See Internet resources 1 and 11–13. Brief descriptions are listed here; consultation with suppliers is recommended to determine which type is best for a given application.

Float Type The buoyant force acting on a float causes it to rise or fall as the fluid level changes. The position of the float can actuate a switch, or a signal can be transmitted to a remote location. Floats are typically used for sensing upper or lower level limits.

Pressure Sensing Placing a pressure sensor in the bottom of a tank senses the depth of fluid using the principle $\Delta p = \gamma h$, where γ is the specific weight of the fluid and h is the depth above the sensor. Care must be exercised when the specific weight can change because of temperature or composition of the material. When the vessel is pressurized, a differential pressure sensor can measure both the ambient pressure above the fluid and the pressure at the bottom of the tank and use the difference to determine the depth.

Capacitance Probe A high-frequency AC electrical signal is delivered to the sensor, and the magnitude of the current that flows through the device is dependent on the capacitance of the material and the depth of submergence of the probe. Although these devices can be used for most types of liquids and solids, calibration for each material is typically required.

Vibration Type This type of sensor is based on the principle that the frequency of vibration of a tuned fork changes with the density of the material with which it is in contact. It is used for point level measurement such as sensing the lowest acceptable level that can trigger the resupply of the tank or shutting down the system. Sensing the maximum level can close a valve to stop the supply of liquid.

Ultrasonic A high-frequency sound pulse is emitted by the sensor and then reflected from the surface of the fluid or solid being sensed because of its higher density compared with the air or other gas above the material. The time for a reflected signal to be detected by the sensor is then related to the distance traveled and thus to the level. The frequency is typically in the range from 12 kHz to 70 kHz. This device is a noncontact type, and can be used for abrasive materials or where the tank configuration does not permit a sensor in the fluid itself. Some disadvantages are sensitivity to dust, foam, ambient noise, turbulent surfaces, and buildup of the material on the emitter. Care must also be exercised when using ultrasonic level sensors with solid materials because the surface tends to take a conical or sloped shape based on the angle of repose of the material. The signal may also be scattered by coarse materials.

Radar Instead of using ultrasonic sound waves, the radar level sensor uses electromagnetic microwaves at a frequency range of 6 GHz to 26 GHz, depending on the design of the transmitter. The signal is directed at the surface by a conical horn and is reflected from the fluid surface because of the change in dielectric constant of the material relative to the medium above the surface. The reflected wave is sensed and the time of flight is related to the distance traveled and thus to the level of the surface.

Guided Radar This type is similar to the radar sensor except a waveguide is attached to the radar unit and is extended down into the material whose level is to be sensed. The waveguide is typically a thin cable or rod that is positioned at approximately one third of the tank diameter from the wall. It can be up to 35 m (115 ft) long in the cable form. Rigid rods range in length from 2 m (6.6 ft) to 4 m (13 ft). The pulsed wave at 100 MHz to 1.5 GHz travels down the guide and is maintained in a focused pattern [within a radius of 200 mm (8 in)], much tighter than is practical with the standard radar unit. It provides a more reliable signal when applied in tall, small-diameter tanks or where there are obstructions in the tank that may send false signals. The reflected wave travels back up the waveguide to the sensor. Guided radar level sensors are relatively insensitive to changes in temperature, pressure, product density, turbulence, foam, obstructions, vessel shape, dust, noise, humidity, and the material from which the tank is made.

15.13 COMPUTER-BASED DATA ACQUISITION AND PROCESSING

Microcomputers, programmable controllers, and other microprocessor-based electronic instrumentation greatly simplify the acquisition, processing, and recording of flow measurement data. As shown in this chapter, many of the flowmeters produce an electrical signal that is proportional to the flow velocity. The signal is either an analog voltage that varies with the velocity or a pulse frequency that can be counted electronically. Analog signals can be converted to digital signals by analog-to-digital converters, often called *A-D converters*, for input to digital computers.

The computers can total the fluid flow rate over time to determine the total quantity of fluid transferred to a given location. A comprehensive measurement and control system can consist of pressure, temperature, level, and flow measurement devices; automatic process controllers; interface units; operator control stations; and large host computers. Video terminals can display the status of several measurements simultaneously for the operator while monitoring the data for values that fall outside prescribed levels. The host computer can acquire data from several places within the plant and maintain the central database for quality control, production data, and inventory control. Wireless systems and cloud computing are frequently used for critical systems that allow monitoring and control from computers, tablets, and smart phones anywhere in the world.

REFERENCES

1. American Society of Mechanical Engineers. 1971. *Fluid Meters: Their Theory and Application*, 6th ed. Howard S. Bean (Ed). New York: Author.

2. _____. 2000. *Method for Establishing Installation Effects on Flow Meters (Standard MFC-10M)*. New York: Author.

3. _____. 2003. *Glossary of Terms Used in the Measurement of Fluid Flow in Pipes (Standard MFC-1M)*. New York: Author.

4. _____. 1998. *Measurement of Liquid Flow in Closed Conduits by Weighing Method (Standard MFC-9M)*. New York: Author.

5. _____. 2004. *Measurement of Fluid Flow in Pipes Using Orifice, Nozzle, & Venturi (Standard MFC-3M)*. New York: Author.

6. _____. 2001. *Measurement of Fluid Flow Using Variable Area Meters (Standard MFC-18M-2001)*. New York: Author.

7. _____. 2007. *Measurement of Liquid by Turbine Flowmeters (Standard MFC-22)*. New York: Author.

8. _____. 1998. *Measurement of Fluid Flow in Pipes Using Vortex Flowmeters (Standard MFC-6M)*. New York: Author.

9. _____. 2007. *Measurement of Fluid Flow in Closed Conduits by Means of Electromagnetic Flowmeters (Standard MFC 16)*. New York: Author.

10. _____. 2011. *Measurement of Liquid Flow in Closed Conduits Using Transit-Time Ultrasonic Flowmeters (Standard MFC-5.1)*. New York: Author.

11. _____. 2006. *Measurement of Fluid Flow in Pipes by Means of Coriolis Mass Flowmeters (Standard MFC-11)*. New York: Author.

12. _____. 2010. *Measurement of Fluid Flow in Closed Conduits by Means of Thermal Mass Dispersion Flowmeters (Standard MFC-21.2)*. New York: Author.

13. _____. 2006. *Measurement of Fluid Flow in Closed Conduits Using Multiport Averaging Pitot Primary Elements (Standard MFC-12M-2006)*. New York: Author.

14. Baker, Roger C. 2005. *Flow Measurement Handbook*. Cambridge: Cambridge University Press.

15. Baker, Roger C. 2004. *An Introductory Guide to Flow Measurement, ASME Edition*. New York: American Society of Mechanical Engineers.

16. International Standards Organization, ISO. 2007. *Measurement of Fluid Flow—Procedures for the Evaluation of Uncertainties (ISO 5168:2005)* West Conshohocken, PA: ASTM International (Distributor).

17. Lee, T. W. 2008. *Thermal and Flow Measurements*. Boca Raton, FL: CRC Press.

18. Miller, Richard W. 1996. *Flow Measurement Engineering Handbook*, 3rd ed. New York: McGraw-Hill.

19. Spitzer, D. W. 2001. *Flow Measurement: Practical Guides for Measurement and Control*, 2nd ed. Research Triangle Park, NC: ISA—The Instrumentation, Systems, and Automation Society.

20. Spitzer, David W. 2004. *Industrial Flow Measurement*, 3rd ed. Research Triangle Park, NC: ISA—The Instrumentation, Systems, and Automation Society.

21. Upp, E. Loy, and Paul J. LaNasa. 2002. *Fluid Flow Measurement*, 2nd ed. Woburn, MA: Gulf Publishing, Butterworh-Heinmann.

INTERNET RESOURCES

1. **Endress+Hauser:** Manufacturer of measurement devices for fluid flow, level, pressure, temperature, and pH. Flowmeter types include orifice, magnetic, Coriolis, ultrasonic, vortex, and pitot tubes.

2. **BadgerMeter, Inc.:** Manufacturer of measurement devices for fluid flow including magnetic, turbine, ultrasonic, variable area, and a variety of positive-displacement designs based on oval gear geometry and oscillating piston styles.

3. **Flow Technology, Inc.:** Manufacturer of turbine, magnetic, and ultrasonic flowmeters and nutating disc and gear type positive-displacement meters for industrial, aerospace and defense, automotive, and oil and gas applications.

4. **ABB, Inc.:** A diversified company offering instrumentation and control products, including flow measurement, through their Automation Technology Products unit. From the home page, select *Products*, then *Measurement*, then *Flow Meters*. Flowmeter designs include, magnetic, vortex, swirl, variable-area, differential pressure, Coriolis mass, and thermal mass.

5. **Alnor Instruments, TSI Incorporated:** Manufacturer of the Alnor AXD Micromanometer, an electronic meter for measuring small differential pressures from pitot-static tubes and small mass flowmeters. From the home page, select *Products*, then *Categories*, then *Flowmeters*.

6. **Invensys Foxboro:** Manufacturer of a variety of flow measurement devices, including Coriolis mass flow, density, vortex, and magnetic. From the home page, select *Flow* to view the complete line of flowmeters. On the *Flow* page, select *View M&I Downloads and Tools* to access sizing software for selecting a suitable meter for a specific application considering flow rate range and fluid properties.

7. **Tri-Flo Tech, Inc.:** Manufacturer of orifice plates, venturi tubes, pitot tubes, nozzles, flow tubes, turbine meters, and other devices for measuring or controlling flow. From the home page, select *Products*, then *Industrial Flow Meters*.

8. **Wyatt Engineering:** Manufacturer of the Wyatt–Badger venturi tube, flow nozzles, Lo-Loss® flow tubes, and orifice plate type flow measurement devices. From the home page, select *Products*.

9. **Racine Federated, Inc.:** Manufacturer of several types of flow measurement devices under several brand names; Blancett turbine flowmeters, Dynasonics ultrasonic flowmeters, Flo-tech turbine flowmeters for hydraulic fluids, Hedland in-line flow meters, Preso differential pressure flowmeters and Racine vortex shedding flowmeters. From the home page, select *Divisions* to then select the desired type of flowmeter.

10. **PRC Flow Measurement & Control, Inc.:** Manufacturer of differential pressure flowmeters including venturi tubes, low-loss flow tubes, ASME flow nozzles, and orifice plates. Selection can be made from the home page.

11. **Omega Engineering, Inc.:** Supplier of flow measurement devices including variable-area, magnetic, turbine, paddle-wheel, vortex, ultrasonic, and thermal mass. Velocity measurement devices include several styles of anemometers and pitot tubes. Also, offers numerous styles of temperature and level sensors. Select the desired type of meter from the home page.

12. **Brooks Instrument:** Manufacturer of several types of flow-meters, including thermal mass, Coriolis, variable-area, and oval gear positive-displacement. Also, offers several level measurement devices.

13. **Dwyer Instruments, Inc.:** Manufacturer of instruments for measuring flow, pressure, temperature, velocity and level, including manometers, digital manometers, pitot tubes, pressure gages, pressure transmitters, and variable-area, thermal mass, turbine, magnetic, orifice plate, and ultrasonic flowmeters.

14. **Cole-Parmer:** Supplier of numerous products for industrial use including flow measurement and fluid handling including variable area, thermal mass, turbine, and paddlewheel flow-meters. From the home page, select *Flow, Level, & Valves*, then select *flowmeters*.

15. **Instrumart:** Supplier of many types of flowmeters including magnetic, thermal mass, Coriolis, ultrasonic, vortex, turbine paddle wheel, variable-area, positive-displacement, and differential pressure. From the home page, select *Products*, then *Flow Meters*.

16. **Flow Control Network:** The online complement to *Flow Control* magazine covering all aspects of fluid-handling systems. The site can be searched for the websites and contact information of numerous companies that advertise in the magazine. An annual *Buyer's Resource* is published.

17. **Dantec Dynamics:** Manufacturer of anemometers and flow-imaging systems using constant-temperature anemometry (CTA) probes, particle image velocimetry (PIV), laser Doppler anemometry (LDA), computational fluid dynamics (CFD), and laser-induced fluorescence (LIF) techniques. From the home page, select *Products & Services*, then *Fluid Mechanics*, then the type of system desired.

18. **R. M. Young Company:** Supplier of a variety of meteorological instruments including wind sensors, anemometers, barometric pressure, rain, temperature and humidity. From the home page, select *Products*, then the type of device desired.

19. **CME Flow:** Manufacturer of gas laminar flowmeters for flow rates from very low 0.01 L/min (0.00035 ft^3/min) to very high 1000 L/min (35 ft^3/min). It is a division of Aerospace Control Products. Select the type of meter from the home page.

20. **Hyspan Precision Products, Inc.:** A manufacturer of a wide variety of HVAC and industrial products including venturi flow measurement systems, expansion compensation systems, vibration eliminators, ball joints, and metal hose assemblies. From the home page, select Hyspan Barco Venturi Flow Measurement Systems for venturi sizes from ½ in to 30 in (13 mm to 760 mm), capable of handling flow rates from 0.2 gal/min (0.8 L/min) to over 40 000 gal/min (155 000 L/min).

REVIEW QUESTIONS

1. List six factors that affect the selection and use of flowmeters.
2. Define *range* as it relates to flowmeters.
3. Describe three methods for calibrating flowmeters.
4. Name four types of variable-head meters.
5. Describe the venturi tube.
6. What is meant by the *throat* of a venturi tube?
7. What is the nominal included angle of the convergent section for a venturi tube?
8. What is the nominal included angle of the divergent section for a venturi tube?
9. Why is there such a difference between the angles of the convergent and the divergent sections for a venturi tube?
10. Describe the term *discharge coefficient* as it relates to variable-head meters.
11. Describe a flow nozzle and how it is used.
12. Describe an orifice meter and how it is used.
13. Describe a flow tube and how it is used.
14. Of the venturi, the flow nozzle, the flow tube, and the orifice, which has the lowest discharge coefficient? Why?
15. Describe *pressure loss* as it relates to flowmeters.
16. Rank the venturi, the flow nozzle, the orifice, and the flow tube on the basis of pressure loss.
17. Describe a rotameter variable-area meter.
18. Describe a turbine flowmeter and how it is used.
19. Describe a vortex flowmeter and how it is used.
20. Describe a magnetic flowmeter and how it is used.
21. Describe how mass flow rate can be measured.
22. Describe a pitot tube and how it is used.
23. Define *stagnation pressure* and show how it can be derived from Bernoulli's equation.
24. Define *static pressure head*.
25. Define *velocity pressure head*.
26. Why is a differential manometer a convenient device for use with a pitot tube?
27. Describe the method used to measure the average velocity of flow in a pipe using a pitot tube.
28. Describe a cup anemometer.
29. Describe a hot-wire anemometer and how it is used.
30. List several types of level measurement devices.

PRACTICE PROBLEMS

15.1 A venturi meter similar to the one in Fig. 15.2 has an inlet diameter of 100 mm and a throat diameter of 50 mm. While it is carrying water at 80°C, a pressure difference of 55 kPa is observed between sections 1 and 2. Calculate the volume flow rate of water.

15.2 Air with a specific weight of 12.7 N/m^3 and a kinematic viscosity of 1.3×10^{-5} m^2/s is flowing through a flow nozzle similar to that shown in Fig. 15.4. A manometer using water as the gage fluid reads 81 mm of deflection. Calculate the volume flow rate if the nozzle diameter is 50 mm. The special inlet tube has an inside diameter = 100 mm.

15.3 The flow of kerosene is being measured with an orifice meter similar to that shown in Fig. 15.6. The pipe is a 2-in Schedule 40 pipe and the orifice diameter is 1.00 in. The kerosene is at 77°F. For a pressure difference of 0.53 psi across the orifice, calculate the volume flow rate of kerosene.

15.4 A sharp-edged orifice is placed in a 10-in-diameter pipe carrying ammonia. If the volume flow rate is 25 gal/min, calculate the deflection of a water manometer (a) if the orifice diameter is 1.0 in and (b) if the orifice diameter is 7.0 in. The ammonia has a specific gravity of 0.83 and a dynamic viscosity of 2.5×10^{-6} lb·s/ft^2.

15.5 A flow nozzle similar to that shown in Fig. 15.4 is used to measure the flow of water at 120°F. The pipe is 6-in Schedule 80 steel. The nozzle diameter is 3.50 in. Determine the pressure difference across the nozzle that would be measured for a flow of 1800 gal/min.

15.6 A venturi meter similar to the one in Fig. 15.2 is attached to a 4-in Schedule 40 steel pipe and has a throat diameter of 1.50 in. Determine the pressure difference across the meter that would be measured for a flow of 600 gal/min of kerosene at 77°F.

15.7 A 50.0 mm, sharp-edge orifice is placed in a DN 100 Schedule 80 steel pipe. Compute the volume flow rate of ethylene glycol at 25°C when a mercury manometer reads a 95-mm deflection.

15.8 An orifice meter is to be used to measure the flow rate of propyl alcohol at 25°C through a PVC plastic pipe, 40 mm OD × 3.0 mm wall. The expected range of flow rates is 1.0 m^3/h to 2.5 m^3/h. Specify the diameter for the orifice to obtain $\beta = 0.40$ and determine the range of readings for a mercury manometer for the given flow rates.

15.9 A flow nozzle is to be installed in a 5-in Type K copper tube carrying linseed oil at 77°F. A mercury manometer is to be used to measure the pressure difference across the nozzle when the expected range of the flow rate is from 700 gal/min to 1000 gal/min. The manometer scale ranges from 0 to 8.0 in of mercury. Determine an appropriate diameter of the nozzle.

15.10 An orifice meter is to be installed in a 12-in ductile iron pipe carrying water at 60°F. A mercury manometer is to be used to measure the pressure difference across the orifice when the expected range of the flow rate is from 1500 gal/min to 4000 gal/min. The manometer scale ranges from 0 to 12.0 in of mercury. Determine the appropriate diameter of the orifice.

15.11 A pitot-static tube is inserted into a pipe carrying methyl alcohol at 25°C. A differential manometer using mercury as the gage fluid is connected to the tube and shows a deflection of 225 mm. Calculate the velocity of flow of the alcohol.

15.12 A pitot-static tube is connected to a differential manometer using water at 40°C as the gage fluid. The velocity of air at 40°C and atmospheric pressure is to be measured, and it is expected that the maximum velocity will be 25 m/s. Calculate the expected manometer deflection.

15.13 A pitot-static tube is inserted in a pipe carrying water at 10°C. A differential manometer using mercury as the gage fluid shows a deflection of 106 mm. Calculate the velocity of flow.

15.14 A pitot-static tube is inserted into a duct carrying air at standard atmospheric pressure and a temperature of 50°C. A differential manometer reads 4.8 mm of water. Calculate the velocity of flow.

15.15 A pitot-static tube is inserted into a duct carrying air at standard atmospheric pressure and a temperature of 80°F. A differential manometer reads 0.24 in of water. Calculate the velocity of flow.

COMPUTER AIDED ENGINEERING ASSIGNMENTS

1. Write a program using Eq. (15–5) for computing the volume flow rate for any variable-head meter. Include the computation of the area in the main pipe, the area in the throat, the diameter ratio β, and the Reynolds number. Ask the user to input a value for C. Use Eq. (15–7) to compute the discharge coefficient for a nozzle. For the orifice, prompt the user to find the value of C from Fig. 15.7 when given the Reynolds number and the diameter ratio. Permit the user to input the differential pressure in SI units (pascals), in U.S. Customary System units (psi), or in terms of the deflection of a differential manometer with a known gage fluid.

2. Write a program for accepting data for the ten measurements required to complete a traverse of a circular pipe using a pitot tube as shown in Fig. 15.19. Compute the velocity of flow for each point by using Eq. (15–12). Then compute the average of ten values to determine the average velocity. Finally, compute the volume flow rate from $Q = Av$.

FORCES DUE TO FLUIDS IN MOTION

Structures that will be exposed to flow need to be designed to withstand the force that results from the flow. Many consider the vertical load of traffic on a bridge, for example, but many historical bridge failures have occurred due to the side loads induced by fluid flow, whether from river water or the wind as shown in Figs. 16.1 and 16.2.

Whenever a fluid stream is deflected from its initial direction or if its velocity is changed, a force is required to accomplish the change. You must be able to determine the magnitude and direction of such forces in order to design the structure to contain the fluid flow

safely. Sometimes the force of the fluid causes a desired motion, such as when a jet of water strikes the blades of a turbine and the rotation of the turbine generates useful power.

Exploration

- How have you experienced forces due to fluids in motion?
- Consider situations in your home, your car, in a factory, or in some public utilities like power plants or water treatment facilities.

FIGURE 16.1 The support structures of this flow monitoring station on a river must withstand huge forces as it diverts the stream flow during a high level, fast moving current event. (*Source:* Daniel Loretto/Fotolia)

FIGURE 16.2 A highway billboard stands in the path of an impending storm. Even though air has a much lower density that the water seen in Figure 16.1, it can still cause major damage when moving at hurricane or tornadic velocities. (*Source:* Vitaly Krivosheev/Fotolia)

- Describe the effect of the forces caused by the fluids in motion as they are deflected from their initial direction or as the velocity of flow is changed.

     List as many situations as you can in which you have observed the effects of forces created when a fluid stream has been deflected or when its velocity has been changed. Consider these examples:

- Have you ever stuck your hands or head outside an open window of a car traveling at highway speed? Or, has your dog done that?

- Have you ever been buffeted by the wind as you try to walk in a storm?

- Have you ever used the spray from a garden hose to knock some dirt loose from the sidewalk?

- Have you ever watched firefighters struggle to control the nozzle of a fire hose that shoots a large stream of water at high velocity? They must exert large forces to hold it steady, and if they lose control, the nozzle thrashes wildly and is very dangerous.

- Winds can also be very damaging. Thunderstorms with winds of 60–100 mi/h (96–160 km/h) can damage roofs, topple signs, and blow over trucks and mobile homes. Tornadoes and hurricanes can generate winds up to 300 mi/h (482 km/h) and can cause great devastation. Have you ever experienced such a storm?

- Useful energy can be derived from the forces due to fluids in motion. High-velocity jets of water impacting on the blades or buckets of a turbine wheel cause it to rotate and enable it to drive a generator to produce electric power.

- Hot combustion gases in a gas turbine engine expand through the turbine wheels to develop very high levels of power to propel an airplane, a helicopter, or a ship.

- The flow of compressed air from a nozzle is often used to move products in a production system or to remove metal chips and other debris.

- Very high velocity narrow streams of water are used to cut fibrous material such as carpet and fabric in water-jet cutting systems.

- Piping systems that carry large volumes of fluids under pressure exert high forces as the fluid passes around elbows or is restricted by a contraction in the flow stream. Thus, any part of the system where the flow direction is changed or where the magnitude of the velocity is changed must be anchored securely.

Your list may have included some items that describe drag forces as a car, truck, or airplane moves through the air, or as a boat or a submarine moves through water. Those are important concepts, but they are covered in the next chapter on Drag and Lift. However, sometimes both the impact type forces discussed in this chapter and the drag or lift forces discussed in Chapter 17 act at the same time. Consider the following examples.

- The winds acting on the sail of a sailboat cause large forces to propel the boat through the water. This can be exhilarating. At the same time, the boat hull experiences drag forces that tend to slow it down because of the relative motion between the hull and the water.

- Drag forces on automobiles, trucks, boats, and aircraft retard their motion and additional, expensive power must be generated by their engines to overcome the drag. But also when a truck, a travel trailer, or a motor home experiences strong side forces due to high wind velocities, they can be blown off the road or even overturned.

     In this chapter, you will learn the fundamental principles that govern the generation of forces due to fluids in motion. Several types of practical problems will be demonstrated. Then, in Chapter 17, you will extend this topic to include drag forces on many shapes of objects and lift on aerodynamic devices.

16.1 OBJECTIVES

After completing this chapter, you should be able to:

1. Use Newton's second law of motion, $F = ma$, to develop the *force equation*, which is used to compute the force exerted by a fluid as its direction of motion or its velocity is changed.

2. Relate the force equation to the *impulse–momentum* principle.

3. Use the force equation to compute the force exerted on a stationary object that causes the change in direction of a fluid flow stream.

4. Use the force equation to compute the force exerted on bends in pipelines.

5. Use the force equation to compute the force on moving objects, such as the vanes of a pump impeller.

16.2 FORCE EQUATION

Whenever the magnitude or direction of the velocity of a body is changed, a force is required to accomplish the change. Newton's second law of motion is often used to express this concept in mathematical form; the most common form is

$$F = ma \qquad (16-1)$$

Force equals mass times acceleration. Acceleration is the time rate of change of velocity. However, because velocity is a vector quantity having both magnitude and direction, changing either the magnitude or the direction will result in acceleration. According to Eq. (16–1), an external force is required to cause the change.

Equation (16–1) is convenient for use with solid bodies because the mass remains constant and the acceleration of the entire body can be determined. In fluid flow problems, a continuous flow is caused to undergo the acceleration, and a different form of Newton's equation is desirable. Because acceleration is the time rate of change of velocity, Eq. (16–1) can be written as

$$F = ma = m\frac{\Delta v}{\Delta t} \qquad \textbf{(16–2)}$$

The term $m/\Delta t$ can be interpreted as the mass flow rate, that is, the amount of mass flowing in a given amount of time. In the discussion of fluid flow in Chapter 6, mass flow rate was indicated by the symbol M. In addition, M is related to the volume flow rate Q by the relationship

$$M = \rho Q \qquad \textbf{(16–3)}$$

where ρ is the density of the fluid. Then Eq. (16–2) becomes

▷ **General form of Force Equation**

$$F = (m/\Delta t)\Delta v = M\,\Delta v = \rho Q\,\Delta v \qquad \textbf{(16–4)}$$

This is the general form of the force equation for use in fluid flow problems because it involves the velocity and volume flow rate, items generally known in a fluid flow system.

16.3 IMPULSE–MOMENTUM EQUATION

The force equation, Eq. (16–4), is related to another principle of fluid dynamics, the *impulse–momentum equation.* Impulse is defined as a force acting on a body for a period of time, and it is indicated by

$$\text{Impulse} = F(\Delta t)$$

This form, relying on the total change in time Δt, is suitable for dealing with steady flow conditions. When conditions vary, the instantaneous form of the equation is used:

$$\text{Impulse} = F(dt)$$

where dt is the differential amount of change in time.

Momentum is defined as the product of the mass of a body and its velocity. The *change* in momentum is

$$\text{Change in momentum} = m(\Delta v)$$

In an instantaneous sense,

$$\text{Change in momentum} = m(dv)$$

Now, Eq. (16–2) can be rearranged to the form

$$F(\Delta t) = m(\Delta v)$$

Here we have shown the impulse–momentum equation for steady flow conditions. In an instantaneous sense,

$$F(dt) = m(dv)$$

16.4 PROBLEM-SOLVING METHOD USING THE FORCE EQUATIONS

We emphasize that problems involving forces must account for the directions in which the forces act. In Eq. (16–4), force and velocity are both vector quantities. The equation is valid only when all terms have the same direction. For this reason, different equations are written for each direction of concern in a particular case. In general, if three perpendicular directions are called x, y, and z, a separate equation can be written for each direction:

▷ **Force Equations in x, y, and z Directions**

$$F_x = \rho Q\,\Delta v_x = \rho Q(v_{2_x} - v_{1_x}) \qquad \textbf{(16–5)}$$
$$F_y = \rho Q\,\Delta v_y = \rho Q(v_{2_y} - v_{1_y}) \qquad \textbf{(16–6)}$$
$$F_z = \rho Q\,\Delta v_z = \rho Q(v_{2_z} - v_{1_z}) \qquad \textbf{(16–7)}$$

This is the form of the force equation that will be used in this book, with the directions chosen according to the physical situation. In a particular direction, say x, the term F_x refers to the net external force that acts on the fluid in that direction. Therefore, it is the algebraic sum of *all* external forces, including that exerted by a solid surface and forces due to fluid pressure. The term Δv_x refers to the change in velocity in the x direction. In addition, v_1 is the velocity as the fluid enters the device and v_2 is the velocity as it leaves. Then v_{1_x} is the component of v_1 in the x direction and v_{2_x} is the component of v_2 in the x direction.

The specific approach to problems using the force equations depends somewhat on the nature of the data given. A general procedure follows.

Procedure for Using the Force Equations

1. Identify a portion of the fluid stream to be considered a free body. This will be the part where the fluid is changing direction or where the geometry of the flow stream is changing.

2. Establish reference axes for directions of forces. Usually one axis is chosen to be parallel to one part of the flow stream. In the example problems to follow, the positive x and y directions are chosen to be in the same direction as the reaction forces.

3. Identify and show on the free-body diagram all external forces acting on the fluid. All solid surfaces that affect the direction of the flow stream exert forces. Also, the fluid pressure acting on the cross-sectional area of the stream exerts a force in a direction parallel to the stream at the boundary of the free body.

4. Show the direction of the velocity of flow as it enters the free body and as it leaves the free body.

5. Using the data thus shown for the free body, write the force equations in the pertinent directions. Use Eq. (16–5), (16–6), or (16–7).

6. Substitute data and solve for the desired quantity.

The example problems presented in the following sections illustrate this procedure.

16.5 FORCES ON STATIONARY OBJECTS

When free streams of fluid are deflected by stationary objects, external forces must be exerted to maintain the object in equilibrium. Some examples follow.

Example Problem 16.1   A 1-in-diameter jet of water having a velocity of 20 ft/s is deflected 90° by a curved vane, as shown in Fig. 16.3. The jet flows freely in the atmosphere in a horizontal plane. Calculate the x and y forces exerted on the water by the vane.

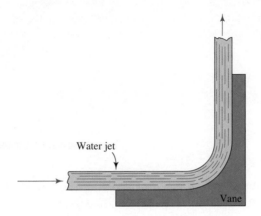

FIGURE 16.3   Water jet deflected by a curved vane.

Water jet

Vane

Solution   Using the force diagram of Fig. 16.4, we can write the force equation for the x direction as

$$F_x = \rho Q(v_{2_x} - v_{1_x})$$
$$R_x = \rho Q[0 - (-v_1)] = \rho Q v_1$$

We know that

$$Q = Av = (0.00545\ \text{ft}^2)(20\ \text{ft/s}) = 0.109\ \text{ft}^3/\text{s}$$

Then, assuming $\rho = 1.94\ \text{slugs/ft}^3 = 1.94\ \text{lb·s}^2/\text{ft}^4$, we write

$$R_x = \rho Q v_1 = \frac{1.94\ \text{lb·s}^2}{\text{ft}^4} \times \frac{0.109\ \text{ft}^3}{\text{s}} \times \frac{20\ \text{ft}}{\text{s}} = 4.23\ \text{lb}$$

For the y direction, assuming $v_2 = v_1$, the force is

$$F_y = \rho Q(v_{2_y} - v_{1_y})$$
$$R_y = \rho Q(v_2 - 0) = (1.94)(0.109)(20)\ \text{lb} = 4.23\ \text{lb}$$

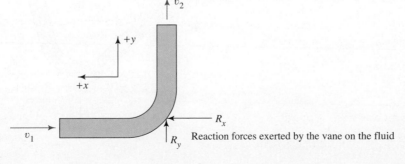

v_2

$+y$

$+x$

R_x

v_1

R_y Reaction forces exerted by the vane on the fluid

FIGURE 16.4   Force diagram for the fluid deflected by the vane.

Example Problem
16.2

In a decorative fountain, 0.05 m³/s of water having a velocity of 8 m/s is being deflected by the angled chute shown in Fig. 16.5. Determine the reactions on the chute in the *x* and *y* directions shown. Also, calculate the total resultant force and the direction in which it acts. Neglect elevation changes.

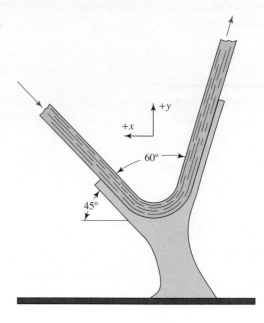

FIGURE 16.5 Decorative fountain deflecting a water jet.

Solution

Figure 16.6 shows the *x* and *y* components of the velocity vectors and the assumed directions for R_x and R_y. The force equation in the *x* direction is

$$F_x = \rho Q(v_{2_x} - v_{1_x})$$

We know that

$$v_{2_x} = -v_2 \sin 15° \quad \text{(toward the right)}$$
$$v_{1_x} = -v_1 \cos 45° \quad \text{(toward the right)}$$

Neglecting friction in the chute, we can assume that $v_2 = v_1$. The only external force is R_x. Then, we have

$$R_x = \rho Q[-v_2\sin 15° - (-v_1\cos 45°)]$$
$$= \rho Qv(-\sin 15° + \cos 45°) = 0.448\,\rho Qv$$

FIGURE 16.6 Force diagram for the fluid deflected by the vane.

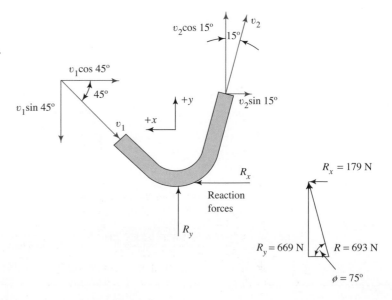

Using $\rho = 1000 \, \text{kg/m}^3$ for water, we get

$$R_x = \frac{(0.448)(1000 \, \text{kg})}{\text{m}^3} \times \frac{0.05 \, \text{m}^3}{\text{s}} \times \frac{8 \, \text{m}}{\text{s}} = \frac{179 \, \text{kg·m}}{\text{s}^2} = 179 \, \text{N}$$

In the y direction, the force equation is

$$F_y = \rho Q(v_{2_y} - v_{1_y})$$

We know that

$$v_{2_y} = v_2 \cos 15° \qquad \text{(upward)}$$
$$v_{1_y} = -v_1 \sin 45° \qquad \text{(downward)}$$

Then, we have

$$R_y = \rho Q[v_2\cos 15° - (-v_1\sin 45°)]$$
$$= \rho Q v(\cos 15° + \sin 45°)$$
$$= (1000)(0.05)(8)(0.966 + 0.707) \, \text{N}$$
$$R_y = 699 \, \text{N}$$

The resultant force R is

$$R = \sqrt{R_x^2 + R_y^2} = \sqrt{179^2 + 669^2} = 693 \, \text{N}$$

For the direction of R, we get

$$\tan \phi = R_y/R_x = 669/179 = (3.74)$$
$$\phi = \text{Tan}^{-1}(3.74) = 75.0°$$

Therefore, the resultant force that the chute must exert on the water is 693 N acting 75° from the horizontal, as shown in Fig. 16.6. The resultant force can be used to ensure the structural safety and rigidity of the fountain.

16.6 FORCES ON BENDS IN PIPELINES

Figure 16.7 shows a typical 90° elbow in a pipe carrying a steady volume flow rate Q. To ensure proper installation, it is important to know how much force is required to hold it in equilibrium. The following problem demonstrates an approach to this type of situation.

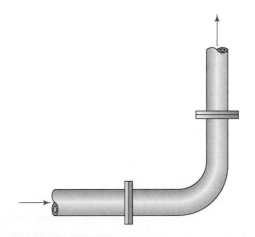

FIGURE 16.7   Pipe elbow.

Example Problem 16.3

Calculate the force that must be exerted on the pipe shown in Fig. 16.7 to hold it in equilibrium. The elbow is in a horizontal plane and is connected to two 4-in Schedule 40 pipes carrying 3000 L/min of water at 15°C. The inlet pressure is 550 kPa.

Solution

The problem may be visualized by considering the fluid within the elbow to be a free body, as shown in Fig. 16.8. Forces are shown in black vectors, and the direction of the velocity of flow is shown by red vectors. A convention must be set for the directions of all vectors. Here we assume that the positive x direction is to the left and the positive y direction is up. The forces R_x and R_y are the external reactions

FIGURE 16.8 Force diagram on the fluid in the elbow.

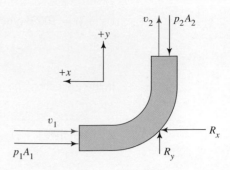

required to maintain equilibrium. The forces $p_1 A_1$ and $p_2 A_2$ are the forces due to the fluid pressure. The two directions will be analyzed separately.

We find the net external force in the x direction by using the equation

$$F_x = \rho Q(v_{2_x} - v_{1_x})$$

We know that

$$F_x = R_x - p_1 A_1$$
$$v_{2_x} = 0$$
$$v_{1_x} = -v_1$$

Then, we have

$$R_x - p_1 A_1 = \rho Q[0 - (-v_1)]$$
$$R_x = \rho Q v_1 + p_1 A_1 \qquad\qquad \textbf{(16–8)}$$

From the given data, $p_1 = 550\,\text{kPa}$, $\rho = 1000\,\text{kg/m}^3$, and $A_1 = 8.213 \times 10^{-3}\,\text{m}^2$. Then,

$$Q = 3000\,\text{L/min} \times \frac{1\,\text{m}^3/\text{s}}{60\,000\,\text{L/min}} = 0.05\,\text{m}^3/\text{s}$$

$$v_1 = \frac{Q}{A_1} = \frac{0.05\,\text{m}^3/\text{s}}{8.213 \times 10^{-3}\,\text{m}^2} = 6.09\,\text{m/s}$$

$$\rho Q v_1 = \frac{1000\,\text{kg}}{\text{m}^3} \times \frac{0.05\,\text{m}^3}{\text{s}} \times \frac{6.09\,\text{m}}{\text{s}} = 305\,\text{kg·m/s}^2 = 305\,\text{N}$$

$$p_1 A_1 = \frac{550 \times 10^3\,\text{N}}{\text{m}^2} \times (8.213 \times 10^{-3}\,\text{m}^2) = 4517\,\text{N}$$

Substituting these values into Eq. (16–8) gives

$$R_x = (305 + 4517)\,\text{N} = 4822\,\text{N}$$

In the y direction, the equation for the net external force is

$$F_y = \rho Q(v_{2_y} - v_{1_y})$$

We know that

$$F_y = R_y - p_2 A_2$$
$$v_{2_y} = +v_2$$
$$v_{1_y} = 0$$

Then, we have

$$R_y - p_2 A_2 = \rho Q v_2$$
$$R_y = \rho Q v_2 + p_2 A_2$$

If energy losses in the elbow are neglected, $v_2 = v_1$, and $p_2 = p_1$ because the sizes of the inlet and outlet are equal. Then,

$$\rho Q v_2 = 305 \text{ N}$$
$$p_2 A_2 = 4517 \text{ N}$$
$$R_y = (305 + 4517) \text{ N} = 4822 \text{ N}$$

The forces R_x and R_y are the reactions caused at the elbow as the fluid turns 90°. They may be supplied by anchors on the elbow or taken up through the flanges into the main pipes.

Example Problem 16.4 Linseed oil with a specific gravity of 0.93 enters the reducing bend shown in Fig. 16.9 with a velocity of 3 m/s and a pressure of 275 kPa. The bend is in a horizontal plane. Calculate the x and y forces required to hold the bend in place. Neglect energy losses in the bend.

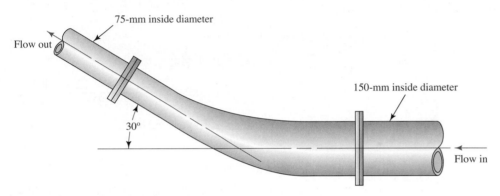

FIGURE 16.9 Reducing bend.

Solution The fluid in the bend is shown as a free body in Fig. 16.10. We must first develop the force equations for the x and y directions shown.

The force equation for the x direction is

$$F_x = \rho Q(v_{2_x} - v_{1_x})$$
$$R_x - p_1 A_1 + p_2 A_2 \cos 30° = \rho Q[-v_2 \cos 30° - (-v_1)] \qquad \textbf{(16–9)}$$
$$R_x = p_1 A_1 - p_2 A_2 \cos 30° - \rho Q v_2 \cos 30° + \rho Q v_1$$

Algebraic signs must be carefully included according to the sign convention shown in Fig. 16.10. Notice that all forces and velocity terms are the components *in the x direction.*

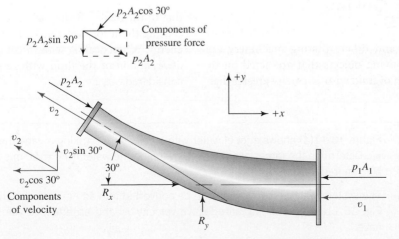

FIGURE 16.10 Force diagram for fluid in the reducing bend.

In the y direction, the force equation is

$$F_y = \rho Q(v_{2_y} - v_{1_y})$$

$$R_y - p_2 A_2 \sin 30° = \rho Q(v_2 \sin 30° - 0) \qquad \textbf{(16–10)}$$

$$R_y = p_2 A_2 \sin 30° + \rho Q v_2 \sin 30°$$

The numerical values of several items must now be calculated. For the entering and leaving pipes, $A_1 = 1.767 \times 10^{-2}\,\text{m}^2$ and $A_2 = 4.418 \times 10^{-3}\,\text{m}^2$. We have

$$\rho = (\text{sg})(\rho_w) = (0.93)(1000\,\text{kg/m}^3) = 930\,\text{kg/m}^3$$

$$\gamma = (\text{sg})(\gamma_w) = (0.93)(9.81\,\text{kN/m}^3) = 9.12\,\text{kN/m}^3$$

$$Q = A_1 v_1 = (1.767 \times 10^{-2}\,\text{m}^2)(3\,\text{m/s}) = 0.053\,\text{m}^3/\text{s}$$

Because of continuity, $A_1 v_1 = A_2 v_2$. Then, we have

$$v_2 = v_1(A_1/A_2) = (3\,\text{m/s})(1.767 \times 10^{-2}/4.418 \times 10^{-3}) = 12\,\text{m/s}$$

Bernoulli's equation can be used to find p_2:

$$\frac{p_1}{\gamma} + z_1 + \frac{v_1^2}{2g} = \frac{p_2}{\gamma} + z_2 + \frac{v_2^2}{2g}$$

But $z_1 = z_2$. Then, we have

$$p_2 = p_1 + \gamma(v_1^2 - v_2^2)/2g$$

$$= 275\,\text{kPa} + \left[\frac{(9.12)(3^2 - 12^2)}{(2)(9.81)} \times \frac{\text{kN}}{\text{m}^3} \times \frac{\text{m}^2}{\text{s}^2} \times \frac{\text{s}^2}{\text{m}}\right]$$

$$= 275\,\text{kPa} - 62.8\,\text{kPa}$$

$$p_2 = 212.2\,\text{kPa}$$

The quantities needed for Eqs. (16–9) and (16–10) are

$$p_1 A_1 = (275\,\text{kN/m}^2)(1.767 \times 10^{-2}\,\text{m}^2) = 4859\,\text{N}$$

$$p_2 A_2 = (212.2\,\text{kN/m}^2)(4.418 \times 10^{-3}\,\text{m}^2) = 938\,\text{N}$$

$$\rho Q v_1 = (930\,\text{kg/m}^3)(0.053\,\text{m}^3/\text{s})(3\,\text{m/s}) = 148\,\text{N}$$

$$\rho Q v_2 = (930\,\text{kg/m}^3)(0.053\,\text{m}^3/\text{s})(12\,\text{m/s}) = 591\,\text{N}$$

From Eq. (16–9), we get

$$R_x = (4859 - 938\cos 30° - 591\cos 30° + 148)\,\text{N} = 3683\,\text{N}$$

From Eq. (16–10), we get

$$R_y = (938\sin 30° + 591\sin 30°)\,\text{N} = 765\,\text{N}$$

16.7 FORCES ON MOVING OBJECTS

The vanes of turbines and other rotating machinery are familiar examples of moving objects that are acted on by high-velocity fluids. A jet of fluid with a velocity greater than that of the blades of the turbine exerts a force on the blades, causing them to accelerate or to generate useful mechanical energy. When dealing with forces on moving bodies, the *relative motion* of the fluid with respect to the body must be considered.

Example Problem 16.5   Figure 16.11(a) shows a jet of water with a velocity v_1 striking a vane that is moving with a velocity v_0. Determine the forces exerted by the vane on the water if $v_1 = 20\,\text{m/s}$ and $v_0 = 8\,\text{m/s}$. The jet is 50 mm in diameter.

Solution   The system with a moving vane can be converted into an equivalent stationary system as shown in Fig. 16.11(b) by defining an effective velocity v_e and an effective volume flow rate Q_e. We then have

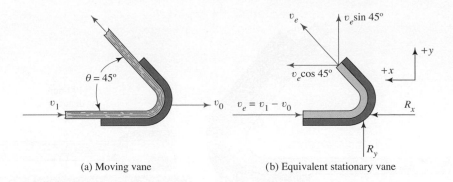

(a) Moving vane   (b) Equivalent stationary vane

FIGURE 16.11 Flow deflected by a moving vane.

◇ **Effective Velocity and Volume Flow Rate**

$$v_e = v_1 - v_0 \tag{16–11}$$

$$Q_e = A_1 v_e \tag{16–12}$$

where A_1 is the area of the jet as it enters the vane. It is only the difference between the jet velocity and the vane velocity that is effective in creating a force on the vane. The force equations can be written in terms of v_e and Q_e. In the *x* direction,

$$R_x = \rho Q_e v_e \cos\theta - (-\rho Q_e v_e)$$
$$= \rho Q_e v_e (1 + \cos\theta) \tag{16–13}$$

In the *y* direction,

$$R_y = \rho Q_e v_e \sin\theta - 0 \tag{16–14}$$

We know that

$$v_e = v_1 - v_0 = (20 - 8)\,\text{m/s} = 12\,\text{m/s}$$
$$Q_e = A_1 v_e = (1.964 \times 10^{-3}\,\text{m}^2)(12\,\text{m/s}) = 0.0236\,\text{m}^3/\text{s}$$

Then the reactions are calculated from Eqs. (16–13) and (16–14):

$$R_x = (1000)(0.0236)(12)(1 + \cos 45°) = 483\,\text{N}$$
$$R_y = (1000)(0.0236)(12)(\sin 45°) = 200\,\text{N}$$

PRACTICE PROBLEMS

16.1 Calculate the force required to hold a flat plate in equilibrium perpendicular to the flow of water at 25 m/s issuing from a 75-mm-diameter nozzle.

16.2 What must be the velocity of flow of water from a 2-in-diameter nozzle to exert a force of 300 lb on a flat wall?

16.3 Calculate the force exerted on a stationary curved vane that deflects a 1-in-diameter stream of water through a 90° angle. The volume flow rate is 150 gal/min.

16.4 A highway sign is being designed to withstand winds of 125 km/h. Calculate the total force on a sign 4 m by 3 m if the wind is flowing perpendicular to the face of the sign. Calculate the equivalent pressure on the sign in Pa. The air is at −10°C. (See Chapter 17 and Problem 17.9 for a more complete analysis of this problem.)

16.5 Compute the forces in the vertical and horizontal directions on the block shown in Fig. 16.12. The fluid stream is a 1.75-in-diameter jet of water at 60°F with a velocity of 25 ft/s. The velocity leaving the block is also 25 ft/s.

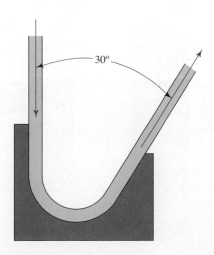

FIGURE 16.12 Problem 16.5.

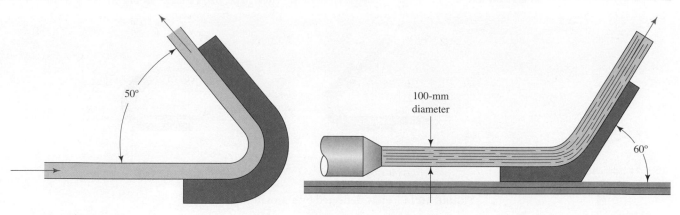

FIGURE 16.13 Problem 16.6.

FIGURE 16.14 Problem 16.7.

16.6 Figure 16.13 shows a free stream of water at 180°F being deflected by a stationary vane through a 130° angle. The entering stream has a velocity of 22.0 ft/s. The cross-sectional area of the stream is constant at 2.95 in² throughout the system. Compute the forces in the horizontal and vertical directions exerted on the water by the vane.

16.7 Compute the horizontal and vertical forces exerted on the vane shown in Fig. 16.14 due to a flow of water at 50°C. The velocity is constant at 15 m/s.

16.8 In a plant where hemispherical cup-shaped parts are made, an automatic washer is being designed to clean the parts prior to shipment. One scheme being evaluated uses a stream of water at 180°F shooting vertically upward into the cup. The stream has a velocity of 30 ft/s and a diameter of 1.00 in. As shown in Fig. 16.15, the water leaves the cup vertically downward in the form of an annular ring having an outside diameter of 4.00 in and an inside diameter of 3.80 in. Compute the external force required to hold the cup down.

16.9 A stream of non-flammable oil (sg = 0.90) is directed onto the center of the underside of a flat metal plate to

keep it cool during a welding operation. The plate weighs 550 N. If the stream is 35 mm in diameter, calculate the velocity of the stream that will lift the plate. The stream strikes the plate perpendicularly.

16.10 A 2-in-diameter stream of water having a velocity of 40 ft/s strikes the edge of a flat plate such that half the stream is deflected downward as shown in Fig. 16.16. Calculate the force on the plate and the moment due to the force at point A.

16.11 Figure 16.17 represents a type of flowmeter in which the flat vane is rotated on a pivot as it deflects the fluid stream. The fluid force is counterbalanced by a spring. Calculate the spring force required to hold the vane in a vertical position when water at 100 gal/min flows from the 1-in Schedule 40 pipe to which the meter is attached.

16.12 Water is piped vertically from below a boat and discharged horizontally in a 4-in diameter jet with a velocity of 60 ft/s. Calculate the force on the boat.

16.13 A 2-in nozzle is attached to a hose with an inside diameter of 4 in. The resistance coefficient K of the nozzle is 0.12 based on the outlet velocity head. If the jet issuing from the nozzle has a velocity of 80 ft/s, calculate the force exerted by the water on the nozzle.

16.14 Seawater (sg = 1.03) enters a heat exchanger through a reducing bend connecting a 4-in Type K copper tube with a 2-in Type K tube. The pressure upstream from the bend is 825 kPa. Calculate the force required to hold the bend in equilibrium. Consider the energy loss in the bend, assuming it has a resistance coefficient K of 3.5 based on the inlet velocity. The flow rate is 0.025 m³/s.

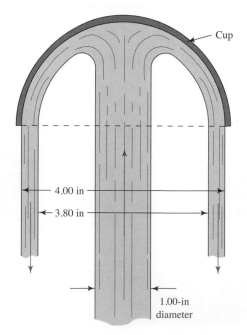

FIGURE 16.15 Problem 16.8.

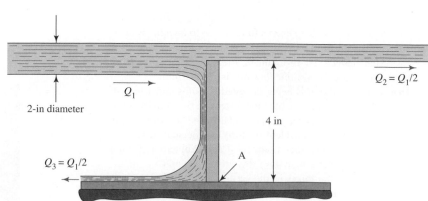

FIGURE 16.16 Problem 16.10.

FIGURE 16.17 Problem 16.11.

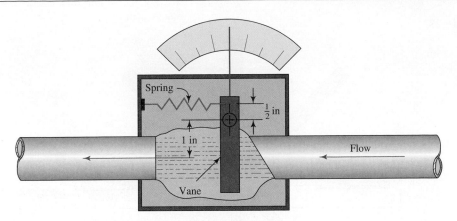

16.15 A reducer connects a standard 6-in Schedule 40 pipe to a 3-in Schedule 40 pipe. The walls of the conical reducer are tapered at an included angle of 40°. The flow rate of water is 500 gal/min and the pressure ahead of the reducer is 125 psig. Considering the energy loss in the reducer, calculate the force exerted on the reducer by the water.

16.16 Calculate the force on a 45° elbow attached to an 8-in Schedule 80 steel pipe carrying water at 80°F at 6.5 ft³/s. The outlet of the elbow discharges into the atmosphere. Consider the energy loss in the elbow.

16.17 Calculate the force required to hold a 90° elbow in place when attached to DN 150 Schedule 40 pipes carrying water at 125 m³/s and 1050 kPa. Neglect energy lost in the elbow.

16.18 Calculate the force required to hold a 180° close return bend in equilibrium. The bend is in a horizontal plane and is attached to a DN 100 Schedule 80 steel pipe carrying 2000 L/min of a hydraulic fluid at 2.0 MPa. The fluid has a specific gravity of 0.89. Neglect energy losses.

16.19 A bend in a tube causes the flow to turn through an angle of 135°. The pressure ahead of the bend is 275 kPa. If the standard hydraulic copper tube, 100 mm OD × 3.5 mm wall, carries 0.12 m³/s of carbon tetrachloride at 25°C, determine the force on the bend. Neglect energy losses.

16.20 A vehicle is to be propelled by a jet of water impinging on a vane as shown in Fig. 16.18. The jet has a velocity of 30 m/s and issues from a nozzle with a diameter of 200 mm. Calculate the force on the vehicle (a) if it is stationary and (b) if it is moving at 12 m/s.

16.21 A part of an inspection system in a packaging operation uses a jet of air to remove imperfect cartons from a conveyor line, as shown in Fig. 16.19. The jet is initiated by a sensor and timed so that the product to be rejected is in front of the jet at the right moment. The product is to be tipped over a ledge on the side of the conveyor as shown in the figure. Compute the required velocity of the air jet to tip the carton off the conveyor. The density of the air is

FIGURE 16.18 Problem 16.20.

FIGURE 16.19 Problem 16.21.

FIGURE 16.20 Problems 16.22 and 16.23.

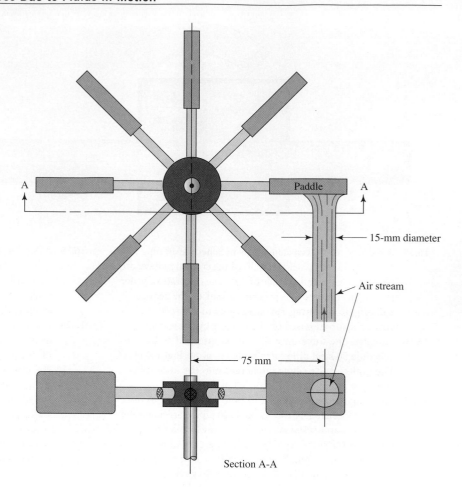

Paddle

15-mm diameter

Air stream

75 mm

Section A-A

1.20 kg/m^3. The carton has a mass of 0.10 kg. The jet has a diameter of 10.0 mm.

16.22 Shown in Fig. 16.20 is a small decorative wheel fitted with flat paddles so the wheel turns about its axis when acted on by a blown stream of air. Assuming that all the air in a 15-mm-diameter stream moving at 0.35 m/s strikes one paddle and is deflected by it at right angles, compute the force exerted on the wheel initially when it is stationary. The air has a density of 1.20 kg/m^3.

16.23 For the wheel described in Problem 16.22, compute the force exerted on the paddle when the wheel rotates at 40 rpm.

16.24 A set of louvers deflects a stream of warm air onto painted parts, as illustrated in Fig. 16.21. The louvers are rotated slowly to distribute the air evenly over the parts. Compute the torque required to rotate the louvers toward the stream of air when it is flowing at a velocity of 10 ft/s. Assume that all the air that approaches a given louver is deflected to the angle of the louver. The air has a density of 2.06×10^{-3} slugs/ft^3. Use $\theta = 45°$.

16.25 For the louvers shown in Fig. 16.21 and described in Problem 16.24, compute the torque required to rotate the louvers when the angle $\theta = 20°$.

16.26 For the louvers shown in Fig. 16.21 and described in Problem 16.24, compute the torque required to rotate the louvers for several settings of the angle θ from 10° to 90°. Plot a graph of torque versus angle.

16.27 Figure 16.22 shows a device for clearing debris using a 1½-in-diameter jet of air issuing from a blower nozzle. As shown, the jet is striking a rectangular box-shaped object

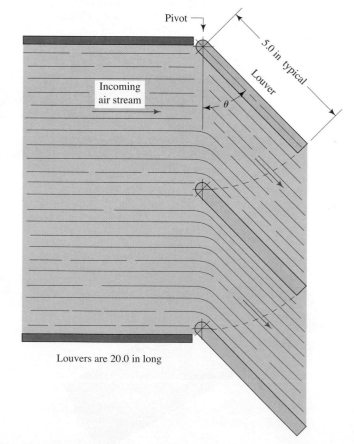

Pivot

5.0 in typical

Louver

Incoming air stream

θ

Louvers are 20.0 in long

FIGURE 16.21 Problems 16.24–16.26.

FIGURE 16.22 Problems 16.27 and 16.28.

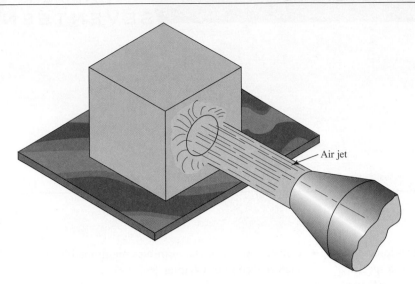

sitting on a floor. If the air velocity is 25 ft/s and the entire jet is deflected by the box, what is the heaviest object that could be moved? Assume that the box slides rather than tumbling over and that the coefficient of friction is 0.60. The air has a density of 2.40×10^{-3} slugs/ft^3.

16.28 Repeat Problem 16.27, except change the jet to water at 50°F and the diameter to 0.75 in.

16.29 Figure 16.23 is a sketch of a turbine in which the incoming stream of water at 15°C has a diameter of 7.50 mm and is moving with a velocity of 25 m/s. Compute the force on one blade of the turbine if the stream is deflected through the angle shown and the blade is stationary.

16.30 Repeat Problem 16.29 with the blade rotating as a part of the wheel at a radius of 200 mm and with a linear tangential velocity of 10 m/s. Also, compute the rotational speed of the wheel in rpm.

16.31 Repeat Problem 16.29, except with the blade rotating as a part of the wheel at a radius of 200 mm and with a linear tangential velocity ranging from 0 to 25 m/s in 5-m/s steps.

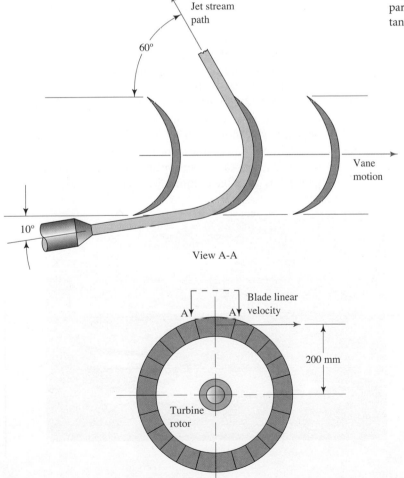

FIGURE 16.23 Problems 16.29–16.31.

DRAG AND LIFT

A moving body immersed in a fluid experiences forces that oppose motion. The force is called *drag*. When a specially shaped body called an *airfoil* moves through air, the flow of air around it causes a net upward force called *lift*. This is the fundamental reason that aircraft can fly. Similarly, specially shaped bodies moving through water are called *hydrofoils*. Both lift and drag occur simultaneously on airfoils and hydrofoils. Consider the ship shown in Fig. 17.1.

While there is also some drag on the hydrofoils, it is much less than it would be if the entire hull were in the water, allowing the ferry to reach high speed and to operate efficiently. You will develop your ability to analyze drag and lift forces as you study this chapter.

Exploration

- Have you experienced the sensation of drag or lift yourself on a bicycle or motorcycle or in a convertible with the top down? Have you participated in bicycle road races? Have you done skydiving or snowboarding? Describe the effects of drag on your body during those events.

- Seek examples of products and equipment where the drag or lift forces have an effect on their behavior or performance.

- Consider some of the examples mentioned in the Big Picture section of Chapter 16.

- As you find examples, document the shape and size with as much detail as you can and describe how they performed.

Introductory Concepts

A moving body immersed in a fluid experiences forces caused by the action of the fluid. The total effect of these forces is quite complex. However, for the purposes of design or for the analysis of the behavior of a body in a fluid, two resultant forces—drag and lift—are the most important. Lift and drag forces are the same regardless of whether the body is moving in the fluid or the fluid is moving over the body.

Drag is the force on a body caused by the fluid that resists motion in the direction of travel of the body. The most familiar applications requiring the study of drag are in the transportation fields. *Wind resistance* is the term often used to describe the effects of drag on aircraft, automobiles, trucks, and trains. The drag force must be opposed by a propulsive force in the opposite direction to maintain or increase the velocity of the vehicle. Because the production of the propulsive force requires added power, it is desirable to minimize drag.

FIGURE 17.1 The hull of this high-speed passenger ferry rises above the water riding on hydrofoils that create lift due to the relative motion between the ship and the water. (*Source:* Tentacle/Fotolia)

Lift is a force caused by the fluid in a direction perpendicular to the direction of travel of the body. Its most important application is the design and analysis of aircraft wings called *airfoils*. The geometry of an airfoil is such that a lift force is produced as air passes over and under it. Of course, the magnitude of the lift must at least equal the weight of the aircraft in order for it to fly. This phenomenon is sometimes called *positive lift*. But *negative lift* is also useful in the case of high speed race cars. The wings are essentially installed upside down from the orientation that would be used on an airplane in order to create a downward force on the race car to provide more traction and improve performance in turns.

The study of the performance of bodies in moving air streams is called *aerodynamics*. Gases other than air could be considered in this field, but due to the obvious importance of the applications in aircraft design, the majority of work has been done with air as the fluid.

Hydrodynamics is the name given to the study of moving bodies immersed in liquids, particularly water. Many concepts concerning lift and drag are similar regardless of whether the fluid is a liquid or a gas. This is not true, however, at high velocities, where the effects of the compressibility of the fluid must be taken into account. Liquids can be considered incompressible in the study of lift and drag. Conversely, a gas such as air is readily compressible.

What kinds of examples have you found of products or equipment where drag or lift forces have an effect on its behavior or performance? Consider the following questions and observations as you describe your examples:

- Are the edges sharp or smooth and well rounded?

- Is the shape flat or does it have a curved surface?

- If the object is cup-shaped, is the open side of the cup facing into the wind (or other fluid) or away from it?

- What attempts have been made to streamline the shape?

- Find two automobiles that have radically different shapes, one that is highly streamlined and another that is more boxy. Perhaps the boxy shape will be an older car, even an antique. How do you think the shape affects drag?

- Describe the shape of high-speed trains such as the *Acela* in the United States, the *TGV* in France, the *ICE* in Germany, or the *shinkansen* (bullet trains) in Japan. What approaches have been used to decrease drag? How do their shapes compare with conventional freight locomotives? Find data for the drag

characteristics of the fast trains on the Internet or from some other information source.

- Compare race cars from different periods of time. Consider Indy-type racers, sports cars, and NASCAR and stock car racers. How are they similar in their approach to reducing drag? How do they differ? In addition to the shape of the cars, notice how race car drivers often follow closely behind the car ahead of them when not in a passing situation. This practice is called *drafting* and it reduces the drag force on the following car dramatically so it uses less fuel during those periods.

- Compare aircraft from different periods of time. What attempts were made in the early days of flight from the Wright brothers through the decade of the 1930s to reduce drag? How did military aircraft change from the advent of World War II through the Korean War and on through more recent conflicts? How do jet aircraft compare with propeller-driven planes with regard to their aerodynamics?

- Watch a bicycle race or participate in one yourself. How do riders work toward reducing drag forces on their bodies and on the bicycle? How do the riders manage drag differently during times when there are several cyclists traveling together? Do they practice drafting as described before for race cars?

Much of the practical data concerning lift and drag have been generated experimentally. We will report some of these data here using relatively simple shapes, such as spheres, cylinders, flat sided objects, cones, and simple streamlined shapes to illustrate the concepts and to help you learn the relationships between the shape of the body and its drag or lift characteristics. One notable example is the familiar golf ball that is featured in Internet resources 1–3. References 3, 7, 10, and 15 provide extensive data on drag exerted on objects having basic shapes. Those fundamental concepts can be applied in principle to the more complex shapes such as race cars, trucks, buses, aircraft, or ships. The complete analyses of such complex systems require much more detail than can be included in this book and entire large texts and reference books are devoted to them. Some examples are:

- For aircraft, consider References 1, 2, 5, and 8 along with Internet resources 7–9.

- For road vehicles, check out References 4, 6, 9, 11, 13, 14, and 16–18 as well as Internet resources 4–6.

- Reference 12 covers the relationships between drag resistance on the hull of ships and the propulsive power required to drive it through the water at design speed.

17.1 OBJECTIVES

After completing this chapter, you should be able to:

1. Define *drag*.
2. Define *lift*.
3. Write the expression for computing the drag force on a body moving relative to a fluid.
4. Define the *drag coefficient*.
5. Define the term *dynamic pressure*.
6. Describe the stagnation point for a body moving relative to a fluid.
7. Distinguish between *pressure drag* and *friction drag*.
8. Discuss the importance of flow separation on pressure drag.
9. Determine the value of the pressure drag coefficient for cylinders, spheres, and other shapes.
10. Discuss the effect of Reynolds number and surface geometry on the drag coefficient.
11. Compute the magnitude of the pressure drag force on bodies moving relative to a fluid.
12. Compute the magnitude of the friction drag force on smooth spheres.
13. Discuss the importance of drag on the performance of ground vehicles.
14. Discuss the effects of compressibility and cavitation on drag and the performance of bodies immersed in fluids.
15. Define the lift coefficient for a body immersed in a fluid.
16. Compute the lift force on a body moving relative to a fluid.
17. Describe the effects of friction drag, pressure drag, and induced drag on airfoils.

17.2 DRAG FORCE EQUATION

Drag forces are usually expressed in the form

Drag Force

$$F_D = \text{drag} = C_D(\rho v^2/2)A \tag{17–1}$$

The terms in this equation are as follows:

- C_D is the *drag coefficient*. It is a dimensionless number that depends on the shape of the body and its orientation relative to the fluid stream. It is usually determined experimentally, but computational fluid dynamic analysis can aid in reducing the value of C_D.
- ρ is the density of the fluid. Because the density of liquids is much greater than that of a gas, the general order of magnitude of the drag forces on objects moving through water is far greater than when objects move through air. The compressibility of the air affects its density somewhat.
- v is the velocity of the free stream of the fluid relative to the body. In general it does not matter whether the body is moving or the fluid is moving. However, the location of other surfaces near the body of interest can affect the drag. For example, when a truck or a car travels on a highway, the interaction of the underside of the vehicle with the roadway affects the drag.
- A is some characteristic area of the body. Be careful to note in later sections just what area is to be used in a given situation. Most often the area of interest is the maximum cross-sectional area of the body, sometimes called the *projected area*. Think of what the largest two-dimensional shape would be if you looked straight into the front of your car. That is the area you would use to compute the drag on a car, called the *form drag* or the *pressure drag*. For very long, smooth shapes such as a passenger train car or a blimp, however, the surface area may be used. Here we are concerned with the *friction drag* as the air flows along the surface of the vehicle.
- The combined term $\rho v^2/2$ is called the *dynamic pressure*, which we define next. Notice that the drag force is proportional to the dynamic pressure and, therefore, it is proportional to the velocity *squared*. This means, for example, that doubling the velocity for a given object will increase the drag force by a factor of four.

You can visualize the influence of the dynamic pressure on drag by referring to Fig. 17.2, which shows a sphere in a fluid stream. The streamlines depict the path of the fluid as it approaches and flows around the sphere. At point s on the surface of the sphere, the fluid stream is at rest or "stagnant." The term *stagnation point* is used to describe this point. The relationship between the pressure p_s and that in the undisturbed stream at point 1 can be found using Bernoulli's equation along a streamline:

$$\frac{p_1}{\gamma} + \frac{v_1^2}{2g} = \frac{p_s}{\gamma} \tag{17–2}$$

Solving for p_s, we get

$$p_s = p_1 + \gamma v_1^2/2g$$

Because $\rho = \gamma/g$, we have

$$p_s = p_1 + \rho v_1^2/2 \tag{17–3}$$

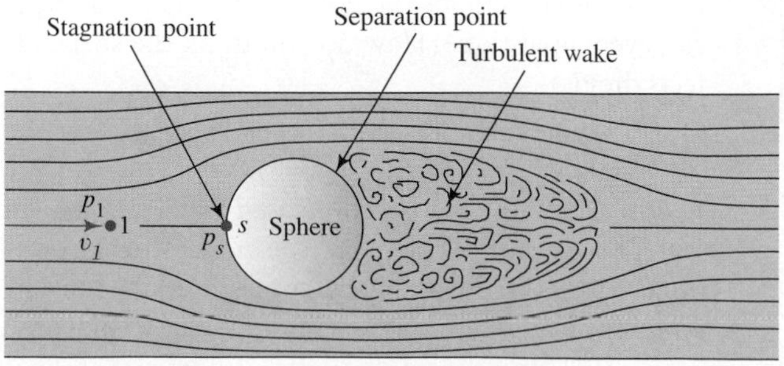

FIGURE 17.2 Sphere in a fluid stream showing the stagnation point on the front surface and the turbulent wake behind.

The stagnation pressure is greater than the static pressure in the free stream by the magnitude of the dynamic pressure $\rho v_1^2/2$. The kinetic energy of the moving stream is transformed into a kind of potential energy in the form of pressure.

The increase in pressure at the stagnation point can be expected to produce a force on the body opposing its motion, that is, a drag force. However, the magnitude of the force is dependent not only on the stagnation pressure, but also on the pressure at the back side of the body. Because it is difficult to predict the actual variation in pressure on the back side, the drag coefficient is typically used.

The total drag on a body is due to two components. (For a lifting body such as an airfoil, a third component exists as described in Section 17.7.) *Pressure drag* (also called *form drag*) is due to the disturbance of the flow stream as it passes the body, creating a turbulent wake. The characteristics of the disturbance are dependent on the form of the body and sometimes on the Reynolds number of flow and the roughness of the surface. *Friction drag* is due to shearing stresses in the thin layer of fluid near the surface of the body called the *boundary layer*. These two types of drag are described in the following sections.

17.3 PRESSURE DRAG

As a fluid stream flows around a body, it tends to adhere to the surface for a portion of the length of the body. Then at a certain point, the thin boundary layer separates from the surface, causing a turbulent wake to be formed (see Fig. 17.2). The pressure in the wake is significantly lower than that at the stagnation point at the front of the body. A net force is thus created that acts in a direction opposite to that of the motion. This force is the pressure drag.

If the point of separation can be caused to occur farther back on the body, the size of the wake can be decreased and the pressure drag will be lower. This is the reasoning for streamlining. Figure 17.3 illustrates the change in the wake caused by the elongation and tapering of the tail of the body. Thus, the amount of pressure drag is dependent on the form of the body, and the term *form drag* is often used.

The pressure drag force is calculated from Eq. (17–1) in which *A is taken to be the maximum cross-sectional area of the*

body perpendicular to the flow. The coefficient C_D is the pressure drag coefficient.

As an illustration of the importance of streamlining, the value of C_D for the drag on a smooth sphere moving through air with a Reynolds number of approximately 10^5 is 0.5. A highly streamlined shape like that used in most airships (blimps) has a C_D of approximately 0.04, a reduction by more than a factor of 10!

17.3.1 Properties of Air

Drag on bodies moving in air is often the goal for drag analysis. To use Eq. (17–1) to calculate the drag forces, we need to know the density of the air. As with all gases, the properties of air change drastically with temperature. In addition, as altitude above sea level increases, the density decreases. Appendix E presents the properties of air at various temperatures and altitudes.

17.4 DRAG COEFFICIENT

The magnitude of the drag coefficient for pressure drag depends on many factors, most notably the shape of the body, the Reynolds number of the flow, the surface roughness, and the influence of other bodies or surfaces in the vicinity. Two of the simpler shapes, the sphere and the cylinder, are discussed first.

17.4.1 Drag Coefficient for Spheres and Cylinders

Data plotted in Fig. 17.4 give the value of the drag coefficient versus Reynolds number for *smooth* spheres and cylinders. For spheres and cylinders, the Reynolds number is computed from the familiar *looking* relation

$$N_R = \frac{\rho v D}{\eta} = \frac{vD}{\nu} \qquad (17\text{–}4)$$

However, the diameter D is the diameter of the body itself, rather than the diameter of a flow conduit, which D represented earlier.

Note the very high values of C_D for low Reynolds numbers, over 100 for a smooth sphere at $N_R = 0.10$. This corresponds to motion through very viscous fluids. It drops rapidly to a value of about 4 for $N_R = 10$ and then to 1.0 for $N_R = 100$. The value of C_D ranges from about 0.38 to 0.46 for the higher Reynolds numbers from 1000 to 10^5.

For cylinders, $C_D \approx 60$ for the very low Reynolds number of 0.10. It drops to a value of 10 for $N_R = 1.0$ and to a value of 1.0 for $N_R = 1000$. In the higher ranges of Reynolds numbers, C_D ranges from about 0.90 to 1.30 for N_R from 1000 to 10^5.

For very small Reynolds numbers ($N_R < 1.0$ approximately), the drag is almost entirely due to friction and will be discussed later. At higher Reynolds numbers, the importance of flow separation and the turbulent wake behind the

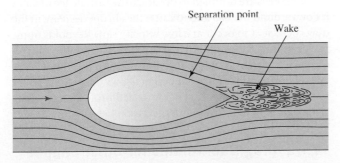

FIGURE 17.3 Effect of streamlining on the wake.

FIGURE 17.4 Drag coefficients for spheres and cylinders.

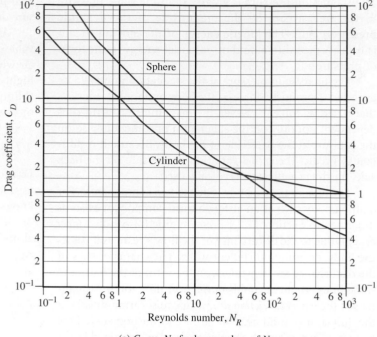

(a) C_D vs. N_R for lower values of N_R

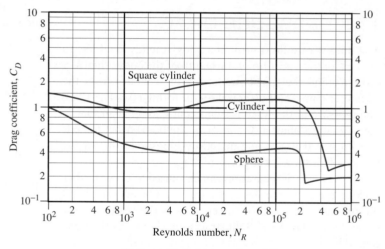

(b) C_D vs. N_R for higher values of N_R

body make pressure drag predominant. The following discussion relates only to pressure drag.

At a value of the Reynolds number of about 2×10^5, the drag coefficient for spheres drops sharply from approximately 0.42 to 0.17. This is caused by the abrupt change in the nature of the boundary layer from laminar to turbulent. Concurrently, the point on the sphere where separation occurs moves farther back, decreasing the size of the wake. For cylinders, a similar phenomenon occurs at approximately $N_R = 4 \times 10^5$, where C_D changes from about 1.2 to 0.30.

Either roughening the surface or increasing the turbulence in the flow stream can decrease the value of the Reynolds number at which the transition from a laminar to a turbulent boundary layer occurs, as illustrated in Fig. 17.5.

This graph is meant to show typical curve shapes only and should not be used for numerical values.

Golf balls are dimpled to optimize the turbulence of air as it flows around the ball and to cause the abrupt decrease in the drag coefficient to occur at a low velocity (low Reynolds number), resulting in longer flights. A perfectly smooth golf ball could be driven only about 100 yards by even the best golfers, whereas the familiar dimpled design allows the average golfer to far exceed this distance. Highly skilled professional golfers can make 300-yard drives (Internet resources 1–3).

17.4.2 Drag Coefficients for Other Shapes

A square cylinder has a uniform square cross section and is relatively long in relation to its height. Shown in Fig. 17.4(b)

FIGURE 17.5 Effect of turbulence and roughness on C_D for spheres.

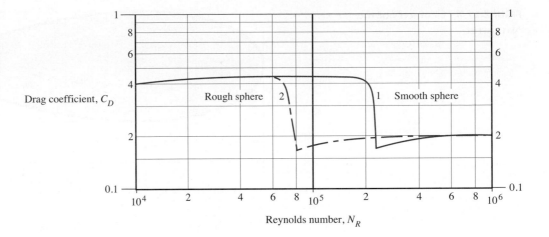

is the drag coefficient for a square cylinder with a flat side perpendicular to the flow for Reynolds numbers from 3.5×10^3 to 8×10^4. The values range from approximately 1.60 to 2.05, somewhat higher than for the circular cylinder. Significant reductions can be obtained by using small to moderate corner radii, bringing values of C_D down to as low as 0.55 at high Reynolds numbers. However, the values tend to be highly affected by changes in Reynolds numbers for such designs. Testing is advised.

Figure 17.6 gives data for C_D for three versions of elliptical cylinders for Reynolds numbers from 3.0×10^4 to 2×10^5. These shapes have an ellipse for a cross section with different ratios of cross-sectional length to maximum thickness, sometimes called *fineness ratio*. Also shown for comparison is the circular cylinder, which can be considered as a special case of the elliptical cylinder with a fineness ratio

of 1:1. Note the dramatic reduction of drag coefficient from about 1.22 for a circular cylinder to about 0.21 for the elliptical cylinders of high fineness ratio of 8:1.

Even more reduction in drag coefficient can be made with the familiar "teardrop" shape, also shown in Fig. 17.6. This is a standard shape called a *Navy strut*, which has values for C_D in the range of 0.07–0.11. Figure 17.7 shows the strut geometry. (See Reference 3.)

Table 17.1 lists values of the drag coefficients for several simple shapes. Note the orientation of the shape relative to the direction of the oncoming flow. The C_D values for such shapes are nearly independent of Reynolds numbers because they have sharp edges that cause the boundary layer to separate at the same place. Most of the testing for these shapes was done in the range of Reynolds numbers from 10^4 to 10^5.

FIGURE 17.6 Drag coefficients for elliptical cylinders and struts.

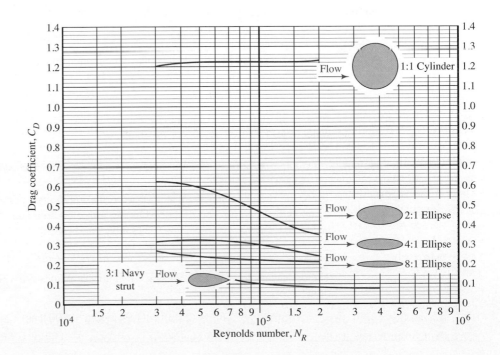

FIGURE 17.7 Geometry of the
Navy strut.

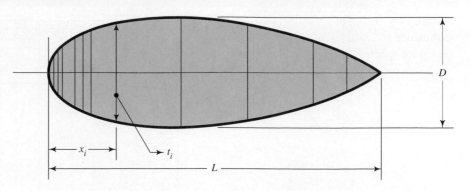

x/L	0.00	.0125	.025	.040	.075	.100	.125	.200
t/D	0.00	.260	.371	.525	.630	.720	.785	.911

x/L	.400	.600	.800	.900	1.00
t/D	.995	.861	.562	.338	0.00

The computation of the Reynolds number for the shapes shown in Table 17.1 uses the *length of the body parallel to the flow* as the characteristic dimension for the body. The formula then becomes

$$N_R = \frac{\rho v L}{\eta} = \frac{vL}{\nu} \qquad (17\text{–}5)$$

For the square cylinder, semitubular cylinders, and triangular cylinders, the data are for models that are long relative to the major thickness dimension. For short cylinders of all shapes, the modified flow around the ends will tend to decrease the values for C_D below those listed in Table 17.1.

**Example Problem
17.1**

Compute the drag force on a 6.00-ft square bar with a cross section of 4.00 in × 4.00 in when the bar is moving at 4.00 ft/s through water at 40°F. The long axis of the bar and a flat face are placed perpendicular to the flow.

Solution

We can use Eq. (17–1) to compute the drag force:

$$F_D = C_D(\rho v^2/2)A$$

Figure 17.4 shows that the drag coefficient depends on the Reynolds number found from Eq. (17–5):

$$N_R = \frac{vL}{\nu}$$

where L is the length of the bar parallel to the flow: 4.0 in or 0.333 ft. The kinematic viscosity of the water at 40°F is 1.67×10^{-5} ft^2/s. Then

$$N_R = \frac{(4.00)(0.333)}{1.67 \times 10^{-5}} = 8.0 \times 10^4$$

Then, the drag coefficient $C_D = 2.05$. The maximum area perpendicular to the flow, A, can now be computed. A can also be described as the projected area seen if you look directly at the bar. In this case, then, the bar is a rectangle 0.333 ft high and 6.00 ft long. That is,

$$A = (0.333 \text{ ft})(6.00 \text{ ft}) = 2.00 \text{ ft}^2$$

The density of the water is 1.94 slugs/ft^3. Equivalent units are 1.94 lb·s^2/ft^4. We can now compute the drag force:

$$F_D = (2.05)(\tfrac{1}{2})(1.94 \text{ lb·s}^2/\text{ft}^4)(4.00 \text{ ft}/\text{s})^2(2.00 \text{ ft}^2) = 63.6 \text{ lb}$$

TABLE 17.1 Typical drag coefficients

Shape of body	Orientation		C_D
Rectangular plate Flow is perpendicular to the flat front face.		*a/b* 1 4 8 12.5 25 50 ∞	1.16 1.17 1.23 1.34 1.57 1.76 2.00
Tandem disks *L* = spacing *d* = diameter		*L/d* 1 1.5 2 3	0.93 0.78 1.04 1.52
One circular disk			1.11
Cylinder *L* = length *d* = diameter		*L/d* 1 2 4 7	0.91 0.85 0.87 0.99
Hemispherical cup, open back			0.41
Hemispherical cup, open front			1.35
Cone, closed base			0.51
			0.34

TABLE 17.1   Typical drag coefficients (*continued*)

Shape of body	Orientation	C_D

Square cylinder
 	
    Flow is perpendicular
    to the flat front face.
		1.60

Semitubular cylinders		1.12
		2.30

Triangular cylinders
 	
    Flow is perpendicular
    to the flat front face - or
    in line with the angled front
    on the axis of the shape.
	30°	1.05

	30°	1.85
	60°	1.39
	60°	2.20
	90°	1.60
	90°	2.15
	120°	1.75
	120°	2.05

Note: Reynolds numbers are typically from 10^4 to 10^5 and are based on the length of the body parallel to the flow direction, except for the semitubular cylinders, for which the characteristic length is the diameter.

Source: Data adapted from Avallone, Eugene A., and Theodore Baumeister III, eds. 1987. *Marks' Standard Handbook for Mechanical Engineers*, 9th ed. New York: McGraw-Hill, Table 4; and Lindsey, W. F. 1938. *Drag of Cylinders of Simple Shapes* (Report No. 619). National Advisory Committee for Aeronautics.

17.5 FRICTION DRAG ON SPHERES IN LAMINAR FLOW

A special method of analysis is used for computing friction drag for spheres moving at low velocities in a viscous fluid, which results in very low Reynolds numbers. An important application of this phenomenon is the *falling-ball viscometer*, discussed in Chapter 2. As a sphere falls through a viscous fluid, no separation occurs, and the boundary layer remains attached to the entire surface. Therefore, virtually all the drag is due to friction rather than to pressure drag.

In Reference 15, George G. Stokes presents important research on spheres moving through viscous fluids. He found that for Reynolds numbers less than about 1.0, the relationship between the drag coefficient and Reynolds number is $C_D = 24/N_R$. Special forms of the drag force equation can then be developed. The general form of the drag force equation is

$$F_D = C_D \left(\frac{\rho v^2}{2} \right) A$$

Letting $C_D = 24/N_R$ and letting $N_R = vD\rho/\eta$, we get

$$C_D = \frac{24}{N_R} = \frac{24\eta}{vD\rho}$$

Then, the drag force becomes

$$F_D = \frac{24\eta}{vD\rho} \left(\frac{\rho v^2}{2} \right) A = \frac{12\eta vA}{D} \qquad (17\text{--}6)$$

When computing friction drag, we use the surface area of the object. For a sphere, the surface area is $A = \pi D^2$. Then,

⇨ **Drag on a Sphere Related to Surface Area**

$$F_D = \frac{12\eta vA}{D} = \frac{12\eta v(\pi D^2)}{D} = 12\pi\eta vD \qquad (17\text{--}7)$$

To correlate drag in the low-Reynolds-number range with that already presented in Section 17.4 dealing with pressure drag, we must redefine the area to be the maximum cross-sectional area of the sphere, $A = \pi D^2/4$. Equation (17–6) then becomes

⇨ **Stokes's Law: Drag on a Sphere Related to Cross-Sectional Area**

$$F_D = \frac{12\eta vA}{D} = \left(\frac{12\eta v}{D} \right) \left(\frac{\pi D^2}{4} \right) = 3\pi\eta vD \qquad (17\text{--}8)$$

This form for the drag on a sphere in a viscous fluid is commonly called *Stokes's law*. As shown in Fig. 17.4, the relation $C_D = 24/N_R$ plots as a straight line on the log-log scales for the low Reynolds numbers.

17.6 VEHICLE DRAG

Decreasing drag is a major goal in designing most kinds of vehicles because a significant amount of energy is required to overcome drag as vehicles move through fluids. You are familiar with the streamlined shapes of aircraft bodies and the hulls of ships. Race cars and sports cars have long had the sleek styling characteristic of low aerodynamic drag. More recently, passenger cars and highway trucks have been redesigned to decrease drag. See also References 3, 6, 9, 11, 13, and 16–18 and Internet resources 4–6 for more data on the aerodynamics of automobiles and other ground vehicles.

Many factors affect the overall drag coefficient for vehicles, including the following:

1. The shape of the forward end, or *nose*, of the vehicle

2. The smoothness of the surfaces of the body

3. Such appendages as mirrors, door handles, antennas, and so forth

4. The shape of the tail section of the vehicle

5. The effect of nearby surfaces, such as the ground beneath an automobile

6. Discontinuities, such as wheels and wheel wells

7. The effect of other vehicles nearby

8. The direction of the vehicle with respect to prevailing winds

9. Air intakes to provide engine cooling or ventilation

10. The ultimate purpose of the vehicle (critical for commercial trucks)

11. The accommodation of passengers

12. Visibility afforded to operators and passengers

13. Stability and control of the vehicle

14. Aesthetics (the attractiveness of the design)

17.6.1 Automobiles

The overall drag coefficient, as defined in Eq. (17–1) based on the maximum projected frontal area, varies widely for passenger cars. Reference 13 lists a nominal mean value of 0.45, with a range of 0.30–0.60. Experimental shapes for cars have shown values as low as 0.137. See Internet resource 4. An approximate value of 0.25 is practical for a "low-drag" design.

The basic principles of drag reduction for automobiles include providing rounded, smooth contours for the forward part; elimination or streamlining of appendages; blending of changes in contour (such as at the hood/windshield interface); and rounding of rear corners.

Example Problem   A prototype automobile has an overall drag coefficient of 0.35. Compute the total drag as it moves at
      17.2   25 m/s through still air at 20°C. The maximum projected frontal area is 2.50 m^2.

Solution   We will use the drag force equation:

$$F_D = C_D\left(\frac{\rho v^2}{2}\right)A$$

From Appendix E, $\rho = 1.204$ kg/m^3. Then,

$$F_D = 0.35\left[\frac{(1.204)(25)^2}{2}\right](2.50) = 329 \text{ kg·m/s}^2 = 329 \text{ N}$$

17.6.2 Power Required to Overcome Drag

Power is defined as the rate of doing work. When a force is
continuously exerted on an object while the object is moving
at a constant velocity, power equals force times velocity.
Then, the power required to overcome drag is

$$P_D = F_D v$$

Using the data from Example Problem 17.2, we get

$$P_D = (329 \text{ N})(25 \text{ m/s}) = 8230 \text{ N·m/s} = 8230 \text{ W}$$
$$= 8.23 \text{ kW}$$

In U.S. Customary System units, this converts to 11.0 hp, a
sizable power loss.

17.6.3 Trucks

The shapes commonly used for trucks fall into the category
called *bluff bodies*. Reference 13 indicates the following
approximate contributions of various parts of a truck to its
total drag:

    70 percent due to the design of the front

    20 percent due to the design of the rear

    10 percent due to friction drag on body surfaces

As with automobiles, rounded smooth contours offer
large improvements. For trucks with box-shaped cargo containers, designing corners with a large radius can assist in keeping the boundary layer from separating at the corners,
consequently reducing the size of the turbulent wake behind
the vehicle and reducing drag. In theory, providing a long,
streamlined tail similar to the shape of an aircraft fuselage will
reduce drag. However, such a vehicle would be too long to be
practical or useful. Newer large highway trucks have drag coefficients in the range from 0.55 to 0.75. Internet resource
6 describes a program in Europe for drag reduction on highway
trucks. Internet resources 16 and 17 provide additional insights.

17.6.4 Trains

Early locomotives had drag coefficients in the range of 0.80–
1.05 (Reference 3). High-speed, streamlined trains can have
values of approximately 0.40. For long passenger and freight
trains, skin friction can be significant.

17.6.5 Aircraft

As with automobiles, wide variations in the overall drag
coefficients of aircraft are to be expected with changes in the
size and shape to accommodate different uses. For subsonic
aircraft, the typical rounded, fairly blunt-nosed design with
smooth blends at wings and tail structures and a long-
tapered tail section results in drag coefficients of approximately 0.12–0.22. At supersonic speeds, the nose is usually
sharp to diminish the effect of the shock wave. Operating at
much lower speeds, the airship (dirigible or blimp) has a
drag coefficient in the range of 0.04. See References 1, 2, and
5 and Internet resources 7–9.

17.6.6 Ships

The total resistance to the motion of floating ships through
water is due to skin friction, pressure or form drag, and
wave-making resistance. Wave-making resistance, a large
contributor to the total resistance, makes analyzing drag
on ships quite different from analyzing drag on ground
vehicles or aircraft. Reference 3 defines the *total ship resistance* R_{ts} as the force required to overcome all forms of
drag. To normalize the value for different sizes of ships
within a given class, values are reported as the ratio R_{ts}/Δ,
where Δ is the displacement of the ship. Representative
values of R_{ts}/Δ are given in Table 17.2. The resistance values can be combined with the speed of the ship v to compute the effective power required to propel it through the
water.

$$P_E = R_{ts}v \qquad (17\text{–}9)$$

See Reference 12 for a comprehensive treatment of this topic.

TABLE 17.2 Resistance of ships	
Ship Type	**R_{ts}/Δ**
Ocean freighter	0.001
Passenger liner	0.004
Tugboat	0.006
Fast naval ship	0.01–0.12

Example Problem 17.3

Assume that a tugboat has a displacement of 625 long tons (1 long ton = 2240 lb) and is moving through water at 35 ft/s. Compute the total ship resistance and the total effective power required to drive the boat.

Solution

From Table 17.2, we find the specific resistance ratio to be $R_{st}/\Delta = 0.006$. Then, the total ship resistance is

$$\Delta = (625 \text{ tons})(2240 \text{ lb/ton}) = 1.4 \times 10^6 \text{ lb}$$

$$R_{ts} = (0.006)(\Delta) = (0.006)(1.4 \times 10^6 \text{ lb}) = 8400 \text{ lb}$$

The power required is

$$P_E = R_{ts}v = (8400 \text{ lb})(35 \text{ ft/s}) = 0.294 \times 10^6 \text{ lb-ft/s}$$

Using 550 lb-ft/s = 1.0 hp, we get

$$P_E = (0.294 \times 10^6)/550 = 535 \text{ hp}$$

17.6.7 Submarines

A floating submarine's resistance can be computed in the same way as can a ship's. However, when completely submerged, none of the submarine's motion causes surface waves, and the computation of resistance is similar to that for an aircraft. The hull shape is similar to the shape of an aircraft fuselage, and skin friction plays a major role in the total resistance. Of course, the total *magnitude* of the drag for a submarine is significantly greater than that for an aircraft because the density of water is far greater than that of air.

17.7 COMPRESSIBILITY EFFECTS AND CAVITATION

The results reported in Section 17.4 are for conditions in which the compressibility of the fluid (usually air) has little effect on the drag coefficient. These data are valid if the velocity of flow is less than about one-half the speed of sound in the fluid. Above that speed for air, the character of the flow changes and the drag coefficient increases rapidly.

When the fluid is a liquid such as water, we do not need to consider compressibility because liquids are very slightly compressible. However, we must consider another phenomenon called *cavitation*. As the liquid flows past a body, the static pressure decreases. If the pressure becomes sufficiently low,

the liquid vaporizes, forming bubbles. Because the region of low pressure is generally small, the bubbles burst when they leave that region. When the collapsing of the vapor bubbles occurs near a surface of the body, rapid erosion or pitting results. Cavitation has other adverse effects when it occurs near control surfaces of boats or on propellers. The bubbles in the water decrease the forces exerted on rudders and control vanes and decrease thrust and the performance of propellers.

17.8 LIFT AND DRAG ON AIRFOILS

We define lift as a force acting on a body in a direction perpendicular to that of the flow of fluid. We will discuss the concepts concerning lift with reference to airfoils. The shape of the airfoil comprising the wings of an airplane determines its performance characteristics. Note that high speed ground vehicles, particularly race cars, also employ airfoils, but with opposite objectives. The force created by the airfoil is directed downward to enhance traction and control of the car. The general design of the airfoil is similar to that used on aircraft, but the orientation is opposite.

The manner in which an airfoil produces lift when placed in a moving air stream (or when moving in still air) is illustrated in Fig. 17.8. As the air flows over the airfoil, it

FIGURE 17.8 Pressure distribution on an airfoil.

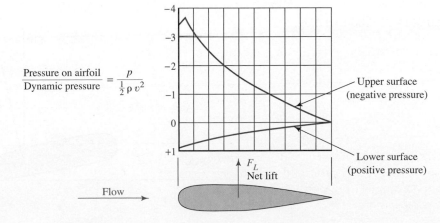

FIGURE 17.9 Span and chord lengths for an airfoil.

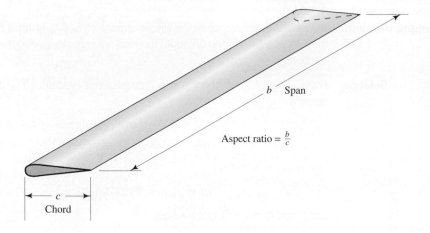

achieves a high velocity on the top surface with a corresponding decrease in pressure. At the same time the pressure on the lower surface is increased. The net result is an upward force called *lift*. We express the *lift force* F_L as a function of a lift coefficient C_L in a manner similar to that presented for drag:

▷ **Lift Force**

$$F_L = C_L(\rho v^2/2)A \qquad (17\text{–}10)$$

The velocity v is the velocity of the free stream of fluid relative to the airfoil. To achieve uniformity in the comparison of one shape with another, we usually define the area A as the product of the span of the wing and the length of the airfoil section called the *chord*. In Fig. 17.9, the span is b and the chord length is c.

The value of the lift coefficient C_L is dependent on the shape of the airfoil and also on the angle of attack. Figure 17.10 shows that the angle of attack is the angle between the chord line of the airfoil and the direction of the fluid velocity. Other factors affecting lift are the Reynolds number, the surface roughness, the turbulence of the air stream, the ratio of the velocity of the fluid stream to the speed of sound, and the aspect ratio. *Aspect ratio* is the name given to the ratio of the span b of the wing to the chord length c. It is important because the characteristics of the flow at the wing tips are different from those toward the center of the span.

The total drag on an airfoil has three components. Friction drag, F_{Df}, and pressure drag, F_{Dp}, occur as described

before in this chapter. The third component is called *induced drag*, F_{Di}, which is a function of the lift produced by the airfoil. At a particular angle of attack, the net resultant force on the airfoil acts essentially perpendicular to the chord line of the section, as shown in Fig. 17.10. Resolving this force into vertical and horizontal components produces the true lift force F_L and the induced drag F_{Di}. Expressing the induced drag as a function of a drag coefficient gives

$$F_{Di} = C_{Di}(\rho v^2/2)A \qquad (17\text{–}11)$$

It can be shown that C_{Di} is related to C_L by the relation

$$C_{Di} = \frac{C_L^2}{\pi(b/c)} \qquad (17\text{–}12)$$

The total drag is then

$$F_D = F_{Df} + F_{Dp} + F_{Di} \qquad (17\text{–}13)$$

Normally, it is the total drag that is of interest when designing airfoils. We determine a single drag coefficient C_D for the airfoil, from which the total drag can be calculated using the relation

$$F_D = C_D(\rho v^2/2)A \qquad (17\text{–}14)$$

As before, the area A is the product of the span b and the chord length c.

We use two methods to present the performance characteristics of airfoil profiles. In Fig. 17.11, the values of C_L, C_D, and the ratio of lift to drag F_L/F_D are all plotted

FIGURE 17.10 Induced drag.

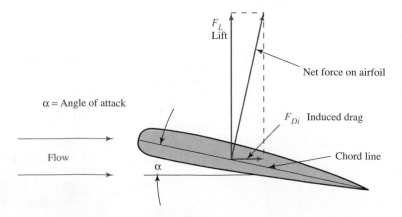

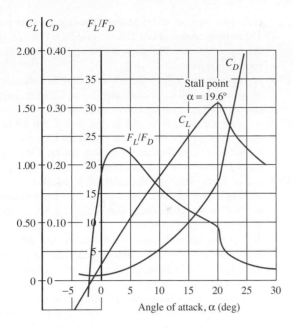

FIGURE 17.11 Airfoil performance curves.

up to a point where it abruptly begins to decrease. This point of maximum lift is called the *stall point*; at this angle of attack, the boundary layer of the air stream separates from the upper side of the airfoil. A large turbulent wake is created, greatly increasing drag and decreasing lift. See References 1, 2, 5, and 8, and Internet resources 7–9 for additional discussion and general data for airfoils, particularly as applied to aircraft. References 4, 6, and 9, and Internet resource 5 address the application of airfoils to race cars.

REFERENCES

1. Anderson, John D. 2010. *Fundamentals of Aerodynamics*, 5th ed. New York: McGraw-Hill.

2. Anderson, John D. 2011. *Introduction to Flight*. New York: McGraw-Hill.

3. Avallone, Eugene A., Theodore Baumeister, and Ali Sadegh, eds. 2006. *Marks' Standard Handbook for Mechanical Engineers*, 11th ed. New York: McGraw-Hill.

4. Barnard, R. H. 2010. *Road Vehicle Aerodynamic Design*, 3rd ed. St. Albans, UK: Mechaero Publishing.

5. Bertin, John J. 2008. *Aerodynamics for Engineers*, 5th ed. Upper Saddle River, NJ: Pearson Education/Prentice-Hall.

6. Edgar, Julian. 2013. *Amateur Car Aerodynamics Sourcebook*. Seattle, WA: CreateSpace Independent Publishing Platform, An Amazon Company.

7. Blevins, R. D. 2003. *Applied Fluid Dynamics Handbook*. Melbourne, FL: Krieger.

8. Houghton, E. L., P. W. Carpenter, Steven Collicott, and Dan Valentine. 2012. *Aerodynamics for Engineering Students*, 6th ed. Burlington, MA: Butterworth-Heinemann.

9. Katz, Joseph. 2003. *Race Car Aerodynamics: Designing for Speed*. Cambridge, MA: Bentley.

10. Lindsey, W. F. 1938. *Drag of Cylinders of Simple Shapes* (Report No. 619). National Advisory Committee for Aeronautics.

11. McBeath, Simon. 2011. *Competition Car Aerodynamics: A Practical Handbook*, 2nd ed. Somerset, UK: Haynes Publishing.

12. Molland, Anthony F., Steven R. Turnock, and Dominic A. Hudson. 2011. *Ship Resistance and Propulsion: Practical Estimation of Propulsive Power*. Cambridge, UK: Cambridge University Press.

13. Morel, T. and C. Dalton, eds. 1979. *Aerodynamics of Transportation*. New York: American Society of Mechanical Engineers.

14. Raveendran, Arun. 2012. *Aerodynamically Efficient Bus Design*. Saarbrücken, Germany: LAP Lambert Academic Publishing.

15. Stokes, George G. 1901. *Mathematical and Physical Papers*, Vol. 3. Cambridge, UK: Cambridge University Press.

16. Williams, Nathan A. 2012. *Drag Optimization of Light Trucks Using Computational Fluid Dynamics*. Seattle, WA: Amazon Digital Services.

17. Wolf-Heinrich, Hucho, ed. 1998. *Aerodynamics of Road Vehicles*, 4th ed. Warrendale, PA: SAE International (SAE).

18. Yakkundi, Vivek. 2011. *Aerodynamics of Cars: An Experimental Investigation – A Synergy of Wind Tunnel and CFD*. Saarbrücken, Germany: LAP Lambert Academic Publishing.

versus the angle of attack as the abscissa. Note that the scale factors are different for each variable. The airfoil to which the data apply has the designation NACA 2409 according to a system established by the National Advisory Committee for Aeronautics. NACA Technical Report 610 explains the code used to describe airfoil profiles. NACA Reports 586, 647, 669, 708, and 824 present the performance characteristics of several airfoil sections. Internet resource 8 provides dimensional data for over 1500 airfoil shapes.

The second method of presenting data for airfoils is shown in Fig. 17.12. This is called the *polar diagram* and is constructed by plotting C_L versus C_D with the angle of attack indicated as points on the curve.

In both Fig. 17.11 and Fig. 17.12, it can be seen that the lift coefficient increases with increasing angle of attack

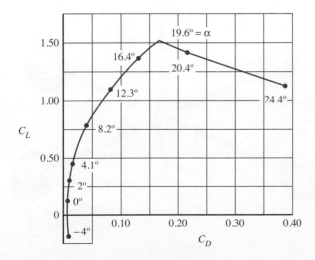

FIGURE 17.12 Airfoil polar diagram.

INTERNET RESOURCES

1. **Golfsmith—The Evolution of Golf Balls:** A discussion on the flight characteristics of golf balls.

2. **GSA Golf Simulators—Physics of Golf:** An overview of the physics of golf, flight characteristics of golf balls, the dynamics of golf clubs.

3. **Golf Ball Aerodynamics—Frankly Golf:** A brief overview of how a golf ball flies, the importance of dimpling, flight conditions, and drag forces.

4. **Curbside Classic—History of Automotive Aerodynamics:** An illustrated history of automotive aerodynamics in three parts: 1. From 1899 to 1939; 2. From 1940 to 1959; 3. From 1960 to the present.

5. **Aerodynamics of Race Cars—Strange Holiday, Joseph Katz:** An overview of the work on race car performance of Joseph Katz from his book listed as Reference 9 above.

6. **Platform for Aerodynamic Road Transport:** A platform for collaboration among scientists, road transport equipment manufacturers, and shippers and carriers to achieve a reduction in fuel consumption and emissions. Based at the Technical University of Delft in Germany.

7. **Allstar Network—FIU:** An online aeronautics learning laboratory for science, technology, and research, hosted by Florida International University. The level 3, Principles of Aerodynamics, section gives introductory descriptions of airfoils and fluid dynamics. Also included is a description of the software FoilSim, developed by NASA Lewis Research Center, that can be downloaded from the website. It analyzes the flow around an airfoil, displays the flow graphically, and computes lift as a function of airspeed and angle of attack.

8. **UIUC Airfoil Data Site:** A database of dimensional data for over 1500 airfoil shapes, hosted by the University of Illinois at Urbana-Champaign, Department of Aerospace Engineering, Applied Aerodynamics Group.

9. **Public Domain Aerodynamic Software:** A site from which public domain software can be downloaded, including programs for computing the coordinates of NACA airfoils.

PRACTICE PROBLEMS

17.1 A cylinder 25 mm in diameter is placed perpendicular to a fluid stream with a velocity of 0.15 m/s. If the cylinder is 1 m long, calculate the total drag force if the fluid is (a) water at 15°C and (b) air at 10°C and atmospheric pressure.

17.2 As part of an advertising sign on the top of a tall building, a 2-m-diameter sphere called a "weather ball" glows different colors if the temperature is predicted to drop, rise, or remain about the same. Calculate the force on the weather ball due to winds of 15, 30, 60, 120, and 160 km/h if the air is at 0°C.

17.3 Determine the terminal velocity (see Section 2.6.4, Chapter 2) of a 75-mm-diameter sphere made of solid aluminum (specific weight = 26.6 kN/m³) in free fall in (a) castor oil at 25°C, (b) water at 25°C, and (c) air at 20°C and standard atmospheric pressure. Consider the effect of buoyancy.

17.4 Calculate the moment at the base of a flagpole caused by a wind of 150 km/h. The pole is made of three sections, each 3 m long, of different-size Schedule 80 steel pipe. The bottom section is DN 150, the middle is DN 125, and the top is DN 100. The air is at 0°C and standard atmospheric pressure.

17.5 A pitcher throws a baseball without spin with a velocity of 20 m/s. If the ball has a circumference of 225 mm, calculate the drag force on the ball in air at 30°C.

17.6 A parachute in the form of a hemispherical cup 1.5 m in diameter is deployed from a car trying for the land speed record. Determine the force exerted on the car if it is moving at 1100 km/h in air at atmospheric pressure and 20°C.

17.7 Calculate the required diameter of a parachute in the form of a hemispherical cup supporting a man weighing 800 N if the terminal velocity (see Section 2.6.4, Chapter 2) in air at 40°C is to be 5 m/s.

17.8 A ship tows an instrument in the form of a 30° cone, point first, at 7.5 m/s in seawater. If the base of the cone has a diameter of 2.20 m, calculate the force in the cable to which the cone is attached.

17.9 A highway sign is being designed to withstand winds of 125 km/h. Calculate the total force on a sign 4 m by 3 m if the wind is flowing perpendicular to the face of the sign. The air is at −10°C. Compare the force calculated for this problem with that for Problem 16.4. Discuss the reasons for the differences.

17.10 Assuming that a semitrailer behaves as a square cylinder, calculate the force exerted if a wind of 20 km/h strikes it broadside. The trailer is 2.5 m by 2.5 m by 12 m. The air is at 0°C and standard atmospheric pressure.

17.11 A type of level indicator incorporates four hemispherical cups with open fronts mounted as shown in Fig. 17.13. Each cup is 25 mm in diameter. A motor drives the cups at a constant rotational speed. Calculate the torque that the motor must produce to maintain the motion at 20 rpm

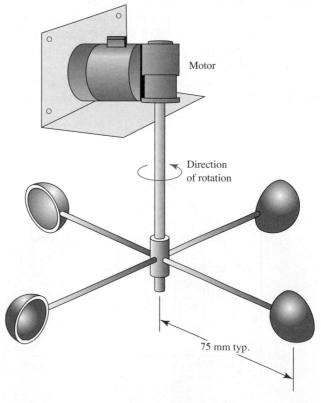

FIGURE 17.13 Problem 17.11.

FIGURE 17.14 Problem 17.12.

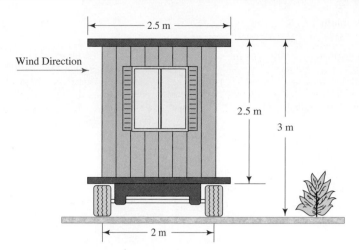

Wind Direction

2.5 m

2.5 m

3 m

2 m

when the cups are in (a) air at 30°C and (b) gasoline at 20°C.

17.12 Determine the wind velocity required to overturn the mobile home sketched in Fig. 17.14 if it is 10 m long and weighs 50 kN. Consider it to be a square cylinder. The width of each tire is 300 mm. The air is at 0°C.

17.13 A bulk liquid transport truck incorporates a cylindrical tank 2 m in diameter and 8 m long. For the tank alone, calculate the pressure drag when the truck is traveling at 100 km/h in still air at 0°C.

17.14 A wing on a race car is supported by two cylindrical rods, as shown in Fig. 17.15. Compute the drag force exerted on the car due to these rods when the car is traveling through still air at −20°F at a speed of 150 mph.

17.15 In an attempt to decrease the drag on the car shown in Fig. 17.15 and described in Problem 17.14, the cylindrical rods are replaced by elongated elliptical cylinders having a length-to-breadth ratio of 8:1. By how much will the drag be reduced? Repeat for the *Navy strut*.

17.16 The four designs shown in Fig. 17.16 for the cross section of an emergency flasher lighting system for police vehicles are being evaluated. Each has a length of 60 in and a width of 9.00 in. Compare the drag force exerted on

each proposed design when the vehicle moves at 100 mph through still air at −20°F.

17.17 A four-wheel drive utility vehicle incorporates a roll bar that extends above the cab and is in the free stream of air. The bar, made from a 3-in Schedule 40 steel pipe, has a total length of 92 in exposed to the wind. Compute the drag exerted on the vehicle by the bar when the vehicle travels at 65 mph through still air at −20°F.

17.18 An advertising sign for the ABC Paper Company is shown in Fig. 17.17. It is made from three flat disks, each with a diameter of 56.0 in. The disks are joined by 4.50-in-diameter tubes measuring 30 in between the disks. Compute the total force on the sign if it is faced into a 100-mph wind. The air is at −20°F.

17.19 An antenna in the shape of a cylindrical rod projects from the top of a locomotive. If the antenna is 42 in long and 0.200 in. in diameter, compute the drag force on it when the locomotive is traveling at 160 mph in still air at −20°F.

17.20 A ship tows an instrument package in the form of a hemisphere with an open back at a velocity of 25.0 ft/s through seawater at 77°F. The diameter of the hemisphere is 7.25 ft. Compute the force in the cable to which the package is attached.

FIGURE 17.15 Problems 17.14, 17.15, and 17.28.

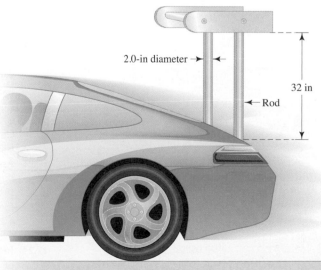

2.0-in diameter

32 in

Rod

FIGURE 17.16 Problem 17.16.

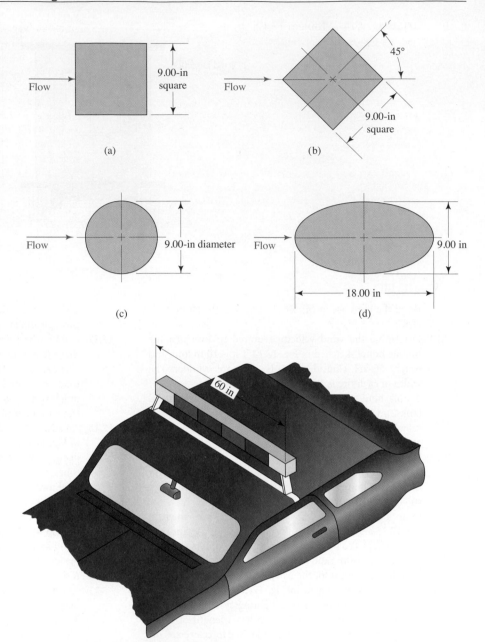

(a)

(b)

(c)

(d)

(e) Pictorial of light assembly mounted on the car

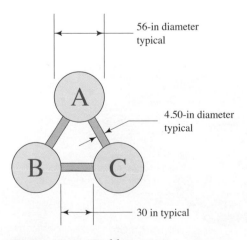

56-in diameter
typical

4.50-in diameter
typical

30 in typical

FIGURE 17.17 Problem 17.18.

17.21 A flat rectangular plate 8.50 × 11.00 in. in size is inserted into lake water at 60°F from a boat moving at 30 mph. What force is required to hold the plate steady relative to the boat with the flat face toward the water?

17.22 The windshield on an antique Stutz Bearcat automobile is a flat circular disk approximately 28 in. in diameter. Compute the drag caused by the windshield when the car travels at 60 mph in still air at 40°F.

17.23 Assume that curve 2 in Fig. 17.5 is a true representation of the performance of a golf ball with a diameter of 1.25 in. If the Reynolds number is 1.5×10^5, compute the drag force on the golf ball and compare it to the drag force on a smooth sphere of the same diameter whose drag coefficient conforms to curve 1. The air is at 40°F.

17.24 In a falling-ball viscometer, a steel sphere with a diameter of 1.200 in drops through a heavy syrup and travels 18.0 in. in 20.40 s at a constant speed. Compute the viscosity of the syrup. The syrup has a specific gravity of 1.18. Note that the free-body diagram of the sphere should include its weight acting down and the buoyant force and the drag force acting up. The steel has a specific gravity of 7.83. See also Chapter 2.

17.25 Compute the power required to overcome drag on a truck with a drag coefficient of 0.75 when the truck moves at 65 mph through still air at 40°F. The maximum cross section of the truck is a rectangle 8.00 ft wide and 12.00 ft high.

17.26 A small, fast boat has a specific resistance ratio of 0.06 (see Table 17.2) and displaces 125 long tons. Compute the total ship resistance and the power required to overcome drag when it is moving at 50 ft/s in seawater at 77°F.

17.27 A passenger liner displaces 8700 long tons. Compute the total ship resistance and the power required to overcome drag when it is moving at 30 ft/s in seawater at 77°F.

17.28 Assume that Fig. 17.11 shows the performance of the wing on the race car shown in Fig. 17.15. Note that it is mounted in the inverted position, so the lift pushes down to aid in skid resistance. Compute the downward force exerted on the car by the wing and the drag when the angle of attack is set at 15° and the speed is 25 m/s. The chord length is 780 mm and the span is 1460 mm.

17.29 Calculate the total drag on an airfoil that has a chord length of 2 m and a span of 10 m. The airfoil is at 3000 m flying at (a) 600 km/h and (b) 150 km/h. Use Fig. 17.11 for C_D and $\alpha = 15°$.

17.30 For the airfoil with the performance characteristics shown in Fig. 17.11, determine the lift and drag at an angle of attack of 10°. The airfoil has a chord length of 1.4 m and a span of 6.8 m. Perform the calculation at a speed of 200 km/h in the standard atmosphere at (a) 200 m and (b) 10 000 m.

17.31 Repeat Problem 17.30 if the angle of attack is the stall point, 19.6°.

17.32 For the airfoil in Problem 17.30, what load could be lifted from the ground at a takeoff speed of 125 km/h when the angle of attack is 15°? The air is at 30°C and standard atmosphere pressure.

17.33 Determine the required wing area for a 1350-kg airplane to cruise at 125 km/h if the airfoil is set at an angle of attack of 2.5°. The airfoil has the characteristics shown in Fig. 17.11. The cruise altitude is 5000 m in the standard atmosphere.

FANS, BLOWERS, COMPRESSORS, AND THE FLOW OF GASES

Fans, blowers, and compressors are used to increase pressure and to cause the flow of air and other gases in ducts and piping systems. The compressibility of gases requires special methods of analysis of the performance of such devices. You should become familiar with recommended ways of evaluating the performance of systems carrying gas flow.

Gases behave much differently than liquids, requiring several different types of equipment to deliver them at the large ranges of pressures and flow rates required for a given application. Care must be taken to account for the compressibility of gases, its light density, and quite different behavior in piping systems.

Exploration

Take a moment to think about where you may have experienced fans or compressors and the piping or ducting systems used to distribute the compressed air or other gases to its final place of use.

- Where are fans, blowers, or compressors used in your home? Consider the kitchen, the laundry room, and the furnace and air conditioning systems.

- Check in your car in the passenger compartment or under the hood. What about the tires?

- Look around stores, offices, or a shopping mall. How are they kept at a comfortable temperature for the customers and those who work there?

- If you have access to a factory, look for places where moving air or air under pressure is used.

Introductory Concepts

Fans, blowers, and compressors are used to increase pressure and to cause the flow of air and other gases in a gas flow system. Their function is similar to that of pumps in a liquid flow system, as discussed in Chapter 13. Some of the principles already developed for the flow of liquids and the application of pumps can be applied also to the flow of gases. However, the compressibility of gases causes some important differences.

As you observed where fans and blowers are used in your home, an obvious example is when you use a fan to circulate air when it is too warm for comfort. The fan draws air from the ambient air in the room, accelerates it by the action of the fan blades, and delivers it at a higher velocity. The moving air tends to create a cooling effect.

What other examples did you think of? How does your list compare with these?

- Does your home have a forced-air heating and air conditioning system? Inspect the unit if it is accessible and discover what causes the flow of air through the ductwork. Perhaps it is a blower that looks like those shown later in Figs. 18.3 and 18.4. The air is drawn from a return air duct into the center of the bladed

FIGURE 18.1 Pneumatics are critical since many operations, including automated assembly equipment like that shown here, depend on the reliable distribution of compressed air. (*Source:* Nataliya Hora/Fotolia)

rotor. The rotation of the rotor imparts energy to the air, throws it outward along the spiral housing, and delivers it at higher pressure and velocity to the discharge ductwork. Chapter 19 covers the topic of the flow of air in ducts at relatively low pressure.

■ For an air conditioner, look for the condensing unit outside the home. A large fan draws ambient air and forces it over the condensing coils to carry heat away from them. The refrigerant in the coils then condenses and passes on to the evaporator of the system. The evaporator fan inside the home performs the important function of enhancing the cooling effect of the air.

■ Many household appliances incorporate fans and blowers: a hair dryer, clothes dryer, your computer, most refrigerators (both inside the cabinet and in the machine compartment), the vacuum cleaner, and power drills and saws.

■ How many fans did you find on your car? The engine cooling fan is under the hood, moving air over the engine to remove heat from the coolant through the radiator and by direct convection from the engine itself. The car's heater and air conditioning system uses a blower in a similar manner to your home's furnace, air conditioner, and refrigerator. The blower is likely mounted within the instrument panel with ductwork to carry the heated or cooled air to the vents near the driver and passengers.

■ Stores, offices, and shopping malls must also use heating systems and air conditioning units to provide a comfortable environment for the people using those spaces. Because the overall demands for heating, cooling, and ventilating air are much larger, the air handling blowers are much larger as well.

■ Now consider the factory. In addition to the heating, ventilating, and air conditioning systems (HVAC), there may be a need for high-pressure air to operate many of the processes in the plant. Compressed air is used to power screwdrivers, drills, air cylinders (also called linear pneumatic actuators), and other pneumatic devices, similar to the automation equipment shown in Fig. 18.1. Larger plants typically employ central compressors to deliver a steady supply of air at approximately 100 psi (690 kPa) to all parts of the factory. Individual work cells can tap into the system as needed. Perhaps the central compressor looks like that in Fig. 18.6.

Special techniques for the design of flow systems carrying gases, such as air, have been developed by professionals based on years of experience. The detailed analysis of the phenomena involved requires knowledge of thermodynamics. Because knowledge of thermodynamics is not expected for the readers of this book, some of the methods in this chapter are presented without extensive development. New terms or concepts required for understanding the methods are, of course, described. See Internet resources 11–13 for specialized information about the movement of air and refrigerants. Internet resource 15 provides a wealth of information about industrial and medical gases, along with air.

18.1 OBJECTIVES

After completing this chapter, you should be able to:

1. Describe the general characteristics of fans, blowers, and compressors.

2. Describe propeller fans, duct fans, and centrifugal fans.

3. Describe blowers and compressors of the centrifugal, axial, vane–axial, reciprocating, lobed, vane, and screw types.

4. Specify suitable sizes for pipes carrying steam, air, and other gases at higher pressures.

5. Compute the flow rate of air and other gases through nozzles.

18.2 GAS FLOW RATES AND PRESSURES

When working in U.S. Customary System units, the flow rate of air or another gas is most frequently expressed in ft³/min, abbreviated cfm (cubic feet per minute). Velocities are typically reported in ft/min. Although these are not the standard units for the U.S. Customary System, they are convenient for the range of flows typically encountered in residential, commercial, and industrial applications

In the SI system, m^3/s for flow rate and m/s for velocity are the units most often used. For systems carrying relatively low flow rates, the unit of L/s is sometimes used. Convenient conversions are listed below:

$$1.0 \text{ ft}^3/\text{s} = 60 \text{ ft}^3/\text{min} = 60 \text{ cfm}$$
$$1.0 \text{ m}^3/\text{s} = 2120 \text{ ft}^3/\text{min} = 2120 \text{ cfm}$$
$$1.0 \text{ ft/s} = 60 \text{ ft/min}$$
$$1.0 \text{ m/s} = 3.28 \text{ ft/s}$$
$$1.0 \text{ m/s} = 197 \text{ ft/min}$$

Pressures can be measured in lb/in² (abbreviated psi) in U.S. Customary System units when relatively large pressures are encountered. However, in most air-handling systems, the pressures are small and are measured in *inches of water gage*, abbreviated as inH_2O or in w.g. This unit is derived from the practice of using a pitot tube and water manometer to measure the pressure in ducts, as illustrated in Figs. 15.17 and 15.18. The equivalent pressure can be derived from the pressure–elevation relation, $\Delta p = \gamma h$. If we use $\gamma = 62.4 \text{ lb/ft}^3$ for water, a pressure of 1.00 inH$_2$O is equivalent to

$$\Delta p = \gamma h = \frac{62.4 \text{ lb}}{\text{ft}^3} \cdot 1.00 \text{ in} \cdot \frac{1 \text{ ft}^3}{1728 \text{ in}^3} = 0.0361 \text{ lb/in}^2$$

Stated differently, 1.0 psi = 27.7 inH$_2$O. In many air flow systems, the pressures involved are only a few inches of water or even a fraction of an inch.

The standard SI unit of pascals (Pa) is quite small and is used directly when designing an air-handling system in SI units. Also used are bars, mmH$_2$O, and mmHg. Large industrial systems may require pressures in the kPa or MPa range. Some useful conversion factors are listed below:

$$1.0 \text{ bar} = 100 \text{ kPa}$$
$$1.0 \text{ psi} = 6895 \text{ Pa}$$
$$1.0 \text{ inH}_2\text{O} = 248.8 \text{ Pa}$$
$$1.0 \text{ mmH}_2\text{O} = 9.81 \text{ Pa}$$
$$1.0 \text{ mmHg} = 132.8 \text{ Pa}$$

FIGURE 18.2   Duct fan.

18.3 CLASSIFICATION OF FANS, BLOWERS, AND COMPRESSORS

Fans, blowers, and compressors are all used to increase the pressure of and move air or other gases. The primary differences among them are their physical construction and the pressures that they are designed to develop. See References 5 and 8.

- A *fan* is designed to operate against small static pressures, up to about 2.0 psi (13.8 kPa) and is used in heating, ventilating, and air-conditioning equipment (HVAC) and similar applications. Typical operating pressures for fans are from 0 to 6 inH$_2$O (0.00 to 0.217 psi or 0.00 to 1500 Pa).

- At pressures from 2.0 psi up to approximately 15.0 psi (103 kPa), the gas mover is called a *blower,* used for high-capacity HVAC systems, to supply combustion air for industrial furnaces, and for some materials processing systems.

- To develop higher pressures, even as high as several thousand psi, *compressors* are used. Supplying compressed air for pneumatic automation equipment in manufacturing systems is a common application.

Fans are used to circulate air within a space, to bring air into or exhaust it from a space, or to move air through ducts in ventilation, heating, or air conditioning systems. Types of fans include propeller fans, duct fans, and centrifugal fans. See Internet resources 1, 3, and 4.

Propeller fans operate at virtually zero static pressure and are composed of two to six blades with the appearance of aircraft propellers. Thus, they draw air in from one side and discharge it from the other side in an approximately axial direction. This type of fan is popular for circulating air in living or working spaces to improve comfort. When mounted in windows or other openings in the walls of a building, they deliver fresh outside air into the building or exhaust air from the building. When mounted in the ceiling or roof, they are often called *ventilators*.

Propeller fans are available from small sizes (a few inches in diameter, delivering a few hundred cfm) to 60 in or more in diameter, delivering over 50 000 cfm at zero static pressure. Operating speeds typically range from about 600 rpm to 1725 rpm. These fans are driven by electric motors, either directly or through belt drives.

Duct fans have a construction similar to that of propeller fans, except that the fan is mounted inside a cylindrical duct, as shown in Fig. 18.2. The duct could be a part of a larger duct system delivering air to or exhausting it from a remote area. Duct fans can operate against static pressures up to about 1.50 inH$_2$O (375 Pa). Sizes range from very small, delivering a few hundred cfm, to about 36 in, delivering over 20 000 cfm.

Two examples of *centrifugal fans* or *centrifugal blowers*, along with their rotors, are shown in Figs. 18.3 and 18.4. Air enters at the center of the rotor, also called the *impeller*, and is thrown outward by the rotating blades, thus, adding kinetic energy. The high-velocity gas is collected by the volute surrounding the rotor, where the kinetic energy is converted into an increased gas pressure for delivery through a duct system for ultimate use. See Internet resources 1–6.

FIGURE 18.3 Centrifugal blower with straight, radial bladed rotor.

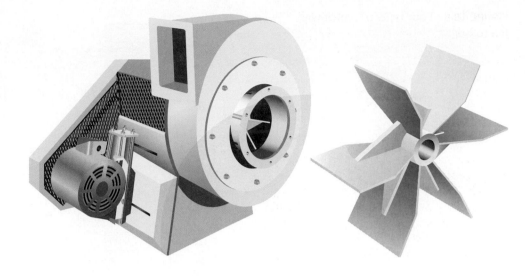

FIGURE 18.4 Centrifugal blower with backward-inclined bladed rotor.

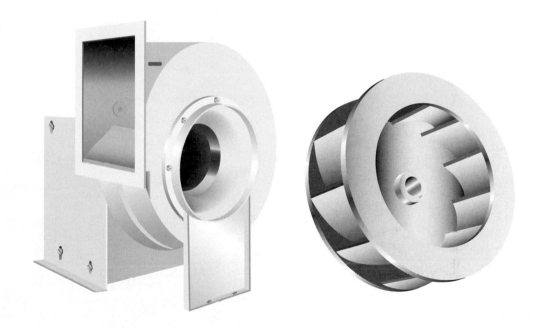

The construction of the rotor for fans and blowers is typically one of four basic designs, as shown in Fig. 18.5. The *backward-inclined blade* is often made with simple flat plates. As the rotor rotates, the air tends to leave parallel to the blade along the vector called v_b in the figure. However, that is added vectorially to the tangential velocity of the blade itself, v_t, which gives the resultant velocity shown as v_R. The *forward-curved blades* yield a generally higher resultant air velocity because the two component vectors are more nearly in the same direction. For this reason, a rotor with forward-curved blades will run at a slower speed than a similar-sized backward-inclined bladed fan for the same air flow and pressure. However, the backward-inclined fan typically requires lower power for the same service (Reference 15). *Airfoil-shaped, backward-inclined fan blades* operate more quietly and efficiently than flat,

backward-inclined blades. All these types of fans are used for ventilation systems and some industrial process uses. *Radial-blade* fans have many applications in industry for supplying large volumes of air at moderate pressures for boilers, cooling towers, material dryers, and bulk material conveying.

Centrifugal compressors employ impellers similar to those in centrifugal pumps (Figs. 13.11 and 13.12). However, the specific geometry is tailored to handle gas rather than liquid. Figure 18.6 shows a large, single-stage, centrifugal compressor. When a single-rotor compressor cannot develop a sufficiently high pressure, a multistage compressor, as shown in Fig. 18.7, is used. Centrifugal compressors are used for flows from about 500 to 100 000 cfm (0.24–47m³/s) at pressures as high as 8000 psi (55 MPa). See Internet resources 2, 9, and 10.

FIGURE 18.5 Four types of centrifugal fan rotors.

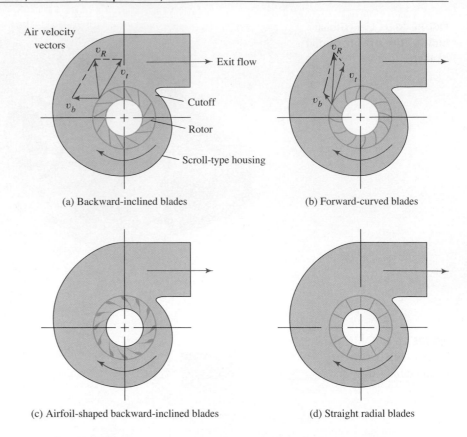

(a) Backward-inclined blades

(b) Forward-curved blades

(c) Airfoil-shaped backward-inclined blades

(d) Straight radial blades

FIGURE 18.6 Single-stage centrifugal compressor. (*Source:* Dresser-Rand, Turbo Products Division)

A multistage *axial compressor* is shown in Fig. 18.8. Only the lower half of the casing, the multistaged rotor, and the shaft assembly are shown. The gas is drawn into the large end, moved axially and compressed by the series of bladed rotors, and discharged from the small end. Axial compressors arc employed to deliver large flow rates, approximately 8000 to 1.0 million cfm (3.8–470 m³/s) at a discharge pressure up to 100 psi (690 kPa).

Vane–axial blowers are similar to duct fans, described earlier, except they typically have blades that are airfoil shaped and include vanes within the cylindrical housing to

redirect the flow axially within the following duct. This results in higher static pressure capability for the blower and reduces swirl of the air.

Positive-displacement blowers and compressors come in a variety of designs:

- Reciprocating—single-acting or double-acting

- Rotary—lobe, vane, or screw

The construction of a reciprocating compressor looks similar to that of an engine. A rotating crank and a connecting

FIGURE 18.7 Lower case of a horizontally split multistage centrifugal compressor with rotor installed. (*Source:* Dresser-Rand, Turbo Products Division)

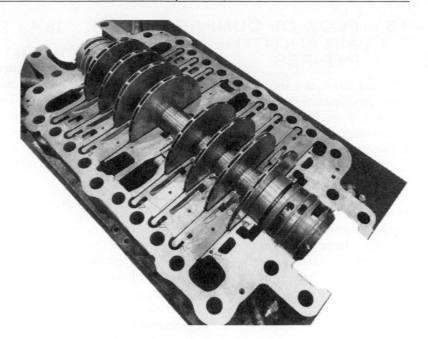

rod drive the piston. The piston reciprocates within its cylinder, drawing in low-pressure gas as it travels away from the cylinder head and then compressing it within the cylinder as it travels toward the head. When the pressure of the gas reaches the desired level, discharge valves open to deliver the compressed gas to the piping system. Figure 13.6 shows the arrangement of both single-acting and double-acting pistons. Small versions of such compressors are seen in small shops and service stations. However, for many industrial users, they can be very large, delivering up to 10 000 cfm (4.7 m³/s) at pressures up to 60 000 psi (413 MPa). See Internet resources 2, 8, and 10.

Lobe and vane compressors appear very similar to the pumps shown in Figs. 13.2 and 13.5. Lobe-type designs can develop up to approximately 15 psi (100 kPa) and are often

called blowers (Reference 14). Vane-type compressors are capable of developing several hundred psi and are often used in pneumatic fluid power systems.

Screw compressors are used in construction and industrial applications requiring compressed air up to 500 psi (3.4 MPa) with delivery up to 20 000 cfm (9.4 m³/s). In the single-screw design, air is trapped between adjacent "threads" rotating inside a close-fitting housing. The axial progression of the threads delivers the air to the outlet. In some designs, the pitch of the threads decreases along the length of the screw, providing compression within the housing as well as delivering it against system resistance. Two or more screws in a mesh can also be employed. See Fig. 13.3 and Internet resources 7, 8, 10, and 14.

FIGURE 18.8 Lower case of an axial compressor with rotor installed. (*Source:* Dresser-Rand, Turbo Products Division)

18.4 FLOW OF COMPRESSED AIR AND OTHER GASES IN PIPES

Many industries use compressed air in fluid power systems to power production equipment, material handling devices, and automation machinery. A common operating pressure for such systems is in the range of 60–125 psig (414–862 kPa gage). Performance and productivity of the equipment are degraded if the pressure drops below the design pressure. Therefore, careful attention must be paid to pressure losses between the compressor and the point of use. A detailed analysis of the piping system should be completed, using the methods outlined in Chapters 6–12, modified for the compressibility of air.

When large changes in pressure or temperature of the compressed air occur along the length of a flow system, the corresponding changes in the specific weight of the air should be taken into account. However, if the change in pressure is less than about 10 percent of the inlet pressure, variations in specific weight will have negligible effect. When the pressure drop is between 10 percent and 40 percent of the inlet pressure, we can use the average of the specific weight for the inlet and outlet conditions to produce results with reasonable accuracy (Reference 9). When the predicted pressure change is greater than 40 percent, one should either redesign the system or consult other references.

18.4.1 Specific Weight for Air

Figure 18.9 shows the variation of the specific weight for air as a function of changes in pressure and temperature. Note the large magnitude of the changes, particularly as pressure changes. The specific weight for any conditions of pressure and temperature can be computed from the *ideal gas law* from thermodynamics, which states

▷ **Ideal Gas Law**

$$\frac{p}{\gamma T} = \text{constant} = R \qquad (18\text{–}1)$$

where

- $p =$ Absolute pressure of the gas
- $\gamma =$ Specific weight of the gas
- $T =$ Absolute temperature of the gas, that is, the temperature above absolute zero
- $R =$ Gas constant for the gas being considered (see Appendix N)

Equation (18–1) also can be solved for the specific weight:

$$\gamma = \frac{p}{RT} \qquad (18\text{–}2)$$

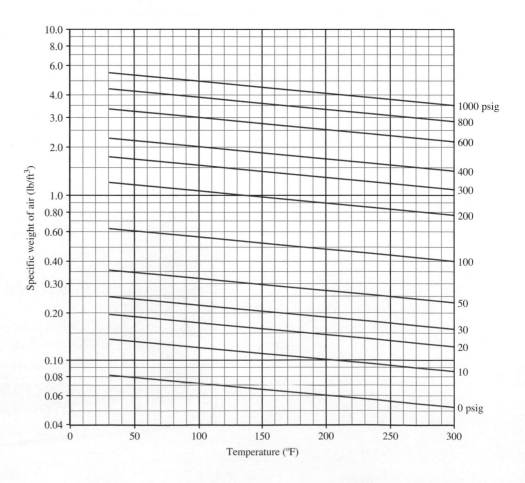

FIGURE 18.9 Specific weight of air versus pressure and temperature.

The absolute temperature is found by adding a constant to the measured temperature. In U.S. Customary System units,

$$T = (t°F + 460)\ °R$$

where °R is degrees Rankine, the standard unit for absolute temperature measured relative to absolute zero. In SI units,

$$T = (t°C + 273)\ K$$

where K (kelvin) is the standard SI unit for absolute temperature.

As presented in Section 3.2 [Eq. (3–2)], absolute pressure is found by adding the prevailing atmospheric pressure to the gage pressure reading. We use $p_{atm} = 14.7$ psia in U.S. Customary System units and $p_{atm} = 101.3$ kPa absolute in SI units unless the true local atmospheric pressure is known.

The value of the gas constant R for air is 53.3 ft·lb/lb·°R in U.S. Customary System units. The numerator of the unit for R is energy in foot pounds (ft·lb), and thus the pound

(lb) unit indicates *force*. The corresponding SI unit for energy is the newton meter (N·m), where

$$1.0\ ft·lb = 1.356\ N·m$$

The pound unit in the denominator for R is the weight of the air, also a force. To convert the U.S. Customary System unit °R to the SI unit K, we use $1.0\ K = 1.8°R$. Using these conversions, we can show that the value of R for air in the SI system is 29.2 N·m/N·K. Appendix N shows values for R for other gases.

In choosing to express the values for the gas constant R in terms of *weight*, we facilitate the computation of the *specific weight* γ, as shown in Example Problem 18.1. Later, we will use γ in calculations for the *weight flow rate* of a gas through a nozzle. It should be noted that the use of the resulting relationships based on weight should be confined to applications near Earth's surface, where the value of the acceleration due to gravity g is fairly constant.

In other types of analysis, particularly those in the field of thermodynamics, R is defined in terms of *mass* rather than weight. In aerospace environments, in which g and the weight can approach zero, the mass basis should also be used.

Example Problem 18.1

Compute the specific weight of air at 100 psig and 80°F.

Solution

Using Eq. (18–2), we find

$$p = p_{atm} + p_{gage} = 14.7\ psia + 100\ psig = 114.7\ psia$$
$$T = t + 460 = 80°F + 460 = 540°R$$

Then,

$$\gamma = \frac{p}{RT} = \frac{114.7\ lb}{in^2} \cdot \frac{lb·°R}{53.3\ ft·lb} \cdot \frac{1}{540°R} \cdot \frac{144\ in^2}{ft^2}$$
$$\gamma = 0.574\ lb/ft^3$$

Note that the quantities 53.3 and 144 will always be used for air in this type of calculation. Then, a convenient, unit-specific equation for specific weight of air can be derived as follows:

$$\gamma = 2.70\ p/T \qquad\qquad (18–3)$$

This gives γ for air directly in pounds per cubic foot (lb/ft³) when the absolute pressure is in psia and the absolute temperature is in °R.

Example Problem 18.2

Compute the specific weight of air at 750 kPa gage and 30°C.

Solution

Using Eq. (18–2), we get

$$p = p_{atm} + p_{gage} = 101.3\ kPa + 750\ kPa = 851.3\ kPa$$
$$T = t + 273 = 30°C + 273 = 303\ K$$

Then,

$$\gamma = \frac{p}{RT} = \frac{851.3 \times 10^3\ N}{m^2} \cdot \frac{N·K}{29.2\ N·m} \cdot \frac{1}{303\ K} = 96.2\ N/m^3$$

18.4.2 Flow Rates for Compressed Air Lines

Ratings for equipment using compressed air and for compressors delivering the air are given in terms of *free air*, sometimes called *free air delivery* (FAD). This gives the quantity of air delivered per unit time assuming that the air is at standard atmospheric pressure (14.7 psia or 101.3 kPa absolute) and at the standard temperature of 60°F or 15°C (absolute temperatures of 520°R or 285 K). To determine the flow rate at other conditions, the following equation can be used:

$$Q_a = Q_s \cdot \frac{p_{atm-s}}{p_{atm} + p_a} \cdot \frac{T_a}{T_s} \qquad (18\text{--}4)$$

where

Q_a = Volume flow rate at actual conditions

Q_s = Volume flow rate at standard conditions

p_{atm-s} = Standard absolute atmospheric pressure

p_{atm} = Actual absolute atmospheric pressure

p_a = Actual gage pressure

T_a = Actual absolute temperature

T_s = Standard absolute temperature = 520°R or 285 K

Using these values and those of the standard atmosphere, we can write Eq. (18–4) as follows.

In U.S. Customary System units:

$$Q_a = Q_s \cdot \frac{14.7 \text{ psia}}{p_{atm} + p_a} \cdot \frac{(t + 460)°R}{520°R} \qquad (18\text{--}4a)$$

In SI units:

$$Q_a = Q_s \cdot \frac{101.3 \text{kPa}}{p_{atm} + p_a} \cdot \frac{(t + 273) \text{ K}}{285 \text{ K}} \qquad (18\text{--}4b)$$

Example Problem 18.3   An air compressor has a rating of 500 cfm free air. Compute the flow rate in a pipeline in which the pressure is 100 psig and the temperature is 80°F.

Solution   Using Eq. (18–4a) and assuming that the local atmospheric pressure is 14.7 psia, we get

$$Q_a = 500 \text{ cfm} \cdot \frac{14.7 \text{ psia}}{14.7 + 100} \cdot \frac{(80 + 460)}{520} = 66.5 \text{ cfm}$$

18.4.3 Pipe Size Selection and Piping System Design

Many factors must be considered when specifying a suitable pipe size for carrying compressed air in industrial plants. Some of those factors and the parameters involved are as follows:

- *Pressure drop* Because friction losses are proportional to the square of the velocity of flow, it is desirable to use as large a pipe as is feasible, to ensure adequate pressure at all points of use in a system.

- *Compressor power requirement* The power required to drive the compressor increases as the pressure drop increases. Therefore, it is desirable to use large pipes to minimize pressure drop.

- *Cost of piping* Large pipes cost more than small pipes, which makes the use of smaller pipes preferable.

- *Cost of the compressor* In general, a compressor designed to operate at a higher pressure will cost more, which makes the use of larger pipes that minimize pressure drop preferable.

- *Installation cost* Small pipes are easier to handle, but that is usually not a major factor.

- *Space required* Small pipes take less space and provide less interference with other equipment or operations.

- *Future expansion* To allow the addition of more air-using equipment in the future, large pipes are desirable.

- *Noise* When air flows at a high velocity through pipes, valves, and fittings, it creates a high noise level. It is desirable to use large pipes to permit lower velocities.

There is no clearly optimum pipe size for all installations, and the designer should evaluate the overall performance of several proposed sizes before making a final specification. To aid in beginning the process, Table 18.1 lists some suggested sizes.

As with pipeline systems studied earlier, compressed air pipe systems typically contain valves and fittings to control the amount and direction of flow. We account for their effects by using the equivalent-length technique, described in Section 10.9. Values for the L_e/D ratio are listed in Table 10.4. See also Reference 9.

Figure 18.10 shows a sketch of a typical layout of a piping system serving an industrial operation. The following are its basic features:

- The compressor draws ambient air and increases its pressure for delivery to the system.

- The aftercooler conditions the air. Some systems may also include an air dryer near the compressor to remove excess moisture before it enters the main distribution piping.

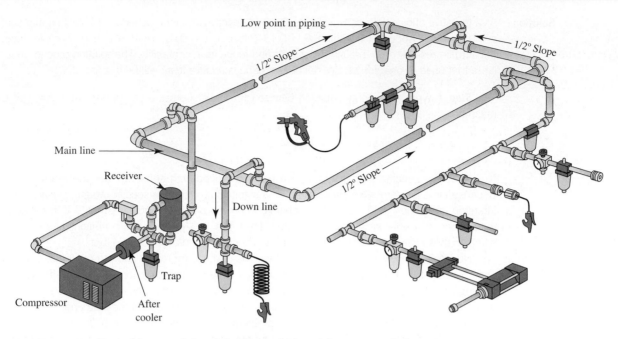

FIGURE 18.10 Typical layout of the piping for an industrial compressed air system.

TABLE 18.1	Suggested compressed air system pipe sizes		
Maximum Flow Rate (cfm)		**Pipe Size (Schedule 40)**	
Free Air	Compressed Air (100 psig, 60°F)	Inch size	DN Metric size (mm)
4	0.513	1/8	6
8	1.025	1/4	8
20	2.563	3/8	10
35	4.486	1/2	15
80	10.25	3/4	20
150	19.22	1	25
300	38.45	1 1/4	32
450	57.67	1 1/2	40
900	115.3	2	50
1400	179.4	2 1/2	65
2500	320.4	3	80
3500	448.6	3 1/2	90
5000	640.8	4	100

Note: The sizes listed are the smallest standard Schedule 40 steel pipes that will carry the given flow rate at a pressure of 100 psig (690 kPa) with no more than 5.0-psi (34.5 kPa) pressure drop in 100 ft (30.5m) of pipe. See Appendix F "Dimensions of Steel Pipe" for pipe dimensions.

- The compressed air goes to a receiver that has a relatively large volume to ensure that an even supply of air is available to the system. Other receivers may be placed closer to points of use in larger, more complex industrial systems.

- A trap is installed before the receiver to remove moisture.

- Piping serving the factory systems is laid out in a loop arrangement.

- Connections to the loop are made at the top of the main loop piping to inhibit the delivery of moisture to branch lines and the tools used there.

- Piping in the loop system slopes away from the compressor with one or more traps to remove moisture at low points in the system.

- Branch lines are sized to carry their given flow rates with the same nominal velocity as in the loop system.

- Pressure regulators are installed in branch lines to enable the adjustment of pressure to suit the tools used in that line.

- The condensate that comes from the air dryer and all of the traps in the system must be removed on a regular basis and treated before being disposed.

- Adequate ventilation and a supply of fresh air must be available in the vicinity of the compressor units to provide adequate input air to the compressor, to carry away heat generated by the drive motors and that generated from the compression process. The compressor room temperature should be within the range of 40°F to 90°F (4.5°C to 32°C).

Example Problem 18.4 Specify a suitable size of pipe for the delivery of 500 cfm (free air) at 100 psig at 80°F to the pneumatic system for an automated machine. The total length of straight pipe required between the compressor and the machine is 140 ft. The line also contains two fully open gate valves, six standard elbows, and two standard tees, in which the flow is through the run of the tee. Then, analyze the pressure required at the compressor to ensure that the pressure at the machine is no less than 100 psig.

Solution   As a tentative choice, let's consult Table 18.1 and specify a 1½-inch Schedule 40 steel pipe to carry the air. Then, from Appendix F we find $D = 0.1342$ ft and $A = 0.01414$ ft². We now check to determine the actual pressure drop through the system and judge its acceptability. The solution procedure is similar to that used in Chapter 11. Special circumstances relative to air will be discussed.

Step 1 Write the energy equation between the outlet from the compressor and the inlet to the machine:

$$\frac{p_1}{\gamma_1} + z_1 + \frac{v_1^2}{2g} - h_L = \frac{p_2}{\gamma_2} + z_2 + \frac{v_2^2}{2g}$$

Note that the specific weight terms have been identified with subscripts for the reference points. Because air is compressible, there could be a significant change in specific weight. However, it is our intention in this design to have a small change in pressure between points 1 and 2. If this is achieved, the change in specific weight can be ignored. Therefore, let $\gamma_1 \approx \gamma_2$. The conditions at point 2 are the same as those used in Example Problem 18.1. Then, we will use $\gamma = 0.574$ lb/ft³.

No information was given about the elevations of the compressor and the machine. Because the specific weight of air and other gases is so small, it is permissible to ignore elevation differences when dealing with the flow of gases, unless these differences are very large. As indicated in Sections 3.3 and 3.4, the pressure change is directly proportional to the specific weight of the fluid and to the change in elevation. For $\gamma = 0.574$ lb/ft³ for air in this problem, an elevation change of 100 ft (about the height of a 10-story building), would change the pressure only 0.40 psi.

The velocity at the two reference points will be the same because we will use the same size of pipe throughout. Then, the velocity head terms can be cancelled from the energy equation.

Step 2 Solve for the pressure at the compressor:

$$p_1 = p_2 + \gamma h_L$$

Step 3 Evaluate the energy loss h_L by using Darcy's equation, and include the effects of minor losses:

$$h_L = f\left(\frac{L}{D}\right)\left(\frac{v^2}{2g}\right) + f_T\left(\frac{L_e}{D}\right)\left(\frac{v^2}{2g}\right)$$

The L/D term is the actual ratio of pipe length to flow diameter:

Pipe: $L/D = (140\text{ ft}/0.1342\text{ ft}) = 1043$

The equivalent L_e/D values for the valves and fittings are found in Table 10.4:

2 valves:	$L_e/D = 2(8) = 16$
6 elbows:	$L_e/D = 6(30) = 180$
2 tees:	$L_e/D = 2(20) = 40$
Total:	$L_e/D = 236$

The flow velocity can be computed from the continuity equation. It was determined in Example Problem 18.3 that the flow rate of 500 cfm of free air at actual conditions of 100 psig and 80°F is 66.5 cfm. Then,

$$v = \frac{Q}{A} = \frac{66.5\text{ ft}^3}{\text{min}} \cdot \frac{1}{0.01414\text{ ft}^2} \cdot \frac{1\text{ min}}{60\text{ s}} = 78.4\text{ ft/s}$$

The velocity head is

$$\frac{v^2}{2g} = \frac{(78.4)^2\text{ ft}^2/\text{s}^2}{2(32.2\text{ ft/s}^2)} = 95.44\text{ ft}$$

To evaluate the friction factor f, we need the density and the viscosity of the air. Knowing the specific weight of the air, we can compute the density from

$$\rho = \frac{\gamma}{g} = \left(\frac{0.574\text{ lb}}{\text{ft}^3}\right)\left(\frac{\text{s}^2}{32.2\text{ ft}}\right) = \frac{0.0178\text{ lb·s}^2}{\text{ft}^4} = 0.0178\text{ slug/ft}^3$$

The dynamic viscosity of a gas does not change much as pressure changes. So, we can use the data from Appendix E, even though they are for standard atmospheric pressure. The dynamic viscosity is found to be $\eta = 3.85 \times 10^{-7}$ lb·s/ft².

It would be incorrect to use the kinematic viscosity listed for air in Appendix E because that value includes the density, which is dramatically different at 100 psig than it is at atmospheric pressure.

We can now compute the Reynolds number:

$$N_R = \frac{vD\rho}{\eta} = \frac{(78.4)(0.1342)(0.0178)}{3.85 \times 10^{-7}} = 4.86 \times 10^5$$

The relative roughness D/ϵ is

$$D/\epsilon = 0.1342/1.5 \times 10^{-4} = 895$$

Then, from the Moody diagram (Fig. 8.7), we read $f = 0.021$. The value for f_T used for the valves and fittings can be found from Table 10.5 to be 0.020 for the 1½-in Schedule 40 pipe. Because this is different from the friction factor for the pipe itself, we can compute the energy losses for the piping and the valves and fittings separately as follows:

For the piping,

$$h_L = f(L/D)(v^2/2g) = (0.021)(1043)(95.44 \text{ ft}) = 2090 \text{ ft}$$

For the valves and fittings,

$$h_L = f_T\left(\frac{L_e}{D}\right)_{total}\left(\frac{v^2}{2g}\right) = (0.020)(236)(95.44) = 450 \text{ ft}$$

The overall total energy loss is then,

$$h_{L-total} = 2090 + 450 \text{ ft} = 2540 \text{ ft}$$

Step 4 Compute the pressure drop in the pipeline:

$$p_1 - p_2 = \gamma h_L = \frac{0.574 \text{ lb}}{\text{ft}^3} \cdot 2540 \text{ ft} \cdot \frac{1 \text{ ft}^2}{144 \text{ in}^2} = 10.12 \text{ psi}$$

Step 5 Compute the pressure at the compressor.

$$p_1 = p_2 + 10.12 \text{ psi} = 100 \text{ psig} + 10.12 \text{ psi} = 110.1 \text{ psig}$$

Step 6 Because the change in pressure is less than 10 percent, the assumption that the specific weight of the air is constant is satisfactory. If a larger pressure drop had occurred, we could either redesign the system with a larger pipe size or adjust the specific weight to the average of those at the beginning and end of the system. This system design appears to be satisfactory with regard to pressure drop.

18.5 FLOW OF AIR AND OTHER GASES THROUGH NOZZLES

The typical design of a nozzle is a converging section through which a fluid flows from a region of higher pressure to a region of lower pressure. Figure 18.11 shows a nozzle installed in the side of a relatively large tank with flow from the tank to the atmosphere. The nozzle shown converges smoothly and gradually, terminating at its smallest section, called the *throat*. Other designs for nozzles, including abrupt orifices and those attached to smaller pipes at the inlet, require special treatment, as we will discuss later.

Some concepts from the field of thermodynamics must now be discussed, along with some additional properties of gases.

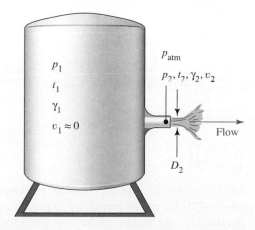

FIGURE 18.11 Discharge of a gas from a tank through a smooth convergent nozzle.

When the flow of a gas occurs very slowly, heat from the surroundings can transfer to or from the gas to maintain its temperature constant. Such flow is called *isothermal*. However, when the flow occurs very rapidly or when the system is very well insulated, very little heat can be transferred to or from the gas. Under ideal conditions with *no* heat transferred, the flow is called *adiabatic*. Real systems behave in some manner between isothermal and adiabatic. However, for rapid flow through a nozzle, we will assume that the flow is adiabatic.

18.5.1 Nozzle Flow for Adiabatic Process

For an adiabatic process, the equation that describes the relationship between the absolute pressure and the specific weight of the gas is

$$\frac{p}{\gamma^k} = \text{Constant} \tag{18–5}$$

The exponent k is called the *adiabatic exponent*, a dimensionless number, and its value for air is 1.40. Appendix N gives the values for k for other gases.

Equation (18–5) can be used to compute the condition of a gas at a point of interest if the condition at some other point is known and if an adiabatic process occurs between the two points. That is,

$$\frac{p}{\gamma^k} = \text{Constant} = \frac{p_1}{\gamma_1^k} = \frac{p_2}{\gamma_2^k} \tag{18–6}$$

Stated differently,

$$\frac{p_2}{p_1} = \left(\frac{\gamma_2}{\gamma_1}\right)^k \tag{18–7}$$

or

$$\frac{\gamma_2}{\gamma_1} = \left(\frac{p_2}{p_1}\right)^{1/k} \tag{18–8}$$

Here, p_2 is in the nozzle and p_1 is in the tank. The pressure outside the nozzle is p_{atm}.

The weight flow rate of gas exiting the tank through the nozzle in Fig. 18.11 is

$$W = \gamma_2 v_2 A_2 \tag{18–9}$$

The principles of thermodynamics can be used to show that the velocity of the flow in the nozzle is

▷ **Velocity of Flow through a Nozzle**

$$v_2 = \left\{\left(\frac{2gp_1}{\gamma_1}\right)\left(\frac{k}{k-1}\right)\left[1 - \left(\frac{p_2}{p_1}\right)^{(k-1)/k}\right]\right\}^{1/2} \tag{18–10}$$

Note that the pressures here are *absolute pressures*. Equations (18–6)–(18–10) can be combined to produce a convenient equation for the weight rate of flow from the tank in terms of the conditions of the gas in the tank and the pressure ratio p_2/p_1:

▷ **Weight Flow Rate when p_2/p_1 > Critical Ratio**

$$W = A_2 \sqrt{\frac{2gk}{k-1}(p_1\gamma_1)\left[\left(\frac{p_2}{p_1}\right)^{2/k} - \left(\frac{p_2}{p_1}\right)^{(k+1)/k}\right]} \tag{18–11}$$

Note that a *decreasing* pressure ratio p_2/p_1 actually indicates an *increasing* pressure difference $(p_1 - p_2)$, and, therefore, it is expected that the weight flow rate W will increase as the pressure ratio is decreased. This is true for the higher values of the pressure ratio. Under these conditions, p_2 in the throat is equal to p_{atm}.

However, it can be shown that the flow rate reaches a maximum at a *critical pressure ratio*, defined as

▷ **Critical Pressure Ratio**

$$\left(\frac{p_2'}{p_1}\right)_c = \left(\frac{2}{k+1}\right)^{k/(k-1)} \tag{18–12}$$

Because the value of the critical pressure ratio is a function only of the adiabatic exponent k, it is a constant for any particular gas.

When the critical pressure ratio is reached, the velocity of flow at the throat of the nozzle is equal to the speed of sound in the gas at the conditions that prevail there. *This velocity of flow remains constant regardless of how much the downstream pressure is lowered.* The speed of sound in the gas is

▷ **Sonic Velocity**

$$c = \sqrt{\frac{kgp_2'}{\gamma_2}} \tag{18–13}$$

Another name for c is the *sonic velocity*, the velocity that a sound wave would travel in the gas. This is the maximum velocity of flow of a gas through a converging nozzle.

Supersonic velocity, velocity greater than the speed of sound, can be obtained only with a nozzle that first converges and then diverges. The analysis of such nozzles is not presented here.

The name *Mach number* is given to the ratio of the actual velocity of flow to the sonic velocity. That is,

▷ **Mach Number**

$$N_M = v/c \tag{18–14}$$

Equation (18–11) must be used to compute the weight flow rate of gas from a tank through a converging nozzle for values of $N_M < 1.0$, for which the pressure ratio p_{atm}/p_1 is greater than the critical pressure ratio. For $N_M = 1$, substituting the critical pressure ratio from Eq. (18–12) into Eq. (18–11) yields

▷ **Maximum Weight Flow Rate when p_2/p_1 Critical**

$$W_{\text{max}} = A_2 \sqrt{\left(\frac{2gk}{k+1}\right)(p_1\gamma_1)\left(\frac{2}{k+1}\right)^{2/(k-1)}} \tag{18–15}$$

This equation must be used when the pressure ratio p_{atm}/p_1 is less than the critical ratio.

Figure 18.12 shows the behavior of gas flow through a nozzle from a relatively large tank, according to Eqs. (18–11) and (18–15). The graph in (a) is for the case in which the pressure in the tank p_1 is held constant and the pressure outside the nozzle p_{atm} is decreased. When $p_{atm} = p_1$, $p_{atm}/p_1 = 1.00$, and there is obviously no flow through the nozzle. As p_{atm} is decreased, the relatively larger difference in pressure $(p_1 - p_{atm})$ causes an increase in the weight flow rate as computed from Eq. (18–11). However, when the critical pressure ratio $(p_2'/p_1)_c$ is reached, the velocity in the throat reaches

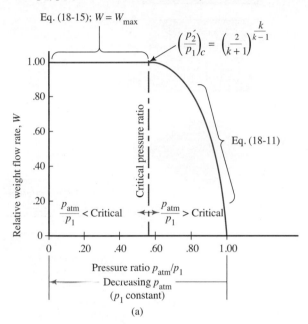

sonic velocity and the pressure remains at the critical pressure p_2' computed from Eq. (18–12). That is,

▷ **Critical Pressure, p_2'**

$$p_2 = p_2' = p_1 \left(\frac{2}{k+1}\right)^{k/(k-1)} \tag{18–16}$$

Decreasing the pressure outside the nozzle any further would *not* increase the rate of flow from the tank.

Figure 18.12(b) shows a different interpretation of the variation of weight flow rate versus pressure ratio. In this case, the pressure outside the nozzle, p_{atm}, is held constant while the pressure in the tank, p_1, is increased. Obviously, when the gage pressure in the tank is zero, no flow occurs because there is no pressure differential. As p_1 is increased, the pressure ratio p_{atm}/p_1 is at first greater than the critical pressure ratio, and Eq. (18–11) applies. When the critical pressure ratio is reached or exceeded, the velocity in the throat will be at sonic velocity *for the condition of the gas in the throat*.

Note, however, that for any given value of p_1, the critical pressure in the throat, p_2', is given by Eq. (18–16). Because p_1 is increasing, p_2' is also increasing. Still, because the pressure ratio between the tank and the throat is at the critical value, Eq. (18–15) must be used to compute the weight flow rate through the nozzle. The weight flow rate is now dependent on p_1 and γ_1. Note also that γ_1 is directly proportional to p_1, as can be seen from Eq. (18–2). Then, after the critical pressure ratio is reached, the weight flow rate increases linearly as the pressure in the tank increases.

If the nozzle has an abrupt reduction rather than the smooth shape shown in Fig. 18.10, the flow will be less than that predicted from Eq. (18–11) or (18–15). A discharge coefficient should be applied, similar to the discharge coefficients described in Chapter 15 for venturi meters, flow nozzles, and orifice meters. In addition, if the upstream section is relatively small, as in a pipe, some correction for the velocity of approach should be applied (see Reference 12).

In summary, the following procedure can be used to compute the weight flow rate of a gas through a nozzle of the type shown in Fig. 18.11, assuming that the flow is adiabatic. Figure 18.13 charts the process.

Computation of Adiabatic Flow of a Gas through a Nozzle

1. Compute the actual pressure ratio between the pressure outside the nozzle and that in the tank, p_{atm}/p_1.

2. Compute the critical pressure ratio by using Eq. (18–12).

3a. If the actual pressure ratio is greater than the critical pressure ratio, use Eq. (18–11) to compute the weight flow rate through the nozzle with $p_2 = p_{atm}$. If desired, the velocity of flow can be computed by using Eq. (18–10).

3b. If the actual pressure ratio is less than the critical pressure ratio, use Eq. (18–15) to compute the weight flow rate through the nozzle. Also, recognize that the velocity of flow in the throat of the nozzle is equal to the sonic velocity, computed from Eq. (18–13), and that the pressure at the throat is that called p_2' in Eq. (18–16). The gas then expands to p_{atm} as it leaves the nozzle.

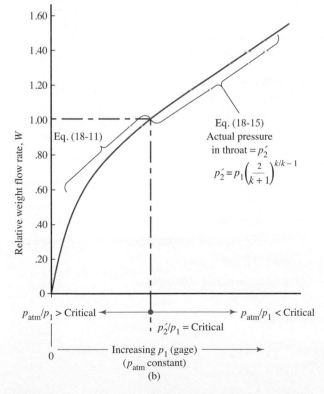

FIGURE 18.12 Weight flow rate of a gas through a nozzle.

FIGURE 18.13 Flow chart for computing
weight flow rate of a gas from a nozzle.

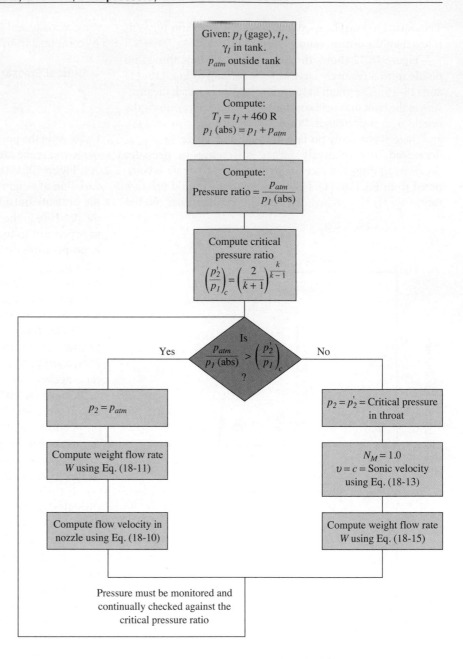

Given: p_1 (gage), t_1,
γ_1 in tank.
p_{atm} outside tank

Compute:
$T_1 = t_1 + 460$ R
p_1 (abs) $= p_1 + p_{atm}$

Compute:
Pressure ratio $= \dfrac{p_{atm}}{p_1 \text{ (abs)}}$

Compute critical
pressure ratio
$\left(\dfrac{p_2}{p_1}\right)_c = \left(\dfrac{2}{k+1}\right)^{\frac{k}{k-1}}$

Is
$\dfrac{p_{atm}}{p_1 \text{ (abs)}} > \left(\dfrac{p_2'}{p_1}\right)_c$
?

Yes

No

$p_2 = p_{atm}$

Compute weight flow rate
W using Eq. (18-11)

Compute flow velocity in
nozzle using Eq. (18-10)

$p_2 = p_2' =$ Critical pressure
in throat

$N_M = 1.0$
$v = c =$ Sonic velocity
using Eq. (18-13)

Compute weight flow rate
W using Eq. (18-15)

Pressure must be monitored and
continually checked against the
critical pressure ratio

**Example Problem
18.5**

For the tank with a nozzle in its side, shown in Fig. 18.11, compute the weight flow rate of air leaving the
tank for the following conditions:

$$p_1 = 10.0 \text{ psig} = \text{Pressure in the tank}$$

$$p_{atm} = 14.2 \text{ psia} = \text{Atmospheric pressure outside the tank}$$

$$t_1 = 80°F = \text{Temperature of the air in the tank}$$

$$D_2 = 0.100 \text{ in} = \text{Diameter of the nozzle at its outlet}$$

Solution Use the procedure outlined above.

1. Actual pressure ratio:

$$\frac{p_{atm}}{p_1} = \frac{14.2 \text{ psia}}{(10.0 + 14.2) \text{ psia}} = 0.587$$

2. Determine the critical pressure ratio from Appendix N. For air, it is 0.528.

3a. Because the actual ratio is greater than the critical ratio, Eq. (18–11) is used for the weight flow rate. We must compute the nozzle throat area A_2:

$$A_2 = \pi (D_2)^2/4 = \pi (0.100 \text{ in})^2/4 = 0.00785 \text{ in}^2$$

Converting to ft^2, we get

$$A_2 = 0.00785 \text{ in}^2 \ (1.0 \text{ ft}^2/144 \text{ in}^2) = 5.45 \times 10^{-5} \text{ ft}^2$$

Equation (18–3) can be used to compute γ_1:

$$\gamma_1 = \frac{2.70 \, p_1}{T_1} = \frac{2.70(24.2 \text{ psia})}{(80 + 460)°R} = 0.121 \text{ lb/ft}^3$$

It is helpful to convert p_1 to the units of lb/ft^2:

$$p_1 = \frac{24.2 \text{ lb}}{\text{in}^2} \frac{144 \text{ in}^2}{\text{ft}^2} = 3485 \text{ lb/ft}^2$$

Then, using Eq. (18–11), with consistent U.S. Customary System units, we find the result for W in lb/s. For these conditions, $p_2 = p_{atm}$:

$$W = A_2 \sqrt{\frac{2gk}{k-1}(p_1 \gamma_1) \left[\left(\frac{p_2}{p_1} \right)^{2/k} - \left(\frac{p_2}{p_1} \right)^{(k+1)/k} \right]} \tag{18–11}$$

$$W = (5.45 \times 10^{-5}) \sqrt{\frac{2(32.2)(1.4)(3485)(0.121)}{(1.4-1)} [(0.587)^{2/1.4} - (0.587)^{2.4/1.4}]}$$

$$W = 4.32 \times 10^{-3} \text{ lb/s}$$

Example Problem 18.6

For the conditions used in Example Problem 18.5, compute the velocity of flow in the throat of the nozzle and the Mach number for the flow.

Solution

Equation (18–10) must be used to compute the velocity in the throat. For consistent U.S. Customary System units, the velocity will be in ft/s:

$$v_2 = \left\{ \frac{2gp_1}{\gamma_1} \left(\frac{k}{k-1} \right) \left[1 - \left(\frac{p_2}{p_1} \right)^{(k-1)/k} \right] \right\}^{1/2} \tag{18–10}$$

$$v_2 = \left\{ \left(\frac{2(32.2)(3485)}{0.121} \right) \left(\frac{1.40}{0.40} \right) \left[1 - (0.587)^{0.4/1.4} \right] \right\}^{1/2}$$

$$v_2 = 957 \text{ ft/s}$$

To compute the Mach number, we need to compute the speed of sound in the air at the conditions existing in the throat by using Eq. (18–13):

$$c = \sqrt{\frac{kgp_2}{\gamma_2}} \tag{18–13}$$

The pressure $p_2 = p_{atm} = 14.2$ psia. Converting to lb/ft^2, we get

$$p_2 = \left(\frac{14.2 \text{ lb}}{\text{in}^2} \right) \left(\frac{144 \text{ in}^2}{1.0 \text{ ft}^2} \right) = 2045 \text{ lb/ft}^2$$

The specific weight γ_2 can be computed from Eq. (18–8):

$$\frac{\gamma_2}{\gamma_1} = \left(\frac{p_2}{p_1} \right)^{1/k} \tag{18–8}$$

Knowing that $\gamma_1 = 0.121$ lb/ft^3, we get

$$\gamma_2 = \gamma_1 \left(\frac{p_2}{p_1} \right)^{1/k}$$

$$\gamma_2 = (0.121)(0.587)^{1/1.4} = 0.0827 \text{ lb/ft}^3$$

Then, the speed of sound is

$$c = \sqrt{\frac{kgp_2}{\gamma_2}}$$

$$c = \sqrt{\frac{(1.4)(32.2)(2045)}{0.0827}} = 1056 \text{ ft/s} \qquad \text{(18–13)}$$

Now we can compute the Mach number:

$$N_M = \frac{v}{c} = \frac{957 \text{ ft/s}}{1056 \text{ ft/s}} = 0.906$$

Example Problem 18.7 Compute the weight flow rate of air from the tank through the nozzle shown in Fig. 18.11 if the pressure in the tank is raised to 20.0 psig. All other conditions are the same as in Example Problem 18.5.

Solution Use the same procedure outlined before.

1. Actual pressure ratio:

$$\frac{p_{atm}}{p_1} = \frac{14.2 \text{ psia}}{(20.0 + 14.2) \text{ psia}} = 0.415$$

2. The critical pressure ratio is again 0.528 for air.

3b. Because the actual pressure ratio is less than the critical pressure ratio, Eq. (18–15) should be used:

$$W_{max} = A_2 \sqrt{\frac{2gk}{k+1} (p_1\gamma_1)\left(\frac{2}{k+1}\right)^{2/(k-1)}} \qquad \text{(18–15)}$$

Computing γ_1 for $p_1 = 34.2$ psia, we get

$$\gamma_1 = \frac{2.70p_1}{T_1} = \frac{(2.70)(34.2 \text{ psia})}{540°R} = 0.171 \text{ lb/ft}^3$$

We need the pressure p_1 in lb/ft²:

$$p_1 = \left(\frac{34.2 \text{ lb}}{\text{in}^2}\right)\left(\frac{144 \text{ in}^2}{\text{ft}^2}\right) = 4925 \text{ lb/ft}^2$$

Then, the weight flow rate is

$$W_{max} = (5.45 \times 10^{-5}) \sqrt{\frac{2(32.2)(1.4)(4925)(0.171)}{2.4}\left(\frac{2}{2.4}\right)^{2/0.4}}$$

$$W_{max} = 6.15 \times 10^{-3} \text{ lb/s}$$

The velocity of the air flow at the throat will be the speed of sound at the throat conditions. However, the pressure at the throat must be determined from the critical pressure ratio, Eq. (18–12):

$$\left(\frac{p_2'}{p_1}\right)_c = \left(\frac{2}{k+1}\right)^{k/(k-1)} = 0.528$$

$$p_2' = p_1(0.528) = (4925 \text{ lb/ft}^2)(0.528) = 2600 \text{ lb/ft}^2 \qquad \text{(18–12)}$$

Knowing that $\gamma_1 = 0.171$ lb/ft³, we find

$$\gamma_2 = \gamma_1\left(\frac{p_2}{p_1}\right)^{1/k}$$

$$\gamma_2 = (0.171)(0.528)^{1/1.4} = 0.1084 \text{ lb/ft}^3$$

Then, the speed of sound and also the velocity in the throat is

$$c = \sqrt{\frac{kgp_2}{\gamma_2}} = \sqrt{\frac{(1.4)(32.2)(2600)}{0.1084}} = 1040 \text{ ft/s}$$

Of course, the Mach number in the throat is then 1.0.

REFERENCES

1. Air Movement and Control Association International. 2011. *Fan and Air Systems Application Handbook: Publications 200-95 (R20110), 201-02 (R2011), 202-98(R2011), and 203.90 (R2011).* Arlington Heights, Illinois: Author.

2. _____. 2008. *Industrial Process/Power Generation Fans: Specification Guidelines, AMCA Publication 801-01 (R2008).* Arlington Heights, Illinois: Author.

3. American Society of Heating, Refrigerating, and Air Conditioning Engineering. 2012. *ASHRAE Handbook.* Atlanta, GA: Author.

4. American Society of Mechanical Engineers. 2003. *Glossary of Terms Used in the Measurement of Fluid Flow in Pipes.* Standard ANSI/ASME MFC-1M. New York: Author.

5. _____. 1996. *Process Fan and Compressor Selection.* New York: Author.

6. Bleier, Frank P. 1997. *Fan Handbook.* New York: McGraw-Hill.

7. Bloch, Heinz and Fred K. Geither. 2012. *Compressors: How to Achieve High Reliability & Availability.* New York: McGraw-Hill.

8. Chopey, Nicholas P., ed., and the staff of *Chemical Engineering.* 1994. *Fluid Movers: Pumps, Compressors, Fans and Blowers,* 2nd ed. New York: McGraw-Hill.

9. Crane Co. 2011. *Flow of Fluids Through Valves, Fittings, and Pipe* (Technical Paper 410). Stamford, CT: Author.

10. Davidson, John and Otto von Bertele. 2011. *Compressor Selection.* New York: Wiley.

11. Hayes, W. H. 2003. *Industrial Exhaust Hood and Fan Piping,* 2nd ed. New York: Merchant Press.

12. Idelchik, I. E. E., N. A. Decker, and M. Steinberg. 1991. *Fluid Dynamics of Industrial Equipment.* New York: Taylor & Francis.

13. Idelchik, I. E. and M. O. Steinberg. 2005. *Handbook of Hydraulic Resistance.* Mumbai, India: Jaico Publishing House.

14. Peng, William W. 2007. *Fundamentals of Turbomachinery.* New York: Wiley.

15. The Trane Company. 1996. *Trane Air Conditioning Manual.* La Crosse, WI: Author.

INTERNET RESOURCES

1. **Hartzell Air Movement, Inc.:** Manufacturer of fans for commercial and industrial service for ventilation, process air supply and exhaust, heating equipment, and original equipment applications.

2. **Dresser-Rand:** Manufacturer of centrifugal and reciprocating compressors, steam turbines, gas turbines, and related products for the oil and gas, chemical, and petrochemical industries.

3. **Lau Industries-Dayton Ohio:** Manufacturer of air-moving components and fan systems for the heating, ventilation, air conditioning, and refrigeration industries.

4. **Continental Fan Manufacturing, Inc.:** Manufacturer of inline fans, blowers, and motorized impellers for the residential, commercial, and industrial markets. Included are duct fans, exhaust fans, dampers, and controls.

5. **New York Blower Company:** Manufacturer of many types and sizes of air-moving equipment for industrial and commercial applications. Included are centrifugal, axial, material handling, and roof ventilating fans, and heating products.

6. **Chicago Blower Corporation:** Manufacturer of fans and blowers for industrial and large commercial applications such as air handling, fume exhaust, HVAC, pneumatic conveying, drying, and supplying combustion air to boilers and incinerators. Included are centrifugal and axial fans and high-pressure blowers.

7. **Sullivan-Palatek:** Manufacturer of electric motor- and engine-driven rotary-screw air compressors for stationary and mobile applications.

8. **Quincy Compressor:** Manufacturer of reciprocating and rotary-screw compressors, vacuum pumps, and related equipment and controls.

9. **Elliott Group-Turbo Machinery:** Manufacturer of steam turbines, centrifugal air and gas compressors, power recovery units, generator systems, and cogeneration systems.

10. **Gardner Denver-Compressors:** Manufacturer of rotary-screw, reciprocating, and centrifugal air compressors and air treatment systems.

11. **Air Movement and Control Association International (AMCA):** An industry association of manufacturers of air system equipment for industrial, commercial, and residential markets.

12. **ASHRAE:** Formerly known as the American Society for Heating, Refrigerating, and Air-Conditioning Engineers, ASHRAE and its members focus on building systems, energy efficiency, indoor air quality, refrigeration and sustainability through research, standards writing, publishing, and continuing education.

13. **Trane® Heating and Air Conditioning:** A brand of the Ingersoll-Rand company, Trane® is a manufacturer of residential and commercial heating and air conditioning equipment. See also their publication, *Trane Air Conditioning Manual,* for technical details of planning and implementing air flow systems for heating and air conditioning systems. See Reference 15.

14. **Kaeser Compressors, Inc.:** Manufacturer of air compressors, blowers, and vacuum systems for industrial and commercial installations. Rotary screw-type, reciprocating, and rotary lobe-type compressors and blowers are provided along with dryers, filters, coolers, and condensate management systems and control systems for regulating compressed air demand.

Their SmartPipe™ aluminum delivery system piping provides low friction losses, high air quality, and easy installation.

15. **Compressed Gas Association (CGA):** An organization that promotes safe, secure, and environmentally responsible manufacture, transportation, storage, transfilling, and disposal of industrial and medical gases and their containers. Included are liquefied, nonliquefied, dissolved, and cryogenic gases.

16. **eCompressedair:** From the bottom of the home page, select *Main Library,* then select *Piping Systems* for an extensive set of documents giving guidelines for the design and installation of piping for compressed air systems for industrial applications.

PRACTICE PROBLEMS

Units and Conversion Factors

18.1 A pipe in a compressed air system is carrying 2650 cfm. Compute the flow rate in ft^3/s.

18.2 A duct in a heating system carries 8320 cfm. Compute the flow rate in ft^3/s.

18.3 A pipe in a compressed air system carries 2650 cfm. Compute the flow rate in m^3/s.

18.4 A duct in a heating system carries 8320 cfm. Compute the flow rate in m^3/s.

18.5 The velocity of flow in a ventilation duct is 1140 ft/min. Compute the velocity in m/s.

18.6 The velocity of flow in an air conditioning system duct is 5.62 m/s. Compute the velocity in ft/s.

18.7 A measurement for the static pressure in a heating duct is 4.38 inH_2O. Express this pressure in psi.

18.8 A fan is rated to deliver 4760 cfm of air at a static pressure of 0.75 inH_2O. Express the flow rate in m^3/s and the pressure in Pa.

18.9 The static pressure in a gas pipe is measured to be 925 Pa. Express this pressure in inH_2O.

18.10 Express the pressure of 925 Pa in psi.

Fans, Blowers, and Compressors

18.11 Describe a centrifugal fan with backward-inclined blades.

18.12 Describe a centrifugal fan with forward-curved blades.

18.13 Describe a duct fan.

18.14 Describe a vane–axial blower, and compare it with a duct fan.

18.15 Name four types of positive-displacement compressors.

18.16 Name a type of compressor often used for pneumatic fluid power systems.

Specific Weight of Air

18.17 Compute the specific weight of air at 80 psig and 75°F.

18.18 Compute the specific weight of air at 25 psig and 105°F.

18.19 Compute the specific weight of natural gas at 4.50 inH_2O and 55°F.

18.20 Compute the specific weight of nitrogen at 32 psig and 120°F.

18.21 Compute the specific weight of air at 1260 Pa(gage) and 25°C.

18.22 Compute the specific weight of propane at 12.6 psig and 85°F.

Flow of Compressed Air in Pipes

18.23 An air compressor delivers 820 cfm of free air. Compute the flow rate of air in a pipe in which the pressure is 80 psig and the temperature is 75°F.

18.24 An air compressor delivers 2880 cfm of free air. Compute the flow rate of air in a pipe in which the pressure is 65 psig and the temperature is 95°F.

18.25 Specify a size of Schedule 40 steel pipe suitable for carrying 750 cfm (FAD) at 100 psig with no more than 5.0-psi pressure drop in 100 ft of pipe.

18.26 Specify a size of Schedule 40 steel pipe suitable for carrying 165 cfm (FAD) at 100 psig with no more than 5.0-psi pressure drop in 100 ft of pipe.

18.27 Specify a size of Schedule 40 steel pipe suitable for carrying 800 cfm (free air) to a reactor vessel in a chemical processing plant in which the pressure must be at least 100 psig at 70°F. The total length of pipe from the compressor to the reactor vessel is 350 ft. The line contains eight standard elbows, two fully open gate valves, and one swing-type check valve. After completing the design, determine the required pressure at the compressor.

18.28 For an aeration process, a sewage treatment plant requires 3000 cfm of compressed air. The pressure must be 80 psig, and the temperature must be 120°F. The compressor is located in a utility building, and 180 ft of pipe is required. The line also contains one fully open butterfly valve, 12 elbows, four tees with the flow through the run, and one ball-type check valve. Specify a suitable size of Schedule 40 steel pipe, and determine the required pressure at the compressor.

Gas Flow through Nozzles

18.29 Air flows from a reservoir in which the pressure is 40.0 psig and the temperature is 80°F to a pipe in which the pressure is 20.0 psig. The flow is to be considered adiabatic. Compute the specific weight of the air both in the reservoir and in the pipe.

18.30 Air flows from a reservoir in which the pressure is 275 kPa and the temperature is 25°C to a pipe in which the pressure is 140 kPa. The flow is to be considered adiabatic. Compute the specific weight of the air both in the reservoir and in the pipe.

18.31 Refrigerant 12 expands adiabatically from 35.0 psig at a temperature of 60°F to 3.6 psig. Compute the specific weight of the refrigerant at both conditions.

18.32 Oxygen is discharged from a tank in which the pressure is 125 psig and the temperature is 75°F through a nozzle with a diameter of 0.120 in. The oxygen flows into the atmosphere, where the pressure is 14.40 psia. Compute the weight flow rate from the tank and the velocity of flow through the nozzle.

18.33 Repeat Problem 18.32, but change the pressure in the tank to 7.50 psig.

18.34 A high-performance racing tire is charged with nitrogen at 50 psig and 70°F. At what weight flow rate would the nitrogen escape through a valve with a diameter of 0.062 in into the atmosphere at a pressure of 14.60 psia?

18.35 Repeat Problem 18.34 at internal pressures of 45 psig through 0 psig in 5.0-psig decrements. Plot a graph of weight flow rate versus the internal pressure in the tire.

18.36 Figure 18.14 shows a two-compartment vessel. The compartments are connected by a smooth convergent nozzle. The left compartment contains propane gas and is maintained at a constant 25.0 psig and 65°F. The right compartment starts at an equal 25.0-psig pressure, and then the pressure is decreased to 0.0 psig. The local atmospheric pressure is 14.28 psia. Compute the weight rate of

FIGURE 18.14 Vessel for Problem 18.36.

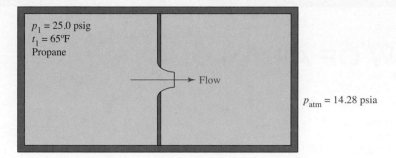

$p_1 = 25.0$ psig
$t_1 = 65°F$
Propane

→ Flow

$p_{atm} = 14.28$ psia

flow of propane through the 0.5-in nozzle as the pressure decreases in 5.0-psi decrements. Plot the weight flow rate versus the pressure in the right compartment.

18.37 Air flows from a large tank through a smooth convergent nozzle into the atmosphere, where the pressure is 98.5 kPa absolute. The temperature in the tank is 95°C. Compute the minimum pressure in the tank required to produce sonic velocity in the nozzle.

18.38 For the conditions of Problem 18.37, compute the magnitude of the sonic velocity in the nozzle.

18.39 For the conditions of Problem 18.37, compute the weight flow rate of air from the tank if the nozzle diameter is 10.0 mm.

18.40 A tank of Refrigerant 12 is at 150 kPa gage and 20°C. At what rate would the refrigerant flow from the tank into the atmosphere, where the pressure is 100.0 kPa absolute, through a smooth nozzle having a throat diameter of 8.0 mm?

18.41 For the tank described in Problem 18.40, compute the weight flow rate through the nozzle for tank gage pressures of 125 kPa, 100 kPa, 75 kPa, 50 kPa, and 25 kPa. Assume that the temperature in the tank is 20°C for all cases. Plot the weight flow rate versus tank pressure.

COMPUTER AIDED ENGINEERING ASSIGNMENTS

1. Write a program or spreadsheet for performing the calculations called for in Eqs. (18–2) and (18–3) for the specific weight of a gas and the correction of the volume flow rate for pressures and temperatures different from standard free air conditions.

2. Write a program or spreadsheet for the analysis of the flow of compressed air in a pipeline system. The program should use a procedure similar to that outlined in Example Problem 18.4. Note that some of the features of the program are similar to those used in earlier chapters for the flow of liquids in pipeline systems.

3. Write a program or spreadsheet for computing the velocity and weight rate of flow of a gas from a tank through a smooth convergent nozzle. The program should use Eqs. (18–6)–(18–16), which involve critical pressure ratio and sonic velocity.

4. Use the program or spreadsheet from Assignment 3 to solve Problems 18.32, 18.33, 18.35, 18.36, and 18.41. These problems call for analysis at several conditions.

FLOW OF AIR IN DUCTS

Heating, ventilation, and air conditioning systems distribute air at relatively low pressure through ducts. The fans or blowers responsible for moving the air, as described in Chapter 18, are generally high-volume, low-pressure devices. Knowledge of the pressures in the duct system is required to properly match a fan to a given system, to ensure the delivery of an adequate amount of air, and to balance the flow to various parts of the system. Often these systems are hidden from the general public or the occupants of a home, hotel, or shopping mall, but Fig. 19.1 shows a good view of a well-designed duct system.

Exploration

Think about all of the different places where you experience the comfortable, conditioned environment provided by heating, ventilating, and air-conditioning systems in any given day.

- Acquire data about a forced-air heating, air conditioning, or ventilation system that you can gain access to. It might be in your home, in your school, in a commercial building, or in an industrial plant.

- Describe the system in as much detail as possible including the size and shape of the ducts, the kind of fan used, the location of the fan, and how the air is distributed to the conditioned space.

Introductory Concepts

Here are some things you should look for when examining a duct system.

- Where is the main fan or blower that forces the air through the system? Describe its physical size and configuration, referring to Figs. 18.2 to 18.4 in Chapter 18 for examples. Can you find ratings for the fan such as its speed, delivery (volume flow rate), and pressure rating? The delivery may be reported in units such as cfm, an abbreviation for cubic feet per minute. The pressure rating might be in psi or inches of water or some other unit. Refer back to Section 18.2 in the early part of Chapter 18 for the definition of common units for measuring flow rate and pressures in air delivery systems.

- How does the air get to the intake of the fan? Where does it come from?

- Where does the air go directly from the discharge section of the fan? Is the fan a part of the burner of a furnace or a space heater? Does it deliver air over the cooling coils of an air conditioning system? Or is the flow delivered directly to ductwork for ventilation without affecting its temperature?

- Follow the ductwork from the fan outlet to each of its discharge points. Try to get measurements for the dimensions of the duct. Is it round, square, or

FIGURE 19.1 Providing fresh, conditioned air to living and working spaces is an important part of building design. This commercial HVAC system includes the large diameter feeder duct and several smaller branch ducts with their diffusers that distribute the air throughout the space. (*Source:* Realchemyst/Fotolia)

rectangular? Are there any bends, reducers, or expansions in the ductwork?

- Are there any control devices such as dampers installed in the ducts that allow the air flow to be partially blocked? This allows the system operator to balance the flow to ensure that an adequate amount of conditioned air is delivered to each destination point.

- Describe the grilles or registers that control the air delivery at each destination. What are their critical dimensions?

- Are there provisions for the air to return back from the conditioned spaces to the fan system to encourage air recirculation? If so, how is that accomplished?

An Example Air Distribution System

Figure 19.2 is a sketch of the layout of an air distribution system. Outside air enters the building at point 1 through louvers that protect the ductwork from wind and rain. The velocity of the air flow through the louvers should be relatively low, approximately 500 ft/min (2.5 m/s), to minimize entrainment of undesirable contaminants. The duct then reduces to a smaller size to deliver the air to the

suction side of a fan. A sudden contraction of the duct is shown, although a more gradual reduction would have a lower pressure loss.

The damper in the intake duct can partially close off the duct to decrease the flow, if desired. The intake duct delivers the air to the fan inlet, where its pressure is increased by the action of the fan wheel. Of course, the fan is responsible for creating the flow, drawing in the outside air and delivering it into the distribution ductwork. It must overcome all of the resistances provided by friction in the ducts and losses due to elbows, branch tees, dampers, and grilles while moving an adequate supply of air.

The outlet from the fan is carried by a main duct from which four branches deliver the air to its points of use. Dampers, shown in each of the branches, permit balancing of the system while in operation. Grilles are used at each outlet to distribute the air to the conditioned spaces (in this example, three offices and a conference room).

The ductwork shown is mostly above the ceiling. Note that an elbow is placed above each outlet grille to direct the air flow downward through the ceiling system.

It is important to understand the basic operating parameters of such an air-handling system. Obviously,

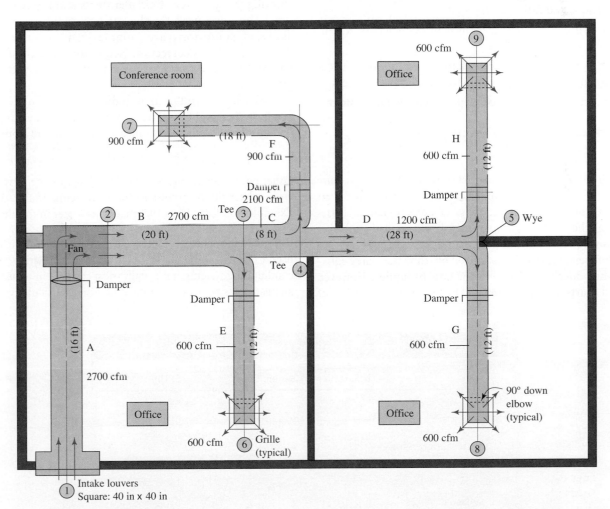

FIGURE 19.2 Air distribution system.

the air outside the building is at the prevailing atmospheric pressure. To cause flow into the duct through the intake louvers, the fan must create a pressure lower than atmospheric in the duct. This is a negative gage pressure. As the air flows through the duct, friction losses cause the pressure to decrease further. In addition, any obstruction to the flow, such as the intake louvers, a damper, and tees or wyes that direct the flow, cause pressure drops. The fan increases the pressure of the air and forces it through the supply ductwork to the outlet grilles.

The air inside the rooms of the building may be slightly above or below atmospheric pressure. Some air-handling-system designers prefer to have a slight positive pressure in the building for better control and to eliminate drafts. However, when designing the ductwork, one usually considers the pressure inside the building to be the same as that outside the building. Consider all of the References and Internet resources listed at the end of the chapter for guidelines on the design of duct systems (References 1–11 and Internet resources 1–6 and 8), for examples of commercially available components (Reference 10 and Internet resource 8), and for software that can aid in the duct design process (Internet resources 7–9).

19.1 OBJECTIVES

After completing this chapter, you should be able to:

1. Describe the basic elements of an air distribution system that may be used for heating, ventilation, or air conditioning.

2. Determine energy losses in ducts, considering straight sections and fittings.

3. Determine the circular equivalent diameters of rectangular ducts.

4. Analyze and design ductwork to carry air to spaces needing conditioning and to achieve balance in the system.

5. Identify the fan selection requirements for the system.

19.2 ENERGY LOSSES IN DUCTS

Two kinds of energy losses in duct systems cause the pressure to drop along the flow path. Friction losses occur as the air flows through straight sections, whereas dynamic losses occur as the air flows through such fittings as tees and wyes and through flow control devices.

Friction losses can be estimated by using the Darcy equation introduced in Chapter 8 for flow of liquids. However, special charts have been prepared by the American Society of Heating, Refrigerating, and Air-Conditioning Engineers (ASHRAE) for the typical conditions found in duct design (Reference 1 and Internet resource 3). Internet resource 5 is the website for the Heating, Refrigeration, and Air Conditioning Institute of Canada (HRAI).

Figures 19.3 and 19.4 show friction loss h_L as a function of volume flow rate, with two sets of diagonal lines showing the diameter of circular ducts and the velocity of flow. The units used for the various quantities and the assumed conditions are summarized in Table 19.1. Reference 2 shows correction factors for other conditions. See Section 18.2 for information on air flow rates and pressures.

We use the symbol h_L to indicate the friction loss per 100 ft of duct read from Fig. 19.3. Then the total energy loss for a given duct length L is called H_L and is found from

$$H_L = h_L(L/100)$$

Other energy losses will also be designated by the H symbol with subscripts pertinent to the unit being analyzed. See Example Problems 19.1–19.4 that follow later in the chapter.

Rectangular Ducts   Although circular ducts are often used for distributing air through heating, ventilating, or air conditioning systems, it is usually most convenient to use rectangular ducts because of space limitations, particularly above

---

TABLE 19.1   Units and assumed conditions for friction charts

	U.S. Customary System Units	SI Units
Flow rate	ft^3/min (cfm)	m^3/s
Friction loss h_L	in of water per 100 ft (inH$_2$O/100 ft)	Pa/m
Velocity	ft/min	m/s
Duct diameter	in	mm
Specific weight of air	0.075 lb/ft^3	11.81 N/m^3
Duct surface roughness	5×10^{-4} ft	1.5×10^{-4} m
Condition of air	14.7 psia; 68°F	101.3 kPa; 20°C

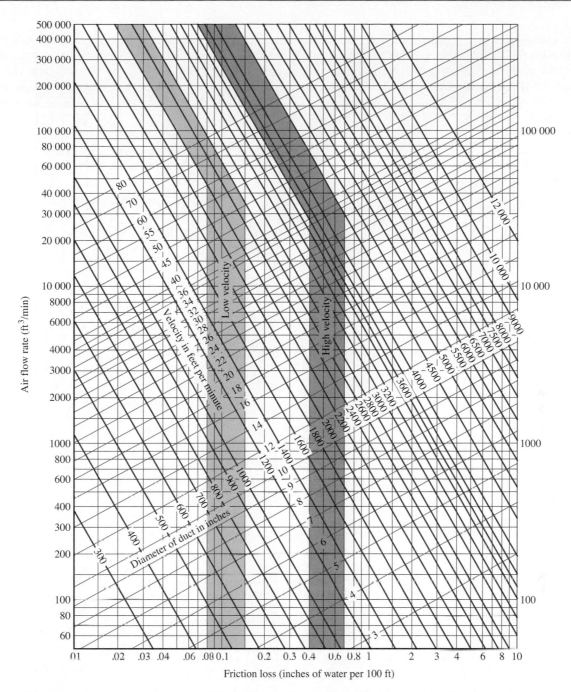

FIGURE 19.3 Friction loss in ducts—U.S. Customary System units. (*Source:* From *ASHRAE Handbook—1981 Fundamentals.* Copyright © 1981 American Society of Heating, Refrigerating, and Air-Conditioning Engineers, Inc. Reprinted with permission.)

ceilings. The hydraulic radius of the rectangular duct can be used to characterize its size (as discussed in Section 9.5). When the necessary substitutions of the hydraulic radius for the diameter are made in relationships for velocity, Reynolds number, relative roughness, and the corresponding friction factor, we see that the *circular equivalent diameter* for a rectangular duct is

Circular Equivalent Diameter for a Rectangular Duct

$$D_e = \frac{1.3\,(ab)^{5/8}}{(a+b)^{1/4}} \qquad (19\text{--}1)$$

where *a* and *b* are sides of the rectangle.

This enables you to use the friction loss charts in Figs. 19.3 and 19.4 for rectangular as well as circular ducts. Table 19.2 shows some results computed using Eq. (19–1).

Flat Oval Ducts Another popular shape for air ducts is the flat oval shown in Fig. 19.5. The cross-sectional area is the sum of a rectangle and a circle, found from

$$A = \pi a^2/4 + a(b - a) \qquad (19\text{--}2)$$

where *a* is the length of the minor axis of the duct and *b* is the length of the major axis.

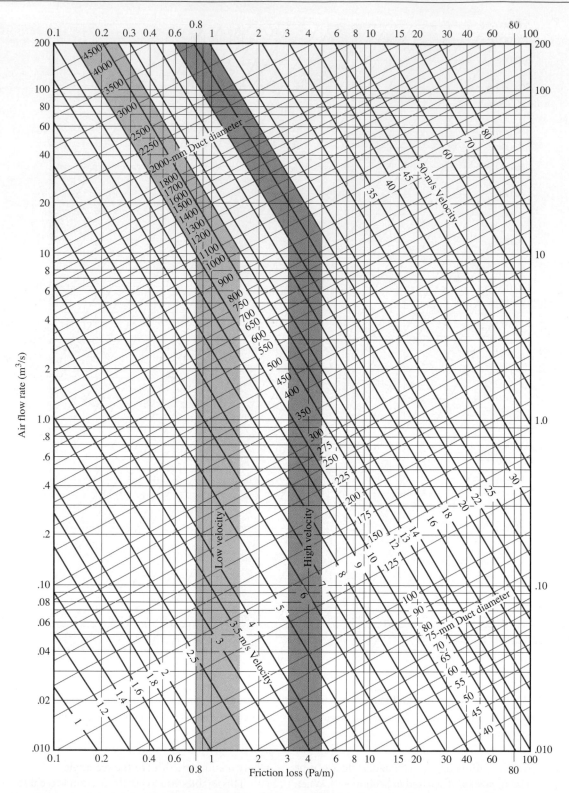

FIGURE 19.4   Friction loss in ducts—SI units. (*Source:* From *ASHRAE Handbook—1981 Fundamentals.* Copyright © 1981 American Society of Heating, Refrigerating, and Air-Conditioning Engineers, Inc. Reprinted with permission.)

The circular equivalent diameter D_e of a flat oval duct is needed to use Figs. 19.3 and 19.4 to determine friction loss:

$$D_e = \frac{1.55A^{0.625}}{WP^{0.250}} \quad (19\text{–}3)$$

where WP is the wetted perimeter as defined in Chapter 9, found from

$$WP = \pi a + 2(b - a) \quad (19\text{–}4)$$

Table 19.3 shows some examples of the circular equivalent diameters for flat oval ducts.

TABLE 19.2 Circular equivalent diameters of rectangular ducts

Side b (in)	6	8	10	12	14	16	Side a (in) 18	20	22	24	26	28	30
6	6.6												
8	7.6	8.7											
10	8.4	9.8	10.9										
12	9.1	10.7	12.0	13.1									
14	9.8	11.5	12.9	14.2	15.3								
16	10.4	12.2	13.7	15.1	16.4	17.5							
18	11.0	12.9	14.5	16.0	17.3	18.5	19.7						
20	11.5	13.5	15.2	16.8	18.2	19.5	20.7	21.9					
22	12.0	14.1	15.9	17.6	19.1	20.4	21.7	22.9	24.0				
24	12.4	14.6	16.5	18.3	19.9	21.3	22.7	23.9	25.1	26.2			
26	12.8	15.1	17.1	19.0	20.6	22.1	23.5	24.9	26.1	27.3	28.4		
28	13.2	15.6	17.7	19.6	21.3	22.9	24.4	25.8	27.1	28.3	29.5	30.6	
30	13.6	16.1	18.3	20.7	22.0	23.7	25.2	26.6	28.0	29.3	30.5	31.7	32.8

FIGURE 19.5 Flat oval duct shape.

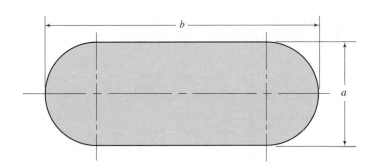

TABLE 19.3 Circular equivalent diameters of flat oval ducts

Minor Axis	8	10	12	14	16	Major Axis 18	20	22	24	26	28	30
6	7.1	8.1	8.9	9.6	10.2	10.8	11.3	11.8	12.3	12.7	13.1	13.5
8		9.2	10.2	11.0	11.8	12.5	13.2	13.8	14.4	14.9	15.4	15.9
10			11.2	12.2	13.2	14.0	14.8	15.5	16.2	16.8	17.4	18.0
12				13.2	14.3	15.3	16.1	17.0	17.7	18.5	19.2	19.8
14					15.2	16.3	17.3	18.3	19.1	19.9	20.7	21.4
16						17.2	18.3	19.4	20.3	21.2	22.1	22.9
18							19.2	20.4	21.4	22.4	23.3	24.2
20								21.2	22.4	23.5	24.5	25.4
22									23.3	24.4	25.5	26.5
24										25.3	26.4	27.5
26											27.3	28.4
28												29.3

Example Problem 19.1 Determine the velocity of flow and the amount of friction loss that would occur as air flows at 3000 cfm through 80 ft of circular duct having a diameter of 22 in.

Solution We can use Fig. 19.3 to determine that the velocity is approximately 1150 ft/min and that the friction loss per 100 ft of duct (h_L) is 0.082 inH$_2$O. Then, by proportion, the loss for 80 ft is

$$H_L = h_L(L/100) = (0.082 \text{ inH}_2\text{O}) \left(\tfrac{80}{100}\right) = 0.066 \text{ inH}_2\text{O}$$

Example Problem 19.2 Specify the dimensions of a rectangular duct that would have the same friction loss as the circular duct described in Example Problem 19.1.

Solution From Table 19.2 we can specify a 14-by-30-in rectangular duct that would have the same loss as the 22.0-in-diameter circular duct. Others that will have approximately the same loss are 16-by-26-in, 18-by-22-in, and 20-by-20-in rectangular ducts. Such a list gives the designer many options when fitting a duct system into given spaces.

Example Problem 19.3 Specify the dimensions of a flat oval duct that would have approximately the same friction loss as the circular duct described in Example Problem 19.1.

Solution From Table 19.3 we can specify a 16-by-28-in flat oval duct that would have approximately the same friction loss as a 22.0-in-diameter circular duct. Others with approximately the same loss are 18-by-26-in and 20-by-24-in flat oval ducts.

Dynamic losses can be estimated by using published data for loss coefficients for air flowing through certain fittings (see References 2 and 5). In addition, manufacturers of special air-handling devices publish extensive data about expected pressure drops. Table 19.4 presents a few examples for use in problems in this book. Note that these data are highly simplified. For example, actual loss coefficients for tees depend on the size of the branches and on the amount of air flow in each branch. As with the minor losses discussed in Chapter 10, changes in flow area or direction of flow should be made as smooth as possible to minimize dynamic losses. The data for round, 90° elbows show the large variations possible.

The dynamic loss for a fitting is calculated from

$$H_L = C(H_v) \tag{19–5}$$

where C is the loss coefficient from Table 19.4 and H_v is the *velocity pressure* or *velocity head*.

In U.S. Customary System units, pressure levels and losses are typically expressed in inches of water, which is actually a measure of pressure head. Then,

$$H_v = \frac{\gamma_a v^2}{2g\gamma_w} \tag{19–6}$$

where γ_a is the specific weight of air, v is the flow velocity, and γ_w is the specific weight of water. When the velocity is expressed in feet per minute and the conditions of standard air are used, Eq. (19–6) reduces to

⇨ **Velocity Pressure for Air Flow (U.S. Units)**

$$H_v = \left(\frac{v}{4005}\right)^2 \tag{19–7}$$

When we use SI units, pressure levels and losses are measured in the pressure unit of Pa. Then,

$$H_v = \frac{\gamma_a v^2}{2g} \tag{19–8}$$

When the velocity is expressed in m/s and the conditions of standard air are used, Eq. (19–8) reduces to

⇨ **Velocity Pressure for Air Flow (SI Units)**

$$H_v = \left(\frac{v}{1.289}\right)^2 \text{Pa} \tag{19–9}$$

TABLE 19.4 Examples of loss factors for duct fittings

Dynamic Loss Coefficient C						
90° elbows						
Smooth, round	0.22					
5-piece, round	0.33					
4-piece, round	0.37					
3-piece, round	0.42					
Mitered, round	1.20					
Smooth, rectangular	0.18					
Tee, branch	1.00					
Tee, flow through main	0.10					
Symmetrical wye	0.30					

Damper Position	0° (wide open)	10°	20°	30°	40°	50°
C	0.20	0.52	1.50	4.5	11.0	29

Outlet grille: Assume total pressure drop through grille is 0.06 inH$_2$O (15 Pa).

Intake louvers: Assume total pressure drop across louver is 0.07 inH$_2$O (17 Pa).

Note: Dynamic loss for fittings is $C(H_v)$, where H_v is the velocity pressure upstream of the fitting. Values shown are examples, for use only in solving problems in this book. Many factors affect the actual values for a given style of fitting. Refer to Reference 2 or manufacturers' catalogs for more complete data.

Example Problem 19.4

Estimate the pressure drop that occurs when 3000 cfm of air flows around a smooth, rectangular, 90° elbow with side dimensions of 14 × 24 in.

Solution

Use Table 19.2 to find that the circular equivalent diameter for the duct is 19.9 in. From Fig. 19.3 we find the velocity of flow to be 1400 ft/min. Then, using Eq. (19–7), we compute

$$H_v = \left(\frac{v}{4005}\right)^2 = \left(\frac{1400}{4005}\right)^2 = 0.122 \text{ inH}_2\text{O}$$

From Table 19.4, we find $C = 0.18$. Then, the pressure drop is

$$H_L = C(H_v) = (0.18)(0.122) = 0.022 \text{ inH}_2\text{O}$$

19.3 DUCT DESIGN

In The Big Picture of this chapter, the general features of ducts for carrying the flow of air were described. Figure 19.2 shows a simple duct system, the operation of which has been described. In this section, we describe one method of designing such a duct system.

The goals of the design process are to specify reasonable dimensions for the various sections of the ductwork, to estimate the air pressure at key points, to determine the requirements to be met by the fan in the system, and to balance the system. Balance requires that the pressure drop from the fan outlet to each outlet grille is the same when the duct sections are carrying their design capacities.

Several different techniques are used by air distribution designers, such as the following:

- *Equal-friction method*   Use Fig. 19.3 or 19.4 to specify a uniform value for the friction loss per unit length of duct. For *low-velocity systems*, the loss is between 0.08 inH$_2$O and 0.16 inH$_2$O (0.8–1.5 Pa/m).

- *Static regain method*   The design of ducts is adjusted to obtain the same static pressure at all junctions. Some iteration is required.

- *T-Method*   This is an optimization procedure that considers system performance along with cost factors. Considered are energy cost, initial system cost, time of operation of the system, efficiencies of the fan and the drive motor, and costs related to financing the investment and inflation.

■ *Industrial exhaust systems for vapors and particulates* Special considerations are discussed in Reference 2 for exhaust systems to ensure that velocities are sufficiently high to entrain and convey particulates. The selection of fitting types is also critical to avoid places where particulates can accumulate.

Only the equal-friction method is demonstrated in this chapter, and a general procedure follows. See Reference 2 and Internet resource 1 for additional information about all four methods. Several commercial software packages are available that assist designers in organizing the system design procedure and completing the numerous calculations. Internet resources 1 and 7–9 give some examples.

Smaller systems for homes and light commercial applications are mostly of the "low-velocity" type, in which ductwork and fittings are relatively simple. Noise is usually not a major problem if the limits shown in Figs. 19.3 and 19.4 are not exceeded. However, the resulting sizes for ducts in a low-velocity system are relatively large.

Space limitations in the design of large office buildings and certain industrial applications make "high-velocity systems" attractive. The name comes from the practice of using smaller ducts to carry a given flow rate. However, several consequences arise:

1. Noise is usually a factor, and special noise-attenuation devices must be employed.

2. Duct construction must be more substantial, and sealing is more critical.

3. Operating costs are generally higher because of greater pressure drops and higher fan total pressures.

High-velocity systems can be justified when lower building costs result or when more efficient use of space can be achieved.

General Procedure for Designing Air Ducts using the Equal-Friction Method

1. Generate a proposed layout of the air distribution system:
   a. Determine the air flow desired into each conditioned space (cfm or m^3/s).
   b. Specify the location of the fan.
   c. Specify the location of the outside air supply inlet.
   d. Propose the layout of the ductwork for the intake duct.

   e. Propose the layout of the air delivery system to each space including fittings such as tees, elbows, dampers, and grilles. Dampers should be included in the final run to each delivery grille to facilitate final balance of the system.

2. For the intake duct and the fan outlet duct, determine the total airflow requirement as the sum of all of the air flows delivered to conditioned spaces.

3. Use Fig. 19.3 or 19.4 to specify the nominal friction loss (inH$_2$O/100 ft or Pa/m). Low-velocity design is recommended for typical commercial or residential systems.

4. Specify the nominal flow velocity for each part of the duct system. For the intake duct and the final runs to occupied spaces, use approximately 600–800 ft/min (3–4 m/s). For main ducts away from occupied spaces, use approximately 1200 ft/min (6 m/s).

5. Specify the size and shape of each part of the duct system. The diameters for circular ducts are found directly from Fig. 19.3 or 19.4. Rectangular ducts can be sized using Table 19.2 and Eq. (19–1). Use Table 19.3 and Eq. (19–3) for flat oval ducts.

6. Compute the energy losses in the intake duct and in each section of the delivery duct.

7. Compute the total energy loss for each path from the fan outlet to each delivery grille.

8. Determine whether the energy losses for all paths are reasonably balanced, that is, the pressure drop from the fan to each outlet grille is approximately equal.

9. If significant unbalance occurs, redesign the ductwork by typically reducing the design velocity in those ducts where high pressure drops occur. This requires using larger ducts.

10. Reasonable balance is achieved when all paths have small differences in pressure drop such that modest adjustment to dampers will achieve a true balance.

11. Determine the pressure at the fan inlet and outlet and the total pressure rise across the fan.

12. Specify a fan that will deliver the total air flow at this pressure rise.

13. Plot or chart the pressure in the duct for each path and inspect for any unusual performance. The following design example illustrates the application of this procedure for a low-velocity system.

Example Problem 19.5

The system shown in Fig. 19.2 is being designed for a small office building. The air is drawn from outside the building by a fan and delivered through four branches to three offices and a conference room. The air flows shown at each outlet grille have been determined by others to provide adequate ventilation to each area. Dampers in each branch permit final adjustment of the system.

Complete the design of the duct system, specifying the size of each section of the ductwork for a low-velocity system. Compute the expected pressure drop for each section and at each fitting. Then, compute the total pressure drop along each branch from the fan to the four outlet grilles and check for system balance. If a major imbalance is predicted, redesign appropriate parts of the system to achieve a more nearly balanced system. Then, determine the total pressure required for the fan. Use Fig. 19.3 for estimating friction losses and Table 19.4 for dynamic loss coefficients.

Solution First, treat each section of the duct and each fitting separately. Then, analyze the branches.

1. Intake duct A: $Q = 2700$ cfm; $L = 16$ ft.
   Let $v \approx 800$ ft / min.

   From Fig. 19.3, required $D = 25.0$ in.

$$h_L = 0.035 \, \text{inH}_2\text{O}/100 \, \text{ft}$$
$$H_L = 0.035(16/100) = 0.0056 \, \text{inH}_2\text{O}$$

2. Damper in duct A: $C = 0.20$ (assume wide open) (Table 19.4).

$$\text{For 800 ft/min, } H_v = (800/4005)^2 = 0.040 \, \text{inH}_2\text{O}$$
$$H_L = 0.20(0.040) = 0.0080 \, \text{inH}_2\text{O}$$

3. Intake louvers: The 40-by-40-in size has been specified to give approximately a velocity of 600 ft/min through the open space of the louvers. Use $H_L = 0.070 \, \text{inH}_2\text{O}$ from Table 19.4.

4. Sudden contraction between louver housing and intake duct: From Fig. 10.8, we know that the resistance coefficient depends on the velocity of flow and the ratio D_1/D_2 for circular conduits. Because the louver housing is a 40-by-40-in square, we can compute its equivalent diameter from Eq. (19–1):

$$D_e = \frac{1.3(ab)^{5/8}}{(a + b)^{1/4}} = \frac{1.3(40 \times 40)^{5/8}}{(40 + 40)^{1/4}} = 43.7 \, \text{in}$$

   Then, in Fig. 10.8,

$$D_1/D_2 = 43.7/25 = 1.75$$

   and $K = C = 0.31$. Then,

$$H_L = C(H_v) = 0.31(0.04) = 0.0124 \, \text{inH}_2\text{O}$$

5. Total loss in intake system:

$$H_L = 0.0056 + 0.0080 + 0.07 + 0.0124 = 0.096 \, \text{inH}_2\text{O}$$

   Because the pressure outside the louvers is atmospheric, the pressure at the inlet to the fan is $-0.096 \, \text{inH}_2\text{O}$, a negative gage pressure. An additional loss may occur at the fan inlet if a geometry change is required to mate the intake duct with the fan. Knowledge of the fan design is required, and this potential loss is ignored in this example.

Note: All ducts on the fan outlet side are rectangular.

6. Fan outlet, duct B: $Q = 2700$ cfm; $L = 20$ ft.
   Let $v \approx 1200$ ft/min; $h_L = 0.110 \, \text{inH}_2\text{O}/100$ ft.

$$D_e = 20.0 \, \text{in; use 12-by-30-in size to minimize overhead space required}$$
$$H_L = 0.110(20/100) = 0.0220 \, \text{inH}_2\text{O}$$
$$H_v = (1200/4005)^2 = 0.090 \, \text{inH}_2\text{O}$$

7. Duct E: $Q = 600$ cfm; $L = 12$ ft.
   Let $v \approx 800$ ft/min; $h_L = 0.085 \, \text{inH}_2\text{O}/100$ ft.

$$D_e = 12.0 \, \text{in; use 12-by-10-in size}$$
$$H_L = 0.085(12/100) = 0.0102 \, \text{inH}_2\text{O}$$
$$H_v = (800/4005)^2 = 0.040 \, \text{inH}_2\text{O}$$

8. Damper in duct E: $C = 0.20$ (assume wide open).

$$H_L = 0.20(0.040) = 0.0080 \, \text{inH}_2\text{O}$$

9. Elbow in duct E: smooth rectangular elbow; $C = 0.18$.

$$H_L = 0.18(0.040) = 0.0072 \, \text{inH}_2\text{O}$$

10. Grille 6 for duct E: $H_L = 0.060 \, \text{inH}_2\text{O}$.

11. Tee 3 from duct B to branch E, flow in branch: $C = 1.00$.
    H_L based on velocity ahead of tee in duct B:

$$H_L = 1.00(0.090) = 0.090 \, \text{inH}_2\text{O}$$

12. Duct C: $Q = 2100 \, \text{cfm}$; $L = 8 \, \text{ft}$.
    Let $v \approx 1200 \, \text{ft/min}$; $h_L = 0.110 \, \text{inH}_2\text{O}/100 \, \text{ft}$.

$$D_e = 18.5 \, \text{in; use 12-by-24-in size}$$
$$H_L = 0.110(8/100) = 0.0088 \, \text{inH}_2\text{O}$$
$$H_v = (1200/4005)^2 = 0.090 \, \text{inH}_2\text{O}$$

13. Tee 3 from duct B to duct C, flow through main: $C = 0.10$.

$$H_L = 0.10(0.090) = 0.009 \, \text{inH}_2\text{O}$$

14. Duct F: $Q = 900 \, \text{cfm}$; $L = 18 \, \text{ft}$.
    Let $v \approx 800 \, \text{ft/min}$; $h_L = 0.068 \, \text{inH}_2\text{O}/100 \, \text{ft}$.

$$D_e = 14.3 \, \text{in; use 12-by-14-in size}$$
$$H_L = 0.068(18/100) = 0.0122 \, \text{inH}_2\text{O}$$
$$H_v = (800/4005)^2 = 0.040 \, \text{inH}_2\text{O}$$

15. Damper in duct F: $C = 0.20$ (assume wide open).

$$H_L = 0.20(0.040) = 0.0080 \, \text{inH}_2\text{O}$$

16. Two elbows in duct F: smooth rectangular elbow; $C = 0.18$.

$$H_L = 2(0.18)(0.040) = 0.0144 \, \text{inH}_2\text{O}$$

17. Grille 7 for duct F: $H_L = 0.060 \, \text{inH}_2\text{O}$.

18. Tee 4 from duct C to branch F, flow in branch: $C = 1.00$.
    H_L based on velocity ahead of tee in duct C:

$$H_L = 1.00(0.090) = 0.090 \, \text{inH}_2\text{O}$$

19. Duct D: $Q = 1200 \, \text{cfm}$; $L = 28 \, \text{ft}$.
    Let $v \approx 1000 \, \text{ft/min}$; $h_L = 0.100 \, \text{inH}_2\text{O}/100 \, \text{ft}$.

$$D_e = 14.7 \, \text{in; use 12-by-16-in size}$$
$$\text{Actual } D_e = 15.1 \, \text{in; new } h_L = 0.087 \, \text{inH}_2\text{O}/100 \, \text{ft}$$
$$H_L = 0.087(28/100) = 0.0244 \, \text{inH}_2\text{O}$$
$$\text{New } v = 960 \, \text{ft/min}$$
$$H_v = (960/4005)^2 = 0.057 \, \text{inH}_2\text{O}$$

20. Tee 4 from duct C to duct D, flow through main: $C = 0.10$.

$$H_L = 0.10(0.090) = 0.009 \, \text{inH}_2\text{O}$$

21. Wye 5 between duct D and ducts G and H: $C = 0.30$.

$$H_L = 0.30(0.057) = 0.017 \, \text{inH}_2\text{O}$$

This loss applies to either duct G or duct H.

22. Ducts G and H are identical to duct E, and losses from Steps 7–10 can be applied to these paths.

This completes the evaluation of the pressure drops through components in the system. Now we can sum the losses through any path from the fan outlet to the outlet grilles.

a. Path to grille 6 in duct E: sum of losses from Steps 6–11:

$$H_6 = 0.0220 + 0.0102 + 0.0080 + 0.0072 + 0.060 + 0.090$$
$$= 0.1974 \, \text{inH}_2\text{O}$$

b. Path to grille 7 in duct F: sum of losses from Steps 6 and 12–18:

$$H_7 = 0.0220 + 0.0088 + 0.0090 + 0.0122 + 0.0080 + 0.0144 + 0.060 + 0.090$$
$$= 0.2244 \, \text{inH}_2\text{O}$$

c. Path to either grille 8 in duct G or grille 9 in duct H: sum of losses from Steps 6, 12, 13, 19–21, and 7–10:

$$H_8 = 0.0220 + 0.0088 + 0.0090 + 0.0244 + 0.0090 + 0.0170$$
$$+ 0.0102 + 0.008 + 0.0072 + 0.06$$
$$= 0.1756 \, \text{inH}_2\text{O}$$

Redesign to Achieve a Balanced System The ideal system design would be one in which the loss along any path, a, b, or c, is the same. Because that is not the case here, some redesign is called for. The loss in path b to grille 7 in duct F is much higher than the others. The component losses from Steps 12, 14–16, and 18 affect this branch, and some reduction can be achieved by reducing the velocity of flow in ducts C and F.

12a. Duct C: $Q = 2100 \, \text{cfm}$; $L = 8 \, \text{ft}$.
Let $v \approx 1000 \, \text{ft/min}$; $h_L = 0.073 \, \text{inH}_2\text{O}/100 \, \text{ft}$.

$$D_e = 19.6 \, \text{in; use 12-by-28-in size}$$
$$H_L = 0.073(8/100) = 0.0058 \, \text{inH}_2\text{O}$$
$$H_v = (1000/4005)^2 = 0.0623 \, \text{inH}_2\text{O}$$

14a. Duct F: $Q = 900 \, \text{cfm}$; $L = 18 \, \text{ft}$.
Let $v \approx 600 \, \text{ft/min}$; $h_L = 0.033 \, \text{inH}_2\text{O}/100 \, \text{ft}$.

$$D_e = 16.5 \, \text{in; use 12-by-18-in size; } D_e = 16.0 \, \text{in}$$
$$\text{Actual } v = 630 \, \text{ft/min; } h_L = 0.038 \, \text{inH}_2\text{O}/100 \, \text{ft}$$
$$H_L = 0.038(18/100) = 0.0068 \, \text{inH}_2\text{O}$$
$$H_v = (630/4005)^2 = 0.0247 \, \text{inH}_2\text{O}$$

15a. Damper in duct F: $C = 0.20$ (assume wide open).

$$H_L = 0.20(0.0247) = 0.0049 \, \text{inH}_2\text{O}$$

16a. Two elbows in duct F: smooth rectangular elbow; $C = 0.18$.

$$H_L = 2(0.18)(0.0247) = 0.0089 \, \text{inH}_2\text{O}$$

18a. Tee 4 from duct C to branch F, flow in branch: $C = 1.00$.
H_L based on velocity ahead of tee in duct C.

$$H_L = 1.00(0.0623) = 0.0623 \, \text{inH}_2\text{O}$$

Now, we can recompute the total loss in path b to grille 7 in duct F. As before, this is the sum of the losses from Steps 6, 12a, 13, 14a, 15a, 16a, 17, and 18a:

$$H_7 = 0.0220 + 0.0058 + 0.009 + 0.0068 + 0.0049 + 0.0089 + 0.06 + 0.0623$$
$$= 0.1797 \, \text{inH}_2\text{O}$$

This is a significant reduction, which results in a total pressure drop less than that of path a. Therefore, let's see if we can reduce the loss in path a by also reducing the velocity of flow in duct E. Steps 7–9 are affected.

7a. Duct E: $Q = 600 \, \text{cfm}$; $L = 12 \, \text{ft}$.
Let $v \approx 600 \, \text{ft/min}$.

$$D_e = 13.8 \, \text{in; use 12-by-14-in size}$$
$$\text{Actual } D_e = 14.2 \, \text{in; } h_L = 0.032 \, \text{inH}_2\text{O; } v = 550 \, \text{ft/min}$$

$$H_L = 0.032(12/100) = 0.0038 \text{ inH}_2\text{O}$$
$$H_v = (550/4005)^2 = 0.0189 \text{ inH}_2\text{O}$$

8a. Damper in duct E: $C = 0.20$ (assume wide open).

$$H_L = 0.20(0.0189) = 0.0038 \text{ inH}_2\text{O}$$

9a. Elbow in duct E: smooth rectangular elbow; $C = 0.18$.

$$H_L = 0.18(0.0189) = 0.0034 \text{ inH}_2\text{O}$$

Now, we can recompute the total loss in path a to grille 6 in duct E. As before, this is the sum of the losses from Steps 6, 7a, 8a, 9a, 10, and 11:

$$H_6 = 0.0220 + 0.0038 + 0.0038 + 0.0034 + 0.060 + 0.090$$
$$= 0.1830 \text{ inH}_2\text{O}$$

This value is very close to that found for the redesigned path b, and the small difference can be adjusted with the dampers.

Now, note that path c to either grille 8 or grille 9 still has a lower total loss than either path a or path b. We could either use a slightly smaller duct size in branches G and H or depend on the adjustment of the dampers here also. To evaluate the suitability of using the dampers, let's estimate how much the dampers would have to be closed to increase the total loss to $0.1830 \text{ inH}_2\text{O}$ (to equal that in path a). The increased loss is

$$H_6 - H_8 = 0.1830 - 0.1756 \text{ inH}_2\text{O} = 0.0074 \text{ inH}_2\text{O}$$

With the damper wide open and with 600 cfm passing at a velocity of approximately 800 ft/min, the loss was $0.0080 \text{ inH}_2\text{O}$, as found in the original Step 8. The loss now should be

$$H_L = 0.0080 + 0.0074 = 0.0154 \text{ inH}_2\text{O}$$

For the damper, however,

$$H_L = C(H_v)$$

Solving for C gives

$$C = \frac{H_L}{H_v} = \frac{0.0154 \text{ inH}_2\text{O}}{0.040 \text{ inH}_2\text{O}} = 0.385$$

Referring to Table 19.4, you can see that a damper setting of less than 10° would produce this value of C, a very feasible setting. Thus, it appears that the duct system could be balanced as redesigned and that the total pressure drop from the fan outlet to any outlet grille will be approximately $0.1830 \text{ inH}_2\text{O}$. This is the pressure that the fan would have to develop.

Summary of the Duct System Design

- Intake duct A: round; $D = 25.0$ in
- Duct B: rectangular; 12 in × 30 in
- Duct C: rectangular; 12 in × 28 in
- Duct D: rectangular; 12 in × 16 in
- Duct E: rectangular; 12 in × 14 in
- Duct F: rectangular; 12 in × 18 in
- Duct G: rectangular; 12 in × 10 in
- Duct H: rectangular; 12 in × 10 in
- Pressure at fan inlet: $-0.096 \text{ inH}_2\text{O}$
- Pressure at fan outlet: $0.1830 \text{ inH}_2\text{O}$
- Total pressure rise by the fan: $0.1830 + 0.096 = 0.279 \text{ inH}_2\text{O}$
- Total delivery by the fan: 2700 cfm

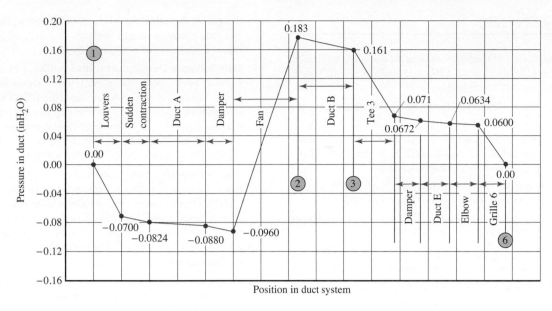

FIGURE 19.6 Pressure in duct (inH_2O) versus position for the system shown in Figure 19.2. Path a to outlet grille 6.

It is helpful to visualize the pressure changes that occur in the system. Figure 19.6 shows a plot of air pressure versus position for the path from the intake louvers, through the fan, through ducts B and E, to outlet grille 6. Similar plots could be made for the other paths.

19.4 ENERGY EFFICIENCY AND PRACTICAL CONSIDERATIONS IN DUCT DESIGN

Additional considerations must be addressed when designing air distribution systems for HVAC systems and industrial exhausts. Internet resources 1–6 and References 1–5 and 8–11 are good sources of guidelines. Listed below are some recommendations.

1. Lower velocities tend to produce lower energy losses in the system, which reduce fan energy usage and may permit using a smaller, less expensive fan. However, ducts will tend to be larger, affecting space requirements and leading to higher installed costs. Total system cost and life-cycle costs should be evaluated.

2. Locating as much of the duct system as practical within the conditioned space will save energy for heating and cooling systems.

3. Ductwork should be well sealed to prevent leaks.

4. Ducts passing through unconditioned spaces should be well insulated.

5. The fan capacity should be well matched to the air supply requirement to avoid excessive control by dampers, which tends to waste energy.

6. When loads vary significantly over time, variable-speed drives should be installed on the fan and connected into the control system to lower the fan speed at times of low demand. The fan laws indicate that lowering speed reduces power required by the cube of the speed reduction ratio. (See Chapter 13.) For example, reducing fan speed by 20 percent will reduce the required power by approximately 50 percent.

7. Ducts can be made from sheet metal, rigid fiberglass duct board, fabric, or flexible nonmetallic duct. Some come with insulation either inside or outside to reduce energy losses and to attenuate noise. Smooth surfaces are preferred for long runs to minimize friction losses.

8. Return air ducts should be provided to maintain consistent flow into and out from each room in the conditioned space.

9. Ducts for most HVAC systems are designed for pressures that range from $-3\,inH_2O$ (-750 Pa) on the intake side of fans to $10\,inH_2O$ (2500 Pa) on the outlet side. However, some large commercial or industrial installations may range from $-10\,inH_2O$ (-2500 Pa) to $100\,inH_2O$ (25 kPa). Structural strength, rigidity, and vibration must be considered.

10. Noise generation in air distribution systems must be considered to ensure that occupants are not annoyed by high noise levels. Special care should be given to fan selection and location and air velocity in ducts and through outlet grilles. Sound insulation, vibration isolators, and mounting techniques should be examined to minimize noise.

REFERENCES

1. American Society of Heating, Refrigerating and Air-Conditioning Engineers (ASHRAE). 2012. *ASHRAE Handbook: HVAC Systems and Equipment.* Atlanta: Author.

2. _____ . 2009. *ASHRAE Handbook: Fundamentals.* Atlanta: Author.

3. Collins, Lane M. and Danielle E. Martinez, eds. 2012. *Guidelines for Improved Duct Design and HVAC Systems in the Home.* Hauppauge, NY: Nova Science Publications.

4. Haines, Roger W. and Michael Myers. 2009. *HVAC Systems Design Handbook,* 5th ed. New York: McGraw-Hill.

5. Hayes, W. H. 2003. *Industrial Exhaust Hood and Fan Piping,* 2nd ed. New York: Merchant Press.

6. Idelchik, I. E. E., N. A. Decker, and M. Steinberg. 1991. *Fluid Dynamics of Industrial Equipment.* New York: Taylor & Francis.

7. Idelchik, I. E. and M. O. Steinberg. 2005. *Handbook of Hydraulic Resistance.* Mumbai, India: Jaico Publishing House.

8. Rutkowski, Hank. 2009. *Residential Duct Systems.* Arlington, VA: Air Conditioning Contractors of America.

9. Sheet Metal and Air-Conditioning Contractors National Association (SMACNA). 1990. *HVAC Systems—Duct Design*, 3rd ed. Chantilly, VA: Author.

10. The Trane Company. 1996. *Air-Conditioning Manual.* La Crosse, WI: Author.

11. Sun, Tseng-Yao. 1994. *Air Handling Systems Design.* New York: McGraw-Hill.

INTERNET RESOURCES

1. **Air Conditioning Contractors of America (ACCA):** An industry association promoting quality design, installation, and operation of air conditioning systems. Producer of many manuals and software products that aid designers of such systems for residential and commercial applications from the ACCA's Online Store. See also Reference 10.

2. **Sheet Metal and Air Conditioning Contractors' National Association (SMACNA):** An international trade association for the sheet metal and air conditioning contractors industry. Publisher of *HVAC Systems—Duct Design* (Reference 9) along with numerous other manuals and references. Calculation aids for duct design are offered in both U.S. and SI units.

3. **ASHRAE:** Formerly known as the American Society for Heating, Refrigerating, and Air Conditioning Engineers, ASHRAE and its members focus on building systems, energy efficiency, indoor air quality, refrigeration and sustainability through research, standards writing, publishing and continuing education.

4. **Air Movement and Control Association International (AMCA):** An industry association of manufacturers of air system equipment for industrial, commercial, and residential markets.

5. **Heating, Refrigeration, and Air Conditioning Institute of Canada (HRAI):** A Canadian national association for heating, ventilation, air conditioning and refrigeration (HVACR) manufacturers, wholesalers, and contractors.

6. **U.S. Department of Energy—Greening Federal Facilities, EERE:** An extensive site offering guides for energy sustainability in buildings including HVAC systems and their duct designs. In the online report, *Greening of Federal Facilities,* 2nd ed., Part V covers Energy Systems, and Section 5.2.2 of the report discusses Air Distribution Systems.

7. **Elite Software Development, Inc.:** Producer of a variety of software products for designing HVAC systems for commercial or residential applications, including DUCTSIZE, an aid to designing optimum air conditioning round, rectangular, or flat oval duct sizes. CAD drawings show system layout, fittings, duct sizes, fans, and outlets. DUCTSIZE links with Autodesk Building Systems and AutoCAD MEP software.

8. **Trane Company:** From the home page, select *Software Downloads.* Numerous software packages are available, including their *VeriTrane*™ *Duct Designer* software based on the *ASHRAE Fundamentals Handbook* (Reference 2). It includes *Duct Configurator* to model duct systems, *Ductulator®* to size duct components, and *Fitting Loss Calculator* to identify optimal fittings considering efficiency and cost.

9. **Wrightsoft®:** Producer of HVAC design software that operates on laptops, tablets, and mobile devices. The software is integrated with publications from the ACCA (Internet resource 1), and the HRAI (Internet resource 5) for system layout, duct design and sizing, load calculations, and operating cost.

PRACTICE PROBLEMS

Energy Losses in Straight Duct Sections

19.1 Determine the velocity of flow and the friction loss as 1000 cfm of air flows through 75 ft of an 18-in-diameter round duct.

19.2 Repeat Problem 19.1 for duct diameters of 16, 14, 12, and 10 in. Then plot the velocity and friction loss versus duct diameter.

19.3 Specify a diameter for a round duct suitable for carrying 1500 cfm of air with a maximum pressure drop of 0.10 inH$_2$O per 100 ft of duct, rounding up to the next inch. For the actual size specified, give the friction loss per 100 ft of duct.

19.4 Determine the velocity of flow and the friction loss as 3.0 m^3/s of air flows through 25 m of a 500-mm-diameter round duct.

19.5 Repeat Problem 19.4 for duct diameters of 600, 700, 800, 900, and 1000 mm. Then plot the velocity and friction loss versus duct diameter.

19.6 Specify a diameter for a round duct suitable for carrying 0.40 m^3/s of air with a maximum pressure drop of 1.00 Pa/m of duct, rounding up to the next 50-mm increment. For the actual size specified, give the friction loss in Pa/m.

19.7 A heating duct for a forced-air furnace measures 10 × 30 in. Compute the circular equivalent diameter. Then determine the maximum flow rate of air that the duct could carry while limiting the friction loss to 0.10 inH$_2$O per 100 ft.

19.8 A branch duct for a heating system measures 3 × 10 in. Compute the circular equivalent diameter. Then determine the maximum flow rate of air that the duct could carry while limiting the friction loss to 0.10 inH$_2$O per 100 ft.

19.9 A ventilation duct in a large industrial warehouse measures 42 × 60 in. Compute the circular equivalent diameter. Then determine the maximum flow rate of air that the duct could carry while limiting the friction loss to 0.10 inH$_2$O per 100 ft.

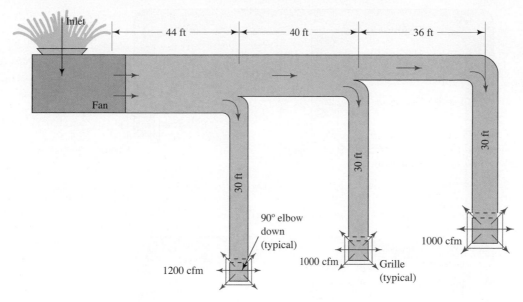

FIGURE 19.7 Duct system for Problem 19.27.

19.10 A heating duct for a forced-air furnace measures 250 × 500 mm. Compute the circular equivalent diameter. Then determine the maximum flow rate of air that the duct could carry while limiting the friction loss to 0.80 Pa/m.

19.11 A branch duct for a heating system measures 75 × 250 mm. Compute the circular equivalent diameter. Then determine the maximum flow rate of air that it could carry while limiting the friction loss to 0.80 Pa/m.

19.12 Specify a size for a rectangular duct suitable for carrying 1500 cfm of air with a maximum pressure drop of 0.10 inH₂O per 100 ft of duct. The maximum vertical height of the duct is 12.0 in.

19.13 Specify a size for a rectangular duct suitable for carrying 300 cfm of air with a maximum pressure drop of 0.10 inH₂O per 100 ft of duct. The maximum vertical height of the duct is 6.0 in.

Energy Losses in Ducts with Fittings

19.14 Compute the pressure drop as 650 cfm of air flows through a three-piece 90° elbow in a 12-in-diameter round duct.

19.15 Repeat Problem 19.14, but use a five-piece elbow.

19.16 Compute the pressure drop as 1500 cfm of air flows through a wide-open damper installed in a 16-in-diameter duct.

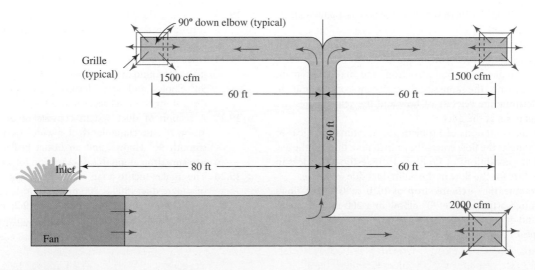

FIGURE 19.8 Duct system for Problem 19.28.

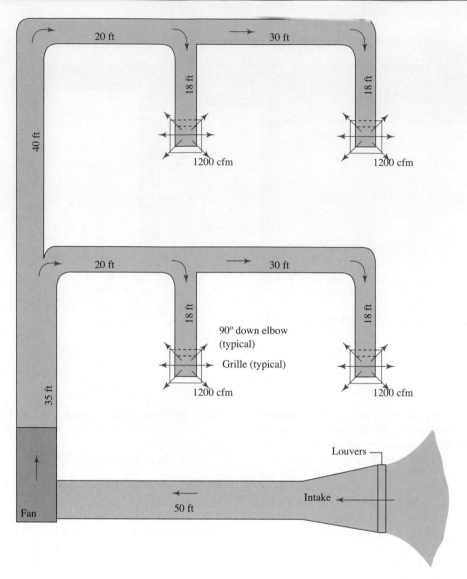

FIGURE 19.9   Duct system for Problem 19.29.

19.17 Repeat Problem 19.16 with the damper partially closed at 10°, 20°, and 30°.

19.18 A part of a duct system is a 10-by-22-in rectangular main duct carrying 1600 cfm of air. A tee to a branch duct, 10 × 10 in, draws 500 cfm from the main. The main duct remains the same size downstream from the branch. Determine the velocity of flow and the velocity pressure in all parts of the duct.

19.19 For the conditions of Problem 19.18, estimate the loss in pressure as the flow enters the branch duct through the tee.

19.20 For the conditions of Problem 19.18, estimate the loss in pressure for the flow in the main duct due to the tee.

19.21 Compute the pressure drop as 0.20 m³/s of air flows through a three-piece 90° elbow in a 200-mm-diameter round duct.

19.22 Repeat Problem 19.21, but use a mitered elbow.

19.23 Compute the pressure drop as 0.85 m³/s of air flows through a damper set at 30° installed in a 400-mm-diameter duct.

19.24 A section of duct system consists of 42 ft of straight 12-in-diameter round duct, a wide-open damper, two three-piece 90° elbows, and an outlet grille. Compute the pressure drop along this duct section for Q = 700 cfm.

19.25 A section of duct system consists of 38 ft of straight 12-by-20-in rectangular duct, a wide-open damper, three smooth 90° elbows, and an outlet grille. Compute the pressure drop along this duct section for Q = 1500 cfm.

19.26 The intake duct to a fan consists of intake louvers, 5.8 m of square duct (800 × 800 mm), a sudden contraction to a 400-mm-diameter round duct, and 9.25 m of the round duct. Estimate the pressure at the fan inlet when the duct carries 0.80 m³/s of air.

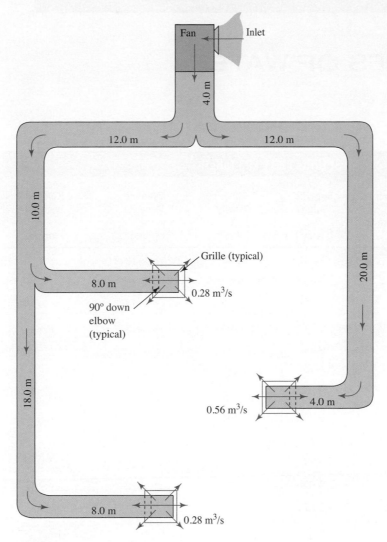

FIGURE 19.10 Duct system for Problem 19.30.

Design of Ducts

For the conditions shown in Figs. 19.7–19.10, complete the design of the duct system by specifying the sizes for all duct sections necessary to achieve a system that is balanced when it carries the flow rates shown. Compute the pressure at the fan outlet, assuming that the final outlets from the duct system are to atmospheric pressure. When an inlet duct section is shown, also complete its design and compute the pressure at the fan inlet. Note that there is no single best solution to these problems, and several design decisions must be made. It may be desirable to change certain features of the suggested system design to improve its operation or to make it simpler to achieve a balanced system.

19.27 Use Fig. 19.7.
19.28 Use Fig. 19.8.
19.29 Use Fig. 19.9.
19.30 Use Fig. 19.10.

PROPERTIES OF WATER

TABLE A.1 SI units [101 kPa (abs)]				
Temperature (°C)	Specific Weight γ (kN/m³)	Density ρ (kg/m³)	Dynamic Viscosity η (Pa·s)	Kinematic Viscosity ν (m²/s)
0	9.81	1000	1.75×10^{-3}	1.75×10^{-6}
5	9.81	1000	1.52×10^{-3}	1.52×10^{-6}
10	9.81	1000	1.30×10^{-3}	1.30×10^{-6}
15	9.81	1000	1.15×10^{-3}	1.15×10^{-6}
20	9.79	998	1.02×10^{-3}	1.02×10^{-6}
25	9.78	997	8.91×10^{-4}	8.94×10^{-7}
30	9.77	996	8.00×10^{-4}	8.03×10^{-7}
35	9.75	994	7.18×10^{-4}	7.22×10^{-7}
40	9.73	992	6.51×10^{-4}	6.56×10^{-7}
45	9.71	990	5.94×10^{-4}	6.00×10^{-7}
50	9.69	988	5.41×10^{-4}	5.48×10^{-7}
55	9.67	986	4.98×10^{-4}	5.05×10^{-7}
60	9.65	984	4.60×10^{-4}	4.67×10^{-7}
65	9.62	981	4.31×10^{-4}	4.39×10^{-7}
70	9.59	978	4.02×10^{-4}	4.11×10^{-7}
75	9.56	975	3.73×10^{-4}	3.83×10^{-7}
80	9.53	971	3.50×10^{-4}	3.60×10^{-7}
85	9.50	968	3.30×10^{-4}	3.41×10^{-7}
90	9.47	965	3.11×10^{-4}	3.22×10^{-7}
95	9.44	962	2.92×10^{-4}	3.04×10^{-7}
100	9.40	958	2.82×10^{-4}	2.94×10^{-7}

TABLE A.2   U.S. Customary System units (14.7 psia)

Temperature (°F)	Specific Weight γ (lb/ft³)	Density ρ (slugs/ft³)	Dynamic Viscosity η (lb-s/ft²)	Kinematic Viscosity ν (ft²/s)
32	62.4	1.94	3.66×10^{-5}	1.89×10^{-5}
40	62.4	1.94	3.23×10^{-5}	1.67×10^{-5}
50	62.4	1.94	2.72×10^{-5}	1.40×10^{-5}
60	62.4	1.94	2.35×10^{-5}	1.21×10^{-5}
70	62.3	1.94	2.04×10^{-5}	1.05×10^{-5}
80	62.2	1.93	1.77×10^{-5}	9.15×10^{-6}
90	62.1	1.93	1.60×10^{-5}	8.29×10^{-6}
100	62.0	1.93	1.42×10^{-5}	7.37×10^{-6}
110	61.9	1.92	1.26×10^{-5}	6.55×10^{-6}
120	61.7	1.92	1.14×10^{-5}	5.94×10^{-6}
130	61.5	1.91	1.05×10^{-5}	5.49×10^{-6}
140	61.4	1.91	9.60×10^{-6}	5.03×10^{-6}
150	61.2	1.90	8.90×10^{-6}	4.68×10^{-6}
160	61.0	1.90	8.30×10^{-6}	4.38×10^{-6}
170	60.8	1.89	7.70×10^{-6}	4.07×10^{-6}
180	60.6	1.88	7.23×10^{-6}	3.84×10^{-6}
190	60.4	1.88	6.80×10^{-6}	3.62×10^{-6}
200	60.1	1.87	6.25×10^{-6}	3.35×10^{-6}
212	59.8	1.86	5.89×10^{-6}	3.17×10^{-6}

PROPERTIES OF COMMON LIQUIDS

TABLE B.1 SI units [101 kPa (abs) and 25°C]

	Specific Gravity sg	Specific Weight γ (kN/m³)	Density ρ (kg/m³)	Dynamic Viscosity η (Pa·s)	Kinematic Viscosity ν (m²/s)
Acetone	0.787	7.72	787	3.16×10^{-4}	4.02×10^{-7}
Alcohol, ethyl	0.787	7.72	787	1.00×10^{-3}	1.27×10^{-6}
Alcohol, methyl	0.789	7.74	789	5.60×10^{-4}	7.10×10^{-7}
Alcohol, propyl	0.802	7.87	802	1.92×10^{-3}	2.39×10^{-6}
Aqua ammonia (25%)	0.910	8.93	910	—	—
Benzene	0.876	8.59	876	6.03×10^{-4}	6.88×10^{-7}
Carbon tetrachloride	1.590	15.60	1 590	9.10×10^{-4}	5.72×10^{-7}
Castor oil	0.960	9.42	960	6.51×10^{-1}	6.78×10^{-4}
Ethylene glycol	1.100	10.79	1 100	1.62×10^{-2}	1.47×10^{-5}
Gasoline	0.68	6.67	680	2.87×10^{-4}	4.22×10^{-7}
Glycerin	1.258	12.34	1 258	9.60×10^{-1}	7.63×10^{-4}
Kerosene	0.823	8.07	823	1.64×10^{-3}	1.99×10^{-6}
Linseed oil	0.930	9.12	930	3.31×10^{-2}	3.56×10^{-5}
Mercury	13.54	132.8	13 540	1.53×10^{-3}	1.13×10^{-7}
Propane	0.495	4.86	495	1.10×10^{-4}	2.22×10^{-7}
Seawater	1.030	10.10	1 030	1.03×10^{-3}	1.00×10^{-6}
Turpentine	0.870	8.53	870	1.37×10^{-3}	1.57×10^{-6}
Fuel oil, medium	0.852	8.36	852	2.99×10^{-3}	3.51×10^{-6}
Fuel oil, heavy	0.906	8.89	906	1.07×10^{-1}	1.18×10^{-4}
Approximate data for selected natural and biological fluids. Values vary significantly with composition.					
Olive oil at 68°F (20°C)	0.92	9.03	920	0.085	9.24×10^{-5}
Honey at 70°F (21°C)	1.42	13.93	1420	10.0	7.04×10^{-3}
Ketchup at 70°F (21°C)	1.48	14.52	1480	50.0	3.38×10^{-2}
Peanut butter at 70°F (21°C)	1.30	12.75	1300	250	1.92×10^{-1}
Blood at 50°F (10°C)	1.06	10.20	1060	0.01	9.43×10^{-6}
Blood at 98.6°F (37°C)	1.06	10.20	1060	3.5×10^{-3}	3.30×10^{-6}

TABLE B.2   U.S. Customary System units (14.7 psia and 77°F)

	Specific Gravity sg	Specific Weight γ (lb/ft^3)	Density ρ (slugs/ ft^3)	Dynamic Viscosity η (lb-s/ft^2)	Kinematic Viscosity ν (ft^2/s)
Acetone	0.787	48.98	1.53	6.60×10^{-6}	4.31×10^{-6}
Alcohol, ethyl	0.787	49.01	1.53	2.10×10^{-5}	1.37×10^{-5}
Alcohol, methyl	0.789	49.10	1.53	1.17×10^{-5}	7.65×10^{-6}
Alcohol, propyl	0.802	49.94	1.56	4.01×10^{-5}	2.57×10^{-5}
Aqua ammonia (25%)	0.910	56.78	1.77	—	—
Benzene	0.876	54.55	1.70	1.26×10^{-5}	7.41×10^{-6}
Carbon tetrachloride	1.590	98.91	3.08	1.90×10^{-5}	6.17×10^{-6}
Castor oil	0.960	59.69	1.86	1.36×10^{-2}	7.31×10^{-3}
Ethylene glycol	1.100	68.47	2.13	3.38×10^{-4}	1.59×10^{-4}
Gasoline	0.68	42.40	1.32	6.00×10^{-6}	4.55×10^{-6}
Glycerin	1.258	78.50	2.44	2.00×10^{-2}	8.20×10^{-3}
Kerosene	0.823	51.20	1.60	3.43×10^{-5}	2.14×10^{-5}
Linseed oil	0.930	58.00	1.80	6.91×10^{-4}	3.84×10^{-4}
Mercury	13.54	844.9	26.26	3.20×10^{-5}	1.22×10^{-6}
Propane	0.495	30.81	0.96	2.30×10^{-6}	2.40×10^{-6}
Seawater	1.030	64.00	2.00	2.15×10^{-5}	1.08×10^{-5}
Turpentine	0.870	54.20	1.69	2.87×10^{-5}	1.70×10^{-5}
Fuel oil, medium	0.852	53.16	1.65	6.25×10^{-5}	3.79×10^{-5}
Fuel oil, heavy	0.906	56.53	1.76	2.24×10^{-3}	1.27×10^{-3}

Approximate data for selected natural and biological fluids. Values vary significantly with composition.

Olive oil at 68°F (20°C)	0.92	57.41	1.78	1.78×10^{-3}	9.98×10^{-4}
Honey at 70°F (21°C)	1.42	88.61	2.75	0.209	7.60×10^{-2}
Ketchup at 70°F (21°C)	1.48	92.35	2.87	1.04	3.62×10^{-1}
Peanut butter at 70°F (21°C)	1.30	81.12	2.52	5.22	2.07
Blood at 50°F (10°C)	1.06	66.14	2.06	2.09×10^{-4}	1.01×10^{-4}
Blood at 98.6°F (37°C)	1.06	66.14	2.06	7.31×10^{-5}	1.51×10^{-4}

TYPICAL PROPERTIES OF PETROLEUM LUBRICATING OILS

Type	Specific Gravity	Kinematic Viscosity ν				Viscosity Index
		At 40°C (104°F)		At 100°C (212°F)		
		(m²/s)	(ft²/s)	(m²/s)	(ft²/s)	
Automotive hydraulic systems	0.887	3.99×10^{-5}	4.30×10^{-4}	7.29×10^{-6}	7.85×10^{-5}	149
Engine oils at 100°C						
Viscosity grade 20	0.880			5.6×10^{-6}	6.03×10^{-5}	
Viscosity grade 40	0.882			12.5×10^{-6}	1.35×10^{-4}	
Viscosity grade 60	0.883			21.9×10^{-6}	2.36×10^{-4}	
Gear lubricants						
Viscosity grade 80	0.890			7.0×10^{-6}	7.53×10^{-5}	
Viscosity grade 140	0.892			24.0×10^{-6}	2.58×10^{-4}	
Machine tool hydraulic systems						
Light	0.887	3.20×10^{-5}	3.44×10^{-4}	4.79×10^{-6}	5.16×10^{-5}	46
Medium	0.895	6.70×10^{-5}	7.21×10^{-4}	7.29×10^{-6}	7.85×10^{-5}	53
Heavy	0.901	1.96×10^{-4}	2.11×10^{-3}	1.40×10^{-5}	1.51×10^{-4}	53
Low temperature	0.844	1.40×10^{-5}	1.51×10^{-4}	5.20×10^{-6}	5.60×10^{-5}	374
Machine tool lubricating oils						
Light	0.881	2.20×10^{-5}	2.37×10^{-4}	3.90×10^{-6}	4.20×10^{-5}	40
Medium	0.915	6.60×10^{-5}	7.10×10^{-4}	7.00×10^{-6}	7.53×10^{-5}	41
Heavy	0.890	2.00×10^{-4}	2.15×10^{-3}	1.55×10^{-5}	1.67×10^{-4}	73

Note: Approximate values for use only in problem solving in this book.

VARIATION OF VISCOSITY WITH TEMPERATURE

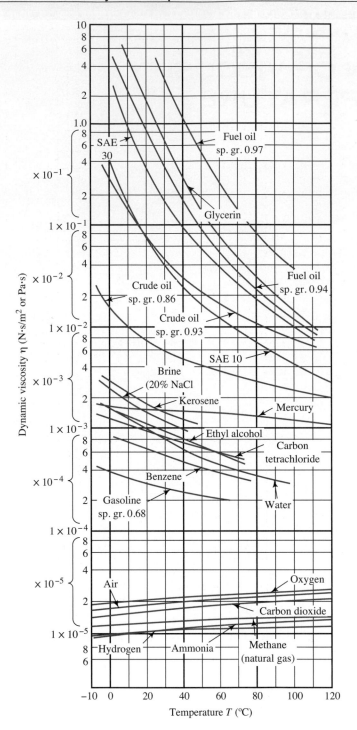

FIGURE D.1 Dynamic viscosity versus temperature—SI units.

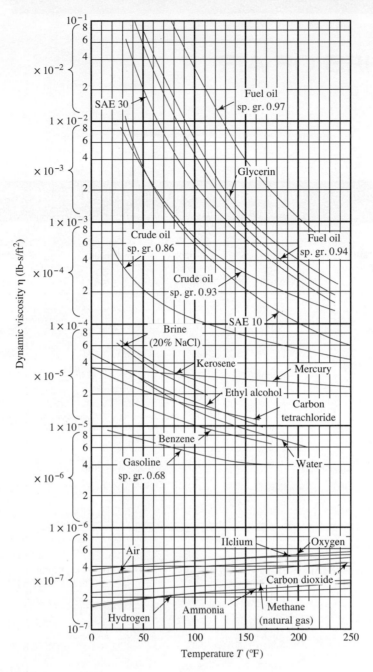

FIGURE D.2 Dynamic viscosity versus temperature—U.S. Customary System units.

PROPERTIES OF AIR

TABLE E.1 Properties of air versus temperature in SI units at standard atmospheric pressure

Temperature T (°C)	Density ρ (kg/m³)	Specific Weight γ (N/m³)	Dynamic Viscosity η (Pa·s)	Kinematic Viscosity ν (m²/s)
−40	1.514	14.85	1.51×10^{-5}	9.98×10^{-6}
−30	1.452	14.24	1.56×10^{-5}	1.08×10^{-5}
−20	1.394	13.67	1.62×10^{-5}	1.16×10^{-5}
−10	1.341	13.15	1.67×10^{-5}	1.24×10^{-5}
0	1.292	12.67	1.72×10^{-5}	1.33×10^{-5}
10	1.247	12.23	1.77×10^{-5}	1.42×10^{-5}
20	1.204	11.81	1.81×10^{-5}	1.51×10^{-5}
30	1.164	11.42	1.86×10^{-5}	1.60×10^{-5}
40	1.127	11.05	1.91×10^{-5}	1.69×10^{-5}
50	1.092	10.71	1.95×10^{-5}	1.79×10^{-5}
60	1.060	10.39	1.99×10^{-5}	1.89×10^{-5}
70	1.029	10.09	2.04×10^{-5}	1.99×10^{-5}
80	0.9995	9.802	2.09×10^{-5}	2.09×10^{-5}
90	0.9720	9.532	2.13×10^{-5}	2.19×10^{-5}
100	0.9459	9.277	2.17×10^{-5}	2.30×10^{-5}
110	0.9213	9.034	2.22×10^{-5}	2.40×10^{-5}
120	0.8978	8.805	2.26×10^{-5}	2.51×10^{-5}

Note: Properties of air for standard conditions at sea level are as follows:

Temperature	15°C
Pressure	101.325 kPa
Density	1.225 kg/m³
Specific weight	12.01 N/m³
Dynamic viscosity	1.789×10^{-5} Pa·s
Kinematic viscosity	1.46×10^{-5} m²/s

TABLE E.2 Properties of air versus temperature in U.S. Customary System units at standard atmospheric pressure

Temperature T (°F)	Density ρ (slugs/ft^3)	Specific Weight γ (lb/ft^3)	Dynamic Viscosity η (lb-s/ft^2)	Kinematic Viscosity ν (ft^2/s)
−40	2.94×10^{-3}	0.0946	3.15×10^{-7}	1.07×10^{-4}
−20	2.80×10^{-3}	0.0903	3.27×10^{-7}	1.17×10^{-4}
0	2.68×10^{-3}	0.0864	3.41×10^{-7}	1.27×10^{-4}
20	2.57×10^{-3}	0.0828	3.52×10^{-7}	1.37×10^{-4}
40	2.47×10^{-3}	0.0795	3.64×10^{-7}	1.47×10^{-4}
60	2.37×10^{-3}	0.0764	3.74×10^{-7}	1.58×10^{-4}
80	2.28×10^{-3}	0.0736	3.85×10^{-7}	1.69×10^{-4}
100	2.20×10^{-3}	0.0709	3.97×10^{-7}	1.80×10^{-4}
120	2.13×10^{-3}	0.0685	4.06×10^{-7}	1.91×10^{-4}
140	2.06×10^{-3}	0.0662	4.16×10^{-7}	2.02×10^{-4}
160	1.99×10^{-3}	0.0641	4.27×10^{-7}	2.15×10^{-4}
180	1.93×10^{-3}	0.0621	4.38×10^{-7}	2.27×10^{-4}
200	1.87×10^{-3}	0.0602	4.48×10^{-7}	2.40×10^{-4}
220	1.81×10^{-3}	0.0584	4.58×10^{-7}	2.52×10^{-4}
240	1.76×10^{-3}	0.0567	4.68×10^{-7}	2.66×10^{-4}

Note: Properties of air for standard conditions at sea level, converted from SI, are as follows:

Temperature	59°F
Pressure	14.696 psi
Density	2.37×10^{-3}
Specific weight	.0764 lb/ft^3
Dynamic viscosity	3.736×10^{-7} lb-s/ft^2
Kinematic viscosity	1.57×10^{-4} ft^2/s

TABLE E.3 Properties of air at various altitudes

SI Units				U.S. Customary System Units			
Altitude (m)	Temperature T (°C)	Pressure P (kPa)	Density ρ (kg/m³)	Altitude (ft)	Temperature T (°F)	Pressure P (psi)	Density ρ (slugs/ft³)
0	15.00	101.3	1.225	0	59.00	14.696	2.38×10^{-3}
200	13.70	98.9	1.202	500	57.22	14.433	2.34×10^{-3}
400	12.40	96.6	1.179	1000	55.43	14.173	2.25×10^{-3}
600	11.10	94.3	1.156	5000	41.17	12.227	2.05×10^{-3}
800	9.80	92.1	1.134	10000	23.34	10.106	1.76×10^{-3}
1000	8.50	89.9	1.112	15000	5.51	8.293	1.50×10^{-3}
2000	2.00	79.5	1.007	20000	−12.62	6.753	1.27×10^{-3}
3000	−4.49	70.1	0.9093	30000	−47.99	4.365	8.89×10^{-4}
4000	−10.98	61.7	0.8194	40000	−69.70	2.720	5.85×10^{-4}
5000	−17.47	54.0	0.7364	50000	−69.70	1.683	3.62×10^{-4}
10000	−49.90	26.5	0.4135	60000	−69.70	1.040	2.24×10^{-4}
15000	−56.50	12.11	0.1948	70000	−67.30	0.644	1.38×10^{-4}
20000	−56.50	5.53	0.0889	80000	−61.81	0.400	8.45×10^{-5}
25000	−51.60	2.55	0.0401	90000	−56.32	0.251	5.22×10^{-5}
30000	−46.64	1.20	0.0184	100000	−50.84	0.158	3.25×10^{-5}

Source: U.S. Standard Atmosphere, 1976, NOAA-S/T76-1562. Washington, DC: National Oceanic and Atmospheric Administration.

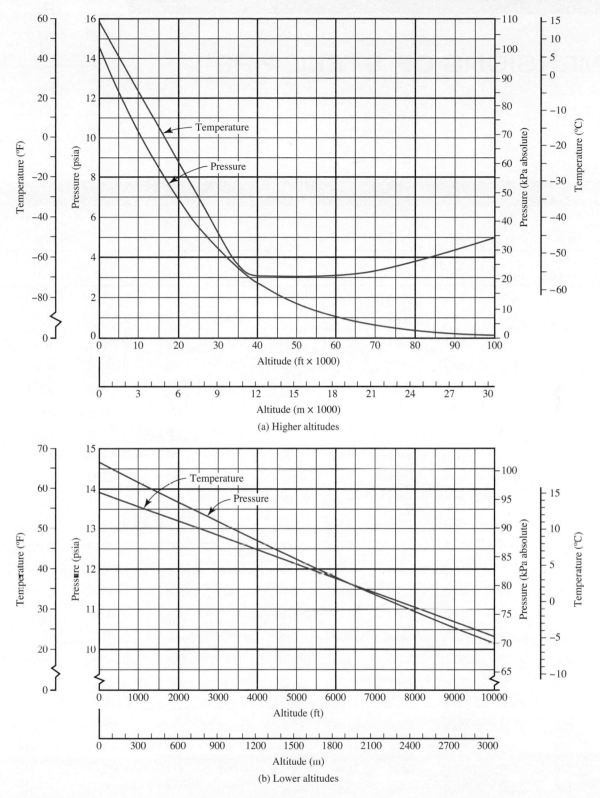

(a) Higher altitudes

(b) Lower altitudes

FIGURE E.1 Properties of the standard atmosphere versus altitude.

DIMENSIONS OF STEEL PIPE

TABLE F.1   Schedule 40

Nominal Pipe Size		Outside Diameter		Wall Thickness		Inside Diameter			Flow Area	
NPS (in)	DN (mm)	(in)	(mm)	(in)	(mm)	(in)	(ft)	(mm)	(ft²)	(m²)
⅛	6	0.405	10.3	0.068	1.73	0.269	0.0224	6.8	0.000 394	3.660×10^{-5}
¼	8	0.540	13.7	0.088	2.24	0.364	0.0303	9.2	0.000 723	6.717×10^{-5}
⅜	10	0.675	17.1	0.091	2.31	0.493	0.0411	12.5	0.001 33	1.236×10^{-4}
½	15	0.840	21.3	0.109	2.77	0.622	0.0518	15.8	0.002 11	1.960×10^{-4}
¾	20	1.050	26.7	0.113	2.87	0.824	0.0687	20.9	0.003 70	3.437×10^{-4}
1	25	1.315	33.4	0.133	3.38	1.049	0.0874	26.6	0.006 00	5.574×10^{-4}
1¼	32	1.660	42.2	0.140	3.56	1.380	0.1150	35.1	0.010 39	9.653×10^{-4}
1½	40	1.900	48.3	0.145	3.68	1.610	0.1342	40.9	0.014 14	1.314×10^{-3}
2	50	2.375	60.3	0.154	3.91	2.067	0.1723	52.5	0.023 33	2.168×10^{-3}
2½	65	2.875	73.0	0.203	5.16	2.469	0.2058	62.7	0.033 26	3.090×10^{-3}
3	80	3.500	88.9	0.216	5.49	3.068	0.2557	77.9	0.051 32	4.768×10^{-3}
3½	90	4.000	101.6	0.226	5.74	3.548	0.2957	90.1	0.068 68	6.381×10^{-3}
4	100	4.500	114.3	0.237	6.02	4.026	0.3355	102.3	0.088 40	8.213×10^{-3}
5	125	5.563	141.3	0.258	6.55	5.047	0.4206	128.2	0.139 0	1.291×10^{-2}
6	150	6.625	168.3	0.280	7.11	6.065	0.5054	154.1	0.200 6	1.864×10^{-2}
8	200	8.625	219.1	0.322	8.18	7.981	0.6651	202.7	0.347 2	3.226×10^{-2}
10	250	10.750	273.1	0.365	9.27	10.020	0.8350	254.5	0.547 9	5.090×10^{-2}
12	300	12.750	323.9	0.406	10.31	11.938	0.9948	303.2	0.777 1	7.219×10^{-2}
14	350	14.000	355.6	0.437	11.10	13.126	1.094	333.4	0.939 6	8.729×10^{-2}
16	400	16.000	406.4	0.500	12.70	15.000	1.250	381.0	1.227	0.1140
18	450	18.000	457.2	0.562	14.27	16.876	1.406	428.7	1.553	0.1443
20	500	20.000	508.0	0.593	15.06	18.814	1.568	477.9	1.931	0.1794
24	600	24.000	609.6	0.687	17.45	22.626	1.886	574.7	2.792	0.2594

TABLE F.2 Schedule 80

Nominal Pipe Size		Outside Diameter		Wall Thickness		Inside Diameter			Flow Area	
NPS (in)	DN (mm)	(in)	(mm)	(in)	(mm)	(in)	(ft)	(mm)	(ft²)	(m²)
⅛	6	0.405	10.3	0.095	2.41	0.215	0.017 92	5.5	0.000 253	2.350×10^{-5}
¼	8	0.540	13.7	0.119	3.02	0.302	0.025 17	7.7	0.000 497	4.617×10^{-5}
⅜	10	0.675	17.1	0.126	3.20	0.423	0.035 25	10.7	0.000 976	9.067×10^{-5}
½	15	0.840	21.3	0.147	3.73	0.546	0.045 50	13.9	0.001 625	1.510×10^{-4}
¾	20	1.050	26.7	0.154	3.91	0.742	0.061 83	18.8	0.003 00	2.787×10^{-4}
1	25	1.315	33.4	0.179	4.55	0.957	0.079 75	24.3	0.004 99	4.636×10^{-4}
1¼	32	1.660	42.2	0.191	4.85	1.278	0.106 5	32.5	0.008 91	8.278×10^{-4}
1½	40	1.900	48.3	0.200	5.08	1.500	0.125 0	38.1	0.012 27	1.140×10^{-3}
2	50	2.375	60.3	0.218	5.54	1.939	0.161 6	49.3	0.020 51	1.905×10^{-3}
2½	65	2.875	73.0	0.276	7.01	2.323	0.193 6	59.0	0.029 44	2.735×10^{-3}
3	80	3.500	88.9	0.300	7.62	2.900	0.241 7	73.7	0.045 90	4.264×10^{-3}
3½	90	4.000	101.6	0.318	8.08	3.364	0.280 3	85.4	0.061 74	5.736×10^{-3}
4	100	4.500	114.3	0.337	8.56	3.826	0.318 8	97.2	0.079 86	7.419×10^{-3}
5	125	5.563	141.3	0.375	9.53	4.813	0.401 1	122.3	0.126 3	1.173×10^{-2}
6	150	6.625	168.3	0.432	10.97	5.761	0.480 1	146.3	0.181 0	1.682×10^{-2}
8	200	8.625	219.1	0.500	12.70	7.625	0.635 4	193.7	0.317 4	2.949×10^{-2}
10	250	10.750	273.1	0.593	15.06	9.564	0.797 0	242.9	0.498 6	4.632×10^{-2}
12	300	12.750	323.9	0.687	17.45	11.376	0.948 0	289.0	0.705 6	6.555×10^{-2}
14	350	14.000	355.6	0.750	19.05	12.500	1.042	317.5	0.852 1	7.916×10^{-2}
16	400	16.000	406.4	0.842	21.39	14.314	1.193	363.6	1.117	0.1038
18	450	18.000	457.2	0.937	23.80	16.126	1.344	409.6	1.418	0.1317
20	500	20.000	508.0	1.031	26.19	17.938	1.495	455.6	1.755	0.1630
24	600	24.000	609.6	1.218	30.94	21.564	1.797	547.7	2.535	0.2344

DIMENSIONS OF STEEL, COPPER, AND PLASTIC TUBING

TABLE G.1	Dimensions of Steel Tubing—Inch-based sizes							
Outside Diameter		Wall Thickness		Inside Diameter			Flow Area	
(in)	(mm)	(in)	(mm)	(in)	(ft)	(mm)	(ft²)	(m²)
⅛	3.18	0.032	0.813	0.061	0.00508	1.549	2.029×10^{-5}	1.885×10^{-6}
		0.035	0.889	0.055	0.00458	1.397	1.650×10^{-5}	1.533×10^{-6}
3/16	4.76	0.032	0.813	0.124	0.01029	3.137	8.319×10^{-5}	7.728×10^{-6}
		0.035	0.889	0.117	0.00979	2.985	7.530×10^{-5}	6.996×10^{-6}
¼	6.35	0.035	0.889	0.180	0.01500	4.572	1.767×10^{-4}	1.642×10^{-5}
		0.049	1.24	0.152	0.01267	3.861	1.260×10^{-4}	1.171×10^{-5}
5/16	7.94	0.035	0.889	0.243	0.02021	6.160	3.207×10^{-4}	2.980×10^{-5}
		0.049	1.24	0.215	0.01788	5.448	2.509×10^{-4}	2.331×10^{-5}
⅜	9.53	0.035	0.889	0.305	0.02542	7.747	5.074×10^{-4}	4.714×10^{-5}
		0.049	1.24	0.277	0.02308	7.036	4.185×10^{-4}	3.888×10^{-5}
½	12.70	0.049	1.24	0.402	0.03350	10.21	8.814×10^{-4}	8.189×10^{-5}
		0.065	1.65	0.370	0.03083	9.40	7.467×10^{-4}	6.937×10^{-5}
⅝	15.88	0.049	1.24	0.527	0.04392	13.39	1.515×10^{-3}	1.407×10^{-4}
		0.065	1.65	0.495	0.04125	12.57	1.336×10^{-3}	1.242×10^{-4}
¾	19.05	0.049	1.24	0.652	0.05433	16.56	2.319×10^{-3}	2.154×10^{-4}
		0.065	1.65	0.620	0.05167	15.75	2.097×10^{-3}	1.948×10^{-4}
⅞	22.23	0.049	1.24	0.777	0.06475	19.74	3.293×10^{-3}	3.059×10^{-4}
		0.065	1.65	0.745	0.06208	18.92	3.027×10^{-3}	2.812×10^{-4}
1	25.40	0.065	1.65	0.870	0.07250	22.10	4.128×10^{-3}	3.835×10^{-4}
		0.083	2.11	0.834	0.06950	21.18	3.794×10^{-3}	3.524×10^{-4}
1¼	31.75	0.065	1.65	1.120	0.09333	28.45	6.842×10^{-3}	6.356×10^{-4}
		0.083	2.11	1.084	0.09033	27.53	6.409×10^{-3}	5.954×10^{-4}
1½	38.10	0.065	1.65	1.370	0.1142	34.80	1.024×10^{-2}	9.510×10^{-4}
		0.083	2.11	1.334	0.1112	33.88	9.706×10^{-3}	9.017×10^{-4}
1¾	44.45	0.065	1.65	1.620	0.1350	41.15	1.431×10^{-2}	1.330×10^{-3}
		0.083	2.11	1.584	0.1320	40.23	1.368×10^{-2}	1.271×10^{-3}
2	50.80	0.065	1.65	1.870	0.1558	47.50	1.907×10^{-2}	1.772×10^{-3}
		0.083	2.11	1.834	0.1528	46.58	1.835×10^{-2}	1.704×10^{-3}

TABLE G.2 Dimensions of Steel and Copper Hydraulic Tubing— Metric-based sizes

			SI Units
Hydraulic Tubing: Steel and copper—selected sizes			
Outside diameter *OD* (mm)	Wall thickness *t* (mm)	Inside diameter *ID* (mm)	Flow area *A* (m²)
6	0.8	4.4	1.521×10^{-5}
6	1.0	4.0	1.257×10^{-5}
8	1.0	6.0	2.827×10^{-5}
8	1.2	5.6	2.463×10^{-5}
15	1.2	12.6	1.247×10^{-4}
15	1.5	12.0	1.131×10^{-4}
20	1.2	17.6	2.433×10^{-4}
20	1.5	17.0	2.270×10^{-4}
25	1.5	22.0	3.801×10^{-4}
25	2.0	21.0	3.464×10^{-4}
32	1.5	29.0	6.605×10^{-4}
32	2.0	28.0	6.158×10^{-4}
40	1.5	37.0	1.075×10^{-3}
40	2.0	36.0	1.018×10^{-3}
50	1.5	47.0	1.735×10^{-3}
50	2.0	46.0	1.662×10^{-3}
60	2.0	56.0	2.463×10^{-3}
60	2.8	54.4	2.324×10^{-3}
80	2.8	74.4	4.347×10^{-3}
100	3.5	93.0	6.793×10^{-3}
120	3.5	113.0	1.003×10^{-2}
140	5.0	130.0	1.327×10^{-2}
160	5.5	149.0	1.744×10^{-2}
180	6.0	168.0	2.217×10^{-2}
200	7.0	186.0	2.717×10^{-2}
220	8.0	204.0	3.269×10^{-2}
240	9.0	222.0	3.871×10^{-2}
260	10.0	240.0	4.524×10^{-2}

Source: Parker Steel Company—Metric Sized Metals, Toledo, Ohio
Note: Numerous other sizes and wall thicknesses available

TABLE G.3 Dimensions of PVC Plastic Pressure Pipe—
Metric-based sizes

SI Units

PVC Plastic Pressure Pipe—selected sizes

Outside diameter OD (mm)	Wall thickness t (mm)	Inside diameter ID (mm)	Flow area A (m^2)	Pressure rating p (bar)
16	1.5	13.0	1.327×10^{-4}	16
20	1.5	17.0	2.270×10^{-4}	16
25	1.9	21.2	3.530×10^{-4}	16
32	2.4	27.2	5.811×10^{-4}	16
40	3.0	34.0	9.079×10^{-4}	16
50	2.4	45.2	1.605×10^{-3}	10
50	3.7	42.6	1.425×10^{-3}	16
63	3.0	57.0	2.552×10^{-3}	10
63	4.7	53.6	2.256×10^{-3}	16
75	3.6	67.8	3.610×10^{-3}	10
75	5.6	63.8	3.197×10^{-3}	16
90	2.8	84.4	5.595×10^{-3}	6
90	4.3	81.4	5.204×10^{-3}	10
90	6.7	76.6	4.608×10^{-3}	16
125	3.1	118.8	1.108×10^{-2}	6
125	4.8	115.4	1.046×10^{-2}	10
125	7.4	110.2	9.538×10^{-3}	16
160	4.0	152.0	1.815×10^{-2}	6
160	6.2	147.6	1.711×10^{-2}	10
160	9.5	141.0	1.561×10^{-2}	16
200	4.9	190.2	2.841×10^{-2}	6
200	7.7	184.6	2.676×10^{-2}	10
200	11.9	176.2	2.438×10^{-2}	16
250	6.2	237.6	4.434×10^{-2}	6
250	9.6	230.8	4.184×10^{-2}	10
250	14.8	220.4	3.815×10^{-2}	16
400	9.8	380.4	1.137×10^{-1}	6
500	12.3	475.4	1.775×10^{-1}	6

Source: epco-plastics.com/pdfs/pvc
Notes: 1. Numerous other sizes and wall thicknesses available

2. Pressure equivalents:
- 6 bar – 600 kPa – 87 psi
- 10 bar = 1000 kPa = 145 psi
- 16 bar = 1600 kPa = 232 psi

DIMENSIONS OF TYPE K COPPER TUBING

Nominal Size	Outside Diameter		Wall Thickness		Inside Diameter			Flow Area	
(in)	(in)	(mm)	(in)	(mm)	(in)	(ft)	(mm)	(ft²)	(m²)
⅛	0.250	6.35	0.035	0.889	0.180	0.0150	4.572	1.767×10^{-4}	1.642×10^{-5}
¼	0.375	9.53	0.049	1.245	0.277	0.0231	7.036	4.185×10^{-4}	3.888×10^{-5}
⅜	0.500	12.70	0.049	1.245	0.402	0.0335	10.21	8.814×10^{-4}	8.189×10^{-5}
½	0.625	15.88	0.049	1.245	0.527	0.0439	13.39	1.515×10^{-3}	1.407×10^{-4}
⅝	0.750	19.05	0.049	1.245	0.652	0.0543	16.56	2.319×10^{-3}	2.154×10^{-4}
¾	0.875	22.23	0.065	1.651	0.745	0.0621	18.92	3.027×10^{-3}	2.812×10^{-4}
1	1.125	28.58	0.065	1.651	0.995	0.0829	25.27	5.400×10^{-3}	5.017×10^{-4}
1¼	1.375	34.93	0.065	1.651	1.245	0.1037	31.62	8.454×10^{-3}	7.854×10^{-4}
1½	1.625	41.28	0.072	1.829	1.481	0.1234	37.62	1.196×10^{-2}	1.111×10^{-3}
2	2.125	53.98	0.083	2.108	1.959	0.1632	49.76	2.093×10^{-2}	1.945×10^{-3}
2½	2.625	66.68	0.095	2.413	2.435	0.2029	61.85	3.234×10^{-2}	3.004×10^{-3}
3	3.125	79.38	0.109	2.769	2.907	0.2423	73.84	4.609×10^{-2}	4.282×10^{-3}
3½	3.625	92.08	0.120	3.048	3.385	0.2821	85.98	6.249×10^{-2}	5.806×10^{-3}
4	4.125	104.8	0.134	3.404	3.857	0.3214	97.97	8.114×10^{-2}	7.538×10^{-3}
5	5.125	130.2	0.160	4.064	4.805	0.4004	122.0	1.259×10^{-1}	1.170×10^{-2}
6	6.125	155.6	0.192	4.877	5.741	0.4784	145.8	1.798×10^{-1}	1.670×10^{-2}
8	8.125	206.4	0.271	6.883	7.583	0.6319	192.6	3.136×10^{-1}	2.914×10^{-2}
10	10.125	257.2	0.338	8.585	9.449	0.7874	240.0	4.870×10^{-1}	4.524×10^{-2}
12	12.125	308.0	0.405	10.287	11.315	0.9429	287.4	6.983×10^{-1}	6.487×10^{-2}

DIMENSIONS OF DUCTILE IRON PIPE

TABLE I.1 Class 150 for 150-psi (1.03-MPa) pressure service Cement lined pipe

Nominal Pipe Size	Outside Diameter		Wall Thickness		Inside Diameter			Flow Area	
(in)	(in)	(mm)	(in)	(mm)	(in)	(ft)	(mm)	(ft²)	(m²)
4	4.80	121.9	0.315	8.001	4.17	0.348	105.9	0.0948	0.00881
6	6.90	175.3	0.315	8.001	6.27	0.523	159.3	0.2144	0.01993
8	9.05	229.9	0.315	8.001	8.42	0.702	213.9	0.3867	0.03594
10	11.10	281.9	0.325	8.255	10.45	0.871	265.4	0.5956	0.05535
12	13.20	335.3	0.345	8.763	12.51	1.043	317.8	0.8536	0.07933
14	15.30	388.6	0.375	9.525	14.55	1.213	369.6	1.155	0.1073
16	17.40	442.0	0.395	10.033	16.61	1.384	421.9	1.505	0.1398
18	19.50	495.3	0.405	10.287	18.69	1.558	474.7	1.905	0.1771
20	21.60	548.6	0.425	10.795	20.75	1.729	527.1	2.348	0.2182
24	25.80	655.3	0.425	10.795	24.95	2.079	633.7	3.395	0.3155
30	32.00	812.8	0.465	11.811	31.07	2.589	789.2	5.265	0.4893
36	38.30	972.8	0.505	12.827	37.29	3.108	947.2	7.584	0.7049
42	44.50	1130.3	0.535	13.589	43.43	3.619	1103.1	10.287	0.9561
48	50.80	1290.3	0.585	14.859	49.63	4.136	1260.6	13.434	1.2485

AREAS OF CIRCLES

TABLE J.1	U.S. Customary System units		
Diameter		**Area**	
(in)	(ft)	(in^2)	(ft^2)
0.25	0.0208	0.0491	3.409×10^{-4}
0.50	0.0417	0.1963	1.364×10^{-3}
0.75	0.0625	0.4418	3.068×10^{-3}
1.00	0.0833	0.7854	5.454×10^{-3}
1.25	0.1042	1.227	8.522×10^{-3}
1.50	0.1250	1.767	1.227×10^{-2}
1.75	0.1458	2.405	1.670×10^{-2}
2.00	0.1667	3.142	2.182×10^{-2}
2.50	0.2083	4.909	3.409×10^{-2}
3.00	0.2500	7.069	4.909×10^{-2}
3.50	0.2917	9.621	6.681×10^{-2}
4.00	0.3333	12.57	8.727×10^{-2}
4.50	0.3750	15.90	0.1104
5.00	0.4167	19.63	0.1364
6.00	0.5000	28.27	0.1963
7.00	0.5833	38.48	0.2673
8.00	0.6667	50.27	0.3491
9.00	0.7500	63.62	0.4418
10.00	0.8333	78.54	0.5454
12.00	1.00	113.1	0.7854
18.00	1.50	254.5	1.767
24.00	2.00	452.4	3.142

TABLE J.2   SI units

Diameter		Area	
(mm)	(m)	(mm^2)	(m^2)
6	0.006	28.27	2.827×10^{-5}
12	0.012	113.1	1.131×10^{-4}
18	0.018	254.5	2.545×10^{-4}
25	0.025	490.9	4.909×10^{-4}
32	0.032	804.2	8.042×10^{-4}
40	0.040	1257	1.257×10^{-3}
45	0.045	1590	1.590×10^{-3}
50	0.050	1963	1.963×10^{-3}
60	0.060	2827	2.827×10^{-3}
75	0.075	4418	4.418×10^{-3}
90	0.090	6362	6.362×10^{-3}
100	0.100	7854	7.854×10^{-3}
115	0.115	1.039×10^4	1.039×10^{-2}
125	0.125	1.227×10^4	1.227×10^{-2}
150	0.150	1.767×10^4	1.767×10^{-2}
175	0.175	2.405×10^4	2.405×10^{-2}
200	0.200	3.142×10^4	3.142×10^{-2}
225	0.225	3.976×10^4	3.976×10^{-2}
250	0.250	4.909×10^4	4.909×10^{-2}
300	0.300	7.069×10^4	7.069×10^{-2}
450	0.450	1.590×10^5	1.590×10^{-1}
600	0.600	2.827×10^5	2.827×10^{-1}

CONVERSION FACTORS

Note: Conversion factors are given here to generally three or four significant figures. More precise values are available in *IEEE/ASTM Standard SI 10-2002* listed as Reference 1 in Chapter 1.

TABLE K.1 Conversion factors

Mass Standard SI unit: kilogram (kg). Equivalent unit: $N \cdot s^2/m$.

$$\frac{14.59 \text{ kg}}{\text{slug}} \qquad \frac{32.174 \text{ lb}_m}{\text{slug}} \qquad \frac{2.205 \text{ lb}_m}{\text{kg}} \qquad \frac{453.6 \text{ grams}}{\text{lb}_m} \qquad \frac{2000 \text{ lb}_m}{\text{ton}_m} \qquad \frac{1000 \text{ kg}}{\text{metric ton}_m}$$

Force Standard SI unit: Newton (N). Equivalent unit: $kg \cdot m/s^2$.

$$\frac{4.448 \text{ N}}{\text{lb}_f} \qquad \frac{10^5 \text{ dynes}}{\text{N}} \qquad \frac{4.448 \times 10^5 \text{ dynes}}{\text{lb}_f} \qquad \frac{224.8 \text{ lb}_f}{\text{kN}}$$

Length

$$\frac{3.281 \text{ ft}}{\text{m}} \qquad \frac{39.37 \text{ in}}{\text{m}} \qquad \frac{12 \text{ in}}{\text{ft}} \qquad \frac{1.609 \text{ km}}{\text{mi}} \qquad \frac{5280 \text{ ft}}{\text{mi}} \qquad \frac{6076 \text{ ft}}{\text{nautical mile}}$$

Area

$$\frac{144 \text{ in}^2}{\text{ft}^2} \qquad \frac{10.76 \text{ ft}^2}{\text{m}^2} \qquad \frac{645.2 \text{ mm}^2}{\text{in}^2} \qquad \frac{10^6 \text{ mm}^2}{\text{m}^2} \qquad \frac{43\,560 \text{ ft}^2}{\text{acre}} \qquad \frac{10^4 \text{ m}^2}{\text{hectare}}$$

Volume

$$\frac{1728 \text{ in}^3}{\text{ft}^3} \qquad \frac{231 \text{ in}^3}{\text{gal}} \qquad \frac{7.48 \text{ gal}}{\text{ft}^3} \qquad \frac{264.2 \text{ gal}}{\text{m}^3} \qquad \frac{3.785 \text{ L}}{\text{gal}} \qquad \frac{35.31 \text{ ft}^3}{\text{m}^3}$$

$$\frac{28.32 \text{ L}}{\text{ft}^3} \qquad \frac{1000 \text{ L}}{\text{m}^3} \qquad \frac{61.02 \text{ in}^3}{\text{L}} \qquad \frac{1000 \text{ cm}^3}{\text{L}} \qquad \frac{1.201 \text{ U.S. gal}}{\text{Imperial gallon}}$$

Volume Flow Rate

$$\frac{449 \text{ gal/min}}{\text{ft}^3/\text{s}} \qquad \frac{35.31 \text{ ft}^3/\text{s}}{\text{m}^3/\text{s}} \qquad \frac{15\,850 \text{ gal/min}}{\text{m}^3/\text{s}} \qquad \frac{3.785 \text{ L/min}}{\text{gal/min}}$$

$$\frac{60\,000 \text{ L/min}}{\text{m}^3/\text{s}} \qquad \frac{2119 \text{ ft}^3/\text{min}}{\text{m}^3/\text{s}} \qquad \frac{16.67 \text{ L/min}}{\text{m}^3/\text{h}} \qquad \frac{101.9 \text{ m}^3/\text{h}}{\text{ft}^3/\text{s}}$$

Density (mass/unit volume)

$$\frac{515.4 \text{ kg/m}^3}{\text{slug/ft}^3} \qquad \frac{1000 \text{ kg/m}^3}{\text{gram/cm}^3} \qquad \frac{32.17 \text{ lb}_m/\text{ft}^3}{\text{slug/ft}^3} \qquad \frac{16.018 \text{ kg/m}^3}{\text{lb}_m/\text{ft}^3}$$

Specific Weight (weight/unit volume)

$$\frac{157.1 \text{ N/m}^3}{\text{lb}_f/\text{ft}^3} \qquad \frac{1728 \text{ lb/ft}^3}{\text{lb/in}^3}$$

Pressure Standard SI unit: pascal (Pa). Equivalent units: N/m^2 or $kg/m \cdot s^2$.

$$\frac{144 \text{ lb/ft}^2}{\text{lb/in}^2} \qquad \frac{47.88 \text{ Pa}}{\text{lb/ft}^2} \qquad \frac{6895 \text{ Pa}}{\text{lb/in}^2} \qquad \frac{1 \text{ Pa}}{\text{N/m}^2} \qquad \frac{100 \text{ kPa}}{\text{bar}} \qquad \frac{14.50 \text{ lb/in}^2}{\text{bar}}$$

$$\frac{27.68 \text{ inH}_2\text{O}}{\text{lb/in}^2} \qquad \frac{249.1 \text{ Pa}}{\text{inH}_2\text{O}} \qquad \frac{2.036 \text{ inHg}}{\text{lb/in}^2} \qquad \frac{3386 \text{ Pa}}{\text{inHg}} \qquad \frac{133.3 \text{ Pa}}{\text{mmHg}} \qquad \frac{51.71 \text{ mmHg}}{\text{lb/in}^2}$$

$$\frac{14.696 \text{ lb/in}^2}{\text{Std. atmosphere}} \qquad \frac{101.325 \text{ kPa}}{\text{Std. atmosphere}} \qquad \frac{29.92 \text{ inHg}}{\text{Std. atmosphere}} \qquad \frac{760.1 \text{ mmHg}}{\text{Std. atmosphere}}$$

TABLE K.1 Conversion factors (*continued*)

Note: Conversion factors based on the height of a column of liquid (e.g., inH_2O and mmHg) are based on a standard gravitational field ($g = 9.806\,65$ m/s^2), a density of water equal to 1000 kg/m^3, and a density of mercury equal to $13\,595.1$ kg/m^3, sometimes called *conventional values* for a temperature at or near 0°C. Actual measurements with such fluids may vary because of differences in local gravity and temperature.

Energy Standard SI unit: joule (J). Equivalent units: N·m or kg·m^2/s^2.

$$\frac{1.356\,\text{J}}{\text{lb-ft}} \qquad \frac{1.0\,\text{J}}{\text{N·m}} \qquad \frac{8.85\,\text{lb-in}}{\text{J}} \qquad \frac{1.055\,\text{kJ}}{\text{Btu}} \qquad \frac{3.600\,\text{kJ}}{\text{W·h}} \qquad \frac{778.17\,\text{ft-lb}}{\text{Btu}}$$

Power Standard SI unit: watt (W). Equivalent unit: J/s or N·m/s.

$$\frac{745.7\,\text{W}}{\text{hp}} \qquad \frac{1.0\,\text{W}}{\text{N·m/s}} \qquad \frac{550\,\text{lb-ft/s}}{\text{hp}} \qquad \frac{1.356\,\text{W}}{\text{lb-ft/s}} \qquad \frac{3.412\,\text{Btu/hr}}{\text{W}} \qquad \frac{1.341\,\text{hp}}{\text{kW}}$$

Dynamic Viscosity Standard SI unit: Pa·s or N·s/m^2 (cP = centipoise)

$$\frac{47.88\,\text{Pa·s}}{\text{lb-s/ft}^2} \qquad \frac{10\,\text{poise}}{\text{Pa·s}} \qquad \frac{1000\,\text{cP}}{\text{Pa·s}} \qquad \frac{100\,\text{cP}}{\text{poise}} \qquad \frac{1\,\text{cP}}{1\,\text{mPa·s}}$$

Kinematic Viscosity Standard SI unit: m^2/s (cSt = centistoke)

$$\frac{10.764\,\text{ft}^2/\text{s}}{\text{m}^2/\text{s}} \qquad \frac{10^4\,\text{stoke}}{\text{m}^2/\text{s}} \qquad \frac{10^6\,\text{cSt}}{\text{m}^2/\text{s}} \qquad \frac{100\,\text{cSt}}{\text{stoke}} \qquad \frac{1\,\text{cSt}}{1\,\text{mm}^2/\text{s}} \qquad \frac{10^6\,\text{mm}^2/\text{s}}{\text{m}^2/\text{s}}$$

Refer to Section 2.6.5 for conversions involving Saybolt Universal seconds.

General Approach to Application of Conversion Factors. Arrange the conversion factor from the table in such a manner that when multiplied by the given quantity, the original units cancel out, leaving the desired units.

Example 1 Convert 0.24 m^3/s to the units of gal/min:

$$(0.24\,\text{m}^3/\text{s})\,\frac{15\,850\,\text{gal/min}}{\text{m}^3/\text{s}} = 3804\,\text{gal/min}$$

Example 2 Convert 150 gal/min to the units of m^3/s:

$$(150\,\text{gal/min})\,\frac{1\,\text{m}^3/\text{s}}{15\,850\,\text{gal/min}} = 9.46 \times 10^{-3}\,\text{m}^3/\text{s}$$

Temperature Conversions (Refer to Section 1.6)

Given the Fahrenheit temperature T_F in °F, the Celsius temperature T_C in °C is

$$T_C = (T_F - 32)/1.8$$

Given the temperature T_C in °C, the Fahrenheit temperature T_F in °F is

$$T_F = 1.8T_C + 32$$

Given the temperature T_C in °C, the absolute temperature T_K in K (kelvin) is

$$T_K = T_C + 273.15$$

Given the temperature T_F in °F, the absolute temperature T_R in °R (degrees Rankine) is

$$T_R = T_F + 459.67$$

Given the temperature T_F in °F, the absolute temperature T_K in K is

$$T_K = (T_F + 459.67)/1.8 = T_R/1.8$$

PROPERTIES OF AREAS

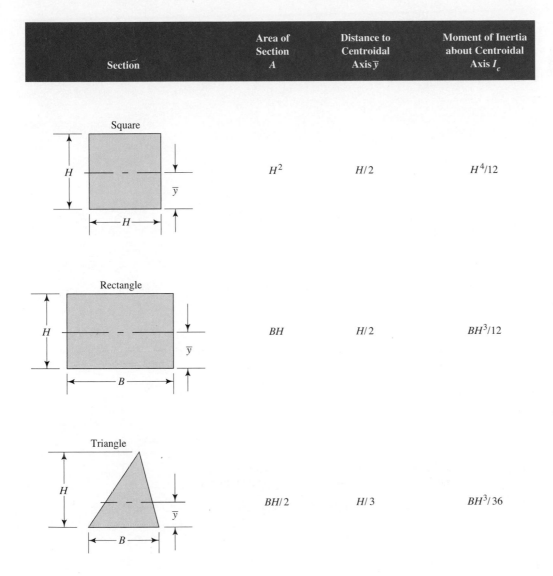

Section	Area of Section A	Distance to Centroidal Axis $\overline{y}$	Moment of Inertia about Centroidal Axis I_c
Square	H^2	$H/2$	$H^4/12$
Rectangle	BH	$H/2$	$BH^3/12$
Triangle	$BH/2$	$H/3$	$BH^3/36$

Properties of Areas (*continued*)

Section	Area of Section A	Distance to Centroidal Axis $\bar{y}$	Moment of Inertia about Centroidal Axis I_c
Circle 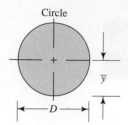	$\pi D^2/4$	$D/2$	$\pi D^4/64$
Ring 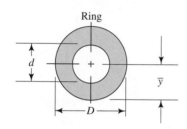	$\dfrac{\pi(D^2-d^2)}{4}$	$D/2$	$\dfrac{\pi(D^4-d^4)}{64}$
Semicircle 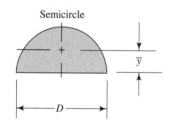	$\pi D^2/8$	$0.212D$	$(6.86\times10^{-3})D^4$
Quadrant 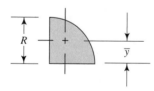	$\pi D^2/16$ $\pi R^2/4$	$0.212D$ $0.424R$	$(3.43\times10^{-3})D^4$ $(5.49\times10^{-2})R^4$
Trapezoid	$\dfrac{H(G+B)}{2}$	$\dfrac{H(G+2B)}{3(G+B)}$	$\dfrac{H^3(G^2+4GB+B^2)}{36(G+B)}$

PROPERTIES OF SOLIDS

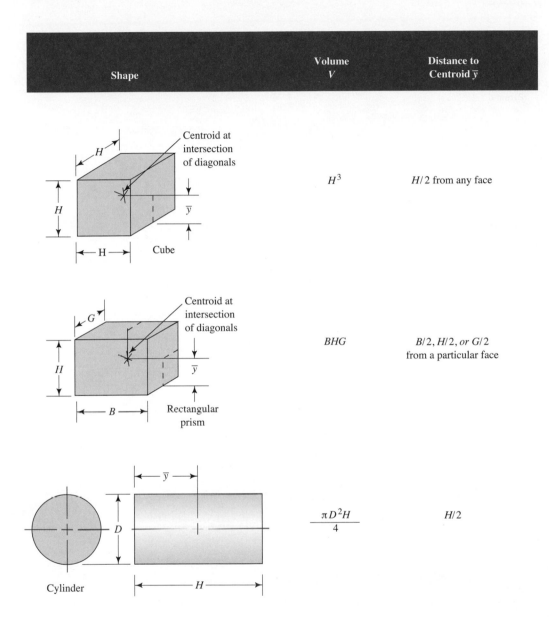

Shape	Volume V	Distance to Centroid $\bar{y}$
Cube	H^3	$H/2$ from any face
Rectangular prism	BHG	$B/2, H/2,$ *or* $G/2$ from a particular face
Cylinder	$\dfrac{\pi D^2 H}{4}$	$H/2$

Properties of Solids (*continued*)

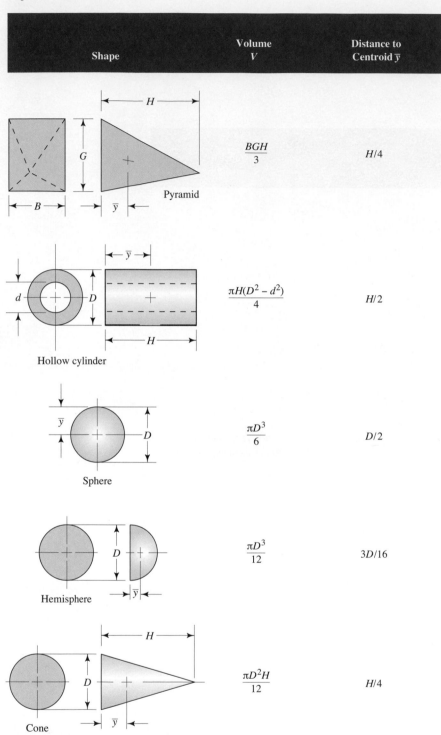

Shape	Volume V	Distance to Centroid $\overline{y}$
Pyramid	$\dfrac{BGH}{3}$	$H/4$
Hollow cylinder	$\dfrac{\pi H(D^2 - d^2)}{4}$	$H/2$
Sphere	$\dfrac{\pi D^3}{6}$	$D/2$
Hemisphere	$\dfrac{\pi D^3}{12}$	$3D/16$
Cone	$\dfrac{\pi D^2 H}{12}$	$H/4$

GAS CONSTANT, ADIABATIC EXPONENT, AND CRITICAL PRESSURE RATIO FOR SELECTED GASES

| Gas | Gas Constant R | | k | Critical Pressure Ratio |
	$\dfrac{\text{ft·lb}}{\text{lb·°R}}$	$\dfrac{\text{N·m}}{\text{N·K}}$		
Air	53.3	29.2	1.40	0.528
Ammonia	91.0	49.9	1.32	0.542
Carbon dioxide	35.1	19.3	1.30	0.546
Natural gas (typical; depends on gas)	79.1	43.4	1.27	0.551
Nitrogen	55.2	30.3	1.41	0.527
Oxygen	48.3	26.5	1.40	0.528
Propane	35.0	19.2	1.15	0.574
Refrigerant 12	12.6	6.91	1.13	0.578

ANSWERS TO SELECTED PROBLEMS

Chapter 1

1.1 1.25 m

1.3 $3.65 \times 10^{-6} \, \text{m}^3$

1.5 $391 \times 10^6 \, \text{mm}^3$

1.7 22.2 m/s

1.9 2993 m

1.11 786 m

1.13 $7.39 \times 10^{-3} \, \text{m}^3$

1.15 1.83 m/s

1.17 47.2 m/s

1.19 48.7 mi/h

1.21 $8.05 \times 10^{-2} \, \text{m/s}^2$

1.23 $0.264 \, \text{ft/s}^2$

1.25 10.8 N·m

1.27 1.76 kN·m

1.29 37.4 g

1.31 1.56 m/s

1.33 26 700 ft·1b

1.35 6.20 slugs

1.37 4.63 ft/s

1.39 2.49 runs/game

1.41 129 innings

1.43 354 psi

1.45 2.72 MPa

1.47 119 psi

1.49 40.25 kN

1.51 2.26 in

1.57 1300 psi   8.96 MPa

1.59 1890 psi   13.03 MPa

1.61 −1.59 percent

1.63 884 lb/in

1.65 14 137 lb/in

1.67 62.2 kg

1.69 8093 N

1.71 0.242 slug

1.73 50.9 lb

1.75 $m = 4.97$ slugs
$w = 712$ N
$m = 72.5$ kg

1.77 9810 N

1.81 $1.225 \, \text{kg/m}^3$

1.83 0.903 at 5°C
0.865 at 50°C

1.85 Density = $883 \, \text{kg/m}^3$
Specific weight = $8.66 \, \text{kN/m}^3$
Specific gravity = 0.883

1.87 634 N

1.89 $2.72 \times 10^{-3} \, \text{m}^3$

1.91 Density = $789 \, \text{kg/m}^3$
Specific weight = $7.74 \, \text{kN/m}^3$

1.93 $w = 3.536$ MN $m = 360.5$ Mg

1.95 3.23 N

1.97 $2.38 \times 10^{-3} \, \text{slugs/ft}^3$

1.99 0.904 at 40°F   0.865 at 120°F

1.101 Specific weight = $56.1 \, \text{lb/ft}^3$
Density = $1.74 \, \text{slugs/ft}^3$
Specific gravity = 0.899

1.103 142 lb

1.105 $2745 \, \text{cm}^3$

1.107 $1.53 \, \text{slugs/ft}^3$; $0.79 \, \text{g/cm}^3$

1.109 Volume = 1.16×10^5 gal
Weight = 6.60×10^5 lb

1.111 0.724 lb

1.113 Required depth = 17.3 in

1.115 Required time = 15.6 min

1.117 Payback time = 2.27 years

1.119 Displacement = 0.442 L

1.121 Rotational speed = 2354 rpm

Chapter 2

2.19 $1.5 \times 10^{-3} \, \text{Pa·s}$

2.23 1.90 Pa·s

2.25 $8.9 \times 10^{-6} \, \text{lb·s/ft}^2$

2.29 $4.1 \times 10^{-3} \, \text{lb·s/ft}^2$

2.31 2.8×10^{-5} lb·s/ft²

2.33 9.5×10^{-5} lb·s/ft²

2.35 2.2×10^{-4} lb·s/ft²

2.55 From Table 2.5: Viscosities at 40°C in mm²/s of cSt

ISO Viscosity grade	Minimum	Nominal	Maximum
10	9.0	10.0	11.0
68	61.2	68.0	74.8
220	198	220	242
1000	900	1000	110

2.57 5.60×10^{-6} m²/s 6.03×10^{-5} ft²/s

2.59 1.36×10^{-4} lb·s/ft²

2.61 0.402 Pa·s

2.63 8.23×10^{-3} lb-s/ft²

2.65 78.0 SUS

2.66 257 SUS

2.67 871 SUS

2.68 1130 SUS

2.69 706 SUS

2.70 955 SUS

2.71 1349 mm²/s

2.72 94.6 mm²/s

2.73 12.5 mm²/s

2.74 37.5 mm²/s

2.75 1018 mm²/s

2.76 113.6 mm²/s

Chapter 3

3.11 12.7 psia

3.13 Zero gage pressure

3.15 56 kPa(gage)

3.17 −23 kPa(gage)

3.19 384 kPa(abs)

3.21 105.4 kPa(abs)

3.23 13 kPa(abs)

3.25 8.1 psig

3.27 −3.2 psig

3.29 55.7 psia

3.31 15.3 psia

3.33 1.9 psia

3.35 1.05

3.37 5.56 psig

3.39 32.37 kPa(gage)

3.41 177.9 psig

3.43 $p_{\text{surface}} = 24.77$ kPa(abs)

   $p_{\text{bottom}} = 67.93$ kPa(abs)

3.45 61.73 psig

3.47 13.36 ft

3.49 6.84 m

3.51 70.6 kPa(gage)

3.53 110 MPa

3.55 −22.47 kPa(gage)

3.63 $p_B - p_A = -0.258$ psi

3.65 $p_A - p_B = 96.03$ kPa

3.67 $p_A = 90.05$ kPa(gage)

3.69 $p_A - p_B = 2.73$ psi

3.71 $p_A = 0.254$ kPa(gage)

3.77 30.06 in

3.81 83.44 kPa

3.83 14.99 psia

3.85 98.94 kPa(abs)

3.87 $p = -0.133$ psi

   $p = -917$ Pa

3.89 $p = 2.88$ kPa

   $p = 0.418$ psi

3.91 $p = -25.7$ inHg

3.93 $p = 4.15$ psi

   $p = 28.6$ kPa

3.95 Required height = 16.3 m

3.97 Pressure = 76.5 kPa

3.99 Depth = 17.0 ft

3.101 Change in pressure = 7.0 kPa

3.103 Change in pressure = 4.38 psi

Chapter 4

4.1 1673 lb

4.3 125 lb

4.5 2.47 kN

4.7 22.0 kN

4.9 6.05 lb

4.11 137 kN

4.13 1.26 MN

4.15 $F_R = 126\,300$ lb

   $h_p = 10.33$ ft vertical depth to center of pressure

   $L_p = 11.93$ ft

4.17 $F_R = 46.8$ kN

   $h_p = 0.933$ m vertical depth to center of pressure

   $L_p = 1.32$ m

4.19 $F_R = 1.09$ kN $L_p = 966$ mm

   $L_p - L_c = 13.3$ mm

4.21 $F_R = 1787$ lb $L_p = 13.51$ ft

   $L_p - L_c = 0.0136$ ft

4.23 $F_R = 1.213$ kN $L_p = 1.122$ m

   $L_p - L_c = 5.98$ mm

4.25 $F_R = 5.79$ kN $\qquad L_p = 1.372$ m
$L_p - L_c = 0.0637$ m

4.27 $F_R = 11.97$ kN $\qquad L_p = 1.693$ m
$L_p - L_c = 0.0235$ m

4.29 $F_R = 329.6$ lb $\qquad L_p = 47.81$ in
$L_p - L_c = 0.469$ in

4.31 $F_R = 247$ N $\qquad L_p = 196.5$ mm
$L_p - L_c = 0.0465$ m

4.33 $F_R = 29\,950$ lb $\qquad L_p = 5.333$ ft

4.35 $F_R = 34\,586$ lb $\qquad L_p = 6.158$ ft

4.37 $F_R = 343$ kN $\qquad L_p = 3.067$ m

4.39 $F_R = 11.92$ kN $\qquad L_p = 1.00$ m

4.41 Force on hinge $= 4.85$ kN to left
Force on stop $= 2.95$ kN to left

4.43 $F_R = 3.29$ kN
$L_{pe} - L_{ce} = 4.42$ mm

4.45 $F_R = 1826$ lb
$L_{pe} - L_{ce} = 1.885$ in

4.47 $F_R = 48.58$ kN $\qquad F_V = 35.89$ kN
$F_H = 32.74$ kN

4.49 $F_R = 120\,550$ lb $\qquad F_V = 99\,925$ lb
$F_H = 67\,437$ lb

4.51 $F_R = 959.1$ kN $\qquad F_V = 927.2$ kN
$F_H = 245.3$ kN

4.53 $F_R = 80.7$ kN $\qquad F_V = 54.0$ kN
$F_H = 60.0$ kN

4.55 $F_R = 64.49$ kN $\qquad F_V = 47.15$ kN
$F_H = 44.00$ kN

4.57 $F_R = 820$ lb $\qquad L_p = 8.327$ ft

4.59 $F_{R1} = 481$ kN on the left side of the door;
$L_p = 1.167$ m from bottom of door

4.61 $F_R = 381$ kN; $L_p = 1.00$ m from bottom of wall

4.63 $F_R = 71.5$ kN; $L_p = 2.14$ m from surface of water

4.65 $F = 669$ lb

Chapter 5

5.1 Buoyant force $= 814$ N $\qquad$ Tension $= 556$ N

5.3 It will sink

5.5 234 mm

5.7 0.217 m^3

5.9 7.515×10^{-3} m^3

5.11 5.055 ft^3

5.13 0.0249 lb

5.15 1.041

5.17 1447 lb

5.19 283.6 m^3

5.21 7.95 kN/m^3

5.23 237 mm

5.25 29 mm

5.27 10.05 kN

5.29 135 mm

5.31 1681 lb

5.33 4.67 in

5.35 300 lb

5.37 14.39 lb

5.39 $y_{mc} = 0.4844$ m (unstable)

5.41 $y_{mc} = 8.256$ ft (stable)

5.43 $y_{mc} = 488.8$ mm (unstable)

5.45 $y_{mc} = 10.55$ in (unstable)

5.47 32.50 ft

5.49 $y_{mc} = 436$ mm (stable)

5.51 $y_{mc} = 90.2$ mm (stable)

5.53 $y_{mc} = 410.3$ mm (stable)

5.55 $y_{mc} = 54.18$ in (stable)

5.57 $y_{mc} = 13.29$ ft (stable)

5.59 $y_{mc} = 467$ mm (stable)

5.61 $y_{mc} = 1.288$ m (stable)

5.63 **a.** 17.09 kN $\qquad$ **b.** 11.85 kN/m^3
c. Unstable; $y_{mc} = 0.822$ m; $y_{cg} = 0.950$ m

5.65 Buoyant force $= 12.8$ lb

5.67 Tension in cable $= F_T$: **a.** $F_T = 72.0$ kN in air;
b. $F_T = 6.3$ kN in sea water; **c.** bell will sink

5.69 Mass of lead $= 6.91$ kg

5.71 Steel will float in mercury; sg $= 13.54$

5.73 $D_{min} = 1.01$ ft; If $D > 1.01$ ft, part of float will be above
the water; If $D < 1.01$ ft, camera and float will sink.

Chapter 6

6.1 1.89×10^{-4} m^3/s

6.3 0.550 m^3/s

6.5 2.08×10^{-3} m^3/s

6.7 0.250 m^3/s

6.9 3.30×10^4 L/min

6.11 2.96×10^{-7} m^3/s

6.13 215 L/min

6.15 1.02 ft^3/s

6.17 5.57 ft^3/s

6.19 561 gal/min

6.21 3368 gal/min

6.23 $Q = 500$ gal/min $= 1.11$ ft^3/s $= 3.15 \times 10^{-2}$ m^3/s
$Q = 2500$ gal/min $= 5.57$ ft^3/s $= 0.158$ m^3/s

6.25 2.77×10^{-2} ft^3/s

6.27 1.76×10^{-5} ft^3/s

6.29 $W = 736$ N/s $\qquad M = 75.0$ kg/s

6.31 $Q = 7.47 \times 10^{-7}$ m^3/s $\qquad M = 8.07 \times 10^{-4}$ kg/s

6.33 $M = 2.48 \times 10^{-2}$ slugs/s $\qquad W = 2878$ lb/h

6.35 5.38 ft^3/s

6.37 3.09 ft

6.39 $v_1 = 0.472$ m/s $v_2 = 1.89$ m/s

6.41 7.88 m/s

6.43 1.97×10^5 lb/h

6.45 $1\frac{1}{4} \times 0.065$ steel tube for $v \geq 8.0$ ft/s min
$\frac{7}{8} \times 0.065$ steel tube for $v \leq 25.0$ ft/s max

6.47 DN 150 Schedule 40 pipe for Q = 1800 L/min
DN 350 Schedule 40 pipe for 9500 L/min

6.49 3.075 m/s

6.51 10.08 ft/s

6.53 20 mm OD $\times$ 1.5 mm wall steel tube

6.55 Suction line: 5-in pipe; $v_s = 12.82$ ft/s
6-in pipe; $v_s = 8.88$ ft/s
Discharge line: $3\frac{1}{2}$-in pipe; $v_d = 25.94$ ft/s
4-in pipe; $v_d = 20.15$ ft/s

6.57 Suction line: DN 80 pipe; $v_s = 3.73$ m/s
DN 90 pipe; $v_s = 2.61$ m/s
Discharge line: DN 50 pipe; $v_d = 7.69$ m/s
DN 65 pipe; $v_d = 5.39$ m/s

6.59 $v_{\text{pipe}} = 7.98$ ft/s $v_{\text{nozzle}} = 65.0$ ft/s

6.61 34.9 kPa

6.63 25.1 psig

6.65 $p_A = 58.1$ kPa   Q = 0.0213 m^3/s

6.67 2.90 ft^3/s

6.69 $Q = 4.66 \times 10^{-3}$ m^3/s $p_A = -3.61$ kPa
$p_B = -12.44$ kPa

6.71 1.30 m

6.73 35.6 ft/s

6.75 3.98×10^{-3} m^3/s

6.77 1.48×10^{-3} m^3/s

6.79 1.035 ft^3/s

6.81 31.94 psig

6.86 6.00×10^{-3} m^3/s

6.90 1.28 ft

6.93 10.18 psig

6.95 296 s

6.97 556 s

6.99 504 s

6.101 1155 s

6.103 252 s

6.105 1.94 s

6.107 2.40 L/s

6.109 2.00 psig

6.111 −33 kPa

6.113 0.708 in, 0.174 in, 0.694 in

6.115 0.038 ft^3/s

Chapter 7

7.1 34.5 lb-ft/lb

7.3 3.33×10^{-2} m^3/s

7.5 15.7 lb-ft/lb

7.7 72.7

7.9 16.2 kW

7.11 0.700 hp   70.0%

7.13 $h_A = 37.46$ m $P_A = 0.390$ kW

7.15 **a.** $p_B = 1.07$ psig   **b.** $p_C = 21.8$ psig
c. $h_A = 48$ ft   **d.** $P_A = 10.9$ hp

7.17 $h_A = 4.68$ m $P_A = 43.6$ W

7.19 2.80 hp

7.21 $P_A = 1.25$ W $P_I = 2.08$ W

7.23 8.59 kW

7.25 $P_R = 16.85$ kW $P_O = 12.64$ kW

7.27 2.00 hp

7.29 35.0 kPa

7.31 12.8 ft

7.33 6.216 m

7.35 21.16 hp

7.37 219.1 psig

7.39 1.01 psig

7.41 5.76 psig

7.43 4.28 hp

7.45 1.26 hp

7.47 47 miles per hour

7.49 11.1 hr

7.51 36.8 W

7.53 0.293 ft/s

Chapter 8

8.1 $N_R = 239$; Laminar flow

8.3 $Q_{max} = 6.82 \times 10^{-3}$ m^3/s

8.5 **a.** 80 mm OD $\times$ 2.8 mm wall copper tube
b. 120 mm OD $\times$ 3.5 mm wall tube
c. 25 mm OD $\times$ 2.0 mm wall tube
d. 6 mm OD $\times$ 1.0 mm wall tube (smallest listed)

8.9 4.76×10^4; Turbulent flow

8.11 $N_R = 9.64 \times 10^5$; Turbulent flow

8.13 $N_R = 33.4$; Laminar flow

8.15 $N_R = 5.22 \times 10^3$; Turbulent flow

8.17 $N_R = 2260$ (critical zone)

8.19 $N_R = 1105$; Laminar flow

8.21 $N_R = 2.12 \times 10^4$; Laminar flow

8.23 $Q_1 = 0.1681$ gal/min
$Q_2 = 0.3362$ gal/min

8.25 $N_R = 1.06 \times 10^4$

8.27 $p_1 - p_2 = -471$ kPa

8.29 $h_L = 1.20$ lb-ft/lb

8.31 $p_1 - p_2 = 25.2$ kPa

8.33 $h = 45.7$ ft

8.35 **a.** $h = 12.60$ ft

    b. $P_A = 113.8$ hp

8.37 $p = 46.9$ psi

8.39 **a.** $\Delta p = 853$ kPa

    b. $P_A = 17.1$ kW

8.41 $p_B = 81.1$ kPa

8.43 $p_1 - p_2 = 39.6$ psi

8.45 2 ½-in Type K copper tube; $p_1 - p_2 = 411$ psi

8.47 $P_A = 151$ hp

8.49 $P_A = 2.64$ hp

8.51 $p_1 - p_2 = 107$ kPa

8.53 $f = 0.0273$

8.55 $f = 0.0155$

8.57 $f = 0.0213$

8.59 $f = 0.0206$

8.61 $f = 0.0175$

8.63 $h_L = 15.2$ ft

8.65 $h_L = 28.5$ ft

8.67 $h_L = 1.78$ m

8.69 **a.** $h_L = 61.4$ ft

    b. $h_L = 25.3$ ft

8.71 $h_L = 14.7$ ft

8.73 $h_L = 252$ m; System cannot deliver water at nearly this rate.

8.75 Four stations required; Schedule 80 or 160 pipe could operate at higher pressure and fewer pumping stations would be required.

8.77 $h_L = 38.0$ ft for 2-in pipe; $h_L = 1273$ ft for 1-in pipe

8.79 $\Delta p = 0.153$ psi; $N_R = 6.6 \times 10^4$; $f = 0.0252$

Chapter 9

9.1 At centerline; $U = 21.44$ ft/s
At $r = 0.20$ in; $U = 20.64$ ft/s
At $r = 0.40$ in; $U = 18.23$ ft/s
At $r = 0.60$ in; $U = 14.22$ ft/s
At $r = 0.80$ in; $U = 8.60$ ft/s
At $r = 1.00$ in; $U = 1.38$ ft/s
At wall; $r = 0.20$ in; $U = 0.00$ ft/s

9.3 At centerline; $U = 0.0133$ m/s $U = 0.01048$ m/s
At $r = 8.00$ mm; $U = 0.01026$ m/s
At $r = 16.00$ mm; $U = 0.00960$ m/s
At $r = 24.00$ mm; $U = 0.00849$ m/s
At $r = 32.00$ mm; $U = 0.00695$ m/s
At $r = 40.00$ mm; $U = 0.00496$ m/s
At $r = 48.00$ mm; $U = 0.00253$ m/s
At wall; $r = 55.1$ mm; $U = 0.0000$ m

9.5 32.4 mm

9.7 At centerline, insertion $= 84.15$ mm
At centerline, $U = 2.00v$
At $r = 5.0$ mm, $U = 1.9907v$; 0.47% low

9.9 1.95 m/s

9.11 Selected values:

y (mm)	U (m/s)
10	0.530
30	0.628
50	0.674
100	0.735
300	0.833
600	$0.895 = U_{max}$

9.13 At $y = 2.44$ in, $U = v_{avg} = 6.00$ ft/s
At $y_1 = 2.94$ in, $U_1 = 6.12$ ft/s
At $y_2 = 1.94$ in, $U_2 = 5.85$ ft/s

9.15.

N_R	f	v/U_{max}
4×10^3	0.041	0.775
1×10^4	0.032	0.796
1×10^5	0.021	0.828
1×10^6	0.0185	0.837

9.17 Selected values: $v = 10.08$ ft/s:

y (in)	U (ft/s)
0.05	5.83
0.15	7.98
0.50	10.35
1.00	11.71
1.50	12.51
2.013	$13.09 = U_{max}$

9.19 $Q_{shell}/Q_{tube} = 2.19$

9.21 $Q_{tube} = 0.3535$ ft³/s $Q_{shell} = 1.998$ ft³/s

9.23 $N_R = 2.77 \times 10^4$

9.25 Tube: $N_R = 1.20 \times 10^5$ Shell: $N_R = 5.027 \times 10^3$

9.27 Pipes: $N_R = 1.00 \times 10^6$
Shell: $N_R = 2.35 \times 10^5$

9.29 $Q = 0.0397$ ft³/s

9.31 $N_R = 552$

9.33 $N_R = 1.112 \times 10^5$

9.35 $R = 0.0471$ ft $Q = 0.0181$ ft³/s

9.37 0.713 psi

9.39 92.0 Pa

9.41 254 kPa

9.43 3.02 kPa

9.45 3.72 psi

9.47 $v = 23.05$ ft/s $Q = 187$ gal/min $h_L = 51.9$ ft

9.49 7.36×10^4

9.51 $Q = 0.0507 \text{ ft}^3/\text{s}$ $h_L = 1.67 \text{ ft}$

9.53 $N_R = 3.30 \times 10^4$ $h_L = 3.149 \text{ m}$

Chapter 10

10.1 $h_L = 0.211 \text{ m}$

10.3 $h_L = 4.55 \text{ ft}$

10.5 $p_1 - p_2 = -0.0891 \text{ psi}$

10.7 $h_L = 0.330 \text{ m}$

10.13 506.98 kPa

10.15 $h_L = 0.235 \text{ m}$

10.17 $h_L = 1.35 \text{ ft}$

10.19 False

10.21 $h_L = 0.224 \text{ m}$

10.23 $h_L = 4.32 \text{ ft}$

10.27 $K = 0.255$

10.29 **a.** $h_L = 0.358 \text{ m}$
   b. $h_L = 0.229 \text{ m}$
   c. $h_L = 0.115 \text{ m}$
   d. $h_L = 0.018 \text{ m}$

10.31 $L_e = 2.04 \text{ m}$

10.33 $\Delta p = 1.47 \text{ psi}$

10.35 7.36 kPa

10.37 $\Delta p = 9.81 \text{ psi}$

10.39 $h_L = 0.321 \text{ ft}$

10.41 $h_L = 3.17 \text{ m}$

10.43 $h_L = 1.177 \text{ m}$ $h_L = 1.572 \text{ m}$

10.45 $h_L = 0.275 \text{ m}$

10.47 $h_L = 0.849 \text{ ft}$

10.49 $K = 9.15$ $h_L = 15.5 \text{ ft}$

10.51 $K = 0.731$ $h_L = 1.25 \text{ ft}$

10.53 $h_L = 175 \text{ psi}$

10.55 $K = 143$

10.57 $C_v = 0.612$

10.59 $\Delta p = 0.764 \text{ psi}$

10.61 $\Delta p = 0.359 \text{ psi}$

10.63 $\Delta p = 0.562 \text{ psi}$

10.65 $\Delta p = 4.952 \text{ psi}$

10.67 $\Delta p = 2.273 \text{ psi}$

10.69 $\Delta p = 0.680 \text{ psi}$

10.71 $\Delta p = 39.13; h_L = 4.939 \text{ m}; N_R = 1.47 \times 10^5; f = 0.01916$

Chapter 11

11.1 $p_B = 85.6 \text{ kPa}$

11.3 $p_A = 212.8 \text{ psig}$

11.5 $p_A = 12.74 \text{ MPa}$

11.7 $p_A - p_B = 18.0 \text{ kPa}$

11.9 $v = 3.36 \text{ m/s}$

11.11 $Q = 1.867 \times 10^{-3} \text{ m}^3/\text{s}$

11.13 **a.** $v = 32.44 \text{ ft/s}$
   b. $v = 81.44 \text{ ft/s}$

11.15 $Q = 0.0273 \text{ m}^3/\text{s}$

11.17 5-in Schedule 80 pipe

11.19 $D_{min} = 1.96 \text{ ft minimum}$

11.21 $h = 3.35 \text{ m}$

11.23 $h_A = 24.17 \text{ ft}$ $P_A = 0.168 \text{ hp}$

11.25 $v_{max} = 6.68 \text{ ft/s}$

11.27 $p_{inlet} = -48.4 \text{ kPa}$

11.29 $h_A = 276.8 \text{ ft}$ $P_A = 16.1 \text{ hp}$ $P_I = 21.2 \text{ hp}$

11.31 $p_A = 319.4 \text{ kPa}$

11.33 $Q = 204 \text{ L/min}$

11.35 $Q = 296.3 \text{ gal/min}$

11.37 $Q = 327.2 \text{ gal/min}$

11.39 $Q = 0.03916 \text{ m}^3/\text{s}$

11.41 $Q = 0.06992 \text{ m}^3/\text{s}$

11.43 4-in Schedule 40 steel pipe

11.45 $1\frac{1}{2}$-in Schedule 40 plastic pipe

11.47 32 mm OD $\times$ 2.0 mm wall steel tube

11.49 $Q = 136.2 \text{ L/min}$

11.51 $p_B = 8.293 \text{ kPa}; h_L = 2.615 \text{ m}; v = 1.623 \text{ m/s};$
   $N_R = 7.303 \times 10^4; f = 0.02105$

Chapter 12

12.1 $Q = 0.0607 \text{ m}^3/\text{s}$

12.3 **a.** $Q_a = 518 \text{ L/min (upper pipe)}$
   $Q_b = 332 \text{ L/min (lower pipe)}$
   b. $p_A - p_B = 95.0 \text{ kPa}$

12.5 $K = 162$

12.6 **a.** $Q = 1.891 \text{ ft}^3/\text{s}$
   b. $Q = 1.403 \text{ ft}^3/\text{s}$
   c. $Q = 0.488 \text{ ft}^3/\text{s}$

12.7 $Q_6 = 2.805 \text{ ft}^3/\text{s in 6-in pipe}$
   $Q_6 = 0.205 \text{ ft}^3/\text{s in 2-in pipe}$

12.11 Flow rates after six iterations:
   $\Delta Q < 0.25\%$ in any pipe
   $Q_1 = 6.942 \text{ ft}^3/\text{s}$ $Q_2 = 4.751 \text{ ft}^3/\text{s}$
   $Q_3 = 8.558 \text{ ft}^3/\text{s}$ $Q_4 = 2.191 \text{ ft}^3/\text{s}$
   $Q_5 = 3.251 \text{ ft}^3/\text{s}$ $Q_6 = 4.170 \text{ ft}^3/\text{s}$
   $Q_7 = 2.151 \text{ ft}^3/\text{s}$ $Q_8 = 4.388 \text{ ft}^3/\text{s}$
   $Q_9 = 3.210 \text{ ft}^3/\text{s}$ $Q_{10} = 1.402 \text{ ft}^3/\text{s}$
   $Q_{11} = 1.388 \text{ ft}^3/\text{s}$ $Q_{12} = 1.598 \text{ ft}^3/\text{s}$

12.13 $Q_{upper} = 1270 \text{ gal/min}; Q_{lower} = 79.7 \text{ gal/min};$
   $v_{upper} = 14.11 \text{ m/s}; v_{lower} = 7.62 \text{ m/s}$
   $N_{R\text{-}upper} = 8.142 \times 10^5; N_{R\text{-}lower} = 1.499 \times 10^5$

Chapter 13

13.16 Capacity is cut in half

13.17 Reduced by a factor of 4

13.18 Reduced by a factor of 8

13.19 Reduced by 25%

13.20 Reduced by 44%

13.21 Reduced by 58%

13.23 $1\frac{1}{2} \times 3 - 10$

13.25 $Q = 234$ gal/min $P = 22$ hp
Efficiency $= 53.5\%$ $NPSH_R = 11.0$ ft

13.26 Head $= 250$ ft $Q = 162$ gal/min
$P = 17$ hp   Efficiency $= 56.7\%$
$NPSH_R = 7.5$ ft

13.35 $Q = 375$ gal/min $N_s = 607$ $D_s = 3.00$

13.37 $n = 795$ rpm

13.39 $N_s = 2659$

13.41 $N_s = 2475$

13.51 $NPSH_R = 4.97$ ft

13.53 $NPSH_A = 20.70$ ft

13.55 $NPSH_A = 3.435$ m

13.57 $NPSH_A = -0.02$ m (incipient cavitation)

13.59 $NPSH_A = 2.63$ ft

13.61 $NPSH_A = -4.42$ ft (cavitation)

13.63 $NPSH_A = 1.02$ m

13.65 Required $p = 1617$ kPa gage

13.66 This is a sample solution using PIPE-FLO® and Pump-Flo™. Many options are available in the Pump-Flo™ catalog. The data shown here are for an end-suction pump (ESP) $4 \times 3 \times 13$.

$h_A = 23.25$ m; $Q = 1813$ L/min;
$NPSH_A = 29.11$ m; $NPSH_R = 2.026$ m

Suction pressure $= -187.5$ kPa; Discharge pressure $= 414.6$ kPa

Power $= 9.398$ kW; Efficiency $= 73.02\%$

Chapter 14

14.1 $R = 75$ mm

14.3 $R = 0.940$ ft

14.5 $R = 40.3$ mm

14.7 $R = 1.606$ in

14.9 $R = 0.909$ m

14.11 $Q = 0.295$ ft³/s

14.13 $S = 0.0125$

14.15 **a.** $Q = 34.7$ ft³/s
    b. $Q = 141.1$ ft³/s

14.17 $y = 1.69$ m

14.19 $Q = 15.89$ m³/s
$N_F = 0.629$ for depth $= 1.50$ m
$y_c = 1.16$ m

14.21 $D_{min} = 1.29$ ft for clay tile

14.23 $y = 0.833$ m $R = 0.270$ m

14.24 $S = 0.0116$

14.25 and 14.26   Selected values:

Width (m)	Depth (m)	R (m)	S
0.50	1.333	0.2105	0.0162
1.00	0.667	0.2857	0.0108
1.50	0.444	0.2791	0.0111
2.00	0.333	0.2500	0.0129

14.27 $A = 7.50$ ft² $R = 0.936$ ft

14.28 $Q = 44.49$ ft³/s

14.29 $Q = 75.63$ ft³/s

14.30 and 14.31   Selected values:

y (in)	A (ft²)	R (ft)	Q (ft³/s)
6.00	1.375	0.3616	4.33
10.00	2.708	0.5412	11.15
18.00	6.375	0.8605	35.75
24.00	10.00	1.086	65.47

14.33 $A = 0.0358$ m² $R = 0.0742$ m

14.35 $Q = 0.0168$ m³/s

14.37

	S
Rectangle	0.00519
Triangle	0.00518
Trapezoid	0.00471
Semicircle	0.00441

14.39 **a.** $y_c = 0.917$ m   **b.** $E_{min} = 1.38$ m
    d. and **e.**
      For $y = 0.50$ m: $E = 2.042$ m,
      $v = 5.50$ m/s, $N_F = 2.48$
      For $y_{alt} = 1.94$ m: $E = 2.042$ m,
      $v = 1.42$ m/s, $N_F = 0.325$

14.41 **a.** $y_c = 0.418$ ft   **b.** $E_{min} = 0.523$ ft
    d. and **e.**
      For $y = 0.25$ ft: $E = 1.067$ ft,
      $v = 7.253$ ft/s, $N_F = 3.615$
      For $y_{alt} = 1.065$ ft: $E = 1.067$ ft,
      $v = 0.400$ ft/s, $N_F = 0.097$

14.43 $Q_{max} = 1.00$ ft³/s

14.45 $H = 0$ in $Q = 0$ ft³/s
$H = 2$ in $Q = 1.35$ ft³/s
$H = 4$ in $Q = 3.84$ ft³/s
$H = 6$ in $Q = 7.14$ ft³/s
$H = 8$ in $Q = 11.1$ ft³/s
$H = 10$ in $Q = 15.7$ ft³/s
$H = 12$ in $Q = 20.8$ ft³/s

14.47 **a.** $Q = 18.8$ ft³/s
    b. $Q = 16.95$ ft³/s
    c. $Q = 6.84$ ft³/s

14.49 For $Q_{min} = 0.09$ ft³/s, $H = 0.100$ ft
For $Q_{max} = 8.9$ ft³/s, $H = 2.01$ ft

14.51 For $L = 4.0$ ft, $H = 2.06$ ft
For $L = 10.0$ ft, $H = 1.155$ ft

14.53 $Q = 7.55 \text{ ft}^3/\text{s}$

14.55 $Q = 1.19 \text{ ft}^3/\text{s}$

14.57 $Q = 0.073 \text{ ft}^3/\text{s}$

14.59 $H = 0.797 \text{ ft}$

14.61 Rectangular design B flume:
For $Q_{min} = 50 \text{ m}^3/\text{h}$, $H_{min} = 0.0863 \text{ m}$
For $H = 0.100 \text{ m}$, $Q = 63.4 \text{ m}^3/\text{h}$
For $H = 0.125 \text{ m}$, $Q = 90.5 \text{ m}^3/\text{h}$
For $H = 0.150 \text{ m}$, $Q = 121.3 \text{ m}^3/\text{h}$
For $H = 0.175 \text{ m}$, $Q = 155.3 \text{ m}^3/\text{h}$
For $Q_{max} = 180 \text{ m}^3/\text{h}$, $H_{max} = 0.1917 \text{ m}$

Chapter 15

15.1 $Q = 2.12 \times 10^{-2} \text{ m}^3/\text{s}$

15.3 $Q = 0.0336 \text{ ft}^3/\text{s}$

15.5 $\Delta p = 21.04 \text{ psi}$

15.7 $Q = 5.824 \times 10^{-3} \text{ m}^3/\text{s}$

15.11 $v = 8.45 \text{ m/s}$

15.13 $v = 5.11 \text{ m/s}$

15.15 $v = 33.0 \text{ ft/s}$

Chapter 16

16.1 2.76 kN

16.3 $R_x = R_y = 39.7 \text{ lb}$ Resultant $= 56.1 \text{ lb}$ $\underline{45°}$↗

16.5 $R_x = 10.13 \text{ lb}$ to the right $R_x = 37.79 \text{ lb}$ up

16.7 $R_x = 873 \text{ N}$ to the left $R_x = 1512 \text{ N}$ up

16.9 25.2 m/s

16.11 Spring force $= 32.0 \text{ lb}$

16.13 368 lb

16.15 2676 lb

16.17 $R_x = R_y = 20.41 \text{ kN}$ Resultant $= 28.9 \text{ kN}$ $\underline{45°}$↗

16.19 $R_x = 8.96 \text{ kN}$ $R_y = 3.71 \text{ kN}$
Resultant $= 9.7 \text{ kN}$ $\underline{22°}$↗

16.21 $v = 45.6 \text{ m/s}$

16.23 $2.72 \times 10^{-7} \text{ N}$

16.25 Moment $= 0.336 \text{ lb-in}$

16.27 0.0307 lb

16.29 $R_s = 41.0 \text{ N}; R_y = 19.1 \text{ N}$

Chapter 17

17.1 **a.** 0.253 N
b. $4.56 \times 10^{-4} \text{ N}$

17.3 1.50 m/s   2.05 m/s   105 m/s

17.5 0.42 N

17.7 7.32 m

17.9 11.2 kN

17.11 **a.** $2.85 \times 10^{-6} \text{ N·m}$
b. $1.67 \times 10^{-3} \text{ N·m}$

17.13 1364 N

17.15 Elliptical cylinder: $F_D = 12.05 \text{ lb}$
Navy strut: $F_D = 4.82 \text{ lb}$

17.17 $F_D = 31.3 \text{ lb}$

17.19 $F_D = 5.86 \text{ lb}$

17.21 $F_D = 1414 \text{ lb}$

17.23 $F_D = 0.080 \text{ lb}$ on golf ball
$F_D = 0.207 \text{ lb}$ on smooth sphere

17.25 $P_D = 140 \text{ hp}$

17.27 $P_E = 4252 \text{ hp}$

17.29 **a.** $F_D = 26.5 \text{ kN}$
b. $F_D = 1.66 \text{ kN}$

17.31 **a.** $F_L = 26.8 \text{ kN}$ $F_D = 2.83 \text{ kN}$
b. $F_L = 9.24 \text{ kN}$ $F_D = 972 \text{ N}$

17.33 $A = 90.4 \text{ m}^2$

Chapter 18

18.1 $44.17 \text{ ft}^3/\text{s}$

18.3 $1.25 \text{ m}^3/\text{s}$

18.5 5.79 m/s

18.7 0.158 psi

18.9 3.72 in H_2O

18.17 0.478 lb/ft^3

18.19 0.0525 lb/ft^3

18.21 11.79 N/m^3

18.23 131 cfm

18.25 2-in Schedule 40 pipe

18.27 $2\frac{1}{2}$-in Schedule 40 pipe, $p = 107 \text{ psig}$

18.29 0.2735 lb/ft^3 in the reservoir
0.198 lb/ft^3 in the pipe

18.31 1.092 lb/ft^3 at 35.0 psig
0.506 lb/ft^3 at 3.6 psig

18.33 $W = 5.76 \times 10^{-3} \text{ lb/s}$
$v = 811 \text{ ft/s}$

18.34 $4.44 \times 10^{-3} \text{ lb/s}$

18.37 186.6 kPa

18.39 $9.58 \times 10^{-3} \text{ N/s}$

18.40 and 18.41

p_1 (kPa gage)	W (N/s)
150	0.555
125	0.500
100	0.444
75	0.389
50	0.326
25	0.238

Chapter 19

19.1 $v = 570 \text{ ft/min}$ $h_L = 0.0203 \text{ inH}_2\text{O}$

19.3 $D = 17.0 \text{ in}$ $h_L = 0.078 \text{ inH}_2\text{O}$

19.6 $D = 350$ mm $h_L = 0.58$ Pa/m

19.8 $D_e = 5.74$ in $Q = 95$ cfm

19.10 $D_e = 381$ mm $Q = 0.60$ m^3/s

19.12 10×24 or 12×20

19.14 0.0180 inH$_2$O

19.16 0.0145 inH$_2$O

19.18 Main duct; 1600 cfm:
      $v = 1160$ ft/min; $H_v = 0.0839$ inH$_2$O

Main duct, 1100 cfm:
$v = 800$ ft/min; $H_v = 0.0399$ inH$_2$O
Branch; 500 cfm:
$v = 720$ ft/min; $H_v = 0.0323$ inH$_2$O

19.20 $H_L = 0.00839$ inH$_2$O

19.22 $H_L = 29.6$ Pa

19.24 $H_L = 0.1629$ inH$_2$O

19.26 $p_{\text{fan}} = -27.6$ Pa

INDEX

Note: Page number followed by f and t indicates figure and table respectively.

A

Absolute pressure, 39, 461
  and gage pressure, 39–40, 39f
Absolute viscosity, 21
Absolute zero, 6
Actively adjustable fluids, 24
A-D converters, 414
Adiabatic exponent, 461
Adiabatic flow, 461
Aerodynamics, 433
Affinity laws, for centrifugal pumps, 332–333
Air, flow of, in ducts, 470–483
  air distribution system, 471–472, 471f
  considerations in duct design, 483
  duct design, 477–483
  energy losses in ducts, 472–477
Airfoils, 432
  lift and drag on, 443–445
Airfoil-shaped, backward-inclined fan blades, 453
Air, properties of, 496–499
American Fire Sprinkler Association (AFSA), 122
American National Standards Institute (ANSI), 342
American Petroleum Institute (API) scale, 13
American Society of Heating, Refrigerating, and Air-Conditioning
    Engineers (ASHRAE), 472
American Society of Mechanical Engineers (ASME), 285
American Water Works Association (AWWA), 122, 285
Aneroid barometer, 51
Angle valves, 245
Apparent viscosity, 23
Areas of circles
  SI units, 508t
  U.S. Customary System units, 507t
Areas, properties of, 511t–512t
Aspect ratio, 444
ASTM International (ASTM), 122
Automated multi-range capillary viscometer, 29f
Axial compressors, 454, 455f

B

Ball valves, 253
Bar, 10
Barometer, 51–52, 51f
Bernoulli's equation, 128, 155
  applications of, 129–137
  fluid elements used in, 128f
  interpretation of, 128–129
  restrictions on, 129
Bingham fluids, 23

Bingham pycnometer, 11
Blower, 452
Bluff bodies, 442
Boundary layer, 435
Bourdon tube pressure gage, 53, 53f
British Standards (BS), 122
Bulk modulus, 10
Bulk storage tanks, 414
Buoyancy, 93
  buoyant force, 95
  concept of, 94–95
  material, 101–102
  procedure for solving buoyancy problems, 95–101
  and stability, 93–108
Butterfly valves, 240f, 246, 253

C

Cannon–Fenske routine viscometer, 29f
Capillary-tube viscometer, 28, 28f
Cavitation, 124, 341
  compressibility effects and, 443
Center of pressure, 63, 67
Centistoke (cSt), 28
Centrifugal blowers, 452–453, 453f
Centrifugal compressors, 453
Centrifugal fan rotors, 454f
Centrifugal fans, 452
Centrifugal grinder pumps, 330, 331f
Centrifugal pumps, 319, 326, 327f
  additional performance charts, 336–341
  affinity laws for, 332–333
  complete performance chart, 335, 336f
  composite rating chart for line of, 333f
  efficiency, 334–335
  impeller size, effect of, 333
  manufacturers' data for, 333–341
  net positive suction head required, 335, 336f
  performance data for, 330–332
  power requirement, 334
  small, 328, 329f
  speed, effect of, 333–334
Check valves, 245–246
  ball type, 240f
  swing type, 239f
Chord for airfoils, 444
Cipolletti weir, 387
Class III series pipeline system, 278–281
  approaches to designing of, 278–279
  spreadsheet for, 279–281

Class II series pipeline system, 272–278
    approaches for designing of, 272–273, 275–278
    iteration approach, 276
    spreadsheet for, 273–275
Class I series pipeline system, 265–268
    critique of, 269
    design changes, 269
    general principles of, 269
    spreadsheet for, 270–272
Cold-cranking simulator, 27
Column pump, 330, 330f
Compressibility, 10
Compressors, 452
Computational fluid dynamics (CFD), 216–218
Cone angle, 236
Conservation of energy, 127–128
Continuity equation, 120–122
    for any fluid, 120
    for liquids, 120
    for parallel systems, 297–298
Contracted weir, 387
Control valves, use of, 356–358
Conversion factors, 509t–510t
Coolant, 118
Copper Development Association (CDA), 123
Copper Tube Sizes (CTS), 123
Copper tubing, 123
Coriolis mass flowmeter, 408, 409f
Critical pressure ratio, 252, 461
Cross iteration technique, 307
Cup anemometer, 413

D

Darcy's equation, 183
    for noncircular sections, 215
Densitometer, 408
Density, 11
    and specific weight, relation between, 11–13
Diaphragm pumps, 319, 323–324
Diaphragm valves, 253
Differential manometer, 48f
Differential pressure meters (dp meters), 397. See also Variable-
        head meters
Diffuser, 233
    ideal, 233
    real, 233
Dilatant fluids, 23
Dimensional analysis, 181
Discharge pipe/discharge line, 226, 346–347
Doppler-type meter, 408
Drafting, 433
Drag and lift, 432–445
    on airfoils, 443–445
    compressibility effects and cavitation, 443
    drag coefficient, 435–440
    drag force equation, 434–435
    friction drag on spheres in laminar flow, 441
    pressure drag, 435
    vehicle drag, 441–443
Drag coefficient, 434–438, 439t–440t
    for other shapes, 436–438
    for spheres and cylinders, 435–436
Duct fan, 452, 452f

Ductile iron pipe, 123
    dimensions of, 506t
Dynamic pressure, 434
Dynamic viscosity, 21
    definition of, 21
    units for, 21–22, 22t

E

Electrorheological fluids (ERF), 24
Elevation, 40
    pressure and, relationship between, 40–43
Elevation head, Bernoulli's equation, 128
Energy losses and additions, in fluid flow systems, 156
    fluid friction and, 157–158
    fluid motors and, 156–157
    pumps and, 156, 156f
    valves and fittings and, 158
Entrance resistance coefficients, 237, 238f
Equal-friction method, for designing air ducts, 477, 478
Equations for friction factor, 194–195
Equivalent length ratio, 241
European Standards (EN) for pipe and tubing, 122
Exit loss, 231
Extensional rheometry, 25

F

Falling-ball viscometer, 29–30, 30f, 441
Fans, blowers, and compressors, 452–455, 453f, 454f
Fineness ratio, 437
Flat oval air ducts, 473–474
Floating bodies, stability of, 103–105
Flow coefficients for valves, 251–252
Flow devices, selection factors, 396–397
    accuracy, 397
    calibration, 397
    fluid types, 397
    pressure loss, 397
    range, 397
    type of flow indication, 397
Flow due to falling head, 140–142
Flow energy, 127, 127f
Flow imaging, 413
Flow, in noncircular sections, 212
    average velocity, 212
    Darcy's equation for, 215
    friction loss, 215–216
    hydraulic radius, 213–214
    Reynolds number, 214–215
Flow measurement, 395–415
    computer-based data acquisition and processing, 414–415
    devices for, considerations in selection of, 396–397
    level measurement, 414
    magnetic flowmeter, 406–407, 407f
    mass flow measurement, 408–410, 409f
    positive-displacement meters, 408
    reasons for, 395–396
    turbine flowmeter, 404, 405f
    turndown in, 397
    ultrasonic flowmeter, 408
    variable-head meters, 397–404
    velocity probes, 410–413
    vortex flowmeter, 404, 406, 406f

Flow of air and gases through nozzles, 461–467
   critical pressure, 463
   critical pressure ratio, 462
   Mach number, 462
   nozzle flow for adiabatic process, 462
   sonic velocity, 462
Flow of compressed air and gases, in pipes, 456–461
   flow rates for compressed air lines, 458
   pipe size selection and piping system design, 458–461
   specific weight for air, 456–457
Flow tubes, 404, 404f
Fluid, 3
   properties, 4
   viscosity, effect of, 359–360
Fluid flow rate, 118–120, 119t
Fluid friction, 157–158
Fluid motors, 156–157
   mechanical efficiency of, 165–167
Fluid power cylinder, 65f, 157f
Fluid power systems, 33, 249–251, 250f
   hydraulic fluids for, 33–34
Flumes
   critical flow, 388
   long-throated, 388, 389t
   Parshall, 388, 388f, 388t
Foams, buoyancy, 101–102
Foot valve with strainer, 246
   hinged disc, 240f
   poppet disc type, 240f
Force equation, 419–420
   procedure for using of, 420–421
   in x, y, and z direction, 420
Forces due to fluids in motion, 418–427
   on bends in pipelines, 423–426
   force equation, 419–420
   impulse–momentum equation, 420
   on moving objects, 426–427
   problems involving forces, 420–421
   on stationary objects, 421–423
Forces due to static fluids, 63–80
   on curved surface with fluid above and below, 79, 80f
   on curved surface with fluid below it, 78–79
   effect of pressure above fluid surface, 78
   gases under pressure, 65–66
   on horizontal flat surfaces under liquids, 66–67
   on rectangular walls, 67–69
   on submerged curved surface, 74–78
   on submerged plane areas, 69–73
Form drag, 434
Free air delivery (FAD), 458
Friction drag, 434, 435
   on spheres in laminar flow, 441
Friction factor, equations for, 194–195
Friction loss
   in laminar flow, 183–184
   in noncircular cross sections, 215–216
   in turbulent flow, 184–190
Froude number, 376

G

Gage fluids, 46
Gage pressure, 53
Gases, 3

Gas flow rates, 451–452
Gas laminar flowmeter, 410
Gate valves, 239f, 245
Gear pumps, 156f, 320–321
General energy equation, 158–162
German Standards (DIN), 122
Gerotor, 157
Globe valve, 239f, 244–245
Gradual enlargement, energy loss for, 231–233
Gradual reducer, 226
Gravitometer, 408

H

Hagen–Poiseuille equation, 183–184
Hazen–Williams coefficient, 195, 196t
Hazen–Williams formula
   nomograph for solution of, 196–198, 197f
   other forms of, 196, 197t
   for SI units, 195
   for U.S. Customary unit system, 195
   for water flow, 195–196
Head loss equation, for parallel systems, 298
Heating, Refrigeration, and Air Conditioning Institute
   of Canada (HRAI), 472
High water-based fluids (HWBF), 34
Hot-wire anemometer, 413
Hydraulic drop, 375
Hydraulic fluids, for fluid power systems, 33–34
Hydraulic hose, 124
Hydraulic Institute (HI), 342
Hydraulic jump, 374, 384–386
Hydraulic motor, 157, 157f
Hydraulic radius, 213–214
   for open-channel flow, 374
Hydrodynamics, 433
Hydrofoils, 432
Hydrometer, 13, 14f

I

Ideal gas law, 456
Impeller, 326, 452
   for kinetic pumps, 328f
Impulse–momentum equation, 420
Inclined well-type manometer, 49, 50f
Induced drag, 444
Industrial exhaust systems, for vapors and particulates, 478
Industrial fluid distribution system, 154–155, 154f
Inherent viscosity, liquid polymers, 25
International Association of Plumbing and Mechanical Officials
   (IAPMO), 122
International Organization for Standardization (ISO), 122
International System of Units, 4. See also Units
   SI unit prefixes, 4, 4t
Intrinsic viscosity, liquid polymers, 25
Isothermal flow, 461
ISO viscosity grades, 33, 33t
Iteration, 272

J

Japanese Standards (JIS), 122
Jet pumps, 319, 326
   deep-well, 328f
   shallow-well, 328f

K

Karman street, 406
Kinematic viscosity
 definition of, 22
 units for, 22, 22t
Kinetic energy, 127
Kinetic pumps, 326
 centrifugal grinder pumps, 330, 331f
 column pump, 330, 330f
 jet pumps, 326
 self-priming pumps, 329–330, 329f
 small centrifugal pumps, 328, 329f
 submersible pumps, 326
Knife gate valve, 245

L

Laminar flow, 178–180, 180f, 182
 friction loss in, 183–184
 velocity profile for, 207–208, 208f
Level measurement devices, 414
 capacitance probe, 414
 float type, 414
 guided radar, 414
 pressure sensing, 414
 radar, 414
 ultrasonic, 414
 vibration type, 414
Life cycle cost (LCC), for pumped fluid systems, 363–364
Lift, 432, 433, 444. *See also* Drag and lift
Linear actuators, 248
Lipkin bicapillary pycnometer, 11
Liquid hydraulic system, 248–249
Liquid polymers, viscosity of, 24–25
Liquids, 3
 bulk modulus for, 10, 10t
 surface tension of, 15t
Liquids, common, properties of
 in SI units, 490t
 in U.S. Customary System units, 491t
Lobe pump, 323, 323f
Long-throated flumes, 388, 389t

M

Mach number, 461
Magnehelic pressure gage, 53, 54f
Magnetic flowmeter, 406–407, 407f
Magnetorheological fluids (MRF), 24
Manning's equation
 in SI units, 376–377
 in U.S. Customary System, 377
Manometer, 45–50
 differential, 48f
 inclined well-type, 49, 50f
 U-tube, 46f, 47f
 well-type, 49, 50f
Mass, 3
 in SI unit system, 5
 in unit lbm, 5–6
 in U.S. Customary System, 5
 and weight, distinction between, 3
Mass flow measurement, 408–410
Mass flow rate, 119, 119t

Mechanical efficiency
 of fluid motors, 165–167
 of pumps, 163–165
Metacenter, 103
Metacentric height, 106
Metric flow coefficient for valves, 252
Minor losses, 225–227
 entrance loss, 237–238
 exit loss and, 231
 gradual contraction and, 236–237
 gradual enlargement and, 231–233
 resistance coefficient and, 227
 sudden contraction and, 233–235
 sudden enlargement and, 228–230
Momentum, 420
Moody diagram, 185–187, 186f, 187f
 laminar zone, 186
 smooth pipes line, 186, 187f
 transition zone, 186, 187f
 use of, 187–190
 zone of complete turbulence, 186

N

Nanofluids, 24
Nappe, 374
National Fire Protection Association (NFPA), 122, 285
Navy strut, 437, 438f
Needle valve, 252
Negative lift, 433
Net positive suction head (NPSH), 341–345
 cavitation and, 341
 NPSH available, 342
 NPSH margin, 342
 NPSH required, 342–345
 vapor pressure and, 341–342
Networks, 298, 307
Neutral buoyancy, 95
Newtonian fluid, 23
 and non-Newtonian fluids, 23–25, 23f
Nominal Pipe Size (NPS), 122
Non-Newtonian fluids, 23. *See also* Newtonian fluids
 time-dependent fluids, 23–24
 time-independent fluids, 23
NPSH. *See* Net positive suction head (NPSH)
NSF International (NSF), 122

O

Open channel, 372
 conveyance, 382
 efficient shapes for, 382, 383f
 geometry of, 380–382, 381f
 uniform steady flow in, 376–380
Open-channel flow, 372–373
 classification of, 374–375
 critical flow and specific energy, 382–384
 cross sections of, 373f
 flume, 90
 Froude number, 376
 hydraulic depth, 376
 hydraulic jump, 374, 384–386
 hydraulic radius for, 374
 irrigation stream, 372, 372f

kinds of, 375–376
measurement, 386–390
Reynolds number for, 374
uniform steady flow, 374
unsteady varied flow, 374
varied steady flow, 374
weir, 386–388 (*see also* Weir)
Orifice discharge coefficient, 403f
Overall efficiency, 163, 326
Overturning couple, 103

P

Parallel and branching pipeline systems, 296–313
    PIPE-FLO®, use of, 304–307
    systems with three or more branches, 307–313
    systems with two branches, 297f, 298–304
Parshall flume, 388, 388f, 388t
Pascals, 452
Pascal's laws, 8
Pascal's paradox, 45, 45f
Peristaltic pumps, 324
Petroleum lubricating oils, 492
Petroleum oils, 34
Piezoelectric effect, 55
Piezometric head, 73–74
Pipe and tubing, 122
    commercially available, 122–124
    copper tubing, 123, 505t
    ductile iron pipe, 123, 506t
    hydraulic hose, 124
    plastic pipe and tubing, 123–124, 504t
    recommended velocity of flow in, 124–126
    steel pipe, 122–123, 500t, 501t
    steel tubing, 123, 502t, 503t
Pipe bends, 246–248
Pipe elbows, 241f
PIPE-FLO® software
    K-factors in, 253–257
    parallel pipeline systems and, 304–307
    for selection of pumps, 352–356
    series pipeline systems and, 281–284
    use of, 190–194, 216
Pipeline design, for structural integrity, 284–286
Pipe roughness, 185f, 185t
Piston pumps, 156f
    for fluid power, 321
    for fluid transfer, 323
Pitot tube, 410–413, 410f, 411f
Plastic pipe and tubing, 123–124, 504t
Plastic valves, 252–253
    ball valves, 253
    butterfly valves, 253
    diaphragm valves, 253
    sediment strainers, 253
    swing check valves, 253
Plug, 252
Plug-flow fluids, 23
Pneumatics, 33
Pneumatic systems, 248
Polar diagram, 445
Positive-displacement blowers and compressors, 454
Positive-displacement meters, 408
Positive-displacement pumps, 319, 320

diaphragm pump, 323–324
gear pumps, 320–321
lobe pump, 323, 323f
performance data for, 324
peristaltic pumps, 324
piston pump, 321, 323
progressing cavity pump, 322–323, 323f
reciprocating pump performance, 324–325
rotary pump performance, 325–326
screw pump, 321–322, 322f
vane pump, 321
Positive lift, 433
Potential energy, 127
Power, 162, 442
    delivered to fluid motors, 165–167
    required by pumps, 162–165
    in U.S. Customary System, 162
Pressure, 3, 8–10, 63, 451–452
    definition of, 8
    drag, 434, 435
    drop in fluid power valves, 248–251
    as height of column of liquid, 52–53
    Pascal's laws, 8, 9f
Pressure–elevation relationship, 40–43
    development of, 43–45
        gases, 44
        liquids, 44
        standard atmosphere, 44–45
Pressure gages, 53, 53f, 54f
Pressure head, Bernoulli's equation, 128
Pressure recovery, 233
Pressure relief valve, 249
Pressure transducers and transmitters, 53–55, 54f
Principle of continuity. *See* Continuity equation
Programmed instruction, 6
Progressing cavity pump, 322–323, 323f
Progressive cavity pump, 319
Projected area, 434
Propeller fans, 452
Pseudoplastic fluids, 23
Pumps, 156, 156f
    alternate system operating modes, 356–361
    centrifugal, 330–341
    considerations in selection of, 320
    discharge line, 346–347
    efficiency, 319
    kinetic, 326–330
    life cycle costs for, 363–364
    mechanical efficiency of, 163–165
    net positive suction head, 341–345
    PIPE-FLO® use for selection of, 352–356
    positive-displacement, 320–326
    power delivered by, to fluid, 319
    power input to, 319
    pump selection and operating point, 350–352
    pump type selection and specific speed, 361–363
    selection and application, 318–364
    suction line, 346
    system resistance curve, 347–350
    total head on, 318–319
    types of, 320
PVC plastic pressure pipe, dimensions of, 504t
Pycnometers, 11

R

Rectangular air ducts, 472–473
Rectangular walls, forces on, 67–69
Rectangular weir, 387
Reduced viscosity, liquid polymers, 25
Relative viscosity, liquid polymers, 25
Resistance coefficients, 158, 227
    for valves and fittings, 238–244
Resistance temperature detectors (RTDs), 410
Reynolds numbers, 178, 181–182, 181f
    for closed noncircular cross sections, 214–215
    critical, 182–183
    for open-channel flow, 374
Rheology, 23
Rheometers. *See* Viscometers
Rheopectic fluids, 24
Righting couple, 103
Rotameter, 404, 405f
Rotary actuators, 248
Rotating-drum viscometer, 27–28, 27f

S

SAE International, 27
SAE viscosity grades, 32–33
Saybolt Universal seconds (SUS), 30–32
Saybolt viscometer, 30, 30f
Screw compressors, 455
Screw pump, 321–322, 322f
    timed multiple, 322
    untimed multiple, 322
Sediment strainers, 253
Self-priming pumps, 329–330, 329f
Series pipeline system, 264–286
    Class III systems, 278–281
    Class II systems, 272–278
    Class I systems, 265–272
    PIPE-FLO® examples for, 281–284
    pipeline design, for structural integrity, 284–286
Shear rate, 21
Shell-and-tube heat exchanger, 206f, 212f
Shut-off valves, 356
Silicone fluids, 34
Single-stage centrifugal compressor, 454f
Sluice gate, 374
Sonic velocity, 461
Specific gravity, 11
    of crude oils, 13
    in degrees Baumé or API, 13
    and density, relation between, 11–13
Specific viscosity, liquid polymers, 25
Specific weight, 11
Stability, 93
    degree of, 107–108
    of floating bodies, 103–107
    of submerged bodies, 102–103
Stabinger viscometer, 28
Staff gage, 387
Stagnation point, 410, 434
Stall point, 445
Standard atmosphere, 44–45
Standard calibrated glass capillary viscometers, 28–29, 29f
Standard Dimension Ratio (SDR) system, 124

Standard Inside Dimension Ratio (SIDR) system, 124
Static regain method, for designing air ducts, 477
Static stability curve, 107–108
    for floating body, 108f
Steady flow, 120
Steel and copper hydraulic tubing, dimensions of, 503t
Steel pipe, 122–123
    dimensions of, 500t–501t
Steel tubing, 123
    dimensions of, 502t
Stokes's law, 441
Strain gage pressure transducer, 53, 54f
Submerged bodies, stability of, 102–103
Submerged curved surface, forces on
    horizontal component, 74–76
    procedure for computing of, 76–78
    resultant force, 76
    vertical component, 76
Submerged plane areas, forces on, 69–72
    center of pressure, 72–73
    development of procedure for, 72–73
    resultant force, 72
Submersible solids handling pump, 329f
Suction pipe/suction line, 226, 341, 346
Sudden contraction, energy loss due to, 233–235
Sudden enlargement, energy loss for, 228–230
Supersonic velocity, 461
Suppressed weir, 387
Surface tension, 14, 15t
Swing check valves, 253
Swirl meter, 406
System resistance curve (SRC), 347–350

T

Temperature, 6
    absolute, 6
    scales for, 6
Tees, standard, 241f
Terminal velocity, 29
Thermohydrometers, 13, 14f
Thixotropic fluids, 24
Time-dependent fluids, 23–24
Time-independent fluids, 23
T method, for designing air ducts, 477–478
Torricelli's theorem, 137–140
Total dynamic head (TDH), 318
Total head, Bernoulli's equation, 128
Total pressure tap, 410
Total ship resistance, 442–443
Triangular weir, 387–388
Turbine flowmeter, 404, 405f
Turbulent flow, 178, 180, 180f, 182
    friction factor for, 195
    friction loss in, 184–190
    velocity profile for, 209–211, 209f

U

Ubbelohde viscometer, 29f
Ultrasonic flowmeter, 408
Uniform steady open-channel flow, 374, 374f, 376–380. *See also*
    Open-channel flow
    Manning's equation, 376

Manning's *n*, value for, 377, 377t
  normal depth, 377
  normal discharge, 377
Unit-cancellation procedure, 6
Units
  in equation, 6–8
  in fluid mechanics, 7t
U.S. Customary System, 4–5. *See also* Units
U-tube manometer, 46f, 47f

V

Valves, standard, application of, 244–246
  angle valves, 245
  butterfly valve, 246
  check valves, 245–246
  foot valves with strainers, 246
  gate valves, 245
  globe valve, 244–245
Vane–axial blowers, 454
Vane pump, 321
Vapor pressure, 341–342
Variable-area meters, 404, 405f
Variable-head meters, 397
  flow nozzle, 401, 401f
  flow tubes, 404
  orifice, 401–404, 402f
  overall pressure loss, 404
  venturi tube, 397–400
Variable-speed drives, 358–359
Varied open-channel flow, 374–375, 374f. *See also*
    Open-channel flow
Vehicle drag, 441
  aircraft, 442
  automobiles, 441–442
  power to overcome drag, 442
  ship, 442–443
  submarine, 443
  trains, 442
  trucks, 442
Velocity gradient, 21
Velocity head, Bernoulli's equation, 128
Velocity probes, 410
  cup anemometer, 413
  flow imaging, 413
  hot-wire anemometer, 413
  pitot tube, 410–413, 410f, 411f
Velocity profile, 207
  for laminar flow, 207–208, 208f
  for turbulent flow, 209–211, 209f

Vena contracta, 142, 233, 234f
Ventilators, 452
Venturi meter, 135
Venturi tube, 397–400
Viscometers, 27
  capillary-tube, 28, 28f
  falling-ball, 29–30, 30f
  rotating-drum, 27–28, 27f
  Saybolt, 30, 30f
  Stabinger, 28
  standard calibrated glass capillary, 28–29, 29f
Viscosity, 4, 34
  definition of, 20
  dynamic, 21–22, 22t
  ISO grades, 33
  of liquid polymers, 24–25
  measurement, 27–32
  SAE grades, 32–33
  variation of, with temperature, 25–26
Viscosity index (VI), 25–26, 26f, 27t
Viscosity, variation of, with temperature, 493–495
Volume flow rate, 118, 119, 119t
  conversion factors for, 119
  definitions and units, 119t
  for variety of types of systems, 119t
Volumetric efficiency, 163, 326
Vortex flowmeter, 404, 406, 406f
Vortex-shedding element, 404

W

Water–glycol fluids, 34
Water, properties of
  in SI units, 488t
  surface tension, 14t
  in U.S. Customary System units, 489t
Weight
  and mass (*see* Mass)
  in SI unit system, 5
  in U.S. Customary System, 5
Weight flow rate, 119, 119t, 162
Weir, 374, 386–387, 386f, 387f
  Cipolletti, 387
  contracted, 387
  rectangular, 387
  triangular, 387–388
Well-type manometer, 49, 50f
Wheatstone bridge, 55
Wind resistance, 432

KEY EQUATIONS

HAZEN-WILLIAMS FORMULA – SI UNITS	$v = 0.85\, C_h\, R^{0.63} s^{0.54}$	(8–9)
HYDRAULIC RADIUS—CLOSED NONCIRCULAR SECTIONS	$R = \dfrac{A}{WP} = \dfrac{\text{area}}{\text{wetted perimeter}}$	(9–5)
REYNOLDS NUMBER FOR NONCIRCULAR SECTIONS	$N_R = \dfrac{v(4R)\rho}{\eta} = \dfrac{v(4R)}{\nu}$	(9–6)
DARCY'S EQUATION FOR NONCIRCULAR SECTIONS	$h_L = f\dfrac{L}{4R}\dfrac{v^2}{2g}$	(9–7)
HYDRAULIC RADIUS—OPEN CHANNELS	$R = \dfrac{A}{WP} = \dfrac{\text{area}}{\text{wetted perimeter}}$	(14–1)
REYNOLDS NUMBER – OPEN CHANNELS	$N_R = \dfrac{vR}{\nu}$	(14–3)
FROUDE NUMBER	$N_F = \dfrac{v}{\sqrt{gy_h}}$	(14–4)
HYDRAULIC DEPTH	$y_h = A/T$	(14–5)
MANNING'S EQUATION—SI UNITS	$v = \dfrac{1.00}{n}R^{2/3}S^{1/2}$	(14–6)
NORMAL DISCHARGE—SI UNITS	$Q = \left(\dfrac{1.00}{n}\right)AR^{2/3}S^{1/2}$	(14–8)
MANNING'S EQUATION – U.S. UNITS	$v = \dfrac{1.49}{n}R^{2/3}S^{1/2}$	(14–10)
NORMAL DISCHARGE – U.S. UNITS	$Q = AV = \left(\dfrac{1.49}{n}\right)AR^{2/3}S^{1/2}$	(14–11)
GENERAL FORM OF FORCE EQUATION	$F = (m/\Delta t)\Delta v = M\,\Delta v = \rho Q\,\Delta v$	(16–4)
FORCE EQUATION IN x-DIRECTION	$F_x = \rho Q\,\Delta v_x = \rho Q(v_{2_x} - v_{1_x})$	(16–5)
FORCE EQUATION IN y-DIRECTION	$F_y = \rho Q\,\Delta v_y = \rho Q(v_{2_y} - v_{1_y})$	(16–6)
FORCE EQUATION IN z-DIRECTION	$F_z = \rho Q\,\Delta v_z = \rho Q(v_{2_z} - v_{1_z})$	(16–7)
EFFECTIVE VELOCITY	$v_e = v_1 - v_0$	(16–11)